Ali Baklouti
Deformation Theory of Discontinuous Groups

De Gruyter Expositions in Mathematics

Volume 72

Ali Baklouti

Deformation Theory of Discontinuous Groups

DE GRUYTER

Mathematics Subject Classification 2020
Primary: 22E25, 22E27, 22E40, 22G15, 32G05, 57S30; Secondary: 81S10, 57M25, 57M27, 57S30

Author
Prof. Ali Baklouti
Faculté des Sciences de Sfax
Département de Mathématiques
Route de Soukra
3038 Sfax
Tunisia
ali.baklouti@usf.tn

ISBN 978-3-11-076529-8
e-ISBN (PDF) 978-3-11-076530-4
e-ISBN (EPUB) 978-3-11-076539-7
ISSN 0938-6572

Library of Congress Control Number: 2022934617

Bibliographic information published by the Deutsche Nationalbibliothek
The Deutsche Nationalbibliothek lists this publication in the Deutsche Nationalbibliografie; detailed bibliographic data are available on the Internet at http://dnb.dnb.de.

Typesetting: VTeX UAB, Lithuania
Printing and Binding: LSC Communications, United States

www.degruyter.com

Preface

Discontinuous actions of groups play an important role in many fields of mathematics, especially in the study of Riemann surfaces. This research axis appears to be a significant and indispensable framework because of its close relationship with so many other fields in mathematics, such as geometry, topology, number theory, algebraic geometry, differential geometry and with different fields, such as physics and other various areas. The study of Kleinian groups (discrete groups of orientation preserving isometries of hyperbolic spaces), Fuchsian groups, and the theory of automorphic forms are all rich areas of mathematics with many deep results. The work of Thurston on 3-manifolds and as a generalization the deformations of Kleinian groups have given additional focus to this very rich field of discontinuous group actions.

When it comes to the setting of solvable groups actions, the literature is somewhat scarce in this area. This book is devoted mainly to studying various geometric and topological concepts related to the deformation and moduli spaces of discontinuous group actions and building some interrelationships between these concepts. It contains the most recent developments of the theory, extending from basic concepts to a comprehensive exposition, and highlighting the newest approaches and methods in deformation theory. It presents full proofs of recent results, computes fundamental examples and serves as an introduction and reference for students and researchers in Lie theory, discontinuous groups and deformation (and moduli) spaces. It also includes the most recent solutions to many open questions over the last decades and brings related newest research results in this area.

The first chapter aims to record some main backgrounds on nilpotent, solvable and exponential solvable Lie groups and some compact extensions. Fundamental and basic examples, such as Heisenberg groups, threadlike groups, Euclidean motion groups and Heisenberg motion groups are treated with extensive details for further developments and use. As a preparation to discontinuous actions, an explicit description of closed and discrete subgroups of these groups is also well developed. The important notion of syndetic hull of closed subgroups is also introduced and many existence and unicity results are proved, including the extensions of homomorphisms of discrete subgroups to their syndetic hulls, which appears to be of major role in the computation of the parameter, deformation and moduli spaces.

The second chapter focuses on the characterization of proper action of closed subgroups on solvmanifolds and on some homogeneous spaces of compact extensions. In the case of m-step nilpotent Lie groups, the proper action of a closed connected subgroup is shown to be equivalent to its free actions for $m \leq 3$. Such a fact fails in general to hold otherwise. We also generate geometric criteria of the proper action of a discontinuous group on an arbitrary homogeneous space, where the group in question stands for the semidirect product group $K \ltimes \mathbb{R}^n$, where K is a compact subgroup of $GL(n, \mathbb{R})$. In the case of Heisenberg motion groups, the same requires the classification into three categories of all discrete subgroups. As shown, this will be a capital

https://doi.org/10.1515/9783110765304-201

role in the study of many geometrical concepts related to corresponding deformation and moduli spaces.

We also define the notions of weak and finite proper actions and substantiate that these are equivalent to free actions of connected closed subgroups operating on special and maximal solvmanifolds.

We pay attention in Chapter 3 to the determination of the parameter, deformation and moduli spaces of the action of a discontinuous group $\Gamma \subset G$ on a homogeneous space G/H in numerous settings, G being a Lie group and H a closed subgroup of G. This issue is of major relevance to understand the local geometric structures of these spaces as many examples reveal. The strategy basically consists in building up accurate cross-sections of adjoint orbits of deformation parameters. Toward such goal, the first step consists in generating an algebraic characterization of the above spaces making use of the results on the existence of syndetic hulls developed in the first chapter. Introducing the Grassmannian topology, we then show that the parameter space is stratified into G-invariant layers, endowed with the structure of a total space of a principal fiber bundle. This allows to explicitly determine (to a certain extent) the parameter and deformation spaces in many fundamental cases. For instance, the setting of Heisenberg groups is extensively pursued in the fourth chapter, where a necessary and sufficient condition for which the deformation space is endowed with a smooth manifold structure is obtained. This further allows to extend the study to the setting of the direct product of Heisenberg groups.

We also deal with the setting of general m-step nilpotent Lie groups in Chapter 4, where a description of the parameter and deformation spaces are derived ($m \leq 3$). A necessary condition for the Hausdorfness of the deformation space is also obtained. The setting of threadlike groups is also studied and an explicit determination of the deformation space is provided. In the case of a non-Abelian discontinuous group of rank k, the deformation space is shown to be endowed with a smooth manifold structure if and only if $k > 3$.

The fifth chapter is devoted to study the local rigidity property of deformations introduced by A. Weil in the Riemannian case and generalized further by T. Kobayashi. We state the local rigidity conjecture in the nilpotent setting, which asserts that the local rigidity fails to hold for any nontrivial discontinuous group acting on nilpotent homogeneous space. We further extend our study to many exponential and solvable settings. Namely, we show that the local rigidity fails when the Lie algebra $\mathfrak{l}$ of the syndetic hull of Γ is not characteristically solvable and in the exponential setting where $\mathfrak{l}$ is Abelian and $\dim(\mathfrak{l}) \geq 2$. Besides, we prove the existence of formal colored discontinuous groups in the general solvable setting. That is, the parameter space admits a mixture of locally rigid and formally nonrigid deformations. In the case where G is the diamond group and Γ a nontrivial finitely generated subgroup of G (not necessarily discrete), then there is no open G-orbits in $\mathrm{Hom}(\Gamma, G)$. In particular, if Γ is a discontinuous group for a homogeneous space G/H, then the strong local rigidity property fails to hold.

We are also concerned with an analogue of the so-called Selberg–Weil–Kobayashi local rigidity theorem in the context of a real exponential group G and H a maximal subgroup of G, where the local rigidity property is shown to hold if and only if the group G is isomorphic to $\mathrm{Aff}(\mathbb{R})$, the group of affine transformations of the real line. For more generality where G is a Lie group and Γ a finite group, we show that the space $\mathrm{Hom}(\Gamma, G)/G$ is discrete and at most countable. This space is finite if in addition G has finitely many connected components. This helps to show an analogue of the local rigidity conjecture holds in both cases where G stands for the compact extension $K \ltimes \mathbb{R}^n$ and for the Heisenberg motion groups.

Chapter 6 deals with the stability property, a different geometrical concept of deformations, which measures in general the fact that in a neighborhood of $\varphi \in \mathrm{Hom}(\Gamma, G)$, the properness property of the action on G/H is preserved. The determination of stable points is a very difficult problem in general, which mainly reduces to describe explicitly the interior of the subset of $\mathrm{Hom}_d^0(\Gamma, G)$ of injective homomorphisms with discrete image. We are then led to investigate about several kinds of questions of geometric nature related to the structure of the deformation space and as a result, many stability theorems will be established in the nilpotent and exponential cases and also in the context of some compact extensions.

On the other hand, it may then happen that there does not exist an infinite discrete subgroup Γ of G, which acts properly discontinuously on G/H. This phenomenon is called the Calabi–Markus phenomenon. Based on several upshots proved in previous chapters, such a phenomenon together with the question of existence of compact Clifford–Klein forms are subject of a study in the context of some compact extensions of nilpotent Lie groups.

The seventh chapter is devoted to resume some of the previous upshots once we remove the assumption on the groups in question to be simply connected. This means that the center may be compact and we show in this case that many previously open questions in the simply connected setting get answered. For instance, in the case of reduced Heisenberg groups H_{2n+1}^r, the deformation space turns out to be a Hausdorff space and even endowed with a smooth manifold structure for any arbitrary connected subgroup H of G and any arbitrary discontinuous group Γ for G/H and that the stability property holds, which is also the case of the product Lie group $G = H_{2n+1}^r \times H_{2n+1}^r$ and $H = \Delta_G$, the diagonal subgroup of G. On the other hand, a (strong) local rigidity theorem is obtained for both H_{2n+1}^r and $H_{2n+1}^r \times H_{2n+1}^r$. That is, the parameter space admits a (strong) locally rigid point if and only if Γ is finite.

The setting of reduced threadlike groups is also considered through similar questions. We show that a local rigidity conjecture holds for Abelian discontinuous groups and that non-Abelian discontinuous groups are stable. We also single out the notion of stability on layers and show that any Abelian discontinuous group is stable on layers.

The purpose of the last chapter is to describe a dequantization procedure for topological modules over a deformed algebra. We define the characteristic variety of a topological module as the common zeroes of the annihilator of the representation obtained by setting the deformation parameter to zero. On the other hand, the Poisson characteristic variety is defined as the common zeroes of the ideal obtained by considering the annihilator of the deformed representation, and then setting the deformation parameter to zero.

We next apply such a dequantization procedure to the case of representations of Lie groups. Let $V = \mathbb{R}^d$ be a linear Poisson manifold. Then the dual V^* of linear forms on V form a Lie subalgebra $\mathfrak{g}$ of the algebra $S(\mathfrak{g}^{\mathbb{C}})$ of polynomials on V endowed with the Poisson bracket. We then regard the Poisson manifold V as the dual $\mathfrak{g}^*$ of the Lie algebra $\mathfrak{g}$.

In the case where $G = \exp \mathfrak{g}$ is an exponential solvable Lie group, the orbit method appears to be a fundamental tool to smoothly link their unitary duals with the space of coadjoint orbits. We first bring explicit computations of the characteristic and the Poisson characteristic varieties in many fundamental Poisson-linear examples. In the nilpotent case, we show that any coadjoint orbit appears as the Poisson characteristic variety of a well-chosen topological module. We then substantiate the Zariski closure conjecture claiming that for an irreducible unitary representation of G, associated to a coadjoint orbit Ω via the Kirillov orbit method, the Poisson characteristic variety associated to a topological module with an adequate way coincides with the Zariski closure in $\mathfrak{g}^*$ of the orbit Ω. We also prove the conjecture in many restrictive cases, notably in the nilpotent setting (with a different approach) and in the case where the representation is induced from a normal polarizing subgoup. We finally investigate the bicontinuity of Kirillov and Dixmier maps in the light of this dequantization process.

Ali Baklouti

Contents

1 Structure theory

The material on solvable Lie groups and Lie algebras is quite standard; the readers may consult [42] and [59] and many references therein. The goal of the chapter is to record some main backgrounds on nilpotent, solvable and exponential solvable Lie groups and some of their compact extensions. We first recall the definitions of real solvable and nilpotent Lie algebras and define their associated Lie groups. We also go through some details with famous and important examples, such as Heisenberg groups, threadlike groups, Euclidean motion groups and Heisenberg motion groups. An explicit description of closed and discrete subgroups is also provided for further developments and use.

We also define the notion of the syndetic hull of a closed subgroup of a Lie group, and prove existence and unicity results in the case of (reduced) completely solvable Lie groups. In the setup of exponential solvable Lie groups, this has major interest and use in the computation of the parameter, deformation and moduli spaces of the action of discontinuous groups on homogeneous spaces as will be dealt with throughout the book.

1.1 Solvable Lie groups

1.1.1 Solvable and exponential solvable Lie groups

Let $\mathfrak{g}_{\mathbb{C}}$ denote the complexified Lie algebra of a given Lie algebra $\mathfrak{g}$, where $\mathbb{R} \subset \mathbb{C}$ denote the fields of complex and real numbers. Let $\mathrm{ad}_X, X \in \mathfrak{g}$ be the adjoint endomorphism defined by

$$\mathrm{ad}_X(Y) = [X, Y],$$

where $[,]$ denotes the Lie bracket. When $[X, Y] = 0$ for any $X, Y \in \mathfrak{g}$, we say that $\mathfrak{g}$ is *commutative* or *Abelian.* A linear subspace $\mathfrak{a}$ of $\mathfrak{g}$ is an *ideal* if $[X, \mathfrak{a}] \subset \mathfrak{a}$ for any $X \in \mathfrak{g}$. Then the quotient space $\mathfrak{g}/\mathfrak{a}$ endowed with the Lie bracket induced from $\mathfrak{g}$, becomes a Lie algebra. Let

$$D\mathfrak{g} = [\mathfrak{g}, \mathfrak{g}] = \mathbb{R}\text{-span}\{[X, Y]; X, Y \in \mathfrak{g}\},$$

then obviously $D\mathfrak{g}$ is an ideal of $\mathfrak{g}$ and the quotient Lie algebra $\mathfrak{g}/D\mathfrak{g}$ is commutative. We define the following sequences of vector spaces:

$$\mathfrak{g} = D^0\mathfrak{g} \supset D^1\mathfrak{g} \supset D^2\mathfrak{g} \supset \cdots \tag{1.1}$$

and

$$\mathfrak{g} = C^0\mathfrak{g} \supset C^1\mathfrak{g} \supset C^2\mathfrak{g} \supset \cdots, \tag{1.2}$$

https://doi.org/10.1515/9783110765304-001

where

$$D^0\mathfrak{g} = \mathfrak{g}, \quad D^k\mathfrak{g} = D(D^{k-1}\mathfrak{g}) \quad (k = 1, 2, \ldots),$$

and

$$C^0\mathfrak{g} = \mathfrak{g}, \quad C^k\mathfrak{g} = [\mathfrak{g}, c^{k-1}\mathfrak{g}] \quad (k = 1, 2, \ldots).$$

In both cases, the decreasing sequences (1.1) and (1.2) turn out to be respectively flags of ideals of $\mathfrak{g}$.

The Lie algebra $\mathfrak{g}$ is said to be *solvable*, if there exists k such that $D^k\mathfrak{g} = \{0\}$ and *nilpotent* if there exists k such that $C^k\mathfrak{g} = \{0\}$. On the other hand, $\mathfrak{g}$ is said to be m-step nilpotent (for some integer m), if $C^m\mathfrak{g} \neq \{0\}$ and $C^{m+1}\mathfrak{g} = \{0\}$.

Besides, a connected Lie group is said to be *solvable* (resp., nilpotent, m-step nilpotent) if its Lie algebra is solvable (resp., nilpotent, m-step nilpotent).

Comparing (1.1) and (1.2), nilpotent Lie groups are solvable Lie groups, but the gap between these two classes is quite wide. Here, we define some classes of solvable non-necessarily nilpotent Lie groups. Let

$$\exp = \exp_G : \mathfrak{g} \to G \tag{1.3}$$

be the exponential mapping of G.

Definition 1.1.1. Let G be a solvable Lie group and $\mathfrak{g}$ its Lie algebra. Then G said to be *exponential solvable* if, the exponential mapping (1.3) is a C^∞-diffeomorphism from $\mathfrak{g}$ onto G. In this case, let $\log = \log_G$ designate the inverse map of $\exp_G$.

Any exponential solvable Lie group is therefore connected and simply connected. The following result provides a characterization of such a class.

Theorem 1.1.2 ([106]). *Let G be a connected, simply connected and solvable Lie group and let $\mathfrak{g}$ be its Lie algebra. Then the following are equivalent:*
(1) *G is exponential solvable.*
(2) *$\exp_G$ is an injective mapping.*
(3) *$\exp_G$ is a surjective mapping.*
(4) *No endomorphism $\mathrm{ad}_X, X \in \mathfrak{g}$ has purely imaginary eigenvalues.*

Definition 1.1.3. Let $\mathfrak{g}$ be a Lie algebra such that $\dim \mathfrak{g} = n$. When there exists a sequence of ideals,

$$\{0\} = \mathfrak{g}_0 \subset \mathfrak{g}_1 \subset \cdots \subset \mathfrak{g}_{n-1} \subset \mathfrak{g}_n = \mathfrak{g}, \quad \dim \mathfrak{g}_j = j \quad (0 \leq j \leq n),$$

we say that $\mathfrak{g}$ is *completely solvable*. Any completely solvable Lie algebra is an exponential solvable Lie algebra for which any endomorphism $\mathrm{ad}_X, X \in \mathfrak{g}$ has only real eigenvalues. A connected Lie group is said to be completely solvable, if its Lie algebra is completely solvable.

Example 1.1.4 (Affine group of the real line). Consider the 2-dimensional Lie group defined by

$$\mathrm{Aff}(\mathbb{R}) = \left\{ \begin{pmatrix} a & b \\ 0 & 1 \end{pmatrix}, \ a, b \in \mathbb{R} \text{ and } a > 0 \right\},$$

known to be the "$ax + b$" group. Let $\mathfrak{g} = \mathbb{R}X \oplus \mathbb{R}Y$ denote its Lie algebra, where

$$X = \begin{pmatrix} 1 & 0 \\ 0 & 0 \end{pmatrix} \quad \text{and} \quad Y = \begin{pmatrix} 0 & 1 \\ 0 & 0 \end{pmatrix}$$

and whose Lie bracket is given by $[X, Y] = Y$. Then $\mathfrak{g}$ is a completely solvable Lie algebra said to be the Lie algebra of the affine group Aff($\mathbb{R}$) of the real line, and will be briefly denoted by aff($\mathbb{R}$).

The following notion of coexponential bases to closed subgroups of connected Lie groups (cf. [106]) plays a capital role in the sequel.

Definition 1.1.5. Let G be a connected and simply connected Lie group and let H be a closed connected subgroup of G. Let $\mathfrak{g}$, $\mathfrak{h}$ be the Lie algebras of G and H, respectively. A basis $\{X_1, \ldots, X_p\}$, $p = \dim(\mathfrak{g}/\mathfrak{h})$, is said to be coexponential to $\mathfrak{h}$ in $\mathfrak{g}$ if the map:

$$\begin{aligned} \varphi_{\mathfrak{g},\mathfrak{h}} : \mathbb{R}^p \times H \quad & \to \quad G \\ ((t_1, \ldots, t_p), h) \quad & \mapsto \quad \exp t_p X_p \cdots \exp t_1 X_1 \cdot h \end{aligned}$$

is a diffeomorphism.

We have the following.

Theorem 1.1.6 ([106, Proposition 2]). *Let G be a connected, simply connected and solvable Lie group. Then every connected closed subgroup of G admits a coexponential basis.*

Remark 1.1.7. A constructive proof of Theorem 1.1.6 is based on the three following assertions:

(i) If $\mathfrak{h}$ is a one-codimensional ideal of $\mathfrak{g}$, then any vector in $\mathfrak{g} \setminus \mathfrak{h}$ is a coexponential basis.

(ii) If $\mathfrak{g} \supset \mathfrak{h}' \supset \mathfrak{h}$ and $\{b_1, \ldots, b_d\}$, respectively, $\{c_1, \ldots, c_r\}$ is a coexponential basis for $\mathfrak{h}'$ in $\mathfrak{g}$, respectively, for $\mathfrak{h}$ in $\mathfrak{h}'$, then $\{b_1, \ldots, b_d, c_1, \ldots, c_r\}$ is a coexponential basis for $\mathfrak{h}$ in $\mathfrak{g}$.

(iii) If $\mathfrak{h}$ is a maximal subalgebra of $\mathfrak{g}$, which is not an ideal of $\mathfrak{g}$, then any coexponential basis for $\mathfrak{h} \cap [\mathfrak{g}, \mathfrak{g}]$ in $[\mathfrak{g}, \mathfrak{g}]$ is also a coexponential basis for $\mathfrak{h}$ in $\mathfrak{g}$.

As a direct consequence from Remark 1.1.7, we get the following lemmas (cf. [106]).

Lemma 1.1.8. *Let* $\mathfrak{h}$ *be a subspace of* $\mathfrak{g}$ *containing* $[\mathfrak{g},\mathfrak{g}]$ *and* V *a linear subspace of* $\mathfrak{g}$ *such that* $\mathfrak{g} = \mathfrak{h} \oplus V$. *Then any basis of* V *is a coexponential basis to* $\mathfrak{h}$ *in* $\mathfrak{g}$.

Proof. Take a basis $\{u_1, \ldots, u_r\}$ of V. As $\mathfrak{h}$ contains $[\mathfrak{g},\mathfrak{g}]$, then the sequence

$$\mathfrak{h}_0 = \mathfrak{h} \subset \cdots \subset \mathfrak{h}_r,$$

where $\mathfrak{h}_i = \mathbb{R}\text{-span}\{u_1, \ldots, u_i\} \oplus \mathfrak{h}$ is a sequence of ideals of $\mathfrak{g}$. Then the result comes directly from assertions ($\imath$) and ($\imath\imath$). □

Lemma 1.1.9. *Let* $\mathfrak{h}$ *be a subalgebra of* $\mathfrak{g}$ *such that* $\mathfrak{h} + [\mathfrak{g},\mathfrak{g}] = \mathfrak{g}$. *Then there exists a coexponential basis* $\{u_1, \ldots, u_r\}$ *to* $\mathfrak{h}$ *in* $\mathfrak{g}$ *such that* $[\mathfrak{g},\mathfrak{g}] = W \oplus \mathfrak{h} \cap [\mathfrak{g},\mathfrak{g}]$, *where* W *is the linear span of the family* $\{u_1, \ldots, u_r\}$.

Definitions 1.1.10.

(1) With the notation above, a strong Malcev (or Jordan–Hölder) basis $\{Z_1, \ldots, Z_m\}$ of $\mathfrak{g}$ is a basis of $\mathfrak{g}$ such that $\mathfrak{g}_i = \mathbb{R}\text{-span}\{Z_1, \ldots, Z_i\}$ is an ideal of $\mathfrak{g}$ for every $i \in \{1, \ldots, m\}$. The obtained flags of ideals

$$\{0\} = \mathfrak{g}_0 \subset \mathfrak{g}_1 \subset \cdots \subset \mathfrak{g}_m = \mathfrak{g}$$

is called strong Malcev (or Jordan–Hölder) sequence of $\mathfrak{g}$. It is well known that such basis exists if $\mathfrak{g}$ is nilpotent.

(2) A family of vectors $\{X_1, \ldots, X_p\}$, $p = \dim(\mathfrak{g}/\mathfrak{h})$ is said to be a Malcev basis of $\mathfrak{g}$ relative to $\mathfrak{h}$ if $\mathfrak{g} = \text{span}\{X_1, \ldots, X_p\} \oplus \mathfrak{h}$ and for all $i \in \{1, \ldots, p\}$, the vector space $\mathbb{R}\text{-span}\{X_1, \ldots, X_i\} \oplus \mathfrak{h}$ is a subalgebra of $\mathfrak{g}$.

Remark 1.1.11.

(1) It is well known that nilpotent Lie groups admit Jordan–Hölder bases.

(2) Let $\mathfrak{g}$ be a nilpotent Lie algebra and let $\mathscr{B} = \{Z_1, \ldots, Z_m\}$ be a strong Malcev basis of $\mathfrak{g}$ and

$$\mathscr{S}_{\mathfrak{g}} : \{0\} = \mathfrak{g}^0 \subset \mathfrak{g}^1 \subset \cdots \subset \mathfrak{g}^{m-1} \subset \mathfrak{g}^m = \mathfrak{g} \tag{1.4}$$

the associated Jordan–Hölder sequence of $\mathfrak{g}$ associated to the basis $\mathscr{B}$. Here, $\mathfrak{g}^i = \mathbb{R}\text{-span}\{Z_1, \ldots, Z_i\}$, $i = 1, \ldots, m$. Such a notion of basis is very important and will be crucially used to adapt our setup with convenient coordinates, which provide an adequate model for the adjoint action. Given a subalgebra $\mathfrak{h}$ of $\mathfrak{g}$, we denote by $\mathscr{I}_{\mathfrak{g}}^{\mathfrak{h}} = \{i_1 < \cdots < i_p\}$ ($p = \dim \mathfrak{h}$) the set of indices i ($1 \leq i \leq m$) such that $\mathfrak{h} + \mathfrak{g}^{i-1} = \mathfrak{h} + \mathfrak{g}^i$. We note for all $i_s \in \mathscr{I}_{\mathfrak{g}}^{\mathfrak{h}}$, $\tilde{Z}_s = Z_{i_s} + \sum_{r<i_s} \alpha_{r,s} Z_r \in \mathfrak{h}$. We get a basis $\mathscr{B}_{\mathfrak{h}} = \{\tilde{Z}_1, \ldots, \tilde{Z}_p\}$, which is a strong Malcev basis of $\mathfrak{h}$ singled out from $\mathscr{B}$. We define the matrix of $\mathfrak{h}$ written in the basis $\mathscr{B}$, denoted by $M_{\mathfrak{h},\mathscr{B}} \in M_{m,p}(\mathbb{R})$ and will be used later.

(3) When $\mathfrak{g}$ is nilpotent, every subalgebra $\mathfrak{h}$ of $\mathfrak{g}$ admits such a Malcev basis, which is obviously a coexponential basis.

1.1.2 Heisenberg Lie groups

Thee are different manners to define the Heisenberg group $\mathbb{H}_n$ for a positive integer n. One complex realization of the group is to define $\mathbb{H}_n = \mathbb{C}^n \times \mathbb{R}$ equipped with the group law

$$(z,t)(w,s) = \left(z + w, t + s - \frac{1}{2}\operatorname{Im}\langle z, w\rangle\right),$$

where $\operatorname{Im}\langle z, w\rangle$ designates the imaginary part of the standard complex scalar product $\langle z, w\rangle$.

Toward a real realization, one can define the Heisenberg group H_{2n+1} as the connected Lie group associated to the 2-step real nilpotent Lie algebra $\mathfrak{g} := \mathfrak{h}_{2n+1}$ of dimension $2n + 1$ admitting a basis $\{Z, X_1, Y_1, \ldots, X_n, Y_n\}$ with the unique nonvanishing brackets:

$$[X_l, Y_l] = Z \quad \text{for all } l = 1, \ldots, n. \tag{1.5}$$

For several upcoming purposes, one can also define $\mathfrak{g}$ as a real vector space, with a skew-symmetric bilinear form b of rank $2n$ and a fixed generator Z of the kernel of b. The center $\mathfrak{z}$ of $\mathfrak{g}$ is then the kernel of b and it is the one-dimensional subspace $[\mathfrak{g}, \mathfrak{g}]$, where for $X, Y \in \mathfrak{g}$, the Lie bracket is given by

$$[X, Y] = b(X, Y)Z. \tag{1.6}$$

There is a way to characterize the subalgebras of $\mathfrak{g}$. Roughly speaking, it consists in building a symplectic basis of $\mathfrak{g}$ constructed from a given subalgebra. We prove the following.

Proposition 1.1.12. *Let $\mathfrak{h}$ be a Lie subalgebra of $\mathfrak{g}$. Then there exists a basis $\mathscr{B}_{\mathfrak{h}} = \{Z, X_1 \ldots, X_n, Y_1, \ldots, Y_n\}$ of $\mathfrak{g}$ with the Lie commutation relations*

$$[X_i, Y_j] = \delta_{i,j} Z, \quad i, j = 1, \ldots, n$$

and satisfying:

(1) *If $\mathfrak{z} \subset \mathfrak{h}$, then there exist two integers $p, q \geq 0$ such that the family*

$$\{Z, X_1, \ldots, X_{p+q}, Y_1, \ldots, Y_p\}$$

constitutes a basis of $\mathfrak{h}$.

(2) *If* $\mathfrak{z} \not\subset \mathfrak{h}$, *then* $\dim \mathfrak{h} \leq n$ *and* $\mathfrak{h}$ *is generated by* $X_1, \ldots, X_s$, *where* $s = \dim \mathfrak{h}$. *The symbol* $\delta_{i,j}$ *designates here the Kronecker index. The basis* $\mathscr{B}_{\mathfrak{h}}$ *is said to be a symplectic basis of* $\mathfrak{g}$ *adapted to* $\mathfrak{h}$.

Proof. (1) Note that the assertion is obviously true if $\mathfrak{h} = \mathfrak{z}$. We can and do assume then that $\mathfrak{h} \supsetneq \mathfrak{z}$, the kernel of the restriction $b_{|\mathfrak{h}}$ is therefore nontrivial, and there exists a subalgebra V_0 such that $\ker b_{|\mathfrak{h}} = \mathfrak{z} \oplus V_0$. For any complementary subspace V_1 of $\mathfrak{z} \oplus V_0$ in $\mathfrak{h}$, the bilinear form $b_{|V_1}$ is nondegenerate. Let $p, q \geq 0$ such that $\dim V_0 = q$ and $\dim V_1 = 2p$. Let now

$$N := \{x \in \mathfrak{g} : b(x, V_1) = 0\}.$$

Remark that $\mathfrak{g} = N \oplus V_1$. Indeed, note first that $N \cap V_1 = \ker b_{|V_1} = \{0\}$. We now consider the map

$$\begin{aligned} f : \mathfrak{g} &\longrightarrow V_1^* \\ x &\mapsto b(x, \cdot), \end{aligned}$$

which is surjective and verifies $\ker f = N$. We deduce therefore for dimension reasons that $\mathfrak{g} = V_1 \oplus N$. Let N_1 be any supplementary subspace of $\mathfrak{z}$ in N, that is, $N = \mathfrak{z} \oplus N_1$. For $x \in \ker b_{|N_1}$, we have $b(v, x) = 0$ for all $v \in \mathfrak{g}$, which means that x is central in $\mathfrak{g}$. But the intersection of N_1 and $\mathfrak{z}$ is trivial, then $\ker b_{|N_1} = \{0\}$, and finally the restriction $b_{|N_1}$ is nondegenerate. Up to this step, one decomposes $\mathfrak{g} = \mathfrak{z} \oplus V_1 \oplus N_1$ as a sum of b-orthogonal subspaces and we can assume that V_0 is an isotropic subspace of N_1. It is therefore well known that any basis of V_0 can be extended to a symplectic basis of N_1 and the result follows by taking any symplectic basis of V_1, a symplectic basis of N_1 passing through V_0 and the generator of $\mathfrak{z}$.

(2) Assume now that $\mathfrak{z} \not\subset \mathfrak{h}$. Let V be a complementary subspace to $\mathfrak{z}$ in $\mathfrak{g}$ containing $\mathfrak{h}$. Then $b_{|V}$ is nondegenerate and $\mathfrak{h}$ is an isotropic subspace of V, in particular, $\dim \mathfrak{h} \leq n$. Take any basis of $\mathfrak{h}$ and extend it to a symplectic basis of V by adding the central vector Z. We obtain a symplectic basis of $\mathfrak{g}$ adapted to $\mathfrak{h}$. □

1.1.3 Threadlike Lie groups

We will be dealing in this book with the actions of discrete groups acting on certain nilpotent homogeneous spaces for which the basis group in question is *m-step threadlike*. A threadlike Lie algebra $\mathfrak{g}_m$ is the $(m+1)$-dimensional real nilpotent Lie algebra with basis $\{X, Y_1, \ldots Y_m\}$ and nontrivial Lie brackets:

$$[X, Y_i] = Y_{i+1}, \quad i \in \{1, \ldots, m-1\}, \quad m \geq 2. \tag{1.7}$$

Let $G_m = \exp(\mathfrak{g}_m)$ be the associated connected and simply connected nilpotent Lie group. Note that $\mathfrak{g}_2$ is the Heisenberg Lie algebra, G_m is m-step nilpotent and is the semidirect product of the one parameter group $\exp(\mathbb{R}X)$ and the Abelian subgroup $G^0 = \exp(\mathfrak{g}^0)$ where $\mathfrak{g}^0 = \text{span}\{Y_1, \dots, Y_m\}$.

As we will be dealing with homogeneous spaces, the following result that describes the structure of a subalgebra $\mathfrak{h}$ of dimension $p \geq 2$, which is not included in $\mathfrak{g}_0$, is of interest and will be used later on.

Lemma 1.1.13. *Let $\mathfrak{g}$ be a threadlike Lie algebra and let $\mathfrak{h}$ be a p-dimensional Lie subalgebra such that $\mathfrak{h} \not\subset \mathfrak{g}_0$ and $p \geq 2$. Then there exists a strong Malcev basis $\mathscr{B} = \{X, X_1, \dots, X_n\}$ of $\mathfrak{g}$ such that*

$$[X, X_i] = X_{i+1} \quad \textit{for all } i = 1, \dots, n-1,$$
$$[X_i, X_j] = 0 \quad \textit{for all } i, j = 1, \dots, n$$

and that $\mathscr{B}_{\mathfrak{h}} = \{X, X_{n-p+2}, \dots, X_n\}$ is a strong Malcev basis of $\mathfrak{h}$.

Proof. First of all, we can and do assume that $X \in \mathfrak{h}$. Let now $\mathscr{S}_{\mathfrak{g}_0}$ be a Jordan–Hölder sequence of ideals of $\mathfrak{g}_0$ associated to $\{Y_n, \dots, Y_1\}$ as in (1.4) such that $\mathfrak{g}^i = \mathbb{R}\text{-span}\{Y_n, \dots, Y_{n-i+1}\}$, $i = 1, \dots, n$. Let $\mathfrak{h}_0 = \mathfrak{h} \cap \mathfrak{g}_0$. For all $i \in \mathscr{S}_{\mathfrak{g}_0}^{\mathfrak{h}_0}$, there exists a vector $X_{n-i+1} = Y_{n-i+1} + \sum_{j=n-i+2}^{n} \alpha_{j,i} Y_j \in \mathfrak{h}$. Even more, we get $[X, X_{n-i+1}] \in \mathfrak{h} \cap \mathfrak{g}^{i-1} \setminus \mathfrak{h} \cap \mathfrak{g}^{i-2}$. Thus $i - 1 \in \mathscr{S}_{\mathfrak{g}_0}^{\mathfrak{h}_0}$, and let $X_{n-i+2} = [X, X_{n-i+1}]$. So, it appears clear that $\sup\{i, i \in \mathscr{S}_{\mathfrak{g}_0}^{\mathfrak{h}_0}\} = p - 1$ and that $\mathscr{S}_{\mathfrak{g}_0}^{\mathfrak{h}_0} = \{1, \dots, p-1\}$, which gives that $\mathfrak{h} = \mathbb{R}\text{-span}\{X, X_{n-p+2}, \dots, X_n\}$. We complete this basis to obtain a basis of $\mathfrak{g}$ by the following induction relation. If $X_j = Y_j + \sum_{s=j+1}^{n} \alpha_{s,n-j+1} Y_s$, $j = n-p+1, \dots, 2$, we get $X_{j-1} = Y_{j-1} + \sum_{s=j+1}^{n} \alpha_{s,n-j+1} Y_{s-1}$. This achieves the proof of the lemma. □

1.1.4 Maximal subgroups of solvable Lie groups

Definition 1.1.14. Let $\mathfrak{g}$ be a Lie algebra. A subalgebra $\mathfrak{h}$ of $\mathfrak{g}$ is said to be maximal in $\mathfrak{g}$, if $\mathfrak{h} \neq \mathfrak{g}$ and for every subalgebra $\mathfrak{l}$ such that $\mathfrak{h} \subset \mathfrak{l} \subset \mathfrak{g}$, then either $\mathfrak{l} = \mathfrak{h}$ or $\mathfrak{l} = \mathfrak{g}$.

The following theorem describes the structure of maximal subalgebras of a solvable Lie algebra. Such result (cf. Theorem 5.3.3 later) is proved in [107] in the case of exponential solvable algebras.

Theorem 1.1.15. *Let $\mathfrak{g}$ be a solvable Lie algebra and $\mathfrak{h}$ a nonnormal maximal subalgebra of $\mathfrak{g}$. Then $\mathfrak{h}$ is of one- or two-codimensional in $\mathfrak{g}$ and we have the following:*

(1) *If $\mathfrak{h}$ is one-codimensional, then there exist a codimension one ideal $\mathfrak{g}_0$ of $\mathfrak{h}$, which is a codimension two ideal in $\mathfrak{g}$, and two elements A, X in $\mathfrak{g}$ such that*

$$\mathfrak{g} = \mathfrak{h} \oplus \mathbb{R}X, \quad \mathfrak{h} = \mathfrak{g}_0 \oplus \mathbb{R}A$$

and

$$[A,X] = X \bmod \mathfrak{g}_0.$$

(2) *If $\mathfrak{h}$ is two-codimensional, then we have one of the following mutually exclusive situations:*

(2.1) *There exists a codimension one ideal $\mathfrak{g}_0$ of $\mathfrak{h}$, which is a three-codimensional ideal of $\mathfrak{g}$, as well as three elements A, X, Y in $\mathfrak{g}$ and a real number α such that*

$$\mathfrak{g} = \mathfrak{h} \oplus \mathbb{R}X \oplus \mathbb{R}Y, \quad \mathfrak{h} = \mathfrak{g}_0 \oplus \mathbb{R}A,$$
$$[A, X + iY] = (\alpha + i)(X + iY) \bmod (\mathfrak{g}_0)_{\mathbb{C}} \tag{1.8}$$

and

$$[X,Y] = 0 \bmod \mathfrak{g}_0. \tag{1.9}$$

(2.2) *There exist a two-codimensional ideal $\mathfrak{g}_0$ of $\mathfrak{h}$, which is a four-codimensional ideal in $\mathfrak{g}$ and four nonzero vectors A, B, X, Y such that*

$$\mathfrak{g} = \mathfrak{h} \oplus \mathbb{R}X \oplus \mathbb{R}Y, \quad \mathfrak{h} = \mathfrak{g}_0 \oplus \mathbb{R}A \oplus \mathbb{R}B,$$
$$[A, X + iY] = X + iY \bmod (\mathfrak{g}_0)_{\mathbb{C}}, \tag{1.10}$$
$$[B, X + iY] = -Y + iX \bmod (\mathfrak{g}_0)_{\mathbb{C}}, \tag{1.11}$$
$$[X,Y] = 0 \bmod \mathfrak{g}_0 \quad \textit{and} \quad [A,B] = 0 \bmod \mathfrak{g}_0. \tag{1.12}$$

Proof. Let $(\mathfrak{g}^i)_{1 \le i \le p}$ be a Jordan–Hölder sequence of $\mathfrak{g}$. Obviously, $\mathfrak{g}^i + \mathfrak{h}$ is a subalgebra containing $\mathfrak{h}$, for $1 \le i \le p$. Let i_0 be the smallest integer such that $\mathfrak{g}^{i_0} + \mathfrak{h} \neq \mathfrak{h}$. Then by maximality of $\mathfrak{h}$, we have $\mathfrak{g}^{i_0-1} + \mathfrak{h} = \mathfrak{h}$ and $\mathfrak{g}^{i_0} + \mathfrak{h} = \mathfrak{g}$ and then $\operatorname{codim} \mathfrak{h} = 1$ or 2.

Assume first that $\operatorname{codim} \mathfrak{h} = 1$. there exist a vector $X \in \mathfrak{g}/\mathfrak{h}$ and a linear form $\psi : \mathfrak{h} \to \mathbb{R}$ such that

$$[T,X] = \psi(T)X \bmod \mathfrak{h}, \quad T \in \mathfrak{h}.$$

Since $\mathfrak{h}$ is not an ideal, the form ψ is not identically zero. Let $\mathfrak{g}_0 = \ker \psi$, then $\mathfrak{g}_0$ is a codimensional 2 ideal of $\mathfrak{g}$. In fact, it is clear that $\mathfrak{g}_0 = \{T \in \mathfrak{h} : [T, \mathfrak{g}] \subset \mathfrak{h}\}$; then $[\mathfrak{g}_0, \mathfrak{g}] \subset \mathfrak{h}$, which merely implies that $\mathfrak{g}_0$ is an ideal of $\mathfrak{g}$.

Let $\mathfrak{n}$ be the nilradical of $\mathfrak{g}$, then $\mathfrak{n} + \mathfrak{h} = \mathfrak{g}$. So, we can suppose that X is in $\mathfrak{n}$. It follows that $[T,X] - \psi(T)X \in \mathfrak{h} \cap \mathfrak{n}$, for all $T \in \mathfrak{h}$, and since $\mathfrak{n}$ is a nilpotent ideal, then for all $T \in \mathfrak{h} \cap \mathfrak{n}$, there exists $n \in \mathbb{N}^*$ such that $0 = \operatorname{ad}_T^n(X) = \psi^n(T)X \bmod \mathfrak{h}$, which entails that $\mathfrak{h} \cap \mathfrak{n} \subset \mathfrak{g}_0$. We obtain therefore

$$[T,X] = \psi(T)X \bmod \mathfrak{g}_0, \quad T \in \mathfrak{h}.$$

Finally, we choose $A \in \mathfrak{h}$ such that $\psi(A) = 1$, then $\mathfrak{h} = \mathfrak{g}_0 \oplus \mathbb{R}A$ and $[A,X] = X \bmod \mathfrak{g}_0$.

Suppose now that codim $\mathfrak{h} = 2$. There exist two vectors X, Y and two linear functionals $\psi_1, \psi_2 : \mathfrak{h} \to \mathbb{R}$ such that

$$[T, X + iY] = (\psi_1(T) + i\psi_2(T))(X + iY) \bmod \mathfrak{h}_{\mathbb{C}}, \quad T \in \mathfrak{h}.$$

Remark first of all, by the maximality of $\mathfrak{h}$, that $\psi_2 \neq 0$ and that $[X, Y] \in \mathfrak{h}$. In fact, suppose that $[X, Y] \notin \mathfrak{h}$, then for $T \in \mathfrak{h}$,

$$[T, [X, Y]] = [[Y, T], X] + [[T, X], Y] = 2\psi_1(T)[X, Y] \bmod \mathfrak{h}.$$

We obtain therefore that $\mathfrak{h} \oplus \mathbb{R}[X, Y]$ is a one-codimensional subalgebra containing $\mathfrak{h}$, which contradicts the maximality of $\mathfrak{h}$. Moreover, if $\mathfrak{n}$ designates the nilradical of $\mathfrak{g}$, then $\mathfrak{n} + \mathfrak{h} = \mathfrak{g}$, so, we can pick the vectors X, Y in a way such that $X \in \mathfrak{n}$ and $Y \in \mathfrak{n}$. We have therefore that $[T, X + iY] - (\psi_1(T) + i\psi_2(T))(X + iY) \in \mathfrak{h}_{\mathbb{C}} \cap \mathfrak{n}_{\mathbb{C}} \subset (\mathfrak{h} \cap \mathfrak{n})_{\mathbb{C}}$, for all $T \in \mathfrak{h}$. Since $\mathfrak{n}$ is a nilpotent ideal, we have

$$\mathfrak{h} \cap \mathfrak{n} \subset \ker \psi_1 \cap \ker \psi_2. \tag{1.13}$$

Let

$$\mathfrak{g}_0 = \ker \psi_1 \cap \ker \psi_2.$$

We observe, as in the previous case, that $\mathfrak{g}_0 = \{T \in \mathfrak{h} : [T, \mathfrak{g}] \subset \mathfrak{h}\}$. Then $[\mathfrak{g}_0, \mathfrak{g}] \subset \mathfrak{h}$, which implies that $\mathfrak{g}_0$ is an ideal of $\mathfrak{g}$.

Assume in a first time that ψ_1 and ψ_2 are linearly dependent, then $\psi_1 = \alpha\psi_2$ for some real number α. In this case, $\mathfrak{g}_0 = \ker \psi_2$ is a two-codimensional ideal of $\mathfrak{g}$. Let $A \in \mathfrak{h}$ satisfying $\psi_2(A) = 1$, and we have $\mathfrak{h} = \mathfrak{g}_0 \oplus \mathbb{R}A$ and

$$[A, X] = \alpha X - Y \bmod \mathfrak{g}_0 \quad \text{and} \quad [A, Y] = \alpha Y + X \bmod \mathfrak{g}_0.$$

We see also have by (1.13) that (1.9) is verified.

Suppose now that ψ_1 and ψ_2 are linearly independent. Therefore, $\mathfrak{g}_0$ is a four-codimensional ideal of $\mathfrak{g}$. So, there exist two linearly independent vectors $A, B \in \mathfrak{h}$ such that

$$\mathfrak{h} = \mathfrak{g}_0 \oplus \mathbb{R}A \oplus \mathbb{R}B,$$
$$(\psi_1(A), \psi_2(A)) = (1, 0) \quad \text{and} \quad (\psi_1(B), \psi_2(B)) = (0, 1).$$

We have then

$$[A, X + iY] = X + iY \bmod \mathfrak{g}_0 \quad \text{and} \quad [B, X + iY] = i(X + iY) \bmod \mathfrak{g}_0.$$

We see then that (1.10) and (1.11) are satisfied. Finally, using the inclusion (1.13), we see that (1.12) is also verified. □

Definition 1.1.16. Let G be connected and simply connected solvable Lie group of Lie algebra $\mathfrak{g}$. A connected subgroup H of G is said to be maximal if its Lie subalgebra is maximal in $\mathfrak{g}$ as in Definition 1.1.14.

Remark 1.1.17.
(1) In other words, H is maximal in G, if $H \neq G$ and for every subgroup L such that $H \subset L \subset G$, then either $L = H$ or $L = G$. As a direct consequence of Theorem 1.1.15, maximal subgroups are of codimension one or two.
(2) On can reformulate the statements of Theorem 1.1.15 in terms of groups, replacing $\mathfrak{g}$ by G and $\mathfrak{h}$ by H and the direct sums by a product of one of the parameter groups.

1.2 Euclidean motion groups

Let $\langle\cdot,\cdot\rangle$ and $\|\cdot\|$ designate respectively the canonical Euclidean scalar product and the related norm on $\mathbb{R}^n$, for a given positive integer n. Let $I(n) := O_n(\mathbb{R}) \ltimes \mathbb{R}^n$ be the semidirect product of the orthogonal group $O_n(\mathbb{R})$ and $\mathbb{R}^n$. Here, $O_n(\mathbb{R})$ merely acts on $\mathbb{R}^n$ naturally. From now on, $I(n)$ will be denoted G unless cited otherwise. Any element γ of G is therefore written down as

$$\gamma = (A, x),$$

where A stands for an orthogonal matrix and $x \in \mathbb{R}^n$. This gives rise that the multiplication law of G is submitted to the following equation:

$$\gamma\gamma' = (A,x)(A',x') = (AA', x + Ax') \tag{1.14}$$

for every $\gamma = (A,x)$ and $\gamma' = (A',x') \in G$. Obviously, the unity element of G equals $(I,0)$ with $I = I_n$ is the unity matrix of $M_n(\mathbb{R})$. Finally, an element of the form (I,t), with $t \in \mathbb{R}^n$ is said to be a translation.

1.2.1 On orthogonal matrices

Let $\perp$ mean the orthogonality symbol with respect to the canonical Euclidean product $\langle\cdot,\cdot\rangle$ on $\mathbb{R}^n$, that is mentioned above. The following are immediate:
(1) For any $A \in O_n(\mathbb{R})$ and any subspace $\mathscr{V}$ of $\mathbb{R}^n$, we have $A(\mathscr{V}) \oplus^{\perp} A(\mathscr{V}^{\perp}) = \mathbb{R}^n$. Here, $A(\mathscr{V})$ denotes the subspace of $\mathbb{R}^n$, image of $\mathscr{V}$ by the linear map associated to A.
(2) A subspace $\mathscr{V}$ of $\mathbb{R}^n$ is fixed by A if and only if $\mathscr{V}^{\perp}$ also is.
(3) Let χ_A denote the characteristic polynomial of a matrix A. Then two matrices $A, B \in O_n(\mathbb{R})$ are similar if and only if $\chi_A = \chi_B$.

(4) Note for any $\theta \in \mathbb{R}$, the orthogonal transformation

$$r(\theta) = \begin{pmatrix} \cos\theta & -\sin\theta \\ \sin\theta & \cos\theta \end{pmatrix}.$$

Then for any $A \in O_n(\mathbb{R})$, there exists $S \in O_n(\mathbb{R})$ such that

$$S^{-1}AS = \begin{pmatrix} I_p & & & & \\ & -I_q & & & \\ & & r(\theta_1) & & \\ & & & \ddots & \\ & & & & r(\theta_l) \end{pmatrix}, \tag{1.15}$$

for some positive integers p, q and l such that $p+q+2l = n$ and some reals $\theta_1, \dots, \theta_l$.

(5) The following lemma is an elementary linear algebra result and will be of use later.

Lemma 1.2.1. *Let $\{O_i\}_{i\in J}$ be a family of commuting orthogonal matrices on $\mathbb{R}^n$. Then there exists $S \in O_n(\mathbb{R})$, some integers $m_+, m_-, m_\pm, l \in \mathbb{N}$ with $m_+ + m_- + m_\pm + 2l = n$ such that for any $i \in J$,*

$$S^{-1}O_iS = \begin{pmatrix} I_{m_+} & & & & & \\ & -I_{m_-} & & & & \\ & & d_i(-1,1) & & & \\ & & & r(\theta_{1,i}) & & \\ & & & & \ddots & \\ & & & & & r(\theta_{l,i}) \end{pmatrix},$$

for some reals $\theta_{1,i}, \dots, \theta_{l,i}$, where

$$d_i(-1,1) = \begin{pmatrix} \varepsilon_{1,i} & & \\ & \ddots & \\ & & \varepsilon_{m_\pm,i} \end{pmatrix} \in M_{m_\pm}(\mathbb{R}) \quad \textit{and} \quad \varepsilon_{k,i} \in \{-1,1\} \quad \textit{for } k \in \{1,\dots,m_\pm\}.$$

Proof. The Motzkin–Taussky theorem (cf. [113]) says that any family of commuting diagonalizable matrices $(O_j)_{j\in J}$ in $M_n(\mathbb{C})$ is simultaneously diagonalizable. The idea uses an induction on the integer n and the reason is that for $i_0 \in J$, and for any $j \in J$, O_j fixes any eigenspace of O_{i_0} and induces on each one a diagonalizable endomorphism.

First, we remark that one can find $i_0 \in J$ and $\lambda \in \mathbb{C}$ depending upon i_0, for which the eigenspace

$$E_\lambda = \ker O_{i_0} - \lambda I$$

of O_{i_0} is of dimension less than n. Otherwise $\lambda \in \mathbb{R}$ and all $O_i, i \in J$ are multiples of the identity and the lemma is trivial. Assume now that $\lambda \notin \mathbb{R}$, then E_λ, $E_{\overline{\lambda}}$ and also

$$F_\lambda := E_\lambda \oplus E_{\overline{\lambda}}$$

are stable by all the O_j's ($j \in J$). For fixed $j \in J$, let $v \in E_\lambda$ be an eigenvector of O_j associated to an eigenvalue λ_j. Then $\overline{v} \in E_{\overline{\lambda}}$ is also an eigenvector of O_j associated to $\overline{\lambda}_j$. This means that if $\lambda' = \pm 1$, is an eigenvalue of the induced endomorphism by $O_{j'}$ on F_λ, $j' \in J$, then ± 1 is of even multiplicity.

We will deploy an induction on n. Assume for awhile that $n = 2$. If $\lambda \in \mathbb{C} \setminus \mathbb{R}$, then any $O_i (i \in J)$ having ± 1 as an eigenvalue coincides, thanks to the above with $\pm I_2$. So no matrix of the family has the set $\{1, -1\}$ as a spectrum and this closes the proof in this case.

Suppose now that $n > 2$ and that the result holds for any integer $k < n$. Let $O_{i_1}, i_1 \in J$ having an eigenvalue $\lambda_{i_1} \neq \pm 1$. Consider

$$F_{\lambda_{i_1}} = E_{\lambda_{i_1}} \oplus E_{\overline{\lambda}_{i_1}}$$

and for any $j \in J$, O_{j,i_1}, the matrix corresponding to the restriction endomorphism associated to O_j on $F_{\lambda_{i_1}}$. The family $\{O_{j,i_1}\}_{j \in J}$, is thus a family of commuting diagonalizable matrices in $M_{2p_1}(\mathbb{C})$, for which $2p_1 = \dim(F_{\lambda_{i_1}})$.

If $2p_1 = n$, then if the spectrum of some $O_j, j \in J$ contains ± 1 as an eigenvalue, then it only does with an even multiplicity. On the other hand, there exists a common unitary basis of eigenvectors $(v_1, \dots, v_{p_1})$ of $E_{\lambda_{i_1}}$ and $(\overline{v}_1, \dots, \overline{v}_{p_1})$ of $E_{\overline{\lambda}_{i_1}}$, which we arrange as $(v_1, \overline{v}_1, \dots, v_{p_1}, \overline{v}_{p_1})$ to obtain a basis of $F_{\lambda_{i_1}}$. This allows to get the result in this case.

More generally, fixing the complex eigenvalues $\lambda_{i_1}, \dots, \lambda_{i_k}$ of some $O_{i_1}, \dots, O_{i_k}$, respectively, for $i_1, \dots, i_k \in J$, one can write

$$\mathbb{C}^n = F_{\lambda_{i_1}} \oplus \cdots \oplus F_{\lambda_{i_k}} \oplus H_k, \tag{1.16}$$

where H_k denotes the orthogonal supplementary subspace of $F_{\lambda_{i_1}} \oplus \cdots \oplus F_{\lambda_{i_k}}$. Now, for any $j \in J$, the restriction of O_j to H_k admits no nonreal eigenvalues. So if $\dim(H_k) = 0$ we are done; otherwise, the spectrum of $O_{j|H_k}$ is sitting inside $\{-1, 1\}$ for any $j \in J$ and, therefore, $O_{j|H_k}$ coincides with

$$A_j(-1,1) = \begin{pmatrix} \pm 1 & & \\ & \ddots & \\ & & \pm 1 \end{pmatrix} \in M_q(\mathbb{R}),$$

where $q = \dim H_k$. Let

$$E(\pm 1) = \bigcap_{j \in J} \ker(O_j \mp I),$$

and $H(-1,1)$ the orthogonal supplementary of $E(1) \oplus E(-1)$ in H_k, according to the decomposition (1.16). This yields the following refined decomposition of $\mathbb{R}^n$ as

$$\mathbb{R}^n = F_1 \oplus \cdots \oplus F_l \oplus \widetilde{H}_k, \tag{1.17}$$

where $2l = \sum_{j=1}^{k} \dim F_{\lambda_{i_j}}$, $\dim F_i = 2$ for $i \in \{1,\dots,l\}$ and $\widetilde{H}_k$ is the orthogonal supplement of $F_1 \oplus \cdots \oplus F_l$ and the spectrum of $O_{j|\widetilde{H}_k}$ is sitting in $\mathbb{R}$. If $m_+ = \dim E(1)$, $m_- = \dim E(-1)$ and $m_\pm = \dim H(-1,1)$, there exists $S \in O_n(\mathbb{R})$ such that for any $j \in J$,

$$S^{-1}O_jS = \begin{pmatrix} I_{m_+} & & & & & \\ & -I_{m_-} & & & & \\ & & d_j(-1,1) & & & \\ & & & r(\theta_{1,j}) & & \\ & & & & \ddots & \\ & & & & & r(\theta_{l,j}) \end{pmatrix}, \tag{1.18}$$

where for any $j \in J$,

$$d_j(-1,1) = \begin{pmatrix} \varepsilon_{1,j} & & \\ & \ddots & \\ & & \varepsilon_{m_\pm,j} \end{pmatrix} \in M_{m_\pm}(\mathbb{R}), \quad \varepsilon_{k,j} \in \{-1,1\} \quad \text{for } k \in \{1,\dots,m_\pm\}.$$

This achieves the proof with the convention that if one of the integers m_+, m_-, $m_\pm$, l is zero, the corresponding block does not show up. □

Note here that according to the matrix form in Lemma 1.2.1 above, $\mathbb{R}^n$ decomposes into direct sums of subspaces as follows:

$$\mathbb{R}^n = E(1) \oplus^\perp E(-1) \oplus^\perp H(-1,1) \oplus^\perp F_1 \oplus^\perp \cdots \oplus^\perp F_l, \tag{1.19}$$

where $E(\pm 1) = \bigcap_{j\in J} \ker(O_j \mp I)$, F_i are two-dimensional subspaces of $\mathbb{R}^n$ and $H(-1,1)$ is an orthogonal supplement for which the restriction $S^{-1}O_iS$ coincides with $d_i(-1,1)$.

1.2.2 Some structure results

For any $g = (A,x) \in G$, denote by $O(g)$ and $O(A) \in \mathbb{N}^* \cup \{\infty\}$, the orders of g and A, respectively. Let also $E_A(1) = \ker(A - I)$ and let $P(A)$ be the orthogonal projection on $E_A(1)$. We have the following.

Lemma 1.2.2. *$P(A)$ is a polynomial upon A. In particular, if A is of the finite order p, say, then*

$$P(A) = \frac{1}{p}(I + A + \cdots + A^{p-1}). \tag{1.20}$$

Proof. The first statement is a well-known fact. Assume that A has a finite order p. If $x \in E_A(1)$, then clearly $P(A)(x) = x$. On the other hand, equation (1.20) says that $P(A)(I - A) = 0$. As such, $I - A$ induces an isomorphism on $E_A(1)^\perp$. Then if $y \in E_A(1)^\perp$, there exists $y' \in E_A(1)^\perp$ such that $y = (I - A)y'$ and, therefore, $P(A)y = P(A)(I - A)y' = 0$. □

Let pr_1 and pr_2 designate the natural projections from $I(n)$ onto $O_n(\mathbb{R})$ and $\mathbb{R}^n$, respectively. Then equation (1.14) says that $\mathrm{pr}_1(\Gamma)$ is a subgroup of $O_n(\mathbb{R})$ for any subgroup Γ of G. Note that $\mathrm{pr}_2(\Gamma)$ is not a subgroup of $\mathbb{R}^n$ in general. Take indeed Γ generated by (A, x) such that $x \neq 0$ and A is of order $k \in \mathbb{N}^*$ such that $P(A)(x) = 0$. Then

$$\mathrm{pr}_2(\Gamma) = \left\{ \sum_{i=0}^{p} A^i x, p \in \{0, \ldots, k-1\} \right\},$$

which is not a subgroup of $\mathbb{R}^n$. For discrete subgroups of G, the following appears to be immediate.

Lemma 1.2.3. *Let Γ be a discrete subgroup of G. Then Γ is infinite if and only if $\mathrm{pr}_2(\Gamma)$ is also.*

Proof. Indeed, let G be a topological group, K a compact subgroup of G and Γ a discrete subgroup of G. Then Γ is finite if and only if it has a finite orbit in G/K, which is enough to conclude. □

Remark 1.2.4. One can develop another elementary proof of Lemma 1.2.3. Indeed, given an infinite discrete subgroup Γ for which $\mathrm{pr}_2(\Gamma) = \{x_1, \ldots, x_k\}$ $(k \in \mathbb{N}^*)$, one can write

$$\Gamma = \bigcup_{i=1}^{k} \mathscr{A}_{x_i} \times \{x_i\},$$

where $\mathscr{A}_{x_i} = \{A \in O_n(\mathbb{R}) \mid (A, x_i) \in \Gamma\}$. There exists therefore $i_0 \in \{1, \ldots, k\}$ such that the set $\mathscr{A}_{x_{i_0}} \times \{x_{i_0}\}$ is an infinite set sitting inside the compact set $O_n(\mathbb{R}) \times \{x_{i_0}\}$, so it cannot be discrete.

Lemma 1.2.5. *Let Γ be a discrete subgroup of G and $y = (A, x) \in \Gamma$. Then $O(y) = p$ if and only if $O(A) = p$ and $P(A)(x) = 0$. That is, for $y \in \Gamma$, $O(y) = +\infty$ if and only if $P(A)(x) \neq 0$.*

Proof. Let $p = O(y)$. Remark first that

$$y^p = \left(A^p, \left(\sum_{s=0}^{p-1} A^s \right) x \right) = (I, 0).$$

Then the order of A divides, that is, of y. If $q = O(A) < p$, then

$$y^q = \left(I, \left(\sum_{s=0}^{q-1} A^s\right)x\right)$$

and, therefore, $O(y) = \infty$ if ever $(\sum_{s=0}^{q-1} A^s)x \neq 0$. Hence, $q = p$ and $(\sum_{s=0}^{p-1} A^s)x = pP(A)(x) = 0$. The converse is trivial. □

The following establishes a necessary and sufficient condition for two elements to be conjugate.

Lemma 1.2.6. *Two elements (A,x) and (A',x') are conjugate in G if and only if there exists $S \in O_n(\mathbb{R})$ such that $S^{-1}AS = A'$ and $Sx' - x \in E_A(1)^\perp$. In particular, if (A,x) and (A',x') are of finite orders, then they are conjugate in G if and only if A and A' are conjugate in $O_n(\mathbb{R})$.*

Proof. Let $g = (S,y) \in G$. A direct computation shows that

$$g^{-1}(A,x)g = (S^{-1}AS, S^{-1}[x + (A - I)y]). \tag{1.21}$$

So $A' = S^{-1}AS$ and $x' = S^{-1}[x + (A - I)y]$. Therefore, $Sx' - x = (A - I)y$, which gives in turn that $P(A)(Sx' - x) = 0$, and finally $Sx' - x \in E_A(1)^\perp$.

Conversely, if $A' = S^{-1}AS$ and $Sx' - x \in E_A(1)^\perp$, then the equation

$$(A - I)z = Sx' - x$$

has at least a solution, say z_0. Take $y = z_0$, then $(A - I)y = Sx' - x$ and, therefore, $x' = S^{-1}[x + (A - I)y]$. This gives that

$$(A',x') = (S^{-1}, -S^{-1}y)(A,x)(S,y).$$

If (A,x) and (A',x') are of finite orders, then by Lemma 1.2.5, $x \in E_A(1)^\perp$ and $x' \in E_{A'}(1)^\perp$. So if there exists $S \in O_n(\mathbb{R})$ such that $A' = S^{-1}AS$, then $Sx' \in E_A(1)^\perp$ and also $Sx' - x \in E_A(1)^\perp$. Then the arguments above allow to conclude. □

1.2.3 Discrete subgroups of $I(n)$

We now prove our first upshot. We have the following.

Proposition 1.2.7. *Let $\{y_i\}_{i\in J}$ be a commuting family of G. There exist some integers m_+, m_-, $m_\pm$ and l satisfying $m_+ + m_- + m_\pm + 2l = n$ and $g \in G$ such that for any $i \in J$,*

$$g^{-1}y_i g = \left(\begin{pmatrix} I_{m_+} & & & & & \\ & -I_{m_-} & & & & \\ & & d_i(-1,1) & & & \\ & & & r(\theta_{1,i}) & & \\ & & & & \ddots & \\ & & & & & r(\theta_{l,i}) \end{pmatrix}, \begin{pmatrix} y_{m_+,i} \\ 0_{m_-} \\ 0_{m_\pm} \\ 0_2 \\ \vdots \\ 0_2 \end{pmatrix}\right), \tag{1.22}$$

where $y_{m_+,i} \in \mathbb{R}^{m_+}$,

$$d_j(-1,1) = \begin{pmatrix} \varepsilon_{1,j} & & \\ & \ddots & \\ & & \varepsilon_{m_\pm,j} \end{pmatrix} \in M_{m_\pm}(\mathbb{R}) \quad (\varepsilon_{k,j} \in \{-1,1\} \textit{ for } k \in \{1,\dots,m_\pm\}),$$

$\theta_{1,i},\dots,\theta_{l,i} \in \mathbb{R}$ and 0_2, 0_{m_-} and $0_{m_\pm}$ are the zeros of $\mathbb{R}^2$, $\mathbb{R}^{m_-}$ and $\mathbb{R}^{m_\pm}$, respectively.

Proof. Let $y_i = (A_i, x_i)$ for $i \in J$. A direct application of Lemma 1.2.1 shows that there exists $S \in O_n(\mathbb{R})$ such that for $i \in J$, $(S,0)^{-1}y_i(S,0) = (S^{-1}A_iS, S^{-1}x_i) := \delta_i$ is of the form

$$\delta_i = \left(\begin{pmatrix} I_{m_+} & & & & & \\ & -I_{m_-} & & & & \\ & & d_i(-1,1) & & & \\ & & & r(\theta_{1,i}) & & \\ & & & & \ddots & \\ & & & & & r(\theta_{l,i}) \end{pmatrix}, \begin{pmatrix} y_{m_+,i} \\ y_{m_-,i} \\ z_{m\pm,i} \\ x_{1,i} \\ \vdots \\ x_{l,i} \end{pmatrix}\right),$$

where the integers m_+, m_-, $m_\pm$ and l are as in Lemma 1.2.1. Trivially, they do not depend upon $i \in J$ and if one of them is zero. This reduces to the fact that the correspondent block of the $\mathbb{R}^n$ side does not show up.

For any $i, j \in J$,

$$\delta_i\delta_j = \left(\begin{pmatrix} I_{m_+} & & & & & \\ & I_{m_-} & & & & \\ & & d_{i,j}(-1,1) & & & \\ & & & r(\theta_{1,i,j}) & & \\ & & & & \ddots & \\ & & & & & r(\theta_{l,i,j}) \end{pmatrix}, \begin{pmatrix} y_{m_+,i} + y_{m_+,j} \\ y_{m_-,i} - y_{m_-,j} \\ z_{m\pm,i} + d_i(-1,1)z_{m\pm,j} \\ x_{1,i} + r(\theta_{1,i})x_{1,j} \\ \vdots \\ x_{l,i} + r(\theta_{l,i})x_{l,j} \end{pmatrix}\right),$$

where $d_{i,j}(-1,1) = d_i(-1,1)d_j(-1,1)$ and $\theta_{s,i,j} = \theta_{s,i} + \theta_{s,j}$ for $1 \leq s \leq l$. The fact that y_i and y_j commute for any $i, j \in J$ says that

$$y_{m_-,i} = y_{m_-,j} := y_{m_-}$$

and that $(d_i(-1,1), z_{m_\pm,i})$ and $(d_j(-1,1), z_{m_\pm,j})$ of $I(m_\pm)$ commute. Let us opt for the notation:

$$(d_i(-1,1), z_{m_\pm,i}) = \left(\begin{pmatrix} \varepsilon_{i,1} & & \\ & \ddots & \\ & & \varepsilon_{i,m_\pm} \end{pmatrix}, \begin{pmatrix} \tau_{i,1} \\ \vdots \\ \tau_{i,m_\pm} \end{pmatrix}\right) \qquad (i \in J),$$

where for $k \in \{1, \ldots, m_\pm\}$, $\varepsilon_{i,k} \in \{-1,1\}$ and $\tau_{i,k} \in \mathbb{R}$. Hence, the commutativity condition says

$$(1 - \varepsilon_{j,k})\tau_{i,k} = (1 - \varepsilon_{i,k})\tau_{j,k}.$$

As $\varepsilon_{i_k,k} = -1$, for some $i_k \in J$, then

$$\frac{(1 - \varepsilon_{j,k})}{2}\tau_{i_k,k} = \tau_{j,k},$$

and $\tau_{j,k} = 0$ whenever $\varepsilon_{j,k} = 1$. Let for $k \in \{1, \ldots, m_\pm\}$, $\tau_k = \tau_{i_k,k}$. Then

$$(d_i(-1,1), z_{m_\pm,i}) = \left(\begin{pmatrix} \varepsilon_{i,1} & & \\ & \ddots & \\ & & \varepsilon_{i,m_\pm} \end{pmatrix}, \begin{pmatrix} \frac{1-\varepsilon_{i,1}}{2}\tau_1 \\ \vdots \\ \frac{1-\varepsilon_{i,m_\pm}}{2}\tau_{m_\pm} \end{pmatrix}\right).$$

Let, on the other hand, for all $k \in \{1, \ldots, l\}$ the integer $i_k \in J$ such that $r(\theta_{k,i_k}) := r(\theta_k) \neq I_2$ and $x'_k := x_{k,i_k}$. So for $j \in J$,

$$x'_k + r(\theta_k)x_{k,j} = x_{k,j} + r(\theta_{k,j})x'_k, \tag{1.23}$$

and equivalently,

$$x_{k,j} = (I_2 - r(\theta_{k,j}))(I_2 - r(\theta_k))^{-1}x'_k,$$

for any $j \in J$. Let $x_k = (I_2 - r(\theta_k))^{-1}x'_k$. Then $(S,0)^{-1}y_i(S,0)$ takes the following form:

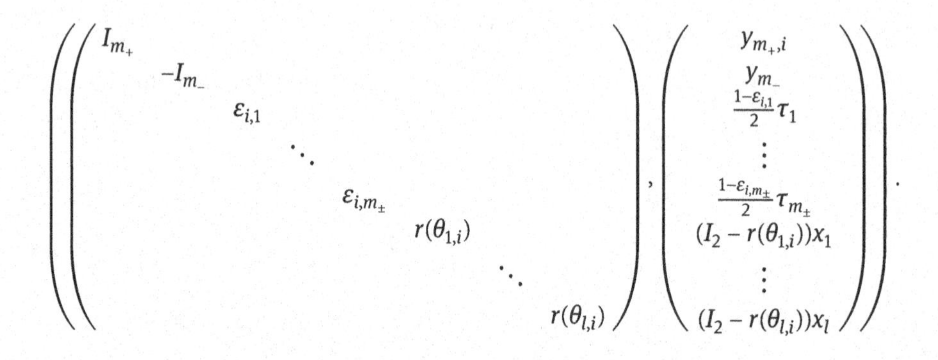

Now, for $t = {}^t({}^t0_{m_+}, \frac{1}{2}{}^ty_{m_-}, \frac{1}{2}\tau_1, \dots, \frac{1}{2}\tau_{m_\pm}, {}^tx_1, \dots, {}^tx_l)$ and $g = (S,0)(I,t) = (S,St)$, one gets

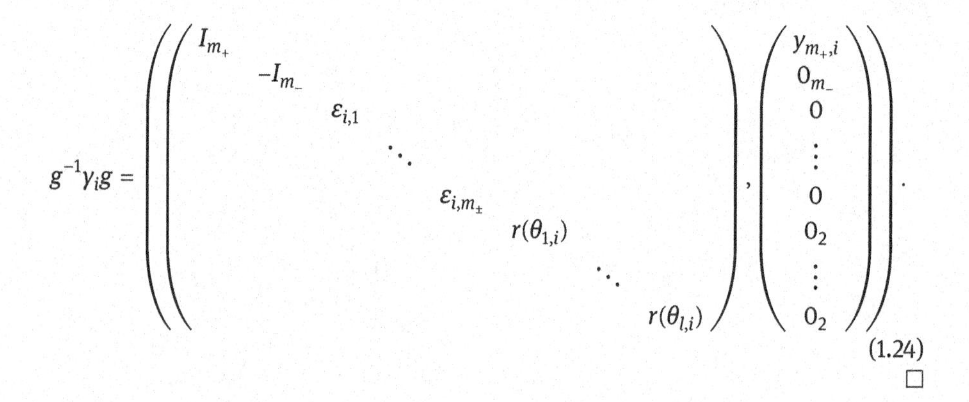

(1.24)

□

Generation families of discrete subgroups of $I(n)$

We first pose the following.

Definition 1.2.8. The rank of a group is the smallest cardinality of a family of its generators.

We next prove the following.

Lemma 1.2.9. *Let Γ be a subgroup of G. Then Γ is finite if and only if $\mathrm{pr}_1(\Gamma)$ is finite and any element in Γ is of finite order. Furthermore, if Γ is compact, then $\Gamma \cap \{I\} \times (\mathbb{R}^n \setminus \{0\}) = \emptyset$.*

Proof. The necessary condition is trivial. For the converse, suppose that $\mathrm{pr}_1(\Gamma) = \{A_1, \dots, A_p\}$ and suppose that Γ is infinite, which is equivalent to the fact that $\mathrm{pr}_2(\Gamma)$ is infinite as in Lemma 1.2.3. Then

$$\Gamma = \bigsqcup_{1 \le i \le p} \{(A_i, x), (A_i, x) \in \Gamma\},$$

and necessarily there exists $i_0 \in \{1, \dots, p\}$ such the set $\{(A_{i_0}, x), (A_{i_0}, x) \in \Gamma\}$ is infinite. So for $x' \neq x$, $(A_{i_0}, x)(A_{i_0}, x')^{-1} = (I, x - x') \neq (I, 0)$ is an element of Γ of infinite order. This is absurd. If now Γ contains (I, y), $y \neq 0$, then $\{(I, my), m \in \mathbb{Z}\} \subset \Gamma$. So Γ cannot be compact. □

The compact subgroups of G are of paramount importance along this work, and we quote the following result (cf. [82]).

Fact 1.2.10 ([82, Lemma 14.1.1]). *Let $G = K \ltimes V$ be the semidirect product Lie group, where V is a finite-dimensional vector space and K is compact. Then for any compact subgroup $U \subseteq G$, there exists $v \in V$ with $vUv^{-1} \subseteq K$.*

Corollary 1.2.11. *Let $G = K \ltimes V$ be the semidirect product Lie group, where V is a finite-dimensional vector space and K is compact. Then for any subgroup $U \subseteq G$ such that $\pi_2(U)$ is compact, there exists $v \in V$ with $vUv^{-1} \subseteq K$. Here, π_2 designates the projection from $K \ltimes V$ into V.*

Proof. Consider a sequence $\{(B_p, z_p)\}_{p \in \mathbb{N}}$ inside U, which converges to some $(B, z) \in \overline{U}$, where $\overline{U}$ is the closure of U. Since $\pi_2(U)$ is compact, then obviously the sequence $\{z_p\}_{p \in \mathbb{N}}$ converges to $z \in \pi_2(U)$. Hence, $\overline{U}$ is compact. By Fact 1.2.10, there exists $v \in V$ such that $vUv^{-1} \subseteq v\overline{U}v^{-1} \subseteq K$. □

Let us announce next the following result, which will be of use.

Lemma 1.2.12 (cf. [115, Lemma 4 and Lemma 5]). *Let $\gamma_1 = (A, x)$ and $\gamma_2 = (B, y)$ be in G such that γ_1 and γ_2 generate a discrete subgroup of G. If $\|A - I\| < \frac{1}{2}$ and $\|B - I\| < \frac{1}{2}$ then γ_1 and γ_2 commute.*

We now prove the following.

Proposition 1.2.13. *Let Γ be a discrete subgroup of G. Then Γ is finite if and only if any element of Γ is of finite order.*

Proof. When Γ is finite, the statement is clear. Thanks to Lemma 1.2.9, it is sufficient to prove that $\mathrm{pr}_1(\Gamma)$ is finite. Assume then that $\mathrm{pr}_1(\Gamma)$ is infinite (so not discrete), there exists an infinite sequence $\{\gamma_p = (A_p, x_p)\}_{\{p \in \mathbb{N}\}}$ in Γ, for which the sequence $\{A_p\}_{\{p \in \mathbb{N}\}}$ converges to I. As above, for a given $p_0 \in \mathbb{N}$ we have $\|A_p - I\| < \frac{1}{2}$ for $p \ge p_0$. Then for any $p > q \ge p_0$, γ_p and γ_q commute thanks to Lemma 1.2.12. Now, apply Proposition 1.2.7

to the commuting family $\{\gamma_p\}_{\{p \geq p_0\}}$ to get that γ_p is conjugate to

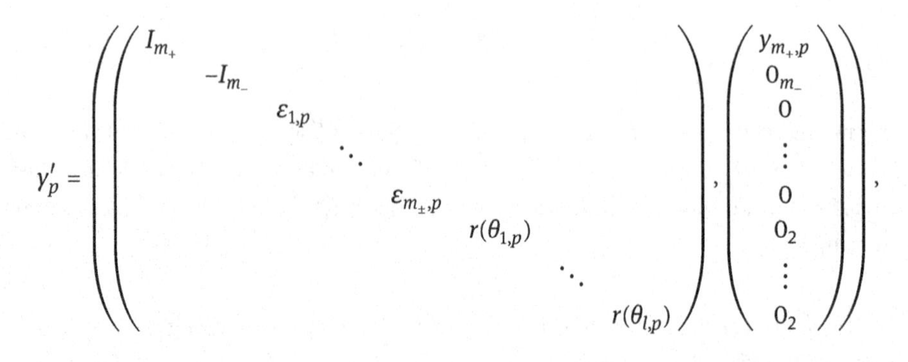

where $r(\theta_{i,p}) \in SO_2(\mathbb{R})$ for $i \in \{1, \ldots, l\}$, and $\varepsilon_{i,p} \in \{-1, 1\}$ for $i \in \{1, \ldots, m_\pm\}$. Thanks to Lemma 1.2.5, we get $y_{m_+,p} = 0$ and then an infinite sequence inside the compact $O_n(\mathbb{R}) \times \{0\}$. We end up with a convergent infinite extract sequence inside Γ. This is absurd and then $\mathrm{pr}_1(\Gamma)$ is finite. □

The following is thus immediate.

Corollary 1.2.14. *A discrete subgroup Γ of G is infinite if and only if it contains an element of infinite order.*

The following result, known as Bieberbach's theorem (cf. [43] and [44]), for which a simpler proof can be found in [115, Theorem 1] will be next utilized.

Theorem 1.2.15. *Any discrete subgroup Γ of G contains an Abelian normal subgroup of finite index in Γ.*

We next record the following.

Theorem 1.2.16 (cf. [84, main theorem]). *A closed solvable subgroup of a locally compact, almost connected group G' (the quotient G'/G'_0 is compact, where G'_0 designates the of G') is compactly generated.*

Now the following is immediate.

Corollary 1.2.17. *A discrete subgroup of an Euclidean motion group is finitely generated.*

Proof. Let Γ be a discrete subgroup of an Euclidean motion group. By Theorem 1.2.15, Γ admits an Abelian normal subgroup Γ_a of finite index q say in Γ. Let then $\Gamma/\Gamma_a = \{\overline{e}, \overline{\delta_1}, \ldots, \overline{\delta_{q-1}}\}$, where e designates the identity of G. By Theorem 1.2.16, the Abelian discrete subgroup Γ_a is finitely generated since G is almost connected. Let then $\{\gamma_1, \ldots, \gamma_k\}$ be a family of generators of Γ_a for some positive integer k, then $\{\gamma_1, \ldots, \gamma_k, \delta_1, \ldots, \delta_{q-1}\}$ generates Γ. □

As above, our strategy is to get a full description of Abelian discrete subgroups of G up to a conjugation. Let Γ be a discrete subgroup of G and Γ_∞ a subgroup of Γ fulfilling Theorem 1.2.15.

Lemma 1.2.18. *There exists $t \in \mathbb{R}^n$ such that for any $(A,x) \in \Gamma_\infty$, we have*

$$(I,t)(A,x)(I,-t) = (A, P_A(x)).$$

In addition, for any (A,x) and (B,y) in Γ_∞, we get

$$(A,P_A(x))(B,P_B(y)) = (AB, P_A(x) + P_B(y)). \tag{1.25}$$

Proof. Since Γ_∞ is Abelian, then it follows from Lemma 1.2.1 that there exist some integers m_+, m_-, $m_\pm$ and l satisfying $m_+ + m_- + m_\pm + 2l = n$ and $g = (S,v) \in I(n)$ such that, for any $y = (A,x) \in \Gamma_\infty$,

$$gyg^{-1} = \left(\begin{pmatrix} I_{m_+} & & & & & \\ & -I_{m_-} & & & & \\ & & d(-1,1) & & & \\ & & & r(\theta_1) & & \\ & & & & \ddots & \\ & & & & & r(\theta_l) \end{pmatrix}, \begin{pmatrix} x_{m_+} \\ 0_{m_-} \\ 0_{m_\pm} \\ 0_2 \\ \vdots \\ 0_2 \end{pmatrix} \right),$$

is as in equation (1.22). Here, $x_{m_+} \in \mathbb{R}^{m_+}$,

$$d(-1,1) = \begin{pmatrix} \varepsilon_1 & & \\ & \ddots & \\ & & \varepsilon_{m_\pm} \end{pmatrix} \in \mathscr{M}_{m_\pm}(\mathbb{R}), \quad \varepsilon_k \in \{-1,1\},$$

for $k \in \{1,\dots,m_\pm\}$,

$$r(\theta_i) = \begin{pmatrix} \cos\theta_i & -\sin\theta_i \\ \sin\theta_i & \cos\theta_i \end{pmatrix},$$

with $\theta_i \in \mathbb{R}$, for $i \in \{1,\dots,l\}$ and 0_2, 0_{m_-} and $0_{m_\pm}$ are the zeros of $\mathbb{R}^2$, $\mathbb{R}^{m_-}$ and $\mathbb{R}^{m_\pm}$, respectively. Set $x' = {}^t(x_{m_+}, 0_{m_-}, 0_{m_\pm}, 0_2, \dots, 0_2)$, then clearly $x' \in \ker SAS^{-1} - I$. Hence, $S^{-1}x' \in \ker A - I$. As $x' = S(x + (I-A)S^{-1}v) = S \cdot P_A(x)$, it is enough to take $t = S^{-1}v$.

Let now (A,x) and (B,y) be two elements of Γ_∞ and set $(A',x') = g(A,x)g^{-1}$ and $(B',y') = g(B,y)g^{-1}$. Then equation (1.22) gives $(A',x')(B',y') = (A'B', x'+y')$. Hence, $A'y' = y'$, which gives in turn $AP_B(y) = P_B(y)$. This completes the proof. □

One important class of discrete subgroups was studied by Bieberbach (cf. [43] and [44]) and is defined as follows.

Definition 1.2.19. A discrete subgroup Γ of a semidirect product $K \ltimes \mathbb{R}^n$ is said to be crystallographic if it contains an Abelian normal subgroup of finite index, which is generated by n free translations.

We now quote the following due to Bieberbach (cf. [43] and [44]).

Lemma 1.2.20 (Bieberbach). *Let Γ_1 and Γ_2 be two isomorphic crystallographic subgroups of G. For any isomorphism φ from Γ_1 into Γ_2, there exists $g \in GL_n(\mathbb{R}) \ltimes \mathbb{R}^n$ such that $\varphi(\gamma) = g\gamma g^{-1}$, for any $\gamma \in \Gamma_1$.*

We now record the following result proved in [121].

Theorem 1.2.21. *If W is a subgroup of a free group F, then W is a free group. Moreover, if W has finite index m in F, the rank of W is precisely $nm + 1 - m$ where n is the rank of F.*

As a consequence, we get the following.

Lemma 1.2.22. *Any discrete free subgroup of an Euclidean motion group is cyclic (and, therefore, Abelian).*

Proof. Let Γ be a discrete free subgroup of an Euclidean motion group G. Due to Theorem 1.2.15, Γ admits an Abelian normal subgroup Γ_a of finite index q say in Γ. By Corollary 1.2.17, Γ and Γ_a are finitely generated, and let p denote the rank of Γ and l that of Γ_a. Theorem 1.2.21 gives that Γ_a is free and, therefore, Γ_a is trivial or isomorphic to $\mathbb{Z}$. Since Γ is torsion-free, then Γ_a is isomorphic to $\mathbb{Z}$, and hence, $l = 1$, which gives in turn that $1 = 1 + q(p-1)$, and conclusively $q(p-1) = 0$. This allows to conclude that $p = 1$ and then Γ is cyclic. □

One important issue to study the deformation of discontinuous groups of Euclidean motion groups is to have an idea about how the generators of a discrete subgroup look like. First, we prove the following.

Proposition 1.2.23. *Let $\{y_1, \ldots, y_k\}$ be a family of vectors of rank k generating a subspace E of $\mathbb{R}^n$. Let also Σ be a subgroup of $O_n(\mathbb{R})$ leaving invariant E. There exists $S \in O_n(\mathbb{R})$ such that:*

(i)

$$S^{-1}y_i = \begin{pmatrix} y_{k,i} \\ 0_{n-k} \end{pmatrix},$$

for some $y_{k,i} \in \mathbb{R}^k$ and where 0_{n-k} is the zero vector of $\mathbb{R}^{n-k}$.

(ii) *For any $A \in \Sigma$,*

$$S^{-1}AS = \begin{pmatrix} C_{k,A} & 0 \\ 0 & C'_{n-k,A} \end{pmatrix}$$

for some $C_{k,A} \in O_k(\mathbb{R})$ and $C'_{n-k,A} \in O_{n-k}(\mathbb{R})$.

Proof. Consider the transition matrix S from the standard basis of $\mathbb{R}^n$ to $B_E \cup B_{E^\perp}$ where B_E and $B_{E^\perp}$ designate two orthonormal bases of E and $E^\perp$, respectively. Immediately, $S \in O_n(\mathbb{R})$ and it satisfies (i) and (ii). □

The following upshot describes to some extent the form of generators of a discrete subgroup.

Theorem 1.2.24. *Let $G = I(n)$ be the Euclidean motion group. For any infinite discrete subgroup Γ of G, there exists $g \in G$ such that the subgroup $\Gamma^g := g^{-1}\Gamma g$ admits generators $\{\gamma_1, \ldots, \gamma_k, \gamma_{k+1}, \ldots, \gamma_{k_0}, \delta_1, \ldots, \delta_{q-1}\}$ meeting the following: For $1 \le i \le k_0$,*

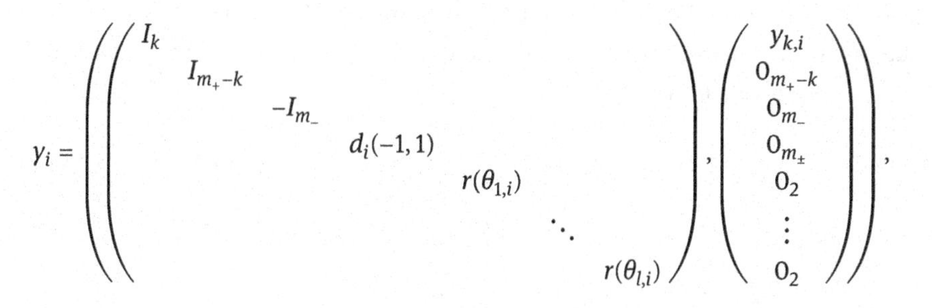

where $y_{k,i} \in \mathbb{R}^k$ satisfying $\{y_{k,1}, \ldots, y_{k,k}\}$ is of rank k, for $k+1 \le i \le k_0$, $y_{k,i} = 0_k$ and where for $1 \le i \le k_0$, $d_i(-1,1)$ is a diagonal matrix with coefficients ± 1, of $O_{m_\pm}(\mathbb{R})$ and for any $1 \le s \le l$, $\theta_{s,i} \in \mathbb{R}$. Furthermore, the family $\{\gamma_i : 1 \le i \le k_0\}$ generates an Abelian normal subgroup Γ^g_a of Γ^g such that $\Gamma^g/\Gamma^g_a = \{\overline{e}, \overline{\delta}_1, \ldots, \overline{\delta}_{q-1}\}$ where for $1 \le i \le q-1$,

$$\delta_i = \left(\begin{pmatrix} C_{k,i} & & & & \\ & C_{m_+-k,i} & & & \\ & & C_{m_-,i} & & \\ & & & C_{m_\pm,i} & \\ & & & & P_i \end{pmatrix}, \begin{pmatrix} z_{k,i} \\ 0_{m_+-k} \\ 0_{m_-} \\ 0_{m_\pm} \\ 0_{2l} \end{pmatrix} \right).$$

Here, $C_{k,i}$, $C_{m_+-k,i}$, $C_{m_-,i}$, $C_{m_\pm,i}$ and P_i belong to $O_k(\mathbb{R})$, $O_{m_+-k}(\mathbb{R})$, $O_{m_-}(\mathbb{R})$, $O_{m_\pm}(\mathbb{R})$ and $O_{2l}(\mathbb{R})$, respectively. The integers m_+, m_-, $m_\pm$ and l satisfy $m_+ + m_- + m_\pm + 2l = n$.

Proof. First, remark that Γ admits a normal Abelian subgroup Γ_a of the finite index q, say, in Γ due to Theorem 1.2.15. Let $\{a_1, \ldots, a_{k_0}\}$ be some generators of Γ_a, then the family $\{a_1, \ldots, a_{k_0}, b_1, \ldots, b_{q-1}\}$ generates Γ, where $\{\overline{e}, \overline{b_1}, \ldots, \overline{b_{q-1}}\}$ is a complete representative of the finite set Γ/Γ_a. Thanks to Lemma 1.2.1, there exists $g \in G$ such that the subgroup $\Gamma^g_a := g^{-1}\Gamma_a g$ is now being generated by the family $\{\gamma_i = g^{-1}a_i g\}_{1 \le i \le k_0}$ as in formula (1.22). If $y_{m_+,i} = 0$ for any $i \in \{1, \ldots, k_0\}$ or $m_+ = 0$, then thanks to Lemma 1.2.5 all the γ_i's are of finite orders and, therefore, due to Lemma 1.2.9, Γ is finite. Otherwise, let $k \le k_0$ be the rank of the family $\{y_{m_+,i}\}_{1 \le i \le k_0}$ and let $\mathscr{L}$ be the subgroup of $\mathbb{R}^{m_+}$ generated by $\{y_{m_+,i}\}_{1 \le i \le k_0}$. Suppose that there exists an infinite sequence $\{y_{m_+}(p)\}_{p \in \mathbb{N}} \subset \mathscr{L}$, which converges to some $y_{m_+} \in \mathbb{R}^{m_+}$. Therefore, by writing for any integer p, $y_{m_+}(p) = \sum_{1\le i\le k_0} n_i(p) y_{m_+,i}$, we define a sequence $\{\gamma(p)\}_{p\in\mathbb{N}} \subset \Gamma$, which is of the general form

$$\gamma(p) = \left(\begin{pmatrix} I_{m_+} & \\ & A(p) \end{pmatrix}, \begin{pmatrix} y_{m_+}(p) \\ 0_{n-m_+} \end{pmatrix} \right).$$

For p large enough, $\gamma(p) \in O_n(\mathbb{R}) \times B(y_{m_+}, 1)$, which is absurd because Γ is discrete. Hence, $\mathscr{L}$ is a discrete subgroup of $\mathbb{R}^{m_+}$. Assume without loss of generality that the first $y_{m_+,i}$ $(1 \le i \le k)$ generate $\mathscr{L}$. Take then $\Gamma_\infty = \langle \gamma_1, \dots, \gamma_k \rangle$, which is the subgroup of Γ_a^g generated by the γ_i's $(1 \le i \le k)$. We now can and do assume that there exists $k' > k$ such that $\gamma_{k'} \notin \Gamma_\infty$. Otherwise, we can take $k_0 = k$. There exist therefore some integers $n_1(k'), \dots, n_k(k')$ such that

$$y_{m_+,k'} = \sum_{i=1}^{k} n_i(k') y_{m_+,i},$$

and then a routine computation gives:

$$\begin{aligned} \gamma'_{k'} &:= \gamma_{k'}^{-1} \prod_{i=1}^{k} \gamma_i^{n_i(k')} \\ &= \left(\begin{pmatrix} I_{m_+} & & & & & \\ & \varepsilon'_{k'} I_{m_-} & & & & \\ & & d'_{k'}(-1,1) & & & \\ & & & r(\theta'_{1,k'}) & & \\ & & & & \ddots & \\ & & & & & r(\theta'_{l,k'}) \end{pmatrix}, \begin{pmatrix} 0_{m_+} \\ 0_{m_-} \\ 0_{m_\pm} \\ 0_2 \\ \vdots \\ 0_2 \end{pmatrix} \right), \end{aligned}$$

where $\varepsilon'_{k'} \in \{-1, 1\}$ and $d'_{k'}(-1,1)$ is a diagonal matrix of diagonal values ± 1, and finally $\gamma'_{k'}$ is of finite order. This entails that Γ_a^g, being Abelian, is generated by $\{\gamma_1, \dots, \gamma_k\} \cup \{\gamma'_{k'}\}_{k+1\le k'\le k_0}$. The family $\{\gamma'_{k'}\}_{k+1\le k'\le k_0}$ generates a discrete subgroup Γ_0 such that all of its elements are of finite order, and finite thanks to Lemma 1.2.9. For the sake of simplicity, let us write $\{\gamma'_{k'}\}_{k+1\le k'\le k_0} = \{\gamma_{k+1}, \dots, \gamma_{k_0}\}$. That is, the subgroup Γ^g being generated by $\{\gamma_1, \dots, \gamma_k, \gamma_{k+1}, \dots, \gamma_{k_0}, \delta_1, \dots, \delta_{q-1}\}$ where $\gamma_i = g^{-1} a_i g$ $(1 \le i \le k_0)$ and $\delta_j = g^{-1} b_j g$ $(1 \le j \le q-1)$ is such that the γ_i's $(1 \le i \le k_0)$ are of the requested form.

On the other hand, $\Gamma^g / \Gamma_a^g = \{\bar{e}, \bar{\delta}_1, \dots, \bar{\delta}_{q-1}\}$. We examine now the elements $\delta_i =: (S_i, z_i)$, for any $i \in \{1, \dots, q-1\}$. As we have $\delta_i^{-1} \Gamma_a^g \delta_i = \Gamma_a^g$, then for any $\gamma = (A_\gamma, x_\gamma) \in \Gamma_a^g$, we get

$$\delta_i^{-1} \gamma \delta_i = (S_i^{-1} A_\gamma S_i, S_i^{-1}[x_\gamma + (A_\gamma - I) z_i]) = (A_{\gamma'}, x_{\gamma'}) \in \Gamma_a^g \tag{1.26}$$

and implies:

$$S_i^{-1}\begin{pmatrix} I_{m_+} & & & \\ & \pm I_{m_-} & & \\ & & d_\gamma(-1,1) & \\ & & & R_\gamma \end{pmatrix} S_i = \begin{pmatrix} I_{m_+} & & & \\ & \pm I_{m_-} & & \\ & & d_{\gamma'}(-1,1) & \\ & & & R_{\gamma'} \end{pmatrix}.$$

Thanks to the arguments of Proposition 1.2.1, this equation gives that for the decomposition (1.19) of $\mathbb{R}^n$ into direct sums is such that $E(1)$, $E(-1)$, $H(-1,1)$ and $F_1 \oplus^\perp \cdots \oplus^\perp F_l$ are stable vector subspaces by all the S_i's. This gives the following writing:

$$(S_i, z_i) = \left(\begin{pmatrix} C_{m_+,i} & & & \\ & C_{m_-,i} & & \\ & & C_{m_\pm,i} & \\ & & & P_i \end{pmatrix}, \begin{pmatrix} z_{m_+,i} \\ z_{m_-,i} \\ z_{m_\pm,i} \\ z_{2l,i} \end{pmatrix}\right),$$

where $C_{m_+,i} \in O_{m_+}(\mathbb{R})$, $C_{m_-,i} \in O_{m_-}(\mathbb{R})$, $C_{m_\pm,i} \in O_{m_\pm}(\mathbb{R})$ and $P_i \in O_{2l}(\mathbb{R})$. Hence, the matrix form of (1.26) reads:

$$\left(\begin{pmatrix} I_{m_+} & & & \\ & \pm I_{m_-} & & \\ & & C_{m_\pm,i}^{-1} d_\gamma(-1,1) C_{m_\pm,i} & \\ & & & P_i^{-1} R_\gamma P_i \end{pmatrix}, \begin{pmatrix} C_{m_+,i}^{-1} y_{m_+,\gamma} \\ (-2C_{m_-,i}^{-1}) z_{m_-,i} \\ (C_{m_\pm,i}^{-1}(d_\gamma - I_{m_\pm})) z_{m_\pm,i} \\ (P_i^{-1}(R_\gamma - I_{2l})) z_{2l,i} \end{pmatrix}\right), \tag{1.27}$$

which is an element of $g^{-1}\Gamma_a g$. For $\gamma \in \Gamma_a$, the following hold:

$$z_{m_-,i} = 0_{m_-}, \quad (d_\gamma - I_{m_\pm}) z_{m_\pm,i} = 0_{m_\pm} \quad \text{and} \quad (R_\gamma - I_{2l}) z_{2l,i} = 0_{2l}.$$

Let

$$z_{m_\pm,i} = \begin{pmatrix} \tau_{1,i} \\ \vdots \\ \tau_{m_\pm,i} \end{pmatrix} \quad \text{and} \quad z_{2l,i} = \begin{pmatrix} u_{1,i} \\ \vdots \\ u_{l,i} \end{pmatrix},$$

where $\tau_{s,i} \in \mathbb{R}$ for $s \in \{1,\dots,m_\pm\}$ and $u_{s,i} \in \mathbb{R}^2$ for $s \in \{1,\dots,l\}$. On the one hand, we get $(\varepsilon_{\gamma,s} - 1)\tau_{s,i} = 0$, for $s \in \{1,\dots,m_\pm\}$ for a given γ such as $\varepsilon_{\gamma,s} = -1$. This implies that $\tau_{s,i} = 0$ and so $z_{m_\pm,i} = 0_{m_\pm}$. On the other hand, $(r(\theta_{\gamma,s}) - I_2)u_{s,i} = 0$ for any $s \in \{1,\dots,l\}$ for a given γ such that $r(\theta_{\gamma,s}) \neq I_2$. This implies that $u_{s,i} = 0$ and so $z_{2l,i} = 0_{2l}$. Hence,

$$\delta_i = \left(\begin{pmatrix} C_{m_+,i} & & & \\ & C_{m_-,i} & & \\ & & C_{m_\pm,i} & \\ & & & P_i \end{pmatrix}, \begin{pmatrix} z_{m_+,i} \\ 0_{m_-} \\ 0_{m_\pm} \\ 0_{2l} \end{pmatrix}\right).$$

Furthermore, equation (1.27) entails that the subspace of $\mathbb{R}^{m_+}$ generated by $\{y_{m_+,i}\}_{i\in\{1,\dots,k\}}$ is stable by the subgroup Σ_{m_+} of $O_{m_+}(\mathbb{R})$ generated by $\{C_{m_+,j}\}_{j\in\{1,\dots,q-1\}}$. Applying Proposition 1.2.23, there exists $S_{m_+} \in O_{m_+}(\mathbb{R})$ such that for any $j \in \{1,\dots,q-1\}$,

$$(S_{m_+}^{-1}, 0_{m_+})(C_{m_+,j}, z_{m_+,j})(S_{m_+}, 0_{m_+}) = \left(\begin{pmatrix} C_{k,j} & \\ & C_{m_+-k,j} \end{pmatrix}, \begin{pmatrix} z_{k,j} \\ z_{m_+-k,j} \end{pmatrix}\right),$$

and for any $i \in \{1,\dots,k\}$,

$$(S_{m_+}^{-1}, 0_{m_+})(I_{m_+}, y_{m_+,i})(S_{m_+}, 0_{m_+}) = \left(\begin{pmatrix} I_k & \\ & I_{m_+-k} \end{pmatrix}, \begin{pmatrix} y_{k,i} \\ 0_{m_+-k} \end{pmatrix}\right).$$

This gives in turn that the subgroup Λ_{m_+-k} of $I(m_+-k)$ generated by the $(C_{m_+-k,j}, z_{m_+-k,j})$'s is finite. Then Fact 1.2.10 asserts that there exists $t_{m_+-k} \in \mathbb{R}^{m_+-k}$ such that

$$(I_{m_+-k}, -t_{m_+-k})(C_{m_+-k,j}, z_{m_+-k,j})(I_{m_+-k}, t_{m_+-k}) = (C_{m_+-k,j}, 0_{m_+-k}),$$

which completes the proof. □

1.2.4 Closed subgroups of $I(n)$

Connected semisimple Lie subgroups of $I(n)$

Let $\mathfrak{a}$ be a Lie algebra. A symmetric bilinear form $b : \mathfrak{a} \times \mathfrak{a} \longrightarrow \mathbb{R}$ is said to be invariant if $b([x,y],z) = b(x,[y,z])$, for any $x,y,z \in \mathfrak{a}$. Further, $\mathfrak{a}$ is said to be compact if there exists a positive definite, invariant and symmetric bilinear form on $\mathfrak{a}$. Recall now the following fact, which will be of use later.

Fact 1.2.25 ([82, Theorem 12.1.17]). *Any connected semisimple Lie group with compact Lie algebra is compact and of finite center.*

Let N be a connected, simply connected nilpotent Lie group, C a compact group of automorphisms of N, and $H \subset N_C := C \ltimes N$ a connected Lie subgroup of N_C. Let $\mathfrak{g} := \mathfrak{c} \oplus \mathfrak{n}$ be the Lie algebra of N_C. Thanks to Levi decomposition (see [132]), any maximal semisimple subgroup S of H (called Levi factor), we have $H = S \cdot R$, where R is the solvable radical of H. Besides, S is unique up to a conjugation and $S \cap R$ is discrete. The following characterizes connected semisimple Lie subgroups of G.

Proposition 1.2.26. *Any connected semisimple Lie subgroup S of N_C is compact.*

Proof. Let S be a connected semisimple Lie subgroup of N_C. We denote by $\mathfrak{s}$ its corresponding Lie algebra. Set $\varphi : \mathfrak{s} \longrightarrow \mathfrak{c}$, $(X,u) \longmapsto X$. It is clear that φ is a homomorphism of Lie algebras. Further, $\ker(\varphi) = \{(0,u) \mid (0,u) \in \mathfrak{s}\}$ is a nilpotent ideal of $\mathfrak{s}$. This means that φ is injective, $\mathfrak{s}$ and $\varphi(\mathfrak{s})$ are isomorphic Lie algebras, and in turn, $\mathfrak{s}$ is a semisimple compact Lie algebra as $\mathfrak{c}$ is a compact Lie algebra. By applying Fact 1.2.25, the proof is complete. □

As a direct consequence, we get the following.

Corollary 1.2.27. *Any closed connected subgroup of N_C contains a normal, closed, connected solvable subgroup cocompactly.*

Connected closed solvable subgroups of $I(n)$

For any matrix A of $O_n(\mathbb{R})$, let P_A denote the orthogonal projection onto $\ker(A - I)$. By Lemma 1.2.2, P_A is a polynomial of A and that $P_A(I - A) = 0$. For an arbitrary subgroup H of $I(n)$, set $V_H := \{t \mid (I, t) \in H\} \subset \mathbb{R}^n$, $E_H := \mathbb{R}\text{-span}\{P_A(x) \mid (A, x) \in H\} \subset \mathbb{R}^n$ and V'_H, the orthogonal complement subspace of $\widetilde{V}_H := \mathbb{R}\text{-span}(V_H)$ in E_H. For any $g \in I(n)$, put $H^g := g^{-1}Hg$. We have the following.

Lemma 1.2.28. *E_H is stable by $\mathrm{pr}_1(H)$.*

Proof. Consider $\{(A_i, x_i)\}_{1\le i\le k}$ a family of H such that $\{P_{A_i}(x_i)\}_{1\le i\le k}$ spans the subspace E_H. For any $(B, t) \in H$, we have

$$(B, t)^{-1}(A_i, x_i)(B, t) = (B^{-1}A_iB, B^{-1}[x_i + (A_i - I)t])$$

and, therefore,

$$P_{B^{-1}A_iB}(B^{-1}[x_i + (A_i - I)t]) = B^{-1}(P_{A_i}[x_i + (A_i - I)t]) = B^{-1}P_{A_i}(x_i),$$

for any $i \in \{1, \dots, k\}$, as P_{A_i} is a polynomial depending upon A_i. This shows that $B^{-1}P_{A_i}(x_i) \in E_H$ and in turn $\mathbb{R}\text{-span}\{B^{-1}P_{A_i}(x_i)\}_{1\le i\le k} \subseteq E_H$, which gives that $B^{-1}E_H = E_H$. □

Throughout this subsection, R denotes a closed connected solvable subgroup of $I(n)$. We record the following.

Fact 1.2.29 ([82, Theorem 14.4.1]). *Let R be a connected solvable Lie group and $T \subseteq R$ be a maximal torus. Then T is maximal compact in R, and there exists a closed submanifold $M \cong \mathbb{R}^q \subseteq R$ (for some $q \in \mathbb{N}$) such that the map $M \times T \to R$, $(m, t) \mapsto mt$ is a diffeomorphism.*

As a direct consequence, we get the following.

Corollary 1.2.30. *$\mathrm{pr}_1(R)$ is a connected Abelian subgroup of $O_n(\mathbb{R})$.*

The following will be used later.

Lemma 1.2.31. *V_R is a closed subgroup of E_R stable by $\mathrm{pr}_1(R)$.*

Proof. Let $\{t_p\}_{p\in\mathbb{N}}$ be a converging sequence of V_R, and denote $t \in \mathbb{R}^n$ its limit. Obviously, $\{(I, t_p)\}_{p\in\mathbb{N}}$ is a converging sequence of R. Since R is closed, then $(I, t) \in R$ and, therefore, $t \in V_R$. Let now $(A, x) \in R$ and $(I, t) \in R$, then

$$(A, x)(I, t)(A, x)^{-1} = (I, At).$$

This shows that $At \in V_R$. □

Lemma 1.2.32. *Let H be a subgroup of $I(n)$, $g \in I(n)$ and $S = \mathrm{pr}_1(g)$. Then $E_{H^g} = S^{-1}\cdot E_H$.*

Proof. For $g = (S, v) \in I(n)$ and $(A, x) \in H$, we have

$$g^{-1}(A, x)g = (S^{-1}AS, S^{-1}[x + (A - I)v]).$$

From $\ker(S^{-1}AS - I) = S^{-1}\ker(A - I)$, we get $P_{S^{-1}AS}(S^{-1}[x + (A - I)v]) = S^{-1}\cdot P_A(x)$. □

Lemma 1.2.33. *For any $x \in \widetilde{V_R}$, there exist $r \in R$ and $y \in \widetilde{V_R}^{\perp}$ such that $r = (A, x + y)$.*

Proof. As V_R is a closed subgroup of $\mathbb{R}^n$, it writes $V_R = F_R \oplus D_R$, where F_R is a linear subspace of $\mathbb{R}^n$ and D_R is a discrete subgroup of $\mathbb{R}^n$. We first prove that F_R is stable by $\mathrm{pr}_1(R)$. For any $A \in \mathrm{pr}_1(R)$ and $x \in A(F_R)$, there exist $x_F \in F_R$ and $x_D \in D_R$ such that $x = x_F + x_D$. For $\alpha \in \mathbb{R}$, $\alpha x = \alpha x_F + \alpha x_D \in A(F_R)$. This shows that $\alpha x_D \in D$, and hence, $x_D = 0$. We deduce that $A(F) \subset F$. As F and $A(F)$ have the same dimension, we get $A(F_R) = F_R$.

Let $\{x_1, \dots, x_q\}$ be a family of generators of D_R, where q designates the rank of D_R. Consider $\widetilde{V}_R = F_R \oplus F'_R$, where F'_R stands for the orthogonal complement of F_R in $\widetilde{V}_R$. For any $i \in \{1, \dots, q\}$, $x_i = v_i + u_i$ where $v_i \in F_R$ and $u_i \in F'_R$. Let D'_R be the subgroup generated by $\{u_1, \dots, u_q\}$. Note that the fact that $V_R = F_R \oplus D_R = F_R \oplus^{\perp} D'_R$ entails

$$q = \dim(\mathbb{R}\text{-span}\, D) = \dim(\mathbb{R}\text{-span}\, D') = \dim(F'),$$

as $\dim(\widetilde{V_R}) = rk(D) + \dim(F)$. Therefore, the rank of D_R and D'_R coincide with q and then D'_R is a lattice of F'_R. As F'_R and V_R are stable by $\mathrm{pr}_1(R)$, it is also the case for D'_R. Write now $\mathbb{R}^n = F'_R \oplus F_R \oplus V_R^{\perp}$, as a direct sum, we easily check that there exists $g := (S, 0) \in G$ such that any $r \in R^g$ writes

$$r = \left(\begin{pmatrix} A'(r) & & \\ & A_F(r) & \\ & & B(r) \end{pmatrix}, \begin{pmatrix} x'(r) \\ x_F(r) \\ y(r) \end{pmatrix} \right). \tag{1.28}$$

This already gives that the subgroup Λ generated by $\{A'(r), r \in R^g\}$ is a connected Abelian subgroup of $O_q(\mathbb{R})$ and leaves the lattice D'_R invariant. This gives in addition that any $A'(r)$, $r \in R^g$ is of finite order. Indeed, the set $\{A'(r)^p.u_i\}_{p\in\mathbb{N}}$ is discrete and lie inside the sphere $S(0, \|u_i\|)$ and then finite. There exists therefore some $p_i(r)$ such that

$A'(r)^{p_i(r)}.u_i = u_i$, and $A'(r)^{(p(r))} = I_q$ for some integer $p(r)$. We conclusively deduce that Λ is trivial, and then equation (1.28) reads

$$r = \left(\begin{pmatrix} I_q & & \\ & A_F(r) & \\ & & B(r) \end{pmatrix}, \begin{pmatrix} x'(r) \\ x_F(r) \\ y(r) \end{pmatrix} \right). \tag{1.29}$$

This implies that $F''_R := \{x'(r), r \in R^g\}$ is a connected subgroup of $\mathbb{R}^q$, and hence, a subspace of $\mathbb{R}^q$. As $D'_R \subset F''$, then $F''_R = \mathbb{R}^q = F'_R$. Write now $g^{-1}rg = (A, x + y)$ with

$$x = S \begin{pmatrix} x'(r) \\ x_F(r) \\ 0_{n-q-\eta_1} \end{pmatrix} \quad \text{and} \quad y = S \begin{pmatrix} 0_q \\ 0_{\eta_1} \\ y(r) \end{pmatrix},$$

where η_1 designates the dimension of F_R, and the proof is complete. □

The following result shows how an element in a closed connected subgroup of $I(n)$ writes up to a conjugation with respect to the orthogonal sum $F'_R \oplus F_R \oplus V'_R \oplus E^{\perp}_R$.

Proposition 1.2.34. *Set $q = \dim(F'_R)$, $\eta_1 := \dim(F_R)$, $\eta_2 := \dim(V'_R)$ and $\eta := \eta_1 + \eta_2 + q$. There exists $g \in I(n)$, such that for any $r \in R^g$,*

$$r = \left(\begin{pmatrix} I_q & & & \\ & A_1(r) & & \\ & & I_{\eta_2} & \\ & & & A_3(r) \end{pmatrix}, \begin{pmatrix} x'(r) \\ x_1(r) \\ u(r) \\ 0_{n-\eta} \end{pmatrix} \right),$$

where $A_1(r) \in M_{\eta_1}(\mathbb{R})$, $A_3(r) \in M_{n-\eta}(\mathbb{R})$, $x'(r) \in F'_R$, $x_1(r) \in F_R$ and $u(r) \in V'_{R^g}$.

Proof. Using Lemma 1.2.1 and Lemma 1.2.33, consider the decomposition

$$\mathbb{R}^n = F'_R \oplus F_R \oplus V'_R \oplus E^{\perp}_R$$

and let S be a transition matrix from the standard basis of $\mathbb{R}^n$ to an orthonormal basis adapted to the above decomposition. For $h := (S, 0)$, any $r \in R^h$ reads

$$r = \left(\begin{pmatrix} I_q & & & \\ & A_1(r) & & \\ & & A_2(r) & \\ & & & A_3(r) \end{pmatrix}, \begin{pmatrix} x'(r) \\ x_1(r) \\ x_2(r) \\ x_3(r) \end{pmatrix} \right).$$

Here, $A_1(r) \in M_{\eta_1}(\mathbb{R})$, $A_2(r) \in M_{\eta_2}(\mathbb{R})$, $A_3(r) \in M_{n-\eta}(\mathbb{R})$ and $x'(r) \in F'_R$, $x_1(r) \in F_R$, $x_2(r) \in V'_{R^h}$, $x_3(r) \in E^{\perp}_{R^h}$.

Clearly,

$$\begin{pmatrix} 0_q \\ x_1(r) \\ 0_{\eta_2} \\ 0_{n-\eta} \end{pmatrix} \in F_R \subset V_R,$$

and then by Lemma 1.2.33,

$$\begin{pmatrix} 0_q \\ -A_1(r)^{-1}x_1(r) \\ 0_{\eta_2} \\ 0_{n-\eta} \end{pmatrix} \in V_{R^h}.$$

We then get

$$\begin{aligned} a(r) &:= r \cdot \left(I, \begin{pmatrix} 0_q \\ -A_1(r)^{-1}x_1(r) \\ 0_{\eta_2} \\ 0_{n-\eta} \end{pmatrix} \right) \\ &= \left(\begin{pmatrix} I_q & & & \\ & A_1(r) & & \\ & & A_2(r) & \\ & & & A_3(r) \end{pmatrix}, \begin{pmatrix} x'(r) \\ 0_{\eta_1} \\ x_2(r) \\ x_3(r) \end{pmatrix} \right) \in R^h. \end{aligned}$$

By Corollary 1.2.30, for any r, r' in R^h, the commutator reads

$$[a(r), a(r')] = \left(I, \begin{pmatrix} 0_q \\ 0_{\eta_1} \\ (I_{\eta_2} - A_2(r'))x_2(r) - (I_{\eta_2} - A_2(r))x_2(r') \\ (I_{n-\eta} - A_3(r'))x_3(r) - (I_{n-\eta} - A_3(r))x_3(r') \end{pmatrix} \right).$$

This means that

$$\begin{pmatrix} 0_q \\ 0_{\eta_1} \\ (I_{\eta_2} - A_2(r'))x_2(r) - (I_{\eta_2} - A_2(r))x_2(r') \\ (I_{n-\eta} - A_3(r'))x_3(r) - (I_{n-\eta} - A_3(r))x_3(r') \end{pmatrix} \in V_{R^h}$$

and, therefore,

$$\begin{aligned} (I_{\eta_2} - A_2(r'))x_2(r) - (I_{\eta_2} - A_2(r))x_2(r') &= 0_{\eta_2}, \\ (I_{n-\eta} - A_3(r'))x_3(r) - (I_{n-\eta} - A_3(r))x_3(r') &= 0_{n-\eta}. \end{aligned}$$

It follows that $(A_2(r), x_2(r))$ and $(A_2(r'), x_2(r'))$ commute and so do $(A_3(r), x_3(r))$ and $(A_3(r'), x_3(r'))$. Applying Lemma 1.2.1, there exist $\rho = (U_2, v_2) \in I(\eta_2)$ and $k \in \mathbb{N}$ such that for any $r \in R^h$, $\rho^{-1}(A_2(r), x_2(r))\rho$ is of the general form:

$$\left(\begin{pmatrix} I_k & \\ & A_2'(r) \end{pmatrix}, \begin{pmatrix} x_k'(r) \\ 0_{\eta_2-k} \end{pmatrix}\right).$$

As

$$\left\{\begin{pmatrix} x'(r) \\ x_1(r) \\ x_2(r) \\ 0_{n-\eta} \end{pmatrix}\right\}_{r\in R^h} \quad \text{and} \quad \left\{\begin{pmatrix} x'(r) \\ x_1(r) \\ 0_{\eta_2} \\ 0_{n-\eta} \end{pmatrix}\right\}_{r\in R^h}$$

span E_{R^h} and $\widetilde{V}_{R^h}$, respectively, the following set spans V'_{R^h},

$$\left\{\begin{pmatrix} 0_q \\ 0_{\eta_1} \\ x_2(r) \\ 0_{n-\eta} \end{pmatrix}\right\}_{r\in R^h}. \tag{1.30}$$

Let

$$U = \begin{pmatrix} I_{\eta_1} & & \\ & U_2 & \\ & & I_{n-\eta} \end{pmatrix}.$$

Then by (1.30), we get

$$U^{-1}V'_{R^h} = \mathbb{R}\text{-span}\left\{\begin{pmatrix} x_k'(r) \\ 0_{\eta_2-k} \end{pmatrix}\right\}_{r\in R^h},$$

and hence, we easily get $k = \eta_2$ and $A_2(r) = I_{\eta_2}$. There exists therefore $t \in \mathbb{R}^n$ such that

$$(I,t)\,a(r)\,(I,-t) = (I,t)\left(\begin{pmatrix} I_q & & & \\ & A_1(r) & & \\ & & I_{\eta_2} & \\ & & & A_3(r) \end{pmatrix}, \begin{pmatrix} 0_q \\ 0_{\eta_1} \\ x_2(r) \\ x_3(r) \end{pmatrix}\right)(I,-t)$$
$$= \left(\begin{pmatrix} I_q & & & \\ & A_1(r) & & \\ & & I_{\eta_2} & \\ & & & A_3(r) \end{pmatrix}, \begin{pmatrix} 0_q \\ 0_{\eta_1} \\ x_2(r) \\ P_{A_3(r)}(x_3(r)) \end{pmatrix}\right).$$

Let now

$$A := \begin{pmatrix} I_q & & & \\ & A_1(r) & & \\ & & I_{\eta_2} & \\ & & & A_3(r) \end{pmatrix} \quad \text{and} \quad z := \begin{pmatrix} 0_q \\ 0_{\eta_1} \\ x_2(r) \\ P_{A_3(r)}(x_3(r)) \end{pmatrix}.$$

Then clearly $z \in \ker(A - I)$, and then $P_A(z) = z$, which implies that $z \in E_{R^h}$ as in Lemma 1.2.32. Hence, $P_{A_3(r)}(x_3(r)) = 0$. □

Corollary 1.2.35. *Let R be a solvable connected subgroup of $I(n)$. Then R is compact if and only if $E_R = \{0\}$.*

Proof. If R is compact, then R is conjugate to a subgroup of $O_n(\mathbb{R})$. It follows from Lemma 1.2.32 that $E_R = \{0\}$. Conversely, $E_R = \{0\}$ entails that $V_R = \{0\}$ and $V_R' = \{0\}$. Proposition 1.2.34 completes the proof. □

We now give a description of the structure of a given, closed, connected, solvable subgroup of $I(n)$.

Proposition 1.2.36. *Any closed, connected, solvable subgroup R of $I(n)$ writes*

$$R = \mathscr{A}_R \cdot F_R,$$

where $\mathscr{A}_R$ is a closed, connected, Abelian subgroup of R and F_R is a subspace of $\mathbb{R}^n$.

Proof. Let

$$r = \left(\begin{pmatrix} I_q & & & \\ & A_1(r) & & \\ & & I_{\eta_2} & \\ & & & A_3(r) \end{pmatrix}, \begin{pmatrix} x'(r) \\ x_1(r) \\ u(r) \\ 0_{n-\eta} \end{pmatrix} \right) \in R^g.$$

Then

$$r = \left(I, \begin{pmatrix} 0_q \\ x_1(r) \\ 0_{\eta_2} \\ 0_{n-\eta} \end{pmatrix} \right) \cdot \left(\begin{pmatrix} I_q & & & \\ & A_1(r) & & \\ & & I_{\eta_2} & \\ & & & A_3(r) \end{pmatrix}, \begin{pmatrix} x'(r) \\ 0_{\eta_1} \\ u(r) \\ 0_{n-\eta} \end{pmatrix} \right),$$

as

$$\left(I, \begin{pmatrix} 0_q \\ x_1(r) \\ 0_{\eta_2} \\ 0_{n-\eta} \end{pmatrix} \right) \in R^g.$$

Let now

$$\mathscr{A}_{R^g} := \left\{ \left(\begin{pmatrix} I_q & & & \\ & A_1(r) & & \\ & & I_{\eta_2} & \\ & & & A_3(r) \end{pmatrix}, \begin{pmatrix} x'(r) \\ 0_{\eta_1} \\ u(r) \\ 0_{n-\eta} \end{pmatrix} \right) \middle| \, r \in R^g \right\},$$

which is Abelian by Corollary 1.2.30. As R^g is connected, then so is $\mathscr{A}_{R^g}$. □

Corollary 1.2.37. *There exists $\tau := (I, v) \in I(n)$ such that $\mathrm{pr}_2(R^\tau) = E_R$.*

Proof. Keep the same notation as in Proposition 1.2.36. Remark that one can write

$$\mathrm{pr}_2(R^g) = F_{R^g} + W,$$

where

$$W := \left\{ \begin{pmatrix} x'(r) \\ 0_{\eta_1} \\ u(r) \\ 0_{n-\eta} \end{pmatrix} \middle| \, r \in R^g \right\} = \mathrm{pr}_2(\mathscr{A}_{R^g}),$$

for some $g := (S, v) \in I(n)$. As $\mathscr{A}_{R^g}$ is a connected Abelian group, then so is W. Let $\{t_p\}_{p \in \mathbb{N}}$ be a sequence of W, which converges to some $t \in \mathbb{R}^n$. Then for some sequence $\{(A_p)\}_{p \in \mathbb{N}}$, we can and do assume that $\{(A_p, t_p)\}_{p \in \mathbb{N}}$ converges to some $(A, t) \in \mathscr{A}_{R^g}$, and then $t \in F$. This justifies that W is a closed, connected subgroup of $\mathbb{R}^n$, and hence, a subspace of $\mathbb{R}^n$. By (1.30), $W = V'_{R^g} \oplus F'_{R^g}$, and then $E_{R^g} = \mathrm{pr}_2(R^g)$. By Lemma 1.2.32, we have $E_{R^g} = S^{-1} E_R$, and a direct computation gives

$$\mathrm{pr}_2(R^g) = S^{-1}\{x + (A - I)v \mid (A, x) \in R\} = S^{-1}\,\mathrm{pr}_2(R^\tau),$$

where $\tau = (I, v)$. Finally, $E_R = \mathrm{pr}_2(R^\tau)$. □

Let H be a closed connected subgroup of $I(n)$. We now prove the following result, which will be of use later.

Proposition 1.2.38. *Let H be a closed connected subgroup of $I(n)$ and R its solvable radical, then $E_H = E_R$.*

Proof. Clearly, $E_R \subset E_H$. Conversely, let $H = R \cdot S$ be a Levi decomposition of H. Since S is compact, then we can assume that S is a subgroup of $O_n(\mathbb{R})$. For any $M \in S$ and $(A, x) \in R$, we have

$$M(A, x)M^{-1} = (MAM^{-1}, Mx) \in R.$$

This gives $P_{MAM^{-1}}(Mx) = MP_A(x)$, which proves that E_R is stable by $\mathrm{pr}_1(H)$. By considering a basis adapted to the decomposition $\mathbb{R}^n = E_R \oplus E_R^\perp$, there exists $Q \in O_n(\mathbb{R})$ such that any element of $Q^{-1}SQ$ writes

$$\begin{pmatrix} M_1 & 0 \\ 0 & M_2 \end{pmatrix}$$

and any element of R writes

$$\left(\begin{pmatrix} A & 0 \\ 0 & B \end{pmatrix}, \begin{pmatrix} x_1 \\ 0 \end{pmatrix}\right),$$

as in Proposition 1.2.34. Let $h \in H$, there exists

$$M := \begin{pmatrix} M_1 & 0 \\ 0 & M_2 \end{pmatrix} \in S \quad \text{and} \quad r := \left(\begin{pmatrix} A_1 & 0 \\ 0 & A_2 \end{pmatrix}, \begin{pmatrix} x_1 \\ 0 \end{pmatrix}\right) \in R$$

such that

$$h = r \cdot M = \left(\begin{pmatrix} A_1M_1 & 0 \\ 0 & A_2M_2 \end{pmatrix}, \begin{pmatrix} x_1 \\ 0 \end{pmatrix}\right).$$

To conclude, it is enough to observe that $P_{A_1M_1}(x_1) \in E_R$, as $P_{A_1M_1}$ depends polynomially upon A_1M_1. ☐

We also get the following.

Corollary 1.2.39. *Let H be a closed, connected subgroup of $I(n)$. Then H is compact if and only if $E_H = \{0\}$.*

We close the chapter by the following, which will be of use.

Proposition 1.2.40. *Let Γ be a discrete subgroup of $I(n)$ and Γ_a a normal Abelian subgroup of finite index in Γ. Then $E_\Gamma = E_{\Gamma_a}$. In particular, Γ is finite if and only if $E_\Gamma = \{0\}$.*

Proof. Take back the same notation as in Theorem 1.2.24. For a given $g := (S, v)$, any $\gamma \in \Gamma^g$ reads

$$\gamma = \left(\begin{pmatrix} C_{k,j(\gamma)} & \\ & \tilde{C}_{n-k} \end{pmatrix}, \begin{pmatrix} w_k(\gamma) \\ 0_{n-k} \end{pmatrix}\right),$$

for some $j(\gamma) \in \{1, \ldots, q-1\}$. As $\delta_j^q \in \Gamma_a^g$, we get $P_{C_{k,j(\gamma)}}(w_k(\gamma)) \in \mathbb{R}\text{-span}\{y_{k,1}, \ldots, y_{k,k}\}$. On the other hand, we easily check that

$$\mathbb{R}\text{-span}\left\{\begin{pmatrix} y_{k,i} \\ 0_{n-k} \end{pmatrix} 1 \le i \le k\right\} = E_{\Gamma_a^g}.$$

Hence, $E_{\Gamma_a^g} = E_{\Gamma^g}$ and by Lemma 1.2.32, we get $E_{\Gamma_a} = E_\Gamma$. ☐

The following is an immediate consequence from Proposition 1.2.40.

Corollary 1.2.41. *Let Γ be a discrete subgroup of $I(n)$. There exists $\tau = (I, v) \in I(n)$ such that* $\mathrm{pr}_2(\Gamma^\tau) \subset E_\Gamma$.

1.3 Heisenberg motion groups

Consider the Heisenberg group $\mathbb{H}_n = \mathbb{C}^n \times \mathbb{R}$ as defined in Subsection 1.1.2. The group $\mathbb{U}_n$ of $n \times n$ complex unitary matrices acts on $\mathbb{H}_n$ by the automorphisms

$$A(z,t) = (Az, t) \quad \text{for each } A \in \mathbb{U}_n, z \in \mathbb{C}^n \text{ and } t \in \mathbb{R}.$$

The Heisenberg motion group is the Lie group defined by the semidirect product $G_n = \mathbb{U}_n \ltimes \mathbb{H}_n$ with the multiplication law:

$$(A,z,t)(B,w,s) = \left(AB, z + Aw, t + s - \frac{1}{2}\operatorname{Im}\langle z, Aw\rangle\right).$$

We deal with the topology structure on the product $\mathbb{U}_n \times \mathbb{C}^n \times \mathbb{R}$ arising from the standard Hermitian inner product $\langle\cdot,\cdot\rangle$ on $\mathbb{C}^n$ with the related norm on $\mathbb{U}_n$. Furthermore, $\mathbb{U}_n$ yields a maximal compact connected subgroup of $\mathrm{Aut}(\mathbb{H}_n)$ and G_n. For the sake of simplicity of notation, G_n will be merely denoted by G, unless explicit mention. Denote $i = \sqrt{-1}$, and for any $g \in G$ and a subgroup H of G, $H^g := gHg^{-1}$. For any $u \in \mathbb{C}^n$, denote $\tau_u := (I, u, 0)$ where I designates the identity matrix of $M_n(\mathbb{C})$ and set $e = \tau_0$.

Define the following:

- $G^1 = \mathbb{U}_n \times \{0\} \times \mathbb{R} \subset G$.
- pr_1, pr_2 and pr_3, the natural projections of G on $\mathbb{U}_n$, $\mathbb{C}^n$ and $\mathbb{R}$, respectively.
- $\mathrm{pr} : G \longrightarrow \mathbb{U}_n \ltimes \mathbb{C}^n$, $(A,z,t) \mapsto (A,z)$ the continuous homomorphism.
- For $(A,z,t) \in G$, $E_A(1) = \ker(A - I)$ and P_A the orthogonal projection on $E_A(1)$. (It is known that P_A depends polynomially upon A.)

1.3.1 First preliminary results

Lemma 1.3.1. *For any $g \in G$, there exists $u \in \mathbb{C}^n$ such that $\tau_u g \tau_{-u} = (A,z,t)$, where $P_A(z) = z$ (i. e., $Az = z$) and $tz = 0$.*

Proof. Let $g = (A, y, \rho) \in G$, write $y = z + (I - A)z'$ where $z \in E_A(1)$ and $z' \in E_A(1)^\perp$. Then $g' = \tau_{-z'} g \tau_{z'} = (A, z, t')$ for some $t' \in \mathbb{R}$. If $z = 0$ or $t' = 0$, we are done. Otherwise, we easily check that $\tau_{\alpha z}(A,z,t')\tau_{-\alpha z} = (A,z,0)$ for $\alpha = -i\frac{t'}{\|z\|^2}$. Hence, for $u = \alpha z - z'$, $\tau_u g \tau_{-u}$ is of the requested form. ☐

Lemma 1.3.2. *Let $(A,z,t), (B,w,s) \in G$. If (A,z,t) and (B,w,s) are conjugate to each other in G, then $P_A(z) = 0$ if and only if $P_B(w) = 0$.*

Proof. Let $g = (S, w, l) \in G$. By a direct computation, we have

$$\begin{aligned}(B, w, s) &= g^{-1}(A, z, t)g \\ &= \left(S^{-1}AS, S^{-1}[z + (A - I)w], \frac{1}{2}\operatorname{Im}\langle w, z + Aw\rangle - \frac{1}{2}\operatorname{Im}\langle z, Aw\rangle + t\right).\end{aligned}$$

Therefore,

$$\begin{aligned}P_{S^{-1}AS}(S^{-1}[z + (A - I)w]) &= S^{-1}P_A(z + (A - I)w) \\ &= S^{-1}P_A(z) + S^{-1}\underbrace{P_A(A - I)}_{=0}w \\ &= S^{-1}P_A(z),\end{aligned}$$

as is to be shown. □

Now we prove the following propositions.

Proposition 1.3.3. *Let $\gamma = (A, z, t) \in G$ be of finite order. Then $P_A(z) = 0$ and γ is conjugate to an element of $U_n \times \{0\} \times \{0\}$ in G.*

Proof. Let $\gamma = (A, z, t) \in G$ be of finite order k_0. By Lemma 1.3.1, there exists $u \in \mathbb{C}^n$ such that $\tau_u\gamma\tau_{-u} = (A, w, s)$, where $P_A(w) = w$ and $sw = 0$. Then

$$\tau_u\gamma^{k_0}\tau_{-u} = (A^{k_0}, k_0w, k_0s) = (I, 0, 0).$$

This implies that $P_A(w) = w = 0$ and $s = 0$. Hence, by Lemma 1.3.2, we have $P_A(z) = 0$ and for $g = \tau_u$, $g\gamma g^{-1} = (A, 0, 0) \in U_n \times \{0\} \times \{0\}$. □

Proposition 1.3.4. *Let $\gamma = (A, z, t) \in G$. If $P_A(z) = 0$, then γ is conjugate to an element of G^1.*

Proof. According to Lemma 1.3.1, there exists $u \in \mathbb{C}^n$ such that $\tau_u\gamma\tau_{-u} = (A, w, s)$, where $P_A(w) = w$ and $sw = 0$. If $P_A(z) = 0$, then Lemma 1.3.2 gives that $P_A(w) = w = 0$. Hence, $g\gamma g^{-1} = (A, 0, s) \in G^1$ for $g = \tau_u$. □

Proposition 1.3.5. *Let $\gamma = (A, z, t) \in G$. If $P_A(z) \neq 0$, then γ is conjugate to an element (B, w, s) of $U_n \times \mathbb{C}^n \times \{0\}$ with $Bw = w$.*

Proof. By Lemma 1.3.1, there exists $u \in \mathbb{C}^n$ such that $\tau_u\gamma\tau_{-u} = (A, w, s)$, where $P_A(w) = w$ and $sw = 0$. If $P_A(z) \neq 0$, then by Lemma 1.3.2, $P_A(w) = w \neq 0$. Hence, $s = 0$ and for $g = \tau_u$ we have $g\gamma g^{-1} = (A, w, 0) \in U_n \times \mathbb{C}^n \times \{0\}$. □

1.3.2 Discrete subgroups of Heisenberg motion groups

Subgroups of the semidirect product group $N_C = C \ltimes N$

We first record the following result, which will be next utilized.

Fact 1.3.6 ([4, Theorem 4.1 and Remark 4.2] and also [9]). *Let N be a connected, simply connected, nilpotent Lie group, C a compact group of automorphisms of N and $\Gamma \subset N_C = C \ltimes N$ a discrete subgroup. Then there exist $b \in N$, a connected Lie subgroup N_Γ of N and a Γ^* of Γ, which is isomorphic to a discrete subgroup of N_Γ satisfying the following conditions:*
(1) *$b\Gamma b^{-1}$ preserves the subset N_Γ of N via the natural action of $C \ltimes N$ on N.*
(2) *The orbit space $(b\Gamma b^{-1})\backslash N_\Gamma$ is compact.*
(3) *The induced $(b\Gamma^* b^{-1})$-action on N_Γ is free.*

It has been shown in [84] that any closed, solvable subgroup of an almost connected Lie group is compactly generated (hence, finitely generated if in addition discrete). Combined with Fact 1.3.6, one gets immediately the following.

Lemma 1.3.7. *Any discrete subgroup of N_C is finitely generated.*

Proof. Let Γ be a discrete subgroup of G, by Fact 1.3.6, Γ contains a subgroup Γ^* of finite index q. Take $\{\overline{e}, \overline{\delta_1}, \dots, \overline{\delta_{q-1}}\}$ for a complete representative of the finite set Γ/Γ^*, where e designates the identity of G. Furthermore, Γ^* is isomorphic to a discrete subgroup of N, which is finitely generated, and then so is Γ^*. Let $\{\gamma_1, \dots, \gamma_k\}$ be a family of generators of Γ^* for some positive integer k, then $\{\gamma_1, \dots, \gamma_k, \delta_1, \dots, \delta_{q-1}\}$ generates Γ. □

We go back now to a Heisenberg motion group G. We take a finite index subgroup Γ^* of Γ as in Fact 1.3.6. Then Γ^* is isomorphic to a discrete subgroup of $\mathbb{H}_n$ and then $D(\Gamma^*)$ is isomorphic to a subgroup of $Z(\mathbb{H}_n)$. Hence, $D(\Gamma^*)$ is a torsion-free subgroup of rank 1 or is trivial. Here, for a group G', $Z(G')$ designates the center of G' and $D(G') = [G', G']$.

The following is also straight to obtain the following.

Proposition 1.3.8. *Let $G = \mathbb{U}_n \ltimes \mathbb{H}_n$ be a Heisenberg motion group and let Γ be a discrete subgroup of G. Then the following are equivalent:*
(1) *Γ is finite.*
(2) *Any element of Γ is of finite order.*
(3) *Γ^* is trivial.*

Proof. As Γ^* is isomorphic to a discrete subgroup of $\mathbb{H}_n$, then it is torsion-free. Hence, Γ^* is trivial whenever any element of Γ is of finite order. The other implications are trivial. □

For any positive integer n, recall the semidirect product $I(n) := O_n(\mathbb{R}) \ltimes \mathbb{R}^n$ of the orthogonal group $O_n(\mathbb{R})$ (with respect to the canonical Euclidean product $\langle \cdot, \cdot \rangle$ on $\mathbb{R}^n$) and $\mathbb{R}^n$.

For $U = A + iB$ and $z = x + iy$, where $A, B \in M_n(\mathbb{R})$ and $x, y \in \mathbb{R}^n$, we easily check that

$$\begin{aligned} \phi_n : \mathbb{U}_n \ltimes \mathbb{C}^n &\hookrightarrow I(2n) \\ (A + iB, x + iy) &\mapsto \left(\begin{pmatrix} A & -B \\ B & A \end{pmatrix}, \begin{pmatrix} x \\ y \end{pmatrix} \right) \end{aligned} \tag{1.31}$$

is a group homomorphism. Note here that for any closed subgroup L of $\mathbb{U}_n \ltimes \mathbb{C}^n$, $\phi_n(L)$ is a closed subgroup of $I(2n)$. By a direct computation, we see that the following two conditions on L are equivalent:
- $P_U(z) = 0$ for any $(U, z) \in L$.
- $P_M(x) = 0$ for any $(M, x) \in \phi_n(L)$.

Corollary 1.3.9. *Let $\{\gamma_j\}_{j \in J}$ be a commuting family of $\mathbb{U}_n \ltimes \mathbb{C}^n$. There exist an integer $0 \le k \le n$ and $g \in \mathbb{U}_n \ltimes \mathbb{C}^n$ such that for any $j \in J$,*

$$g^{-1}\gamma_j g = \left(\begin{pmatrix} I_k & \\ & A_{n-k,j} \end{pmatrix}, \begin{pmatrix} z_{k,j} \\ 0_{n-k} \end{pmatrix} \right), \tag{1.32}$$

where $z_{k,j} \in \mathbb{C}^k$, 0_{n-k} is the zero of $\mathbb{C}^{n-k}$ and $A_{n-k,j} \in \mathbb{U}_{n-k}$.

Proof. For any $j \in J$, let $\gamma_j = (A_j, z_j)$ and $\delta_j := (B_j, x_j) = \phi_n(\gamma_j)$, where ϕ_n is in (1.31). By Lemma 1.2.1, there exist $h = (C, y) \in E(2n)$ and $0 \le s \le 2n$ such that for any $j \in J$,

$$\delta_j' := (C, y)\delta_j(C, y)^{-1} = \left(\begin{pmatrix} I_s & \\ & B_{2n-s,j} \end{pmatrix}, \begin{pmatrix} y_{s,j} \\ 0_{2n-s} \end{pmatrix} \right),$$

for some $B_{2n-s,j} \in O_{2n-s}(\mathbb{R})$ and $y_{s,j} \in \mathbb{R}^s$ $(j \in J)$. Therefore, for $j_0 \in J$,

$$\delta_{j_0}'' := (C^{-1}, 0)\delta_{j_0}'(C, 0) = (B_{j_0}, x_{j_0}'),$$

for some

$$x_{j_0}' = C^{-1}\begin{pmatrix} y_{s,j_0} \\ O_{2n-s} \end{pmatrix} \in \bigcap_{j \in J} E_{B_j}(1).$$

On the other hand, $\delta_{j_0}'' = (I_{2n}, C^{-1}y)\delta_{j_0}(I_{2n}, -C^{-1}y)$ by a simple computation. Set now $C^{-1}y = \binom{a_1}{a_2}$ for $a_1, a_2 \in \mathbb{R}^n$. For $(I, u) = (I, a_1 + ia_2) = \phi_n^{-1}((I_{2n}, C^{-1}y))$, one gets

$$\phi_n^{-1}(\delta_{j_0}'') = (I, u)\phi_n^{-1}(\delta_{j_0})(I, -u) = (A_{j_0}, z_{j_0}'),$$

for some $z_{j_0}' \in \mathbb{C}^n$. We easily check that $z_{j_0}' \in \bigcap_{j \in J} E_{A_j}(1)$. Let Q be the transition matrix from the standard basis of $\mathbb{C}^n$ into an orthonormal basis adapted to the orthogonal direct sum $[\bigcap_{j\in J} E_{A_j}(1)] \oplus [\bigcap_{j\in J} E_{A_j}(1)]^\perp$. Then for $g = (Q, Qu)$, $g\gamma_j g^{-1}$ is of the requested form (1.32), where $k = \dim \bigcap_{j\in J} E_{A_j}(1)$. □

We next have the following.

Proposition 1.3.10. *Let Γ be a discrete subgroup of G such that for any $\gamma = (B, z, t) \in \Gamma$, $P_B(z) = 0$. Then there exists $u \in \mathbb{C}^n$ such that $\Gamma^{\tau_u} \subset G^1$.*

Proof. Making use of Fact 1.3.6, Γ contains a 2-step nilpotent torsion-free subgroup Γ^* and of finite index in Γ. Assume that $D(\Gamma^*)$ is of rank 1 and has a generator $\gamma_0 = (A_0, u_0, s_0)$, say. By Proposition 1.3.4, there exists $w \in \mathbb{C}^n$ such that $\tau_w \gamma_0 \tau_{-w} = (A_0, 0, t_0) \in G^1$. For any $(B, z) \in \mathrm{pr}[(\Gamma^*)^{\tau_w}]$, we have

$$(B,z)^{-1}(A_0,0)(B,z) = (B^{-1}A_0B, B^{-1}(A_0 - I)z) \in \mathrm{pr}[D(\Gamma^*)^{\tau_w}]$$

and then $z \in E_{A_0}(1)$. Let Q be the transition matrix from the standard basis of $\mathbb{C}^n$ to an orthonormal basis of $E_{A_0}(1) \oplus E_{A_0}(1)^\perp$ and consider $\{\overline{(B_j, z_j)}\}_{j\in J}$ to be a complete representative set of $\mathrm{pr}((\Gamma^*)^{\tau_w})/\,\mathrm{pr}(D[(\Gamma^*)^{\tau_w}])$ where $\overline{(B_j, z_j)}$ designates the equivalent class of (B_j, z_j) modulo $\mathrm{pr}(D[(\Gamma^*)^{\tau_w}])$. For $\omega_1 = (Q, 0, 0) \in \mathbb{U}_n$, any $(C, v) \in \mathrm{pr}((\Gamma^*)^{\tau_w \omega_1})$ reads

$$(C,v) = \underbrace{\left(\begin{pmatrix} B_{1,j} & \\ & B_{2,j}\end{pmatrix}, \begin{pmatrix} v_j \\ 0\end{pmatrix}\right)}_{=\,\omega_1(B_j,z_j)\omega_1^{-1}} \underbrace{\left(\begin{pmatrix} I_k & \\ & C_{n-k}\end{pmatrix}, \begin{pmatrix} 0_k \\ 0_{n-k}\end{pmatrix}\right)}_{\in \mathrm{pr}(D((\Gamma^*)^{\tau_w \omega_1}))}, \tag{1.33}$$

where $k = \dim E_{A_0}(1)$. It turns out that $\{(B_{1,j}, v_j)\}_{j\in J}$ generates an Abelian subgroup Λ_k of $\mathbb{U}_k \ltimes \mathbb{C}^k$, say. By applying Corollary 1.3.9, there exist an integer $k' \le k$ and $\omega_k' = (S_k, u_k') \in \mathbb{U}_k \ltimes \mathbb{C}^k$ such that for any j,

$$\omega_k'(B_{1,j}, v_j)\omega_k'^{-1} = \left(\begin{pmatrix} I_{k'} & \\ & \widetilde{B}_j\end{pmatrix}, \begin{pmatrix} \widetilde{v}_j \\ 0_{k-k'}\end{pmatrix}\right),$$

and where all the $\widetilde{v}_j$'s belong to $\mathbb{C}^{k'}$ and the $\widetilde{C}_j$'s to $\mathbb{U}_{k-k'}$. Since for all $\gamma = (B, z, t) \in \Gamma$, $P_B(z) = 0$, $\widetilde{v}_j = 0_{k'}$ for any $j \in J$, then one can easily justify that for

$$w' = Q^{-1}\begin{pmatrix} S_k^{-1}u_k' \\ 0_{n-k}\end{pmatrix},$$

and $v = w' + w$, $(\Gamma^*)^{\tau_v} \subset G^1$. Take a complete representative set $\{\overline{\delta}_1, \dots, \overline{\delta}_q\}$ of $\Gamma^{\tau_v}/_{(\Gamma^*)^{\tau_v}}$. For any $j \in \{1, \dots, q\}$, denote $\delta_j = (N_j, x_j, t_j)$. Then

$$\Gamma^{\tau_v} = \bigcup_{j=1}^{q} \delta_j \cdot (\Gamma^*)^{\tau_v} = \{(N_jA, x_j, t_j + t) : (A, 0, t) \in (\Gamma^*)^{\tau_v}, j \in \{1, \dots, q\}\}. \tag{1.34}$$

On the other hand, remark that $\mathrm{pr}_2(\Gamma^{\tau_v}) = \{x_1, \dots, x_q\}$ and, therefore, by Corollary 1.2.11 there exist some $u' \in \mathbb{C}^n$ such that

$$(I, u')(\mathrm{pr}(\Gamma^{\tau_v}))(I, -u') \subset \mathbb{U}_n \times \{0\}.$$

Conclusively, for $u = u' + v$, $\Gamma^{\tau_u} \subset G^1$.

To complete the proof, just remark that when $D(\Gamma^*)$ is trivial, and it is sufficient to take $C_{n-k} = I_{n-k}$, the identity matrix, in equation (1.33). □

We now look at the analogue of Proposition 1.3.10 in the case where H is connected. We first record the following (cf. [82]).

Fact 1.3.11 ([82, Theorem 12.2.2]). *For a compact connected Lie group C with Lie algebra $\mathfrak{c}$, the following assertions hold:*
(i) *A subalgebra $\mathfrak{t} \subseteq \mathfrak{c}$ is maximal Abelian if and only if it is the Lie algebra of a maximal torus of C.*
(ii) *For two maximal tori T and T', there exists $c \in C$ with $cTc^{-1} = T'$.*
(iii) *Every element of C is contained in a maximal torus.*

We get conclusively the following.

Corollary 1.3.12. *For a closed, connected, solvable subgroup R of G, $\mathrm{pr}_1(R)$ is a connected Abelian subgroup of $\mathbb{U}_n$.*

Proof. Consider the compact connected Lie group $\overline{\mathrm{pr}_1(R)}$, the closure of the solvable subgroup $\mathrm{pr}_1(R)$ of $\mathbb{U}_n$. By Fact 1.3.11, $\overline{\mathrm{pr}_1(R)}$ contains a maximal torus T. Since $\overline{\mathrm{pr}_1(R)}$ is solvable (cf. [48, Corollary 2, p. 342]), then by Fact 1.2.29, for some submanifold $M \cong \mathbb{R}^n \subseteq \overline{\mathrm{pr}_1(R)}$ the map $M \times T \to \overline{\mathrm{pr}_1(R)}$, $(m, t) \mapsto mt$ is a diffeomorphism. By the compacity of $\overline{\mathrm{pr}_1(R)}$, M is trivial, and hence, $\overline{\mathrm{pr}_1(R)} = T$. This completes the proof. □

Lemma 1.3.13. *Let H be a closed, connected subgroup of G such that for any $h = (B, z, t) \in H$, $P_B(z) = 0$. Then there exists $u \in \mathbb{C}^n$ such that $H^{\tau_u} \subset G^1$.*

Proof. Let $H = S \cdot R$ be a Levi decomposition, where S is a maximal semisimple subgroup of H and R is the solvable radical of H. By Proposition 1.2.26, S is compact. We easily see that $\mathrm{pr}(R)$ is Abelian. Indeed, for any $(B, x, t), (B', x', t') \in R$, B and B' commute as in Corollary 1.3.12, and then the commutator $[(B, x, t)(B', x', t')]$ is of the form

$$[(B, x, t)(B', x', t')] = (I, a, s),$$

for some $a \in \mathbb{C}^n$ and $s \in \mathbb{R}$. Since $(I, a, s) \in H$, by the assumption of H, we have $a = P_I(a) = 0$ as it is to be shown. Corollary 1.3.9 asserts now that there exist an integer $0 \leq k \leq n$ and $g = (Q, w) \in \mathbb{U}_n \ltimes \mathbb{C}^n$ such that for any $y = (A, z) \in \mathrm{pr}(R)$,

$$g^{-1} y g = \left(\begin{pmatrix} I_k & \\ & A_{n-k,y} \end{pmatrix}, \begin{pmatrix} z_{k,y} \\ 0_{n-k} \end{pmatrix} \right),$$

where $z_{k,y} \in \mathbb{C}^k$, 0_{n-k} is the zero of $\mathbb{C}^{n-k}$ and $A_{n-k,y} \in \mathbb{U}_{n-k}$. Hence, for any $r = (A,z,t) \in R$,

$$\begin{aligned}(Q,w,0)^{-1}r(Q,w,0) &= (Q^{-1}AQ, Q^{-1}(z+(A-I)w), t'_r)\\ &= \left(\begin{pmatrix} I_k & \\ & A_{n-k,\mathrm{pr}(r)}\end{pmatrix}, \begin{pmatrix} z_{k,\mathrm{pr}(r)} \\ 0_{n-k}\end{pmatrix}, t'_r\right),\end{aligned}$$

for some $t'_r \in \mathbb{R}$. Finally, $z_{k,\mathrm{pr}(r)} = 0_k$ as $P_{Q^{-1}AQ}(Q^{-1}(z+(A-I)w)) = 0$. Therefore, for any $r \in R$, $(Q,w,0)^{-1}r(Q,w,0) = (Q^{-1}AQ, 0, t'_r)$, which entails that $R^{\tau_{-w}} \subset G^1$. Conclusively, $\mathrm{pr}_2(H^{\tau_{-w}}) = \mathrm{pr}_2(S^{\tau_{-w}})$, which is compact and then by Corollary 1.2.11 there exists $v \in \mathbb{C}^n$ such that $(I,v)\,\mathrm{pr}(H^{\tau_{-w}})(I,-v) \subset \mathbb{U}_n \times \{0\}$. Thus, $H^{\tau_u} \subset G^1$ for $u = -w+v$. □

Proposition 1.3.14. *Let H be a closed subgroup of G such that for any $h = (B,z,t) \in H$, $P_B(z) = 0$. Then there exists $u \in \mathbb{C}^n$ such that $H^{\tau_u} \subset G^1$.*

Proof. Consider H_0 to be the connected component of H, thanks to Lemma 1.3.13, we can assume that $H_0 \subset G^1$. Take $\{(B_j, z_j, t_j)\}_{j\in\mathscr{J}}$ a complete representative of the group H/H_0. $H/H_0 = \{\delta_j = \overline{(B_j, z_j, t_j)}\}_{j\in\mathscr{J}}$ is a discrete group. Without loss of generality, we can let $j \in \mathscr{J}$ and $(S,0,t) \in H_0$, and we have

$$(B_j^{-1}, -B_j^{-1}z_j, -t_j)(S,0,t)(B_j,z_j,t_j) = \left(B_j^{-1}AB_j, B_j^{-1}(S-I)z_j, t + \frac{1}{2}\operatorname{Im}\langle z_j, Sz_j\rangle\right) \in H_0.$$

This entails that for any $j \in \mathscr{J}$,

$$z_j \in V := \bigcap_{S\in\mathrm{pr}_1(H_0)} E_S(1).$$

With respect to the decomposition $\mathbb{C}^n = V \oplus V^\perp$, there exists $\omega = (Q,0,0) \in G$ such that for any $h \in H^\omega = \omega H \omega^{-1}$,

$$h = \underbrace{\left(\begin{pmatrix} B_{j,k} & 0 \\ 0 & B_{j,n-k}\end{pmatrix}, \begin{pmatrix} z_{j,k} \\ 0\end{pmatrix}, t_j\right)}_{\omega\delta_j\omega^{-1}} \underbrace{\left(\begin{pmatrix} I_k & 0 \\ 0 & A_{n-k}\end{pmatrix}, 0, t\right)}_{\in\, \omega H_0 \omega^{-1}}, \tag{1.35}$$

where $k = \dim(V)$, $B_{j,k} \in \mathbb{U}_k$, $B_{j,n-k}$ and A_{n-k} belong to $\mathbb{U}_{n-k}$ and $z_{j,k} \in \mathbb{C}^k$.

If $\mathrm{pr}_3(H_0) \neq \{0\}$, then for any $j \in \mathscr{J}$, there exists $(S_j, 0, -t_j) \in \omega H_0 \omega^{-1}$ such that

$$(S_j, 0, -t_j)(B_j, z_j, t_j) = (S_jB_j, S_jz_j, 0) = (B'_j, z'_j, 0).$$

Equation (1.35) then reads

$$h = \left(\begin{pmatrix} B'_{j,k} & 0 \\ 0 & B'_{j,n-k}\end{pmatrix}, \begin{pmatrix} z'_{j,k} \\ 0\end{pmatrix}, 0\right)\left(\begin{pmatrix} I_k & 0 \\ 0 & A'_{n-k}\end{pmatrix}, 0, t'\right),$$

and we easily verify that $\Lambda_k := \{(B'_{j,k}, z'_{j,k}, 0)\}_{j\in\mathcal{J}}$ is a discrete subgroup of $\mathbb{U}_k \ltimes \mathbb{H}_k$. Making use of Proposition 1.3.10, there exists $v \in \mathbb{C}^k$ such that $\tau_v \Lambda_k \tau_v^{-1} \subset \mathbb{U}_k \times \{0\} \times \mathbb{R}$. This accomplishes the proof in this case.

If $\mathrm{pr}_3(H_0) = \{0\}$, then $F := \{(B_{j,k}, z_{j,k}, t_j)\}_{j\in\mathcal{J}}$ is a discrete subgroup of $\mathbb{U}_k \ltimes \mathbb{H}_k$. Again by Proposition 1.3.10, F is conjugate to a subgroup of $\mathbb{U}_k \times \{0\} \times \mathbb{R}$ and so is H. □

The following observation will be of use later.

Lemma 1.3.15. *Let H be a closed subgroup of G such that $H \cap (\mathbb{U}_n \times B(0,r) \times \mathbb{R})$ is compact for any $r > 0$. Then $\mathrm{pr}(H)$ is a closed subgroup of $\mathbb{U}_n \ltimes \mathbb{C}^n$.*

Proof. Let $\{(B_p, z_p)\}_{p\in\mathbb{N}}$ be a sequence of $\mathrm{pr}(H)$ converging to $(B, z) \in \mathbb{U}_n \ltimes \mathbb{C}^n$. One can define a sequence $\{(B_p, z_p, t_p)\}_{p\in\mathbb{N}} \subset H$. As for p large enough, $(B_p, z_p) \in \mathbb{U}_n \times B(0, r)$ for some $r > 0$, then $\{(B_p, z_p, t_p)\}_{p\in\mathbb{N}} \subset H \cap (\mathbb{U}_n \times B(0, r) \times \mathbb{R})$. Thus one can find a subsequence $\{(B_{\alpha(p)}, z_{\alpha(p)}, t_{\alpha(p)})\}_{p\in\mathbb{N}}$, which converges to $(B, z, t) \in H$ for some $t \in \mathbb{R}$ and this implies that $(B, z) \in \mathrm{pr}(H)$. □

Taking all the above mentioned into account, one can conclusively categorize discrete subgroups of G as follows.

Definition 1.3.16. Let Γ be an infinite discrete subgroup of G. Γ is said to be:
(1) of type (A), if Γ is conjugate to a subgroup of $G^1 = \mathbb{U}_n \times \{0\} \times \mathbb{R}$,
(2) of type (B), if any element of Γ is conjugate to an element of $\mathbb{U}_n \times \mathbb{C}^n \times \{0\}$,
(3) of type (C), if Γ contains at least a mixture of two infinite subgroups of type (A) and type (B).

We now prove the following.

Proposition 1.3.17. *Any infinite discrete subgroup of G_n is either of type (A) or (B) or (C). Furthermore, two conjugate infinite discrete subgroups are of the same type.*

Proof. Let Γ be an infinite discrete subgroup of G. Suppose that Γ is neither of type (A) nor (B). We shall prove that Γ is of type (C). By combining with the assumption for Γ and Propositions 1.3.10, 1.3.3 and 1.3.5, we can find an element

$$\gamma_1 = (A_1, z_1, t_1), \gamma_2 = (A_2, z_2, t_2) \in \Gamma$$

such that γ_1, γ_2 are both of infinite order, $P_{A_1}(z_1) \neq 0$ and $P_{A_2}(z_2) = 0$. In particular, by applying Propositions 1.3.4 and 1.3.5, we have that γ_1 (resp., γ_2) is conjugate to an element of $U_n \times \mathbb{C}^n \times \{0\}$ (resp., of G^1). Therefore, Γ contains infinite discrete subgroups of type (B) as $\langle\gamma_1\rangle$ and type (A) as $\langle\gamma_2\rangle$. This means that Γ itself is of type (C). □

1.4 Syndetic hulls

Definition 1.4.1. Let G be a Lie group, Γ a closed subgroup of G and $Z^c(G)$ the maximal compact subgroup of $Z(G)$, the center of G. A *syndetic hull* of Γ is any connected Lie subgroup L of G, which contains $\Gamma \cdot Z^c(G)$ cocompactly. Then obviously, L contains Γ cocompactly.

When G is simply connected and solvable, then $Z^c(G)$ is trivial. In this case, by a syndetic hull of Γ, we mean any connected Lie subgroup of G, which contains Γ cocompactly.

1.4.1 Existence results for completely solvable Lie groups

In [122], Saito proved that any closed subgroup of a completely solvable Lie group admits a unique syndetic hull. Here, we give a simpler proof.

Theorem 1.4.2. *Let G be a simply connected completely solvable Lie group. Then any closed subgroup of G admits a unique syndetic hull.*

Proof. Let Γ be a closed subgroup of G, being completely solvable. We first show that the syndetic hull is unique provided its existence. Let L_1 and L_2 be two connected, closed subgroups such that L_1/Γ and L_2/Γ are compact. Note first that $L_1 \cap L_2$ is connected as being the Lie group associated to the Lie algebra $\mathfrak{l}_1 \cap \mathfrak{l}_2$, where $\mathfrak{l}_i$ denotes the Lie algebra of $L_i, i = 1, 2$. We claim that $L_i/(L_1 \cap L_2)$, $i = 1, 2$ are also compact. To see that, consider for $i = 1,\ 2$ the canonical surjection

$$\pi_i : L_i \longrightarrow L_i/(L_1 \cap L_2),$$

which factors through the canonical surjection $\rho_i : L_i \to L_i/\Gamma$ to a surjection $\tilde{\pi}_i : L_i/\Gamma \to L_i/(L_1 \cap L_2)$ such that $\pi_i = \tilde{\pi}_i \circ \rho_i$. The map $\tilde{\pi}_i$ is surjective and continuous, and thus its image $L_i/(L_1 \cap L_2)$ is compact. Now, G is connected simply connected, solvable Lie group, the quotient $L_i/(L_1 \cap L_2)$ is diffeomorphic to $\mathbb{R}^d$, where $d = \dim L_i - \dim(L_1 \cap L_2)$. It follows therefore that $d = 0$ and that $L_1 \cap L_2 = L_1 = L_2$, as it is to be shown.

We tackle now the proof of existence, we proceed by induction on the dimension of G. Let now G_0 be a one-codimensional, closed, normal subgroup of G. Provided that $\Gamma \subset G_0$, we are done using the induction hypothesis. We assume henceforth that $\Gamma \not\subset G_0$ and we consider $\Gamma_0 = \Gamma \cap G_0$, which is a closed subgroup of G_0. Let $\tilde{G} = G/G_0 \simeq \mathbb{R}$ and $\pi : G \to \tilde{G}$ the canonical projection. Provided that Γ_0 is trivial, the homomorphism $\tilde{\pi} = \pi|\Gamma : \Gamma \to \tilde{\Gamma} := \tilde{\pi}(\Gamma)$ appears to be a groups isomorphism. This gives then that Γ is therefore Abelian and the result follows. We assume from now on that Γ_0 is not trivial. There exists by the induction hypothesis, a closed, connected subgroup L_0 of G_0, which contains Γ_0 cocompactly. As such, there exists a compact set C in G contained in G_0 and fulfills the following identity:

$$L_0 = C\Gamma_0. \tag{1.36}$$

Assume for a while that L_0 is normal in G, then $L_0\Gamma$ is a subgroup of G and by equation (1.36), we get that

$$L_0\Gamma = C\Gamma_0\Gamma = C\Gamma, \tag{1.37}$$

which is closed in G as Γ is. Let $G' = G/L_0$ and $\pi : G \to G'$ the associated canonical surjection. Then $\Gamma' = \pi(L_0\Gamma)$ is a closed subgroup of G'. Using the induction hypothesis, there exists a connected closed subgroup S' of G' such that S'/Γ' is compact, in particular, there exists by Lemma 1 in [74] a compact set C' of G such that $S' = \overline{C'}\Gamma'$, where $\overline{C'}$ is the image of C' by π. Let S be the preimage of S' by π, then S is closed subgroup of G, which contains L_0. Moreover, we have

$$S = C'L_0\Gamma L_0 = C'L_0\Gamma = C'C\Gamma$$

by (1.36) and (1.37), which merely entails that S/Γ is compact. On the other hand, we have $S' = S/L_0$ with S' and L_0 are connected. This shows that S is connected.

We finally treat the case where L_0 is not normal. Let $N_G(L_0)$ be the normalizer of L_0 in G. The subgroup G_0 being normal in G, we get that for any $\gamma \in \Gamma$, $\gamma\Gamma_0\gamma^{-1} \subset G_0 \cap \Gamma = \Gamma_0$, and then Γ_0 is normal in Γ. Therefore, for any $\gamma \in \Gamma$ the subgroup $\gamma L_0\gamma^{-1}$ is connected and closed in G_0 such that $\gamma L_0\gamma^{-1}/\Gamma_0$ is compact. By the uniqueness of L_0, we get that $\gamma L_0\gamma^{-1} = L_0$ and that $\Gamma \subset N_G(L_0)$. Recall that $N_G(L_0)$ is a connected, closed subgroup of G and $\dim N_G(L_0) < \dim G$. The result follows using again the induction hypothesis. □

1.4.2 Case of exponential Lie groups

In the case where G is more generally an exponential solvable Lie group, the result of Theorem 1.4.2 fails to hold in general (a counterexample is given in [122]), except when Γ turns out to be Abelian. In this case, we have the following.

Proposition 1.4.3. *Assume that G is an exponential solvable Lie group and Γ is an Abelian closed subgroup of G. Then Γ admits a unique syndetic hull.*

Proof. We first prove the following.

Lemma 1.4.4. *Let G be an exponential Lie group and $\mathfrak{g}$ its Lie algebra. For all $X, Y \in \mathfrak{g}$ the following properties are equivalent:*

($\imath$) $\exp(X)\exp(Y) = \exp(Y)\exp(X)$

($\imath\imath$) $[X, Y] = 0$.

Proof. From the injectivity of the exponential map, the first property (ι) is equivalent to

$$\mathrm{Ad}(\exp(X))(Y) = Y.$$

This gives that for all $t \in \mathbb{R}$, we have $\mathrm{Ad}(\exp(X))(tY) = tY$. Exponentiating back, we obtain

$$\exp(tY)\exp(X)\exp(-tY) = \exp(X),$$

which merely entails that

$$\mathrm{Ad}(\exp(tY))(X) = X$$

for any $t \in \mathbb{R}$. Therefore, the analytic map

$$a(t) = \sum_{p=1}^{\infty} \frac{t^p}{p!}(\mathrm{ad}(Y))^p(X)$$

vanishes on $\mathbb{R}$ and so do its coefficients. We get conclusively that $[X, Y] = 0$. The converse is trivial. □

We now go back to the proof of Proposition 1.4.3. Let $\mathfrak{g}$ be the Lie algebra of G, $\Lambda = \log \Gamma$ and $\mathfrak{l} = \mathbb{R}\text{-span}(\Lambda)$. We get from (1.4.4) that $\mathfrak{l}$ is an Abelian subalgebra of $\mathfrak{g}$ and then $L = \exp(\mathfrak{l}) = \exp(\mathbb{R}X_1)\cdots\exp(\mathbb{R}X_k)$ is isomorphic to $\mathbb{R}^k$, where $X_1, \ldots, X_k$ belong to Λ. It is then clear that $\exp(\mathbb{Z}X_1)\cdots\exp(\mathbb{Z}X_k) \subset \Gamma$, which means that L/Γ is compact. To see that L is unique, let L' be another connected subgroup of G, which contains Γ cocompactly. Then obviously $L \subset L'$ and then L'/L is compact. By choosing an coexponential basis of L' in L, we see that L/L' is trivial, which achieves the proof of the proposition. □

A straight application of Theorem 1.4.2 and Proposition 1.4.3 is to establish a homeomorphism between the spaces $\mathrm{Hom}(\Gamma, G)$ and $\mathrm{Hom}(\mathfrak{l}, \mathfrak{g})$ where $\mathfrak{l}$ stands for the Lie algebra of the syndetic hull L of Γ. Recall that in [122], it is proved that any homomorphism from Γ to G can be uniquely extended to a homomorphism from the syndetic hull of Γ to G. We now give an alternative simpler proof of this upshot and even develop this aspect later (cf. Proposition 1.4.7 below). We first have the following.

Theorem 1.4.5. *Let G be a completely solvable Lie group, Γ a closed subgroup of G and L its syndetic hull in G. Then any continuous homomorphism from Γ to G uniquely extends to a continuous homomorphism from L to G. This also applies to the case where G is exponentially solvable and Γ is Abelian.*

Proof. First, we provide a proof of existence. We start with a continuous homomorphism φ from Γ to G, and we denote by $\hat{\Gamma} \subset L \times G$ its graph. Clearly, $\hat{\Gamma}$ is a closed subgroup of $L \times G$. By Theorem 1.4.2, $\hat{\Gamma}$ has a unique syndetic hull K. Let

$$p : L \times G \to L, \quad q : L \times G \to G$$

be the natural homomorphisms projections. We now proceed to prove that $q_{|K} \circ p_{|K}^{-1}$ is a well-defined continuous homomorphism from L to G and that its restriction to Γ coincides with φ. We first show that $p_{|K}$ realizes an isomorphism between K and L. Indeed, p is a continuous homomorphism and K is a connected exponential Lie subgroup of $L \times G$. Let ψ be the differential of $p_{|K}$ at e and $\mathfrak{k} = \log K$, the Lie algebra associated to K. Then

$$p_{|K} = \exp_L \circ \psi \circ \log_K \quad \text{and} \quad p(K) = \exp_L(\psi(\mathfrak{k})), \tag{1.38}$$

where $\psi(\mathfrak{k})$ is a Lie subalgebra of $\mathfrak{l} = \log L$, in particular, $p(K)$ is closed, connected subgroup of L. Let π be the canonical surjection $L \to L/\Gamma$. The subgroup Γ is included in $p(K)$ being a saturated (with respect to π) closed set in L. This means that $\pi(p(K))$ is closed in L/Γ and then $p(K)/\Gamma$ is compact. From the uniqueness of the syndetic hull of Γ, we get that $p(K) = L$, and that $p_{|K}$ is a surjective homomorphism from K to L. To prove the injectivity, we need the following lemma.

Lemma 1.4.6. *The maps $p_{|\hat{\Gamma}}$ and $p_{|K}$ are proper. That is, the inverse image of any compact set of L by $p_{|\hat{\Gamma}}$ and $p_{|K}$ is compact.*

Proof. Let C be a compact set in L, then

$$\begin{aligned} p^{-1}(C) \cap \hat{\Gamma} &= (C \times G) \cap \hat{\Gamma} \\ &= \{(x, \varphi(x)), x \in C \cap \Gamma\}, \end{aligned}$$

which is compact as Γ is closed in G. This shows that $p_{|\hat{\Gamma}}$ is proper.

Let now C be a compact set of L and C_1 a compact set of K such that $K = C_1\hat{\Gamma}$. Clearly, $p^{-1}(C) \cap K$ is closed in K and we have

$$\begin{aligned} p^{-1}(C) \cap K = p^{-1}(C) \cap C_1\hat{\Gamma} &\subset C_1(C_1^{-1}p^{-1}(C) \cap \hat{\Gamma}) \\ &\subset C_1(p^{-1}(p(C_1^{-1})C) \cap \hat{\Gamma}) = C_1 p_{|\hat{\Gamma}}^{-1}(p(C_1^{-1})C). \end{aligned}$$

To conclude that $p^{-1}(C) \cap K$ is compact, it is thus sufficient to prove that $C_1 p_{|\hat{\Gamma}}^{-1}(p(C_1^{-1})C)$ is a compact in K. By the continuity of p, we see that $p(C_1^{-1})C$ is a compact set in L. Now $p_{|\hat{\Gamma}}$ is proper. Then $p_{|\hat{\Gamma}}^{-1}(p(C_1^{-1})C)$ is compact in $\hat{\Gamma}$ and the result follows. □

Now the map $p_{|K}$ is a continuous surjective homomorphism and from its properness, we see that its kernel is a compact subgroup of K. Up to this step, $p_{|K}$ is shown to be a continuous bijective homomorphism. Using the first equality in (1.38), we can see that

$$p_{|K}^{-1} = \exp_K \circ \psi^{-1} \circ \log_L,$$

which is also continuous. This conclusively shows that $p_{|K}$ is an isomorphism. Furthermore, $q_{|K}$ is a continuous homomorphism, then $q_{|K} \circ p_{|K}^{-1}$ is a continuous homomorphism from L to G and for $x \in \Gamma$, we have

$$q_{|K} \circ p_{|K}^{-1}(x) = q_{|K}(x, \varphi(x)) = \varphi(x),$$

which entails that the restriction of $q_{|K} \circ p_{|K}^{-1}$ to Γ is φ. This achieves the proof of existence. As for the uniqueness, let L' be the subgroup of G generated by the family of the one parameter subgroups $(\gamma(t), t \in \mathbb{R})$ for $\gamma \in \Gamma$ defined by the derivative at $t = 0$,

$$\frac{d\gamma(t)}{dt}(0) = \log(\gamma).$$

Then L' is a connected subgroup of G, contained in L and contains Γ, which entails that $L = \overline{L'}$. Let ψ be a continuous homomorphism from L to G and $\gamma \in \Gamma$. By the continuity of ψ, we get

$$\psi(\gamma(t)) = \exp(t \log(\psi(\gamma))), \quad t \in \mathbb{R}. \tag{1.39}$$

Let ψ and ψ' be a two continuous homomorphisms from L to G. By continuity, $\psi = \psi'$ if and only if $\psi_{|L'} = \psi'_{|L'}$. But L' is generated by the subgroups $\gamma(t)$, $\gamma \in \Gamma$. This means therefore that $\psi = \psi'$ if and only if $\psi(\gamma(t)) = \psi'(\gamma(t))$ for all $\gamma \in \Gamma$, which is equivalent by means of (1.39) to $\psi_{|\Gamma} = \psi'_{|\Gamma}$. The same proof is adopted to the case where G is exponential solvable and Γ is Abelian. This achieves the proof of the theorem. □

We finally prove the following result, which will be of great interest in the sequel.

Proposition 1.4.7. *Let G be a completely solvable Lie group, Γ a closed subgroup of G and L its syndetic hull in G. Let $\mathrm{Hom}_c(L, G)$ and $\mathrm{Hom}_c(\Gamma, G)$ designate the sets of continuous homomorphisms from L to G and Γ to G, respectively, endowed with the pointwise convergence topology. Then the restriction natural map $R : \mathrm{Hom}_c(L, G) \to \mathrm{Hom}_c(\Gamma, G) : \psi \mapsto \psi_{|\Gamma}$ is a homeomorphism. This also applies to the case where G is exponentially solvable and Γ is Abelian.*

Proof. We obviously get from Theorem 1.4.5 that R is bijective, and it is clear that R is continuous. To prove the continuity of its inverse, let $(\varphi_n)_n$ be a sequence in $\mathrm{Hom}(\Gamma, G)$, which converges to some element φ. We denote by $(\psi_n)_n$ (resp., ψ) the extensions of $(\varphi_n)_n$ (resp., φ). To prove that the sequence $(\psi_n)_n$ converges to ψ, it is sufficient to show that $(\psi_n(\gamma(t)))_n$ converges to $\psi(\gamma(t))$ for every $\gamma \in \Gamma$. Finally, by (1.39) we have for all $\gamma \in \Gamma$,

$$\begin{aligned}\lim_{n\to\infty} \psi_n(\gamma(t)) &= \lim_{n\to\infty} \exp(t\log(\varphi_n(\gamma))) \\ &= \exp(t\log(\varphi(\gamma))) \\ &= \psi(\gamma(t)).\end{aligned}$$

□

1.4.3 Case of reduced exponential Lie groups

Let us first recall the notion of the universal covering of a Lie group. Here, we record some results, which will be of interest in our study and prove the existence of a unique syndetic hull for any closed subgroup of a reduced completely solvable Lie group. We first record the following results.

Theorem 1.4.8 ([56, Theorem XII.10]). *Let G be a connected Lie group. Then there exists a connected simply connected Lie group $\widetilde{G}$ and a Lie group homomorphism $\pi : \widetilde{G} \to G$, which is a covering. The kernel of π is a discrete, normal subgroup, so central in $\widetilde{G}$. In addition, up to isomorphism, $\widetilde{G}$ is unique. The set is called the universal covering of G.*

Theorem 1.4.9 ([80, Proposition C.8]). *Let G and H be two Lie groups associated to Lie algebras $\mathfrak{g}$ and $\mathfrak{h}$, respectively. Let F be a continuous group homomorphism from G to H. Then there exists an algebra homomorphism f from $\mathfrak{g}$ to $\mathfrak{h}$ such that $F \circ \exp_G = \exp_H \circ f$ where as in* (1.3), *$\exp_G$ and $\exp_H$ are the exponential mappings of G and H, respectively.*

Definition 1.4.10. A reduced solvable Lie group G (resp., exponential solvable, completely solvable, nilpotent) is such that $\widetilde{G}$ is solvable (resp., exponential solvable, completely solvable, nilpotent).

Throughout this section, $\mathfrak{g}$ will denote a real exponential solvable Lie algebra and $\widetilde{G}$ its simply connected associated Lie group. That is, $\widetilde{G}$ is connected and simply connected and it is the universal covering of a connected Lie group G, for which the exponential mapping may fail to be injective. Besides, G and $\widetilde{G}$ have the same Lie algebra. Furthermore, the following is true.

Proposition 1.4.11. *Let G be a reduced connected exponential solvable Lie group. Then the exponential mapping $\exp_G$ is surjective.*

Proof. Let $\widetilde{G}$ be the universal covering of G and $\pi : \widetilde{G} \to G$ the associated covering as in Theorem 1.4.8. We have $\exp_{\widetilde{G}} : \mathfrak{g} \to \widetilde{G}$ is a diffeomorphism. Therefore, according to Theorem 1.4.9, there exists a Lie algebra endomorphism f of $\mathfrak{g}$ such that $\pi \circ \exp_{\widetilde{G}} = \exp_G \circ f$. As $\pi \circ \exp_{\widetilde{G}}$ is surjective, then so is $\exp_G \circ f$. Thus, the map $\exp_G$ is surjective.

□

The following result generalizes Theorem 1.4.2 and establishes the existence of the syndetic hull of any closed subgroup of a reduced connected completely solvable Lie group. More precisely, we have the following.

Theorem 1.4.12. *Any closed subgroup* Γ *of a reduced connected completely solvable Lie group admits a unique syndetic hull* L*, where* $L = \exp_G \mathfrak{l}$, $\mathfrak{l} = \mathbb{R}\text{-span}(\log_{\widetilde{G}} \pi^{-1}(\Gamma))$ *and* $\log_{\widetilde{G}}$ *is as in Definition* 1.1.1.

Proof. Let Γ be a closed subgroup of a reduced completely solvable Lie group G. First, we show that there exists a connected Lie subgroup of G, which contains Γ cocompactly. Let $\widetilde{G}$ be the universal covering of G, and π be the associative covering $\pi : \widetilde{G} \longrightarrow G$. We denote by $\Lambda = \ker \pi$. As π is continuous, $\widetilde{\Gamma} = \pi^{-1}(\Gamma)$ is a closed subgroup of $\widetilde{G}$. Since $\widetilde{G}$ is a connected, simply connected and completely solvable, then according to Theorem 1.4.2, $\widetilde{\Gamma}$ has unique syndetic hull, say, $\widetilde{L}$. Let us prove that $L = \pi(\widetilde{L})$ is a connected Lie subgroup of G, which contains Γ cocompactly. We have $\widetilde{\Gamma} \subset \widetilde{L}$, so $\Gamma \subset \pi(\pi^{-1}(\Gamma)) \subset L$. In addition, we have $\widetilde{L} = \widetilde{C}\widetilde{\Gamma}$, for some compact set $\widetilde{C}$ of $\widetilde{G}$, then

$$\pi(\widetilde{L}) = \pi(\widetilde{C})\pi(\widetilde{\Gamma}) = \pi(\widetilde{C})\Gamma.$$

We must show that L is closed in G, which means that $\widetilde{L}\Lambda$ is closed in $\widetilde{G}$. Let $\log_{\widetilde{G}}\Lambda = \mathbb{Z}\text{-span}(Z_1, \ldots, Z_d)$ for some $d \in \mathbb{N}$, $\mathfrak{a} = \mathbb{R}\text{-span}(Z_1, \ldots, Z_d)$ and $\widetilde{A} = \exp_{\widetilde{G}} \mathfrak{a}$. Then $\widetilde{L}\widetilde{A}$ is closed in $\widetilde{G}$ as $\widetilde{L}\widetilde{A}$ is a connected subgroup in a simply connected solvable Lie group $\widetilde{G}$. Then by Theorem 1.1.6, there exits a coexponential basis of $\widetilde{L}$ in $\widetilde{L}\widetilde{A}$, which means that $\widetilde{L}\widetilde{A}$ is diffeomorphic to $\widetilde{L} \times \mathbb{R}^s$ and conclusively $\widetilde{L}\Lambda$ to $\widetilde{L} \times \mathbb{Z}^s$ for some $s \leqslant d$. Therefore, $\widetilde{L}\Lambda$ is closed in $\widetilde{G}$. Let $\Gamma' = \Gamma Z^c(G)$, which is a closed subgroup of G. Then Γ' admits at least one connected Lie subgroup L of G containing it cocompactly. We now show that L is unique. Indeed, if $L_1 = \exp_G \mathfrak{l}_1$ and $L_2 = \exp_G \mathfrak{l}_2$ are two such Lie groups. We claim that $L_i/(L_1 \cap L_2)$, $i = 1, 2$ are compact. To see that, consider for $i = 1, 2$ the canonical surjection

$$s_i : L_i \longrightarrow L_i/(L_1 \cap L_2),$$

which factors through the canonical surjection $\rho_i : L_i \to L_i/\Gamma$ to a surjection $\overline{s_i} : L_i/\Gamma \to L_i/(L_1 \cap L_2)$ such that $s_i = \overline{s_i} \circ \rho_i$. The map $\overline{s_i}$ is surjective and continuous, and thus its image $L_i/(L_1 \cap L_2)$ is compact. Moreover, it is obvious that $L_i/(L_1 \cap L_2)$ is homeomorphic to $(L_i/Z^c(G))/((L_1 \cap L_2)/Z^c(G))$, which is homeomorphic to $\mathbb{R}^p$ for some $p \in \mathbb{N}$. Indeed, $G/Z^c(G)$ turns out to be a connected simply connected completely solvable Lie group and the existence of the coexponential basis of $(L_1 \cap L_2)/Z^c(G)$ in $L_i/Z^c(G)$ allows us to conclude. Finally, as this quotient is compact, we get conclusively that $p = 0$. Hence, $L_1 \cap L_2 = L_1 = L_2$, as was to be shown. □

Let as above $\widetilde{G}$ be the universal covering of a connected Lie group G and $\pi : \widetilde{G} \to G$ the covering map. A pre-Abelian subgroup Γ of G is a subgroup such that $\widetilde{\Gamma} = \pi^{-1}(\Gamma)$ is

Abelian. When more generally G is exponential solvable and connected, the following could also be seen.

Proposition 1.4.13. *Any pre-Abelian closed subgroup of a reduced connected exponential solvable Lie group admits a unique syndetic hull.*

Proof. Keep the same notation as in the proof of Theorem 1.4.12. In this situation, $\widetilde{\Gamma}$ is Abelian and $\widetilde{L}$ exists by Proposition 1.4.3. Then $\mathfrak{l} = \log_{\widetilde{G}}(\widetilde{L})$ is an Abelian Lie subalgebra of $\mathfrak{g}$. Finally, $\exp_G(\mathfrak{l})$ is a syndetic hull of Γ and the unicity is immediate. □

2 Proper actions on homogeneous spaces

The topological property of being *proper* of an action of a group G on a locally compact space M assures a good behavior for a topological group action. Such actions particularly admit nice properties (the stabilizers are compact, the G-orbits are closed, etc.) when G and M are submitted to some conditions, which has important consequences for the structure of the G-space and the orbit space M/G. On the other hand, the determination of a criterion of the proper action appears to be a key ingredient point in the computation of the parameter, deformation and moduli spaces of the action of a discontinuous group, as we shall see throughout the next coming chapters.

We first give basic definitions of proper and free actions and also define weak and finite proper actions, discontinuous groups and Clifford–Klein forms. We shall then focus on the characterization of the proper (and weak and finite) action of connected, closed subgroups on special and maximal solvmanifolds, stating that it is equivalent to free actions (cf. [35] and [36]). Passing to discrete actions, many phenomena happen and some criteria of properness are also obtained. In the case where G is n-step nilpotent, the proper action of a closed, connected subgroup is shown (with detailed proofs) to be equivalent to its free actions whenever $n \leq 3$.

We also generate a geometric criteria of the proper action of a discontinuous group on an arbitrary homogeneous space, where the group in question is the semidirect product group $K \ltimes \mathbb{R}^n$ (K a compact subgroup of $GL(n, \mathbb{R})$) and in the case of Heisenberg motion groups. As shown, this will be a capital role in the study of many geometrical concepts related to corresponding deformation and moduli spaces.

2.1 Proper and fixed-point actions

Let $\mathscr{M}$ be a locally compact space and K a locally compact topological group. The continuous action of the group K on $\mathscr{M}$ is said to be:

(1) Proper if, for each compact subset $S \subset \mathscr{M}$ the set $K_S = \{k \in K : k \cdot S \cap S \neq \emptyset\}$ is compact (cf. [116]).
(2) Fixed point free (or free) if, for each $m \in \mathscr{M}$, the isotropy group $K_m = \{k \in K : k \cdot m = m\}$ is trivial.
(3) The action has the compact intersection property, denoted (CI), if for each $x \in X$, the isotropy group K_x is compact (see [93]).

In the case where G is a Lie group and H and K are closed subgroups of G, the action of K on the homogeneous space $\mathscr{M} = G/H$ is proper if $SHS^{-1} \cap K$ is compact for any compact set S in G. Likewise the action of K on $\mathscr{M}$ is (CI) (resp., free), if for every $g \in G$, $K \cap gHg^{-1}$ is compact (resp., $K \cap gHg^{-1} = \{e\}$). For the sake of brevity, one says that the triple (G, H, K) is proper, (CI) or free, respectively.

https://doi.org/10.1515/9783110765304-002

Here, for two sets A and B of the locally compact topological group G, the product AB is the subset $\{ab : a \in A, b \in B\}$.

2.1.1 Discontinuous groups

Let $\mathscr{M}$ be a locally compact space and K a locally compact topological group. The action of the group K on $\mathscr{M}$ is said to be properly discontinuous if K is discrete, and for each compact subset $S \subset \mathscr{M}$, the set K_S is finite. The group K is said to be a discontinuous group for the homogeneous space $\mathscr{M}$ if K is discrete and acts properly and fixed point free on $\mathscr{M}$.

2.1.2 Clifford–Klein forms

For any given discontinuous group Γ for a homogeneous space G/H, the quotient space $\Gamma\backslash G/H$ is said to be a *Clifford–Klein form* for the homogeneous space G/H. It is then well known that any Clifford–Klein form is endowed through the action of Γ with a manifold structure for which the quotient canonical surjection

$$\pi : G/H \to \Gamma\backslash G/H \tag{2.1}$$

turns out to be an open covering and particularly a local diffeomorphism. On the other hand, any Clifford–Klein form $\Gamma\backslash G/H$ inherits any G-invariant geometric structure (e. g., complex structure, pseudo-Riemanian structure, conformal structure, symplectic structure, etc.) on the homogeneous space G/H through the covering map π defined as in equation (2.1) below.

Definition 2.1.1 (cf. [95]). Let H and K be subsets of a locally compact topological group G. We denote by $H \pitchfork K$ in G if the set $SHS^{-1} \cap K$ is relatively compact for any compact set S in G. Here, $SHS^{-1} = \{ahb^{-1}, a, b \in S, h \in H\}$. We denote by $H \sim K$ in G if there exists a compact set S of G such that $K \subset SHS^{-1}$ and $H \subset SKS^{-1}$.

Let also $\pitchfork_{gp} (\Gamma : G)$ be the set of closed subgroups belonging to $\pitchfork (\Gamma : G)$.

With the notation above, we have the following results.

Fact 2.1.2 (cf. [95]). *Let G be a locally compact topological group, H, H' and K be three subsets of G. Then:*

(i) *$H \pitchfork K$ in G if and only if H acts on G/K properly.*
(ii) *If $H \sim H'$ and if $H \pitchfork K$ in G, then $H' \pitchfork K$ in G.*
(iii) *$H \pitchfork K$ in G if and only if $K \pitchfork H$ in G.*
(iv) *$H \pitchfork K$ in G if and only if $H \pitchfork \overline{K}$ in G. Here, $\overline{K}$ denotes the closure of K in G.*

Fact 2.1.3 (cf. [95]). *Suppose G' is a closed subgroup of a locally compact topological group G. Let H and K be subsets of G'.*
(i) *If $H \sim K$ in G', then $H \sim K$ in G.*
(ii) *If $H \pitchfork K$ in G, then $H \pitchfork K$ in G'.*

2.1.3 Weak and finite proper actions

In this section, we single out the definitions of weak and finite proper actions. We shall show that these notions of actions are equivalent.

Definition 2.1.4. Let K be a locally compact group and X be a K-locally compact space. We say that the action of K on X is:
(1) weakly proper if for every compact set S in X; the set $K_{x,S} = \{k \in K : k \cdot x \in S\}$ is compact for every $x \in X$,
(2) finitely proper if for every finite set S in X; the set $K_S = \{k \in K : k \cdot S \cap S \neq \emptyset\}$ is finite.

We have the following.

Lemma 2.1.5. *Let G be a locally compact topological group, and let H and K be closed subgroups of G. Then:*
(i) *K acts weakly properly on G/H if and only if for every compact set S in G and every $g \in G$, the set $K \cap SHg$ is compact.*
(ii) *K acts weakly properly on G/H if and only if H acts weakly properly on G/K.*
(iii) *K acts finitely properly on G/H if and only if for every finite set S in G the set $K \cap SHS^{-1}$ is finite.*
(iv) *K acts finitely properly on G/H if and only if H acts finitely properly on G/K.*
(v) *Both finite proper action and weak proper action imply (CI)-action.*
(vi) *Proper action implies weak proper action.*

Proof. Let S be a compact of G then $\bar{S} = SH$ is a compact of G/H. Let $g \in G$, $x = gH$ and $k \in K_{\bar{x},\bar{S}}$. Then $kgH \subset SH$ and, therefore, $k \in SHg^{-1}$. Finally,

$$K_{\bar{x},\bar{S}} = K \cap SHg^{-1},$$

which proves assertion (i).

Remark now that for every compact set S in G and every $g \in G$, we have

$$K \cap SHg \subset S(S^{-1}Kg^{-1} \cap H)g,$$

which proves (ii).

To prove (iii), it is then sufficient to see that for every finite set S in G,

$$K_{\bar{S}} = K \cap SHS^{-1},$$

where $\bar{S} = SH$ is also a finite set of G/H.

The assertions (iv), (v) and (vi) are clear. □

In [101], the authors define the notion of the right strongly similar of two nonempty subsets L and H in a locally compact group G denoted by $L \sim_s H$. We point out here that one can similarly define the notion of the left strongly similar denoted by $L \sim_s^\ell H$, which means that there exist $g_0 \in G$ and a compact S_0 in G such that $L \subset S_0 H g_0$ and $H \subset S_0 L g_0$. In the setting where the sets L and H are subgroups of G, the notions $\sim_s$ and $\sim_s^\ell$ are evidently equivalent. Moreover, the following result stems directly from the definition above.

Proposition 2.1.6. *Let G be a locally compact topological group and let H, L and K be closed subgroups of G. Assume that $L \sim_s^\ell H$. Then the triple (G, H, K) is weakly proper if and only if the triple (G, L, K) is.*

We will need throughout the book to use an induction on the dimension of the group G descending to lower-dimensional subgroups. The following elementary fact due to T. Kobayashi dealing with proper actions under an equivariant map will be used.

Fact 2.1.7 ([94]). *Let G_i $(i = 1, 2)$ be locally compact groups and $H_i, K_i \subset G_i$ be closed subgroups. Suppose that $f : G_1 \to G_2$ is a continuous homomorphism such that $f(H_1) \subset H_2$ and $f(K_1) \subset K_2$. Assume that $f(K_1)$ is closed in G_2 and $K_1 \cap \ker f$ is compact. If K_2 acts properly on G_2/H_2, then K_1 acts properly on G_1/H_1.*

As remarked by T. Kobayashi, there is a strong relationship between the proper action of a cocompact discrete subgroup and the proper action of its syndetic hull on a locally compact Hausdorff space. More precisely, one has the following.

Fact 2.1.8 (cf. [92, Lemma 2.3]). *Suppose a locally compact group L acts on a Hausdorff, locally compact space X. Let Γ be a cocompact discrete subgroup of L. Then the L-action on X is proper if and only if the Γ-action on X is properly discontinuous.*

In [92], T. Kobayashi made a bridge between the action of a discrete group and that of a connected group by noticing that if Γ is a cocompact discrete subgroup of a connected subgroup K, then the action of K on X is proper if and only if the action of Γ on X is properly discontinuous. In this context, and as an analogy of fundamental questions on discontinuous groups, he poses the following problems in the continuous setting.

Problem 1: *Find a criterion on the triple* (G, H, K) *such that the action of* K *on* G/H *is proper.*

Problem 2: *Find a criterion on the triple* (G, H, K) *such that the double coset* $K \setminus G/H$ *is compact for the quotient topology.*

The following problem due to T. Kobayashi is strongly linked to Problem 1 and somehow worth being emphasized:

Problem 3: *When is it true that the property (CI) implies the properness of the action of* K *on* X?

In [95], an affirmative answer was given by T. Kobayashi himself in the case where each of the triple (G, H, K) is reductive. Moreover, the last question may have a negative answer even in the case where G is reductive or Abelian (see [92] and [114]). For $n \geq 1$, let $G^n = N_n(\mathbb{R}) \ltimes \mathbb{R}^n$, where $N_n(\mathbb{R})$ designates the group of n-upper real triangular matrices. Several authors considered the action given by an affine transformation subgroup contained in G^n. For $n = 2$, more in fact is true. T. Kobayashi shows in [93] that for $G = GL(2, \mathbb{R}) \ltimes \mathbb{R}^2$ and $H = GL(2, \mathbb{R})$, the triple (G, H, K) is (CI) if and only if K acts properly on G/H for any connected closed subgroup K in G.

In [108], R. Lipsman established the last result for $n = 3$ taking $N_3(\mathbb{R})$ for one of the subgroups in question. He conjectured the following.

Conjecture 2.1.9. *Let* G *be a simply connected nilpotent Lie group,* H *and* K *be connected subgroups of* G. *Then the triple* (G, H, K) *has the (CI) property if and only if* K *acts properly on* G/H.

T. Yoshino proved in [136] that the result holds for $n = 4$ as well, but it fails for every $n \geq 5$ (see [134]). This implies that Conjecture 2.1.9 may fail for n-step nilpotent Lie groups ($n \geq 4$). We here aim to partially answer the above questions in some situations of exponential solvable homogeneous spaces. More precisely, we shall prove Conjecture 2.1.9 below for some classes of nilpotent Lie groups and go beyond the nilpotent for some restricted contexts as well. We first introduce the following.

Definition 2.1.10. Let G be an exponential solvable Lie group and H a connected and closed subgroup of G. A pair (G, H) is said to have the Lipsman property, if for any closed, connected subgroup L of G, there is equivalence between proper and fixed point- free for the triple (G, H, L).

Exponential solvable Lie groups admit no nontrivial connected compact subgroups, which merely entails that properties (2) and (3) in Definitions 2.1 are equivalent as the isotropy group K_x is $K \cap gHg^{-1}$ for $x = gH$. On the other hand, it is obvious that Property (1) implies both Properties (2) and (3) in this case; so, it appears natural to seek the converse. With the above in mind, the following lemma is immediate.

Lemma 2.1.11. *Let G be an exponential solvable Lie group; H and K are closed connected subgroups of G. Then the following conditions are equivalent:*

(i) *The triple (G, H, K) has the (CI) property.*
(ii) *The action of K on G/H is free, that is, $K \cap gHg^{-1} = \{e\}$ for any $g \in G$.*
(iii) $\mathfrak{k} \cap \mathrm{Ad}_g \mathfrak{h} = \{0\}$ *for any $g \in G$. Here, $\mathfrak{h}$ and $\mathfrak{k}$ are the Lie algebras respectively of H and K.*

2.1.4 Campbell–Baker–Hausdorff series

Let G be an exponential solvable Lie group. In this context, the Campbell–Baker–Hausdorff formula allows to reconstruct the group G with its multiplication law knowing only the structure of $\mathfrak{g}$. Furthermore, for every A, $B \in \mathfrak{g}$, we have

$$\exp A \exp B = \exp(A * B) = \exp C(A, B),$$

where $C(A, B) = \sum_{n\geq 1} C_n(A, B)$ and $C_n(A, B)$ is determined by the recursion formula (see [128]): $C_1(A, B) = A + B$ and for $n \geq 1$,

$$\begin{aligned}(n+1)C_{n+1}(A, B) &= \frac{1}{2}[A - B, C_n(A, B)] \\ &+ \sum_{p\geq 1, 2p\leq n} K_{2p} \sum_{\substack{k_1,\dots,k_{2p}>0 \\ k_1+\dots+k_{2p}=n}} [C_{k_1}(A, B), [\cdots [C_{k_{2p}}(A, B), A + B] \cdots]],\end{aligned}$$

where the rational numbers K_{2p}, $p \geq 1$ are given by

$$\frac{z}{1 - e^{-z}} - \frac{1}{2}z = 1 + \sum_{p=1}^{+\infty} K_{2p} z^{2p}.$$

This series is absolutely convergent. In particular,

$$C_2(A, B) = \frac{1}{2}[A, B]$$

and

$$C_3(A, B) = \frac{1}{12}[A, [A, B]] + \frac{1}{12}[[A, B], B].$$

Furthermore, if G is nilpotent, then $C(A, B)$ is a polynomial function in both the variables A and B.

2.1.5 Proper actions and coexponential bases

We have then the following.

Proposition 2.1.12. *Let G be a connected simply connected solvable Lie group, H a connected subgroup and let $\{X_1, \ldots, X_p\}$ be a coexponential basis of H in G as in Theorem 1.1.6. Let S be a compact set of G, then there exist two compact sets S_H and $S_1 = \prod_{i=p}^{1} \exp(I_i X_i)$ such that $(I_i)_{1\leq i\leq p}$ are compact sets of $\mathbb{R}$, $S_H \subset H$ and $S \subset S_1 S_H$.*

Proof. Let S be a compact set in G, then $(\varphi_{\mathfrak{g},\mathfrak{h}})^{-1}(S)$ is a compact set in $\mathbb{R}^p \times H$. So, there exist some compact sets $(I_i)_{1\leq i\leq p}$ of $\mathbb{R}$ and a compact set S_H of H such that $(\varphi_{\mathfrak{g},\mathfrak{h}})^{-1}(S) \subset I_1 \times \cdots \times I_p \times S_H$. Then $S \subset \prod_{i=p}^{1} \exp(I_i X_i) S_H$. □

Corollary 2.1.13. *Let G be a connected simply connected solvable Lie group, and H and K be connected subgroups of G. Then the action of K on G/H is proper if and only if for every compact set S of G of the form $S = \prod_{i=p}^{1} \exp(I_i X_i)$, where $(I_i)_{1\leq i\leq p}$ are compact sets of $\mathbb{R}$, and the set $SHS^{-1} \cap K$ is relatively compact in G.*

Proof. Let S be a compact set in G, using Proposition 2.1.12, we have that $S \subset S_1 S_H$ where $S_1 = \prod_{i=p}^{1} \exp(I_i X_i)$, and S_H is a compact set in H. Then

$$SHS^{-1} \cap K \subset S_1 S_H H S_H^{-1} S_1^{-1} \cap K = S_1 H S_1^{-1} \cap K,$$

which is relatively compact in G according to our assumption. □

The following result will be used later and deals with proper action when one of the subgroups is normal in G.

Proposition 2.1.14. *Let G be a connected, simply connected, solvable Lie group and H, K connected subgroups of G. Assume that one of the subgroups H or K is normal in G, then K acts properly on G/H if and only if triple (G, H, K) has the (CI) property.*

Proof. Without loss of generality, we may and do assume that H is normal. Then the result follows from Fact 2.1.7 with f being the natural quotient map $f : G \to G/H$. We must show that $f(K)$ is closed in G/H, which is equivalent to the fact that HK is closed in G. To see that, we engage an induction on the dimension of G. This is obvious for small dimensions. Let G_0 be a normal subgroup of G of codimension one containing H. If $K \subset G_0$, we are done. Otherwise, let $X \in \mathfrak{k} \setminus \mathfrak{g}$ and $K_0 = K \cap G_0$, which is a closed subgroup of G_0. Obviously, thanks to Remark 1.1.7, $\{X\}$ is a coexponential basis of G_0 in G, and we have

$$HK = HK_0 \cdot \exp(\mathbb{R}X),$$

which closes the proof as HK_0 is a closed in G. □

2.2 Proper actions for 3-step nilpotent Lie groups

We consider hereafter the case of 3-step nilpotent Lie groups. In this context, the ascending central sequence is defined by

$$\mathfrak{g}_{(0)} = \{0\},\ \mathfrak{g}_{(j)} = \{X \in \mathfrak{g} : X \text{ is central mod } \mathfrak{g}_{(j-1)}\}.$$

Then each $\mathfrak{g}_{(j)}$ is an ideal of $\mathfrak{g}$, $\mathfrak{g}_{(1)} = \mathfrak{z}(\mathfrak{g})$ is the center of $\mathfrak{g}$, $\mathfrak{g}_{(2)} = \mathfrak{d} = [\mathfrak{g}, \mathfrak{g}]$ and we have

$$\mathfrak{g}_{(0)} \subsetneq \mathfrak{g}_{(1)} = \mathfrak{z}(\mathfrak{g}) \subsetneq \mathfrak{g}_{(2)} \subsetneq \mathfrak{g}_{(3)} = \mathfrak{g}.$$

Furthermore, for every $A,\ B \in \mathfrak{g}$, we have (cf. Section 2.1.4)

$$\exp A \exp B = \exp(A * B) = \exp\left(A + B + \frac{1}{2}[A, B] + \frac{1}{12}[A, [A, B]] + \frac{1}{12}[B, [B, A]]\right).$$

In the case of 2-step nilpotent Lie groups (i. e., $\mathfrak{g}_{(2)} = \mathfrak{g}$), we prove the following result. An alternative proof is also provided in [114].

Theorem 2.2.1. *Let G be a connected, simply connected at most 2-step nilpotent Lie group, H and K connected subgroups of G. Then K acts properly on G/H if and only if the triple (G, H, K) has the (CI) property.*

Proof. Suppose that the action of K on G/H is not proper, then there exists a compact set S in G such that $K \cap SHS^{-1}$ is not relatively compact. Hence, one can find sequences $V_n \in \mathfrak{h}$, $W_n \in \mathfrak{k}$, $A_n, B_n \in \mathfrak{g}$, such that:

(2.2.1.1) $\exp A_n \in S$, $\exp B_n \in S$, $\lim_{n\to+\infty} A_n = A$, $\lim_{n\to+\infty} B_n = B$,
(2.2.1.2) $\lim_{n\to+\infty} \|V_n\| = \lim_{n\to+\infty} \|W_n\| = +\infty$,
(2.2.1.3) $\lim_{n\to+\infty} \frac{V_n}{\|V_n\|} = V$, $\lim_{n\to+\infty} \frac{W_n}{\|W_n\|} = W$, where $V \in \mathfrak{h}$, $W \in \mathfrak{k}$ and $\|V\| = \|W\| = 1$,
(2.2.1.4) $\exp(W_n) = \exp(A_n)\exp(V_n)\exp(-B_n)$.

The last equation (2.2.1.4) gives

$$W_n = A_n * (-B_n) + V_n + \frac{1}{2}[A_n + B_n, V_n] = \mathrm{Ad}_{\exp(\frac{A_n+B_n}{2})} V_n + A_n * (-B_n).$$

Let $\alpha_n = \frac{\|V_n\|}{\|W_n\|}$, then $\frac{W_n}{\|W_n\|} = \alpha_n \mathrm{Ad}_{\exp(\frac{A_n+B_n}{2})} \frac{V_n}{\|V_n\|} + \frac{1}{\|W_n\|}(A_n * (-B_n))$. Taking the limit as n tends to $+\infty$, we obtain

$$W = \lim_{n\to+\infty} \alpha_n \mathrm{Ad}_{\exp(\frac{A_n+B_n}{2})} \frac{V_n}{\|V_n\|}.$$

Since $V \neq 0$, the limit $\lim_{n\to+\infty} \mathrm{Ad}_{\exp(\frac{A_n+B_n}{2})} \frac{V_n}{\|V_n\|} = \mathrm{Ad}_{\exp(\frac{A+B}{2})} V$ is not zero, which implies that the sequence α_n converges to some $\alpha \in \mathbb{R}_+^*$, and finally $W = \mathrm{Ad}_{\exp(\frac{A+B}{2})}(\alpha V)$, which is impossible by the (CI) property of (G, H, K). □

We generalize now the above result. We prove the following.

Theorem 2.2.2. *Let G be a connected simply connected at most 3-step nilpotent Lie group, and H and K be connected subgroups of G. Then K acts properly on G/H if and only if the triple (G, H, K) has the (CI) property.*

Proof. Let us drop for the moment the assumption that G is 3-step. Let

$$\mathscr{S} : \{0\} = \mathfrak{g}_0 \subset \mathfrak{g}_1 \subset \cdots \subset \mathfrak{g}_m = \mathfrak{g} \tag{2.2}$$

be a strong Malcev sequence and $G_i = \exp(\mathfrak{g}_i)$, $i = 0, \dots, m$. Choosing for every $j \in \{1, \dots, m\}$ a vector Z_j in $\mathfrak{g}_j \setminus \mathfrak{g}_{j-1}$, we obtain a strong Malcev basis $\mathscr{B} = \{Z_1, \dots, Z_m\}$ of $\mathfrak{g}$. We denote by

$$\mathscr{I}^{\mathfrak{h}} = \mathscr{I} = \{i_1 < \cdots < i_a\} \quad (a = \dim \mathfrak{h})$$

the set of indices i $(1 \le i \le m)$ such that $\mathfrak{h} \cap \mathfrak{g}_i \neq \mathfrak{h} \cap \mathfrak{g}_{i-1}$. Let us put

$$\mathscr{J}^{\mathfrak{g}/\mathfrak{h}} = \mathscr{J} = \{j_1 < \cdots < j_p\} = \{1, \dots, m\} \setminus \mathscr{I},$$

with $p = \dim(\mathfrak{g}/\mathfrak{h}) = m - a$. So, it appears clear that we can pick the vectors $Z_k, k \in \mathscr{I}$ of $\mathscr{B}$ in such that $Z_k \in \mathfrak{h}$. This implies that the basis $\{X_1 = Z_{j_1}, \dots, X_p = Z_{j_p}\}$ is a Malcev basis of $\mathfrak{g}$ relative to $\mathfrak{h}$, which is a coexponential basis of $\mathfrak{h}$ in $\mathfrak{g}$. So by Corollary 2.1.13, the action of K on G/H is proper if and only if for every compact set S of G of the form $S = \prod_{i=p}^{1} \exp(I_i X_i)$, where $(I_i)_{1\le i\le p}$ are compact sets of $\mathbb{R}$; the set $SHS^{-1} \cap K$ is relatively compact in G. Even more, the following more general fact holds.

Proposition 2.2.3. *We keep the same notation and hypotheses. The subgroup K acts on G/H properly if and only if for every $q \in \{1, \dots, m\}$ and for every compact set S^q of G included in G_q, the set $SHS^{-1} \cap K$ is relatively compact in G, where*

$$S = \prod_{k\in\mathscr{J},\, k>q} \exp(I_k Z_k) S^q$$

and $(I_k)_{k\in\mathscr{J},\, k>q}$ are compact sets in $\mathbb{R}$.

We come back now to 3-step nilpotent Lie groups. Recall the notation $\mathfrak{d} = [\mathfrak{g}, \mathfrak{g}]$ and $D = \exp(\mathfrak{d})$. We prove first the following.

Lemma 2.2.4. *Let G be a connected, simply connected at most 3-step nilpotent Lie group, and H and K be connected subgroups of G such that (G, H, K) has the (CI) property. Let S be a compact set of G. Then:*

(1) *If S is included in D, the set $SHS^{-1} \cap K$ is relatively compact in G.*
(2) *If H or K is included in D, the action of K on G/H is proper.*

Proof. Let S be a compact set in G such that $SHS^{-1} \cap K$ is not relatively compact. Then there exist some sequences $A_n, B_n \in \mathfrak{g}$, $V_n \in \mathfrak{h}$ and $W_n \in \mathfrak{k}$ meeting conditions (2.2.1.1), ..., (2.2.1.4) of Theorem 2.2.1. So, if $S \subset D$ according to (2.2.4.1) or if one of the subgroups H or K (assume, e. g., H) is included in D according to (2.2.4.2), then equation (2.2.1.4) gives

$$W_n = A_n * (-B_n) + V_n + \frac{1}{2}[A_n + B_n, V_n] = A_n * (-B_n) + \mathrm{Ad}_{\exp(\frac{A_n+B_n}{2})} V_n.$$

Hence, the same procedure as in the proof of Theorem 2.2.1 gives us a contradiction. □

We suppose from now on that both $\mathfrak{h}$ and $\mathfrak{k}$ are not included in $\mathfrak{d}$. Take by the way $\{Y_1, \dots, Y_q\}$, a Malcev basis of $\mathfrak{h}$ relative to $\mathfrak{d} \cap \mathfrak{h}$ and $\{Y'_1, \dots, Y'_r\}$, a Malcev basis of $\mathfrak{k}$ relative to $\mathfrak{d} \cap \mathfrak{k}$. Let $\{i_1, \dots, i_s\}$ be a maximal set of indices in $\{1, \dots, r\}$ such that the vectors $\{Y_1, \dots, Y_q, Y'_{i_1}, \dots, Y'_{i_s}\}$ are linearly independent modulo $\mathfrak{d}$. We note $T_k = Y'_{i_k}$, $k = 1, \dots, s$. Remark that

$$\mathfrak{d} \oplus \left(\bigoplus_{k=1}^{q} \mathbb{R} Y_k \right) = \mathfrak{d} + \mathfrak{h}$$

and that

$$\mathfrak{d} \oplus \left(\bigoplus_{k=1}^{q} \mathbb{R} Y_k \right) \oplus \left(\bigoplus_{k=1}^{s} \mathbb{R} T_k \right) = \mathfrak{d} + \mathfrak{h} + \mathfrak{k}.$$

In the case where $\mathfrak{d} + \mathfrak{h} + \mathfrak{k} = \mathfrak{g}$, the action of K on G/H is proper. In fact, let $\mathscr{S}$ be a flag of ideals as in (2.2) such that $\mathfrak{g}_d = \mathfrak{d}$ and constructed by means of the vectors $\{Y_1, \dots, Y_q, T_1, \dots, T_s\}$. Let also $S = \prod_{k=s}^{1} \exp(I_k T_k) S^d$ be a compact set of G where $(I_k)_{1 \le k \le s}$ are compact sets of $\mathbb{R}$ and S^d is a compact set of D. We have to show according to Proposition 2.2.3 that $K \cap SHS^{-1}$ is relatively compact. In fact,

$$\begin{aligned} K \cap SHS^{-1} &= K \cap \prod_{k=s}^{1} \exp(I_k T_k)(S^d H (S^d)^{-1}) \prod_{k=1}^{s} \exp(-I_k T_k) \\ &= \prod_{k=s}^{1} \exp(I_k T_k)(K \cap S^d H (S^d)^{-1}) \prod_{k=1}^{s} \exp(-I_k T_k). \end{aligned}$$

But $K \cap S^d H (S^d)^{-1}$ is relatively compact using the assertion (1) of Lemma 2.2.4, which completes the proof in this case. Suppose finally that $\mathfrak{d} + \mathfrak{h} + \mathfrak{k} \subsetneq \mathfrak{g}$. Let $\{X_1, \dots, X_l\}$ be a

Malcev basis of $\mathfrak{g}$ relative to $\mathfrak{d}+\mathfrak{h}+\mathfrak{k}$, and $\{Z_1,\dots,Z_d\}$ a strong Malcev basis of $\mathfrak{d}$ passing through $\mathfrak{d}\cap\mathfrak{h}$. The basis

$$\mathscr{B}=\{Z_1,\dots,Z_d,X_1,\dots,X_l,T_1,\dots,T_s,Y_1,\dots,Y_q\},$$

is a strong Malcev basis of $\mathfrak{g}$. We take a corresponding flag of ideals of $\mathfrak{g}$ such that $\mathfrak{g}_d=\mathfrak{d}$. We prove now the following.

Lemma 2.2.5. *Let G be a connected, simply connected at most 3-step nilpotent Lie group, and H and K connected subgroups of G such that (G,H,K) has the (CI) property. With the same notation as above, the action of K on G/H is proper if and only if for every compact set of G of the form $S=\prod_{k=l}^{1}\exp(I_kX_k)S^d$, where $(I_k)_{1\le k\le l}$, S^d are respectively compact sets of $\mathbb{R}$ and D; the set $SHS^{-1}\cap K$ is relatively compact in G.*

Proof. In order to prove that the action of K on G/H is proper, it is clear that according to Proposition 2.2.3, we merely have to show that $S_1HS_1^{-1}\cap K$ is relatively compact in G for every compact S_1 of G of the form $S_1=\prod_{k=s}^{1}\exp(J_kT_k)\prod_{k=l}^{1}\exp(I_kX_k)S^d$, where $(I_k)_{1\le k\le l}$ and $(J_k)_{1\le k\le s}$ are compact sets in $\mathbb{R}$ and S^d is a compact set of D. Write $S_K=\prod_{k=s}^{1}\exp(J_kT_k)$ and $S=\prod_{k=l}^{1}\exp(I_kX_k)S^d$. We then have

$$S_1HS_1^{-1}\cap K\subset S_KSHS^{-1}S_K^{-1}\cap K=S_K(SHS^{-1}\cap K)S_K^{-1},$$

which is relatively compact in G using our hypothesis. □

Let us go back to the proof of Theorem 2.2.2. Suppose that the action of K on G/H is not proper, then there exists by Lemma 2.2.5 a compact subset $S=\prod_{k=l}^{1}\exp(I_kX_k)S^d$ of G such that $(I_k)_{1\le k\le l}$ are compact sets of $\mathbb{R}$, S^d is a compact set of D and $K\cap SHS^{-1}$ is not relatively compact in G. Thus, we can find as above some sequences $V_n\in\mathfrak{h}$, $W_n\in\mathfrak{k}$, A_n, $B_n\in\mathfrak{g}$ meeting conditions (2.2.1.1), …, (2.2.1.4) of Theorem 2.2.1. By equation (2.2.1.4), we have that

$$\exp(W_n)\exp(C_n)=\exp(A_n)\exp(V_n)\exp(-A_n)=\mathrm{Ad}_{\exp A_n}V_n, \tag{2.3}$$

where $C_n=B_n*(-A_n)$. Then we have by (2.3) that

$$W_n+C_n=V_n\operatorname{mod}(\mathfrak{d}). \tag{2.4}$$

It follows according to our choice of the basis $\mathscr{B}$ and (2.4), that $W_n=V_n\operatorname{mod}(\mathfrak{d})$ and that $C_n\in\mathfrak{d}$. Then equation (2.3) gives

$$C_n+\mathrm{Ad}_{\exp(-\frac{1}{2}C_n)}W_n=\mathrm{Ad}_{\exp A_n}V_n$$

as $[C_n, \mathfrak{g}] \subset \mathfrak{z}(\mathfrak{g})$. Writing as above $\alpha_n = \frac{\|V_n\|}{\|W_n\|}$, we get

$$\frac{C_n}{\|W_n\|} + \mathrm{Ad}_{\exp(-\frac{1}{2}C_n)} \frac{W_n}{\|W_n\|} = \alpha_n \,\mathrm{Ad}_{\exp A_n} \frac{V_n}{\|V_n\|}.$$

Taking the limit as n tends to $+\infty$, we obtain

$$\mathrm{Ad}_{\exp(-\frac{1}{2}C)} W = \alpha \,\mathrm{Ad}_{\exp A} V,$$

where $C = \lim_{n\to+\infty} C_n$, $\alpha = \lim_{n\to+\infty} \alpha_n$ and $A = \lim_{n\to+\infty} A_n$. This contradicts the (CI) property of the triple (G, H, K), which achieves the proof of the theorem. An alternative proof of Theorem 2.2.2 is subject of the paper [133]. □

2.3 Special nilpotent Lie groups

We consider in this section Conjecture 2.1.9 in another setting. A solvable Lie algebra $\mathfrak{g}$ is said to be special, if it admits a codimensional one Abelian ideal. In terms of groups, the Lie group G associated to $\mathfrak{g}$ turns out to be a semidirect product $\mathbb{R} \ltimes \mathbb{R}^m$ being noncommutative. Let $G_m = \exp(\mathfrak{g}_m)$ be the threadlike nilpotent Lie group as defined in Subsection 1.1.3. We begin this section by proving Conjecture 2.1.9 for $G_m, m \geq 2$.

Proposition 2.3.1. *Conjecture 2.1.9 holds for $G_m, m \geq 2$.*

Proof. Recall the notation, $K = \exp(\mathfrak{k})$, $H = \exp(\mathfrak{h})$ and $G^0 = \exp(\mathfrak{g}^0)$. We proceed by induction on the step m of G. If $m = 2$, then G is 2-step and we are already done by Theorem 2.2.1. We suppose then that $m > 2$ and we shall discuss several cases. Suppose in a first time that both $\mathfrak{h}$ and $\mathfrak{k}$ are included in $\mathfrak{g}^0$ and suppose that the K action on G/H is not proper. There exists a compact set S in G such that $K \cap SHS^{-1}$ is not relatively compact. There exist by Proposition 2.2.3 a compact set S_0 in $\mathfrak{g}^0$ and a compact set I in $\mathbb{R}$ such that $S \subset \exp(IX)S_0$. Thus, we can find sequences A_n, $B_n \in \mathfrak{g}^0$, $t_n, s_n \in I$, $V_n \in \mathfrak{h}$ and $W_n \in \mathfrak{k}$, such that:

(2.3.1.1) $\exp(A_n) \in S_0$, $\exp(B_n) \in S_0$,
(2.3.1.2) $\lim_{n\to+\infty} t_n = t$, $t \in I$ and $\lim_{n\to+\infty} s_n = s$, $s \in I$,
(2.3.1.3) $\lim_{n\to+\infty} \|V_n\| = \lim_{n\to+\infty} \|W_n\| = +\infty$,
(2.3.1.4) $\lim_{n\to+\infty} \frac{V_n}{\|V_n\|} = V$, $\lim_{n\to+\infty} \frac{W_n}{\|W_n\|} = W$, where $V \in \mathfrak{h}$, $W \in \mathfrak{k}$ and $\|V\| = \|W\| = 1$,
(2.3.1.5) $\exp(W_n) = \exp(t_n X)\exp(A_n)\exp(V_n)\exp(-B_n)\exp(-s_n X)$.

The last equation gives

$$\exp(W_n) = \exp(t_n X)\exp(A_n + V_n - B_n)\exp(-t_n X)\exp((t_n - s_n)X).$$

Thus, $s_n = t_n$ and $W_n = \mathrm{Ad}_{\exp t_n X}(A_n - B_n + V_n)$, as $\mathfrak{g}^0$ is Abelian. Denoting $\alpha_n = \frac{\|V_n\|}{\|W_n\|}$, $n \in \mathbb{N}$, we get

$$\frac{W_n}{\|W_n\|} = \alpha_n \,\mathrm{Ad}_{\exp t_n X}\left(\frac{A_n - B_n + V_n}{\|V_n\|}\right).$$

Since $V \neq 0$ and $\lim_{n\to+\infty} \mathrm{Ad}_{\exp(t_n X)}(\frac{V_n + A_n - B_n}{\|V_n\|}) = \mathrm{Ad}_{\exp(tX)}\, V \neq 0$, α_n converges to some $\alpha \in \mathbb{R}_+^*$, and finally $W = \alpha\, \mathrm{Ad}_{\exp tX}\, V$, which is impossible by the (CI) property.

Suppose now that both $\mathfrak{h}$ and $\mathfrak{k}$ are not contained in $\mathfrak{g}^0$. We can then assume that $X \in \mathfrak{h}$. So, there exists a one-codimensional ideal $\mathfrak{g}^1$ of $\mathfrak{g}$, which contains $\mathfrak{h}$. As $[X, Y_i] \in \mathfrak{g}^1$, $i = 1, \ldots, m-1$, it follows that $\mathfrak{g}^1 = \mathrm{span}\{X, Y_2, \ldots Y_m\}$. When $\mathfrak{k} \subset \mathfrak{g}^1$, we write $\mathfrak{k} = \mathbb{R}Y \oplus \mathfrak{k}^0$ where $\mathfrak{k}^0 = \mathfrak{k} \cap \mathfrak{g}^0$ and $Y = X + \sum_{i=2}^m y_i Y_i$, for some $y_i \in \mathbb{R}$. Let $T = \sum_{i=1}^{m-1} y_{i+1} Y_i$. Then

$$\begin{aligned}
\mathrm{Ad}_{\exp T}\, Y &= Y + \left[\sum_{i=1}^{m-1} y_{i+1} Y_i, Y\right] \\
&= X + \sum_{i=2}^{m} y_i Y_i + \sum_{i=1}^{m-1} y_{i+1} [Y_i, X] \\
&= X + \sum_{i=2}^{m} y_i Y_i - \sum_{i=1}^{m-1} y_{i+1} Y_{i+1} = X,
\end{aligned}$$

which contradicts the (CI) property of (G, H, K). Suppose now that $\mathfrak{k} \not\subset \mathfrak{g}^1$. Write likewise $\mathfrak{k} = \mathbb{R}X_1 \oplus \mathfrak{k}^1$ where $\mathfrak{k}^1 = \mathfrak{k} \cap \mathfrak{g}^1$. It is clear that $\{X_1\}$ is a coexponential basis of $\mathfrak{g}^1$ in $\mathfrak{g}$. Let S a compact set of G, there exist a compact set J in $\mathbb{R}$ and a compact set S_1 in $G^1 = \exp(\mathfrak{g}^1)$ such that $S \subset \exp(JX_1)S_1$. Then, noting $K^1 = \exp(\mathfrak{k}^1)$, we have

$$\begin{aligned}
K \cap SHS^{-1} &\subset \exp(JX_1)(K \cap S_1 H S_1^{-1}) \exp(-JX_1) \\
&= \exp(JX_1)(K^1 \cap S_1 H S_1^{-1}) \exp(-JX_1).
\end{aligned} \tag{2.5}$$

Since $\mathfrak{g}^1$ is $(m-1)$-step threadlike Lie algebra and the triple (G^1, H, K^1) fulfills the (CI) property, the set $K^1 \cap S_1 H S_1^{-1}$ turns out to be relatively compact in G^1 by the induction hypothesis. We conclude then that K acts properly on G/H by (2.5). Finally, when only one of the subalgebras $\mathfrak{h}$ or $\mathfrak{k}$ is included in $\mathfrak{g}^0$, we argue similarly as above replacing G^1 by G^0, K^1 by $K^0 = \exp(\mathfrak{k}^0)$ and using the fact that G^0 is Abelian. □

We extend now the above result to an arbitrary nilpotent special Lie group, and we have the following.

Theorem 2.3.2. *Conjecture 2.1.9 holds for special nilpotent Lie groups.*

Proof. Let G be a special nilpotent Lie group and $\mathfrak{g}$ its Lie algebra. We write $\mathfrak{g} = \mathfrak{g}_0 \oplus \mathbb{R}X$ where $\mathfrak{g}_0$ is the one-codimensional Abelian ideal of $\mathfrak{g}$ and $G_0 = \exp(\mathfrak{g}_0)$ and $X \in \mathfrak{g} \setminus \mathfrak{g}_0$. Recall the notation $\mathfrak{z}(\mathfrak{g})$ of the center of $\mathfrak{g}$. We proceed by induction on $\dim G$.

We suppose in a first time that $\mathfrak{h}$ or $\mathfrak{k}$ contains a nonzero central vector of $\mathfrak{g}$; take, for example, $\mathfrak{h} \cap \mathfrak{z}(\mathfrak{g}) \neq \{0\}$. Let then $Z \in \mathfrak{h} \cap \mathfrak{z}(\mathfrak{g})$ and $\mathfrak{i} = \operatorname{span}\{Z\}$. We note $I = \exp(\mathfrak{i})$ and $\overline{G} = G/I$. We consider the canonical surjection:

$$f : G \to \overline{G}.$$

Then f is a continuous homomorphism, and let $f(H) = \overline{H}$ and $f(K) = \overline{K}$. As $\mathfrak{i} \subset \mathfrak{g}_0$, it is clear that $\overline{G}$ is a special Lie group and if the triple (G, H, K) has the (CI) property, then the triple $(\overline{G}, \overline{H}, \overline{K})$ does. In fact, let $\overline{B} \in \overline{\mathfrak{h}}$, $\overline{A} \in \overline{\mathfrak{k}}$ and $\overline{T} \in \overline{\mathfrak{g}}$ such that $\overline{A} = \operatorname{Ad}_{\exp \overline{T}} \overline{B}$, then there exists $\alpha \in \mathbb{R}$ such that $B = \operatorname{Ad}_{\exp T} A + \alpha Z$. This implies that $B - \alpha Z = \operatorname{Ad}_{\exp T} A$. Finally, $B = \alpha Z$ as (G, H, K) has the (CI) property and, therefore, $\overline{B} = \overline{0}$. We apply the induction hypothesis to $\overline{G}$ to obtain that the action of $\overline{K}$ on $\overline{G}/\overline{H}$ is proper. Moreover, $f(K)$ is closed in $\overline{G}$ and $K \cap \ker f = K \cap I = \{e\}$ is compact. Using Lemma 2.1.7, we conclude that the action of K on G/H is proper. Hence, we can suppose from now on that

$$\mathfrak{h} \cap \mathfrak{z}(\mathfrak{g}) = \mathfrak{k} \cap \mathfrak{z}(\mathfrak{g}) = \{0\}. \tag{2.6}$$

The case where $\mathfrak{h}$ and $\mathfrak{k}$ are included in $\mathfrak{g}_0$ is settled exactly with the same way as in Proposition 2.3.1. We tackle now the remaining cases. Suppose for a while that $H \subset G_0$ and $K \not\subset G_0$, then we can assume that $X \in \mathfrak{k}$ and then that $\dim \mathfrak{k} = 1$ by our assumption (2.6) above. In fact, suppose that $\dim K > 1$, then there exists $Y \in \mathfrak{k} \setminus \mathfrak{z}(\mathfrak{g})$ such that X and Y are linearly independent. Let n_0 be the largest integer such that $\operatorname{ad}_X^{n_0}(Y) \neq 0$ and $\operatorname{ad}_X^{n_0+1}(Y) = 0$. Then $\operatorname{ad}_X^{n_0}(Y) \in \mathfrak{k} \cap \mathfrak{z}(\mathfrak{g})$, which is absurd. Let now S be a compact set of G. Then there exist by Proposition 2.1.12 a compact set I of $\mathbb{R}$ and a compact set S_0 of G_0, such that $S \subset \exp(IX)S_0$. we have then that

$$\begin{aligned} K \cap SHS^{-1} &\subset K \cap \exp(IX)S_0HS_0^{-1}\exp(-IX) \\ &= \exp(IX)(K \cap S_0HS_0^{-1})\exp(-IX) = \exp(IX)\exp(-IX), \end{aligned}$$

which is compact in G. Hence, $K \cap SHS^{-1}$ is relatively compact.

Finally, if $H \not\subset G_0$ and $K \not\subset G_0$, then as in the previous case, we have that $\dim H = \dim K = 1$. Write $\mathfrak{h} = \operatorname{span}\{X\}$ and then $\mathfrak{k} = \operatorname{span}\{X + U_0\}$ for some nonzero $U_0 \in \mathfrak{g}_0$. If the action of K on G/H is not proper, then choosing a flag of ideals $\mathscr{S}$ as in (2.2) passing through $\mathfrak{g}_0$ and $\mathfrak{h}$, according to Proposition 2.2.3, we can suppose that there exists a compact subset S_0 of G_0 such that $K \cap S_0HS_0^{-1}$ is not relatively compact. Thus, we can find sequences $x_n, y_n \in \mathbb{R}$, $A_n, B_n \in \mathfrak{g}_0$, such that:

(2.3.2.1) $\exp(A_n) \in S_0$, $\exp(B_n) \in S_0$,
(2.3.2.2) $\lim_{n\to+\infty} |x_n| = +\infty$, $\lim_{n\to+\infty} |y_n| = +\infty$,
(2.3.2.3) $\exp(y_n(X + U_0)) = \exp(A_n)\exp(x_nX)\exp(-B_n) = \exp(x_nX + A_n - B_n + C(x_n, A_n, B_n))$

for some $C(x_n, A_n, B_n) \in \mathfrak{d} = [X, \mathfrak{g}] \subset \mathfrak{g}_0$. It follows in view of the last equation that $x_n = y_n$ for every $n \in \mathbb{N}$ and that

$$x_n U_0 = A_n - B_n \operatorname{mod}(\mathfrak{d}). \tag{2.7}$$

If $U_0 \in \mathfrak{d}$, then there exists $T_0 \in \mathfrak{g}_0 \backslash \{0\}$ such that $U_0 = [T_0, X]$. Whence,

$$\mathrm{Ad}_{\exp(T_0)} X = X + U_0 \in \mathfrak{k},$$

which contradicts the (CI) property of the triple (G, H, K). Thus, $U_0 \notin \mathfrak{d}$, which means that x_n is a bounded sequence by (2.7). We get then an absurdity taking into account our assumption (2.3.2.3). This completes the proof of the theorem. □

2.4 Proper actions on solvable homogeneous spaces

2.4.1 Proper actions on special solvmanifolds

The following result has been obtained earlier in Theorem 2.3.2 in the setup of nilpotent Lie groups. We prove here that it still remains true for general connected and simply connected solvable Lie groups.

Theorem 2.4.1. *Let H, K be closed, connected subgroups of a connected and simply connected, special, solvable Lie group G. Then K acts properly on G/H if and only if the triple (G, H, K) has the (CI) property. In other words, the statement of Conjecture 2.1.9 holds.*

Proof. We write $\mathfrak{g} = \mathfrak{g}_0 \oplus \mathbb{R}X$ where $\mathfrak{g}_0$ is the one-codimensional Abelian ideal of $\mathfrak{g}$ and $G_0 = \exp \mathfrak{g}_0$. We can adopt the proof of the nilpotent situation to assume that none of the subalgebras $\mathfrak{h}$ and $\mathfrak{k}$ contain a nonzero ideal of $\mathfrak{g}$. Likewise, we can argue as in the proof of Theorem 2.3.2 to settle the situation when both $\mathfrak{h}$ and $\mathfrak{k}$ are included in $\mathfrak{g}_0$ and also if merely one of them is.

So we have only to tackle the case where $\mathfrak{h} \not\subset \mathfrak{g}_0$ and $\mathfrak{k} \not\subset \mathfrak{g}_0$, then we can assume that $\dim \mathfrak{h} = \dim \mathfrak{k} = 1$ by our assumption above. Write $\mathfrak{h} = \operatorname{span}\{X\}$ and then $\mathfrak{k} = \operatorname{span}\{X + U_0\}$ for some nonzero $U_0 \in \mathfrak{g}_0$. If the action of K on G/H is not proper, we can suppose that there exists a compact subset S_0 of G_0 such that $H \cap S_0 K S_0^{-1}$ is not relatively compact. In fact, for every compact set S of G, there exist by Proposition 2.1.12 a compact set S_0 in $\mathfrak{g}_0$ and a compact set I in $\mathbb{R}$ such that $S \subset \exp(IX)S_0$, and we have

$$\begin{aligned} H \cap SKS^{-1} &\subset H \cap \exp(IX) S_0 K S_0^{-1} \exp(-IX) \\ &= \exp(IX)(H \cap S_0 K S_0^{-1}) \exp(-IX). \end{aligned}$$

Thus, we can find sequences x_n, $y_n \in \mathbb{R}$, A_n, $B_n \in \mathfrak{g}_0$, such that:

(2.4.2.1) $\exp(A_n) \in S_0$, $\exp(B_n) \in S_0$,
(2.4.2.2) $\lim_{n\to+\infty} |x_n| = +\infty$, $\lim_{n\to+\infty} |y_n| = +\infty$,
(2.4.2.3) $\exp(x_n X) = \exp(A_n)\exp(y_n(X + U_0))\exp(-B_n)$.

Denote $\mathfrak{d} = [\mathfrak{g}, \mathfrak{g}]$ and D the Lie group associated to $\mathfrak{d}$. We show now that for all $t \in \mathbb{R}$,

$$\exp(t(X + U_0)) = \exp(tX)\exp(tU_0) \bmod D. \tag{2.8}$$

In fact, thanks to the Campbell–Baker–Hausdorff formula in a neighborhood $\mathscr{V}_0$ of 0, there exists an analytic map ϱ from $\mathscr{V}_0$ to D such that

$$\exp(-tX)\exp(t(X + U_0)) = \exp(tU_0) \cdot \varrho(t). \tag{2.9}$$

We consider the analytic map $\psi(t) = \exp(-tU_0)\exp(-tX)\exp(t(X + U_0))$ and let $\{X_1, \dots, X_p\}$ be a coexponential basis of $\mathfrak{d}$ in $\mathfrak{g}$. There exist then $p + 1$ analytic functions $t \in \mathbb{R} \mapsto \alpha_i(t) \in \mathbb{R}$, $i \in \{1, \dots, p\}$ and $t \in \mathbb{R} \mapsto \rho(t) \in D$ such that

$$\psi(t) = \exp(\alpha_1(t)X_1) \cdots \exp(\alpha_p(t)X_p) \cdot \rho(t).$$

By (2.9), the functions $\alpha_i (i \in \{1, \dots, p\})$ vanish on $\mathscr{V}_0$, and then on $\mathbb{R}$ by analyticity, which completes the proof of (2.8).

It follows in view of equations (2.4.2.3) and (2.8) that $x_n = y_n$ for every $n \in \mathbb{N}$ and that

$$\exp(x_n X) = \exp(x_n X)\exp(-x_n X)\exp(A_n)\exp(x_n X)\exp(x_n U_0)\exp(C_n)\exp(-B_n)$$

for some $C_n \in \mathfrak{d}$. Thus,

$$\mathrm{Ad}_{\exp(-x_n X)}(A_n) - B_n + x_n U_0 + C_n = 0. \tag{2.10}$$

Hence,

$$x_n U_0 = B_n - A_n \bmod \mathfrak{d}. \tag{2.11}$$

If $U_0 \in \mathfrak{d}$, then there exists $T_0 \in \mathfrak{g}_0 \setminus \{0\}$ such that $U_0 = [T_0, X]$. Whence,

$$\mathrm{Ad}_{\exp(T_0)} X = X + U_0 \in \mathfrak{k},$$

which contradicts the (CI) property of the triple (G, H, K). Thus, we do assume $U_0 \notin \mathfrak{d}$, which means that x_n is a bounded sequence by (2.11), and we get then an absurdity taking into account our assumption (2.4.2.2). This completes the proof of the theorem. □

2.4.2 Weak and finite proper actions on solvmanifolds

In this section, we provide a criterion of weak and finite proper actions as defined in Subsection 2.1.3, for a solvable triplet (G, H, K). Our main result in this section is the following.

Theorem 2.4.2. *Let G be a connected, simply connected, solvable Lie group, and H and K be closed connected subgroups of G. Then the following assertions are equivalent:*
(i) *The action of K on G/H is finitely proper.*
(ii) *The action of K on G/H is weakly proper.*
(iii) *The action of K on G/H is free.*
(iv) *The triple (G, H, K) has the (CI) property.*

Proof. First of all, we remark that as G is connected, simply connected and solvable, it admits no compact nontrivial subgroups. Indeed, compact subgroups are central (and hence Abelian). It turns out that the action of K on G/H is free if and only if the triple (G, H, K) is (CI) as for every $x \in X$ the isotropy group K_x is given by

$$K_x = K \cap gHg^{-1}, \quad x = gH.$$

Whence, we have only to prove by means of Lemma 2.1.5 that (CI)-action implies both properties (i) and (ii).

Assume that the triple (G, H, K) has the (CI) property and let us show that K acts finitely properly on G/H. We have only to show that for every s, g in G, the set $H \cap gKs$ is finite, and by replacing K by gKg^{-1}, that $H \cap Ks$ is finite for every $s \in G$. Let then $s \in G$ and h_1, h_2 be in $H \cap Ks$, then there exist k_1, k_2 in K such that $h_1 = k_1 s$ and $h_2 = k_2 s$. Hence, $k_2 k_1^{-1} = h_2 h_1^{-1} \in H \cap K = \{e\}$. Finally, $h_1 = h_2$, which entails that $H \cap Ks$ consists at most of one single point, which completes the proof in this case.

We prove now that if the triple (G, H, K) has the (CI) property, then K acts weakly properly on G/H. Likewise, we have only to show that $K \cap SH$ is compact in G for every compact S in G. We use an induction on the dimension of G. The result is obviously true when $\dim G = 1$. Let $\mathfrak{g}$, $\mathfrak{h}$ and $\mathfrak{k}$ be the Lie algebras of G, H and K, respectively. Assume in a first time that one of the subalgebras $\mathfrak{h}$ or $\mathfrak{k}$ is included in a proper ideal of $\mathfrak{g}$, and then by extension, in a one-codimensional ideal $\mathfrak{g}'$ of $\mathfrak{g}$. This is actually what typically happens in the nilpotent context. Let G' be the Lie group associated to $\mathfrak{g}'$ and write $\mathfrak{g} = \mathfrak{g}' \oplus \mathbb{R}X$ for some $X \in \mathfrak{g}$. Let S be a compact set of G, then there exist a compact set I of $\mathbb{R}$ and a compact set S' of G', such that $S \subset \exp(IX)S'$. Suppose in a first time that $\mathfrak{k} \not\subset \mathfrak{g}'$, we can then pick the vector X in a way such that $X \in \mathfrak{k}$. We get therefore that

$$K \cap SH \subset K \cap \exp(IX)S'H = \exp(IX)(K' \cap S'H), \tag{2.12}$$

where $K' = K \cap G'$. We apply the induction hypothesis for the (CI)-triple (G', K', H) to obtain the fact that $K \cap SH$ is compact, which implies that K acts weakly properly on G/H.

Suppose now that $\mathfrak{k} \subset \mathfrak{g}'$. Then

$$H \cap SK \subset H \cap \exp(IX)S'K \subseteq H \cap S'K, \tag{2.13}$$

which is compact using again the induction procedure for the (CI)-triple (G', H, K). This completes the proof in this case.

We suppose from now on that none of the subalgebras $\mathfrak{h}$ and $\mathfrak{k}$ is included in a proper ideal in $\mathfrak{g}$. So, there exists a maximal nonnormal subalgebra $\mathfrak{g}_1$ containing $\mathfrak{h}$. As $\mathfrak{g}$ is solvable, $\mathfrak{g}_1$ is one- or two-codimensional. We keep the same notation as in Theorem 1.1.15 and we tackle separately these two situations. Suppose in a first time that $\mathfrak{g}_1$ is one- codimensional. Let $\mathfrak{g}_0$ be the one-codimensional ideal of $\mathfrak{g}_1$ and let A and X be the vectors of $\mathfrak{g}$ such that

$$\mathfrak{g} = \mathfrak{g}_1 \oplus \mathbb{R}X, \quad \mathfrak{g}_1 = \mathfrak{g}_0 \oplus \mathbb{R}A \quad \text{and} \quad [A, X] = X \bmod \mathfrak{g}_0.$$

If $\mathfrak{k} \subset \mathfrak{g}_1$, then we argue similarly as to use the induction hypothesis on the triple (G_1, H, K) making use of the coexponential basis $\{X\}$ of $\mathfrak{g}_1$ in $\mathfrak{g}$. Here, G_1 is the Lie group associated to $\mathfrak{g}_1$. Otherwise, there exists in $\mathfrak{k}$ a nonzero vector of the form

$$T = X + aA + u_0$$

for some real number a and $u_0 \in \mathfrak{g}_0$. If $a = 0$, then $\{T\}$ is a coexponential basis of $\mathfrak{g}_1$ in $\mathfrak{g}$. Then every compact set S of G is included by means of Proposition 2.1.12, in some compact of G of the form $\exp(IT)S_1$ where I is a compact set of $\mathbb{R}$ and S_1 is compact set of G_1 and we argue as in (2.12). Assume then that $a \neq 0$. If $\mathfrak{k}_1 = \mathfrak{k} \cap \mathfrak{g}_1 \not\subset \mathfrak{g}_0$, then there exists in $\mathfrak{k}$ a vector of the form $A + v_0$ for some $v_0 \in \mathfrak{g}_0$. It follows that

$$T' = [A + v_0, T] = X + w_0 \in \mathfrak{k}$$

for some $w_0 \in \mathfrak{g}_0$, and then we are led to the previous case. Assume now that $\mathfrak{k}_1 \subset \mathfrak{g}_0$. Let $b = \frac{1}{a}$ and consider the subalgebra $\mathfrak{k}' = \mathrm{Ad}_{\exp(bX)}\, \mathfrak{k}$. Then

$$\mathfrak{k}' = \mathbb{R}(A + u_0') \oplus (\mathfrak{k}' \cap \mathfrak{g}_0)$$

for some $u_0' \in \mathfrak{g}_0$. As $\mathfrak{k}'$ is included in $\mathfrak{g}_1$, the action of the subgroup K' associated to $\mathfrak{k}'$ on G/H is weakly proper as the triple (G, H, K') is (CI), which already entails that the action of K on G/H is weakly proper.

We tackle now the case where $\mathfrak{g}_1$ is two-codimensional. If $\mathfrak{k} \subset \mathfrak{g}_1$, then if $\{X', Y'\}$ is the coexponential basis of $\mathfrak{g}_1$ in $\mathfrak{g}$, every compact set can be included by means of

Proposition 2.1.12, in a compact set of the form $\exp(IX')\exp(JY')S_1$ for some compact sets I, J in $\mathbb{R}$ and S_1 in G_1. Then obviously,

$$H \cap SK \subset H \cap \exp(IX')\exp(JY')S_1K \subset H \cap S_1K,$$

and we can use the induction hypothesis for the (CI)-triplet (G_1, H, K). So, we assume from now on that $\mathfrak{k}$ is not included in $\mathfrak{g}_1$. Let $\mathfrak{g}_0$ be the ideal defined as in Theorem 1.1.15. We study first the case where $\mathfrak{g}_0$ is three-codimensional. Let A, X, Y be the vectors of $\mathfrak{g}$ and $\alpha \in \mathbb{R}$ given as in Theorem 1.1.15 such that

$$\mathfrak{g}_1 = \mathfrak{g}_0 \oplus \mathbb{R}A, \quad [A, X + iY] = (\alpha + i)(X + iY) \bmod \mathfrak{g}_0$$

and

$$[X, Y] = 0 \bmod \mathfrak{g}_0.$$

If $\dim \mathfrak{k}/\mathfrak{k}_1 = 1$ (here as above $\mathfrak{k}_1 = \mathfrak{k} \cap \mathfrak{g}_1$ and K_1 the Lie group associated to $\mathfrak{k}_1$), then there exists in $\mathfrak{k}$ a vector of the form

$$T = xX + yY + aA + u_0, \quad u_0 \in \mathfrak{g}_0, \quad x, y, a \in \mathbb{R}$$

such that $x^2 + y^2 \neq 0$. If $a = 0$, then $\{T\}$ is a part of a coexponential basis of $\mathfrak{g}_1$ in $\mathfrak{g}$. Let T' be another vector such that $\{T, T'\}$ forms a coexponential basis of $\mathfrak{g}_1$ in $\mathfrak{g}$. Then every compact set is included in a compact set of the form $\exp(IT)\exp(JT')S_1$ for some compact sets I, J in $\mathbb{R}$ and S_1 in G_1. Then obviously,

$$\begin{aligned} K \cap SH &\subset K \cap \exp(IT)\exp(JT')S_1H = \exp(IT)(K \cap \exp(JT')S_1H) \\ &\subset \exp(IT)(K \cap S_1H) = \exp(IT)(K_1 \cap S_1H). \end{aligned}$$

We can use the induction hypothesis for the (CI)-triplet (G_1, H, K_1), so we are done in this case. Assume then that $a \neq 0$. If $\mathfrak{k}_1 \not\subset \mathfrak{g}_0$, then a vector of the form $A + v_0$ belongs to $\mathfrak{k}$ for some $v_0 \in \mathfrak{g}_0$, and then $\mathfrak{k}$ contains consequently a vector of the form

$$T' = x'X + y'Y + w_0,$$

for some $w_0 \in \mathfrak{g}_0$ and $x', y' \in \mathbb{R}$ such that $x'^2 + y'^2 \neq 0$, which takes us back to the previous case. We look now at the case where $\mathfrak{k}_1 \subset \mathfrak{g}_0$. Take $u = \frac{\alpha x - y}{a(1+\alpha^2)}$ and $v = \frac{\alpha y + x}{a(1+\alpha^2)}$, and consider the algebra $\mathfrak{k}' = \mathrm{Ad}_{\exp(uX+vY)}\mathfrak{k}$, which is included in $\mathfrak{g}_1$, as

$$\mathfrak{k}' = \mathbb{R}(A + z_0) \oplus (\mathfrak{k}' \cap \mathfrak{g}_0)$$

for some $z_0 \in \mathfrak{g}_0$ making use of (2.16). We are also done in this case.

We treat now the case where $\dim(\mathfrak{k}/\mathfrak{k}_1) = 2$, then there exists two vectors in $\mathfrak{k}$ of the form

$$T_1 = x_1X + y_1Y + a_1A + u_0, \quad T_2 = x_2X + y_2Y + a_2A + v_0,$$

where $u_0, v_0 \in \mathfrak{g}_0$, $x_1, x_2, y_1, y_2, a_1, a_2 \in \mathbb{R}$ such that the vectors $x_1X + y_1Y$ and $x_2X + y_2Y$ are linearly independent. If $\mathfrak{k}_1 \not\subset \mathfrak{g}_0$, then as above, a vector $A + v_0$ belongs to $\mathfrak{k}$ for some $v_0 \in \mathfrak{g}_0$ and, therefore,

$$T_i' = [A + v_0, T_i] = (\alpha x_i + y_i)X + (\alpha y_i - x_i)Y + t_0^i \in \mathfrak{k}$$

for some $t_0^i \in \mathfrak{g}_0$, $i = 1, 2$. It is not hard to check that the vectors T_1' and T_2' are linearly independent in $\mathfrak{k}$, and obviously constitute a coexponential basis of $\mathfrak{g}_1$ in $\mathfrak{g}$. As above, we write a compact set S of G as included in a compact of the form $\exp(IT_1')\exp(JT_2')S_1$ for some compact sets I, J in $\mathbb{R}$ and S_1 in G_1. Then

$$\begin{aligned} K \cap SH &\subset K \cap \exp(IT_1')\exp(JT_2')S_1H = \exp(IT_1')\exp(JT_2')(K \cap S_1H) \\ &= \exp(IT_1')\exp(JT_2')(K_1 \cap S_1H). \end{aligned}$$

So, we are done by the induction hypothesis in this case.

Suppose now that $\mathfrak{k}_1 \subset \mathfrak{g}_0$. If $a_1 = a_2 = 0$, we are done as above. Otherwise, suppose for instance that $a_2 \neq 0$ and consider the Lie subalgebra $\mathfrak{k}' = \mathrm{Ad}_{\exp(uX+vY)}\,\mathfrak{k}$ where as above $u = \frac{\alpha x_2 - y_2}{a_2(1+\alpha^2)}$ and $v = \frac{\alpha y_2 + x_2}{a_2(1+\alpha^2)}$. By (2.16), one has that

$$\mathfrak{k}' = \mathbb{R}\left(T_1 - \frac{a_1}{a_2}T_2 + u_0'\right) \oplus \mathbb{R}(A + w_0') \oplus (\mathfrak{k}' \cap \mathfrak{g}_0)$$

for some u_0' and w_0' in $\mathfrak{g}_0$. So, we are led back to the case where $\dim(\mathfrak{k}/\mathfrak{k}_1) = 1$.

We study now the situation where $\mathfrak{g}_0$ is four-codimensional. Let A, B, X, Y be as in Theorem 1.1.15 such that

$$\begin{gathered} \mathfrak{g} = \mathfrak{g}_1 \oplus \mathbb{R}X \oplus \mathbb{R}Y, \quad \mathfrak{g}_1 = \mathfrak{g}_0 \oplus \mathbb{R}A \oplus \mathbb{R}B, \\ [A, X + iY] = X + iY \bmod \mathfrak{g}_0, \quad [B, X + iY] = -Y + iX \bmod \mathfrak{g}_0, \\ [X, Y] = 0 \bmod \mathfrak{g}_0 \quad \text{and} \quad [A, B] = 0 \bmod \mathfrak{g}_0. \end{gathered}$$

We can and do assume that $\mathfrak{k} \not\subset \mathfrak{g}_1$. We shall discuss as above several cases. Suppose first that $\dim(\mathfrak{k}/\mathfrak{k}_1) = 1$. There exists then in $\mathfrak{k}$ a vector of the form

$$T = xX + yY + aA + bB + u_0, \quad u_0 \in \mathfrak{g}_0, \quad x, y, a, b \in \mathbb{R}$$

such that $x^2 + y^2 \neq 0$. We can also assume that $a^2 + b^2 \neq 0$; otherwise $\{T\}$ is a part of a coexponential basis of $\mathfrak{g}_1$ in $\mathfrak{g}$ and we can use the argument made earlier. Assume for a while that $\mathfrak{k}_1 \subset \mathfrak{g}_0$. Take $u = \frac{ax-by}{a^2+b^2}$ and $v = \frac{ay+bx}{a^2+b^2}$, and consider the Lie algebra $\mathfrak{k}' = \mathrm{Ad}_{\exp(uX+vY)}\,\mathfrak{k}$. Then by (2.17), we deduce that

$$\mathfrak{k}' = \mathbb{R}(aA + bB + w_0) \oplus (\mathfrak{k}' \cap \mathfrak{g}_0) \subset \mathfrak{g}_1,$$

for some $w_0 \in \mathfrak{g}_0$. So, we can argue as above to have the answer in this case. We are led then to study the case where $\mathfrak{k}_1 \not\subset \mathfrak{g}_0$, then as above, a vector $a'A+b'B+v_0$ belongs to $\mathfrak{k}$ such that $a'^2 + b'^2 \neq 0$ and $v_0 \in \mathfrak{g}_0$. Therefore,

$$T' = [a'A + b'B + v_0, T] = (a'x + b'y)X + (a'y - b'x)Y + t_0 \in \mathfrak{k}\backslash\{0\},$$

for some $t_0 \in \mathfrak{g}_0$, which completes the proof in this case.

We look now at the case where $\dim(\mathfrak{k}/\mathfrak{k}_1) = 2$, then there exists in $\mathfrak{k}$ two vectors of the form:

$$T_1 = x_1X + y_1Y + a_1A + b_1B + u_0, \quad T_2 = x_2X + y_2Y + a_2A + b_2B + v_0,$$

where $u_0, v_0 \in \mathfrak{g}_0$, $x_1, x_2, y_1, y_2, a_1, a_2, b_1, b_2 \in \mathbb{R}$ such that the vectors $x_1X + y_1Y$ and $x_2X + y_2Y$ are linearly independent. Suppose first that $\mathfrak{k}_1 \not\subset \mathfrak{g}_0$. As $\mathfrak{k}$ is a subalgebra, there exists a vector $aA + bB + w_0 \in \mathfrak{k}_1$ such that $a^2 + b^2 \neq 0$. This gives rise to the fact that the pair $\{T_i' = [aA + bB + w_0, T_i], i = 1, 2\}$ is a coexponential basis of $\mathfrak{g}_1$ in $\mathfrak{g}$ contained in $\mathfrak{k}$, which takes us back to a previous situation. Suppose finally that $\mathfrak{k}_1 \subset \mathfrak{g}_0$, if $a_1^2 + b_1^2 = a_2^2 + b_2^2 = 0$. Then as above, $\{T_1, T_2\}$ is a coexponential basis of $\mathfrak{g}_1$ in $\mathfrak{g}$ contained in $\mathfrak{k}$ and the result holds as drawn up earlier. Thus, we can assume for instance that $a_1^2 + b_1^2 \neq 0$. Let $u = \frac{a_1x_1-b_1y_1}{a_1^2+b_1^2}$, $v = \frac{a_1y_1+b_1x_1}{a_1^2+b_1^2}$ and $\mathfrak{k}' = \mathrm{Ad}_{\exp(uX+vY)}\,\mathfrak{k}$. So, we are led back to the case where $\dim(\mathfrak{k}'/\mathfrak{k}_1') \leq 1$. This completes the proof of the theorem. □

Corollary 2.4.3. *Assume the same hypotheses as in Theorem 2.4.2. Then the action of K on $X = G/H$ is weakly proper if and only if the action of K on X is (CI) and all K-orbits of X are closed.*

Proof. We have only to prove by means of Theorem 2.4.2 that the weak proper action of K on X implies that all K-orbits are closed in X. Let $x, y \in X$ such that $x \notin Ky$ and let U be a compact neighborhood of x in X such that

$$Ky \cap U \neq \emptyset. \tag{2.14}$$

So, there exist $y' \in U$ and $k \in K$ such that $y' = ky$, and then $Ky = Ky'$. We can therefore admit that $y \in U$. As $K_{y,U} = \{k \in K : ky \in U\}$ is a compact set of X, the set $K_{y,U} \cdot y$ turns out to be compact in X and $x \notin K_{y,U} \cdot y \subset U$. We can therefore use similar arguments developed in [74, Théorème 3, Section 1] to achieve the proof. □

2.4.3 Proper actions on maximal solvmanifolds

2.4.4 Connected subgroups acting properly on maximal solvmanifolds

We now prove our first upshot on proper actions on maximal solvable homogeneous spaces, which basically makes use of the above classification result of maximal subalgebras in a solvable Lie algebra.

Theorem 2.4.4. *Let H, K be closed connected subgroups of a connected simply connected solvable Lie group G. Assume that one of the subgroups H or K is maximal in G (as in Definition* 1.1.16*), then K acts properly on G/H if and only if the triple (G,H,K) has the (CI) property.*

Proof. Recall the notation $\mathfrak{g}$, $\mathfrak{h}$ and $\mathfrak{k}$ for the Lie algebras associated respectively to G, H and K. We assume, for example, that $\mathfrak{h}$ is maximal in $\mathfrak{g}$. If $\mathfrak{h}$ is an ideal in $\mathfrak{g}$, then the result is not difficult to obtain. So, we assume that none of the subalgebras $\mathfrak{h}$ and $\mathfrak{k}$ is an ideal of $\mathfrak{g}$. We denote by $\mathfrak{g}_0$ the ideal defined as in Theorem 1.1.15 and G_0 its corresponding Lie group. We consider the quotient group $\overline{G} = G/G_0$ and we designate by

$$f : G \to \overline{G},$$

the canonical projection homomorphism. Likewise, we denote $\overline{H} = f(H)$ and $\overline{K} = f(K)$. So, it appears clear that $K \cap \ker f = K \cap G_0 \subset K \cap H = \{e\}$, and then $K \cap \ker f$ is compact. On the other hand, it is not difficult to show that $(\overline{G},\overline{H},\overline{K})$ has the (CI) property. In fact, let $\overline{h} \in \overline{H}$, $\overline{k} \in \overline{K}$ and $\overline{t} \in \overline{G}$ such that $\overline{h} = \overline{tht}^{-1}$. Then $h = tkt^{-1}t_0$ for some $t_0 \in G_0$, which implies that $ht_0^{-1} = tkt^{-1} \in H \cap tKt^{-1}$. Finally, $h = t_0$ as (G,H,K) has the (CI) property and, therefore, $\overline{h} = \overline{e}$.

To prove that K acts properly on G/H, it is then sufficient to prove that $\overline{K}$ acts properly on $\overline{G}/\overline{H}$ according to Lemma 2.1.7. Suppose in a first time that $\mathfrak{h}$ is one-codimensional. We denote by $\overline{\mathfrak{g}}$ the Lie algebra associated to $\overline{G}$. It is clear that $\overline{\mathfrak{g}}$ is the 2-dimensional Lie algebra spanned by the vectors fields A, X satisfying the bracket relation $[A,X] = X$ and that $\overline{\mathfrak{h}} = \mathbb{R}A$. If $\overline{\mathfrak{k}}$ is an ideal of $\overline{\mathfrak{g}}$, we are already done as in the proof of Theorem 2.3.2. If not, there exists $x \in \mathbb{R}^*$ such that $\overline{\mathfrak{k}} = \operatorname{span}(A + xX)$, but this gives

$$\operatorname{Ad}_{\exp(-xX)} A = A + xX, \tag{2.15}$$

which contradicts the (CI) property of the triple $(\overline{G},\overline{H},\overline{K})$. This completes the proof in this case. Suppose now that $\mathfrak{h}$ is two-codimensional and we assume in a first time that $\operatorname{codim} \mathfrak{g}_0 = 3$. Then $\overline{\mathfrak{h}} = \mathbb{R}A$. We note $\mathfrak{g}_1 = \operatorname{span}\{X, Y\}$. If $\overline{\mathfrak{k}} \not\subset \mathfrak{g}_1$, then there exists a nonzero $(x,y) \in \mathbb{R}^2$ such that $A + xX + yY \in \overline{\mathfrak{k}}$. It comes up by a simple computation that

$$\mathrm{Ad}_{\exp(\frac{-ax+y}{1+a^2}X-\frac{x+ay}{1+a^2}Y)}(A) = A + xX + yY \tag{2.16}$$

which is impossible by the (CI) property. So, we are led to study the case where $\overline{\mathfrak{k}} \subset \mathfrak{g}_1$. Let S be a compact set in $\overline{G}$. Then there exists by Proposition 2.1.12 a compact set S_1 in G_1, the analytic Lie group associated to $\mathfrak{g}_1$ and a compact set I in $\mathbb{R}$ such that $S \subset \exp(IA)S_1$. Then

$$\begin{aligned}\overline{H} \cap S\overline{K}S^{-1} &\subset \overline{H} \cap \exp(IA)S_1\overline{K}S_1^{-1}\exp(-IA)\\ &= \exp(IA)(\overline{H} \cap S_1\overline{K}S_1^{-1})\exp(-IA)\\ &= \exp(IA)\exp(-IA)\end{aligned}$$

which is compact in $\overline{G}$. Suppose now that $\mathfrak{g}_0$ is four-codimensional. Then $\overline{\mathfrak{h}} = \mathbb{R}A \oplus \mathbb{R}B$. We note as above $\mathfrak{g}_1 = \mathrm{span}\{X, Y\}$ and G_1 the Lie group associated to $\mathfrak{g}_1$. If $\overline{\mathfrak{k}} \not\subset \mathfrak{g}_1$, then there exists a vector $aA+bB+xX+yY \in \overline{\mathfrak{k}}$ such that $a^2+b^2 \neq 0$ and $x^2+y^2 \neq 0$. A routine computation shows that

$$\mathrm{Ad}_{\exp(\frac{-ax+by}{a^2+b^2}X-\frac{bx+ay}{a^2+b^2}Y)}(aA + bB) = aA + bB + xX + yY \tag{2.17}$$

which is impossible by the (CI) property. Suppose finally that $\overline{\mathfrak{k}} \subset \mathfrak{g}_1$. Let S be a compact set in $\overline{G}$. Then there exists by Proposition 2.1.12 a compact set S_1 in G_1, two compact sets I and J in $\mathbb{R}$ such that $S \subset \exp(IA)\exp(JB)S_1$. Then

$$\begin{aligned}\overline{H} \cap S\overline{K}S^{-1} &\subset \overline{H} \cap \exp(IA)\exp(JB)S_1\overline{K}S_1^{-1}\exp(-JB)\exp(-IA)\\ &= \exp(IA)\exp(JB)(\overline{H} \cap S_1\overline{K}S_1^{-1})\exp(-JB)\exp(-IA)\\ &= \exp(IA)\exp(JB)\exp(-JB)\exp(-IA),\end{aligned}$$

which is compact in $\overline{G}$. This completes the proof of the theorem. □

The following corollary stems directly from Propositions 2.1.14 and 2.4.4.

Corollary 2.4.5. *Assume that G is a connected and simply connected, solvable Lie group and one of the subgroups in question is normal or maximal. Then the following statements are equivalent:*
(i) *K acts weakly properly on G/H.*
(ii) *K acts properly on G/H.*
(iii) *The triple (G, H, K) is (CI).*

2.4.5 From continuous to discrete actions

Throughout this subsection, we consider a connected nonnormal and maximal subgroup H of a connected solvable Lie group G as in Definition 1.1.16. Let $\mathfrak{h} = \mathrm{Lie}(H)$ and $\mathfrak{g}_0 = \mathrm{Lie}(G_0)$ the ideal contained in $\mathfrak{h}$ (accordingly, $G_0 \subset H$) as in Theorem 1.1.15.

In case 3, the family $\{A, B, X, Y\}$ constitutes a coexponential basis of G_0 in G. Thus, any element g in G can be written as

$$g = (a, b, v, g_0) := e^{aA} e^{bB} e^{xX+yY} g_0,$$

where $a, b \in \mathbb{R}$, $v = (x, y) \in \mathbb{R}^2$ and $g_0 \in G_0$.

As such, Theorem 1.1.15 gives that the multiplication law group of G is submitted to the following equation:

$$(a, b, v, g_0)(a', b', v', g_0') = (a + a', b + b', e^{-a'} r(-b')v + v') \operatorname{mod}(G_0), \tag{2.18}$$

where for $t \in \mathbb{R}$,

$$r(t) = \begin{pmatrix} \cos(t) & \sin(-t) \\ \sin(t) & \cos(t) \end{pmatrix}$$

and

$$r(t)v = \exp(r(t)(xX + yY)).$$

Accordingly, the law group of G in case 2 of Theorem 1.1.15 can be written down as

$$(a, v, g_0)(a', v', g_0') = (a + a', e^{-\alpha a'} r(-a')v + v') \operatorname{mod}(G_0), \tag{2.19}$$

where the coexponential basis of G_0 in G here is $\{A, X, Y\}$.

Finally in case 1, the law group of G is described as follows:

$$(a, x, g_0)(a', x', g_0') = (a + a', e^{-a'} x + x') \operatorname{mod}(G_0), \tag{2.20}$$

where $\{A, X\}$ is the coexponential basis of G_0 in G as in Theorem 1.1.15.

Our first result is the following.

Theorem 2.4.6. *Let G be a connected, simply connected, solvable Lie group and H a nonnormal connected maximal subgroup of G. Then any nontrivial discrete subgroup of G acting properly and freely on G/H is Abelian of rank 1 in case 1 and of rank ≤ 2 otherwise.*

Proof. We keep the same notation as before and we consider the quotient group $\overline{G} = G/G_0$ and $\pi : G \longrightarrow \overline{G}$, the canonical projection. Let $\overline{H} = \pi(H)$ and $\overline{\Gamma} = \pi(\Gamma)$. We first need to prove the following lemmas.

Lemma 2.4.7. *Let Γ be a discrete subgroup of G and H a nonnormal maximal subgroup of G. Then:*
(i) *If Γ acts properly on G/H, then $\overline{\Gamma}$ acts properly on $\overline{G}/\overline{H}$.*
(ii) *If Γ acts freely on G/H, then $\overline{\Gamma}$ acts freely on $\overline{G}/\overline{H}$ and the restriction $\pi_{|\Gamma}$ is injective.*

Proof. (i) Suppose that the action of $\overline{\Gamma}$ on $\overline{G}/\overline{H}$ is not proper, then there exist an infinite sequence $(\gamma_n)_n \in \Gamma$ and a convergent sequence $(s_n)_n$ in G such that $(\overline{\gamma}_n\overline{s}_n\overline{H})_n$ converges in $\overline{G}$. On the other hand, $\overline{\gamma}_n\overline{s}_n\overline{H} = \gamma_n G_0 s_n G_0 H G_0 = \gamma_n s_n H$ and $\overline{G}/\overline{H} = G/G_0/H/G_0 \simeq G/H$, then $(\gamma_n s_n)_n$ converges in G/H. As $(\gamma_n)_n$ is not convergent, this is a contradiction with the properness of the action of Γ.

(ii) First remark that $\ker(\pi_{|\Gamma}) = \Gamma \cap G_0 \subset \Gamma \cap H = \{e\}$. Let $\overline{h} \in \overline{H}$, $\overline{\gamma} \in \overline{\Gamma}$ and $\overline{t} \in \overline{G}$ such that $\overline{h} = \overline{t\gamma t}^{-1}$. Then there is a $t_0 \in G_0$ such that $h = t\gamma t^{-1}t_0$ and $ht_0^{-1} = t\gamma t^{-1} \in H \cap t\Gamma t^{-1} = \{e\}$, which implies that $h = t_0$ so $\overline{h} = \overline{e}$. Then we obtain $\overline{H} \cap \overline{t\Gamma t}^{-1} = \{\overline{e}\}$ for all $\overline{t} \in \overline{G}$. □

Lemma 2.4.8. *If Γ acts properly on G/H, then $\overline{\Gamma}$ is discrete.*

Proof. Suppose that $\overline{\Gamma}$ is not discrete, then there exist a nonstationary sequence $(\overline{\gamma}_n)_n \in \overline{\Gamma}$ such that $(\overline{\gamma}_n)_n$ converges to $\overline{e}$, which means that $(\gamma_n G_0)_n$ converges to G_0. Let S be a compact neighborhood of e in G, then $\overline{S}$ is a compact neighborhood of $\overline{e}$ in $\overline{G}$. We have for sufficiently large n, $\overline{\gamma}_n \in \overline{S}$, then $\overline{\gamma}_n \in \overline{\gamma}_n\overline{S} \cap \overline{S}$ and so $\overline{\gamma}_n \in \overline{\gamma}_n\overline{SH} \cap \overline{SH}$, which implies that $\gamma_n \in \gamma_n SHG_0 \cap SHG_0 = \gamma_n SH \cap SH$. This is a contradiction with the fact that Γ acts properly on G/H. □

Lemma 2.4.9. *Assume the situation of case 1 and let Γ be a subgroup of G. Then the action of Γ on G/H is free if and only if any element of Γ is written as $\gamma = (0, x, g)$ with $x \neq 0$. In particular, $\Gamma \subset \{0\} \times \mathbb{R} \times G_0$.*

Proof. Let $\gamma = (a, x, g)$ be an element of Γ. From equation (2.20), for $g = (a', x', g_0') \in G$ we have

$$g\gamma g^{-1} = (a, e^{a'}((e^{-a} - 1)x' + x)) \bmod (G_0).$$

The action is free if and only if $g\gamma g^{-1} \notin H$ for any $g \in G$ whenever γ is nontrivial, which is equivalent to $a = 0$ and $x \neq 0$. □

Lemma 2.4.10. *Under the assumptions of case 2 and let Γ be a subgroup of G. Then the action of Γ on G/H is free if and only if:*
(i) *Any element of Γ is written as $\gamma = (0, v, g)$ with $v \in \mathbb{R}^2 \setminus \{(0,0)\}$ and $g \in G_0$, in the case where $\alpha \neq 0$. In such case, we have*

$$\Gamma \subset \{0\} \times \mathbb{R}^2 \times G_0.$$

(ii) *Any element of* Γ *is written as* $\gamma = (2k\pi, v, g)$ *with* $k \in \mathbb{Z}$, $v \in \mathbb{R}^2 \setminus \{(0,0)\}$ *and* $g \in G_0$ *in the case where* $\alpha = 0$. *We have in this case*

$$\Gamma \subset 2\pi\mathbb{Z} \times \mathbb{R}^2 \times G_0.$$

Proof. Let $\gamma = (a, v, g)$ be an element of Γ. From equation (2.19), for $g = (a', v', g_0') \in G$, we have

$$\begin{aligned} g\gamma g^{-1} &= (a, e^{\alpha a'} r(a')((e^{-\alpha a} r(-a) - Id)v' + v)) \bmod(G_0) \\ &= (0, e^{\alpha a'} r(a')((I - e^{\alpha a} r(a))v' + e^{\alpha a} r(a)v)) \bmod(H). \end{aligned}$$

The action of Γ on G/H is free if and only if $g\gamma g^{-1} \notin H$ for any $g \in G$ whenever γ is nontrivial. Thus, if $\alpha \neq 0$, the free action of Γ on G/H is equivalent to $a = 0$ and $v \neq 0$. If $\alpha = 0$, the free action is equivalent to $a \in 2\pi\mathbb{Z}$ and $v \neq 0$. □

Lemma 2.4.11. *Let G and H be as in case* 3. *For any subgroup* Γ *of* G, *the action of* Γ *on* G/H *is free if and only any element of* Γ *is written as* $(0, 2k\pi, v, g)$ *such that* $k \in \mathbb{Z}$, $v \in \mathbb{R}^2 \setminus \{(0,0)\}$ *and* $g \in G_0$. *In particular,*

$$\Gamma \subset \{0\} \times 2\pi\mathbb{Z} \times \mathbb{R}^2 \times G_0.$$

Proof. Let $\gamma = (a, b, v, g)$ be an element of Γ. For $g = (a', b', v', g_0') \in G$, we have

$$g\gamma g^{-1} = (a, b, e^{a'} r(b')((e^{-a} r(-b) - Id)v' + v)) \bmod(G_0). \tag{2.21}$$

As before, the action of Γ on G/H is free if and only if $g\gamma g^{-1} \notin H$ for any $g \in G$ whenever γ is nontrivial, which equivalent to $a = 0$, $b \in 2\pi\mathbb{Z}$ and $v \neq 0$. □

We now go back to the proof of Theorem 2.4.6. By Lemma 2.4.7, the map $\bar{\pi} = \pi_{|\Gamma} : \Gamma \longrightarrow \overline{\Gamma}$ is a group isomorphism. To prove therefore that Γ is Abelian, it is sufficient to show that $\overline{\Gamma}$ is. We take back the three cases enumerated in Theorem 1.1.15.

Case 1: $\mathfrak{g} = \mathfrak{h} \oplus \mathbb{R}X$ and $\mathfrak{h} = \mathfrak{g}_0 \oplus \mathbb{R}A$, then $\bar{\mathfrak{g}} = \mathrm{Lie}(\overline{G}) = \bar{\mathfrak{h}} \oplus \mathbb{R}\overline{X}$ and $\bar{\mathfrak{h}} = \mathbb{R}\overline{A}$. By Lemma 2.4.9, any element of $\overline{\Gamma}$ is written as $(0, x)$. Thus, $\overline{\Gamma}$ is identified to a subgroup of $\mathbb{R}$. Furthermore, as Γ acts properly on G/H, $\overline{\Gamma}$ turns out to be discrete by Lemma 2.4.8. Then Γ is isomorphic to $\mathbb{Z}$.

Case 2: $\mathfrak{g} = \mathfrak{h} \oplus \mathbb{R}X \oplus \mathbb{R}Y$ and $\mathfrak{h} = \mathfrak{g}_0 \oplus \mathbb{R}A$, then $\bar{\mathfrak{g}} = \mathrm{Lie}(\overline{G}) = \bar{\mathfrak{h}} \oplus \mathbb{R}\overline{X} \oplus \mathbb{R}\overline{Y}$ and $\bar{\mathfrak{h}} = \mathbb{R}\overline{A}$.

If $\alpha \neq 0$, then by Lemma 2.4.10, the free action of $\overline{\Gamma}$ on $\overline{G}/\overline{H}$ implies that any element of $\overline{\Gamma}$ is written as $(0, v)$. In particular, $\overline{\Gamma}$ is identified to a subgroup of $\mathbb{R}^2$. As the action of Γ is proper, it comes out that $\overline{\Gamma}$ is a discrete and in particular $\mathrm{rk}\,\Gamma \leq 2$.

Now, if $\alpha = 0$ then

$$g\gamma g^{-1} = (a', r(a)((r(-a') - Id)v + v')).$$

The action of Γ on G/H is free if and only if $a' \in 2\pi\mathbb{Z}$ and $v' \neq 0$. Thus, $\overline{\Gamma}$ is identified to a discrete subgroup of $G_1 = 2\pi\mathbb{Z}\times\mathbb{R}^2 \subset \overline{G}$ thanks to Lemma 2.4.10. Now G_1 can be viewed as a closed subgroup of the Abelian group $\mathbb{R}^3$. In particular, $\overline{\Gamma}$ is identified to a discrete subgroup of $\mathbb{R}^3$, so $\mathrm{rk}(\overline{\Gamma}) \leq 3$. Assume that $\mathrm{rk}(\overline{\Gamma}) = 3$ and let $\gamma_i = (a_i, v_i)$, $i = 1, 2, 3$ be some generators of Γ. Then the vectors v_1, v_2 and v_3 are not linearly independent. Then we can choose a sequence $(u_{1,n}, u_{2,n}, u_{3,n})_n \subset \mathbb{Z}^3$ such that

$$\lim_{n\to\infty} u_{1,n}v_1 + u_{2,n}v_2 + u_{3,n}v_3 = 0.$$

It comes out that for $\gamma_n = \gamma_1^{u_{1,n}}\gamma_2^{u_{2,n}}\gamma_3^{u_{3,n}}$ we have

$$\gamma_n = (0, u_{1,n}v_1 + u_{2,n}v_2 + u_{3,n}v_3) \bmod(\overline{H}).$$

In particular, the sequence of general term $(\gamma_n\overline{H})_n$ converges to $\overline{e}$. Thus, the action of $\overline{\Gamma}$ on $\overline{G}/\overline{H}$ is not proper.

Case 3: $\overline{\mathfrak{g}} = \mathrm{Lie}(\overline{G}) = \overline{\mathfrak{h}} \oplus \mathbb{R}\overline{X} \oplus \mathbb{R}\overline{Y}$ and $\overline{\mathfrak{h}} = \mathbb{R}\overline{A} \oplus \mathbb{R}\overline{B}$. As before, from Lemma 2.4.11 the subgroup $\overline{\Gamma}$ is identified to a discrete subgroup of $2\pi\mathbb{Z} \times \mathbb{R}^2$, which is enough to conclude. □

We keep the same hypotheses and notation. Our next upshot consists in providing a simple criterion of the proper action of a rank one discontinuous group for a maximal solvable homogeneous space. This is actually a first step toward the study of the local rigidity problem. We prove the following.

Theorem 2.4.12. *Let G be a connected, simply connected, solvable Lie group, H a non-normal maximal subgroup of G and Γ a rank one subgroup of G. Then the following properties are equivalent:*
(i) *The action of Γ on G/H is free.*
(ii) *The action of Γ on G/H is CI.*
(iii) *The action of Γ on G/H is proper.*

Proof. The group G is simply connected, then Γ is torsion-free and any compact subgroup of G is trivial. In this case, we have obviously (i) $\Longleftrightarrow$ (ii) and (iii) $\Longrightarrow$ (ii), then we have to prove (i) $\Longrightarrow$ (iii). Suppose that the action of Γ on G/H is not proper, then there exist a sequence $(\gamma_n)_n \in \Gamma$ such that $\{\gamma_n, n \in \mathbb{N}\}$ is not a compact set and a convergent sequence $(s_n)_n$ in G such that $(\gamma_n s_n)_n$ converges modulo H.

Let γ be a generator of Γ and (λ_n) a sequence of integers such that $\gamma_n = \gamma^{\lambda_n}$. In case 1, for γ as in Lemma 2.4.9, we have $\gamma_n = (0, \lambda_n x, 0) \bmod(G_0)$. For $s_n = (t_n, y_n, 0) \bmod(G_0)$, from equation (2.20) we get

$$\gamma_n s_n = (t_n, e^{-t_n}\lambda_n x + y_n, 0) \bmod(G_0).$$

Therefore, $(\gamma_n s_n)_n$ converges modulo H if and only if $x = 0$, then the action is not free by Lemma 2.4.9.

We now look at case 2. For y as in Lemma 2.4.10, we have

$$y_0^{\lambda_n} = \begin{cases} (0, \lambda_n v, 0) \operatorname{mod}(G_0) & \text{if } \alpha \neq 0, \\ (2k\lambda_n \pi, \lambda_n v, 0) & \text{if } \alpha = 0. \end{cases}$$

Thus, independently from α, for $s_n = (t_n, w_n, 0) \operatorname{mod}(G_0)$ we have

$$y_n s_n = (0, \lambda_n v + e^{\alpha t_n} r(t_n) w_n, 0) \operatorname{mod}(H).$$

As $(t_n)_n$ is a convergent sequence, it comes out that $(y_n s_n)_n$ converges modulo H if and only if $v = 0$ and then the action is not free by Lemma 2.4.10.

We finally tackle case 3. We have $y_n = (0, 2k\lambda_n \pi, \lambda_n v, 0) \operatorname{mod}(G_0)$ and for $s_n = (t_n, f_n, w_n, 0) \operatorname{mod}(G_0)$, the sequence

$$y_n s_n = (0, 0, \lambda_n v + e^{t_n} r(f_n) w_n) \operatorname{mod}(H),$$

converges modulo H if and only if $v = 0$. □

Proposition 2.4.13. *Let Γ be a subgroup of G of rank ≤ 2 and φ a homomorphism from Γ to G. Then:*
(i) *If Γ is a rank one subgroup, then $\varphi(\Gamma)$ is nontrivial if and only if φ is injective and $\varphi(\Gamma)$ is discrete.*
(ii) *If $\varphi(\Gamma)$ is a rank two subgroup acting properly on G/H, then φ is injective and $\varphi(\Gamma)$ is discrete.*

Proof. Note that any rank one subgroup of connected simply connected solvable Lie group is a torsion-free discrete subgroup. Thus, any nontrivial homomorphism is injective and its image is discrete. For the rank two case and keeping the same notation as before, if φ is noninjective, there exist two integers n_1, n_2, such that $y_1^{n_1} y_2^{n_2} = e$. If $\varphi(\Gamma)$ is not discrete, then there exists an infinite sequence $(y_n = y_1^{\lambda_n} y_2^{\mu_n})_n$ in $\varphi(\Gamma)$, which converges to the unit element. In the first situation, we have $n_1 v_1 + n_2 v_2 = 0$ and in the second, one obtains that $(\lambda_n v_1 + \mu_n v_2)_n$ converges to zero. In both of the situations, we get $\det(v_1, v_2) = 0$. □

2.5 Proper action for the compact extension $K \ltimes \mathbb{R}^n$

Let K be a compact subgroup of $GL(n, \mathbb{R})$ and $G := K \ltimes \mathbb{R}^n \subset I(n)$ the semidirect product group, where $I(n) := O_n(\mathbb{R}) \ltimes \mathbb{R}^n$ denotes the semidirect product of the orthogonal group $O_n(\mathbb{R})$ as in Section 1.2. Let H be a closed subgroup of G and Γ a discontinuous group for the homogeneous space $\mathscr{X} = G/H$. We establish a geometrical criterion of the proper action of Γ on $\mathscr{X}$, which requires an accurate description of the structure of closed connected subgroups of Euclidean motion groups as developed in Subsection 1.2.4. We first prove the following, which concerns the case where $K = O_n(\mathbb{R})$.

Proposition 2.5.1. *Let $G = I(n)$ be the Euclidean motion group, H is a closed subgroup of G and Γ an infinite discrete subgroup of G. Then Γ acts properly on G/H if and only if H is compact.*

Proof. We remark first that Γ acts properly on G/H if and only if $g\Gamma g^{-1}$ acts properly on $G/g'Hg'^{-1}$ for any $g, g' \in G$. Suppose that H is not compact. We can then find a sequence $\{(h_p, y_p)\}_{p\in\mathbb{N}}$ of H such that $\lim_{p\to\infty}\|y_p\| = \infty$. As in Corollary 1.2.14, take $\gamma = (A, x) \in \Gamma$ of infinite order as given in (1.24), then $x \neq 0$ by Lemma 1.2.5 and up to conjugation, $\gamma^p = (A^p, px)$. For $p \in \mathbb{N}$, there exists $\alpha(p) \in \mathbb{N}$ such that

$$\alpha(p)\|x\| \leq \|y_p\| < (\alpha(p) + 1)\|x\|.$$

Let $z_p^+ = \lambda x \in S(0, \|y_p\|) \cap \mathbb{R}x$ where $\alpha(p) \leq \lambda < \alpha(p) + 1$. Consider

$$t_p = z_p^+ - \alpha(p)x = (\lambda - \alpha(p))x,$$

and then $\|t_p\| < \|x\|$. Furthermore, $O_n(\mathbb{R})$ acts transitively on $S(0, \|y_p\|)$; hence, there exists $O_p \in O_n(\mathbb{R})$ such that $O_p y_p = z_p^+$. Let $B'(0, \|x\|)$ be the closed ball and $K = O_n(\mathbb{R}) \times B'(0, \|x\|)$. Then

$$\begin{aligned}(O_p, -t_p)(h_p, y_p)(h_p^{-1}O_p^{-1}A^{\alpha(p)}, 0) &= (A^{\alpha(p)}, -t_p + O_p y_p)\\ &= (A^{\alpha(p)}, \alpha(p)x) = \gamma^{\alpha(p)} \in \Gamma \cap KHK^{-1}.\end{aligned}$$

As the set $\{\gamma^{\alpha(p)}\}_{p\in\mathbb{N}}$ is infinite, we meet a contradiction and then H must be compact. The converse is trivial. □

Remark 2.5.2. In the setting where H, H', L, L' are closed subgroups of G, Fact 2.1.2 gives immediately that if L acts properly on G/H, then so does L on G/H' whenever $H \sim H'$. Furthermore, if L acts properly on G/H, then so does L' on G/H whenever $L \sim L'$.

The following is then immediate.

Proposition 2.5.3. *Let Γ be a discrete subgroup of the Lie group N_C (as in Subsection 1.2.4). Let also H be a closed subgroup of N_C and R its solvable radical. Then the following are equivalent:*
(1) *Γ acts properly on N_C/H.*
(2) *Γ acts properly on N_C/R.*
(3) *For any subgroup Γ' of finite index in Γ, Γ' acts properly on N_C/R.*

Proof. Let $H = S \cdot R$ be a Levi decomposition of H, where S is semisimple, and hence compact. Clearly, $R \subset SHS^{-1} = H$ and $H = SRS^{-1}$, then $H \sim R$. This establishes the equivalence between (1) and (2) since the fact that $\Gamma \pitchfork H$ in N_C is equivalent to that Γ acts properly on N_C/R, as described in Remark 2.5.2. Now $\Gamma = \Gamma' D$ where D is a finite set of Γ and Γ' is a subgroup of Γ. As $\Gamma = D^{-1}\Gamma' D$ and $\Gamma' \subset D^{-1}\Gamma D = \Gamma$, then $\Gamma \sim \Gamma'$. Hence, similar arguments allow to establish the equivalence between (2) and (3), which closes the proof. □

2.5.1 Criterion for proper action

Let H be a closed connected subgroup of G and R its solvable radical. We designate by $B(0,r)$ the closed ball of radius $r > 0$ centered at the origin. For a subset X of $\mathbb{R}^n$ and a subgroup L of $GL(n,\mathbb{R})$, set $L \cdot X = \{A.x \mid A \in L, x \in X\}$. Note also $\mathscr{L}_\Gamma = \{I\} \times E_\Gamma$ and $\mathscr{L}_H = \{I\} \times E_H$.

We first prove the following results.

Lemma 2.5.4. *Let Γ be a discrete subgroup of G. There exists $r > 0$ such that, for any $x \in E_\Gamma$, $B(x,r) \cap pr_2(\Gamma) \neq \emptyset$.*

Proof. Keep the same notation as in Theorem 1.2.24. In virtue of Proposition 1.2.40, $E_{\Gamma^g} = E_{\Gamma^g_a}$. For some $g := (I,t) \in G$, one can take $(A_1,y_1),\ldots,(A_k,y_k) \in \Gamma^g_a$ such that $\{y_1,\ldots,y_k\}$ is a basis of $E_{\Gamma^g_a}$ and $y_i = P_{A_i}(y_i)$, for $i = 1\ldots k$, thanks to Proposition 1.2.18. For any $x \in E_{\Gamma^g}$, there exist some reals $a_1,\ldots,a_k$ such that $x = \sum_{i=1}^k a_i y_i$. Put $\overline{x} = \sum_{i=1}^k [a_i] y_i$, where $[a]$ designates the floor of a real number a. This yields by (1.25):

$$\prod_{i=1}^k (A_i,y_i)^{[a_i]} = \left(\prod_{i=1}^k A_i^{[a_i]}, \sum_{i=1}^k [a_i] y_i\right) = \left(\prod_{i=1}^k A_i^{[a_i]}, \overline{x}\right).$$

This proves that $\overline{x} \in pr_2(\Gamma^g)$ and satisfies

$$\|x - \overline{x}\| \leq r := \sum_{i=1}^k \|y_i\|.$$

□

The following clear facts give some simple characterizations of the proper action that will be discussed later.

Lemma 2.5.5. *Let Γ and H be two closed subgroups of G. The following assertions are equivalent:*

(i) *Γ acts properly on G/H.*
(ii) *For any $r > 0$, $[C_r \cdot H \cdot C_r] \cap \Gamma$ is compact, where $C_r := K \times B(0,r)$.*
(iii) *For any compact sets C and C' of G, $[C \cdot H \cdot C'] \cap \Gamma$ is compact.*

We now prove the following.

Lemma 2.5.6. *Let E and F be two linear subspaces of $\mathbb{R}^n$. Then $E \cap (K \cdot F) = \{0\}$ if and only if $E \cap [K \cdot F + B(0,r)]$ is compact for all $r > 0$.*

Proof. We denote by $S(0,r)$ the sphere centered at the origin and of radius $r > 0$. It is clear that $E \cap [K \cdot F + B(0,r)]$ is closed and that $S_r := (K \cdot F) \cap S(0,r)$ is a compact set of $\mathbb{R}^n$. For all $x \in S_r$, consider the continuous map $\varepsilon(x) = d(x,E)$, where $d(x,E)$ designates the distance from x to E. Set $\varepsilon_0 := \min\{\varepsilon(x) \mid x \in S_r\}$. There exists then $x_0 \in S_r$ such that $\varepsilon_0 = d(x_0,E) > 0$, as $E \cap (K \cdot F) = \{0\}$. For $a \in K \cdot F$, set $a' \in \mathbb{R}_+ a \cap S(0,r)$, and $\widetilde{a'} := \frac{r}{\varepsilon(a')} a'$. Note that $d(\widetilde{a'},E) = r$ and $\|\widetilde{a'}\| \leq \frac{r^2}{\varepsilon_0}$. Let then $y = a + v \in E \cap [K \cdot F + B(0,r)] \neq \emptyset$, with $a \in K \cdot F$ and $v \in B(0,r)$. There exists $t \geq 0$ such that $a = t \cdot \widetilde{a'}$. If $t > 1$, then $d(a,E) > r$, and hence $B(a,r) \cap E = \emptyset$, which is absurd. Therefore, $t \in [0,1]$,

$$\|y\| \leq \|a\| + \|v\| \leq \|\widetilde{a'}\| + r \leq c_{\varepsilon_0,r} := \frac{r^2}{\varepsilon_0} + r,$$

and conclusively $[K \cdot F + B(0,r)] \cap E \subset B(0, c_{\varepsilon_0,r})$.

Conversely, note that $E \cap (K \cdot F)$ is compact as $E \cap (K \cdot F) \subset E \cap [K \cdot F + B(0,r)]$. If $0 \neq x \in E \cap (K \cdot F)$, then $\mathbb{R}x \subset E \cap (K \cdot F)$, which is a contradiction. □

We now move to prove the main result of this section.

Theorem 2.5.7. *Let $G = K \ltimes \mathbb{R}^n$ where K is a compact subgroup of $GL(n,\mathbb{R})$, Γ a discrete subgroup of G and H a closed connected subgroup of G. Then the following are equivalent:*
(i) *Γ acts properly on G/H.*
(ii) $[K \cdot E_H] \cap E_\Gamma = \{0\}$.

Proof. We first prove that Γ acts properly on G/H if and only if $[K \cdot E_R + B(0,r)] \cap E_\Gamma$ is compact, for any $r > 0$. The required result follows then from Lemma 2.5.6.

Assume first that Γ acts properly on G/H and let $r > 0$ for which $W_r := [K \cdot E_R + B(0,r)] \cap E_\Gamma$ is not compact. There exists a sequence $\{x_p\}_{p \in \mathbb{N}}$ of W_r such that $\|x_p\| > p$, for sufficiently large $p \in \mathbb{N}$. Set $x_p = k_p y_p + u_p$, where $k_p \in K$, $y_p \in E_R$ and $u_p \in B(0,r)$. By Lemma 2.5.4, there exists $r' > 0$ such that for any $p \in \mathbb{N}$, there exists $\overline{x}_p \in pr_2(\Gamma)$ verifying $\|x_p - \overline{x}_p\| \leq r'$. This means that there exist $v_p \in B(0,r')$ such that $x_p = \overline{x}_p + v_p$. In addition, Corollary 1.2.37 and Lemma 1.2.38 show that there exists $\tau = (I,v)$ such that $y_p \in E_R = pr_2(R^\tau) \subset pr_2(H^\tau)$, for any $p \in \mathbb{N}$. Clearly, we get $\overline{x}_p = k_p y_p + z_p$, with $z_p = u_p - v_p$ satisfying $\|z_p\| \leq r + r'$, for any $p \in \mathbb{N}$. Let A_p and B_p be two matrices of K such that $(A_p, \overline{x}_p) \in \Gamma$ and $(B_p, y_p) \in H^\tau$ and set $C_{r+r'} = K \times B(0, r+r')$. It is straightforward that

$$(k_p, z_p)(B_p, y_p)(B_p^{-1} k_p^{-1} A_p, 0) = (A_p, \overline{x}_p),$$

which means that $[C_{r+r'} \cdot H^\tau \cdot C_{r+r'}] \cap \Gamma$ is infinite, and thus the action of Γ on G/H^τ is not proper by Lemma 2.5.5, which is enough to conclude.

Conversely, assume that $[K \cdot E_R + B(0,r)] \cap E_\Gamma$ is compact for any $r > 0$. Let Γ_a be a normal Abelian subgroup of Γ of finite index. By means of Lemma 2.5.5 and Corollary 2.5.3, it is enough to show that $[C_r \cdot R \cdot C_r] \cap \Gamma_a$ is finite, where $C_r = K \times B(0,r)$.

From Lemma 1.2.40 and Corollary 1.2.37, there exist $\tau_1 := (I, v_1)$ and $\tau_2 := (I, v_2)$ such that $pr_2(\Gamma_a^{\tau_1}) \subset E_{\Gamma_a} = E_\Gamma$ and $pr_2(R^{\tau_2}) = E_R$. In order to show that $[C_r \cdot R \cdot C_r] \cap \Gamma_a$ is finite, it is enough to show that $[C_r \cdot R^{\tau_2} \cdot C_r] \cap \Gamma_a^{\tau_1}$ is finite. Let $(A,x) \in T_r$, there exist (M,t), $(N,s) \in C_r$ and $(B,y) \in R$ such that

$$(A,x) = (M,t)(B,y)(N,s),$$

which gives $x = My + u$, with $u := t + MBs$. Clearly,we have $x \in pr_2(\Gamma_a^{\tau_1}) \subset E_{\Gamma_a} = E_\Gamma$ and $y \in pr_2(R^{\tau_2}) = E_R$. This gives that

$$x \in [K \cdot E_R + B(0,2r)] \cap E_\Gamma,$$

which is compact, and finally $pr_2(T_r)$ is compact, for any $r > 0$. It follows that T_r is compact as $T_r \subset K \times pr_2(T_r)$, and hence finite, which shows that the action of Γ on G/R is proper, and hence the result by Proposition 2.5.3. □

The following is then immediate.

Corollary 2.5.8. *Let $G = K \ltimes \mathbb{R}^n$ where K is a compact subgroup of $GL(n,\mathbb{R})$, Γ be a discrete subgroup and H a closed connected subgroup of G. Then the following are equivalent:*

(i) *Γ acts properly on G/H.*

(ii) *$\mathscr{L}_\Gamma$ acts properly on $G/\mathscr{L}_H$.*

Proof. Assume first that Γ acts properly on G/H. For any $r > 0$, set $C_r = K \times B(0,r)$. We shall prove that $\mathscr{L}_\Gamma \cap [C_r \cdot \mathscr{L}_H \cdot C_{r'}]$ is compact, for any $r, r' > 0$. Let $(I,x) \in \mathscr{L}_\Gamma \cap [C_r \cdot \mathscr{L}_H \cdot C_{r'}]$. There exist $(A,u) \in C_r$, $(I,y) \in \mathscr{L}_H$ and $(B,v) \in C_{r'}$, such that $(I,x) = (A,u)(I,y)(B,v)$. An easy computation shows that

$$B = A^{-1} \quad \text{and} \quad x = Ay + u + Av.$$

This gives $x \in [K \cdot E_H + B(0,2r)] \cap E_\Gamma$, which is compact by Theorem 2.5.7 and Lemma 2.5.6. This yields $\mathscr{L}_\Gamma \cap [C_r \cdot \mathscr{L}_H \cdot C_{r'}]$ is compact, and hence the action of $\mathscr{L}_\Gamma$ on $G/\mathscr{L}_H$ is proper.

Conversely, assume that $\mathscr{L}_\Gamma$ acts properly on $G/\mathscr{L}_H$. Then we get $\mathscr{L}_\Gamma \cap [K \cdot \mathscr{L}_H]$ is compact. This means that $\mathscr{L}_\Gamma \cap [K \cdot \mathscr{L}_H] = \{0\}$ and as a consequence $E_\Gamma \cap [K \cdot E_H] = \{0\}$. By Theorem 2.5.7, we conclude that Γ acts properly on G/H. □

As a further consequence of Theorem 2.5.7, we can reestablish the following more general result than that of Proposition 2.5.1.

Corollary 2.5.9. *Let Γ be a discrete subgroup and H a closed connected subgroup of G, where K acts transitively on the sphere S^{n-1}. Then Γ acts properly on G/H if and only if H is compact or Γ is finite.*

Proof. Assume that Γ acts properly on G/H. If $E_H = \{0\}$, then H is compact by Corollary 1.2.39. Otherwise, let $0 \neq x \in E_H$. For any $0 \neq y \in \mathbb{R}^n$, there exists $A \in K$ such that $\frac{y}{\|y\|} = A\frac{x}{\|x\|}$, which gives that $y \in [K \cdot E_H]$, and hence $[K \cdot E_H] = \mathbb{R}^n$. It follows that $E_\Gamma = \{0\}$ and that Γ is finite by Proposition 1.2.40. □

Example 2.5.10. Recall the unitary group $U_n(\mathbb{C}) = \{u \in M_n(\mathbb{C}) \mid uu^* = I_n\}$. One can identify U_n with a compact real group through the homomorphism

$$\begin{aligned} \phi_n : U_n &\hookrightarrow O_{2n}(\mathbb{R}) \\ A + iB &\mapsto \begin{pmatrix} A & -B \\ B & A \end{pmatrix}; \end{aligned}$$

here, $A, B \in M_n(\mathbb{R})$. Clearly, $\phi_n(U_n)$ is a compact subgroup, which acts transitively on the sphere S^{2n-1} of $\mathbb{R}^{2n}$. Hence, for $K = \phi_n(U_n)$, the proper action of a discrete subgroup Γ on G/H is equivalent to the fact that Γ is finite or H is compact, thanks to Corollary 2.5.9.

Remark 2.5.11. The problem of finding a criterion of proper action is studied by T. Kobabyashi and T. Yoshino (cf. [93]) in the context of Cartan motion groups. Let G be a linear reductive Lie group and K a maximal compact subgroup of G. Let θ be the corresponding involution and write $G := K\exp(\mathfrak{p})$ a Cartan decomposition of G. The semidirect product $G_\theta := K \ltimes \mathfrak{p}$ is called the Cartan motion group associated to the pair (G, K). The multiplication law of G_θ is given by

$$(k, X)(k', X') := (kk', X + \mathrm{Ad}(k)X').$$

Let $\mathfrak{a}$ be maximal Abelian subspace of $\mathfrak{p}$. For a subset L of G_θ, define $\mathfrak{a}(L) := KLK \cap \mathfrak{a}$. If L is a subgroup of G_θ set $\mathfrak{p}_L := L \cap \mathfrak{p}$ and $d(L) = \dim \mathfrak{p}_L$. Finally, define an involutive automorphism

$$\overline{\theta} : G_\theta \longrightarrow G_\theta, \quad (k, X) \longmapsto (k, -X).$$

Under the assumptions that L and H are $\overline{\theta}$-stable subgroups with at most finitely many connected components and that L acts properly on G_θ/H, it was shown that the Clifford–Klein form $L\backslash G_\theta/H$ is compact if and only if $d(L) + d(H) = d(G_\theta)$ (cf. [93, Lemma 5.3.5]). Further, a criterion of proper action has been established. It states that for any subsets L and H of G_θ, L acts properly on G_θ/H if and only if $\mathfrak{a}(L)$ acts properly on $\mathfrak{a}/\mathfrak{a}(H)$ (cf. [93, Lemma 5.3.6]).

2.6 Proper actions for Heisenberg motion groups

This section aims to generate a criterion of the proper action of a discontinuous group $\Gamma \subset G = \mathbb{U}_n \ltimes \mathbb{H}_n$, the Heisenberg motion group, acting on a homogeneous space G/H. We keep all the notation and definitions of Section 1.3. Remark first the following immediate result.

Lemma 2.6.1. *Let Γ be a discrete subgroup of G and H a closed subgroup of G.*
The following assertions are equivalent:
(i) *Γ acts properly on G/H.*
(ii) *For any $r > 0$, $[C_r \cdot H \cdot C_r] \cap \Gamma$ is finite, where $C_r := \mathbb{U}_n \times B(0,r) \times [-r,r]$.*
(iii) *For any compact sets C and C' of G, $[C \cdot H \cdot C'] \cap \Gamma$ is finite.*

We refer back to Proposition 1.3.17 for the classification of discrete subgroups of G. We first show the following.

Proposition 2.6.2. *Let H be a closed subgroup of G and Γ a discrete subgroup of G of type (A). Then Γ acts properly on G/H if and only if $H \cap (\mathbb{U}_n \times B(0,r) \times \mathbb{R})$ is compact for any $r > 0$.*

Proof. Assume that Γ is a discrete subgroup of G^1. For any $\gamma \in [C_r \cdot \Gamma \cdot C_r] \cap H$ $(r > 0)$, there exist $\lambda := (A_\lambda, 0, t_\lambda) \in \Gamma$ and $(S,u,x), (S',v,y) \in C_r$ such that

$$\gamma = (S,u,s)\lambda(S',u',s') = \left(SA_\lambda S', u + SA_\lambda u', s + s' + t_\lambda - \frac{1}{2}\operatorname{Im}\langle u, SA_\lambda u'\rangle\right).$$

Hence, $\gamma \in H \cap (\mathbb{U}_n \times B(0,2r) \times \mathbb{R})$. Therefore, $[C_r \cdot \Gamma \cdot C_r] \cap H \subset H \cap (\mathbb{U}_n \times B(0,2r) \times \mathbb{R})$.

Conversely, as Γ is infinite, it contains necessarily some $\gamma = (A,0,t)$ with $t \neq 0$. Suppose that there exists $r > 0$ such that $H \cap (\mathbb{U}_n \times B(0,r) \times \mathbb{R})$ is not compact, we can find a sequence $\{(h_p, w_p, s_p)\}_{p\in\mathbb{N}}$ of $H \cap (\mathbb{U}_n \times B(0,r) \times \mathbb{R})$ such that $\lim_{p\to\infty} s_p = \infty$. For $p \in \mathbb{N}$, we have $\gamma^p = (A^p, 0, pt)$ and let $\alpha(p)$ be the floor of $\frac{s_p}{t}$. As $\alpha(p) \leq \frac{s_p}{t} < (\alpha(p)+1)$. Then $l_p := s_p - \alpha(p)t$ satisfies $|l_p| \leq |t|$. Set the compact set $K := \mathbb{U}_n \times B(0,r) \times [-|t|, |t|]$, and we have

$$\gamma^{\alpha(p)} = (I_n, -w_p, -l_p)(h_p, w_p, s_p)(h_p^{-1}A^{\alpha(p)}, 0, 0) \in KHK^{-1} \cap \Gamma.$$

As the set $\{\gamma^{\alpha(p)}\}_{p\in\mathbb{N}}$ is infinite, we meet a contradiction and this completes the proof. □

We next show the following.

Proposition 2.6.3. *Let H be a closed subgroup of G and Γ a discrete subgroup of G of type (B).*

(a) *If* $\mathrm{pr}(\Gamma)$ *is discrete, then* Γ *acts properly on* G/H *if and only if* H *is conjugate to a subgroup of* G^1.
(b) *Otherwise,* Γ *acts properly on* G/H *if and only if* H *is compact.*

Proof. We first prove that if a type (B) discrete subgroup acts properly on G/H, then H is conjugate to a subgroup of G^1. Indeed, assume that one can find some $(A, z, 0) \in \Gamma$ such that $Az = z \neq 0$. If there exists $h = (B, w, s) \in H$ such that $P_B(w) \neq 0$, then by Lemma 1.3.1 there exists $u \in \mathbb{C}^n$ such that $\tau_{-u} h \tau_u = (B, w', 0) \in H^{\tau_u} := \tau_{-u} H \tau_u$. Here, $Bw' = w' \neq 0$. For $p \in \mathbb{N}$, take $\alpha(p)$ to be the floor of $\frac{\|pw'\|}{\|z\|}$. Then

$$\alpha(p)\|z\| \leq \|pw'\| < (\alpha(p) + 1)\|z\|.$$

Let $z_p = \rho_p z \in S(0, \|pw'\|) \cap \mathbb{R}z$, where $\alpha(p) \leq \rho_p < \alpha(p) + 1$, and $S(0, \|pw'\|)$ designates the sphere of center 0 and radius $\|pw'\|$. Then

$$y_p = z_p - \alpha(p)z = (\rho_p - \alpha(p))z$$

satisfies $\|y_p\| < \|z\|$. Since $\mathbb{U}_n$ acts transitively on $S(0, \|pw'\|)$, there exists $C_p \in \mathbb{U}_n$ such that $C_p(pw') = z_p$. Let $K = \mathbb{U}_n \times B(0, \|z\|) \times \{0\}$. Then we easily check that

$$(C_p, -y_p, 0)(B_p, pw', 0)(B_p^{-1} C_p^{-1} A^{\alpha(p)}, 0, 0) = (A^{\alpha(p)}, \alpha(p)z, 0) \in \Gamma \cap KHK^{-1}.$$

As the set $\{(A^{\alpha(p)}, \alpha(p)z, 0)\}_{p \in \mathbb{N}}$ is infinite. Then the proper action of Γ on G/H entails that any $(B, w, s) \in H$ is such that $P_B(w) = 0$. By Proposition 1.3.14, H is conjugate to a subgroup of G^1.

Let us go back now to the proof of the proposition. For statement (a), assume that H is a subgroup of G^1. Let $r > 0$ and $\gamma \in \Gamma \cap [C_r \cdot H \cdot C_r]$, then there exist $(S, u, s), (S', u', s') \in C_r$ and $(B, 0, t) \in H$ such that

$$\gamma = (S, u, s)(B, 0, t)(S', u', s') = \left(SBS', u + SBu', s + s' + t - \frac{1}{2} \operatorname{Im}\langle u, SBu' \rangle \right).$$

Then $\mathrm{pr}(\gamma) \in \mathbb{U}_n \times B(0, 2r)$. As $\mathrm{pr}(\Gamma)$ is discrete, then $\mathrm{pr}(\Gamma) \cap \mathbb{U}_n \times B(0, 2r)$ coincides with a finite set $\{(A_1, z_1), \dots, (A_k, z_k)\}$. Suppose in addition that $\Gamma \cap [C_r \cdot H \cdot C_r]$ is infinite, then obviously there exists $s_0 \in \{1, \dots, k\}$ such as the set $\{(A_{s_0}, z_{s_0}, t) \in \Gamma \cap [C_r \cdot H \cdot C_r]\}$ is infinite. Hence, there exist (A_{s_0}, z_{s_0}, t) and $(A_{s_0}, z_{s_0}, t') \in \Gamma$ such that $t \neq t'$, which gives that

$$(A_{s_0}, z_{s_0}, t)(A_{s_0}, z_{s_0}, t')^{-1} = (I, 0, t - t') \in \Gamma,$$

and this is absurd.

For statement (b), assume that $\mathrm{pr}(\Gamma)$ is not discrete. There exists an infinite sequence $\{(A_p, z_p)\}_{p \in \mathbb{N}} \subset (\mathbb{U}_n \times B(0, \rho)) \cap \mathrm{pr}(\Gamma)$ for some positive real ρ. For any $p \in \mathbb{N}$, one can find some $t_p \in \mathbb{R}$ such as $(A_p, z_p, t_p) \in \Gamma$. Since Γ is discrete, then necessarily the sequence $\{t_p\}_{p\in\mathbb{N}}$ is not bounded. We can assume that $|t_p|$ tends to ∞. As H is conjugate to a subgroup of G^1, we may assume that there exists some $(C, 0, t) \in H$ with $t \neq 0$, and let $\beta(p)$ be the floor of $\frac{t_p}{t}$. Remark that $|t_p - \beta(p)t| \leq |t|$, for $R := \max(\rho, |t|)$, $\Gamma \cap C_R H C_R$ contains the infinite set $\{(A_{\beta(p)}, z_{\beta(p)}, t_{\beta(p)})\}$, which is absurd. Hence, H is compact. The converse is trivial. □

Proposition 2.6.4. *Let Γ be an infinite discrete subgroup of G of type (C). For a closed subgroup H of G, Γ acts properly on G/H if and only if H is compact.*

Proof. By Proposition 2.6.2, Γ acts properly on G/H, and thus $H \cap (\mathbb{U}_n \times B(0, r) \times \mathbb{R})$ is compact for any $r > 0$. Hence, H does not contain an element conjugate to some $(B, 0, t)$ with $t \neq 0$. Proposition 2.6.3 shows that H is conjugate to a subgroup of G^1. Conclusively, H is compact. □

3 Deformation and moduli spaces

Our attention in this chapter is focused on the explicit determination (to a certain extent) of the deformation and moduli spaces of discontinuous groups acting on homogeneous spaces. This issue is of major importance to understand the local geometric structures of these spaces as many examples reveal. Toward such a purpose, many people have been interested in studying the deformation spaces of some geometric structures of closed surfaces, such as projective structures, hyperbolic structures (the Teichmüller space), complex structures and so on. In some restrictive cases, these deformation spaces are explicitly computed and some of their topological features are studied. For broader details, the reader could consult [40, 57, 67, 73, 76, 77, 92, 103–105] and some references therein. For instance, when $G/H = SL(2,\mathbb{R})/SO(2)$ is the Poincaré disk, the deformation space consists of the deformation of complex structures on a Riemann surface M_g of genus $g \geq 2$ and when $G/H = G' \times G'/\operatorname{diag}(G')$ for $G' = SL(2,\mathbb{R})$, the deformation space is nothing but the deformation of complex structures on a 3-dimensional manifold. We refer the reader to the above papers where many settings have been considered.

Furthermore, the explicit determination of the deformation space helps to understand many of its geometrical features, such as the local rigidity, the stability and the Calabi–Markus phenomenon, which will be subject of study later on (cf. Chapters 5 and 6).

3.1 Deformation and moduli spaces of discontinuous actions

3.1.1 Parameter, deformation and moduli spaces

The parameter space

The problem of describing deformations is advocated by T. Kobayashi in [97] where he formalized the study of the deformation of Clifford–Klein forms from a theoretic point of view. (See *Problem C* in [95] for further perspectives and basic examples.) When it comes to the solvable context, the strategy basically consists in building up an accurate cross-section of the adjoint orbits of elements of the parameter space, which defines all of the possible deformations of the discrete subgroup in question, in such a way that the deformed subgroups preserve the property of being discontinuous groups for the homogeneous space.

Let G be a Lie group and Γ be a finitely generated discrete group. We designate by $\operatorname{Hom}(\Gamma, G)$ of group homomorphisms from Γ to G endowed with the pointwise convergence topology. The same topology is obtained by taking generators $\gamma_1, \dots, \gamma_k$ of Γ. Then using the injective map

$$\operatorname{Hom}(\Gamma, G) \hookrightarrow G \times \cdots \times G, \quad \varphi \mapsto (\varphi(\gamma_1), \dots, \varphi(\gamma_k))$$

https://doi.org/10.1515/9783110765304-003

to equip $\mathrm{Hom}(\Gamma, G)$ with the relative topology induced from the direct product $G\times\cdots\times G$. Let H be a closed subgroup of G. If H is not compact, then the discrete subgroup Γ does not necessarily act properly discontinuously on G/H. We consider then the parameter space $\mathscr{R}(\Gamma, G, H)$ of $\mathrm{Hom}(\Gamma, G)$ defined by

$$R(\Gamma, G, H) = \{\varphi \in \mathrm{Hom}(\Gamma, G) : \varphi \text{ is injective and } \varphi(\Gamma) \text{ acts properly discontinuously and freely on } G/H\}. \tag{3.1}$$

This set plays an important role as we will see later. According to this definition, for each $\varphi \in \mathscr{R}(\Gamma, G, H)$, the space $\varphi(\Gamma)\backslash G/H$ is a Clifford–Klein form, which is a Hausdorff topological space and even equipped with a structure of a manifold for which the quotient canonical map is an open covering.

The deformation and moduli space

Let now $\varphi \in \mathscr{R}(\Gamma, G, H)$ and $g \in G$. We consider the element $\varphi^g := g^{-1}\cdot\varphi\cdot g$ of $\mathrm{Hom}(\Gamma, G)$ defined by

$$\varphi^g(\gamma) = g^{-1}\varphi(\gamma)g, \quad \gamma \in \Gamma.$$

It is then clear that the element $\varphi^g \in \mathscr{R}(\Gamma, G, H)$ and that the map

$$\varphi(\Gamma)\backslash G/H \longrightarrow \varphi^g(\Gamma)\backslash G/H, \quad \varphi(\Gamma)xH \mapsto \varphi^g(\Gamma)g^{-1}xH$$

is a natural diffeomorphism. We consider then the orbits space

$$\mathscr{T}(\Gamma, G, H) = \mathscr{R}(\Gamma, G, H)/G$$

instead of $\mathscr{R}(\Gamma, G, H)$ in order to avoid the unessential part of deformations arising inner automorphisms and to be quite precise on parameters. We call the set $\mathscr{T}(\Gamma, G, H)$ as the space of the deformation of the action of Γ on the homogeneous space G/H.

On the other hand, let the group $\mathrm{Aut}(\Gamma)$ act on $\mathrm{Hom}(\Gamma, G)$ by

$$T\cdot\varphi(\gamma) := \varphi(T^{-1}(\gamma)), \quad \varphi \in \mathrm{Hom}(\Gamma, G), \quad T \in \mathrm{Aut}(\Gamma), \quad \gamma \in \Gamma.$$

It is then easy to check that the group $\mathrm{Aut}(\Gamma)$ leaves the parameter space $\mathscr{R}(\Gamma, G, H)$ invariant and its action on it is G-equivariant. We define then (to avoid this unessential part, too) the moduli space as the double coset space

$$\mathscr{M}(\Gamma, G, H) := \mathrm{Aut}(\Gamma)\backslash\mathscr{R}(\Gamma, G, H)/G.$$

With the above in mind, the following facts are immediate.

Lemma 3.1.1. *Let G be a locally compact group and let H and Γ be subgroups of G. Then the following assertions are equivalent:*

(i) Γ *is a discontinuous group for* G/H.
(ii) Γ *is a discontinuous group for* G/gHg^{-1}, *for any* $g \in G$.

In particular, $\mathscr{R}(\Gamma, G, H) = \mathscr{R}(\Gamma, G, gHg^{-1})$, *for all* $g \in G$.

Lemma 3.1.2. *Let* Γ *be a discontinuous group for* G/H, *assume that* H *is contained co-compactly in* K *and* Γ *is torsion-free. Then* $\mathscr{R}(\Gamma, G, H) = \mathscr{R}(\Gamma, G, K)$.

Proof. As Γ is torsion-free, the discontinuous action of Γ on G/H is equivalent to its proper action. Then by definition of the parameter space, the result comes from Fact 2.1.7. □

Remark 3.1.3. Let Γ be generated by $\{y_1, \dots, y_k\}$. Let $F(k)$ be the non-Abelian free group with a k generators $x_1, \dots, x_k$ and φ a group homomorphism from $F(k)$ to G, then φ is completely determined by the image of the generators. As $F(k)$ is a group without any nontrivial relation, the map

$$\mathrm{Hom}(F(k), G) \to G^k, \quad \varphi \mapsto (\varphi(x_1), \dots, \varphi(x_k)),$$

is a bijection and when $\mathrm{Hom}(F(k), G)$ is endowed with the pointwise convergence topology and G^k with the product topology. this map becomes a homeomorphism. Now the natural map

$$\pi_k : F(k) \to \Gamma, \quad x_{i_1}^{n_1} \cdots x_{i_r}^{n_r} \mapsto y_{i_1}^{n_1} \cdots y_{i_r}^{n_r},$$

is a group homomorphism. This allows to conclude that $\Gamma \cong F(k)/\ker \pi_k$ and

$$\mathrm{Hom}(\Gamma, G) = \{\varphi \in \mathrm{Hom}(F(k), G), \ker \varphi \supset \ker \pi_k\}.$$

If we consider the identification of $\mathrm{Hom}(F(k), G)$ with G^k, then we can write

$$\mathrm{Hom}(\Gamma, G) = \{(g_1, \dots, g_k) \in G^k, g_{i_1} \cdots g_{i_r} = e \text{ for all } x_{i_1} \cdots x_{i_r} \in \ker \pi_k\},$$

which is clearly a closed subset. As G is a real analytic manifold and the multiplication map is also analytic, it comes out that $\mathrm{Hom}(\Gamma, G)$ is an analytic subset of G^k whenever Γ is finitely presented, i. e., the kernel of π_k is finitely generated; see [98, Section 5.2].

3.1.2 Case of effective actions

We here refer to Subsection 2.1.2. We first introduce the Clifford–Klein space

$$\mathrm{CK}(\Gamma, G, H) = \{\Gamma' \backslash G/H, \Gamma' \text{ is isomorphic to } \Gamma \text{ and } \Gamma' \text{ is discontinuous for } G/H\}.$$

By a deformation of the Clifford–Klein form $\Gamma\backslash G/H$, we mean any element of the related Clifford–Klein space $\mathrm{CK}(\Gamma, G, H)$. The next section is devoted to describe in the general context how to interpret these deformations and the space $\mathrm{CK}(\Gamma, G, H)$.

We now introduce the kernel of effectiveness related to the action of G on G/H, which equals the normal subgroup $N = \bigcap_{g\in G} gHg^{-1}$. For any $\varphi \in \mathscr{R}(\Gamma, G, H)$, $\varphi(\Gamma) \cap N = \{e\}$, which means that $\varphi(\Gamma)$ acts effectively on G/H. There is a natural surjective map $\Psi : \mathscr{R}(\Gamma, G, H) \to \mathrm{CK}(\Gamma, G, H)$, $\varphi \mapsto \varphi(\Gamma)\backslash G/H$. Now the group G itself acts on $\mathscr{R}(\Gamma, G, H)$ and $\mathrm{CK}(\Gamma, G, H)$ and the map Ψ is G-equivariant. In sum, $\Psi(\mathscr{T}(\Gamma, G, H)) = \mathrm{CK}(\Gamma, G, H)/G$ and the deformation space determines therefore all possible deformations of the related Clifford–Klein form modulo the G-action. Let Γ_i, $i = 1, 2$ be discontinuous groups for G/H, then $\Gamma_1\backslash G/H = \Gamma_2\backslash G/H$ if and only if $\Gamma_1 gHg^{-1} = \Gamma_2 gHg^{-1}$ for all $g \in G$. We define an equivalence relation $\curlywedge$ on $\mathscr{R}(\Gamma, G, H)$ as follows: $\varphi \curlywedge \varphi'$ if and only if $\varphi(\Gamma)gHg^{-1} = \varphi'(\Gamma)gHg^{-1}$ for any $g \in G$. We then introduce the space $\widehat{\mathscr{R}}(\Gamma, G, H)$ as being the quotient subsequent space. Clearly, the map Ψ factors to a bijection from $\widehat{\mathscr{R}}(\Gamma, G, H)$ to $\mathrm{CK}(\Gamma, G, H)$ and to a topological homeomorphism when these spaces are endowed with adequate topologies. Now the action of G on $\mathscr{R}(\Gamma, G, H)$ induces an associated action on $\widehat{\mathscr{R}}(\Gamma, G, H)$, which commutes with the relation $\curlywedge$. The space $\widehat{\mathscr{T}}(\Gamma, G, H) := \widehat{\mathscr{R}}(\Gamma, G, H)/G$ is called the *refined deformation space* and is identified to the space $\mathrm{CK}(\Gamma, G, H)/G$.

3.1.3 Deformation of (G, X)-structures

Our interest to these spaces comes from the deformation theory of the (G,X)-structures, where G is a Lie group and X is a homogeneous space. Let M be a smooth manifold such that $\dim X = \dim M$. A (G,X)-atlas on M is a collection $(U_\alpha, \phi_\alpha)_{\alpha\in I}$, where $\{U_\alpha, \alpha \in I\}$ is an open covering of M and $\{\phi_\alpha : U_\alpha \to X, \alpha \in I\}$ is a family of local coordinates charts such that, on a connected component C of $U_\alpha \cap U_\beta$, there exists $g_{C,\alpha,\beta} \in G$ satisfying

$$g_{C,\alpha,\beta} \circ \phi_\alpha = \phi_\beta.$$

A (G, X)-structure on M is a maximal (G, X)-atlas on M and a (G, X)-manifold is a manifold endowed with a (G, X)-structure. Let Σ be a smooth manifold, a marked (G, X)-structure on Σ is a pair (M, f) where M is a (G, X)-manifold and $f : \Sigma \to M$ is a diffeomorphism. Let $D_{(G,X)}(\Sigma)$ be the space of the all marked (G, X)-structures on Σ. The group $\mathrm{Diff}_0(\Sigma)$ (the subgroup of the group of diffeomorphisms of Σ isotopic to the identity) acts on $D_{(G,X)}(\Sigma)$ through the law:

$$\psi \star (M, f) = (M, f \circ \psi^{-1}), \quad \psi \in \mathrm{Diff}_0(\Sigma).$$

The deformation space of the (G,X)-structures on Σ is the quotient space

$$\mathrm{Def}_{(G,X)}(\Sigma) = D_{(G,X)}(\Sigma)/\operatorname{Diff}_0(\Sigma).$$

Assume Σ is compact. By the deformation Theorem of Thurston, the *holonomy* map is a local homeomorphism between the deformation space of marked (G,X)-structures on Σ and the quotient space $\mathrm{Hom}(\pi_1(\Sigma), G)/G$, (cf. [78]). If Γ is a discontinuous subgroup for X, the Clifford–Klein form $\Gamma\backslash X$ is a (G,X)-manifold. If there is a diffeomorphism $f : \Sigma \to \Gamma\backslash X$, then the marked (G,X)-structure $(\Gamma\backslash X, f)$ is said to be complete. The set $D^c_{(G,X)}(\Sigma)$ of the complete (G,X)-structures on Σ, is invariant under the action of $\operatorname{Diff}_0(\Sigma)$. The deformation space of complete (G,X)-structures on Σ is defined as

$$\mathrm{Def}^c_{(G,X)}(\Sigma) = D^c_{(G,X)}(\Sigma)/\operatorname{Diff}_0(\Sigma).$$

Then the deformation space $\mathscr{T}(\Gamma, G, H)$ of the discontinuous actions of Γ on $X = G/H$ contains the image of $\mathrm{Def}^c_{(G,X)}(\Sigma)$ by the holonomy map. Furthermore, if all the forms $\varphi(\Gamma)\backslash G/H$ are diffeomorphic for $\varphi \in \mathscr{R}(\Gamma, G, H)$, then the deformation space $\mathscr{T}(\Gamma, G, H)$ coincides with the image of $\mathrm{Def}^c_{(G,X)}(\Sigma)$.

Remark 3.1.4. We close this section with the following important remark. Assume that the deformation space $\mathscr{T}(\Gamma, G, H)$ of the discontinuous actions of Γ on $X = G/H$ coincides with the image of $\mathrm{Def}^c_{(G,X)}(\Sigma)$ by the holonomy map *hol* and that the stability holds. Then the restriction

$$hol : \mathrm{Def}^c_{(G,X)}(\Sigma) \to \mathscr{T}(\Gamma, G, H)$$

is a local homeomorphism. Indeed, $\mathscr{T}(\Gamma, G, H)$ is an open set of $\mathrm{Hom}(\Gamma, G)/G$, therefore, $\mathrm{Def}^c_{(G,X)}(\Sigma)$ is an open set of $\mathrm{Def}_{(G,X)}(\Sigma)$. As *hol* is a local homeomorphism, we are done.

3.2 Algebraic characterization of the deformation space

3.2.1 The deformation and moduli spaces in the exponential setting

Let $\mathfrak{g}$ denote a n-dimensional real exponential solvable Lie algebra, G will be the associated connected and simply connected exponential Lie group. The Lie algebra $\mathfrak{g}$ acts on $\mathfrak{g}$ by the adjoint representation ad, and the group G acts on $\mathfrak{g}$ by the adjoint representation Ad, defined by

$$\mathrm{Ad}_g = \exp \circ \mathrm{ad}_T, \quad g = \exp T \in G. \tag{3.2}$$

Let $H = \exp \mathfrak{h}$ be a closed, connected subgroup of G. Let Γ be an Abelian discrete subgroup of G of rank k and define the parameter space $\mathscr{R}(\Gamma, G, H)$ as given in (3.1). The aim now is to generate a characterization of the parameter and the deformation spaces is derived as follows. Let L be the syndetic hull of Γ, which is the smallest (and hence the unique) connected Lie subgroup of G, which contains Γ cocompactly (cf. Theorem 1.4.2). Recall that the Lie subalgebra $\mathfrak{l}$ of L is the real span of the Abelian lattice $\log \Gamma$, which is generated by $\{\log \gamma_1, \dots, \log \gamma_k\}$ where $\{\gamma_1, \dots, \gamma_k\}$ is a set of generators of Γ. The group G also acts on $\mathrm{Hom}(\mathfrak{l}, \mathfrak{g})$ by

$$\psi \cdot g = \mathrm{Ad}_{g^{-1}} \circ \psi. \tag{3.3}$$

Our first observation is that the parameter space defined above, only depends on the structure of the syndetic hull of Γ when the basis group G is completely solvable. Recall that any continuous homomorphism of a connected Lie groups is smooth and its derivative is a homomorphism of Lie algebras. We consider the smooth map d : $\mathrm{Hom}_c(L, G) \longrightarrow \mathrm{Hom}(\mathfrak{l}, \mathfrak{g})$, $\varphi \mapsto d\varphi_{|_e}$ where $\mathfrak{l}$ is the Lie algebras of L. In the case of exponential Lie groups, $d\varphi_{|e}(X) = \log \circ \varphi \circ \exp(X)$ for any $X \in \mathfrak{g}$. The group G acts on the spaces $\mathrm{Hom}(\Gamma, G)$, $\mathrm{Hom}(L, G)$ and $\mathrm{Hom}(\mathfrak{l}, \mathfrak{g})$, respectively, through the following laws:

$$(g \cdot \varphi)(\gamma) = g\varphi(\gamma)g^{-1}, \quad \gamma \in \Gamma \text{ (or } L), \quad \varphi \in \mathrm{Hom}(\Gamma, G) \text{ (or } \mathrm{Hom}(L, G)), \quad g \in G,$$
$$g \cdot \psi = \mathrm{Ad}_g \circ \psi, \quad \psi \in \mathrm{Hom}(\mathfrak{l}, \mathfrak{g}), \quad g \in G.$$

As L is exponential, the differential map d from $\mathrm{Aut}(L)$ to $\mathrm{Aut}(\mathfrak{l})$ is a topological isomorphism. Using the natural injection i of $\mathrm{Aut}(\Gamma)$ in $\mathrm{Aut}(L)$, we see that $\mathrm{Aut}(\Gamma)$ acts on $\mathrm{Hom}(L, G)$ and $\mathrm{Hom}(\mathfrak{l}, \mathfrak{g})$ by

$$a \cdot \varphi = \varphi \circ i(a)^{-1} \quad \text{and} \quad a \cdot \psi = \psi \circ d(i(a))_{|_e}^{-1}, \tag{3.4}$$

for $\varphi \in \mathrm{Hom}(L, G)$, $\psi \in \mathrm{Hom}(\mathfrak{l}, \mathfrak{g})$ and $a \in \mathrm{Aut}(\Gamma)$. Let

$$\mathscr{R}(\mathfrak{l}, \mathfrak{g}, \mathfrak{h}) = \{\psi \in \mathrm{Hom}(\mathfrak{l}, \mathfrak{g}) : \dim \psi(\mathfrak{l}) = \dim \mathfrak{l}, \exp(\psi(\mathfrak{l})) \text{ acts properly on } G/H\}. \tag{3.5}$$

The following result provides a preliminary algebraic interpretation of the parameter, deformation and moduli spaces and of capital importance in the sequel.

Theorem 3.2.1. *Let G be a completely solvable Lie group, H a connected subgroup of G and Γ a discrete subgroup of G acting properly discontinuously on G/H. Then up to a homeomorphism,*

$$\mathscr{R}(\Gamma, G, H) = \mathscr{R}(\mathfrak{l}, \mathfrak{g}, \mathfrak{h}).$$

In particular, if Γ and Γ' have the same syndetic hull, then $\mathscr{R}(\Gamma, G, H)$ and $\mathscr{R}(\Gamma', G, H)$ are homeomorphic. Furthermore, up to homeomorphism, the deformation space $\mathscr{T}(\Gamma, G, H)$ coincides with $\mathscr{R}(\mathfrak{l}, \mathfrak{g}, \mathfrak{h})/G$ and the moduli space is identified to $\mathrm{Aut}(\Gamma)\backslash\mathscr{R}(\mathfrak{l}, \mathfrak{g}, \mathfrak{h})/G$.

Proof. The composition map $d \circ \xi$ is a homeomorphism from

$$\{\varphi \in \mathrm{Hom}(\Gamma, G),\ \varphi(\Gamma) \text{ isomorphic to } \Gamma\}$$

to its image, where ξ is the inverse of the restriction map as in Proposition 1.4.7. We now show that this image is precisely the set

$$\{\varphi \in \mathrm{Hom}(\mathfrak{l}, \mathfrak{g}),\ \dim \varphi(\mathfrak{l}) = \dim \mathfrak{l}\}.$$

Indeed, let $\varphi \in \mathrm{Hom}(\Gamma, G)$ such that Γ is isomorphic to $\Gamma' = \varphi(\Gamma)$. The homomorphism $\xi(\varphi)$ is continuous and L is connected, then $\xi(\varphi)(L)$ is a connected subgroup containing $\varphi(\Gamma)$. This means that $\xi(\varphi)(L)$ contains the syndetic hull of Γ' denoted by L'. In particular, $\dim L \geq \dim \xi(\varphi)(L) \geq \dim L'$. Let φ^{-1} be the inverse of $\varphi : \Gamma \to \Gamma' \subset G$. The composition of φ^{-1} and the natural injection of Γ in G is a homomorphism from Γ' to G and Γ' is isomorphic to $\Gamma = \varphi^{-1}(\Gamma')$. Then for ξ' the extension map from $\mathrm{Hom}(\Gamma', G)$ to $\mathrm{Hom}(L', G)$, we have $\xi'(\varphi^{-1})(L') \supset L$. Then $\dim L' \geq \dim \xi'(\varphi^{-1})(L') \geq \dim L$. This entails that $\dim L = \dim \xi(\varphi)(L)$, which is equivalent to $\dim \mathfrak{l} = \dim d \circ \xi(\varphi)(\mathfrak{l})$. Conversely, if ψ is a Lie algebras homomorphism, such that $\dim \mathfrak{l} = \dim \psi(\mathfrak{l})$, then $\exp \circ \psi \circ \log \in \mathrm{Hom}(L, G)$, is an isomorphism from L to its image. Its restriction to Γ denoted by φ say, is an isomorphism from Γ to its image and satisfies $d \circ \xi(\varphi) = \psi$. Now, from Fact 2.1.8, the proper action of $\varphi(\Gamma)$ is equivalent to the proper action of its syndetic hull $L' = \xi(\varphi)(L) = \exp \circ d \circ \xi(\varphi)(\mathfrak{l})$. We conclude therefore that $d \circ \xi(\mathscr{R}(\Gamma, G, H)) = \mathscr{R}(\mathfrak{l}, \mathfrak{g}, \mathfrak{h})$. To achieve the proof of the theorem, we need the following.

Lemma 3.2.2. *With the same hypotheses as in Theorem* 3.2.1. *The maps d and ξ are G and* $\mathrm{Aut}(\Gamma)$*-equivariant.*

Proof. For $g \in G$ and $\varphi \in \mathrm{Hom}(\Gamma, G)$, we have $g \cdot \xi(\varphi) = \xi(g \cdot \varphi)$ if and only if $(g \cdot \xi(\varphi))_{|\Gamma} = g \cdot \varphi$. Furthermore, for any $\gamma \in \Gamma$, we have $(g \cdot \xi(\varphi))_{|\Gamma}(\gamma) = g\xi(\varphi)(\gamma)g^{-1} = (g \cdot \varphi)(\gamma)$. Let $\tau_g : G \to G, t \mapsto gtg^{-1}$ be the conjugation. For any $\varphi \in \mathrm{Hom}(L, G)$ and any $g \in G$, we have

$$d(g \cdot \varphi)_{|_e} = d(\tau_g \circ \varphi)_{|_e} = d(\tau_g)_{|_e} \circ d(\varphi)_{|_e} = \mathrm{Ad}_g \circ d(\varphi)_{|_e} = g \cdot d(\varphi)_{|_e},$$

where Ad_g is the adjoint representation defined as in formula (3.2). Likewise, for $a \in \mathrm{Aut}(\Gamma)$ and $\varphi \in \mathrm{Hom}(\Gamma, G)$, we have $a \cdot \xi(\varphi) = \xi(a \cdot \varphi)$ if and only if $(a \cdot \xi(\varphi))_{|\Gamma} = a \cdot \varphi$, and as before we have $(a \cdot \xi(\varphi))_{|\Gamma}(\gamma) = \xi(\varphi) \circ i(a)^{-1}(\gamma) = \varphi \circ a^{-1}(\gamma) = (a \cdot \varphi)(\gamma)$. For $\varphi \in \mathrm{Hom}(L, G)$, we have $d(a \cdot \varphi)_{|_e} = d(\varphi \circ i(a)^{-1})_{|_e} = d(\varphi)_{|_e} \circ d(i(a)^{-1})_{|_e} = a \cdot d(\varphi)_{|_e}$. □

Our theorem is thus proved. □

Using Lemma 3.2.2 and Theorem 3.2.1, the following is immediate.

Corollary 3.2.3. *Retain the same hypotheses as in Theorem 3.2.1. If Γ is uniform in G, then H is trivial and we have $\mathscr{R}(\Gamma, G, H) = \mathrm{Aut}(\mathfrak{g})$,*

$$\mathscr{T}(\Gamma, G, H) = \mathrm{Aut}(\mathfrak{g})/\operatorname{Ad}(G) \quad \textit{and} \quad \mathscr{M}(\Gamma, G, H) = \mathrm{Aut}(\Gamma)\backslash \mathrm{Aut}(\mathfrak{g})/\operatorname{Ad}(G).$$

In the exponential setting, we can even get a similar result as in the preceding theorem under the condition that Γ is Abelian.

Theorem 3.2.4. *Let $G = \exp \mathfrak{g}$ be an exponential solvable Lie group, $H = \exp \mathfrak{h}$ a closed connected subgroup of G, $\Gamma \simeq \mathbb{Z}^k$ a discrete subgroup of G and $L = \exp \mathfrak{l}$ its syndetic hull. Then the parameter space $\mathscr{R}(\Gamma, G, H)$ is homeomorphic to $\mathscr{R}(\mathfrak{l}, \mathfrak{g}, \mathfrak{h})$. Likewise, the deformation space can equivalently be described as $\mathscr{T}(\Gamma, G, H) = \mathscr{R}(\Gamma, G, H)/\operatorname{Ad}$, where the action Ad of G is given as in (3.3).*

Proof. The proof will be divided into some separate results. We start by proving the following.

Proposition 3.2.5. *Let G be an exponential solvable Lie group, $\Gamma \cong \mathbb{Z}^k$ a discrete subgroup of G and L its syndetic hull and $\mathfrak{l} = \log L$. Then the map*

$$\begin{aligned} \xi : \mathrm{Hom}(\mathfrak{l}, \mathfrak{g}) &\longrightarrow \mathrm{Hom}(\Gamma, G) \\ \psi &\longmapsto \exp \circ \psi \circ \log \end{aligned}$$

is a G and $\mathrm{Aut}(\Gamma)$-equivariant homeomorphism.

Proof. The group G acts on $\mathfrak{g}^k$ by

$$g \cdot (X_1, \ldots, X_k) = (\operatorname{Ad}_g(X_1), \ldots, \operatorname{Ad}_g(X_k))$$

and on G^k by

$$g \cdot (x_1, \ldots, x_k) = (g x_1 g^{-1}, \ldots, g x_k g^{-1}).$$

The map

$$\exp \mathfrak{g}^k \to G^k, \quad (X_1, \ldots, X_k) \mapsto (\exp X_1, \ldots, \exp X_k)$$

is then a G-equivariant diffeomorphism. For a system of generators $\{\gamma_1, \ldots, \gamma_k\}$ of Γ, we define a natural G-equivariant injections as follows:

$$\begin{aligned} \mathrm{Hom}(\mathfrak{l}, \mathfrak{g}) \to \mathfrak{g}^k, &\quad \psi \mapsto (\psi(\log \gamma_1), \ldots \psi(\log \gamma_k)), \\ \mathrm{Hom}(\Gamma, G) \to G^k, &\quad \varphi \mapsto (\varphi(\gamma_1, \ldots, \varphi(\gamma_k)). \end{aligned}$$

Thanks to these inclusions, we regard $\mathrm{Hom}(\mathfrak{l}, \mathfrak{g})$ as a subset of $\mathfrak{g}^k$ and $\mathrm{Hom}(\Gamma, G)$ as a subset of G^k. We have then $\exp(\mathrm{Hom}(\mathfrak{l}, \mathfrak{g})) = \mathrm{Hom}(\Gamma, G)$ and $\xi = \exp_{|\mathrm{Hom}(\mathfrak{l}, \mathfrak{g})}$, so ξ is a

G-equivariant homeomorphism. To complete the proof, we only need to prove that ξ is $\mathrm{Aut}(\Gamma)$-equivariant. Indeed, Let $T \in \mathrm{Aut}(\Gamma)$ and $\psi \in \mathrm{Hom}(\mathfrak{l}, \mathfrak{g})$ then for any γ in L we have

$$\begin{aligned}\xi(T \cdot \psi)(\gamma) &= \exp \circ (T \cdot \psi) \circ \log \gamma \\ &= \exp \circ \psi \circ \log T^{-1}(\gamma) \\ &= \xi(\psi)(T^{-1}(\gamma)) \\ &= T \cdot \xi(\psi)(\gamma).\end{aligned}$$

□

We now prove the following.

Lemma 3.2.6. *Assume that G is an exponential solvable Lie group, $\Gamma \simeq \mathbb{Z}^k$ a discrete subgroup of G and $\mathfrak{l} = \log L$, where L is the syndetic hull of Γ. Then for $\psi \in \mathrm{Hom}(\mathfrak{l}, \mathfrak{g})$, the following two conditions are equivalent:*
(i) *ψ is injective.*
(ii) *$\xi(\psi)$ is injective and $\xi(\psi)(\Gamma)$ is discrete.*

Proof. Let $\gamma, \gamma' \in \Gamma$ such that $\xi(\psi)(\gamma) = \xi(\psi)(\gamma')$, then clearly $\psi(\log \gamma) = \psi(\log \gamma')$ suppose that ψ is injective then $\log \gamma = \log \gamma'$, which is equivalent to $\gamma = \gamma'$. Now the exponential map is an isomorphism from $\log \Gamma$ to Γ and ψ is an isomorphism between $\mathfrak{l}$ and $\psi(\mathfrak{l})$. If we restrict ψ to $\log \Gamma$, we obtain an isomorphism from $\log \Gamma$ to $\psi(\log \Gamma)$. But $\psi(\log \Gamma) = \log \xi(\psi)(\Gamma)$, which is isomorphic to its exponential, then Γ is isomorphic to $\xi(\psi)(\Gamma)$, which is enough to show the first part of the proposition. Conversely, if (ii) is satisfied then $\xi(\psi)$ is a topological isomorphism between Γ and $\xi(\psi)(\Gamma)$ and then $\log \circ \xi(\psi) \circ \exp$ is an isomorphism between the Abelian, discrete subgroups $\log \Gamma$ and $\psi(\log \Gamma)$. Then the linear subspaces $\mathfrak{l}$ and $\psi(\mathfrak{l})$ are isomorphic, which means in particular that $\dim \psi(\mathfrak{l}) = \dim \mathfrak{l}$ as was to be shown. □

Back now to the proof of Theorem 3.2.4. Using Proposition 3.2.5, we can identify $\mathscr{R}(\Gamma, G, H)$ to its inverse image by ξ. We get then

$$\mathscr{R}(\Gamma, G, H) = \left\{ \psi \in \mathrm{Hom}(\mathfrak{l}, \mathfrak{g}) \;\middle|\; \begin{array}{l} \xi(\psi) \text{ injective and } \xi(\psi)(\Gamma) \text{ acts} \\ \text{properly discontinuously and freely on } G/H \end{array} \right\}.$$

Now the group Γ is torsion free, if therefore the action of Γ is proper, then it is free. Then the first equality is a direct consequence of Lemma 3.2.6 and Fact 2.1.8. We now consider the map

$$\begin{array}{rccc} \tilde{\xi} : & \mathrm{Hom}(\mathfrak{l}, \mathfrak{g})/G & \longrightarrow & \mathrm{Hom}(\Gamma, G)/G \\ & [\psi] & \mapsto & [\xi(\psi)], \end{array}$$

which is clearly a homeomorphism. Then the result follows through the identification of $\mathscr{T}(\Gamma, G, H)$ with its inverse image $\tilde{\xi}^{-1}$. This achieves the proof of the theorem. □

3.2.2 On pairs (G, H) having Lipsman's property

Any information concerning the spaces $\mathrm{Hom}(\Gamma, G)$ and $\mathscr{R}(\Gamma, G, H)$ may help to understand the properties and the structure of the deformation space $\mathscr{T}(\Gamma, G, H)$. The sets $\mathrm{Hom}(\Gamma, G)$ and $\mathscr{R}(\Gamma, G, H)$ may have some singularities and there is no clear reason, to say that the parameter space $\mathscr{R}(\Gamma, G, H)$ is an analytic or algebraic or smooth manifold. For instance, when the parameter space is a semialgebraic set (in the sense of Definition 3.2.7 below), it has certainly a finite number of connected components, which means in turn, that the deformation space itself enjoys this feature. To figure out such an issue, we treat here one situation when the basis group in question is nipotent.

Definition 3.2.7 (cf. [41]).

(1) A subset V of $\mathbb{R}^n$ is called semialgebraic if it admits some representation of the form

$$V = \bigcup_{i=1}^{s} \bigcap_{j=1}^{r_i} \{x \in \mathbb{R}^n : P_{i,j}(x) \quad s_{ij} \quad 0\},$$

where for each $i = 1, \dots, s$ and $j = 1, \dots, r_i$, $P_{i,j}$ are some polynomials on $\mathbb{R}^n$ and $s_{ij} \in \{>, =, <\}$.

(2) Let $X \subset \mathbb{R}^n$ and $Y \subset \mathbb{R}^m$ be semialgebraic sets. A map $f : X \to Y$ is called semialgebraic if its graph is a semialgebraic set of $\mathbb{R}^{n+m}$.

The following proposition will be of use in the sequel and shows that the parameter space is semialgebraic whenever the pair (G, H) has the Lipsman property with G a connected simply connected nilpotent Lie group.

Proposition 3.2.8. *Let (G, H) be a pair having the Lipsman property (in the sense of Definition 2.1.10) with G a connected simply connected nilpotent Lie group, Γ a discontinuous group for G/H and $\mathfrak{l}$ the Lie algebra of the syndetic hull of Γ. Then the parameter space $\mathscr{R}(\mathfrak{l}, \mathfrak{g}, \mathfrak{h})$ is semialgebraic.*

Proof. Note that the action of $\exp(\varphi(\mathfrak{l}))$ on G/H is free if and only if $\mathrm{Ad}_g\, \varphi(\mathfrak{l}) \cap \mathfrak{h} = \{0\}$ for all $g \in G$. Let $\mathscr{L}(\mathfrak{l}, \mathfrak{g})$ be the set of linear map of $\mathfrak{l}$ into $\mathfrak{g}$. Consider the maps

$$\begin{array}{rcl} i : \mathfrak{g} \times \mathscr{L}(\mathfrak{l}, \mathfrak{g}) & \longrightarrow & \mathscr{L}(\mathfrak{l}, \mathfrak{g}) \\ (X, \varphi) & \longmapsto & \mathrm{Ad}_{\exp(X)}\, \varphi \end{array}$$

and define the semialgebraic set

$$S = \left\{ (X, \varphi) \,\middle|\, \begin{array}{l} \varphi \in \mathrm{Hom}^0(\mathfrak{l}, \mathfrak{g}) \\ \mathrm{Ad}_{\exp(X)}\, \varphi(\mathfrak{l}) \cap \mathfrak{h} \neq \{0\} \end{array} \right\} \subset \mathfrak{g} \times \mathscr{L}(\mathfrak{l}, \mathfrak{g}).$$

The map $P_{r_2} : \mathfrak{g} \times \mathscr{L}(\mathfrak{l}, \mathfrak{g}) \longrightarrow \mathscr{L}(\mathfrak{l}, \mathfrak{g});\ (X, \varphi) \longmapsto \varphi$ is semialgebraic, then the set

$$P_{r_2}(S) = \{\varphi \in \mathrm{Hom}^0(\mathfrak{l}, \mathfrak{g}) : \text{there exists } X \in \mathfrak{g} \text{ such that } \mathrm{Ad}_{\exp(X)}\, \varphi(\mathfrak{l}) \cap \mathfrak{h} \neq \{0\}\}$$

is semialgebraic and its complement

$$\mathscr{R}(\mathfrak{l}, \mathfrak{g}, \mathfrak{h}) = P_{r_2}(S)^c = \{\varphi \in \mathrm{Hom}^0(\mathfrak{l}, \mathfrak{g}) :\ \mathrm{Ad}_{\exp(X)}\, \varphi(\mathfrak{l}) \cap \mathfrak{h} = \{0\},\ \forall X \in \mathfrak{g}\}$$

is also semialgebraic. □

3.3 Case of Abelian discontinuous groups

Let $M_{n,k}(\mathbb{R})$ be the real vector space of $n \times k$ matrices with real entries and $M^{\circ}_{n,k}(\mathbb{R})$ be the open set of $M_{n,k}(\mathbb{R})$ consisting of rank k matrices in $M_{n,k}(\mathbb{R})$. For a positive integer $s < n$, we define the set

$$I_s(n, k) = \{(i_1, \dots, i_k) \in \mathbb{N}^k : s < i_1 < \cdots < i_k \leq n\}. \tag{3.6}$$

For

$$M = \begin{pmatrix} L_1 \\ \vdots \\ L_n \end{pmatrix} \in M_{n,k}(\mathbb{R}) \quad \text{and} \quad \alpha = (i_1, \dots, i_k) \in I_s(n, k),$$

we denote by M_α the $k \times k$ relative minor

$$\begin{pmatrix} L_{i_1} \\ \vdots \\ L_{i_k} \end{pmatrix}$$

and set

$$U_\alpha = \{M \in M^{\circ}_{n,k}(\mathbb{R}) :\ M_\alpha = I_k\} \cong M_{n-k,k}(\mathbb{R}),$$

where I_k designates the identity element of $M_k(\mathbb{R})$. Let $\mathscr{B} = \{X_1, \dots, X_l\}$ be a basis of $[\mathfrak{g}, \mathfrak{g}]$. Write for $X, Y \in \mathfrak{g}$,

$$[X, Y] = \sum_{i=1}^{l} b_i(X, Y) X_i$$

for some l alternated bilinear forms $b_1, \dots, b_l$. We also designate by $J_{b_1}, \dots, J_{b_l}$ the matrices of $b_1, \dots, b_l$ written through a basis of $\mathfrak{g}$ completed from $\mathscr{B}$ and passing through $\mathfrak{h}$.

We consider the space

$$\mathscr{V} = \{M \in M_{n,k}(\mathbb{R}) : {}^tMJ_{b_i}M = 0, i = 1, \dots, l\}$$

and also for $\alpha \in I_s(n,k)$,

$$\mathscr{V}_\alpha = \mathscr{V} \cap U_\alpha. \tag{3.7}$$

For $\alpha \in I_s(n,k)$, the group G acts on $\mathscr{V}_\alpha$ by left-side multiplication once we regard Ad_g as a matrix operator multiplication for every $g \in G$. As it will be shown, the layers $(\mathscr{V}_\alpha)_{\alpha \in I_s(n,k)}$ turn out to be G-invariant and this plays important role for the description of the structure of the deformation and moduli spaces. We have the following.

Theorem 3.3.1. *Let G be an exponential solvable Lie group of dimension n, H a connected subgroup of dimension s, which contains $[G,G]$ and Γ a discrete subgroup of G of rank k. Then*

$$\mathscr{T}(\Gamma, G, H) = \bigcup_{\alpha \in I_s(n,k)} \mathscr{T}_\alpha \quad \text{and} \quad \mathscr{M}(\Gamma, G, H) = \bigcup_{\alpha \in I_s(n,k)} \mathscr{M}_\alpha,$$

where for every $\alpha \in I_s(n,k)$, the set $\mathscr{T}_\alpha$ is an open subset of $\mathscr{T}(\Gamma, G, H)$ homeomorphic to the product $GL_k(\mathbb{R}) \times (\mathscr{V}_\alpha/G)$ and $\mathscr{M}_\alpha$ is an open subset of $\mathscr{M}(\Gamma, G, H)$ homeomorphic to the product $GL_k(\mathbb{R})/GL_k(\mathbb{Z}) \times (\mathscr{V}_\alpha/G)$.

As we will see later, this result follows from the fact that the parameters space is the total space of a topological $GL_k(\mathbb{R})$-principal bundle.

In the case of compact Clifford–Klein forms, we get the following.

Corollary 3.3.2. *Let G be an exponential solvable Lie group, H a connected subgroup which contains $[G,G]$ and Γ a discrete subgroup of G such that the Clifford–Klein form $\Gamma\backslash G/H$ is compact. Let $\alpha_0 = (s+1, \dots, n)$, then the deformation space $\mathscr{T}(\Gamma, G, H)$ is homeomorphic to the space $GL_k(\mathbb{R}) \times (\mathscr{V}_{\alpha_0}/G)$ and $\mathscr{M}(\Gamma, G, H)$ is homeomorphic to the space $GL_k(\mathbb{R})/GL_k(\mathbb{Z}) \times (\mathscr{V}_{\alpha_0}/G)$.*

3.3.1 Analysis on Grassmannians

Let $G_{n,k}(\mathbb{R})$ be the Grassmannian of k dimensional linear subspaces of $\mathbb{R}^n$. We recall how $G_{n,k}(\mathbb{R})$ is endowed with a topological structure. The map

$$\begin{array}{rccc} \eta : & M^\circ_{n,k}(\mathbb{R}) & \longrightarrow & G_{n,k}(\mathbb{R}) \\ & M & \mapsto & M(\mathbb{R}^k) \end{array}$$

is surjective. The linear group $GL_k(\mathbb{R})$ acts on $M^\circ_{n,k}(\mathbb{R})$ by right-side multiplication and $\eta(M) = \eta(M')$ if and only if there exists $A \in GL_k(\mathbb{R})$ such that $M' = MA$. This means that the column vectors of an element of $M^\circ_{n,k}(\mathbb{R})$ generate an element of $G_{n,k}(\mathbb{R})$ and the column vectors of two elements of $M^\circ_{n,k}(\mathbb{R})$ generate the same element of $G_{n,k}(\mathbb{R})$ if and only if one of them is a multiple of the other by an element of $GL_k(\mathbb{R})$. It follows therefore that $G_{n,k}(\mathbb{R})$ is identified with $M^\circ_{n,k}(\mathbb{R})$ via the equivalence relation:

$$M \sim M' \quad \text{in } M^\circ_{n,k}(\mathbb{R}) \quad \text{if and only if} \quad M' = MA \quad \text{for some } A \in GL_k(\mathbb{R}).$$

So, we regard the space $G_{n,k}(\mathbb{R})$ as the quotient space $M^\circ_{n,k}(\mathbb{R})/_\sim := M^\circ_{n,k}(\mathbb{R})/GL_k(\mathbb{R})$, endowed with the quotient topology. Let

$$I(n,k) = \{(i_1,\ldots,i_k) : 1 \le i_1 < \cdots < i_k \le n\}.$$

For $M \in M_{n,k}(\mathbb{R})$ and $\alpha = (i_1,\ldots,i_k) \in I(n,k)$, we denote as in the previous section by M_α its $k \times k$ relative minor and $U_\alpha = \{M \in M^\circ_{n,k}(\mathbb{R}) : M_\alpha = I_k\}$. It is not hard to check that the restriction η_α of η on U_α is injective and that with the above in mind one gets quite easily that

$$G_{n,k}(\mathbb{R}) = \bigcup_{\alpha\in I(n,k)} \eta_\alpha(U_\alpha). \tag{3.8}$$

It is well known that these bijections define a compatible affine charts on $G_{n,k}(\mathbb{R})$, which endow this space with a structure of a manifold of dimension $k(n-k)$. In order to make this precise, we have the following.

Lemma 3.3.3. *For any $\alpha, \beta \in I(n,k)$, we have:*
(i) *$\eta(U_\alpha)$ is an open subset of $G_{n,k}(\mathbb{R})$.*
(ii) *The restriction map η_α from U_α to $W_\alpha := \eta(U_\alpha)$ is a homeomorphism.*
(iii) *The transition function*

$$\eta_\beta^{-1} \circ \eta_\alpha : \eta_\alpha^{-1}(W_\alpha \cap W_\beta) \longrightarrow \eta_\beta^{-1}(W_\alpha \cap W_\beta)$$

are defined by $\eta_\beta^{-1} \circ \eta_\alpha(M) = MM_\beta^{-1}$ for every $M \in \eta_\alpha^{-1}(W_\alpha \cap W_\beta)$ and is a diffeomorphism.

Definition 3.3.4. A topological fiber bundle is a data (π, E, B, F, G) where E, B and F are topological spaces called respectively the total space, the base space and the fiber (or typical fiber), π is a continuous surjection from E to B and G is a topological subgroup of the group of homeomorphism of F called the structure group, such that there exists:
(1) A family $\{U_\alpha\}$ of open sets which recovers B,
(2) A family $\chi_\alpha : F \times U_\alpha \to \pi^{-1}(U_\alpha)$ of homeomorphisms called local trivializations, satisfying the following conditions:
 (i) $\pi \circ \chi_\alpha(f,p) = p$.
 (ii) For $\chi_{\alpha,p}(f) = \chi_\alpha(f,p)$, the map $\chi_{\alpha,p} : F \to \pi^{-1}(p)$ is a homeomorphism,

(*iii*) For $p \in U_\alpha \cap U_\beta$, the map $t_{\alpha,\beta}(p) = \chi_{\alpha,p}^{-1} \circ \chi_{\beta,p} : F \to F$ is an element of G and the transition function $t_{\alpha,\beta} : U_\alpha \cap U_\beta \to G$ is continuous.

A G-principal bundle is a fiber bundle such that the action of G on the fibers is free and transitive, which merely means that the fibers are homeomorphic to the structure group (cf. [30]).

It is easily seen that the open subset $M_{n,k}^0(\mathbb{R})$ can be interpreted as a total space of a $GL_k(\mathbb{R})$-principal bundle, with the base space equal to the Grassmannian $G_{n,k}(\mathbb{R})$ and the following local trivialization:

$$\begin{aligned} \chi_\alpha : GL_k(\mathbb{R}) \times \eta_\alpha(U_\alpha) &\to \eta^{-1}(\eta_\alpha(U_\alpha)) \\ (A, W) &\mapsto \eta_\alpha^{-1}(W)A, \end{aligned} \tag{3.9}$$

which is obviously a homeomorphism, its inverse is the map $M \mapsto (M_\alpha, \eta(M))$. In addition, we have

$$\eta \circ \chi_\alpha(A, W) = \eta_\alpha^{-1}(W)A(\mathbb{R}^k) = \eta_\alpha^{-1}(W)(\mathbb{R}^k) = W.$$

Furthermore, the map $\chi_{\alpha,W} : GL_k(\mathbb{R}) \to F_W$ is given by

$$\chi_{\alpha,W}(A) = \eta_\alpha^{-1}(W)A,$$

which is a continuous bijection and its inverse is the continuous map

$$\chi_{\alpha,W}^{-1}(M) = M_\alpha.$$

It is clear that the set $I_s(n,k)$ defined in equation (3.6) is empty if and only if $k + s > n$, and when $k + s = n$, it only consists of one single point. This will correspond later on, to the case of compact Clifford–Klein forms. Let now $\mathscr{X}$ be a subspace of $\mathbb{R}^n$ of dimension s, write it for instance as

$$\mathscr{X} = \left\{ \begin{pmatrix} x_1 \\ \vdots \\ x_s \\ 0 \\ \vdots \\ 0 \end{pmatrix} ; x_1, \dots, x_s \in \mathbb{R} \right\}. \tag{3.10}$$

Lemma 3.3.5. *Let $W \in G_{n,k}(\mathbb{R})$, $W = M(\mathbb{R}^k)$ say. Then $W \cap \mathscr{X} = \{0\}$ if and only if*

$$\eta(M) \in \bigcup_{\alpha \in I_s(n,k)} \eta_\alpha(U_\alpha).$$

Proof. Write the matrix M as

$$M = \begin{pmatrix} A \\ B \end{pmatrix}$$

for some $A \in M_{s,k}(\mathbb{R})$ and $B \in M_{n-s,k}(\mathbb{R})$. Then it is quite clear that $W \cap \mathscr{X} = \{0\}$ if and only if $\operatorname{rank}(B) = k$, which is also equivalent to the fact that there exists $\alpha \in I_s(n,k)$ such that $\det(M_\alpha) \neq 0$, and then the existence of $\alpha \in I_s(n,k)$ with the property that $\eta(M) \in \eta_\alpha(U_\alpha)$. □

We now consider some bilinear forms $b_1, \dots, b_l : \mathbb{R}^n \times \mathbb{R}^n \to \mathbb{R}$. Having fixed a canonical basis for $\mathbb{R}^n$, we designate by $J_{b_1}, \dots J_{b_l}$ the matrices of $b_1, \dots, b_l$ written through. The set

$$\mathscr{V} = \{M \in M_{n,k}(\mathbb{R}) : {}^tMJ_{b_i}M = 0,\ i = 1, \dots, l\}$$

is clearly $GL_k(\mathbb{R})$-invariant. Let also for $\alpha \in I(n,k)$, $\mathscr{V}_\alpha = \mathscr{V} \cap U_\alpha$. We pose now the set

$$U_k(\mathbb{R}^n, \mathscr{X}) = \left\{ W \in G_{n,k}(\mathbb{R}) \;\middle|\; \begin{array}{l} W \cap \mathscr{X} = \{0\} \text{ and} \\ b_i(W,W) = 0,\ i = 1, \dots, l \end{array} \right\} \subset G_{n,k}(\mathbb{R}), \tag{3.11}$$

endowed with the trace topology induced by the Grassmannian topology. The following lemma is then immediate.

Lemma 3.3.6. *We keep the previous notation and hypotheses. The sets $\eta_\alpha(\mathscr{V}_\alpha)$ where α in $I_s(n,k)$, constitute an open covering of $U_k(\mathbb{R}^n, \mathscr{X})$.*

Proof. We prove first of all that the sets $\eta_\alpha(\mathscr{V}_\alpha)$, $\alpha \in I_s(n,k)$ form a covering of $U_k(\mathbb{R}^n, \mathscr{X})$. Let W be a subspace of $U_k(\mathbb{R}^n, \mathscr{X})$. As $W \cap \mathscr{X} = \{0\}$, we get by lemma (3.3.5) that $W \in \bigcup_{\alpha \in I_s(n,k)} \eta_\alpha(U_\alpha)$. Let $\alpha \in I_s(n,k)$ such that $W \in \eta_\alpha(U_\alpha)$. Then from $b_i(W,W) = 0$ for all $i = 1, \dots, l$, we obtain $\eta_\alpha^{-1}(W) \in \mathscr{V}$ and, therefore, $W \in \eta_\alpha(\mathscr{V}_\alpha)$. To see that these sets are opens, it is sufficient to show the following set equality:

$$\eta_\alpha(\mathscr{V}_\alpha) = \eta_\alpha(U_\alpha) \cap U_k(\mathbb{R}^n, \mathscr{X}).$$

Indeed, the direct inclusion is immediate. Let W be in the intersection, then there exists $\beta \in I_s(n,k)$ such that $W \in \eta_\beta(\mathscr{V}_\beta)$. This means that $\eta_\beta^{-1}(W) \in \mathscr{V}$, which is invariant by the action of $GL_k(\mathbb{R})$. Now $W \in \eta_\alpha(U_\alpha)$ implies that $\eta_\alpha^{-1}(W) \in U_\alpha$. Furthermore, $\eta_\alpha^{-1}(W)$ and $\eta_\beta^{-1}(W)$ are in the same $GL_k(\mathbb{R})$-orbit, then we get $\eta_\alpha^{-1}(W) \in \mathscr{V}$. We obtain conclusively that $\eta_\alpha^{-1}(W) \in \mathscr{V} \cap U_\alpha = \mathscr{V}_\alpha$. □

Let

$$R_k(\mathbb{R}^n, \mathscr{X}) := \eta^{-1}(U_k(\mathbb{R}^n, \mathscr{X})).$$

Then it is easy to see that $R_k(\mathbb{R}^n, \mathscr{X})$ is the semialgebraic set given by

$$R_k(\mathbb{R}^n, \mathscr{X}) = \left\{ M \in M_{n,k}(\mathbb{R}) \;\middle|\; \begin{array}{l} \sum_{\alpha \in I_s(n,k)} (\det M_\alpha)^2 \neq 0 \text{ and} \\ {}^t M J_{b_i} M = 0,\ i = 1, \dots, l \end{array} \right\}. \tag{3.12}$$

From Lemma 3.11, the restriction of η to $R_k(\mathbb{R}^n, \mathscr{X})$ and the restrictions of the trivializations χ_α to $GL_k(\mathbb{R}) \times \eta_\alpha(\mathscr{V}_\alpha)$, give a $GL_k(\mathbb{R})$-principal bundle, whose total space is $R_k(\mathbb{R}^n, \mathscr{X})$ and the base space is $U_k(\mathbb{R}^n, \mathscr{X})$. In particular,

$$R_k(\mathbb{R}^n, \mathscr{X}) = \bigcup_{\alpha \in I_s(n,k)} \chi_\alpha(GL_k(\mathbb{R}) \times \eta_\alpha(\mathscr{V}_\alpha)). \tag{3.13}$$

3.3.2 The parameter space for normal subgroups

We assume from now on that H is a normal subgroup of the exponential solvable Lie group G. Let as earlier Γ be an Abelian discrete subgroup of G and L its syndetic hull. We shall identify our group G with $\mathbb{R}^n$, which allows us to make use of the results of the previous section, specially concerning Grassmannians. We begin by remarking the following fact.

Proposition 3.3.7. *Retain the same hypotheses and notation. For any Abelian discrete subgroup Γ acting on G/H properly discontinuously, we have*

$$\mathscr{R}(\Gamma, G, H) = \left\{ \psi \in \operatorname{Hom}(\mathfrak{l}, \mathfrak{g}) \;\middle|\; \begin{array}{l} \dim \psi(\mathfrak{l}) = \dim \mathfrak{l} \text{ and} \\ \psi(\mathfrak{l}) \cap \mathfrak{h} = \{0\} \end{array} \right\}.$$

Proof. We know already that for a subalgebra $\mathfrak{l}$ of $\mathfrak{g}$, $\exp \mathfrak{l}$ acts properly on G/H if and only if $\mathfrak{h} \cap \mathfrak{l} = \{0\}$. Then the result follows from Theorem 3.2.4. □

We choose a good sequence of subalgebras of $\mathfrak{g}$ passing through $[\mathfrak{g}, \mathfrak{g}]$ and $\mathfrak{h}$ assumed to contain $[\mathfrak{g}, \mathfrak{g}]$, from which we extract a basis $\{X_1, \dots, X_n\}$ of $\mathfrak{g}$ such that $\{X_1, \dots, X_s\}$ is a basis of $\mathfrak{h}$ and $\{X_1, \dots, X_l\}$ a basis of $[\mathfrak{g}, \mathfrak{g}]$ with $l \leq s \leq n = \dim \mathfrak{g}$. As earlier, write for $X, Y \in \mathfrak{g}$,

$$[X, Y] = \sum_{i=1}^{l} b_i(X, Y) X_i$$

for some l alternated bilinear forms $b_1, \dots, b_l$. Let $\mathscr{B} = \{e_1, \dots, e_n\}$ be the canonical basis of $\mathbb{R}^n$. We identify $\mathfrak{g}$, $\mathfrak{h}$ and the space of the linear maps $\mathscr{L}(\mathfrak{l}, \mathfrak{g})$ with $\mathbb{R}^n$, the subspace $\mathscr{X}$ defined in (3.10) and the space of matrices $M_{n,k}(\mathbb{R})$, respectively, via the isomorphism $e_i \mapsto X_i$. We designate by the way by $U_k(\mathfrak{g}, \mathfrak{h})$ the corresponding set once we implement the above identifications. We characterize first $\operatorname{Hom}(\mathfrak{l}, \mathfrak{g})$ as a quadratic cone of $M_{n,k}(\mathbb{R})$.

Lemma 3.3.8. *For any Abelian discrete subgroup* Γ *of* G, *we have*

$$\mathrm{Hom}(\mathfrak{l},\mathfrak{g}) = \{M \in M_{n,k}(\mathbb{R}) : {}^tMJ_{b_i}M = 0, i = 1,\dots,l\},$$

where $k = \dim \mathfrak{l}$.

Proof. Since Γ is Abelian, its syndetic hull algebra $\mathfrak{l}$ is obviously so. Thus, $\psi \in \mathscr{L}(\mathfrak{l},\mathfrak{g})$ is an algebras homomorphism if and only if ψ is a linear map satisfying $[\psi(X),\psi(Y)] = 0$, for all $X, Y \in \mathfrak{l}$, which gives rise to the following writing:

$$\mathrm{Hom}(\mathfrak{l},\mathfrak{g}) = \{\psi \in \mathscr{L}(\mathfrak{l},\mathfrak{g}),\ b_i(\psi(X),\psi(Y)) = 0, \text{for all } X, Y \in \mathfrak{l},\ i = 1,\dots,l\}.$$

If now M is the matrix of ψ written through $\mathscr{B}$, then $b_i(\psi(X),\psi(Y)) = 0$ for all $X, Y \in \mathfrak{l}$ if and only if ${}^tMJ_{b_i}M = 0$. □

Therefore, the following result is immediate.

Lemma 3.3.9. *With the same hypotheses and notation as above we have,* $\mathscr{R}(\Gamma, G, H) = R_k(\mathfrak{g},\mathfrak{h}) := \eta^{-1}(U_k(\mathfrak{g},\mathfrak{h}))$ *where* k *designates the rank of* Γ.

Proof. Proposition 3.3.7 and Lemma 3.3.8 enable us to get that

$$\mathscr{R}(\Gamma, G, H) = \left\{ M \in M_{n,k}(\mathbb{R}) \;\middle|\; \begin{array}{l} \dim M(\mathbb{R}^k) = k, \\ M(\mathbb{R}^k) \cap \mathfrak{h} = \{0\} \text{ and} \\ {}^tMJ_{b_i}M = 0, i = 1,\dots,l \end{array} \right\}$$

as Γ is Abelian. We get therefore using the set equality given in equation (3.11) that $\mathscr{R}(\Gamma, G, H) = \eta^{-1}(U_k(\mathfrak{g},\mathfrak{h}))$. □

The following result stems directly from the equality (3.13) and the last lemma.

Corollary 3.3.10. *Assume that* G *is exponential solvable,* H *is a normal subgroup of* G *and* Γ *is an Abelian discrete subgroup of* G. *Let* $(\mathscr{V}_\alpha)_{\alpha \in I_s(n,k)}$ *be the net of layers defined by (3.7) and denote for every* $\alpha \in I_s(n,k)$ *by* $\widetilde{\chi}_\alpha$ *the restriction of the trivialization map given in (3.9) to the set* $GL_k(\mathbb{R}) \times \eta_\alpha(\mathscr{V}_\alpha)$. *Then*

$$\mathscr{R}(\Gamma, G, H) = \bigcup_{\alpha \in I_s(n,k)} \widetilde{\chi}_\alpha(GL_k(\mathbb{R}) \times \eta_\alpha(\mathscr{V}_\alpha)).$$

3.3.3 The deformation space for normal subgroups

We proceed in this section to the proof of our results. From now on, we assume that our group G is exponentially solvable and unless specific mention that the subgroup H contains $[G, G]$. We remark first of all of the following.

Lemma 3.3.11. *Every discrete subgroup Γ of G, acting on G/H properly discontinuously is isomorphic to $\mathbb{Z}^k$ for some integer k.*

Proof. As H is a normal subgroup of G and Γ acts on G/H properly discontinuously, we get that $\Gamma \cap H = \{e\}$. As such, we also that

$$[\Gamma,\Gamma] \subset [G,G] \cap \Gamma \subset H \cap \Gamma,$$

as $\mathfrak{h}$ contains $[\mathfrak{g},\mathfrak{g}]$. This shows then that Γ is Abelian. □

We consider now the actions of G on $M_{n,k}(\mathbb{R}) \cong \mathscr{L}(\mathfrak{l},\mathfrak{g})$ and $G_{n,k}(\mathbb{R})$ given by $M \cdot g = \mathrm{Ad}_{g^{-1}} M$ and $W \cdot g = \mathrm{Ad}_{g^{-1}} W$. Then we have the following.

Lemma 3.3.12. *For every $\alpha \in I_s(n,k)$, the set $\mathscr{V}_\alpha$ is G-stable and η_α is G-equivariant. In particular, $\eta^{-1}(\eta_\alpha(\mathscr{V}_\alpha))$ is an open G-stable subset of $\mathscr{R}(\Gamma, G, H)$.*

Proof. Let $X \in \mathfrak{g}$, using the hypothesis $\mathfrak{h} \supseteq [\mathfrak{g},\mathfrak{g}]$, and we can write

$$\mathrm{Ad}_{\exp X} = \begin{pmatrix} C & D \\ (0) & I_{n-s} \end{pmatrix}$$

for some $C \in M_s(\mathbb{R})$ and $D \in M_{s,n-s}(\mathbb{R})$. Let $W \in \eta_\alpha(\mathscr{V}_\alpha)$, $M = \eta_\alpha^{-1}(W)$ and write M as

$$M = \begin{pmatrix} A \\ B \end{pmatrix}$$

for some $A \in M_{s,k}(\mathbb{R})$ and $B \in M_{n-s,k}(\mathbb{R})$ and $M_\alpha = I_k$. We get therefore that

$$\mathrm{Ad}_{\exp X} M = \begin{pmatrix} CA + DB \\ B \end{pmatrix},$$

and that $(\mathrm{Ad}_{\exp X} M)_\alpha = M_\alpha = I_k$, because M_α depends only on B, for all $\alpha \in I_s(n,k)$. Moreover, and with the above in mind, the hypothesis $[W,W] = \{0\}$ is equivalent to $M^t J_{b_i} M = 0$, $i = 1,\dots,l$, which leads to the fact that $[\mathrm{Ad}_{\exp X}(W), \mathrm{Ad}_{\exp X}(W)] = \{0\}$ as $\mathrm{Ad}_{\exp X}$ is an automorphism of G. This is also equivalent to

$${}^t(\mathrm{Ad}_{\exp X} M) J_{b_i} \mathrm{Ad}_{\exp X} M = 0, \quad i = 1,\dots,l.$$

We obtain finally that for every $\alpha \in I_s(n,k)$ the set $\mathscr{V}_\alpha$ is invariant through the right-side action of G on $M^\circ_{n,k}(\mathbb{R})$ by left multiplication. On the other hand, for $g \in G$ and $M \in \mathscr{V}_\alpha$, we have that

$$\begin{aligned} \eta_\alpha(M \cdot g) &= \eta_\alpha(\mathrm{Ad}_{g^{-1}} M) \\ &= (\mathrm{Ad}_{g^{-1}} M)(\mathbb{R}^k) \\ &= \mathrm{Ad}_{g^{-1}}(M(\mathbb{R}^k)) = \eta_\alpha(M) \cdot g, \end{aligned}$$

which shows that the map η_α is G-equivariant. □

For any $g \in G$ and any $(f, W) \in GL_k(\mathbb{R}) \times \eta_\alpha(\mathscr{V}_\alpha)$, define the product

$$(f, W) \cdot g = (\chi_{\alpha,W\cdot g}^{-1} \circ \mathrm{Ad}_{g^{-1}} \circ \chi_{\alpha,W} \circ f, W \cdot g). \tag{3.14}$$

Proposition 3.3.13. *For every $\alpha \in I_s(n,k)$, the set $GL_k(\mathbb{R}) \times \eta_\alpha(\mathscr{V}_\alpha)$ is a G-space through the right-side action defined as in (3.14). Furthermore,*

$$\begin{array}{rcl} \tilde{\chi}_\alpha : GL_k(\mathbb{R}) \times \eta_\alpha(\mathscr{V}_\alpha) & \longrightarrow & \eta^{-1}(\eta_\alpha(\mathscr{V}_\alpha)) \\ (f, W) & \mapsto & \chi_{\alpha,W} \circ f \end{array}$$

defines a G-equivariant homeomorphism.

Proof. Note first of all that

$$\tilde{\chi}_\alpha(GL_k(\mathbb{R}) \times \eta_\alpha(\mathscr{V}_\alpha)) = \eta^{-1}(\eta_\alpha(\mathscr{V}_\alpha)).$$

As the trivalization χ_α is a homeomorphism, so is its restriction $\tilde{\chi}_\alpha$. For $(f, W) \in GL_k(\mathbb{R}) \times \eta_\alpha(\mathscr{V}_\alpha)$ and $g, g' \in G$, we get easily the following:

$$\begin{aligned} ((f, W) \cdot g) \cdot g' &= (\chi_{\alpha,W\cdot g}^{-1} \circ \mathrm{Ad}_{g^{-1}} \circ \chi_{\alpha,W} \circ f, W \cdot g) \cdot g' \\ &= (\chi_{\alpha,(W\cdot g)\cdot g'}^{-1} \circ \mathrm{Ad}_{g'^{-1}} \circ \chi_{\alpha,W\cdot g} \circ \chi_{\alpha,W\cdot g}^{-1} \circ \mathrm{Ad}_{g^{-1}} \circ \chi_{\alpha,W} \circ f, (W \cdot g) \cdot g') \\ &= (\chi_{\alpha,W\cdot gg'}^{-1} \circ \mathrm{Ad}_{g'^{-1}} \circ \mathrm{Ad}_{g^{-1}} \circ \chi_W \circ f, W \cdot gg') \\ &= (f, W) \cdot gg', \end{aligned}$$

which shows that G acts on the space $GL_k(\mathbb{R}) \times \eta_\alpha(\mathscr{V}_\alpha)$. On the other hand, for $g \in G$ and $\varphi \in \mathscr{R}(\Gamma, G, H)$, we have

$$\begin{aligned} \chi_\alpha((f, W) \cdot g) &= \chi_\alpha(\chi_{\alpha,W\cdot g}^{-1} \circ \mathrm{Ad}_{g^{-1}} \circ \chi_{\alpha,W} \circ f, W \cdot g) \\ &= \chi_{\alpha,W\cdot g} \circ (\chi_{\alpha,W\cdot g}^{-1} \circ \mathrm{Ad}_{g^{-1}} \circ \chi_{\alpha,W} \circ f) \\ &= \mathrm{Ad}_{g^{-1}} \circ \chi_\alpha(f, W) \\ &= \chi_\alpha(f, W) \cdot g, \end{aligned}$$

which shows that $\tilde{\chi}_\alpha$ is G-equivariant. This completes the proof of the proposition. □

Proof of Theorem 3.3.1. From the continuity of η, Lemma 3.3.6 and Lemma 3.3.9, we get that the sets $\eta^{-1}(\eta_\alpha(\mathscr{V}_\alpha))$, $\alpha \in I_s(n,k)$ form an open covering of $\mathscr{R}(\Gamma, G, H)$. Using Proposition 3.3.13, we can see that, for any $\alpha \in I_s(n,k)$ the set $\eta^{-1}(\eta_\alpha(\mathscr{V}_\alpha))$ is G-stable. Denote by

$$\mathscr{T}_\alpha = \eta^{-1}(\eta_\alpha(\mathscr{V}_\alpha))/G.$$

It follows that

$$\mathscr{T}(\Gamma, G, H) = \bigcup_{\alpha \in I_s(n,k)} \mathscr{T}_\alpha.$$

Furthermore, Proposition 3.3.13 gives the identification

$$\mathscr{T}_\alpha = (GL_k(\mathbb{R}) \times \eta_\alpha(\mathscr{V}_\alpha))/G.$$

For $\alpha \in I_s(n,k)$, one has that

$$\begin{aligned} \chi_{\alpha,W\cdot g}^{-1} \circ \mathrm{Ad}_g \circ \chi_{\alpha,W} &= (\mathrm{Ad}_g \circ \eta_\alpha^{-1}(W))_\alpha \\ &= (\eta_\alpha^{-1}(W))_\alpha \\ &= \mathrm{I}_k. \end{aligned}$$

The action of G on $GL_k(\mathbb{R}) \times \eta_\alpha(\mathscr{V}_\alpha)$ does not affect therefore the factor $GL_k(\mathbb{R})$. Furthermore, two pairs (f, W) and (f', W') of the product $GL_k(\mathbb{R}) \times \eta_\alpha(\mathscr{V}_\alpha)$ are in the same orbit if and only if W and W' are in the same orbit of the action on $\eta_\alpha(\mathscr{V}_\alpha)$. Then up to a homeomorphism, and using the G-equivariance of the homeomorphism η_α, we obtain

$$\begin{aligned} \mathscr{T}_\alpha &= GL_k(\mathbb{R}) \times (\eta_\alpha(\mathscr{V}_\alpha)/G) \\ &= GL_k(\mathbb{R}) \times \eta_\alpha(\mathscr{V}_\alpha/G) \\ &= GL_k(\mathbb{R}) \times \mathscr{V}_\alpha/G. \end{aligned}$$

This completes the proof of the assertion concerning the deformation space. Now the group $\mathrm{Aut}(\Gamma)$ can be identified to the subgroup of $\mathrm{Aut}(\mathfrak{l}) = GL_k(\mathbb{R})$, which leaves the Abelian subgroup $\log \Gamma = \mathbb{Z}^k$ stable in $\mathfrak{l} = \mathbb{R}^k$. Then $\mathrm{Aut}(\Gamma) = GL_k(\mathbb{Z})$ and acts on $M_{n,k}(\mathbb{R})$ by the same way as $GL_k(\mathbb{R})$ does. This action leaves the sets $\eta^{-1}(\eta_\alpha(\mathscr{V}_\alpha))$ stable and commutes with the action of G. Then, for

$$\mathscr{M}_\alpha := \mathrm{Aut}(\Gamma)\backslash\eta^{-1}(\eta_\alpha(\mathscr{V}_\alpha))/G,$$

we have

$$\mathscr{M}(\Gamma, G, H) = \bigcup_{\alpha \in I_s(n,k)} \mathscr{M}_\alpha.$$

For a given $\alpha \in I_s(n,k)$, the trivialization $\tilde{\chi}_\alpha$ transfers the action of $\mathrm{Aut}(\Gamma)$ on $\eta^{-1}(\eta_\alpha(\mathscr{V}_\alpha))$ to the action of $\mathrm{Aut}(\Gamma)$ on $GL_k(\mathbb{R}) \times \eta_\alpha(\mathscr{V}_\alpha)$ given by $a\cdot(f, W) = (f \circ a^{-1}, W)$. Let now (f, W) and (f', W') belong to the same double coset $\mathrm{Aut}(\Gamma) \cdot x \cdot G$, which means equivalently that W and W' are in the same G-orbit in $\eta_\alpha(\mathscr{V}_\alpha)$ and f and f' are in the same orbit of the action of $\mathrm{Aut}(\Gamma) = GL_k(\mathbb{Z})$ on $GL_k(\mathbb{R})$, given by $a \cdot f = f \circ a^{-1}$. This means that the canonical quotient map

$$GL_k(\mathbb{R}) \times \eta_\alpha(\mathscr{V}_\alpha) \to \mathrm{Aut}(\Gamma)\backslash(GL_k(\mathbb{R}) \times \eta_\alpha(\mathscr{V}_\alpha))/G$$

factors to a homeomorphism from

$$GL_k(\mathbb{R})/GL_k(\mathbb{Z}) \times \eta_\alpha(\mathscr{V}_\alpha/G) \cong GL_k(\mathbb{R})/GL_k(\mathbb{Z}) \times (\mathscr{V}_\alpha/G)$$

to $\mathrm{Aut}(\Gamma)\backslash(GL_k(\mathbb{R}) \times \eta_\alpha(\mathscr{V}_\alpha))/G = \mathscr{M}_\alpha$, which is an open set of $\mathscr{M}(\Gamma, G, H)$. This completes the proof of the theorem. □

3.3.4 Examples

We study in the section some concrete examples of exponential and nilpotent Lie groups. We put the emphasis on the case where $\dim G \leq 3$.

Example 3.3.14. Let $G = \mathrm{Aff}(\mathbb{R})$ with the Lie algebra $\mathfrak{g} = \mathrm{aff}(\mathbb{R})$ and a basis $\{X, Y\}$ such that $[X, Y] = Y$. So, obviously $[\mathfrak{g}, \mathfrak{g}] = \mathbb{R}Y$. If $H = \exp(\mathfrak{h})$ and Γ are as before, then $\mathfrak{h} = \mathbb{R}Y$ and the Clifford–Klein form $\Gamma\backslash G/H$ is compact (otherwise, $\mathfrak{h} = \mathfrak{g}$ and Γ is trivial). It turns out that the deformation space $\mathscr{T}(\Gamma, G, H)$ is homeomorphic to the space $GL_1(\mathbb{R}) \times (\mathscr{V}_1/G)$, where $\mathscr{V}_1$ is the set $X + \mathbb{R}Y$. Let now $g = \exp(tX)\exp(sY)$ for some $s, t \in \mathbb{R}$, then a routine computation shows that

$$\mathrm{Ad}(g)(X + \alpha Y) = X + (\alpha - s)e^t Y, \quad \alpha \in \mathbb{R}.$$

It turns out that the deformation space is homeomorphic to the space $GL_1(\mathbb{R}) \times \{X\}$. Finally, we get that

$$\mathscr{T}(\Gamma, G, H) \simeq \mathbb{R}^\times := \mathbb{R} \setminus \{0\}$$

and

$$\mathscr{M}(\Gamma, G, H) \simeq \mathbb{R}_{>0}.$$

Example 3.3.15. Suppose now that G is exponential solvable of dimension 3. Then up to an isomorphism, one can assume that $\mathfrak{g}$ admits a basis $\{A, X, Y\}$ with nontrivial brackets:

$$[A, X] = X - \alpha Y, \quad [A, Y] = \alpha X + Y$$

for some $\alpha \in \mathbb{R}^\times$ (see [106]). In this situation, $[\mathfrak{g}, \mathfrak{g}] = \mathbb{R}X \oplus \mathbb{R}Y$ and similar to the setting above, $\mathfrak{h} = [\mathfrak{g}, \mathfrak{g}]$ and the Clifford–Klein form $\Gamma\backslash G/H$ is compact. We get then that $\mathscr{T}(\Gamma, G, H)$ is homeomorphic to the space $GL_1(\mathbb{R}) \times (\mathscr{V}_1/G)$, where $\mathscr{V}_1$ is the set $A + \mathbb{R}Y + \mathbb{R}X$. By an easy computation, we get

$$\begin{aligned} \mathrm{Ad}(g)(A + aX + bY) = A &+ e^s X((a - x - \alpha y)\cos(\alpha s) + (b - y + \alpha x)\sin(\alpha s)) \\ &+ e^s Y((b - y + \alpha x)\cos(\alpha s) + ((-a + x + \alpha y)\sin(\alpha s)). \end{aligned}$$

Here, $a, b \in \mathbb{R}$ and $g = \exp(sA)\exp(xX)\exp(yY)$. This gives rise to the fact that the deformation space is homeomorphic to the space $GL_1(\mathbb{R}) \times \{A\}$. Finally, we have again that

$$\mathscr{T}(\Gamma, G, H) \simeq \mathbb{R}^\times$$

and

$$\mathscr{M}(\Gamma, G, H) \simeq \mathbb{R}_{>0}.$$

Example 3.3.16. Assume now that G is nilpotent and non-Abelian. Then up to an isomorphism, G is the three-dimensional Heisenberg group whose Lie algebra admits a basis (X, Y, Z) such that $[X, Y] = Z$. According to our circumstances, we may assume for instance that $\mathfrak{h} = \mathbb{R}Z \oplus \mathbb{R}Y$ and Γ is a rank one discrete subgroup generated, for example, by $\exp(X)$. Note here that this choice of $\mathfrak{h}$ is irrelevant once we agree that it is two-dimensional and that the Clifford–Klein form in question is compact. We obtain as above that $\mathscr{T}(\Gamma, G, H)$ is homeomorphic to the space $GL_1(\mathbb{R}) \times ((X + \mathbb{R}Y + \mathbb{R}Z)/G) \simeq \mathbb{R}^\times \times \{X + \mathbb{R}Y\} \simeq \mathbb{R}^\times \times \mathbb{R}$ and similarly that

$$\mathscr{M}(\Gamma, G, H) \simeq \mathbb{R}_{>0} \times \mathbb{R}.$$

If $\mathfrak{h}$ is one-dimensional, there does not exist any cocompact discrete subgroup for G/H. We are then submitted to look at the situation where the rank of Γ is one for which the Clifford–Klein form in question is not compact. As such, the deformation space $\mathscr{T}(\Gamma, G, H)$ is the union of the open sets

$$GL_1(\mathbb{R}) \times ((X + \mathbb{R}Y + \mathbb{R}Z)/G)$$

and

$$GL_1(\mathbb{R}) \times ((Y + \mathbb{R}X + \mathbb{R}Z)/G).$$

Finally,

$$\mathscr{T}(\Gamma, G, H) \simeq \mathbb{R}^\times \times (Y + \mathbb{R}X) \cup \mathbb{R}^\times \times (X + \mathbb{R}Y).$$

On the other hand, when we look at the trivialization corresponding to the layers $Y + \mathbb{R}X + \mathbb{R}Z$ and $X + \mathbb{R}Y + \mathbb{R}Z$, one easily gets that $\mathbb{R}^\times \times (Y + \mathbb{R}X)$ is homeomorphic to

$\mathbb{R}^{\times}Y + \mathbb{R}X$ and likewise $\mathbb{R}^{\times} \times (X + \mathbb{R}Y)$ is homeomorphic to $\mathbb{R}^{\times}X + \mathbb{R}Y$. Finally, we get that $\mathscr{T}(\Gamma, G, H) \simeq \mathbb{R}^2 \setminus \{(0,0)\}$ and

$$\begin{aligned}\mathscr{M}(\Gamma, G, H) &\simeq \{(x,y) \in \mathbb{R}^2 : x > 0 \text{ or } y > 0\}\\ &\simeq \mathbb{R}^2 \setminus (\mathbb{R}_{\leq 0} \times \mathbb{R}_{\leq 0}).\end{aligned}$$

3.4 Non-Abelian discontinuous groups

3.4.1 Structure of a principal fiber bundle

We retain all our hypotheses and notation. Let $k = \dim \mathfrak{l}$, $s = \dim \mathfrak{h}$ and $n = \dim \mathfrak{g}$. We fix a basis $\{X_1, \dots, X_n\}$ of $\mathfrak{g}$ passing through $\mathfrak{h}$. We identify the vector spaces $\mathfrak{g}$ to $\mathbb{R}^n$, $\mathfrak{l}$ to $\mathbb{R}^k$, $\mathfrak{h}$ to the s-dimensional subspace $\mathbb{R}^s \times 0_{\mathbb{R}^{n-s}}$ of $\mathbb{R}^n$, $\mathscr{L}(\mathfrak{l}, \mathfrak{g})$ to $M_{n,k}(\mathbb{R})$ the real vector space of $n \times k$ matrices with real entries and $\operatorname{Hom}(\mathfrak{l}, \mathfrak{g})$ to a closed subset of $M_{n,k}(\mathbb{R})$. Let $M^{\circ}_{n,k}(\mathbb{R})$ be the open set of $M_{n,k}(\mathbb{R})$ consisting of rank k matrices in $M_{n,k}(\mathbb{R})$, which is also identified to the set $\{\varphi \in \mathscr{L}(\mathfrak{l}, \mathfrak{g}),\ \varphi \text{ injective}\}$. We define the set

$$I(n,k) = \{(i_1, \dots, i_k) \in \mathbb{N}^k,\ 1 \leq i_1 < \cdots < i_k \leq n\}. \tag{3.15}$$

For

$$M = \begin{pmatrix} L_1 \\ \vdots \\ L_n \end{pmatrix} \in M_{n,k}(\mathbb{R}) \quad \text{and} \quad \alpha = (i_1, \dots, i_k) \in I(n,k),$$

we denote by M_α the $k \times k$ relative minor

$$\begin{pmatrix} L_{i_1} \\ \vdots \\ L_{i_k} \end{pmatrix}.$$

Let

$$U_\alpha = \{M \in M^{\circ}_{n,k}(\mathbb{R}) : M_\alpha = I_k\} \cong M_{n-k,k}(\mathbb{R})$$

and

$$\mathscr{U}_\alpha = \{M \in M^{\circ}_{n,k}(\mathbb{R}),\ \det M_\alpha \neq 0\} \cong M_{n-k,k}(\mathbb{R}) \times GL_k(\mathbb{R}),$$

where I_k designates the identity element of $M_k(\mathbb{R})$. Then clearly

$$M^{\circ}_{n,k}(\mathbb{R}) = \bigcup_{\alpha \in I(n,k)} \mathscr{U}_\alpha.$$

The group $GL_k(\mathbb{R})$ acts on $M_{n,k}(\mathbb{R})$ through the right multiplication and the Grassmannian $G_{n,k}(\mathbb{R})$ of the k-dimensional subspaces of $\mathbb{R}^n$ is identified to the quotient topological space $M^\circ_{n,k}(\mathbb{R})/GL_k(\mathbb{R})$. Let $\eta : M^\circ_{n,k}(\mathbb{R}) \to G_{n,k}(\mathbb{R})$, $M \mapsto M(\mathbb{R}^k)$ be the canonical surjection. It is easy to see that the restriction η_α of η to the set U_α is a homeomorphism between U_α and its image. The group $\operatorname{Aut}(\mathfrak{l})$ of automorphisms of $\mathfrak{l}$ is a closed subgroup of $GL_k(\mathbb{R})$, then the homogeneous space $GL_k(\mathbb{R})/\operatorname{Aut}(\mathfrak{l})$ is endowed with a manifold structure and the quotient map $p : GL_k(\mathbb{R}) \to GL_k(\mathbb{R})/\operatorname{Aut}(\mathfrak{l})$ admits local sections. Consider an open covering $\{V_\beta\}_{\beta\in I}$ of $GL_k(\mathbb{R})/\operatorname{Aut}(\mathfrak{l})$ such that for any $\beta \in I$, there is a section $s_\beta : V_\beta \to GL_k(\mathbb{R})$ satisfying $p \circ s_\beta = Id_{|V_\beta}$. For every $\alpha \in I(n,k)$, we consider the map

$$\begin{array}{rcl} \pi_\alpha : \mathscr{U}_\alpha & \longrightarrow & GL_k(\mathbb{R})/\operatorname{Aut}(\mathfrak{l}) \times \eta(U_\alpha) \\ M & \mapsto & (p(M_\alpha), \eta(M)) \end{array}$$

and let $\mathscr{U}_{\alpha\beta} = \pi_\alpha^{-1}(V_\beta \times \eta(U_\alpha))$. Clearly, π_α is continuous and surjective, $\mathscr{U}_{\alpha\beta}$ is open in $M_{n,k}(\mathbb{R})$ and the collection $(\mathscr{U}_{\alpha\beta})_{\beta\in I}$ constitutes an open covering of $\mathscr{U}_\alpha$. We finally consider the map

$$\begin{array}{rcl} \xi_{\alpha\beta} : \operatorname{Aut}(\mathfrak{l}) \times V_\beta \times \eta(U_\alpha) & \longrightarrow & \mathscr{U}_{\alpha\beta} \\ (A, x, W) & \mapsto & \eta_\alpha^{-1}(W)s_\beta(x)A. \end{array}$$

This map is well-defined as for any $(A, x, W) \in \operatorname{Aut}(\mathfrak{l}) \times V_\beta \times \eta(U_\alpha)$; the conclusion $\pi_\alpha \circ \xi_{\alpha\beta}(A, x, W) = (x, W)$ holds. Indeed, it is clear that the matrix $M = \eta_\alpha^{-1}(W)s_\beta(x)$ is in $\pi_\alpha^{-1}(x, W)$ and the set $M \operatorname{Aut}(\mathfrak{l})$ is a subset of $\pi_\alpha^{-1}(x, W)$. We further have the following useful properties.

Lemma 3.4.1. *For every $\alpha \in I(n,k)$, $\beta, \beta' \in I$ and $(A, x, W) \in \operatorname{Aut}(\mathfrak{l}) \times V_\beta \times \eta(U_\alpha)$, we have:*

(1) *$\xi_{\alpha\beta}$ is a homeomorphism.*
(2) *$\pi_\alpha^{-1}(x, W) = \eta_\alpha^{-1}(W)s_\beta(x) \operatorname{Aut}(\mathfrak{l})$.*
(3) *The map $\xi_{\alpha\beta,x,W} : \operatorname{Aut}(\mathfrak{l}) \to \pi_\alpha^{-1}(x, W)$ given by $\xi_{\alpha\beta,x,W}(A) = \eta_\alpha^{-1}(W)s_\beta(x)A$ is a homeomorphism.*
(4) *The map $t_{\alpha\beta\beta'} : (V_\beta \cap V_{\beta'}) \times \eta(U_\alpha) \to \operatorname{Aut}(\mathfrak{l})$ given by $t_{\alpha\beta\beta'}(x, W) = \xi^{-1}_{\alpha\beta,x,W} \circ \xi_{\alpha\beta',x,W}$ is continuous.*

Proof. For the first statement, it is easy to see that the map $\xi_{\alpha\beta}$ is continuous. Note that for every $M \in \mathscr{U}_{\alpha\beta}$, we have $p(s_\beta(p(M_\alpha))) = p(M_\alpha)$ and then $s_\beta(p(M_\alpha))^{-1}M_\alpha \in \operatorname{Aut}(\mathfrak{l})$. Let

$$\begin{array}{rcl} \xi'_{\alpha\beta} : \mathscr{U}_{\alpha\beta} & \longrightarrow & \operatorname{Aut}(\mathfrak{l}) \times V_\beta \times \eta(U_\alpha) \\ M & \mapsto & (s_\beta(p(M_\alpha))^{-1}M_\alpha, p(M_\alpha), \eta(M)), \end{array} \tag{3.16}$$

then $\xi'_{\alpha\beta}$ is continuous and $\xi_{\alpha\beta} \circ \xi'_{\alpha\beta} = Id_{|\mathscr{U}_{\alpha\beta}}$. Concerning the second statement, let $M \in \pi_\alpha^{-1}(x, W)$, then $\eta(M) = W$ and $p(M_\alpha) = x$, which is equivalent to

$$M = \eta_\alpha^{-1}(W)M_\alpha \quad \text{and} \quad s_\beta(x)^{-1}M_\alpha \in \operatorname{Aut}(\mathfrak{l}).$$

It follows that $M = \eta_\alpha^{-1}(W)s_\beta(x)B$ for some $B \in \operatorname{Aut}(\mathfrak{l})$. This proves the first direct inclusion and the converse is straight immediate. To prove (3), we consider the map $\xi'_{\alpha\beta,x,W} : \pi_\alpha^{-1}(x, W) \to \operatorname{Aut}(\mathfrak{l})$ given by $\xi'_{\alpha\beta,x,W}(M) = s_\beta(x)^{-1}M_\alpha$. Then clearly $\xi_{\alpha\beta,x,W}$ and $\xi'_{\alpha\beta,x,W}$ are continuous and $\xi'_{\alpha\beta,x,W} \circ \xi_{\alpha\beta,x,W}$ is the identity map. For the last statement, it is pretty clear that

$$\xi^{-1}_{\alpha\beta,x,W} \circ \xi_{\alpha\beta',x,W}(A) = s_\beta(x)^{-1}s_{\beta'}(x)A.$$

The composition $\xi^{-1}_{\alpha\beta,x,W} \circ \xi_{\alpha\beta',x,W}$ is then the left translation identified to the element $s_\beta(x)^{-1}s_{\beta'}(x) \in \operatorname{Aut}(\mathfrak{l})$. Furthermore, the continuity of $t_{\alpha\beta\beta'}$ is a direct consequence of the continuity of s_β and $s_{\beta'}$. □

Consider the set

$$\mathscr{V}_\alpha = \operatorname{Hom}(\mathfrak{l}, \mathfrak{g}) \cap \mathscr{U}_\alpha. \tag{3.17}$$

Note that for any $\alpha \in I$, the set $\mathscr{V}_\alpha$ is closed and stable by the action of $\operatorname{Aut}\mathfrak{l}$ on $\mathscr{U}_\alpha$, (regarded as a subgroup of $GL_k(\mathbb{R})$). Let us define the sets

$$W_\alpha = \pi_\alpha(\mathscr{V}_\alpha), \quad W_{\alpha\beta} = W_\alpha \cap (V_\beta \times \eta(U_\alpha)) \quad \text{and} \quad \mathscr{W}_{\alpha\beta} = \pi_\alpha^{-1}(W_{\alpha\beta}) = \mathscr{V}_\alpha \cap \mathscr{U}_{\alpha\beta}. \tag{3.18}$$

Then obviously

$$\mathscr{V}_\alpha = \bigcup_{\beta\in I} \mathscr{W}_{\alpha\beta} \quad \text{and} \quad W_\alpha = \bigcup_{\beta\in I} W_{\alpha\beta}. \tag{3.19}$$

The following lemma is immediate.

Lemma 3.4.2. *The set W_α is closed in $(GL_k(\mathbb{R})/\operatorname{Aut}(\mathfrak{l})) \times \eta(U_\alpha)$, $W_{\alpha\beta}$ is open in W_α and $\mathscr{W}_{\alpha\beta}$ is open in $\mathscr{V}_\alpha$.*

Proof. The set $\mathscr{V}_\alpha$ is a closed in $\mathscr{U}_\alpha$ and $\operatorname{Aut}(\mathfrak{l})$-stable, then $\pi_\alpha(\mathscr{V}_\alpha) = W_\alpha$ is closed. The set $W_{\alpha\beta}$ (resp., $\mathscr{W}_{\alpha\beta}$) is an intersection of W_α (resp., $\mathscr{V}_\alpha$) with an open set, as it is to be shown. □

With the above in mind, we get the following.

Corollary 3.4.3. *For any $\alpha \in I(n, k)$ and any $\beta \in I$, the map $\xi_{\alpha\beta}$ realizes a homeomorphism between $\operatorname{Aut}(\mathfrak{l}) \times W_{\alpha\beta}$ and $\mathscr{W}_{\alpha\beta}$.*

Proof. Toward the proof, it is sufficient to see that $\xi_{\alpha\beta}(\mathrm{Aut}(\mathfrak{l}) \times W_{\alpha\beta}) = \mathscr{W}_{\alpha\beta}$. Let $(A, x, W) \in (\mathrm{Aut}(\mathfrak{l}) \times W_{\alpha\beta})$, then $\xi_{\alpha\beta}(A, x, W) \in \pi_\alpha^{-1}(x, W) \subset \pi_\alpha^{-1}(W_{\alpha\beta}) = \mathscr{W}_{\alpha\beta}$. Conversely, let $M \in \mathscr{W}_{\alpha\beta}$, then $\xi'_{\alpha\beta}(M) \in \mathrm{Aut}(\mathfrak{l}) \times W_{\alpha\beta}$ and $M = \xi_{\alpha\beta}(\xi'_{\alpha\beta}(M))$. □

With the above in mind, we get the following.

Theorem 3.4.4. *The collection $(\pi_\alpha, \mathscr{U}_\alpha, (GL_k(\mathbb{R})/\mathrm{Aut}(\mathfrak{l})) \times \eta(U_\alpha), \mathrm{Aut}(\mathfrak{l}))$ defines a principal bundle; the maps $(\xi_{\alpha\beta})_{\beta\in I}$ are the local trivializations and the maps $(t_{\alpha\beta\beta'})_{\beta'\in I}$ are the transition functions. Furthermore, the data $(\pi_\alpha, \mathscr{V}_\alpha, W_\alpha, \mathrm{Aut}(\mathfrak{l}))$ defines a principal bundle.*

We now make the following remark. For any α and β, the group $\mathrm{Aut}(\Gamma)$ acts on $\mathrm{Aut}(\mathfrak{l}) \times W_{\alpha\beta}$ by

$$a \cdot (A, x, W) = (A \circ d(i(a))_{|_{\mathfrak{l}_e}}^{-1}, x, W) \tag{3.20}$$

and the map $\xi_{\alpha\beta}$ is $\mathrm{Aut}(\Gamma)$-equivariant. If the set $\mathscr{W}_{\alpha\beta}$ is G-stable, then we define a map $c_{W,\alpha}$ from G in $GL_k(\mathbb{R})$ given by

$$\mathrm{Ad}_g\, \eta_\alpha^{-1}(W) = \eta_\alpha^{-1}(g \cdot W) c_{W,\alpha}(g). \tag{3.21}$$

Using the identity $\mathrm{Ad}_{g'g}\, \eta_\alpha^{-1}(W) = \mathrm{Ad}_{g'}\, \mathrm{Ad}_g\, \eta_\alpha^{-1}(W)$ and the definition of $c_{W,\alpha}$, we get

$$c_{W,\alpha}(g'g) = c_{g\cdot W,\alpha}(g') c_{W,\alpha}(g). \tag{3.22}$$

As a direct consequence, we obtain the following.

Proposition 3.4.5. *For any $\alpha \in I(n,k)$ and any $\beta \in I$, if $\mathscr{W}_{\alpha\beta}$ is stable by the G-action, then G acts on $\mathrm{Aut}(\mathfrak{l}) \times W_{\alpha\beta}$ through the following law:*

$$g \cdot (A, x, W) := (s_\beta(p(c_{W,\alpha}(g)s_\beta(x)))^{-1} c_{W,\alpha}(g)s_\beta(x)A, p(c_{W,\alpha}(g)s_\beta(x)), g \cdot W)$$

for any $(A, x, W) \in \mathrm{Aut}(\mathfrak{l}) \times W_{\alpha\beta}$ and any $g \in G$. Moreover, this action commutes with the action of $\mathrm{Aut}(\Gamma)$ and the trivialization map $\xi_{\alpha\beta}$ is G-equivariant.

Proof. Let $M = \eta_\alpha^{-1}(W)s_\beta(x)A \in \mathscr{W}_{\alpha\beta}$. As $\mathscr{W}_{\alpha\beta}$ is G-stable, $\mathrm{Ad}_g M \in \mathscr{W}_{\alpha\beta}$ and, therefore, $\xi'_{\alpha\beta}(\mathrm{Ad}_g M) \in \mathrm{Aut}(\mathfrak{l}) \times W_{\alpha\beta}$ for any $g \in G$. It is then easy to see that $\xi'_{\alpha\beta}(\mathrm{Ad}_g M) = g \cdot (A, x, W)$, where $\xi'_{\alpha\beta}$ is given as in (3.16), which proves that the action is well-defined. Using (3.22), we directly get

$$\begin{aligned}
&g' \cdot (g \cdot (A, x, W)) \\
&\quad = (s_\beta(p(c_{g\cdot W,\alpha}(g')s_\beta(p(c_{W,\alpha}(g)s_\beta(x)))))^{-1} c_{g\cdot W,\alpha}(g')s_\beta(p(c_{W,\alpha}(g)s_\beta(x))) \\
&\quad\quad \times s_\beta(p(c_{W,\alpha}(g)s_\beta(x)))^{-1} c_{W,\alpha}(g)s_\beta(x)A, \\
&\quad\quad p(c_{g\cdot W,\alpha}(g')s_\beta(p(c_{W,\alpha}(g)s_\beta(x)))), g'g \cdot W),
\end{aligned}$$

for any $g, g' \in G$. Besides, $p(c_{g\cdot W,\alpha}(g')s_\beta(p(c_{W,\alpha}(g)s_\beta(x)))) = p(c_{g\cdot W,\alpha}(g')c_{W,\alpha}(g)s_\beta(x))$. This conclusively gives

$$\begin{aligned} & g' \cdot (g \cdot (A, x, W)) \\ & = (s_\beta(p(c_{g\cdot W,\alpha}(g')c_{W,\alpha}(g)s_\beta(x)))^{-1} c_{g\cdot W,\alpha}(g')c_{W,\alpha}(g)s_\beta(x)A, \\ & \quad p(c_{g\cdot W,\alpha}(g')c_{W,\alpha}(g)s_\beta(x)), g'g \cdot W) \\ & = (s_\beta(p(c_{W,\alpha}(g'g)s_\beta(x)))^{-1} c_{W,\alpha}(g'g)s_\beta(x)A, p(c_{W,\alpha}(g'g)s_\beta(x)), g'g \cdot W) \\ & = g'g \cdot (A, x, W). \end{aligned}$$

The commutation properties is a consequence of (3.20). For the G-equivariance equation, let $g \in G$ and $(A, x, W) \in \mathrm{Aut}(\mathfrak{l}) \times W_{\alpha\beta}$. We have

$$\begin{aligned} g \cdot \xi_{\alpha\beta}(A, x, W) &= \mathrm{Ad}_g\, \eta_\alpha^{-1}(W)s_\beta(x)A = \eta_\alpha^{-1}(g \cdot W)c_{W,\alpha}(g)s_\beta(x)A \\ &= \xi_{\alpha\beta}(g \cdot (A, x, W)). \end{aligned}$$

□

3.4.2 The context where $[\Gamma, \Gamma]$ is uniform in $[G, G]$

Throughout this section, H denotes a connected Lie subgroup of a completely solvable Lie group G and Γ a discontinuous group for the homogeneous space G/H whose syndetic hull L fulfills $[L, L] = [G, G]$ (or equivalently $[\Gamma, \Gamma]$ is uniform in $[G, G]$). The aim here is to give a comprehensive description of the associated deformation space. It is worth mentioning that the techniques used here also allow us to treat the context where $[G, G] \subset H$ as it was the case in [31]. In such a situation and unlike the present setup, the discontinuous group is easily shown to be Abelian.

The parameter space

This subsection is devoted to produce a description of the subsequent parameter space $R(\Gamma, G, H)$. This is important to a major extent to see that such a space smoothly splits into some G-invariant constituents. We first see the following.

Proposition 3.4.6. *Retain the same hypotheses and notation and assume that H is normal or $[L, L] = [G, G]$. Then*

$$R(\Gamma, G, H) = \left\{ \varphi \in \mathrm{Hom}(\mathfrak{l}, \mathfrak{g}) \;\middle|\; \begin{array}{l} \dim \varphi(\mathfrak{l}) = \dim \mathfrak{l} \\ \varphi(\mathfrak{l}) \cap \mathfrak{h} = \{0\} \end{array} \right\}.$$

In particular, $R(\Gamma, G, H)$ is open in $\mathrm{Hom}(\mathfrak{l}, \mathfrak{g})$ *and semialgebraic.*

Proof. Our assumption says that $[\mathfrak{l}, \mathfrak{l}] = [\mathfrak{g}, \mathfrak{g}]$. Let $\varphi \in R(\mathfrak{l}, \mathfrak{g}, \mathfrak{h})$. Clearly, $\varphi([\mathfrak{g}, \mathfrak{g}]) = [\varphi(\mathfrak{l}), \varphi(\mathfrak{l})] \subset [\mathfrak{g}, \mathfrak{g}]$. From the injectivity of φ, we deduce that $[\varphi(\mathfrak{l}), \varphi(\mathfrak{l})] = [\mathfrak{g}, \mathfrak{g}]$. In particular, $\varphi(\mathfrak{l}) \supset [\mathfrak{g}, \mathfrak{g}]$, which means that $\varphi(\mathfrak{l})$ is an ideal of $\mathfrak{g}$. Using Lemma 2.1.11, the proper action of $\exp \varphi(\mathfrak{l})$ on G/H is equivalent to $\varphi(\mathfrak{l}) \cap \mathfrak{h} = \{0\}$, and the result follows. □

With the above in mind, we now consider the decomposition,

$$\{\varphi \in \mathrm{Hom}(\mathfrak{l}, \mathfrak{g}), \dim \varphi(\mathfrak{l}) = \dim \mathfrak{l}\} = \mathrm{Hom}(\mathfrak{l}, \mathfrak{g}) \cap M^{\circ}_{n,k}(\mathbb{R}) = \bigcup_{\alpha \in I(n,k)} \mathscr{V}_{\alpha}, \tag{3.23}$$

which is open in $\mathrm{Hom}(\mathfrak{l}, \mathfrak{g})$ and closed in $\mathscr{U}_{\alpha}$. Now, the following fact is proved in Lemma 3.3.5: For all $\varphi \in \mathscr{L}(\mathfrak{l}, \mathfrak{g})$, such that $\dim \varphi(\mathfrak{l}) = \dim \mathfrak{l}$, we have $\varphi(\mathfrak{l}) \cap \mathfrak{h} = \{0\}$ if and only if there exists α in $I_s(n, k)$ such that $\varphi \in \mathscr{U}_{\alpha}$, where

$$I_s(n, k) = \{(i_1, \ldots, i_k), s < i_1 < \cdots < i_k \leq n\}.$$

Up to this step, Lemma 3.23 and the decomposition (3.19) of $\mathscr{V}_{\alpha}$, we get the following.

Corollary 3.4.7. *Suppose that H is normal or $[L, L] = [G, G]$. Then up to the homeomorphism we have*

$$\mathscr{R}(\Gamma, G, H) = \bigcup_{\substack{\alpha \in I_S(n,k), \\ \beta \in I}} \mathscr{W}_{\alpha\beta},$$

as a union of open sets in $\mathrm{Hom}(\mathfrak{l}, \mathfrak{g})$.

The deformation space

Our main upshot in this section is the following.

Theorem 3.4.8. *Let G be a completely solvable Lie group, H a connected subgroup of G and Γ a discontinuous group for G/H such that $[L, L] = [G, G]$. Then for any $\alpha \in I_s(n, k)$ and any $\beta \in I$, the set $\mathscr{W}_{\alpha\beta}$ is G-stable and the deformation and moduli spaces read*

$$\mathscr{T}(\Gamma, G, H) = \bigcup_{\substack{\alpha \in I_S(n,k), \\ \beta \in I}} \mathscr{W}_{\alpha\beta}/G, \quad \mathscr{M}(\Gamma, G, H) = \bigcup_{\substack{\alpha \in I_S(n,k), \\ \beta \in I}} \mathrm{Aut}(\Gamma)\backslash \mathscr{W}_{\alpha\beta}/G,$$

as a union of open sets. Furthermore, G acts on $\mathrm{Aut}(\mathfrak{l}) \times W_{\alpha\beta}$, the set $\mathscr{W}_{\alpha\beta}/G$ is homeomorphic to $(\mathrm{Aut}(\mathfrak{l}) \times W_{\alpha\beta})/G$ and the set $\mathrm{Aut}(\Gamma)\backslash \mathscr{W}_{\alpha\beta}/G$ is homeomorphic to $\mathrm{Aut}(\Gamma)\backslash(\mathrm{Aut}(\mathfrak{l}) \times W_{\alpha\beta})/G$.

Proof. Note first from Corollary 3.4.7 that

$$\mathscr{R}(\Gamma, G, H) = \bigcup_{\substack{\alpha \in I_S(n,k), \\ \beta \in I}} \mathscr{W}_{\alpha\beta}.$$

From Corollary 3.4.3 and Proposition 3.4.5, we are done, provided that for any $\alpha \in I_s(n, k)$ and $\beta \in I$, the set $\mathscr{W}_{\alpha\beta}$ is G-stable, which remains to be proved. Indeed, let

$M \in \mathscr{W}_{\alpha\beta}$. For $W = \eta(M)$ and $x = p(M_\alpha)$, there is $A \in \mathrm{Aut}(\mathfrak{l})$ such that $M = \eta_\alpha^{-1}(W)s_\beta(x)A$. When $[\mathfrak{l},\mathfrak{l}] = [\mathfrak{g},\mathfrak{g}]$, for any $M \in R(\mathfrak{l},\mathfrak{g},\mathfrak{h})$, the subspace $M(\mathfrak{l})$ is an ideal of $\mathfrak{g}$. It follows therefore that $g \cdot W = W$ for any $g \in G$ and from (3.21) we obtain

$$\mathrm{Ad}_g M = \eta_\alpha^{-1}(W)c_{W,\alpha}(g)s_\beta(x)A = Mc_{A,x,W,\alpha}(g), \tag{3.24}$$

where $c_{A,x,W,\alpha}(g) = A^{-1}s_\beta(x)^{-1}c_{W,\alpha}(g)s_\beta(x)A$. The set $\mathscr{W}_{\alpha\beta}$ is stable by the action of $\mathrm{Aut}(\mathfrak{l})$, then to prove that $\mathrm{Ad}_g M \in \mathscr{W}_{\alpha\beta}$, it is sufficient to show the following lemma.

Lemma 3.4.9. *The map $c_{A,x,W,\alpha}$ is a continuous homomorphism from G to* $\mathrm{Aut}(\mathfrak{l})$.

Proof. The map $c_{A,x,W,\alpha} : G \to GL_k(\mathbb{R})$ is nothing but the composition of $c_{W,\alpha}$ and the conjugation by the element $A^{-1}s_\beta(x)^{-1}$. From the fact that $g \cdot W = W$ for any $g \in G$ and equation (3.22), we deduce that the map $c_{W,\alpha}$ is a homomorphism and so is $c_{A,x,W,\alpha}$. The continuity is also a direct consequence from the continuity of the action of G on $M_{n,k}(\mathbb{R})$ and the expression

$$c_{W,\alpha}(g) = (g \cdot \eta_\alpha^{-1}(W))_\alpha, \quad g \in G. \tag{3.25}$$

To conclude, we have to prove that $c_{A,x,W}(g) \in \mathrm{Aut}(\mathfrak{l})$. Let $X, Y \in \mathfrak{l}$, then

$$\begin{aligned} M([c_{A,x,W,\alpha}(g)(X), c_{A,x,W,\alpha}(g)(Y)]) &= [Mc_{A,x,W,\alpha}(g)(X), Mc_{A,x,W,\alpha}(g)(Y)] \\ &= \mathrm{Ad}_g M[X,Y] \\ &= Mc_{A,x,W,\alpha}(g)[X,Y]. \end{aligned}$$

This entails that $M([c_{A,x,W,\alpha}(g)(X), c_{A,x,W,\alpha}(g)(Y)] - c_{A,x,W,\alpha}(g)[X,Y]) = 0$. As M is of maximal rank, we are done. □ □

We close this section by the following remark, which will be of interest later on. Retain the same hypotheses and notation, especially that G is completely solvable and that $[\mathfrak{l},\mathfrak{l}] = [\mathfrak{g},\mathfrak{g}]$. Then as in Proposition 3.4.5, $\xi_{\alpha\beta}$ is a G-equivariant homeomorphism from $\mathscr{W}_{\alpha\beta}$ and $\mathrm{Aut}(\mathfrak{l}) \times W_{\alpha\beta}$ for any $\alpha \in I_s(n,k)$ and any $\beta \in I$. Under these circumstances, the action of G is defined by

$$g \cdot (A, x, W) = (c_{x,W,\alpha}(g)A, x, W), \quad (A, x, W) \in \mathrm{Aut}(\mathfrak{l}) \times W_{\alpha\beta}, \quad g \in G, \tag{3.26}$$

where $c_{x,W,\alpha}(g) = s_\beta(x)^{-1}c_{W,\alpha}(g)s_\beta(x)$. Then $\mathrm{Aut}(\mathfrak{l}) \times W_{\alpha\beta}$ splits into a union of fibers of the natural projection on $W_{\alpha\beta}$ and each fiber is G-stable, or equivalently, the orbits of $\mathrm{Aut}(\mathfrak{l})$ in $\mathscr{W}_{\alpha\beta}$ are G-stable. From Lemma 3.4.9, for $(x, W) \in W_{\alpha\beta}$ we can see that the map $c_{x,W,\alpha} : G \to \mathrm{Aut}(\mathfrak{l})$ is a group homomorphism. This entails in fact that $c_{x,W,\alpha}(G)$ is a Lie subgroup of $\mathrm{Aut}(\mathfrak{l})$. To get that, it is sufficient to see the following.

Lemma 3.4.10. *Let $c : G \to G'$ be a continuous Lie groups homomorphism. Suppose that G is connected, then the image of G is connected Lie subgroup of G'.*

Proof. Let $\mathfrak{g}$ be the Lie algebra of G and $\mathfrak{k} = dc_{|e}(\mathfrak{g})$, then $\mathfrak{k}$ is a Lie subalgebra of $\mathfrak{g}'$, the Lie algebra of G'. Now G is connected, then it is generated by $\exp(\mathfrak{g})$ and

$$c \circ \exp_G(X) = \exp_{G'} \circ dc_{|e}(X),$$

for any $X \in \mathfrak{g}$. This conclusively shows that $c(G)$ is the connected Lie subgroup generated by $\exp_{G'}(\mathfrak{k})$. □

Remark 3.4.11. From (3.26), we can see that the natural projection of $\mathrm{Aut}(\mathfrak{l}) \times W_{\alpha\beta}$ on $W_{\alpha\beta}$ factors through the action of G to a continuous surjection from $\mathscr{W}_{\alpha\beta}/G$ to $W_{\alpha\beta}$ such that the fiber of (x, W) is homeomorphic to the homogeneous space $\mathrm{Aut}(\mathfrak{l})/c_{x,W,\alpha}(G)$.

4 The deformation space for nilpotent Lie groups

The present chapter aims to provide a comprehensive description of the deformation and moduli spaces of the action of a discontinuous group Γ for a homogeneous space G/H, where H stands for an arbitrary, connected, closed subgroup of a connected and simply connected, nilpotent Lie group G. We shall figure out how complicated the explicit calculations could be, given the complexity of Lie brackets of Lie algebras in question and the lack of complete characterization of discrete subgroups of a given nilpotent Lie group.

We treat first the setting of Heisenberg groups answering several questions. The case where the center of G sits inside the subgroup H plays a capital role in the computation. The case where the Clifford–Klein forms are compact presents itself major technical difficulties. We also consider the settings of general 2-step and 3-step nilpotent Lie groups and also the threadlike case (cf. Subsection 1.1.3).

Let $P := G \times G$ the direct product Lie group and $\Delta := \Delta(G)$ the diagonal Lie subgroup of P. The emphasis in also focused on the explicit determination of the deformation space of discontinuous groups acting on the nilpotent homogeneous space P/Δ for which the group G is the Heisenberg group. One motivation to seek such a setting comes up from the test case $G = SL(2, \mathbb{R})$, where all of the related objects are nicely determined, and specifically, the deformation space is nothing but the deformation of complex structures on 3-dimensional manifolds. Though this case does not overlap in nature with the classes of groups we are interested in, it arouses our interest to begin the study of such a problem in some nilpotent contexts.

4.1 Deformation and moduli spaces for Heisenberg groups

4.1.1 A criterion of the proper action, continued

Let $\mathfrak{g}$ be a Heisenberg Lie algebra and G its associated Lie group as defined in Subsection 1.1.2. Let $\mathscr{B}_\mathfrak{h}$ a symplectic basis of $\mathfrak{g}$ adapted to $\mathfrak{h}$ as in Proposition 1.1.12. Remark first that the matrix J_b of the bilinear form b defined by equation (1.6) written in $\mathscr{B}_\mathfrak{h}$ is as follows:

$$J_b := \mathscr{M}(b, \mathscr{B}_\mathfrak{h}) = \begin{pmatrix} 0 & \cdots & 0 \\ \vdots & (0) & (-I_n) \\ 0 & (I_n) & (0) \end{pmatrix}.$$

https://doi.org/10.1515/9783110765304-004

Using Theorem 1.1.6, one can view G as the direct product of $\mathscr{D} = \mathbb{R}^{2n}$ and $\mathbb{R}$ with the following pointwise multiplication:

$$g_1 g_2 = \left(v + w, s + t + \frac{1}{2} b(v, w)\right), \quad g_1 = (v, s), \quad g_2 = (w, t),$$

where b is explicitly given on $\mathscr{D}$ by

$$b(v, w) = \langle v_1, w_2 \rangle - \langle v_2, w_1 \rangle, \quad v = (v_1, v_2), \quad w = (w_1, w_2),$$

where v_1, w_1 designate the coordinates of v and w, respectively, through the basis vectors $(X_1, \dots, X_n)$ and v_2, w_2 their coordinates through the vectors $(Y_1, \dots, Y_n)$. We get now the following characterization (which will be of multiple use later on) of the proper action in the Heisenberg setting. We have the following.

Lemma 4.1.1. *Let $\mathfrak{h}$, $\mathfrak{l}$ be two subalgebras of $\mathfrak{g}$ and $H = \exp \mathfrak{h}$. Then $\exp \mathfrak{l}$ acts properly on G/H if and only if one of these two properties is satisfied:*
(*i*) $\mathfrak{z} \subset \mathfrak{h}$ *and* $\mathfrak{l} \cap \mathfrak{h} = \{0\}$.
(*ii*) $\mathfrak{z} \not\subset \mathfrak{h}$, $\mathfrak{l} \cap \mathfrak{h} = \{0\}$ *and* $\mathfrak{z} \cap (\mathfrak{h} \oplus \mathfrak{l}) = \mathfrak{l} \cap \mathfrak{z}$.

Proof. The Heisenberg Lie algebra is a 2-step nilpotent Lie algebra. Using Theorem 2.2.2, we get that the proper action is equivalent to the property $\mathrm{Ad}_g \mathfrak{h} \cap \mathfrak{l} = \{0\}$ for all $g \in G$ (or equivalently $\mathrm{Ad}_g \mathfrak{l} \cap \mathfrak{h} = \{0\}$ for all $g \in G$). If $\mathfrak{z} \subset \mathfrak{h}$, it is then clear that $\mathrm{Ad}_g \mathfrak{h} = \mathfrak{h}$ and that the proper action is equivalent to $\mathfrak{l} \cap \mathfrak{h} = \{0\}$. Assume that the action is proper and $\mathfrak{z} \not\subset \mathfrak{h}$, so obviously $\mathfrak{h} \cap \mathfrak{l} = \{0\}$. Toward the equality $\mathfrak{z} \cap (\mathfrak{h} \oplus \mathfrak{l}) = \mathfrak{l} \cap \mathfrak{z}$, it is sufficient to show that $\mathfrak{z} \cap (\mathfrak{l} \oplus \mathfrak{h}) \subset (\mathfrak{z} \cap \mathfrak{l})$. Let $x \in \mathfrak{z} \cap (\mathfrak{l} \oplus \mathfrak{h})$. There exist then $l \in \mathfrak{l}$ and $h \in \mathfrak{h}$ such that $x = l + h$. We have to show that $x = l$. Suppose that $l \notin \mathfrak{z}$. Then there exist $X \in \mathfrak{g}$ such that $[X, l] = -x$ as the center is one-dimensional. As such, the nontrivial element $\mathrm{Ad}_{\exp X} l = l - x = -h$ belongs to the intersection $\mathrm{Ad}_{\exp X} \mathfrak{l} \cap \mathfrak{h}$, which is impossible. This leads to the fact that $x - l$ is a central element and belongs to $\mathfrak{h}$, which also means that it is trivial. Conversely, let $t \in \mathrm{Ad}_g \mathfrak{l} \cap \mathfrak{h}$. We have $t = l + x$ with $x \in \mathfrak{z}$ and $l \in \mathfrak{l}$. Then $x \in \mathfrak{z} \cap (\mathfrak{l} \oplus \mathfrak{h})$, which means that $x \in \mathfrak{l}$, and finally $t \in \mathfrak{h} \cap \mathfrak{l} = \{0\}$. This consideration shows conclusively that $\exp \mathfrak{l}$ acts properly on G/H. □

4.1.2 The deformation space for non-Abelian actions

Let Γ be a non-Abelian discontinuous subgroup of G for G/H. Let $L = \exp(\mathfrak{l})$ be the syndetic hull of Γ and $\mathscr{B}_{\mathfrak{l}}$ a symplectic basis of $\mathfrak{g}$ adapted to $\mathfrak{l}$ as in Proposition 1.1.12. We single out from $\mathscr{B}_{\mathfrak{l}}$ a new basis $\mathscr{B}'_{\mathfrak{l}}$ of $\mathfrak{l}$ as follows:

$$\mathscr{B}'_{\mathfrak{l}} := \{Z, X_{p+1}, \dots, X_{p+q}, X_1, \dots, X_p, Y_1, \dots, Y_p\}. \tag{4.1}$$

We also fix a basis of $\mathfrak{g}$ passing through $\mathfrak{h}$.

We refer back to Subsection 3.4 for notation and definitions. Let us denote by $I_s^1(2n+1,k)$ the set of elements (as defined by equation (3.15)) of the form $\alpha = (s+1, i_2, \dots, i_k)$. Pick first a finite open covering $\{V_\beta\}_{\beta\in I}$ of $GL_k(\mathbb{R})/\operatorname{Aut}(\mathfrak{l})$ and its local sections $s_\beta : V_\beta \to GL_k(\mathbb{R}), \beta \in I$. For $\alpha \in I_s^1(2n+1,k)$ and $\beta \in I$, we consider the set

$$\mathscr{A}_{\alpha,\beta} := \{(A,M) \in s_\beta(V_\beta) \times U_\alpha : {}^tA\,{}^tMJ_{\mathfrak{g}}MA = J_{\mathfrak{l}}\},$$

where $J_{\mathfrak{g}}$ and $J_{\mathfrak{l}}$ designates the matrices of b and of $b_{|\mathfrak{l}}$, respectively, written in the basis $\mathscr{B}_{\mathfrak{l}}$. Our main result in this section is the following.

Theorem 4.1.2. *Let G be the $(2n+1)$-dimensional Heisenberg group, H a connected Lie subgroup of G and Γ a non-Abelian discontinuous subgroup of G for G/H. Let $L = \exp(\mathfrak{l})$ be the syndetic hull of Γ. There exists a finite set of local sections $(s_\beta)_{\beta\in I}$ for the canonical surjection $GL_k(\mathbb{R}) \to GL_k(\mathbb{R})/\operatorname{Aut}(\mathfrak{l})$, such that the deformation space of Γ acting on G/H reads*

$$\mathscr{T}(\Gamma, G, H) = \bigcup_{\substack{\beta\in I \\ \alpha\in I_s^1(2n+1,k)}} \mathscr{T}_{\alpha\beta},$$

where for $\beta \in I$ and $\alpha \in I_s^1(2n+1,k)$ the set $\mathscr{T}_{\alpha\beta}$ is open in $\mathscr{T}(\Gamma, G, H)$ and homeomorphic to the set $\mathbb{R}^* \times \mathbb{R}^{2pq} \times \operatorname{Sp}(2p) \times GL_q(\mathbb{R}) \times \mathscr{A}_{\alpha,\beta}$. Here, $1 + 2p + q = k$ and $q + 1 = \dim \mathfrak{z}(\mathfrak{l})$, where $\mathfrak{z}(\mathfrak{l})$ is the center of $\mathfrak{l}$.

Proof. We first start by proving a series of useful results on which the proof of the main upshot rests. Let f be an automorphism of $\mathfrak{l}$. Then clearly $f(Z) \in \mathfrak{z}$ and $f(\mathfrak{z}(\mathfrak{l})) = \mathfrak{z}(\mathfrak{l})$. Then with respect to the basis $\mathscr{B}'_{\mathfrak{l}}$, the matrix M of f is of the form

$$M = \begin{pmatrix} a & c & d \\ 0 & G & F \\ 0 & 0 & E \end{pmatrix}, \quad \begin{array}{lll} \text{with } a \in \mathbb{R}^*, & c \in M_{1,q}(\mathbb{R}), & d \in M_{1,2p}(\mathbb{R}), \\ G \in GL_q(\mathbb{R}), & F \in M_{q,2p}(\mathbb{R}), & E \in GL_{2p}(\mathbb{R}). \end{array} \tag{4.2}$$

Let us denote by a_M, the coefficient a of the matrix M. As f is a homomorphism, it comes out that

$${}^tMJ_{\mathfrak{l}}M = a_MJ_{\mathfrak{l}}, \quad \text{where } J_{\mathfrak{l}} = \begin{pmatrix} 0_{\mathbb{R}} & 0 & 0 \\ 0 & 0_{\mathbb{R}^q} & 0 \\ 0 & 0 & J_{2p} \end{pmatrix} \in M_k(\mathbb{R}), \quad \text{with } J_{2p} = \begin{pmatrix} 0 & I_p \\ -I_p & 0 \end{pmatrix}.$$

By a direct calculation, this condition is equivalent to ${}^tEJ_{2p}E = a_MJ_{2p}$.

Lemma 4.1.3. *Let $\mathfrak{l}$ be a non-Abelian Lie subalgebra of $\mathfrak{g}$,*

$$K = \begin{pmatrix} 1 & M_{1,q}(\mathbb{R}) & M_{1,2p}(\mathbb{R}) \\ 0 & GL_q(\mathbb{R}) & M_{q,2p}(\mathbb{R}) \\ 0 & 0 & \mathrm{Sp}(2p) \end{pmatrix} \quad \text{and} \quad Q = \left\{ \begin{pmatrix} a & 0 & 0 \\ 0 & I_{q+p} & 0 \\ 0 & 0 & aI_p \end{pmatrix}, a \in \mathbb{R}^* \right\}.$$

Then $\mathrm{Aut}(\mathfrak{l}) = KQ$. *Here, p and q are given as in Proposition* 1.1.12.

Proof. Clearly, any matrix in K or Q is an automorphism of $\mathfrak{l}$. Conversely, let f be an element of $\mathrm{Aut}(\mathfrak{l})$ and M its matrix written as in (4.2). Let us write

$$E = \begin{pmatrix} D_1 & D_2 \\ D_3 & D_4 \end{pmatrix} \quad \text{and} \quad E_M = \begin{pmatrix} D_1 & a^{-1}D_2 \\ D_3 & a^{-1}D_4 \end{pmatrix},$$

then ${}^tEJ_{2p}E = aJ_{2p}$ if and only if $E_M \in \mathrm{Sp}(2p)$ and

$$M = \begin{pmatrix} 1 & c & d' \\ 0 & G & F' \\ 0 & 0 & E_M \end{pmatrix} \begin{pmatrix} a & 0 & 0 \\ 0 & I_{p+q} & 0 \\ 0 & 0 & aI_p \end{pmatrix},$$

where for $d = (d_1, \ldots, d_{2p})$, $d' = (d_1, \ldots, d_p, a^{-1}d_{p+1}, \ldots, a^{-1}d_{2p})$ and for $F = (F_1, F_2)$ with F_1, F_2 in $M_{q,p}(\mathbb{R})$, $F' = (F_1, a^{-1}F_2)$. □

The condition $[\mathfrak{l}, \mathfrak{l}] = [\mathfrak{g}, \mathfrak{g}]$ implies that any homomorphism from $\mathfrak{l}$ to $\mathfrak{g}$ stabilizes the center of $\mathfrak{g}$. The following lemma then is immediate.

Lemma 4.1.4. *If $\mathfrak{l}$ is a non-Abelian Lie subalgebra of $\mathfrak{g}$. Then $\mathfrak{l}$ contains the center $\mathfrak{z}$, and any homomorphism from $\mathfrak{l}$ to $\mathfrak{g}$ stabilizes $\mathfrak{z}$.*

We consider a basis $\mathscr{B}_{\mathfrak{h}} = \{Z, X'_1 \ldots, X'_n, Y'_1, \ldots, Y'_n\}$ of $\mathfrak{g}$ adapted to $\mathfrak{h}$ as in Proposition 1.1.12 with the following ordering:

$$X'_1 \ldots, X'_s, Z, Y'_1, \ldots, Y'_s, X'_{s+1}, \ldots, X'_n, Y'_{s+1}, \ldots, Y'_n.$$

and for $\mathfrak{l}$ we keep fixing the basis (4.1). As a direct consequence, the set $\mathrm{Hom}(\mathfrak{l}, \mathfrak{g})$ is identified with a subset of matrices in $M_{2n+1,k}(\mathbb{R})$ written through the bases $\mathscr{B}'_{\mathfrak{l}}$ and $\mathscr{B}_{\mathfrak{h}}$ as

$$M = \begin{pmatrix} 0 & M_1 \\ a & d \\ 0 & M_2 \end{pmatrix}, \tag{4.3}$$

where $a \in \mathbb{R}$, $d \in \mathbb{R}^{k-1}$, $M_1 \in M_{s,k-1}(\mathbb{R})$, $M_2 \in M_{2n-s,k-1}(\mathbb{R})$. As a direct consequence of (4.3), we have the following.

Lemma 4.1.5. *For any $\alpha \in I_s(2n+1,k)$, the set $\mathscr{V}_\alpha$ is empty whenever $\alpha \notin I^1_s(2n+1,k)$.*

Proof. From (4.3), for any M in $\mathrm{Hom}(\mathfrak{l}, \mathfrak{g})$ and $\alpha \notin I_s^1(2n+1, k)$, we get $\det(M_\alpha) = 0$. Then our result is a direct consequence of (3.17). □

Lemma 4.1.6. *For $W = \eta_\alpha(M)$ and $\alpha \in I_s^1(2n+1, k)$, the image of the group homomorphism $c_{W,\alpha}$ is independent from the choice of W and α. More precisely, $c_{W,\alpha}(G)$ equals the matrix group*

$$c(G) := \begin{pmatrix} 1 & M_{1,k-1}(\mathbb{R}) \\ 0 & I_{k-1} \end{pmatrix}.$$

Proof. As a direct consequence of (3.25), we get for $X \in \mathfrak{g}$:

$$c_{W,\alpha}(e^X) = \begin{pmatrix} 1 & b(X, c_2) \cdots b(X, c_k) \\ \cdot & \\ \cdot & \\ \cdot & \\ 0 & (I_{2p+q}) \end{pmatrix},$$

where the c_i are the column of $\eta_\alpha^{-1}(W)$. Furthermore, the column vectors $c_2, \ldots, c_k$ are linearly independent and the center of $\mathfrak{g}$ is not contained in the space generated by these vectors. This means that the linear forms on $\mathfrak{g}$, $b(\cdot, c_2), \ldots, b(\cdot, c_k)$ are independent. Then the system $b(X, c_2) = x_1, \ldots, b(X, c_k) = x_{k-1}$ has a solution for any $(x_1, \ldots, x_{k-1}) \in \mathbb{R}^{k-1}$ and the result follows. □

The following result is immediate.

Lemma 4.1.7. *The normalizer of $c(G)$ in $GL_k(\mathbb{R})$ is the parabolic subgroup P of matrices of the form*

$$\begin{pmatrix} a & b \\ 0 & B \end{pmatrix},$$

where $a \in \mathbb{R}^*$, $b \in \mathbb{R}^{k-1}$ and $B \in GL_{k-1}(\mathbb{R})$.

Lemma 4.1.8. *Let $\alpha \in I_s(2n+1, k)$ and $\beta \in I$. For any $(x, W) \in V_\beta \times \eta(U_\alpha)$, we have $\eta_\alpha^{-1}(W)s_\beta(x) \in \mathscr{V}_\alpha$ only if $s_\beta(x) \in P$. In other words, $c_{x,W,\alpha}(G) = c(G)$. In particular, $c(G)$ is normal in* $\mathrm{Aut}(\mathfrak{l})$.

Proof. By Lemma 4.1.7, the parabolic subgroup P coincides with the normalizer $N(c(G))$ and it is clear that $\mathrm{Aut}(\mathfrak{l})$ is a subgroup of P. Now $s_\beta(x) = (\eta_\alpha^{-1}(W)s_\beta(x))_\alpha$ and $\eta_\alpha^{-1}(W)s_\beta(x) \in \mathscr{V}_\alpha$ only if $\eta_\alpha^{-1}(W)s_\beta(x)$ is written as in (4.3). Then $s_\beta(x) \in P$. □

Lemma 4.1.9. *The topological group* $\mathrm{Aut}(\mathfrak{l})/c(G)$ *is isomorphic to* $K'Q$, *where*

$$K' = \begin{pmatrix} 1 & 0 & 0 \\ 0 & GL_q(\mathbb{R}) & M_{q,2p}(\mathbb{R}) \\ 0 & 0 & \mathrm{Sp}(2p) \end{pmatrix}.$$

In particular, $\mathrm{Aut}(\mathfrak{l})/c(G)$ *is homeomorphic to* $\mathbb{R}^* \times \mathbb{R}^{2pq} \times \mathrm{Sp}(2p) \times GL_q(\mathbb{R})$.

Proof. Consider the projection $p : \mathrm{Aut}(\mathfrak{l}) \to K'Q$, where $p(A)$ is the matrix obtained from A by vanishing the $k-1$ last coefficients of the first line of A. Then p is a surjective homomorphism and its kernel is $c(G)$. Now K' is normal in $K'Q$ and $K' \cap Q$ is trivial; then $K'Q$ is isomorphic to the semidirect product $K \times Q$. The question of the related topology is straight clear. □

Let us go back now to the proof of Theorem 4.1.2. Take a family of continuous local sections $(s_\beta)_{\beta \in I}$ of the canonical surjection $GL_k(\mathbb{R}) \to GL_k(\mathbb{R})/\mathrm{Aut}(\mathfrak{l})$. From Theorem 3.4.8, the set $\mathscr{T}_{\alpha\beta}$ is homeomorphic to the quotient space $(\mathrm{Aut}(\mathfrak{l}) \times W_{\alpha\beta})/G$. By Lemmas 4.1.6 and 4.1.8, we have $c_{x,W,\alpha}(G) = c(G)$ and then the map

$$\begin{array}{rccc} i: & \mathrm{Aut}(\mathfrak{l})/c(G) \times W_{\alpha\beta} & \to & (\mathrm{Aut}(\mathfrak{l}) \times W_{\alpha\beta})/G \\ & (\overline{A}, x, W) & \mapsto & \overline{(A, x, W)} \end{array}$$

is a well-defined bijection and the canonical surjection $p_1 : \mathrm{Aut}(\mathfrak{l}) \times W_{\alpha\beta} \to (\mathrm{Aut}(\mathfrak{l}) \times W_{\alpha\beta})/G$ factors through the canonical surjection $p_2 : \mathrm{Aut}(\mathfrak{l}) \times W_{\alpha\beta} \to \mathrm{Aut}(\mathfrak{l})/c(G) \times W_{\alpha\beta}$ and i. Now p_1 and p_2 are continuous and open; then i is bicontinuous. We now focus attention on the set $W_{\alpha\beta}$. By (3.18) and Lemma 3.4.1, we can write

$$W_{\alpha\beta} = \{(x, W) \in V_\beta \times \eta(U_\alpha) : \eta_\alpha^{-1}(W) s_\beta(x) \in \mathscr{V}_\alpha\}. \tag{4.4}$$

Using Lemma 4.1.5, we can see that $W_{\alpha\beta}$ is empty for $\alpha \notin I_s^1(2n+1, k)$. Let us denote by $a_{s_\beta(x)}$ the coefficient of the first line and column in the matrix $s_\beta(x)$. Then $\eta_\alpha^{-1}(W) s_\beta(x)(Z) = a_{s_\beta(x)} Z$ and we can state that

$$W_{\alpha\beta} = \{(x, W) \in V_\beta \times \eta(U_\alpha) \mid {}^t s_\beta(x) {}^t \eta_\alpha^{-1}(W) J_{\mathfrak{g}} \eta_\alpha^{-1}(W) s_\beta(x) = a_{s_\beta(x)} J_{\mathfrak{l}}\}. \tag{4.5}$$

To complete the proof, we have to replace the sections $(s_\beta)_\beta$ by new sections satisfying $W_{\alpha\beta}$ is empty or $a_{s_\beta(x)} = 1$ for all $(x, W) \in W_{\alpha\beta}$. For $i = 1, \dots, k$, consider the open covering of $GL_k(\mathbb{R})$,

$$G_i = \{A = (a_{ij}) \in GL_k(\mathbb{R}),\ a_{i1} \neq 0\}.$$

Then all the G_i's are $\mathrm{Aut}(\mathfrak{l})$-stable and we can actually replace the sections $(s_\beta)_\beta$ and the covering $(V_\beta)_\beta$, by the covering $(V_{\beta,i})_{\beta,i} = (V_\beta \cap G_i)_{\beta,i}$ and the sections $(s_{\beta,i})_{\beta,i} = (s_{\beta|V_{\beta,i}})_{\beta,i}$.

For $i \neq 1$, $s_{\beta,i}(x) \notin P$ for all $x \in V_{\beta,i}$ and then $W_{\alpha,\beta,i}$ is empty by Lemma 4.1.8. For $i = 1$, let

$$G_1 = \{A = (a_{ij}) \in GL_k(\mathbb{R}),\ a_{11} = 1\}$$

and consider the map

$$\begin{aligned} \delta_\beta : s_\beta(V_{\beta,1}) &\rightarrow G_1 \\ s_\beta(x) &\mapsto s_\beta(x)Q(a_{s_\beta(x)}), \end{aligned}$$

where

$$Q(a_{s_\beta(x)}) = \begin{pmatrix} a_{s_\beta(x)}^{-1} & 0 & 0 \\ 0 & I_{q+p} & 0 \\ 0 & 0 & a_{s_\beta(x)}^{-1} I_p \end{pmatrix}.$$

Lemma 4.1.10. *The map δ_β is a homeomorphism from $s_\beta(V_{\beta,1})$ on its image.*

Proof. Clearly, δ_β is continuous. Note that $s_\beta(x)$ and its image have the same coset class modulo $\mathrm{Aut}(\mathfrak{l})$. Then δ_β is injective and the map $\delta'_\beta : \delta_\beta(s_\beta(V_{\beta,1})) \rightarrow s_\beta(V_{\beta,1})$ given by $\delta'_\beta(M) = s_\beta(p(M))$ is well-defined, continuous and turns out to be the inverse of δ_β. □

We finally end up with new family of continuous local sections $(s'_{\beta,i})_{\beta,i}$ given by

$$s'_{\beta,i} = \begin{cases} \delta_\beta \circ s_{\beta,1} & \text{if } i = 1, \\ s_{\beta,i} & \text{if } i \neq 1 \end{cases}$$

defined on the open sets $V_{\beta,i}$ such that $a_{s'_{\beta,i}}(x) = 1$. This completes the proof of the theorem. □

4.1.3 Deformation and moduli spaces when *H* contains the center

We assume in this section that the subalgebra $\mathfrak{h}$ of $\mathfrak{g}$ contains the center $\mathfrak{z} = [\mathfrak{g}, \mathfrak{g}]$ and that $\mathfrak{l}$ is a subalgebra of $\mathfrak{g}$ such that $\mathfrak{l} \cap \mathfrak{h} = \{0\}$. Then $\mathfrak{l}$ is an Abelian subalgebra and if $\mathscr{L}(\mathfrak{l}, \mathfrak{g})$ designates the vector space of the linear maps from $\mathfrak{l}$ to $\mathfrak{g}$ the set $\mathrm{Hom}(\mathfrak{l}, \mathfrak{g})$ of Lie algebras homomorphisms can be regarded as the set

$$\mathrm{Hom}(\mathfrak{l}, \mathfrak{g}) := \{\psi \in \mathscr{L}(\mathfrak{l}, \mathfrak{g}),\ [\psi(x), \psi(y)] = 0 \text{ for all } x, y \in \mathfrak{l}\}.$$

We fix by the way a symplectic basis $\mathscr{B}_\mathfrak{h}$ of $\mathfrak{g}$ adapted to $\mathfrak{h}$ as provided by Proposition 1.1.12. We identify $\mathfrak{g}$ to $\mathbb{R}^{2n+1}$, $\mathfrak{h}$ to a subspace of $\mathbb{R}^{2n+1}$ and $\mathscr{L}(\mathfrak{l}, \mathfrak{g})$ to a subset of real matrices $M_{2n+1,k}(\mathbb{R})$, where $k = \dim \mathfrak{l}$. Let as usual $s = \dim \mathfrak{h}$. For any $\alpha \in I_s(n,k)$, we consider the set

$$\mathscr{V}'_\alpha := \left\{ M = \begin{pmatrix} 0 \\ A \end{pmatrix}, A \in M_{2n,k}(\mathbb{R}),\ M_\alpha = \mathrm{I}_k \text{ and } {}^tMJ_bM = 0 \right\} \subset \mathscr{V}_\alpha, \tag{4.6}$$

where J_b is the matrix of b in $\mathscr{B}_\mathfrak{h}$. The following theorem provides a description of the deformation and the moduli space in this context.

Theorem 4.1.11. *Let G be the Heisenberg Lie group of dimension $2n + 1$, H a connected Lie subgroup of dimension s, which contains the center of G and Γ a discontinuous group for G/H of rank k. Then*

$$\mathscr{T}(\Gamma, G, H) = \bigcup_{\alpha \in I_s(2n+1,k)} \mathscr{T}_\alpha \quad \text{and} \quad \mathscr{M}(\Gamma, G, H) = \bigcup_{\alpha \in I_s(2n+1,k)} \mathscr{M}_\alpha,$$

where for every $\alpha \in I_s(2n+1, k)$, the set $\mathscr{T}_\alpha$ is an open subset of $\mathscr{T}(\Gamma, G, H)$ homeomorphic to the product $GL_k(\mathbb{R}) \times \mathscr{V}'_\alpha$ and $\mathscr{M}_\alpha$ is an open subset of $\mathscr{M}(\Gamma, G, H)$ homeomorphic to the product $GL_k(\mathbb{R})/GL_k(\mathbb{Z}) \times \mathscr{V}'_\alpha$.

Proof. We will make use of Theorem 3.3.1 as $\mathfrak{h}$ contains the first derivative group of $\mathfrak{g}$. As it stands there, we just have to prove that the quotient space $\mathscr{V}_\alpha/G$ is homeomorphic to $\mathscr{V}'_\alpha$ for any $\alpha \in I_s(2n + 1, k)$. We first fix all of the symplectic basis $\mathscr{B}_\mathfrak{h} = (Z, X_1, \ldots, X_n, Y_1, \ldots, Y_n)$ adapted to $\mathfrak{h}$. Take any $\alpha \in I_s(2n + 1, k)$ and $M \in \mathscr{V}_\alpha$, we can then write

$$M = \begin{pmatrix} a \\ A \end{pmatrix} \quad \text{with } a = (a_1, \ldots, a_k) \in \mathbb{R}^k \text{ and } A \in M_{2n,k}(\mathbb{R}).$$

Note that for $X \in \mathfrak{g}$ we have

$$\mathrm{Ad}_{\exp X} = \begin{pmatrix} 1 & b(X, X_1) & \cdots & b(X, Y_n) \\ \cdot & & & \\ \cdot & & & \\ \cdot & & & \\ 0 & & (I_{2n}) & \end{pmatrix}.$$

Then the action of $\mathrm{Ad}_{\exp X}$ on M affects only the first line. More precisely, if we identify the columns c_i of A to a vector of $\mathfrak{g}$, then the first line of the product is

$$(a_1 + b(X, c_1), \ldots, a_k + b(X, c_k)).$$

The following result, the subject of Lemma 3.3.5, will next be used. Let W_M denote the subspace of $\mathfrak{g}$ generated by the columns of M. If $M \in \mathscr{V}_\alpha$ for $\alpha \in I_s(2n + 1, k)$, then

$W_M \cap \mathfrak{h} = \{0\}$. It turns out as the center $\mathfrak{z}$ is not contained in W_M that the map $\mathfrak{g} \to W_M^*$, $x \mapsto b(x,.)$ is surjective and there exists therefore $X \in \mathfrak{g}$ such that

$$b(X, c_1) = -a_1, \dots, b(X, c_k) = -a_k.$$

It follows then that any M in $\mathscr{V}_\alpha$ is G-equivalent to the matrix obtained from M by vanishing the first line of M. Conversely, if the first lines of M and M' are zero, then $M = \mathrm{Ad}_{\exp X} M'$ only if $M = M'$.

Let π be the continuous map from $\mathscr{V}_\alpha$ to $\mathscr{V}_\alpha'$, which sends the matrix

$$M = \begin{pmatrix} a \\ A \end{pmatrix}$$

to the matrix

$$\pi(M) = \begin{pmatrix} 0 \\ A \end{pmatrix},$$

where we consider the trace topology on $\mathscr{V}_\alpha'$. Then the canonical surjection $p : \mathscr{V}_\alpha \to \mathscr{V}_\alpha/G$ factors through π to a continuous bijection f between $\mathscr{V}_\alpha'$ and $\mathscr{V}_\alpha/G$ defined by $p = f \circ \pi$. Now G acts continuously on $\mathscr{V}_\alpha$, then p is open and we can easily see that f^{-1} is continuous. This achieves the proof of the theorem. □

4.1.4 The case when *H* does not meet the center

We now tackle the case where the center of $\mathfrak{g}$ does not meet $\mathfrak{h}$. In such a situation, H is an Abelian subgroup of G. We still need some other results. The following lemma describes the structure of the parameter space in this case.

Lemma 4.1.12. *Let G be the Heisenberg Lie group, H a connected subgroup, which does not contain the center, Γ a rank k Abelian discontinuous group for G/H and $L = \exp(\mathfrak{l})$ its syndetic hull in G. Then the parameter space $\mathscr{R}(\Gamma, G, H)$ is the disjoint union of the two G-invariant sets*

$$R_1(\Gamma, G, H) = \left\{ \psi \in \mathrm{Hom}(\mathfrak{l}, \mathfrak{g}) \;\middle|\; \begin{array}{l} \dim \psi(\mathfrak{l}) = k, \\ \mathfrak{h} \cap \psi(\mathfrak{l}) = \{0\} \\ \textit{and } \mathfrak{z} \subset \psi(\mathfrak{l}) \end{array} \right\}$$

and

$$R_2(\Gamma, G, H) = \left\{ \psi \in \mathrm{Hom}(\mathfrak{l}, \mathfrak{g}) \;\middle|\; \begin{array}{l} \dim \psi(\mathfrak{l}) = k \textit{ and} \\ (\mathfrak{h} \oplus \mathfrak{z}) \cap \psi(\mathfrak{l}) = \{0\} \end{array} \right\}.$$

Proof. From Theorem 3.2.4 and Lemma 4.1.1, we can easily see that $\mathscr{R}(\Gamma, G, H)$ is the union of the following sets:

$$R_1(\Gamma, G, H) = \left\{ \psi \in \operatorname{Hom}(\mathfrak{l}, \mathfrak{g}) \;\middle|\; \begin{array}{l} \dim \psi(\mathfrak{l}) = k, \\ \mathfrak{h} \cap \psi(\mathfrak{l}) = \{0\}, \\ \mathfrak{z} \cap (\mathfrak{h} \oplus \psi(\mathfrak{l})) = \psi(\mathfrak{l}) \cap \mathfrak{z} \\ \text{and } \mathfrak{z} \subset \psi(\mathfrak{l}) \end{array} \right\}$$

and

$$\begin{aligned} R_2(\Gamma, G, H) &= \left\{ \psi \in \operatorname{Hom}(\mathfrak{l}, \mathfrak{g}) \;\middle|\; \begin{array}{l} \dim \psi(\mathfrak{l}) = k, \\ \mathfrak{h} \cap \psi(\mathfrak{l}) = \{0\}, \\ \mathfrak{z} \cap (\mathfrak{h} \oplus \psi(\mathfrak{l})) = \psi(\mathfrak{l}) \cap \mathfrak{z} \\ \text{and } \mathfrak{z} \not\subset \psi(\mathfrak{l}) \end{array} \right\} \\ &= \left\{ \psi \in \operatorname{Hom}(\mathfrak{l}, \mathfrak{g}) \;\middle|\; \begin{array}{l} \dim \psi(\mathfrak{l}) = k, \\ \mathfrak{h} \cap \psi(\mathfrak{l}) = \{0\}, \\ \mathfrak{z} \cap (\mathfrak{h} \oplus \psi(\mathfrak{l})) = \{0\} \end{array} \right\}. \end{aligned}$$

To conclude, note that the third condition $\mathfrak{z} \cap (\mathfrak{h} \oplus \psi(\mathfrak{l})) = \psi(\mathfrak{l}) \cap \mathfrak{z}$ involved in the set $R_1(\Gamma, G, H)$ is trivial as $\mathfrak{z} \subset \psi(\mathfrak{l})$. Likewise, it is easily seen that $\mathfrak{z} \cap (\mathfrak{h} \oplus \psi(\mathfrak{l})) = \{0\}$ if and only if $(\mathfrak{z} \oplus \mathfrak{h}) \cap \psi(\mathfrak{l}) = \{0\}$, and then the three last set equations of $R_2(\Gamma, G, H)$ together are equivalent to $(\mathfrak{h} \oplus \mathfrak{z}) \cap \psi(\mathfrak{l}) = \{0\}$. On the other hand, for any $g \in G$, $\mathfrak{z} \subset \operatorname{Ad}_{g^{-1}} \circ \psi(\mathfrak{l})$ if and only if $\mathfrak{z} \subset \psi(\mathfrak{l})$, which proves the G-invariance of $R_1(\Gamma, G, H)$. Furthermore, for any $\psi \in R_2(\Gamma, G, H)$ and any $g \in G$, one has

$$(\mathfrak{z} \oplus \mathfrak{h}) \cap \operatorname{Ad}_{g^{-1}}(\psi(\mathfrak{l})) = (\mathfrak{z} \oplus \mathfrak{h}) \cap (\psi(\mathfrak{l})) = \{0\},$$

which shows the G-invariance of the set $R_2(\Gamma, G, H)$. □

We now fix a basis $\mathscr{B}_{\mathfrak{h}} = \{Z, X_1 \dots, X_n, Y_1, \dots, Y_n\}$ of $\mathfrak{g}$ adapted to $\mathfrak{h}$. We consider the decomposition

$$\mathfrak{g} = \mathfrak{z} \oplus \mathfrak{h} \oplus \mathfrak{h}' \oplus \mathfrak{k} \oplus \mathfrak{k}',$$

where

$$\begin{aligned} \mathfrak{h} &= \langle X_1, \dots, X_s \rangle, \quad & \mathfrak{h}' &= \langle Y_1, \dots, Y_s \rangle, \\ \mathfrak{k} &= \langle X_{s+1}, \dots, X_n \rangle \quad \text{and} \quad & \mathfrak{k}' &= \langle Y_{s+1}, \dots, Y_n \rangle. \end{aligned}$$

We identify as previously $\mathfrak{g}$ to $\mathbb{R}^{2n+1} = \mathbb{R} \oplus \mathbb{R}^s \oplus \mathbb{R}^s \oplus \mathbb{R}^{n-s} \oplus \mathbb{R}^{n-s}$ and $\operatorname{Hom}(\mathfrak{l}, \mathfrak{g})$ to the set of matrices given in (4.3), with $l = 1$ and $b_1 = b$. Then with respect to this decomposition, any element of $\mathfrak{g}$,

$$x = a_0 Z + \sum_{i=1}^{n} a_i X_i + \sum_{i=1}^{n} b_i Y_i,$$

is identified to the column vector

$$^t(a_0\, a_1 \cdots a_s\, b_1 \cdots b_s\, a_{s+1} \cdots a_n\, b_{s+1} \cdots b_n)$$

and every homomorphism $\psi \in \mathrm{Hom}(\mathfrak{l}, \mathfrak{g})$, can be written as a matrix

$$M = \begin{pmatrix} A_0 \\ A_1 \\ B_1 \\ A_2 \\ B_2 \end{pmatrix},$$

where $A_0 \in M_{1,k}(\mathbb{R})$, $A_1, B_1 \in M_{s,k}(\mathbb{R})$ and $A_2, B_2 \in M_{n-s,k}(\mathbb{R})$. Then from Lemma (4.1.12), we get

$$R_1(\Gamma, G, H) = \left\{ M \in M_{2n+1,k}(\mathbb{R}) \;\middle|\; \begin{array}{l} \dim M(\mathbb{R}^k) = k, \\ \mathfrak{h} \cap M(\mathbb{R}^k) = \{0\}, \\ \mathfrak{z} \subset M(\mathbb{R}^k) \text{ and} \\ {}^t M J_b M = 0 \end{array} \right\}.$$

Up to this step, we consider the set

$$I_s^1(2n+1, k) = \{(i_1, \ldots, i_k),\ i_1 = 1 \text{ and } i_2 > s+1\}. \tag{4.7}$$

Now we can state the following.

Lemma 4.1.13. *The set $R_1(\Gamma, G, H)$ is open in* $\mathrm{Hom}(\mathfrak{l}, \mathfrak{g})$ *and the sets $\eta^{-1}(\eta_\alpha(\mathscr{V}_\alpha))$, $\alpha \in I_s^1(2n+1, k)$ constitutes an open G-invariant covering of $R_1(\Gamma, G, H)$.*

Proof. The condition $\mathfrak{z} \subset M(\mathbb{R}^k)$ equivalent to the existence of a matrix

$$M' = \begin{pmatrix} 1 & 0 \\ 0 & A_1' \\ 0 & B_1' \\ 0 & A_2' \\ 0 & B_2' \end{pmatrix},$$

with $A'_1, B'_1 \in M_{s,k-1}(\mathbb{R})$, $A'_2, B'_2 \in M_{n-s,k-1}(\mathbb{R})$ and $M(\mathbb{R}^k) = M'(\mathbb{R}^k)$. The conditions $M(\mathbb{R}^k) \cap \mathfrak{h} = \{0\}$ and $\dim M(\mathbb{R}^k) = k$ are equivalent to

$$\operatorname{rank}\begin{pmatrix} B'_1 \\ A'_2 \\ B'_2 \end{pmatrix} = k-1,$$

which is also equivalent to the existence of $\alpha \in I^1_s(2n+1,k)$, such that $M(\mathbb{R}^k) \in \eta_\alpha(U_\alpha)$. Now, if $M(\mathbb{R}^k) \in \eta_\alpha(U_\alpha)$, then ${}^tMJ_bM = 0$ if and only if

$${}^t\{\eta_\alpha^{-1}(M(\mathbb{R}^k))\}J_b\{\eta_\alpha^{-1}(M(\mathbb{R}^k))\} = 0,$$

or equivalently $\eta_\alpha^{-1}(M(\mathbb{R}^k)) \in \mathscr{V}_\alpha$. Then $M \in R_1(\Gamma, G, H)$ if and only if

$$M(\mathbb{R}^k) \in \bigcup_{\alpha \in I^1_s(2n+1,k)} \eta_\alpha(\mathscr{V}_\alpha).$$

This means that

$$R_1(\Gamma, G, H) = \bigcup_{\alpha \in I^1_s(2n+1,k)} \eta^{-1}(\eta_\alpha(\mathscr{V}_\alpha)).$$

Furthermore,

$$\eta^{-1}(\eta_\alpha(\mathscr{V}_\alpha)) = \left\{ M \in M_{2n+1,k}(\mathbb{R}) \;\middle|\; \begin{matrix} (\det M_\alpha) \neq 0 \\ \text{and } {}^tMJ_bM = 0 \end{matrix} \right\},$$

which is an open set of $\mathscr{V}$, and then $R_1(\Gamma, G, H)$ is also open in $\operatorname{Hom}(\mathfrak{l}, \mathfrak{g})$. Let $X \in \mathfrak{g}$ and $M \in \eta^{-1}(\eta_\alpha(\mathscr{V}_\alpha))$. Then there exist $A \in GL_k(\mathbb{R})$ and $M' \in M_{2n+1,k}(\mathbb{R})$ such that

$$\mathrm{Ad}_{\exp X} = \begin{pmatrix} 1 & b(X,X_1) & \cdots & b(X,Y_n) \\ \cdot & & & \\ \cdot & & & \\ \cdot & & & \\ 0 & & (I_{2n}) & \end{pmatrix}$$

and $M = M'A$, with

$$M' = \begin{pmatrix} 1 & 0 \\ 0 & A'_1 \\ 0 & B'_1 \\ 0 & A'_2 \\ 0 & B'_2 \end{pmatrix}.$$

Therefore,

$$\mathrm{Ad}_{\exp X} M = M' \begin{pmatrix} 1 & b(X,c_2) & \cdots & b(X,c_n) \\ \cdot & & & \\ \cdot & & & \\ \cdot & & & \\ 0 & & (I_{k-1}) & \end{pmatrix} A,$$

where $c_2, \dots, c_k$ are the $k-1$ last columns vectors of M' and we can see that $\eta(M) = \eta(\mathrm{Ad}_{\exp X} M)$, which proves the G-invariance of $\eta^{-1}(\eta_\alpha(\mathscr{V}_\alpha))$ for any $\alpha \in I_s^1(2n+1,k)$. □

Now we are ready to state our main result in this section concerning the deformation and the moduli space in the case where $\mathfrak{h}$ does not meet the center of $\mathfrak{g}$. We have the following.

Theorem 4.1.14. *Let G be the Heisenberg Lie group, H a connected subgroup, which does not meet the center of G, Γ a rank k Abelian discontinuous group for G/H and $L = \exp(\mathfrak{l})$ its syndetic hull in G. Then*

$$\mathscr{T}(\Gamma,G,H) = \bigcup_{\alpha\in I_{s+1}(2n+1,k)} \mathscr{T}_\alpha \bigcup_{\alpha\in I_s^1(2n+1,k)} \mathscr{T}_\alpha$$

and

$$\mathscr{M}(\Gamma,G,H) = \bigcup_{\alpha\in I_{s+1}(2n+1,k)} \mathscr{M}_\alpha \bigcup_{\alpha\in I_s^1(2n+1,k)} \mathscr{M}_\alpha,$$

where

(1) *For every $\alpha \in I_{s+1}(2n+1,k)$, the set $\mathscr{T}_\alpha$ is open in $\mathscr{T}(\Gamma,G,H)$ and homeomorphic to the product $GL_k(\mathbb{R}) \times \mathscr{V}'_\alpha$ and the set $\mathscr{M}_\alpha$ is open in $\mathscr{M}(\Gamma,G,H)$ and homeomorphic to $GL_k(\mathbb{Z})\backslash GL_k(\mathbb{R}) \times \mathscr{V}'_\alpha$.*
(2) *For every $\alpha \in I_s^1(2n+1,k)$, the set $\mathscr{T}_\alpha$ is open in $\mathscr{T}(\Gamma,G,H)$ and is homeomorphic to the product $O_k \times \mathbb{R}^k \times N_k \times \mathscr{V}_\alpha$. N_k designates here the set of upper triangular unipotent matrices. Likewise, the set $\mathscr{M}_\alpha$ is open in $\mathscr{M}(\Gamma,G,H)$ and homeomorphic to the product $(GL_k(\mathbb{Z})\backslash GL_k(\mathbb{R})/\mathbb{R}^{k-1}) \times \mathscr{V}_\alpha$.*

Proof. We use Lemma 4.1.12 to write the following decomposition of the deformation space:

$$R_1(\Gamma,G,H)/G \cup R_2(\Gamma,G,H)/G.$$

The set $R_2(\Gamma,G,H)$ can be identified to the parameter space $R(\Gamma,G,K)$, where $K = Z(G)H$ and $Z(G)$ is the center of G. Then by Theorem 4.1.11, we get the following description of the quotient set:

$$R_2(\Gamma, G, H)/G = \bigcup_{\alpha \in I_{s+1}(2n+1,k)} \mathscr{T}_\alpha,$$

where for any $\alpha \in I_{s+1}(2n+1,k)$, the set $\mathscr{T}_\alpha$ is open in $\mathscr{T}(\Gamma, G, H)$ and homeomorphic to the product $GL_k(\mathbb{R}) \times \mathscr{V}'_\alpha$. On the other hand, thanks to Lemma 4.1.13, one can write

$$R_1(\Gamma, G, H)/G = \bigcup_{\alpha \in I_s^1(2n+1,k)} \eta^{-1}(\eta_\alpha(\mathscr{V}_\alpha))/G$$

as union of open sets. Let then $\mathscr{T}_\alpha = \eta^{-1}(\eta_\alpha(\mathscr{V}_\alpha))/G$. Recall that the map χ_α is a homeomorphism between $GL_k(\mathbb{R}) \times \eta_\alpha(\mathscr{V}_\alpha)$ and $\eta^{-1}(\eta_\alpha(\mathscr{V}_\alpha))$. Consider the G-action on $GL_k(\mathbb{R}) \times \eta_\alpha(\mathscr{V}_\alpha)$ given by

$$(A, W) \cdot g = (\chi_{\alpha,W}^{-1} \operatorname{Ad}_{g^{-1}} \chi_{\alpha,W} A, W).$$

Then the map χ_α is G-equivariant. Indeed,

$$\begin{aligned} \chi_\alpha((A, W) \cdot g) &= \chi_\alpha(\chi_{\alpha,W}^{-1} \circ \operatorname{Ad}_{g^{-1}} \circ \chi_{\alpha,W} \circ A, W) \\ &= \chi_{\alpha,W} \circ (\chi_{\alpha,W}^{-1} \circ \operatorname{Ad}_{g^{-1}} \circ \chi_{\alpha,W} \circ A) \\ &= \operatorname{Ad}_{g^{-1}} \circ \chi_\alpha(A, W) \\ &= \chi_\alpha(A, W) \cdot g. \end{aligned}$$

For every $\alpha \in I_s^1(2n+1,k)$ and $W \in \eta_\alpha(\mathscr{V}_\alpha)$, we can easily see that there is $A_1, B_1 \in M_{s,k}(\mathbb{R})$, $A_2, B_2 \in M_{n-s,k}(\mathbb{R})$ such that

$$\eta_\alpha^{-1}(W) = \begin{pmatrix} 1 & 0 \\ 0 & A_1 \\ 0 & B_1 \\ 0 & A_2 \\ 0 & B_2 \end{pmatrix}.$$

Then, for $g^{-1} = \exp X$ we have

$$\chi_{\alpha,W}^{-1} \operatorname{Ad}_{g^{-1}} \chi_{\alpha,W} A = (\operatorname{Ad}_{g^{-1}} \eta_\alpha^{-1}(W) A)_\alpha = \begin{pmatrix} 1 & b(X, c_2) \cdots b(X, c_n) \\ \cdot & \\ \cdot & \\ \cdot & \\ 0 & (I_{k-1}) \end{pmatrix} A,$$

where $c_2, \dots, c_k$ are the $k-1$ last columns of $\eta_\alpha^{-1}(W)$. We now consider the free action of $\mathbb{R}^{k-1}$ on $GL_k(\mathbb{R})$ defined by

$$(x_1, \dots, x_{k-1}) \cdot A = \begin{pmatrix} 1 & x_1 \cdots x_{k-1} \\ \cdot & \\ \cdot & \\ \cdot & \\ 0 & (I_{k-1}) \end{pmatrix} A.$$

The subspace W' of W generated by $c_2, \dots, c_k$ is an Abelian subalgebra of dimension $k-1$, that does not meet the center, which means that the map $\mathfrak{g} \to {W'}^*, X \mapsto b(X, \cdot)_{|W'}$ is surjective. It follows therefore that for any $(x_1, \dots, x_{k-1}) \in \mathbb{R}^{k-1}$ there is $X \in \mathfrak{g}$ such that $b(X, c_i) = x_{i-1}$ for all $i = 2, \dots, k$. Therefore, the quotient map

$$\pi : GL_k(\mathbb{R}) \times \eta_\alpha(\mathscr{V}_\alpha) \longrightarrow (GL_k(\mathbb{R}) \times \eta_\alpha(\mathscr{V}_\alpha))/G,$$

factors through the canonical surjection

$$p : GL_k(\mathbb{R}) \times \eta_\alpha(\mathscr{V}_\alpha) \longrightarrow (GL_k(\mathbb{R})/\mathbb{R}^{k-1}) \times \eta_\alpha(\mathscr{V}_\alpha)$$

to give a continuous surjective map

$$f : (GL_k(\mathbb{R})/\mathbb{R}^{k-1}) \times \eta_\alpha(\mathscr{V}_\alpha) \longrightarrow (GL_k(\mathbb{R}) \times \eta_\alpha(\mathscr{V}_\alpha))/G$$

defined by $\pi = f \circ p$ and we can easily see that f is injective. Now G acts continuously on $GL_k(\mathbb{R}) \times \eta_\alpha(\mathscr{V}_\alpha)$, which entails that π is open and that f is a homeomorphism. The following lemma enables us to achieve the proof of the assertion concerning the deformation space.

Lemma 4.1.15. *Fix a positive integer p and regard $\mathbb{R}^p$ as a subgroup of $GL_{p+1}(\mathbb{R})$ through the writing*

$$R_p := \left\{ \begin{pmatrix} 1 & {}^t x \\ 0 & I_p \end{pmatrix} : x \in \mathbb{R}^p \right\}.$$

Then

$$GL_{p+1}(\mathbb{R})/\mathbb{R}^p \simeq O_{p+1} \times \mathbb{R}^{p+1} \times N_p, \tag{4.8}$$

where N_p denotes the totality of upper triangular unipotent matrices.

Proof. Using the Iwasawa decomposition, we have

$$GL_{p+1}(\mathbb{R}) \simeq O_{p+1} \times A_{p+1} \times N_{p+1},$$

where A_{p+1} ($\simeq \mathbb{R}^{p+1}$) denotes the totality of diagonal matrices with positive entries. Thus, we obtain (4.8) because of the decomposition $N_{p+1} \simeq N_p \times R_p$. □

As for the moduli space, recall that $\mathrm{Aut}(\Gamma) = GL_k(\mathbb{Z})$ and if we consider the action of $\mathrm{Aut}(\Gamma)$ on $GL_k(\mathbb{R}) \times \eta_\alpha(\mathscr{V}_\alpha)$ given by $T \cdot (A, W) = (AT^{-1}, W)$, then χ_α is $\mathrm{Aut}(\Gamma)$-equivariant and the result follows immediately. □

The following result provides accurate layering of the deformation space in the context.

Theorem 4.1.16. *Let G be the Heisenberg Lie group, H a connected subgroup, which does not meet the center of G, Γ a rank k Abelian discontinuous group for G/H and $L = \exp(\mathfrak{l})$ its syndetic hull in G. Then*

$$\mathscr{T}(\Gamma, G, H) = \bigcup_{\alpha \in I_{s+1}(2n+1,k)} \mathscr{T}_\alpha \bigcup_{\alpha \in I_s^1(2n+1,k)} \bigcup_{j=1}^{k} \mathscr{T}_{\alpha,j},$$

where for every $\alpha \in I_{s+1}(2n+1, k)$, the set $\mathscr{T}_\alpha$ is open in $\mathscr{T}(\Gamma, G, H)$ and homeomorphic to the product $GL_k(\mathbb{R}) \times \mathscr{V}'_\alpha$. Furthermore, for any $\alpha \in I_s^1(2n+1, k)$ and $j \in \{1, \dots, k\}$, the set $\mathscr{T}_{\alpha,j}$ is open in $\mathscr{T}(\Gamma, G, H)$ and is homeomorphic to the multiple direct product $\mathbb{R}^* \times \mathbb{R}^{k-1} \times GL_{k-1}(\mathbb{R}) \times \mathscr{V}_\alpha$.

Proof. We only need to show that

$$GL_k(\mathbb{R})/\mathbb{R}^{k-1} = \bigcup_{j=1}^{k} \mathscr{U}_i,$$

where for any $j = 1, \dots, k$, $\mathscr{U}_j$ is homeomorphic to $\mathbb{R}^* \times \mathbb{R}^{k-1} \times GL_{k-1}(\mathbb{R})$. Indeed, let $A \in GL_k(\mathbb{R})$ and denote by A_i the matrix obtained from A by deleting the first line and the ith column of A. Then the union of the open sets

$$U_i = \{A \in GL_k(\mathbb{R}), \det A_i \neq 0\}, \quad 1 \leq i \leq k$$

is equal to $GL_k(\mathbb{R})$ and each of them is $\mathbb{R}^{k-1}$-stable. Therefore,

$$GL_k(\mathbb{R})/\mathbb{R}^{k-1} = \bigcup_{i=1}^{k} U_i/\mathbb{R}^{k-1}.$$

The following lemma enables us to achieve the proof.

Lemma 4.1.17. *For $1 \leq i \leq k$, we have $U_i/\mathbb{R}^{k-1} \cong \mathbb{R}^* \times \mathbb{R}^{k-1} \times GL_{k-1}(\mathbb{R})$.*

Proof. Let $A \in U_i$, write

$$A = \begin{pmatrix} a_{11} & \cdots & a_{1k} \\ (a_1) & \cdots & (a_k) \end{pmatrix},$$

where $a_{11}, \dots, a_{1k} \in \mathbb{R}$ and $a_1, \dots, a_k \in \mathbb{R}^{k-1}$. So for $x \in \mathbb{R}^{k-1}$ we have

$$x \cdot A = \begin{pmatrix} a_{11} + \langle x, a_1 \rangle & \cdots & a_{1k} + \langle x, a_k \rangle \\ (a_1) & \cdots & (a_k) \end{pmatrix},$$

where $\langle \cdot, \cdot \rangle$ designates the natural scalar product on $\mathbb{R}^{k-1}$. For $x_0 = -b_i A_i^{-1}$, where $b_i \in \mathbb{R}^{k-1}$ obtained from $(a_{11}, \dots, a_{1k})$ by eliminating of the ith coordinate, we have

$$a_{1j} + \langle x_0, a_j \rangle = 0 \quad \text{for all } j \neq i.$$

This means that A is equivalent (modulo $\mathbb{R}^{k-1}$) to a certain matrix in the set

$$K_i := \left\{ A = \begin{pmatrix} a_{11} & \cdots & a_{1k} \\ (a_1) & \cdots & (a_k) \end{pmatrix} \in GL_k(\mathbb{R}),\ a_{1j} = 0 \text{ for all } j \neq i \right\}.$$

Note that

$$K_i \cong \mathbb{R}^* \times \mathbb{R}^{k-1} \times GL_{k-1}(\mathbb{R}).$$

Let $\pi : U_i \longrightarrow U_i/\mathbb{R}^{k-1}$ be the canonical surjection and $p : U_i \longrightarrow K_i$ the continuous surjection defined by

$$p(A) = \begin{pmatrix} p_1(A) & \cdots & p_k(A) \\ (a_1) & \cdots & (a_k) \end{pmatrix},$$

where $p_j(A) = 0$, if $j \neq i$ and $p_i(A) = a_{1i} - \langle b_i A_i^{-1}, a_i \rangle \neq 0$. Then clearly the map $f : K_i \longrightarrow U_i/\mathbb{R}^{k-1}$ defined by $f(p(A)) = \pi(A)$ is surjective. For the injectivity, let $A, A' \in K_i$ such that $f(A) = f(A')$, which means that there is $x_0 \in \mathbb{R}^{k-1}$ such that $x_0 \cdot A = A'$. But $x \cdot A \in K_i$ only if $x = 0$. Thus, $x_0 = 0$ and $A = A'$. Using the continuity of π and p with the fact that π is open, we obtain the bicontinuity of f. This achieves the proof of the lemma and also of the theorem. □

4.1.5 Case of compact Clifford–Klein forms

We finally describe the deformation and the moduli space for compact Clifford–Klein forms. We have the following.

Theorem 4.1.18. *Let G be the Heisenberg Lie group of dimension $2n + 1$, H a connected Lie subgroup of dimension s and Γ a rank k discontinuous group for G/H. Assume in addition that the Clifford–Klein form $\Gamma \backslash G/H$ is compact. Then:*

(1) *If H contains the center of G, then $k < s$ and*

$$\mathscr{T}(\Gamma, G, H) = GL_k(\mathbb{R}) \times M_{p,q}(\mathbb{R})^2 \times \mathrm{Sym}(\mathbb{R}^q) \times \mathrm{Sp}(p, \mathbb{R})/\,\mathrm{Sp}(p-r, \mathbb{R})$$

and equivalently

$$\mathscr{M}(\Gamma, G, H) = GL_k(\mathbb{R})/GL_k(\mathbb{Z}) \times M_{p,q}(\mathbb{R})^2 \times \mathrm{Sym}(\mathbb{R}^q) \times \mathrm{Sp}(p, \mathbb{R})/\,\mathrm{Sp}(p-r, \mathbb{R}),$$

where for $\mathfrak{h} = \log H$, $q + 1 = \dim(\ker b_{|\mathfrak{h}})$, $2p + q + 1 = \dim \mathfrak{h}$ and $p + q + r = n$.

(2) *If H does not contain the center of G then*

$$\mathscr{T}(\Gamma, G, H) = O_{n+1} \times \mathbb{R}^{n+1} \times N_n \times \mathrm{Sym}(\mathbb{R}^n)$$

and

$$\mathscr{M}(\Gamma, G, H) = (GL_{n+1}(\mathbb{Z})\backslash GL_{n+1}(\mathbb{R})/\mathbb{R}^n) \times \mathrm{Sym}(\mathbb{R}^n).$$

Proof. Note first of all that if Γ is a discontinuous group for G/H and H contains the center of G, then Γ is Abelian and so is its syndetic hull L. By Proposition 1.1.12, we get $k < n + 1$ where k designates the rank of Γ. If $k = 2n + 1 - s$ and $k > s$, then obviously $k > n$, which means that either Γ is not a discontinuous group for G/H or Γ is not Abelian. So, if $k > s$ and $2n + 1 - s = k$, then the parameters space is empty.

Assume now that $k < s$ and $2n + 1 - s = k$ then the set $I_s(2n + 1, k)$ is reduced to the element $\alpha_0 = (s + 1, \ldots, 2n + 1)$. Using Theorem 4.1.11, we get

$$\mathscr{T}(\Gamma, G, H) = GL_k(\mathbb{R}) \times \mathscr{V}'_{\alpha_0}.$$

To conclude, we just have to prove that

$$\mathscr{V}'_{\alpha_0} = M_{p,q}(\mathbb{R})^2 \times \mathrm{Sym}(\mathbb{R}^q) \times \mathrm{Sp}(p, \mathbb{R})/\,\mathrm{Sp}(p-r, \mathbb{R}).$$

Having fixed an adapted basis $\mathscr{B}_{\mathfrak{h}} = \{Z, X_1 \ldots, X_n, Y_1, \ldots, Y_n\}$ of $\mathfrak{g}$ adapted to $\mathfrak{h}$, we consider the vector subspaces:

$$\begin{aligned} V_1' &= \mathbb{R}\text{-span}\{X_1, \ldots, X_p\}, \quad V_1'' = \mathbb{R}\text{-span}\{Y_1, \ldots, Y_p\}, \\ V_0 &= \mathbb{R}\text{-span}\{X_{p+1}, \ldots, X_{p+q}\}, \quad N_0 = \mathbb{R}\text{-span}\{Y_{p+1}, \ldots, Y_{p+q}\}, \\ N_1' &= \mathbb{R}\text{-span}\{X_{p+q+1}, \ldots, X_n\} \quad \text{and} \quad N_1'' = \mathbb{R}\text{-span}\{Y_{p+q+1}, \ldots, Y_n\}. \end{aligned}$$

So we have the following decompositions:

$$\mathfrak{g} = \mathfrak{z} \oplus V_1' \oplus V_1'' \oplus V_0 \oplus N_0 \oplus N_1' \oplus N_1'' \quad \text{and} \quad \mathfrak{h} = \mathfrak{z} \oplus V_1' \oplus V_1'' \oplus V_0. \tag{4.9}$$

Any matrix $M \in \mathscr{V}'_\alpha$ can be written as

$$M = \begin{pmatrix} 0 & 0 & 0 \\ A_1 & A_2 & A_3 \\ B_1 & B_2 & B_3 \\ C_1 & C_2 & C_3 \\ I & 0 & 0 \\ 0 & I & 0 \\ 0 & 0 & I \end{pmatrix} \begin{matrix} \mathfrak{z} \\ V_1' \\ V_1'' \\ V_0, \\ N_0 \\ N_1' \\ N_1'' \end{matrix}$$

where $A_1, B_1 \in M_{p,q}(\mathbb{R})$, $C_1 \in M_{q,q}(\mathbb{R})$, $A_2, A_3, B_2, B_3 \in M_{p,r}(\mathbb{R})$ and $C_2, C_3 \in M_{q,r}(\mathbb{R})$, for $r = n - p - q$. The matrix of b is

$$J_b = \begin{pmatrix} 0 & 0 & 0 & 0 & 0 & 0 & 0 \\ 0 & 0 & -I & 0 & 0 & 0 & 0 \\ 0 & I & 0 & 0 & 0 & 0 & 0 \\ 0 & 0 & 0 & 0 & -I & 0 & 0 \\ 0 & 0 & 0 & I & 0 & 0 & 0 \\ 0 & 0 & 0 & 0 & 0 & 0 & -I \\ 0 & 0 & 0 & 0 & 0 & I & 0 \end{pmatrix}$$

and the condition ${}^tMJ_bM = 0$ is equivalent to the following system:

$$\begin{pmatrix} {}^tB_1A_1 - {}^tA_1B_1 + C_1 - {}^tC_1 & {}^tB_1A_2 - {}^tA_1B_2 + C_2 & {}^tB_1A_3 - {}^tA_1B_3 + C_3 \\ {}^tB_2A_1 - {}^tA_2B_1 - {}^tC_2 & {}^tB_2A_2 - {}^tA_2B_2 & {}^tB_2A_3 - {}^tA_2B_3 - I \\ {}^tB_3A_1 - {}^tA_3B_1 - {}^tC_3 & {}^tB_3A_2 - {}^tA_3B_2 + I & {}^tB_3A_3 - {}^tA_3B_3 \end{pmatrix} = 0.$$

This is in turn equivalent to

$$C_2 = {}^tA_1B_2 - {}^tB_1A_2, \quad C_3 = {}^tA_1B_3 - {}^tB_1A_3, \quad C_1 = \frac{1}{2}({}^tA_1B_1 - {}^tB_1A_1) + D$$

and

$$\begin{pmatrix} {}^tB_2A_2 - {}^tA_2B_2 & {}^tB_2A_3 - {}^tA_2B_3 \\ {}^tB_3A_2 - {}^tA_3B_2 & {}^tB_3A_3 - {}^tA_3B_3 \end{pmatrix} = \begin{pmatrix} 0 & I \\ -I & 0 \end{pmatrix}, \tag{4.10}$$

where $D \in \mathrm{Sym}(\mathbb{R}^q)$ and $A_1, B_1 \in M_{p,q}(\mathbb{R})$. Let

$$Y = \begin{pmatrix} B_2 & B_3 \\ A_2 & A_3 \end{pmatrix} \in M_{2p,2r}(\mathbb{R})$$

and

$$J_m = \begin{pmatrix} 0 & I_m \\ -I_m & 0 \end{pmatrix} \in M_{2m,2m}(\mathbb{R}).$$

Then the condition (4.10) can be written as ${}^tYJ_pY = J_r$ and for

$$U = \{Y \in M_{2p,2r}(\mathbb{R}),\ {}^tYJ_pY = J_r\}$$

we easily see that

$$\mathscr{V}_\alpha' \cong M_{p,q}(\mathbb{R})^2 \times \mathrm{Sym}(\mathbb{R}^q) \times U.$$

To conclude, we finally prove the following lemma.

Lemma 4.1.19. $U \cong \mathrm{Sp}(p,\mathbb{R})/\mathrm{Sp}(p-r,\mathbb{R})$.

Proof. Note first that the symplectic group $\mathrm{Sp}(p,\mathbb{R})$ acts on U by multiplication on the left and its action is transitive. The matrix

$$Y = \begin{pmatrix} I_r & 0 \\ 0 & 0 \\ 0 & I_r \\ 0 & 0 \end{pmatrix}$$

belongs to U and with a direct verification, we get

$$\mathrm{Stab}(Y) = \left\{ P = \begin{pmatrix} I_r & 0 & 0 & 0 \\ 0 & A & 0 & B \\ 0 & 0 & I_r & 0 \\ 0 & C & 0 & D \end{pmatrix},\ \begin{pmatrix} A & B \\ C & D \end{pmatrix} \in \mathrm{Sp}(p-r) \right\} \cong \mathrm{Sp}(p-r).$$

□

We pay attention finally to the case where H does not meet $Z(G)$, the center of G. As we are dealing with compact Clifford–Klein forms, we are obviously submitted to write that $s + k = 2n + 1$, which entails that $I_{s+1}(2n+1,k)$ is empty and $I_s^1(2n+1,k)$ is merely reduced to the single element $\alpha_0 = (1, s+2, \ldots, 2n+1)$. As it stands here, the subgroups H and Γ are Abelian, and we get by Proposition 1.1.12 that $\dim \mathfrak{h} = n$ and $\operatorname{rank}\Gamma = n+1$. Then Theorem 4.1.14 enables us to write that

$$\mathscr{T}(\Gamma, G, H) = GL_{n+1}(\mathbb{R})/\mathbb{R}^n \times \mathscr{V}_{\alpha_0}.$$

Now every matrix M in $\mathscr{V}_{\alpha_0}$ can be written as

$$M = \begin{pmatrix} 1 & 0 \\ 0 & A \\ 0 & I_n \end{pmatrix}$$

for some $A \in M_n(\mathbb{R})$. The relation ${}^tMJ_bM = 0$ is then equivalent to ${}^tA - A = 0$ and the result follows from Lemma 4.1.15. □

A straight consequence of the last theorem, is the following.

Corollary 4.1.20. *Let G be the Heisenberg Lie group, H a connected Lie subgroup of G and Γ an Abelian discontinuous group of G for G/H. Assume in addition that the Clifford–Klein form $\Gamma\backslash G/H$ is compact. Then the deformation space $\mathscr{T}(\Gamma, G, H)$ is endowed with a structure of a smooth manifold.*

4.1.6 Examples

To end this section, we present some enriching examples for which we carry out explicit computations of some chosen layers $\mathscr{T}_\alpha$ and $\mathscr{M}_\alpha$ involved in the description of the deformation and moduli space as we did in the case of compact Clifford–Klein forms where only one single strate occurs. We precise that our computations take into account the precise basis of $\mathfrak{g}$ adapted to $\mathfrak{h}$ and utterly rely on the position of $\mathfrak{h}$ inside $\mathfrak{g}$. All of the matrices considered in the following examples are written in a basis $\mathscr{B}_\mathfrak{h}$ of $\mathfrak{g}$ adapted to $\mathfrak{h}$.

Example 4.1.21. We assume in this first example that $\mathfrak{h}$ does not contain the center of $\mathfrak{g}$ and that $\dim \mathfrak{h} = s$ according to Proposition 1.1.12. Take for instance $k = s + 1$ and $\alpha = (1, s + 2, \ldots, 2s + 1)$. Then any $M \in \mathscr{V}_\alpha$ can be written as

$$M = \begin{pmatrix} 1 & 0 \\ 0 & A \\ 0 & I_s \\ 0 & B \\ 0 & C \end{pmatrix},$$

where $A \in M_s(\mathbb{R})$, $B, C \in M_{n-s,s}(\mathbb{R})$. The matrix J_b of the bilinear form b is then given by

$$J_b = \begin{pmatrix} 0 & 0 & 0 & 0 & 0 \\ 0 & 0 & -I_s & 0 & 0 \\ 0 & I_s & 0 & 0 & 0 \\ 0 & 0 & 0 & 0 & -I_{n-s} \\ 0 & 0 & 0 & I_{n-s} & 0 \end{pmatrix}.$$

By a routine computation, we can easily see that the condition ${}^tMJ_bM = 0$ gives rise to the following equation:

$$-{}^tA + A - {}^tBC + {}^tCB = 0.$$

So $A = -\frac{1}{2}(-{}^tBC + {}^tCB) + D$, for some $D \in \mathrm{Sym}(\mathbb{R}^s)$, and we finally get that

$$\begin{aligned}\mathscr{T}_\alpha &= GL_k(\mathbb{R})/\mathbb{R}^{k-1} \times M^2_{n-s,s}(\mathbb{R}) \times \mathrm{Sym}(\mathbb{R}^s)\\ &\simeq O_k \times \mathbb{R}^k \times N_k \times M^2_{n-s,s}(\mathbb{R}) \times \mathrm{Sym}(\mathbb{R}^s)\end{aligned}$$

and

$$\mathscr{M}_\alpha = GL_k(\mathbb{Z})\backslash GL_k(\mathbb{R})/\mathbb{R}^{k-1} \times M^2_{n-s,s}(\mathbb{R}) \times \mathrm{Sym}(\mathbb{R}^s).$$

We assume henceforth that $\mathfrak{h}$ contains the center of $\mathfrak{g}$ and, therefore, $\dim \mathfrak{h} = 1 + 2p + q$ according to the notation of Proposition 1.1.12.

Example 4.1.22. Take for instance $p + q + k = n$ and $\alpha = (1 + 2p + 2q + k + 1, \dots, 2n + 1)$. Let $M \in \mathscr{V}'_\alpha$. Then

$$M = \begin{pmatrix} 0 \\ A_1 \\ A_2 \\ A_3 \\ A_4 \\ A_5 \\ I_k \end{pmatrix},$$

where A_1, $A_2 \in M_{p,k}(\mathbb{R})$, A_3, $A_4 \in M_{q,k}(\mathbb{R})$ and $A_5 \in M_k(\mathbb{R})$. The matrix of the bilinear form b is

$$J_b = \begin{pmatrix} 0 & 0 & 0 & 0 & 0 & 0 & 0 \\ 0 & 0 & -I_p & 0 & 0 & 0 & 0 \\ 0 & I_p & 0 & 0 & 0 & 0 & 0 \\ 0 & 0 & 0 & 0 & -I_q & 0 & 0 \\ 0 & 0 & 0 & I_q & 0 & 0 & 0 \\ 0 & 0 & 0 & 0 & 0 & 0 & -I_k \\ 0 & 0 & 0 & 0 & 0 & I_k & 0 \end{pmatrix}.$$

So, the condition ${}^tMJ_bM = 0$ is equivalent to the equation

$$-{}^tA_1A_2 + {}^tA_2A_1 - {}^tA_3A_4 + {}^tA_4A_3 + A_5 - {}^tA_5 = 0.$$

Therefore, for A_1, $A_2 \in M_{p,k}(\mathbb{R})$, A_3, $A_4 \in M_{q,k}(\mathbb{R})$ we can take

$$A_5 = -\frac{1}{2}(-A_1^tA_2 + {}^tA_2A_1 - {}^tA_3A_4 + {}^tA_4A_3) + D, \quad \text{with } D \in \mathrm{Sym}(\mathbb{R}^k).$$

and then

$$\mathscr{T}_\alpha \cong GL_k(\mathbb{R}) \times M_{p,k}(\mathbb{R})^2 \times M_{q,k}(\mathbb{R})^2 \times \mathrm{Sym}(\mathbb{R}^k).$$

Example 4.1.23. We still take $p+q+k = n$ and consider $\alpha = (2p+2q+2, \dots, 2p+2q+k+1)$ and let $M \in \mathscr{V}'_\alpha$. Then

$$M = \begin{pmatrix} 0 \\ A_1 \\ A_2 \\ A_3 \\ A_4 \\ I_k \\ A_5 \end{pmatrix},$$

where $A_1, A_2 \in M_{p,k}(\mathbb{R})$, $A_3, A_4 \in M_{q,k}(\mathbb{R})$ and $A_5 \in M_k(\mathbb{R})$. Then the same calculation as in the first example gives

$$\mathscr{T}_\alpha \cong GL_k(\mathbb{R}) \times M_{p,k}(\mathbb{R})^2 \times M_{q,k}(\mathbb{R})^2 \times \mathrm{Sym}(\mathbb{R}^k).$$

Example 4.1.24. Assume now that $k = q$ and take $\alpha = (2p + q + 2, \dots, 2p + 2q + 1)$ and let $M \in \mathscr{V}'_\alpha$. Then

$$M = \begin{pmatrix} 0 \\ A_1 \\ A_2 \\ A_3 \\ I_q \\ A_5 \\ A_6 \end{pmatrix},$$

where $A_1, A_2 \in M_{p,q}(\mathbb{R})$, $A_3 \in M_q(\mathbb{R})$ and $A_5, A_6 \in M_{r,q}(\mathbb{R})$ with $r = n - p - q$. So, the condition ${}^tMJ_bM = 0$ is equivalent to the equation

$$-{}^tA_1A_2 + {}^tA_2A_1 - {}^tA_3 + A_3 - {}^tA_5A_6 + {}^tA_6A_5 = 0.$$

Then as above we have

$$A_3 = -\frac{1}{2}(-{}^tA_1A_2 + {}^tA_2A_1 - {}^tA_5A_6 + {}^tA_6A_5) + D, \quad \text{with } D \in \mathrm{Sym}(\mathbb{R}^q).$$

We get then that

$$\mathscr{T}_\alpha \cong GL_q(\mathbb{R}) \times M_{p,q}(\mathbb{R})^2 \times M_{r,q}(\mathbb{R})^2 \times \mathrm{Sym}(\mathbb{R}^q).$$

Example 4.1.25. Assume finally that $k = q + r$ where $p + q + r = n$ and take $\alpha = (2p + q + 2, \ldots, 2p + 2q + r + 1)$. For $M \in \mathscr{V}'_\alpha$, we have

$$M = \begin{pmatrix} 0 & 0 \\ A_1 & B_1 \\ A_2 & B_2 \\ A_3 & B_3 \\ I_q & 0 \\ 0 & I_r \\ A_4 & B_4 \end{pmatrix},$$

where $A_1, A_2 \in M_{p,q}(\mathbb{R})$, $B_1, B_2 \in M_{p,r}(\mathbb{R})$, $A_3 \in M_q(\mathbb{R})$, $B_3 \in M_{q,r}(\mathbb{R})$, $A_4 \in M_{r,q}(\mathbb{R})$ and $B_4 \in M_r(\mathbb{R})$. Then

$${}^tMJ_bM = \begin{pmatrix} -{}^tA_1A_2 + {}^tA_2A_1 - {}^tA_3 + A_3 & -{}^tA_1B_2 + {}^tA_2B_1 + B_3 + {}^tA_4 \\ -{}^tB_1A_2 + {}^tB_2A_1 - {}^tB_3 - A_4 & -{}^tB_1B_2 + {}^tB_2B_1 + {}^tB_4 - B_4 \end{pmatrix}.$$

Then the condition ${}^tMJ_bM = 0$ is equivalent to

$$\begin{aligned} A_3 &= -\frac{1}{2}(-{}^tA_1A_2 + {}^tA_2A_1) + D, \quad D \in \mathrm{Sym}(\mathbb{R}^q), \\ B_4 &= -\frac{1}{2}({}^tB_1B_2 - {}^tB_2B_1) + D', \quad D' \in \mathrm{Sym}(\mathbb{R}^r), \\ -A_4 &= {}^tB_3 + {}^tB_1A_2 - {}^tB_2A_1. \end{aligned}$$

We obtain therefore that

$$\mathscr{T}_\alpha = GL_k(\mathbb{R}) \times M_{p,q}(\mathbb{R})^2 \times M_{p,r}(\mathbb{R})^2 \times M_{q,r}(\mathbb{R}) \times \mathrm{Sym}(\mathbb{R}^q) \times \mathrm{Sym}(\mathbb{R}^r).$$

4.1.7 A smooth manifold structure on $\mathscr{T}(\Gamma, H_{2n+1}, H)$

Recall the matrix J_b of the bilinear form b written through the basis $\mathscr{B}$. Having fixed a basis $\mathscr{B}_\mathfrak{l}$ of $\mathfrak{l}$, it appears clear that the map

$$\Psi : \mathrm{Hom}(\mathfrak{l}, \mathfrak{g}) \longrightarrow \mathscr{M}_{2n+1,k}(\mathbb{R}), \tag{4.11}$$

which associates to any element of $\mathrm{Hom}(\mathfrak{l}, \mathfrak{g})$ its matrix written through the bases $\mathscr{B}_\mathfrak{l}$ and the basis $\mathscr{B}$ of $\mathfrak{g}$ is a homeomorphism on its range. Throughout the whole text, the set $\mathrm{Hom}(\mathfrak{l}, \mathfrak{g})$ is therefore homeomorphically identified to a set $\mathscr{U}$ of $\mathscr{M}_{2n+1,k}(\mathbb{R})$ and $\mathrm{Hom}^\circ(\mathfrak{l}, \mathfrak{g})$ of all injective homomorphisms to the subset $\mathscr{U}^\circ$ of $\mathscr{U}$ consisting of the totality of matrices in $\mathscr{U}$ of maximal rank. Obviously, the set $\mathscr{U}$ is closed and algebraic in $\mathscr{M}_{2n+1,k}(\mathbb{R})$. In addition, the set $\mathscr{U}^\circ$ is semialgebraic (a difference of two Zariski open sets) and open in $\mathscr{U}$. More on that can be found in [24]. In general, the set $\mathrm{Hom}^\circ(\mathfrak{l}, \mathfrak{g})$

fails in most of the cases to be equipped with a smooth manifold structure. (In the whole text, we only speak about C^∞ smooth structures). When restricted to the Heisenberg setup, the following result will be proved throughout the next coming sections.

Theorem 4.1.26. *Let $\mathfrak{g}$ be the Heisenberg algebra and $\mathfrak{l}$ a subalgebra of $\mathfrak{g}$. Then the semialgebraic set* $\mathrm{Hom}^\circ(\mathfrak{l},\mathfrak{g})$ *is endowed with a smooth manifold structure of dimension* $\frac{k(4n-k+3)}{2}$ *if $\mathfrak{l}$ is Abelian and* $\frac{(k-1)(4n-k+2)}{2}+k$ *otherwise.*

One first pace toward that aim is the following. Let G act on $\mathscr{M}_{2n+1,k}(\mathbb{R})$ by

$$g\cdot M=\mathrm{Ad}_g\cdot M,\quad M\in\mathscr{M}_{2n+1,k}(\mathbb{R}),\quad g\in G.$$

Here, we view Ad_g as a real valued matrix for any $g\in G$. More precisely, for X in $\mathfrak{g}$ with coordinates ${}^t(\gamma,\alpha,\beta)$, $\gamma\in\mathbb{R}$, $\alpha,\beta\in\mathscr{M}_{1,n}(\mathbb{R})$ in the basis $\mathscr{B}$ and

$$M=M(x,A,B)=\begin{pmatrix}x\\A\\B\end{pmatrix}\in\mathscr{M}_{2n+1,k}(\mathbb{R}),$$

$$\mathrm{Ad}_{\exp X}\cdot M=\begin{pmatrix}x-\beta A+\alpha B\\A\\B\end{pmatrix}. \tag{4.12}$$

Taking into account the action of G on $\mathrm{Hom}(\mathfrak{l},\mathfrak{g})$ defined in (3.3), the following lemma is immediate.

Lemma 4.1.27. *The map Ψ defined in (4.11) is G-equivariant. That is, for any $\psi\in$* $\mathrm{Hom}(\mathfrak{l},\mathfrak{g})$ *and $g\in G$, we have $\Psi(g\cdot\psi)=g\cdot\Psi(\psi)$.*

Let $H=\exp\mathfrak{h}$ be a connected subgroup of G of dimension s and Γ a discontinuous group of G for G/H of rank k with a syndetic hull $L=\exp\mathfrak{l}$. For a given matrix M, we adopt henceforth and unless a specific mention M_i and M^i to denote the rows and respectively the columns of M. Our first result concerns the case where Γ is not Abelian. We will prove the following.

Theorem 4.1.28. *Let H be a connected subgroup of the Heisenberg group G and Γ a non-Abelian discontinuous group of G for the homogeneous space G/H. Then the spaces* $\mathrm{Hom}^\circ(\mathfrak{l},\mathfrak{g})$ *and $R(\mathfrak{l},\mathfrak{g},\mathfrak{h})$ (resp.,* $\mathrm{Hom}^\circ(\mathfrak{l},\mathfrak{g})/G$ *and $\mathscr{T}(\mathfrak{l},\mathfrak{g},\mathfrak{h})$) are endowed with a smooth manifold structure of dimension* $\frac{(k-1)(4n-k+2)}{2}+k$ *(resp., of dimension* $\frac{(k-1)(4n-k+2)}{2}+1$*).*

Proof. It is clear that L contains the center of G as it is non-Abelian. Since Γ acts properly on G/H, $\mathfrak{h}\cap\mathfrak{l}=\{0\}$ and then $\mathfrak{h}$ is Abelian. Thus, $\dim\mathfrak{h}\le n$. We can therefore choose a symplectic basis $\mathscr{B}=\{Z,X_1,\dots,X_n,Y_1,\dots,Y_n\}$ of $\mathfrak{g}$ adapted to $\mathfrak{h}$ (according to Proposition 1.1.12 above), being generated by $\mathscr{B}_\mathfrak{h}=\{X_1,\dots,X_s\}$. Let also $\mathscr{B}_\mathfrak{l}=\{e_1=Z,e_2,\dots,e_k\}$ be a basis of $\mathfrak{l}$. As $\mathfrak{l}$ is non-Abelian, there exist $X_0,Y_0\in\mathfrak{l}$ such that $[X_0,Y_0]=Z$. So, for ψ in $\mathrm{Hom}(\mathfrak{l},\mathfrak{g})$, $\psi(Z)=\psi([X_0,Y_0])=[\psi(X_0),\psi(Y_0)]\in\mathfrak{z}(\mathfrak{g})$. Its matrix written through the bases $\mathscr{B}_{\mathfrak{l}}$ and $\mathscr{B}$ is therefore of the form

$$M(a_0, a, A, B) = \begin{pmatrix} a_0 & a \\ 0 & C \end{pmatrix}, \quad a_0 \in \mathbb{R}, \quad a \in \mathbb{R}^{k-1} \tag{4.13}$$

and

$$C(A, B) = \begin{pmatrix} A \\ B \end{pmatrix} \quad \text{and} \quad A, B \in \mathscr{M}_{n,k-1}(\mathbb{R}). \tag{4.14}$$

We opt also for the notation (4.14) for any pair of matrices having a common number of colums. Let E denote the subspace of $\mathscr{M}_{2n+1,k}(\mathbb{R})$ consisting of the totality of matrices M as in equation (4.13). For any $i, j \in \{1, \dots, k\}$, we have

$$[\psi(e_i), \psi(e_j)] = \psi([e_i, e_j]) = b(e_i, e_j)\psi(Z) = a_0 b(e_i, e_j)Z.$$

Denote then by

$$L = \begin{pmatrix} 0 & 0 \\ 0 & L_0 \end{pmatrix}$$

the matrix of the restriction of the bilinear form b to $\mathfrak{l}$. So, $\psi \in \operatorname{Hom}(\mathfrak{l}, \mathfrak{g})$ if and only if ${}^tMJ_bM = a_0L$. The set $\operatorname{Hom}(\mathfrak{l}, \mathfrak{g})$ is therefore homeomorphic to the set $\mathscr{U}$ of all $M(a_0, a, A, B) \in E$ for which

$${}^tAB - {}^tBA = a_0L_0.$$

As such, the set $\operatorname{Hom}^\circ(\mathfrak{l}, \mathfrak{g})$ is homeomorphically identified to the set

$$\mathscr{U}^\circ = \{M(a_0, a, A, B) \in \mathscr{U} : a_0 \neq 0 \text{ and } \operatorname{rk}(C) = k - 1\},$$

where the symbol rk merely designates the rank. Let $\mathscr{A}(p, \mathbb{R})$ denote the subspace of skew-symmetric matrices. We now establish the following elementary result.

Lemma 4.1.29. *Let $M \in \mathscr{M}_{m,p}(\mathbb{R})$ $(p \leq m)$ of maximal rank. Then the map*

$$\varphi_M : \mathscr{M}_{m,p}(\mathbb{R}) \to \mathscr{A}(p, \mathbb{R}), \quad H \mapsto {}^tMH - {}^tHM$$

is surjective.

Proof. Let M_i, $i = 1, \dots, m$ be the rows of the matrix M. Since M is of rank p, there exist $1 \leq i_1 < \cdots < i_p \leq m$ such that the matrix $\tilde{M}$ constituted of the rows $M_{i_1}, \dots, M_{i_p}$ belongs to $GL_p(\mathbb{R})$. Let $\hat{M}$ denote the matrix in $\mathscr{M}_{m-p,p}(\mathbb{R})$ obtained by subtracting from M the rows $M_{i_1}, \dots, M_{i_p}$. Opting for the same notation for a given matrix H in $\mathscr{M}_{m,p}(\mathbb{R})$, we get

$$\varphi_M(H) = {}^t\tilde{M}\tilde{H} - {}^t\tilde{H}\tilde{M} + {}^t\hat{M}\hat{H} - {}^t\hat{H}\hat{M}.$$

The map $\varphi_{\tilde{M}} : \mathscr{M}_p(\mathbb{R}) \to \mathscr{A}(p,\mathbb{R})$, $K \mapsto {}^t\tilde{M}K - {}^tK\tilde{M}$ is surjective as $\ker(\varphi_{\tilde{M}}) = {}^t\tilde{M}^{-1}\mathscr{S}(p,\mathbb{R})$, where $\mathscr{S}(p,\mathbb{R})$ denotes the space of symmetric matrices. Finally, if $S \in \mathscr{A}(p,\mathbb{R})$ and $K \in \mathscr{M}_p(\mathbb{R})$ with $\varphi_{\tilde{M}}(K) = S$, the matrix $H \in \mathscr{M}_{m,p}(\mathbb{R})$ such that $\tilde{H} = K$ and $\hat{H} = 0$ satisfies the equation $\varphi_M(H) = S$. The lemma is thus proved. □

Let $\mathscr{V} = \{M(x,a,A,B) \in E : x \neq 0, \text{and } \operatorname{rk}(C) = k-1\}$ be the open subset of E and Ψ the smooth map

$$\Psi : \mathscr{V} \to \mathscr{A}(k-1,\mathbb{R}), \quad M \mapsto {}^tAB - {}^tBA - xL_0.$$

Clearly, $\mathscr{U}^\circ = \Psi^{-1}(\{0\})$. The goal now is to show that zero is a regular value of the map Ψ. That is, for all $M = M(a_0,a,A,B)$ in $\mathscr{U}^\circ$, the derivative

$$d\Psi_M : E \to \mathscr{A}(k-1,\mathbb{R}), \quad X = M(h_0,h,H,K) \mapsto {}^tHB - {}^tBH + {}^tAK - {}^tKA - h_0L_0$$

is surjective. But

$$d\Psi_M(X) + h_0L_0 = {}^t\begin{pmatrix} -K \\ H \end{pmatrix} C - {}^tC\begin{pmatrix} -K \\ H \end{pmatrix}$$

which is enough to conclude thanks to Lemma 4.1.29. The following lemma conclusively enables us to conclude that $\operatorname{Hom}^\circ(\mathfrak{l},\mathfrak{g})$ is endowed with a smooth manifold structure with the mentioned dimension.

Lemma 4.1.30. *Let $\mathscr{X}$ and $\mathscr{Y}$ be two Hausdorff topological spaces and $h : \mathscr{X} \to \mathscr{Y}$ a homeomorphism. Assume that one of these spaces is endowed with a smooth manifold structure, then so is the second.*

We now focus attention to the space $\operatorname{Hom}^\circ(\mathfrak{l},\mathfrak{g})/G$. For any $X = {}^t(\gamma,\alpha,\beta) \in \mathfrak{g}$ and $M(a_0,a,A,B) \in \mathscr{U}^\circ$, we have as in equation (4.12),

$$\operatorname{Ad}_{\exp X} \cdot M(a_0,a,A,B) = M(a_0, a + \beta A - \alpha B, A, B).$$

Here, $\gamma \in \mathbb{R}$, α and β are in $\mathbb{R}^n$. Now we can easily see that the matrix through the canonical basis of $\mathbb{R}^{2n}$ and $\mathbb{R}^{k-1}$ of the map $\Phi_{A,B} : \mathbb{R}^n \times \mathbb{R}^n \to \mathbb{R}^{k-1}$, $(\alpha,\beta) \mapsto \beta A - \alpha B$ is $M(\Phi_{A,B}) = (-{}^tB, {}^tA)$, which means that $\operatorname{rk}(M(\Phi_{A,B})) = k-1$ and that $\Phi_{A,B}$ is surjective. Let $\widetilde{\mathscr{U}^\circ} = \{M(a_0,a,A,B) \in \mathscr{U}^\circ : a = 0\}$, then the mapping

$$\tilde{\pi} : \mathscr{U}^\circ/G \to \widetilde{\mathscr{U}^\circ}, \quad [M(a_0,a,A,B)] \mapsto M(a_0,0,A,B)$$

is a continuous bijection. In addition, its inverse coincides with the restriction of the canonical quotient surjection to $\widetilde{\mathscr{U}^\circ}$ regarded as a subset of $\mathscr{U}^\circ$. This shows that $\operatorname{Hom}^\circ(\mathfrak{l},\mathfrak{g})/G$ is also endowed with a smooth manifold structure using the same technique. To achieve the proof of the theorem, it is sufficient to show that $R(\mathfrak{l},\mathfrak{g},\mathfrak{h})$ is open in $\operatorname{Hom}^\circ(\mathfrak{l},\mathfrak{g})$. Indeed, let $K = {}^t(0, I_s, 0) \in \mathscr{M}_{2n+1,s}(\mathbb{R})$ be the matrix of the canonical injection from $\mathfrak{h}$ to $\mathfrak{g}$ through the bases $\mathscr{B}_{\mathfrak{h}}$ and $\mathscr{B}$ of $\mathfrak{h}$ and $\mathfrak{g}$, respectively. As G is 2-step, it is immediate that the action of $\exp(\psi(\mathfrak{l}))$ on G/H is proper if and only if

$$\mathrm{rk}(M(a_0, a, A, B) ⋒ K) = \mathrm{rk}(M(a_0, a, C(0, A_{(n-s)}), B) ⋒ K) = k + s, \quad A = \begin{pmatrix} A_{(s)} \\ A_{(n-s)} \end{pmatrix},$$

where the symbol ⋒ merely means the concatenation of the matrices written through $\mathscr{B}$. This leads to the fact that $R(\mathfrak{l}, \mathfrak{g}, \mathfrak{h})$ is homeomorphic to the set

$$\mathscr{W} = \{M(a_0, a, A, B) \in \mathscr{U}^\circ : \mathrm{rk}(C(A_{(n-s)}, B)) = k - 1\},$$

which is semialgebraic and open in $\mathscr{U}^\circ$ as was to be shown. □

The following result is then immediate.

Corollary 4.1.31. *Retain the same assumptions as in Theorem 4.1.28. Then the deformation space $\mathscr{T}(\Gamma, G, H)$ is semialgebraic and homeomorphic to the set*

$$\mathscr{W} \cap \widetilde{\mathscr{U}^\circ} = \{M(a_0, 0, A, B) \in \mathscr{U}^\circ : \mathrm{rk}(C(A_{(n-s)}, B)) = k - 1\}.$$

We also get the following upshot making use of Theorem 4.1.18. As mentioned in the first section, such a result may fail in higher steps nilpotent Lie groups.

Corollary 4.1.32. *Let H be a connected subgroup of the Heisenberg group G and Γ a discontinuous group of G for the homogeneous space G/H such that the Clifford–Klein form $\Gamma\backslash G/H$ is compact. Then the deformation space $\mathscr{T}(\mathfrak{l}, \mathfrak{g}, \mathfrak{h})$ is endowed with a smooth manifold structure.*

The case where *H* meets the center of *G*

We assume henceforth that Γ is Abelian. Our first upshot concerns the case when $\mathfrak{h}$ contains the center of $\mathfrak{g}$. More precisely, we prove the following.

Theorem 4.1.33. *Let H be a connected subgroup of the Heisenberg group G containing the center of G and Γ a discontinuous subgroup of G for the homogeneous space G/H. Then the parameter and the deformation spaces are semialgebraic sets and both of them are endowed with a smooth manifold structure of dimension $\frac{k(4n-k+3)}{2}$ and $\frac{k(4n-k+1)}{2}$, respectively.*

Proof. We keep the same notation as in Theorem 4.1.28. Note first that $\mathfrak{l}$ is Abelian as $\mathfrak{h}$ contains the center of $\mathfrak{g}$. Let $\mathscr{B} = \{Z, X_1, \dots, X_n, Y_1, \dots, Y_n\}$ be a basis of $\mathfrak{g}$ such that the subalgebra $\mathfrak{h}$ is generated by $\mathscr{B}_{\mathfrak{h}} = \{Z, X_1, \dots, X_p, Y_1, \dots, Y_q\}$ with $\dim \mathfrak{h} = s = p + q + 1 =$

$r+1$. We also fix a basis $\mathscr{B}_{\mathfrak{l}} = \{e_1, \ldots, e_k\}$ of $\mathfrak{l}$. Let ψ be a linear map from $\mathfrak{l}$ to $\mathfrak{g}$ and

$$M(x,A,B) = \begin{pmatrix} x \\ C \end{pmatrix} \in \mathscr{M}_{2n+1,k}(\mathbb{R}); \quad x \in \mathbb{R}^k, \quad C = \begin{pmatrix} A \\ B \end{pmatrix} \quad \text{and} \quad A,B \in \mathscr{M}_{n,k}(\mathbb{R})$$

its matrix through the basis $\mathscr{B}_{\mathfrak{l}}$ and $\mathscr{B}$. The space of all matrices of this form is noted by E. Then $\mathrm{Hom}(\mathfrak{l},\mathfrak{g})$ is clearly homeomorphic to the set $\mathscr{U} = \{M(x,A,B) \in E : {}^tBA - {}^tAB = 0\}$. This entails therefore that $R(\mathfrak{l},\mathfrak{g},\mathfrak{h})$ is open in $\mathrm{Hom}^\circ(\mathfrak{l},\mathfrak{g})$ and homeomorphic to the semialgebraic set

$$\mathscr{W} = \{M(x,A,B) \in \mathscr{U} : \mathrm{rk}(C_{(2n-r)}) = k\},$$

according to the writing

$$C = \begin{pmatrix} C_{(r)} \\ C_{(2n-r)} \end{pmatrix},$$

where $C_{(i)} \in \mathscr{M}_{i,k}(\mathbb{R})$ for $i \in \{r, 2n-r\}$. Looking at the action of G, which only affects the central variables and following the same arguments as in Theorem 4.1.28. One immediately gets that the deformation space $\mathscr{T}(\mathfrak{l},\mathfrak{g},\mathfrak{h})$ is homeomorphic to the semialgebraic set $\{M(x,A,B) \in \mathscr{W} : x = 0\}$. These spaces are endowed with smooth manifold structures making use of the same technique of the proof of Theorem 4.1.28. □

The case where *H* does not meet the center of *G*

We now pay attention to the case where $\mathfrak{h}$ does not meet the center of $\mathfrak{g}$. As we shall remark, here is the single case where the deformation space may fail to be endowed with a smooth manifold structure, even with a Hausdorff topology. Our first result in this direction is the following.

Theorem 4.1.34. *Let H be a connected subgroup of the Heisenberg group G, which does not meet the center of G and Γ an Abelian discontinuous group of G for the homogeneous space G/H. Then the parameter space is described by a union of finitely many semialgebraic and G-invariant smooth manifolds. Accordingly, the deformation space is described by a union of finitely many semialgebraic smooth manifolds.*

Proof. We fix a basis $\mathscr{B} = \{Z, X_1, \ldots, X_n, Y_1, \ldots, Y_n\}$ of $\mathfrak{g}$ such that $\mathfrak{h} = \mathbb{R}\text{-span}\{X_1, \ldots, X_s\}$. Denote by $K = {}^t(0\ I_s\ 0)$ the matrix of the canonical injection from $\mathfrak{h}$ to $\mathfrak{g}$. As earlier, if ψ denotes a linear map from $\mathfrak{l}$ to $\mathfrak{g}$ with a matrix M, the action of $\exp\psi(\mathfrak{l})$ on G/H is proper if and only if

$$\mathrm{rk}(\mathrm{Ad}_{\exp X} \cdot M ⋒ K) = k + s \quad \text{for any } X \in \mathfrak{g}.$$

Keeping the same notation as previously, the set $\mathrm{Hom}(\mathfrak{l},\mathfrak{g})$ is clearly homeomorphic to the set $\mathscr{U} = \{M(x,A,B) \in E : {}^tBA - {}^tAB = 0\}$ and $\mathrm{Hom}^\circ(\mathfrak{l},\mathfrak{g})$ to the set $\mathscr{U}^\circ$ of matrices in $\mathscr{U}$ of maximal rank. For any $X = {}^t(y, \alpha, \beta) \in \mathfrak{g}$ and $M(x, A, B) \in \mathscr{U}$, we have as in equation (4.12), $\mathrm{Ad}_{\exp X} \cdot M(x, A, B) = M(x - \beta A + \alpha B, A, B)$. Here, $y \in \mathbb{R}$, α and β are in $\mathbb{R}^n$. We adopt for the notation

$$\begin{pmatrix} A \\ B \end{pmatrix} = \begin{pmatrix} A_{(s)} \\ C \end{pmatrix},$$

where $A_{(s)} \in \mathscr{M}_{s,k}(\mathbb{R})$ and $C \in \mathscr{M}_{2n-s,k}(\mathbb{R})$ and also $M(x, A_{(s)}, C)$ instead of $M(x, A, B)$. Then the fact that

$$\mathrm{rk}(\mathrm{Ad}_{\exp X} \cdot M \Cap K) = k + s \quad \text{for all } X \in \mathfrak{g}$$

is equivalent to say that

$$\mathrm{rk}(M(x - \beta A + \alpha B, A_{(s)}, C) \Cap K) = k + s, \tag{4.15}$$

for all α and β in $\mathbb{R}^n$, which is also equivalent to the fact that

$$\mathrm{rk}(M(x + wA_{(s)}, 0, C)) = k \tag{4.16}$$

for all $w \in \mathbb{R}^s$. Indeed, this follows from simple manipulations of the rows of the matrix in equation (4.15). Assume for a while that $\mathrm{rk}(M(0, A, B)) = k$, then condition (4.16) is equivalently expressed to mean that $\mathrm{rk}(M(y, 0, C)) = k$, for all $y \in \mathbb{R}^k$ which in turn gives $\mathrm{rk}(C) = k$. Otherwise, the matrix $M(0, A, B)$ is found to be of rank $k-1$ and so is C. Let

$$I_{2n-s}^{k-1} = \{(i_1, \ldots, i_{k-1}) \in \mathbb{N}^{k-1} : 1 \leq i_1 < \cdots < i_{k-1} \leq 2n - s\}.$$

There exists therefore $\theta \in I_{2n-s}^{k-1}$ such that the matrix

$$\begin{pmatrix} x_1 + wA_{(s)}^1 & \cdots & x_k + \omega A_{(s)}^k \\ & C_\theta & \end{pmatrix}$$

is regular for all $w \in \mathbb{R}^s$, where $A_{(s)}^i$, $i = 1, \ldots, k$ designate the columns of $A_{(s)}$ and C_θ is the submatrix of C formed by the rows C_i, $i \in \{i_1, \ldots, i_{k-1}\}$. This fact is also equivalent to

$$\sum_{i=1}^{s} w_i \sum_{j=1}^{k} (-1)^{j+1} a_{ij} \det(C_\theta^{(j)}) + \sum_{j=1}^{k} (-1)^{j+1} x_j \det(C_\theta^{(j)}) \neq 0, \tag{4.17}$$

for any reals $w_1, \ldots, w_s$. Here, $C_\theta^{(j)}$ merely designates the resulting matrix after the substitution of the j-column. Obviously, (4.17) is equivalent to

$$\sum_{j=1}^{k} (-1)^{j+1} x_j \det(C_\theta^{(j)}) \neq 0 \quad \text{and} \quad \sum_{j=1}^{k} (-1)^{j+1} a_{ij} \det(C_\theta^{(j)}) = 0 \quad (i \in \{1, \ldots, s\}),$$

which gives in turn that

$$\det\begin{pmatrix} x \\ C_\theta \end{pmatrix} \neq 0 \quad \text{and} \quad \det\begin{pmatrix} A_i \\ C_\theta \end{pmatrix} = 0$$

for any $i \in \{1,\dots,s\}$. According to our circumstances, the last condition obviously holds as $\mathrm{rk}(C) = k-1$. Let now

$$R_0(\mathfrak{l},\mathfrak{g},\mathfrak{h}) = \{M(x,A,B) \in \mathscr{U}^0 : \mathrm{rk}(M(0,A,B)) = k\}$$

and for $\theta \in I_{2n-s}^{k-1}$, $j \in \{1,\dots,k\}$,

$$\begin{aligned} R_\theta^j(\mathfrak{l},\mathfrak{g},\mathfrak{h}) = \{M(x,A,B) \in \mathscr{U}^0 : \mathrm{rk}(M(x,0,C_\theta)) = \mathrm{rk}(M(0,A,B)) + 1 = k \\ \text{and } \det(C_\theta^{(j)}) \neq 0\}. \end{aligned}$$

Let $I(s,k) = I_{2n-s}^{k-1} \cup \{0\}$. The parameter space splits as follows:

$$\mathscr{R}(\mathfrak{l},\mathfrak{g},\mathfrak{h}) = \bigcup_{\substack{\theta\in I(s,k)\\ j\in\{1,\dots,k\}}} R_\theta^j(\mathfrak{l},\mathfrak{g},\mathfrak{h}).$$

When $\theta = 0$, we also opt for the notation $R_0^j(\mathfrak{l},\mathfrak{g},\mathfrak{h}) = R_0(\mathfrak{l},\mathfrak{g},\mathfrak{h})$ for any $j \in \{1,\dots,k\}$. Furthermore, the layers $R_\theta^j(\mathfrak{l},\mathfrak{g},\mathfrak{h}), \theta \in I(s,k)$ are G-invariant and semialgebraic sets. Indeed, the point is obviously clear for $R_0(\mathfrak{l},\mathfrak{g},\mathfrak{h})$. The G-invariance of $R_\theta^j(\mathfrak{l},\mathfrak{g},\mathfrak{h})$ comes from the conditions $\mathrm{rk}(C) = \mathrm{rk}(C_\theta^{(j)}) = k-1$ and $\mathrm{rk}(M(x-\beta A+\alpha B,0,C)) = k$, for all $\alpha,\beta \in \mathbb{R}^n$. This gives in turn that

$$\mathscr{T}(\mathfrak{l},\mathfrak{g},\mathfrak{h}) = \bigcup_{\substack{\theta\in I(s,k)\\ j\in\{1,\dots,k\}}} \mathscr{T}_\theta^j(\mathfrak{l},\mathfrak{g},\mathfrak{h}),$$

where $\mathscr{T}_\theta^j(\mathfrak{l},\mathfrak{g},\mathfrak{h}) = R_\theta^j(\mathfrak{l},\mathfrak{g},\mathfrak{h})/G$. We now show that $\mathscr{T}_\theta^j(\mathfrak{l},\mathfrak{g},\mathfrak{h})$ is homeomorphic to

$$\mathscr{T}_\theta^j = \{M(x,A,B) \in R_\theta^j(\mathfrak{l},\mathfrak{g},\mathfrak{h}) : x_i = 0 \text{ for all } i \neq j\} \tag{4.18}$$

if $\theta \in I_{2n-s}^{k-1}$ and $j \in \{1,\dots,k\}$ and to $\mathscr{T}_0 = \{M(0,A,B) \in R_0(\mathfrak{l},\mathfrak{g},\mathfrak{h})\}$ when $\theta = 0$. The last fact holds similarly as it was the situation in Theorems 4.1.28 and 4.1.33. Let now $\theta = (i_1,\dots,i_{k-1}) \in I_{2n-s}^{k-1}$, $j \in \{1,\dots,k\}$ and $M(x,A,B) \in R_\theta^j$. We pick

$$Y = (-y_1,\dots,-y_n,y_{n+1},\dots,y_{2n}) = (-u,v) \in \mathbb{R}^{2n}$$

regarded as a vector of $\mathfrak{g} \simeq \mathbb{R}^{2n+1}$ such that

$$Y_\theta = (y_{i_1},\dots,y_{i_{k-1}}) = -(x_1,\dots,\check{x}_j,\dots,x_k)(C_\theta^{(j)})^{-1}$$

and $y_i = 0$ otherwise. Let also $\xi_j = x_j + vA^j - uB^j$ and $\xi = (0,\dots,0,\xi_j,0,\dots,0) \in \mathbb{R}^k$. Clearly, $\mathrm{Ad}_{\exp(Y)} M(x,A,B) = M(\xi,A,B)$ and the map

$$\psi_\theta^j : R_\theta^j(\mathfrak{l},\mathfrak{g},\mathfrak{h}) \to \mathscr{T}_\theta^j; \quad M(x,A,B) \mapsto M(\xi,A,B)$$

is continuous. Furthermore, $G \cdot M \cap \mathscr{T}_\theta^j$ consists of one single point. Indeed, let $M = M(\xi,A,B)$ and $M' = M(\xi',A,B)$; $\xi = (0,\dots,0,\xi_j,0,\dots,0)$, $\xi' = (0,\dots,0,\xi_j',0,\dots,0)$ be points in $G \cdot M \cap \mathscr{T}_\theta^j$. There exists $X = (u,v) \in \mathbb{R}^{2n}$ regarded as earlier as a vector of $\mathfrak{g}$ such that $M = \mathrm{Ad}_{\exp X} M'$. This entails that $\xi_j' = \xi_j + vA^j - uB^j$ and $vA^i - uB^i = 0$ for any $i \neq j$. Since

$$\mathrm{rk}(M(0,A,B)) = \mathrm{rk}(C) = \mathrm{rk}(C_\theta^{(j)}) = k-1,$$

$M^j(0,A,B)$ turns out to be a linear combination of $M^i(0,A,B)$ $(i \neq j)$. So, $\xi_j' = \xi_j$ and, therefore, $M = M'$. It follows that the quotient map

$$\widetilde{\psi}_\theta^j : \mathscr{T}_\theta^j(\mathfrak{l},\mathfrak{g},\mathfrak{h}) \to \mathscr{T}_\theta^j; \quad [M] \mapsto M(\xi,A,B) \tag{4.19}$$

is a continuous bijection and its inverse coincides with the restriction of the canonical surjective map $\mathscr{R}(\mathfrak{l},\mathfrak{g},\mathfrak{h}) \to \mathscr{T}(\mathfrak{l},\mathfrak{g},\mathfrak{h})$, as was to be shown. Using similar details as in the proof of Theorem 4.1.26, we get right away that $R_\theta^j(\mathfrak{l},\mathfrak{g},\mathfrak{h})$ and $\mathscr{T}_\theta^j$ are smooth manifolds for $\theta \in I_{2n-s}^{k-1}$ and $j \in \{1,\dots,k\}$. Then Lemma 4.1.30 enables us to conclude. □

We next prove the following.

Proposition 4.1.35. *Let H be a connected subgroup of the Heisenberg group G, which does not meet the center of G and Γ a maximal discontinuous group of G for the homogeneous space G/H. Then the parameter space $\mathscr{R}(\mathfrak{l},\mathfrak{g},\mathfrak{h})$ and the deformation space $\mathscr{T}(\mathfrak{l},\mathfrak{g},\mathfrak{h})$ are endowed with smooth manifold structures of dimension $\frac{(n+1)(3n+2)}{2}$ and $\frac{3n(n+1)}{2}+1$, respectively.*

Proof. The assertion concerning the parameter space $\mathscr{R}(\mathfrak{l},\mathfrak{g},\mathfrak{h})$ is a direct consequence from above, as it is open in $\mathrm{Hom}^0(\mathfrak{l},\mathfrak{g})$. As in Subsection 4.1.7, $R_0(\mathfrak{l},\mathfrak{g},\mathfrak{h})$ is empty and

$$\mathscr{T}(\mathfrak{l},\mathfrak{g},\mathfrak{h}) = \bigcup_{\substack{\theta \in I_{2n-s}^n \\ j \in \{1,\dots,n+1\}}} \mathscr{T}_\theta^j(\mathfrak{l},\mathfrak{g},\mathfrak{h})$$

with $\mathscr{T}_\theta^j(\mathfrak{l},\mathfrak{g},\mathfrak{h}) = R_\theta^j(\mathfrak{l},\mathfrak{g},\mathfrak{h})/G$. Note that

$$R_\theta^j(\mathfrak{l},\mathfrak{g},\mathfrak{h}) = \{M = M(x,A,B) \in R(\mathfrak{l},\mathfrak{g},\mathfrak{h}) : P_{j,\theta}(M) := \det(C_\theta^{(j)}) \neq 0\},$$

which is open in $R(\mathfrak{l},\mathfrak{g},\mathfrak{h})$ for any $\theta \in I_{2n-s}^n$ and $j \in \{1,\dots,n+1\}$. We first need to show that $\mathscr{T}(\mathfrak{l},\mathfrak{g},\mathfrak{h})$ is a Hausdorff space. Let then $[M_1] \neq [M_2]$ be points of $\mathscr{T}(\mathfrak{l},\mathfrak{g},\mathfrak{h})$ representing two orbits $G \cdot M_1$ and $G \cdot M_1$ of $\mathscr{R}(\mathfrak{l}, \mathfrak{g}, \mathfrak{h})$. When both the points belong to the same stratum $\mathscr{T}_\theta^j(\mathfrak{l}, \mathfrak{g}, \mathfrak{h})$, we are done as being open and endowed with a smooth manifold structure. Assume then $[M_i] \in \mathscr{T}_{\theta_i}^{j_i}(\mathfrak{l}, \mathfrak{g}, \mathfrak{h}), i = 1, 2$ and $(\theta_1, j_1) \neq (\theta_2, j_2)$ with $P_{j_1,\theta_1}(M_1) \neq 0$, $P_{j_1,\theta_1}(M_2) = 0$ and $P_{j_2,\theta_2}(M_2) \neq 0$. Let $\varepsilon = \min\{|P_{j_1,\theta_1}(M_1)|, |P_{j_2,\theta_2}(M_2)|\} > 0$,

$$\mathscr{U}_1 = \left\{M = M(x, A, B) \in R(\mathfrak{l}, \mathfrak{g}, \mathfrak{h}) : |P_{j_1,\theta_1}(M) - P_{j_1,\theta_1}(M_1)| < \frac{\varepsilon}{2}\right\}$$

and

$$\mathscr{U}_2 = \left\{M = M(x, A, B) \in R(\mathfrak{l}, \mathfrak{g}, \mathfrak{h}) : |P_{j_1,\theta_1}(M) - P_{j_1,\theta_1}(M_1)| > \frac{\varepsilon}{2}\right\}.$$

Then $\mathscr{U}_1$ and $\mathscr{U}_2$ are open and G-invariant disjoint sets containing M_1 and M_2, respectively, which is enough to conclude. Up to this step, $\mathscr{T}_\theta^j(\mathfrak{l}, \mathfrak{g}, \mathfrak{h})$ being open in $\mathscr{T}(\mathfrak{l}, \mathfrak{g}, \mathfrak{h})$ for any $\theta \in I_{2n-s}^n$ and $j \in \{1, \ldots, n+1\}$, we also have to prove that $\mathscr{T}(\mathfrak{l}, \mathfrak{g}, \mathfrak{h})$ is equipped with a manifold structure. As in Theorem 4.1.34, the map $\widetilde{\psi}_\theta^j$ defined as in equation (4.19) is a homeomorphism for all θ and j. We need to show that for all θ, α in I_{2n-s}^n and j, l in $\{1, \ldots, n+1\}$ such that $\mathscr{T}_\theta^j(\mathfrak{l}, \mathfrak{g}, \mathfrak{h}) \cap \mathscr{T}_\alpha^l(\mathfrak{l}, \mathfrak{g}, \mathfrak{h}) \neq \emptyset$, the transition map

$$\widetilde{\psi}_\theta^j \circ (\widetilde{\psi}_\alpha^l)^{-1} : U_{\theta,\alpha}^{j,l} = \widetilde{\psi}_\alpha^l(\mathscr{T}_\theta^j(\mathfrak{l}, \mathfrak{g}, \mathfrak{h}) \cap \mathscr{T}_\alpha^l(\mathfrak{l}, \mathfrak{g}, \mathfrak{h})) \to V_{\theta,\alpha}^{j,l} = \widetilde{\psi}_\theta^j(\mathscr{T}_\theta^j(\mathfrak{l}, \mathfrak{g}, \mathfrak{h}) \cap \mathscr{T}_\alpha^l(\mathfrak{l}, \mathfrak{g}, \mathfrak{h}))$$

is a C^∞ map. Indeed, when $j = l$, it is easy to see that $U_{\theta,\alpha}^{j,l} = V_{\theta,\alpha}^{j,l} = \mathscr{T}_\theta^j \cap \mathscr{T}_\alpha^j$ and that $\widetilde{\psi}_\theta^j \circ (\widetilde{\psi}_\alpha^j)^{-1}$ is the identity map. Otherwise, if $j \neq l$ ($j < l$, e. g.), then

$$U_{\theta,\alpha}^{j,l} = \{M(\xi, A, B) \in \mathscr{T}_\alpha^l : \mathrm{rk}(C_\theta^{(j)}) = n\} \quad \text{and} \quad V_{\theta,\alpha}^{j,l} = \{M(\xi, A, B) \in \mathscr{T}_\theta^j : \mathrm{rk}(C_\alpha^{(l)}) = n\}$$

and for $M = M(\xi, A, B) \in U_{\theta,\alpha}^{j,l}$, $\xi = (0, \ldots, 0, \xi_l, 0, \ldots, 0)$

$$\widetilde{\psi}_\theta^j \circ (\widetilde{\psi}_\alpha^l)^{-1}(M) = M(\eta, A, B),$$

where $\eta = (0, \ldots, 0, \eta_j, 0, \ldots, 0)$ with $\eta_j = vA^j - uB^j$ and

$$(-u, v) = Y = (-y_1, \ldots, -y_n, y_{n+1}, \ldots, y_{2n})$$

such that $Y_\theta = (y_{i_1}, \ldots, y_{i_n}) = -(0, \ldots, \check{0}, \ldots, 0, \xi_l, 0, \ldots, 0)(C_\theta^j)^{-1}$ and $y_i = 0$ otherwise. This achieves the proof of the proposition. □

Remark 4.1.36. We close this section with the following example, which treats the case where Γ is of rank one. In this case, we clearly have

$$\mathscr{R}(\mathfrak{l}, \mathfrak{g}, \mathfrak{h}) = R_1(\mathfrak{l}, \mathfrak{g}, \mathfrak{h}) \coprod R_2(\mathfrak{l}, \mathfrak{g}, \mathfrak{h}),$$

where

$$R_1(\mathfrak{l},\mathfrak{g},\mathfrak{h}) = \{M(x,0,0) \in \mathscr{M}_{2n+1,1}(\mathbb{R}) : x \in \mathbb{R}^\times\}$$

and

$$R_2(\mathfrak{l},\mathfrak{g},\mathfrak{h}) = \left\{M(x,a,b) \in \mathscr{M}_{2n+1,1} : a,b \in \mathscr{M}_{1,n}(\mathbb{R}) \text{ and } \|b\|^2 + \sum_{i=s+1}^{n} a_i^2 \neq 0\right\}.$$

Here, a_i, $i = 1,\dots,n$ designate the coordinates of a and $\|\cdot\|$ the Euclidian norm. This description shows that the G-orbits in $\mathscr{R}(\mathfrak{l},\mathfrak{g},\mathfrak{h})$ are not separated by open neighborhoods, which entails that $\mathscr{T}(\mathfrak{l},\mathfrak{g},\mathfrak{h})$ fails to be a Hausdorff space. As in the proof of Theorem 4.1.34, $\mathscr{T}_2(\mathfrak{l},\mathfrak{g},\mathfrak{h}) = R_2(\mathfrak{l},\mathfrak{g},\mathfrak{h})/G$ is homeomorphic to

$$\mathscr{T}_2 = \left\{M(0,a,b) \in \mathscr{M}_{2n+1,1} : a,b \in \mathscr{M}_{1,n}(\mathbb{R}) \text{ and } \|b\|^2 + \sum_{i=s+1}^{n} a_i^2 \neq 0\right\}$$

and $\mathscr{T}_1(\mathfrak{l},\mathfrak{g},\mathfrak{h})$ to $\mathscr{T}_1 = \{M(x,0,0) \in \mathscr{M}_{2n+1,1}(\mathbb{R}) : x \in \mathbb{R}^\times\}$. Clearly, $\mathscr{T}_i(\mathfrak{l},\mathfrak{g},\mathfrak{h})$, $i = 1,2$ are equipped with smooth manifold structures and that $\mathscr{T}_2(\mathfrak{l},\mathfrak{g},\mathfrak{h})$ is open and dense in $\mathscr{T}(\mathfrak{l},\mathfrak{g},\mathfrak{h})$.

Case where Γ is Abelian and maximal

We assume in this subsection that Γ is maximal. We first pose the following.

Definition 4.1.37. Let $\mathfrak{g}$ be a Lie algebra. A maximal Abelian subalgebra of $\mathfrak{g}$ is an Abelian subalgebra of $\mathfrak{g}$ of maximal dimension. Maximal subalgebras are not unique and obviously contain the center of $\mathfrak{g}$.

Definition 4.1.38. Let $G = \exp\mathfrak{g}$ be an exponential solvable Lie group. An Abelian discrete subgroup Γ is said to be maximal, if its rank is maximal (hence equals $n+1$ in the Heisenberg case). The Lie algebra of its syndetic hull is therefore maximal according to Definition 4.1.37, and thus contains the center of $\mathfrak{g}$.

We now show the following elementary result which entails that $R_0(\mathfrak{l},\mathfrak{g},\mathfrak{h})$ is empty in this case.

Lemma 4.1.39. *Let A and B be matrices in $\mathscr{M}_{n,n+1}(\mathbb{R})$ such that ${}^tBA - {}^tAB = 0$. Then* $\operatorname{rk}(M(0,A,B)) \leq n$.

Proof. Equation ${}^tBA - {}^tAB = 0$ is equivalent to ${}^tB^iA^j - {}^tA^iB^j = 0$ for all $i,j = 1,\dots,n+1$ or equivalently the matrix product ${}^tM^i(0,B,-A)M^j(0,A,B)$ vanishes for all $i,j = 1,\dots,n+1$. Let $\mathfrak{a}$ and $\mathfrak{b}$ be the subspace of $\mathbb{R}^{2n}$ spanned with the vectors $\{M^j(0,A,B),\ j = 1,\dots,n\}$ and $\{M^j(0,-B,A),\ j = 1,\dots,n+1\}$, respectively. Then clearly $\mathfrak{b} \subset \mathfrak{a}^\perp$, where $\perp$ means the orthogonality symbol with respect to the Euclidian scalar product of $\mathbb{R}^{2n}$. Then

$\dim(\mathfrak{b}) = \dim(\mathfrak{a}) \leq \dim(\mathfrak{a}^{\perp}) = 2n - \dim(\mathfrak{a})$. This entails that $\dim(\mathfrak{a}) \leq n$ as was to be shown. □

This gives in turn that

$$\mathscr{T}(\mathfrak{l}, \mathfrak{g}, \mathfrak{h}) = \bigcup_{\substack{\theta \in I_{2n-s}^{n} \\ j \in \{1,\dots,n+1\}}} \mathscr{T}_{\theta}^{j}(\mathfrak{l}, \mathfrak{g}, \mathfrak{h})$$

with $\mathscr{T}_{\theta}^{j}(\mathfrak{l}, \mathfrak{g}, \mathfrak{h}) = R_{\theta}^{j}(\mathfrak{l}, \mathfrak{g}, \mathfrak{h})/G$ which is homeomorphic to the set $\mathscr{T}_{\theta}^{j}$ as defined in equation (4.18).

The following is a direct consequence of Theorem 4.1.34.

Corollary 4.1.40. *Retain the same assumptions as in Theorem 4.1.34. Then G acts on the parameter space $\mathscr{R}(\mathfrak{l}, \mathfrak{g}, \mathfrak{h})$ with constant dimension orbits if and only if Γ is maximal.*

The following corollary stems from Theorems 4.1.28, 4.1.33 and 4.1.34.

Corollary 4.1.41. *Let H be a connected subgroup of the Heisenberg group G and Γ a discontinuous group of G for the homogeneous space G/H. Then the parameter space $\mathscr{R}(\mathfrak{l}, \mathfrak{g}, \mathfrak{h})$ is semialgebraic (as in Definition 3.2.7).*

4.1.8 Proof of Theorem 4.1.26

When Γ is non-Abelian, the theorem is already proved in Theorem 4.1.28. We now focus attention to the case where Γ is Abelian. Let U be the open subset of $\mathscr{M}_{2n+1,k}(\mathbb{R})$ of matrices of rank k and consider the smooth map $\varphi : U \to \mathscr{A}(k, \mathbb{R})$, $M \mapsto {}^{t}AB - {}^{t}BA$. One has $\mathrm{Hom}^{\circ}(\mathfrak{l}, \mathfrak{g}) = \varphi^{-1}(\{0\})$. So, it is sufficient to show that 0 is a regular value of φ. Let $M = M(x, A, B) \in U$. The derivative $d\varphi_M : \mathscr{M}_{2n+1,k}(\mathbb{R}) \longrightarrow \mathscr{A}(k, \mathbb{R})$ can be easily written as

$$d\varphi_M(M(a, H, K)) = {}^{t}HB - {}^{t}BH + {}^{t}AK - {}^{t}KA.$$

Just in case the matrix $M_0 = M(0, A, B)$ is of maximal rank, we are already done thanks to Lemma 4.1.29. Assume otherwise that M_0 has $k-1$ as its rank. Let $\widetilde{M_0}$ be the resulting matrix from M_0 after subtracting its first column. We also adopt here the same notation for any given matrix. We can and do admit that $\widetilde{M_0}$ is of rank $k - 1$. Then

$${}^{t}AB - {}^{t}BA = \begin{pmatrix} 0 & u \\ -{}^{t}u & {}^{t}\tilde{A}\tilde{B} - {}^{t}\tilde{B}\tilde{A} \end{pmatrix},$$

where $u = {}^{t}A^{1}\tilde{B} - {}^{t}B^{1}\tilde{A}$. So,

$$d\varphi_M(M(a, H, K)) = \begin{pmatrix} 0 & v \\ -{}^{t}v & {}^{t}\tilde{H}\tilde{B} - {}^{t}\tilde{B}\tilde{H} + {}^{t}\tilde{A}\tilde{K} - {}^{t}\tilde{K}\tilde{A} \end{pmatrix},$$

where $v = {}^tH^1\tilde{B} - {}^tB^1\tilde{H} + {}^tA^1\tilde{K} - {}^tK^1\tilde{A}$. Let

$$S = \begin{pmatrix} 0 & S_1 \\ -{}^tS_1 & \tilde{S} \end{pmatrix} \in \mathscr{A}(k, \mathbb{R}),$$

where $\tilde{S}$ belongs to $\mathscr{A}(k-1, \mathbb{R})$ and S_1 to $\mathscr{M}_{1,k-1}(\mathbb{R})$. As

$$\operatorname{rk}\begin{pmatrix} \tilde{A} \\ \tilde{B} \end{pmatrix} = k - 1,$$

there exist by Lemma 4.1.29 some matrices $\tilde{H}$ and $\tilde{K}$ in $\mathscr{M}_{n,k-1}(\mathbb{R})$ such that $\tilde{S} = {}^t\tilde{H}\tilde{B} - {}^t\tilde{B}\tilde{H} + {}^t\tilde{A}\tilde{K} - {}^t\tilde{K}\tilde{A}$. On the other hand, as

$$\operatorname{rk}\begin{pmatrix} \tilde{A} \\ \tilde{B} \end{pmatrix} = k - 1,$$

there exist H^1 and K^1 such that

$${}^t\begin{pmatrix} -K^1 \\ H^1 \end{pmatrix}\begin{pmatrix} \tilde{A} \\ \tilde{B} \end{pmatrix} = S_1 + {}^t\begin{pmatrix} B^1 \\ -A^1 \end{pmatrix}\begin{pmatrix} \tilde{H} \\ \tilde{K} \end{pmatrix}.$$

This proves that $d\varphi_M$ is surjective. □

4.1.9 From H_{2n+1} to the product group $H_{2n+1} \times H_{2n+1}$

Let $\mathfrak{p} = \mathfrak{h}_{2n+1} \times \mathfrak{h}_{2n+1}$ be the Lie algebra of P. For $(X, Y), (X', Y') \in \mathfrak{p}$, the Lie bracket $[(X, Y), (X', Y')]$ is given by

$$[(X, Y), (X', Y')] = ([X, X'], [Y, Y']).$$

It is then obvious that the group P is a 2-step nilpotent Lie group. If

$$\exp : \mathfrak{h}_{2n+1} \to H_{2n+1}$$

is the exponential map, we denote also by exp the exponential map from $\mathfrak{p}$ on P, which is simply defined by

$$\exp(X, Y) = (\exp X, \exp Y), \quad (X, Y) \in \mathfrak{p}.$$

The center $\mathfrak{z}$ of the Lie algebra $\mathfrak{p}$ is two-dimensional and is generated by the vectors $\{(Z, 0), (Z, Z)\}$. We denote by Δ the diagonal subgroup of P:

$$\Delta = \{(x, x),\ x \in H_{2n+1}\},$$

which is a $(2n+1)$-dimensional subgroup of P. Its Lie algebra

$$\mathfrak{D} = \{(X, X),\ X \in \mathfrak{h}_{2n+1}\}$$

is generated by $\{(Z, Z), (X_i, X_i), (Y_i, Y_i), \ 1 \leq i \leq n\}$ so that $\mathfrak{p}$ is decomposed as

$$\mathfrak{p} = \mathbb{R}(Z, 0) \oplus \mathfrak{D} \oplus \mathfrak{k},$$

where $\mathfrak{k}$ is the vector space spanned by $\{(X_i, 0), (Y_i, 0),\ 1 \leq i \leq n\}$. The group P acts on $\mathfrak{p}$ by the adjoint action Ad_P such that for all $(X, X'), (Y, Y') \in \mathfrak{p}$,

$$\begin{aligned}\mathrm{Ad}_{\exp(X,X')}(Y, Y') &= e^{\mathrm{ad}_{(X,X')}}(Y, Y')\\ &= (Y, Y') + \mathrm{ad}_{(X,X')}(Y, Y')\\ &= (Y, Y') + ([X, Y], [X', Y']).\end{aligned}$$

From now on, we fix the following basis of $\mathfrak{p}$:

$$\mathscr{B} = \{(Z, 0), (Z, Z), (X_i, X_i), (Y_i, Y_i), (X_i, 0), (Y_i, 0),\ 1 \leq i \leq n\}.$$

For $(X, Y), (X', Y') \in \mathfrak{p}$, one has

$$\begin{aligned}[(X, Y), (X', Y')] &= ([X, X'], [Y, Y'])\\ &= b_1((X, Y), (X', Y'))(Z, 0) + b_2((X, Y), (X', Y'))(Z, Z),\end{aligned}$$

where b_1 and b_2 are the skew symmetric bilinear forms on $\mathfrak{p}$ defining its Lie bracket. By a routine computation, the matrices J_{b_1} and J_{b_2} of b_1 and b_2 written through the basis $\mathscr{B}$ are given by

$$J_{b_1} = \begin{pmatrix} 0_{\mathscr{M}_2(\mathbb{R})} & 0 & 0\\ 0 & 0 & J\\ 0 & J & J\end{pmatrix}, \quad J_{b_2} = \begin{pmatrix} 0_{\mathscr{M}_2(\mathbb{R})} & 0 & 0\\ 0 & J & 0\\ 0 & 0 & 0\end{pmatrix} \quad \text{and} \quad J = \begin{pmatrix} 0 & I_n\\ -I_n & 0\end{pmatrix}.$$

Let now Γ be a discontinuous group for P/Δ and $L = \exp \mathfrak{l}$ its syndetic hull. Since $\mathfrak{l} \cap \mathfrak{D} = \{0\}$, then $(Z, Z) \notin \mathfrak{l}$ that is $\dim(\mathfrak{l} \cap \mathfrak{z}) \leq 1$. We fix then a basis $\mathscr{B}_0 = \{e_1, \ldots, e_k\}$ of $\mathfrak{l}$ where e_1 is a generator of $\mathfrak{l} \cap \mathfrak{z}$ whenever this space is not trivial. For all $1 \leq i, j \leq k$, we have

$$[e_i, e_j] = \alpha_{ij} e_1.$$

Denote by K the matrix $(\alpha_{ij})_{i,j=\{1,\ldots,k\}}$. This matrix equals zero when $\mathfrak{l}$ is Abelian. Otherwise, K has the form

$$K = \begin{pmatrix} 0 & 0\\ 0 & K_0\end{pmatrix}, \tag{4.20}$$

where $K_0 \in \mathscr{M}_{k-1}(\mathbb{R})$ is a skew symmetric matrix. Let ψ be a linear map from $\mathfrak{l}$ to $\mathfrak{p}$. Its matrix written through the bases $\mathscr{B}_0$ and $\mathscr{B}$ is

$$M = M(x,y,A,B) = \begin{pmatrix} x \\ y \\ A \\ B \end{pmatrix}, \quad x,y \in \mathscr{M}_{1,k}(\mathbb{R}), \quad A = \begin{pmatrix} A_1 \\ A_2 \end{pmatrix}, \quad B = \begin{pmatrix} B_1 \\ B_2 \end{pmatrix}, \tag{4.21}$$

where A_1, A_2, B_1, B_2 are in $\mathscr{M}_{n,k}(\mathbb{R})$.

If $\psi \in \mathrm{Hom}(\mathfrak{l}, \mathfrak{p})$, we have

$$[\psi(e_i), \psi(e_j)] = \psi([e_i, e_j]) \quad \text{for all } 1 \le i,j \le k. \tag{4.22}$$

When $\mathfrak{l}$ is not Abelian, there exist $V_1, V_2 \in \mathfrak{l}$ such that $[V_1, V_2] = e_1$. So,

$$\psi(e_1) = \psi([V_1, V_2]) = [\psi(V_1), \psi(V_2)] \in \mathfrak{z}.$$

There exist therefore x_0 and y_0 in $\mathbb{R}$ such that $\psi(e_1) = x_0(Z,0) + b_0(Z,Z)$. Hence,

$$\begin{aligned} \psi([e_i, e_j]) &= \alpha_{ij}\psi(e_1) \\ &= \alpha_{ij}(x_0(Z,0) + y_0(Z,Z)) \\ &= x_0\alpha_{ij}(Z,0) + y_0\alpha_{ij}(Z,Z). \end{aligned}$$

Then equations (4.22) read in term of matrices:

$${}^tMJ_{b_1}M = x_0K \quad \text{and} \quad {}^tMJ_{b_2}M = y_0K,$$

which is equivalent to

$${}^tAJB + {}^tBJA + {}^tBJB = x_0K \quad \text{and} \quad {}^tAJA = y_0K,$$

or also

$$\begin{cases} {}^tB_2A_1 - {}^tA_1B_2 + {}^tA_2B_1 - {}^tB_1A_2 + {}^tB_2B_1 - {}^tB_1B_2 = x_0K, \\ {}^tA_2A_1 - {}^tA_1A_2 = y_0K. \end{cases} \tag{4.23}$$

If $\mathfrak{l}$ is Abelian, we have the same equations but with $K = 0$ as it was mentioned above.

Let

$$\mathscr{E} = \{M = M(x,y,A,B) \in \mathscr{M}_{4n+2,k}(\mathbb{R}) \text{ satisfying (4.23)}\}.$$

Since the spaces $\mathrm{Hom}(\Gamma, P)$ and $\mathrm{Hom}(\mathfrak{l}, \mathfrak{p})$ are homeomorphic, the following proposition is straightforward.

Proposition 4.1.42. *The space* $\operatorname{Hom}(\Gamma, P)$ is homeomorphic to $\mathscr{E}$.

Recall that P acts on $\operatorname{Hom}(\mathfrak{l}, \mathfrak{p})$ by

$$\psi^g = \operatorname{Ad}_{g^{-1}} \circ \psi, \quad g \in P, \quad \psi \in \operatorname{Hom}(\mathfrak{l}, \mathfrak{p})$$

and P acts on $\mathscr{E}$ by

$$M^g = \operatorname{Ad}_{g^{-1}} \cdot M, \quad g \in P \quad \text{and} \quad M \in \mathscr{E}.$$

The identification $\Phi : \psi \longmapsto M(\psi, \mathscr{B}_0, \mathscr{B})$ is a homeomorphism, which is P-equivariant. That is,

$$(\Phi(\psi))^g = \Phi(\psi^g), \quad g \in P.$$

Let $X \in \mathfrak{p}$ with coordinates ${}^t(a, b, \alpha, \beta, \gamma, \delta)$ through the basis $\mathscr{B}$, where $a, b \in \mathbb{R}$ and $\alpha, \beta, \gamma, \delta \in \mathbb{R}^{2n}$. For a matrix $M = M(x, y, A, B)$ as in (4.21), we have

$$\operatorname{Ad}_{\exp X} \cdot M = M(x - \delta A_1 + \gamma A_2 - (\beta + \delta)B_1 + (\alpha + \gamma)B_2, y - \beta A_1 + \alpha A_2, A, B).$$

Putting $u = (-(\beta + \delta), \alpha + \gamma)$ and $v = (-\beta, \alpha)$, we get

$$\operatorname{Ad}_{\exp X} \cdot M = M(x + (u - v)A + uB, y + vA, A, B). \tag{4.24}$$

This allows us to give the following description of the parameter space. We then have the following.

Proposition 4.1.43. *The parameter space* $\mathscr{R}(\mathfrak{l}, \mathfrak{p}, \mathfrak{D})$ *is homeomorphic to the space*

$$\left\{ M \in \mathscr{E} : \operatorname{rk}\begin{pmatrix} x - \omega A \\ B \end{pmatrix} = k \text{ for any } \omega \in \mathbb{R}^{2n} \right\}. \tag{4.25}$$

Proof. Let $\psi \in \operatorname{Hom}(\mathfrak{l}, \mathfrak{p})$ and $M = M(x, y, A, B)$ its associated matrix. The action of $\exp \psi(\mathfrak{l})$ on P/Δ is proper if and only if for all $X \in \mathfrak{p}$,

$$\operatorname{Ad}_{\exp X}(\psi(\mathfrak{l})) \cap \mathfrak{D} = \{0\}.$$

That is, $\operatorname{Ad}_{\exp X} M(Y) \notin \mathfrak{D}$ for all $Y \in \mathfrak{l} \setminus \{0\}$ and for all $X \in \mathfrak{p}$. According to (4.24), this means that for all $u, v \in \mathbb{R}^{2n}$ and $Y \in \mathfrak{l} \setminus \{0\}$,

$$\begin{pmatrix} x + (u - v)A + uB \\ B \end{pmatrix} Y \neq 0,$$

which is in turn equivalent to

$$\mathrm{rk}\begin{pmatrix} x + (u - v)A + uB \\ B \end{pmatrix} = \mathrm{rk}\begin{pmatrix} x + (u - v)A \\ B \end{pmatrix} = k$$

for all $u, v \in \mathbb{R}^{2n}$. □

Assume now that the subgroup Γ is non-Abelian, which means that $\mathfrak{l} \cap \mathfrak{z} = \mathbb{R}e_1$. Note that if $\psi \in \mathrm{Hom}(\mathfrak{l}, \mathfrak{p})$, $\psi(e_1)$ belongs to $\mathfrak{z}$ and so its associated matrix is $M = M(x, y, A, B)$ such that $x = (x_0, \tilde{x})$, $y = (y_0, \tilde{y})$, $A = (0, A_0)$ and $B = (0, B_0)$ where $x_0, y_0 \in \mathbb{R}$, $\tilde{x}, \tilde{y} \in \mathbb{R}^{k-1}$ and $A_0, B_0 \in \mathscr{M}_{2n,k-1}(\mathbb{R})$. So, we have the following.

Corollary 4.1.44. *If the subgroup Γ is non-Abelian, the parameter space $\mathscr{R}(\mathfrak{l}, \mathfrak{p}, \mathfrak{D})$ is homeomorphic to the space*

$$\{M = M((x_0, \tilde{x}), (y_0, \tilde{y}), (0, A_0), (0, B_0)) \in \mathscr{E} : x_0 \neq 0 \text{ and } \mathrm{rk}(B_0) = k - 1\}. \tag{4.26}$$

Proof. It is sufficient to notice that

$$\begin{aligned} \mathrm{rk}(M) = k &\Longleftrightarrow \mathrm{rk}\begin{pmatrix} x_0 & x - \omega A_0 \\ 0 & B_0 \end{pmatrix} = k \quad \text{for all } \omega \in \mathbb{R}^{2n} \\ &\Longleftrightarrow x_0 \neq 0 \quad \text{and} \quad \mathrm{rk}(B_0) = k - 1. \end{aligned}$$

□

4.2 Case of 2-step nilpotent Lie groups

We assume henceforth that $\mathfrak{g}$ is a 2-step nilpotent Lie algebra, $\mathfrak{l}$ a subalgebra of $\mathfrak{g}$ and $\mathfrak{z}$ the (nontrivial) center of $\mathfrak{g}$. We consider the decompositions

$$\mathfrak{g} = \mathfrak{z} \oplus \mathfrak{g}' \quad \text{and} \quad \mathfrak{l} = [\mathfrak{l}, \mathfrak{l}] \oplus \mathfrak{l}', \tag{4.27}$$

where $\mathfrak{g}'$ (resp., $\mathfrak{l}'$) designates a subspace of $\mathfrak{g}$ (of $\mathfrak{l}$, resp.). Any $\varphi \in \mathscr{L}(\mathfrak{l}, \mathfrak{g})$ can be written as

$$\varphi = \begin{pmatrix} A_\varphi & B_\varphi \\ C_\varphi & D_\varphi \end{pmatrix}, \tag{4.28}$$

where $A_\varphi \in \mathscr{L}([\mathfrak{l}, \mathfrak{l}], \mathfrak{z})$, $B_\varphi \in \mathscr{L}(\mathfrak{l}', \mathfrak{z})$, $C_\varphi \in \mathscr{L}([\mathfrak{l}, \mathfrak{l}], \mathfrak{g}')$ and $D_\varphi \in \mathscr{L}(\mathfrak{l}', \mathfrak{g}')$. Let

$$\varphi' = \begin{pmatrix} A_\varphi & 0 \\ 0 & D_\varphi \end{pmatrix}. \tag{4.29}$$

We first remark the following assertion.

Lemma 4.2.1. *Any element $\varphi \in \mathscr{L}(\mathfrak{l}, \mathfrak{g})$ is a Lie algebra homomorphism if and only if $C_\varphi = 0$ and $\varphi' \in \mathrm{Hom}(\mathfrak{l}, \mathfrak{g})$.*

Proof. We point out first that if $\varphi \in \mathrm{Hom}(\mathfrak{l}, \mathfrak{g})$, then $\varphi([\mathfrak{l}, \mathfrak{l}]) = [\varphi(\mathfrak{l}), \varphi(\mathfrak{l})] \subset \mathfrak{z}$, in particular $C_\varphi = 0$ and

$$\varphi = \begin{pmatrix} A_\varphi & B_\varphi \\ 0 & D_\varphi \end{pmatrix}.$$

Let $x = x_1 + y_1$ and $x' = x_2 + y_2 \in \mathfrak{l}$ where $x_i \in [\mathfrak{l}, \mathfrak{l}]$ and $y_i \in \mathfrak{l}'$, then

$$\begin{aligned}
\varphi \in \mathrm{Hom}(\mathfrak{l}, \mathfrak{g}) &\Leftrightarrow [\varphi(x), \varphi(x')] = \varphi([x, x']) \\
&\Leftrightarrow [A_\varphi(x_1) + B_\varphi(y_1) + D_\varphi(y_1), A_\varphi(x_2) + B_\varphi(y_2) + D_\varphi(y_2)] \\
&\qquad = \varphi([x_1 + y_1, x_2 + y_2]) \\
&\Leftrightarrow [D_\varphi(y_1), D_\varphi(y_2)] = \varphi([y_1, y_2]) = A_\varphi([y_1, y_2]) \\
&\Leftrightarrow [D_\varphi(y_1), D_\varphi(y_2)] = A_\varphi([y_1, y_2]) \\
&\Leftrightarrow \varphi' \in \mathrm{Hom}(\mathfrak{l}, \mathfrak{g}).
\end{aligned}$$

□

For any $g \in G$, let as earlier Ad_g designate the adjoint representation, which can be written making use the decomposition (4.37) as

$$\mathrm{Ad}_g = \begin{pmatrix} I_\mathfrak{z} & \sigma(g) \\ 0 & I_{\mathfrak{g}'} \end{pmatrix} \tag{4.30}$$

for some map $\sigma : G \to \mathscr{L}(\mathfrak{g}', \mathfrak{z})$. Here, $I_\mathfrak{z}$ and $I_{\mathfrak{g}'}$ denote the identity maps of $\mathfrak{z}$ and $\mathfrak{g}'$, respectively.

Lemma 4.2.2. *The map σ is a group homomorphism. In particular, the range of σ is a linear subspace of $\mathscr{L}(\mathfrak{g}', \mathfrak{z})$.*

Proof. Clearly, $\mathrm{Ad} : G \to GL(\mathfrak{g})$ is a group homomorphism. Then for g and $g' \in G$,

$$\begin{aligned}
\mathrm{Ad}_{gg'} = \mathrm{Ad}_g \mathrm{Ad}_{g'} &\Leftrightarrow \begin{pmatrix} I_\mathfrak{z} & \sigma(gg') \\ 0 & I_{\mathfrak{g}'} \end{pmatrix} = \begin{pmatrix} I_\mathfrak{z} & \sigma(g) \\ 0 & I_{\mathfrak{g}'} \end{pmatrix} \begin{pmatrix} I_\mathfrak{z} & \sigma(g') \\ 0 & I_{\mathfrak{g}'} \end{pmatrix} \\
&\Leftrightarrow \begin{pmatrix} I_\mathfrak{z} & \sigma(gg') \\ 0 & I_{\mathfrak{g}'} \end{pmatrix} = \begin{pmatrix} I_\mathfrak{z} & \sigma(g) + \sigma(g') \\ 0 & I_{\mathfrak{g}'} \end{pmatrix} \\
&\Leftrightarrow \sigma(gg') = \sigma(g) + \sigma(g').
\end{aligned}$$

In particular, σ is continuous and G is connected which means therefore that $\mathrm{Im}(\sigma)$ is a connected (linear) subgroup of the linear space $\mathscr{L}(\mathfrak{g}', \mathfrak{z})$. □

Let $\mathfrak{h}$ be a subalgebra of $\mathfrak{g}$ and consider the decompositions

$$\mathfrak{g} = (\mathfrak{z} \cap \mathfrak{h}) \oplus \mathfrak{z}' \oplus \mathfrak{h}' \oplus V \quad \text{and} \quad \mathfrak{l} = [\mathfrak{l}, \mathfrak{l}] \oplus \mathfrak{l}', \tag{4.31}$$

where $\mathfrak{z} = (\mathfrak{z} \cap \mathfrak{h}) \oplus \mathfrak{z}'$, $\mathfrak{h} = (\mathfrak{z} \cap \mathfrak{h}) \oplus \mathfrak{h}'$ and V is a linear subspace supplementary to $(\mathfrak{z} \cap \mathfrak{h}) \oplus \mathfrak{z}' \oplus \mathfrak{h}'$ in $\mathfrak{g}$. Then with respect to these decompositions, the adjoint representation Ad_g, $g \in G$ can once again written down as

$$\mathrm{Ad}_g = \begin{pmatrix} I_1 & 0 & \sigma_1(g) & \delta_1(g) \\ 0 & I_2 & \sigma_2(g) & \delta_2(g) \\ 0 & 0 & I_3 & 0 \\ 0 & 0 & 0 & I_4 \end{pmatrix},$$

where

$$\sigma(g) = \begin{pmatrix} \sigma_1(g) & \delta_1(g) \\ \sigma_2(g) & \delta_2(g) \end{pmatrix},$$

$\sigma_1(g) \in \mathscr{L}(\mathfrak{h}', \mathfrak{z} \cap \mathfrak{h})$, $\delta_1(g) \in \mathscr{L}(V, \mathfrak{z} \cap \mathfrak{h})$, $\sigma_2(g) \in \mathscr{L}(\mathfrak{h}', \mathfrak{z}')$ and $\delta_2(g) \in \mathscr{L}(V, \mathfrak{z}')$ (here I_1, I_2, I_3 and I_4 designate the identity maps of $\mathfrak{z} \cap \mathfrak{h}$, $\mathfrak{z}'$, $\mathfrak{h}'$ and V, resp.). This leads to the fact that any element of $\mathrm{Hom}(\mathfrak{l}, \mathfrak{g})$ can be written accordingly, as a matrix

$$\varphi(A, B) = \begin{pmatrix} A_1 & B_1 \\ A_2 & B_2 \\ 0 & B_3 \\ 0 & B_4 \end{pmatrix},$$

where

$$A = \begin{pmatrix} A_1 \\ A_2 \end{pmatrix} \quad \text{and} \quad B = \begin{pmatrix} B_1 \\ B_2 \\ B_3 \\ B_4 \end{pmatrix}. \tag{4.32}$$

Here, $A_1 \in \mathscr{L}([\mathfrak{l}, \mathfrak{l}], \mathfrak{z} \cap \mathfrak{h})$, $A_2 \in \mathscr{L}([\mathfrak{l}, \mathfrak{l}], \mathfrak{z}')$, $B_1 \in \mathscr{L}(\mathfrak{l}', \mathfrak{z} \cap \mathfrak{h})$, $B_2 \in \mathscr{L}(\mathfrak{l}', \mathfrak{z}')$, $B_3 \in \mathscr{L}(\mathfrak{l}', \mathfrak{h}')$ and $B_4 \in \mathscr{L}(\mathfrak{l}', V)$.

We can now state our first result.

Theorem 4.2.3. *Let G be a 2-step nilpotent Lie group, H a connected subgroup of G and Γ a discontinuous group for G/H. Then the parameter space $\mathscr{R}(\mathfrak{l}, \mathfrak{g}, \mathfrak{h})$ splits into the disjoint union $\mathscr{R}_1 \sqcup \mathscr{R}_2$, where*

$$\mathscr{R}_1 := \left\{ \varphi(A, B) \in \mathrm{Hom}(\mathfrak{l}, \mathfrak{g}) \;\middle|\; \begin{array}{l} \mathrm{rk}(B_4) = \dim(\mathfrak{l}') \\ \textit{and } \mathrm{rk}(A_2) = \dim([\mathfrak{l}, \mathfrak{l}]) \end{array} \right\}$$

and

$$\mathscr{R}_2 := \left\{ \varphi(A,B) \in \operatorname{Hom}(\mathfrak{l},\mathfrak{g}) \;\middle|\; \begin{array}{l} \operatorname{rk}(B_4) < \dim(\mathfrak{l}') \textit{ and for all } g \in G, \\ \operatorname{rk}\begin{pmatrix} A_2 & B_2 + \sigma_2(g)B_3 + \delta_2(g)B_4 \\ 0 & B_4 \end{pmatrix} = \dim(\mathfrak{l}) \end{array} \right\}.$$

Proof. As the pair (G, H) has the Lipsman property (cf. Theorem 2.2.2), the equality set (3.5) enables us to state that

$$\mathscr{R}(\mathfrak{l},\mathfrak{g},\mathfrak{h}) = \left\{ \varphi(A,B) \in \operatorname{Hom}(\mathfrak{l},\mathfrak{g}) \;\middle|\; \begin{array}{l} \dim \varphi(A,B)(\mathfrak{l}) = \dim(\mathfrak{l}), \\ \operatorname{Ad}_g \varphi(A,B)(\mathfrak{l}) \cap \mathfrak{h} = \{0\} \text{ for all } g \in G \end{array} \right\}.$$

Now

$$\operatorname{Ad}_g \varphi(A,B) = \begin{pmatrix} A_1 & B_1 + \sigma_1(g)B_3 + \delta_1(g)B_4 \\ A_2 & B_2 + \sigma_2(g)B_3 + \delta_2(g)B_4 \\ 0 & B_3 \\ 0 & B_4 \end{pmatrix}, \tag{4.33}$$

which means that the condition $[\operatorname{Ad}_g \varphi(A,B)](\mathfrak{l}) \cap \mathfrak{h} = \{0\}$ is equivalent to the fact that

$$\operatorname{rk}\begin{pmatrix} A_2 & B_2 + \sigma_2(g)B_3 + \delta_2(g)B_4 \\ 0 & B_4 \end{pmatrix} = \dim(\mathfrak{l}),$$

which is in turn equivalent to

$$\operatorname{rk}(B_4) = \dim(\mathfrak{l}') \quad \text{and} \quad \operatorname{rk}(A_2) = \dim([\mathfrak{l},\mathfrak{l}])$$

or

$$\operatorname{rk}(B_4) < \dim(\mathfrak{l}') \quad \text{and} \quad \operatorname{rk}\begin{pmatrix} A_2 & B_2 + \sigma_2(g)B_3 + \delta_2(g)B_4 \\ 0 & B_4 \end{pmatrix} = \dim(\mathfrak{l}). \qquad \square$$

4.2.1 Description of the deformation space $\mathscr{T}(\mathfrak{l},\mathfrak{g},\mathfrak{h})$

We fix in this section our objects and we keep the same notation. Let $\mathfrak{g}$ and $\mathfrak{l}$ be as above and

$$\operatorname{Hom}_1(\mathfrak{l},\mathfrak{g}) := \{\varphi' : \varphi \in \operatorname{Hom}(\mathfrak{l},\mathfrak{g})\},$$

where φ' is as in (4.29). The group G acts on $\operatorname{Hom}_1(\mathfrak{l},\mathfrak{g}) \times \mathscr{L}(\mathfrak{l},\mathfrak{g})$ through the following law:

$$g \cdot (\varphi', B_\varphi) = (\varphi', B_\varphi + \sigma(g)D_\varphi),$$

where B_φ and D_φ are as in formula (4.28) and σ is as in (4.30).

Lemma 4.2.4. *The map*

$$\begin{array}{rcl}\psi: \mathrm{Hom}(\mathfrak{l}, \mathfrak{g}) & \longrightarrow & \mathrm{Hom}_1(\mathfrak{l}, \mathfrak{g}) \times \mathscr{L}(\mathfrak{l}', \mathfrak{z}) \\ \varphi & \longmapsto & (\varphi', B_\varphi)\end{array}$$

is a G-equivariant homeomorphism.

Proof. The fact that ψ is a well-defined homeomorphism comes directly from Lemma 4.2.1. Let $g \in G$ and $\varphi \in \mathrm{Hom}(\mathfrak{l}, \mathfrak{g})$, then

$$\begin{aligned}\psi(g \cdot \varphi) &= \psi(\mathrm{Ad}_g \varphi) \\ &= (\varphi', B_\varphi + \sigma(g) D_\varphi) \\ &= g \cdot \psi(\varphi),\end{aligned}$$

which proves the lemma. □

4.2.2 Decomposition of $\mathrm{Hom}_1(\mathfrak{l}, \mathfrak{g})$

As in Lemma 4.2.2, the group $\sigma(G)$ is a linear space. For any $\varphi \in \mathrm{Hom}(\mathfrak{l}, \mathfrak{g})$, let l_φ be the linear map defined by

$$\begin{array}{rcl}l_\varphi : \sigma(G) & \longrightarrow & \mathscr{L}(\mathfrak{l}', \mathfrak{z}) \\ \sigma(g) & \longmapsto & \sigma(g) D_\varphi.\end{array}$$

The range of l_φ is a linear subspace of $\mathscr{L}(\mathfrak{l}', \mathfrak{z})$ and the orbit $G \cdot (\varphi', B_\varphi) = (\varphi', B_\varphi + \mathrm{Im}(l_\varphi))$. Let $m = \dim(\mathscr{L}(\mathfrak{l}', \mathfrak{z}))$ and $q = \dim(\sigma(G))$. For $t = 0, 1, \ldots, q$, we define the sets

$$\mathrm{Hom}_1^t(\mathfrak{l}, \mathfrak{g}) := \{\varphi' \in \mathrm{Hom}_1(\mathfrak{l}, \mathfrak{g}) : \mathrm{rk}(l_\varphi) = t\}.$$

Then clearly,

$$\mathrm{Hom}_1(\mathfrak{l}, \mathfrak{g}) = \bigsqcup_{t=0}^{q} \mathrm{Hom}_1^t(\mathfrak{l}, \mathfrak{g}).$$

We fix a basis $\{e_1, \ldots, e_m\}$ of $\mathscr{L}(\mathfrak{l}', \mathfrak{z})$ and let

$$I(m, m-t) = \{(i_1, \ldots, i_{m-t}); 1 \le i_1 < \cdots < i_{m-t} \le m\}.$$

For $\beta = (i_1, \ldots, i_{m-t}) \in I(m, m-t)$, we consider the subspace $V_\beta := \bigoplus_{j=1}^{m-t} \mathbb{R} e_{i_j}$.

Proposition 4.2.5. *For any $\varphi \in \mathrm{Hom}_1^t(\mathfrak{l},\mathfrak{g})$, let $P_\varphi : \mathscr{L}(\mathfrak{l}',\mathfrak{z}) \to \mathscr{L}(\mathfrak{l}',\mathfrak{z})/\operatorname{Im}(l_\varphi)$ and $\mathrm{Hom}_{1,\beta}^t(\mathfrak{l},\mathfrak{g}) := \{\varphi \in \mathrm{Hom}_1^t(\mathfrak{l},\mathfrak{g}) : \det(P_\varphi(e_{i_1}),\dots,P_\varphi(e_{i_{m-t}})) \neq 0\}$. Then*

$$\mathrm{Hom}_1^t(\mathfrak{l},\mathfrak{g}) = \bigcup_{\beta\in I(m,m-t)} \mathrm{Hom}_{1,\beta}^t(\mathfrak{l},\mathfrak{g})$$

as a union of open subsets.

Proof. We know that for all $\varphi \in \mathrm{Hom}_1^t(\mathfrak{l},\mathfrak{g})$, the set $\operatorname{Im}(l_\varphi)$ is a linear subspace of $\mathscr{L}(\mathfrak{l}',\mathfrak{z})$ of dimension t. There exists therefore $(i_1,\dots,i_{m-t}) \in I(m,m-t)$ such that the family $\{P_\varphi(e_{i_1}),\dots,P_\varphi(e_{i_{m-t}})\}$ forms a basis of $\mathscr{L}(\mathfrak{l}',\mathfrak{z})/\operatorname{Im}(l_\varphi)$, and consequently

$$\det(P_\varphi(e_{i_1}),\dots,P_\varphi(e_{i_{m-t}})) \neq 0. \qquad \square$$

We are now ready to prove our main result in this section.

Theorem 4.2.6. *Let G, H and $\mathfrak{l}$ be as before. The deformation space reads*

$$\mathscr{T}(\mathfrak{l},\mathfrak{g},\mathfrak{h}) = \bigsqcup_{i=1}^{2}\bigsqcup_{t=0}^{q}\bigcup_{\beta\in I(m,m-t)} \mathscr{T}_{t,\beta,i},$$

where for $\beta = (i_1,\dots,i_{m-t})$, the set $\mathscr{T}_{t,\beta,1}$ is homeomorphic to the semialgebraic set

$$\mathscr{T}_{t,\beta,1} := \left\{ \varphi(A,B) \in \mathrm{Hom}(\mathfrak{l},\mathfrak{g}) \;\middle|\; \begin{array}{l} \begin{pmatrix} B_1 \\ B_2 \end{pmatrix} \in V_\beta,\ \mathrm{rk}(l_{\varphi(A,B)}) = t, \\ \det(P_{\varphi(A,B)}(e_{i_1}),\dots,P_{\varphi(A,B)}(e_{i_{m-t}})) \neq 0, \\ \mathrm{rk}\begin{pmatrix} A_2 & 0 \\ 0 & B_4 \end{pmatrix} = \dim(\mathfrak{l}) \end{array} \right\}$$

and $\mathscr{T}_{t,\beta,2}$ is homeomorphic to

$$\mathscr{T}_{t,\beta,2} := \left\{ \varphi(A,B) \in \mathrm{Hom}(\mathfrak{l},\mathfrak{g}) \;\middle|\; \begin{array}{l} \begin{pmatrix} B_1 \\ B_2 \end{pmatrix} \in V_\beta,\ \mathrm{rk}(l_{\varphi(A,B)}) = t, \\ \det(P_{\varphi(A,B)}(e_{i_1}),\dots,P_{\varphi(A,B)}(e_{i_{m-t}})) \neq 0, \\ \mathrm{rk}(B_4) < \dim(\mathfrak{l}'), \\ \mathrm{rk}\begin{pmatrix} A_2 & B_2 + \sigma_2(g)B_3 + \delta_2(g)B_4 \\ 0 & B_4 \end{pmatrix} = \dim(\mathfrak{l}) \\ \text{for all } g \in G \end{array} \right\},$$

which is also semialgebraic.

Proof. It is easy to see that $D_\varphi = D_{g\cdot\varphi}$, which means that $l_\varphi = l_{g.\varphi}$ and $P_{\varphi'} = P_{g\cdot\varphi'}$ for all $\varphi \in \mathrm{Hom}(\mathfrak{l}, \mathfrak{g})$ and $g \in G$. Then for all $\beta \in I(m, m-t)$ and $0 \le t \le q$, the set $\mathrm{Hom}^t_{1,\beta}(\mathfrak{l}, \mathfrak{g}) \times \mathscr{L}(\mathfrak{l}', \mathfrak{z})$ is G-stable. Let $\mathrm{Hom}^t_\beta(\mathfrak{l}, \mathfrak{g}) = \psi^{-1}(\mathrm{Hom}^t_{1,\beta}(\mathfrak{l}, \mathfrak{g}) \times \mathscr{L}(\mathfrak{l}', \mathfrak{z}))$, then

$$\mathrm{Hom}(\mathfrak{l}, \mathfrak{g}) = \bigsqcup_{t=0}^{q} \mathrm{Hom}^t(\mathfrak{l}, \mathfrak{g}),$$

where

$$\mathrm{Hom}^t(\mathfrak{l}, \mathfrak{g}) := \bigcup_{\beta\in I(m,m-t)} \mathrm{Hom}^t_\beta(\mathfrak{l}, \mathfrak{g}). \tag{4.34}$$

Since ψ is G-equivariant, it is clear that for all $t \in \{0, \dots, q\}$ and $\beta \in I(m, m-t)$, the set $\mathrm{Hom}^t_\beta(\mathfrak{l}, \mathfrak{g})$ is G-stable and by (4.34), we get

$$\mathrm{Hom}^t(\mathfrak{l}, \mathfrak{g})/G = \bigcup_{\beta\in I(m,m-t)} \mathrm{Hom}^t_\beta(\mathfrak{l}, \mathfrak{g})/G, \tag{4.35}$$

and then

$$\mathrm{Hom}(\mathfrak{l}, \mathfrak{g})/G = \bigsqcup_{t=0}^{q} \bigcup_{\beta\in I(m,m-t)} \mathrm{Hom}^t_\beta(\mathfrak{l}, \mathfrak{g})/G.$$

Whence,

$$\mathscr{R}(\mathfrak{l}, \mathfrak{g}, \mathfrak{h}) = \bigsqcup_{i=1}^{2} \bigsqcup_{t=0}^{q} \bigcup_{\beta\in I(m,m-t)} \mathrm{Hom}^t_\beta(\mathfrak{l}, \mathfrak{g}) \cap \mathscr{R}_i.$$

We now consider the maps

$$\begin{aligned} \overline{\psi} : \mathrm{Hom}^t_\beta(\mathfrak{l}, \mathfrak{g})/G &\to (\mathrm{Hom}^t_{1,\beta}(\mathfrak{l}, \mathfrak{g}) \times \mathscr{L}(\mathfrak{l}', \mathfrak{z}))/G \\ G \cdot \varphi &\mapsto G \cdot \psi(\varphi), \\ \pi^t_\beta : (\mathrm{Hom}^t_{1,\beta}(\mathfrak{l}, \mathfrak{g}) \times \mathscr{L}(\mathfrak{l}', \mathfrak{z}))/G &\to \mathrm{Hom}^t_{1,\beta}(\mathfrak{l}, \mathfrak{g}) \times V_\beta \\ (\varphi', B_\varphi + \mathrm{Im}(l_\varphi)) &\mapsto (\varphi', P^{-1}_{\varphi|_{V_\beta}}(B_\varphi + \mathrm{Im}(l_\varphi))) \end{aligned}$$

and $\varepsilon^t_\beta = \psi^{-1} \circ \pi^t_\beta \circ \overline{\psi}$.

We now show that for all $t \in \{0, \dots, q\}$, the collection

$$(\varepsilon^t_\beta, \mathrm{Hom}^t_\beta(\mathfrak{l}, \mathfrak{g})/G)_{\beta\in I(m,m-t)}$$

is a family of local sections of the canonical surjection $\pi^t : \mathrm{Hom}^t(\mathfrak{l}, \mathfrak{g}) \longrightarrow \mathrm{Hom}^t(\mathfrak{l}, \mathfrak{g})/G$. Indeed, we have to show that $\pi^t \circ \varepsilon^t_\beta = \mathrm{Id}_{\mathrm{Hom}^t_\beta(\mathfrak{l}, \mathfrak{g})/G}$. Let $\varphi \in \mathrm{Hom}(\mathfrak{l}, \mathfrak{g})$ such that the orbit

$$\begin{pmatrix} A_\varphi & B_\varphi + \mathrm{Im}(l_\varphi) \\ 0 & D_\varphi \end{pmatrix} \in \mathrm{Hom}^t_\beta(\mathfrak{l}, \mathfrak{g})/G,$$

then

$$\begin{aligned} \pi^t \circ \varepsilon^t_\beta \begin{pmatrix} A_\varphi & B_\varphi + \mathrm{Im}(l_\varphi) \\ 0 & D_\varphi \end{pmatrix} &= \pi^t \begin{pmatrix} A_\varphi & P^{-1}_{\varphi'|V_\beta}(B_\varphi + \mathrm{Im}(l_\varphi)) \\ 0 & D_\varphi \end{pmatrix} \\ &= \begin{pmatrix} A_\varphi & P^{-1}_{\varphi'|V_\beta}(B_\varphi + \mathrm{Im}(l_\varphi)) + \mathrm{Im}(l_\varphi) \\ 0 & D_\varphi \end{pmatrix} \\ &= \begin{pmatrix} A_\varphi & B_\varphi + \mathrm{Im}(l_\varphi) \\ 0 & D_\varphi \end{pmatrix}. \end{aligned}$$

In particular,

$$\varepsilon^t_\beta(\mathrm{Hom}^t_\beta(\mathfrak{l}, \mathfrak{g})/G) = \left\{ \varphi \in \mathrm{Hom}(\mathfrak{l}, \mathfrak{g}) \;\middle|\; \begin{array}{l} \varphi' \in \mathrm{Hom}^t_{1,\beta}(\mathfrak{l}, \mathfrak{g}), \\ B_\varphi \in V_\beta \end{array} \right\}. \tag{4.36}$$

Now

$$\begin{aligned} \mathscr{T}(\mathfrak{l}, \mathfrak{g}, \mathfrak{h}) &= \{[\varphi] \in \mathrm{Hom}(\mathfrak{l}, \mathfrak{g})/G : \exp(\varphi(\mathfrak{l})) \text{ acts properly on } G/H\} \\ &= \bigsqcup_{t=0}^{q} \bigcup_{\beta \in I(m,m-t)} \{[\varphi] \in \mathrm{Hom}^t_\beta(\mathfrak{l}, \mathfrak{g})/G\} \cap \mathscr{R}(\mathfrak{l}, \mathfrak{g}, \mathfrak{h}). \end{aligned}$$

□

4.2.3 Hausdorffness of the deformation space

This section aims to study the Hausdorffness of the deformation space $\mathscr{T}(\mathfrak{l}, \mathfrak{g}, \mathfrak{h})$ in the setting where $\mathfrak{g}$ is 2-step nilpotent. We first prove the following.

Lemma 4.2.7. *For all $\varphi \in \mathrm{Hom}(\mathfrak{l}, \mathfrak{g})$, we have*

$$\dim(G \cdot \varphi) = \dim(\mathfrak{g}) - \dim(\varphi(\mathfrak{l})^\perp),$$

where $\varphi(\mathfrak{l})^\perp = \{X \in \mathfrak{g} : [X, Y] = 0, \forall\, Y \in \varphi(\mathfrak{l})\}$.

Proof. Recall that the map $\sigma : G \to \mathscr{L}(\mathfrak{l}', \mathfrak{z})$ is a group homomorphism and it is clear that $\ker(\sigma) = Z(G)$ where $Z(G)$ is the center of G. Then σ factors via the projection map $G \to G/Z(G)$ to obtain an injective homomorphism $\tilde{\sigma} : G/Z(G) \to \mathscr{L}(\mathfrak{l}', \mathfrak{z})$ such that $\tilde{\sigma}(\overline{G}) = \sigma(G)$.

Let now $\varphi \in \operatorname{Hom}(\mathfrak{l}, \mathfrak{g})$, then $\ker(l_\varphi) = \{\tilde{\sigma}(\overline{e^X}) : [X, Y] = 0 \,\forall\, Y \in \varphi(\mathfrak{l})\}$. Thus,

$$\begin{aligned}\dim(G \cdot \varphi) &= \dim(\sigma(G)) - \dim(\overline{e^{\varphi(\mathfrak{l})^\perp}})\\ &= \dim(\mathfrak{g}) - \dim(\mathfrak{z}) - (\dim(\varphi(\mathfrak{l})^\perp) - \dim(\mathfrak{z}))\\ &= \dim(\mathfrak{g}) - \dim(\varphi(\mathfrak{l})^\perp).\end{aligned}$$ □

We now prove the main result of this section.

Theorem 4.2.8. *Let $\mathfrak{g}$ be a 2-step nilpotent Lie algebra, if all G-orbits in $\mathscr{R}(\mathfrak{l}, \mathfrak{g}, \mathfrak{h})$ have a common dimension, then the deformation space $\mathscr{T}(\mathfrak{l}, \mathfrak{g}, \mathfrak{h})$ is a Hausdorff space.*

Proof. In such a situation, there is $t \in \{0, \dots, q\}$ such that $\mathscr{R}(\mathfrak{l}, \mathfrak{g}, \mathfrak{h}) \subset \operatorname{Hom}^t(\mathfrak{l}, \mathfrak{g})$, where $\operatorname{Hom}^t(\mathfrak{l}, \mathfrak{g})$ is as in equation (4.34) and $q = \dim \sigma(G)$. Indeed, let $\varphi \in \operatorname{Hom}^t(\mathfrak{l}, \mathfrak{g})$ then $\operatorname{rk} l_\varphi = t$ and so

$$\begin{aligned}\dim G \cdot \varphi &= \dim \psi(G \cdot \varphi)\\ &= \dim G \cdot \psi(\varphi)\\ &= \dim(\varphi', B_\varphi + \sigma(G) D_\varphi)\\ &= \dim(\varphi', B_\varphi + \operatorname{Im} l_\varphi)\\ &= \operatorname{rk} l_\varphi = t.\end{aligned}$$

The deformation space $\mathscr{T}(\mathfrak{l}, \mathfrak{g}, \mathfrak{h})$ is therefore contained in $\operatorname{Hom}^t(\mathfrak{l}, \mathfrak{g})/G$, and it is sufficient to show that $\operatorname{Hom}^t(\mathfrak{l}, \mathfrak{g})/G$ is a Hausdorff space. Let $[\varphi] \neq [\xi]$ be some points in $\operatorname{Hom}^t(\mathfrak{l}, \mathfrak{g})/G$. Suppose that $[\varphi]$ and $[\xi]$ are not separated, then there exist $(\varphi_n)_n \subset \operatorname{Hom}^t(\mathfrak{l}, \mathfrak{g})$ and $g_n \in G$ such that φ_n converges to φ and $g_n \cdot \varphi_n$ converges to ξ in $\operatorname{Hom}^t(\mathfrak{l}, \mathfrak{g})$. Using the map ψ defined in Lemma 4.2.4, we can see that the sequence $(\varphi'_n, B_{\varphi_n})_n$ converges to (φ', B_φ), $(\varphi'_n, B_{\varphi_n} + \sigma(g_n) D_{\varphi_n})_n$ converges to (φ', B_φ) and $(\varphi'_n, B_{\varphi_n} + \sigma(g_n) D_{\varphi_n})_n$ converges to (ξ', B_ξ). This means that $\varphi' = \xi'$ and in particular $D_\varphi = D_\xi$ and $P_\varphi = P_\xi$. Finally, $[\varphi]$ and $[\xi]$ belong to the open set $\operatorname{Hom}^t_\beta(\mathfrak{l}, \mathfrak{g})/G$ for some $\beta \in I(m, m-t)$. From (4.36), $\operatorname{Hom}^t_\beta(\mathfrak{l}, \mathfrak{g})/G$ is homeomorphic to a semialgebraic set. Therefore, $\operatorname{Hom}^t_\beta(\mathfrak{l}, \mathfrak{g})/G$ is a Hausdorff space. This leads thus to a contradiction. □

Corollary 4.2.9. *Let $\mathfrak{g}$ be a 2-step nilpotent Lie algebra. If $\mathfrak{l}$ is a maximal Abelian subalgebra of $\mathfrak{g}$, then the deformation space $\mathscr{T}(\mathfrak{l}, \mathfrak{g}, \mathfrak{h})$ is a Hausdorff space.*

Proof. If $\mathfrak{l}$ is a maximal Abelian subalgebra, then so is $\varphi(\mathfrak{l})$ and we have $\varphi(\mathfrak{l})^\perp = \varphi(\mathfrak{l})$. Hence, from Lemma 4.2.7

$$\dim(G \cdot \varphi) = \dim(\mathfrak{g}) - \dim(\varphi(\mathfrak{l})) = \dim(\mathfrak{g}) - \dim(\mathfrak{l}),$$

which is constant. This achieves the proof. □

Proposition 4.2.10. *Let $\mathfrak{l}$ and $\mathfrak{h}$ be two subalgebras of $\mathfrak{g}$ and suppose that:*

(1) *There is decomposition* $\mathfrak{g} = [\mathfrak{l},\mathfrak{l}] \oplus \mathfrak{z}_1 \oplus \mathfrak{z}_2 \oplus \mathfrak{z}_3 \oplus \mathfrak{h}' \oplus \mathfrak{l}' \oplus V$, *where* $\mathfrak{z} \cap \mathfrak{l} = [\mathfrak{l},\mathfrak{l}] \oplus \mathfrak{z}_1$, $\mathfrak{l} = [\mathfrak{l},\mathfrak{l}] \oplus \mathfrak{z}_1 \oplus \mathfrak{l}'$, $\mathfrak{z} \cap \mathfrak{h} = \mathfrak{z}_2$, $\mathfrak{h} = \mathfrak{h}' \oplus \mathfrak{z}_2$ *and* $\mathfrak{z} = [\mathfrak{l},\mathfrak{l}] \oplus \mathfrak{z}_1 \oplus \mathfrak{z}_2 \oplus \mathfrak{z}_3$.

(2) *There is a codimension one subalgebra* $\mathfrak{l}_1$ *of* $\mathfrak{l}$ *such that* $[\mathfrak{l}_1^{\perp},\mathfrak{l}] \not\subset \mathfrak{l}\oplus\mathfrak{h}$ *and* $\mathfrak{z}(\mathfrak{l}) + \mathfrak{l}_1 = \mathfrak{l}$.

Then the deformation space $\mathscr{T}(\mathfrak{l},\mathfrak{g},\mathfrak{h})$ *fails to be a Hausdorff space.*

Proof. Recall that $\dim([\mathfrak{l},\mathfrak{l}]) = s$ and set $\dim(\mathfrak{z}_1) = q$, $\dim(\mathfrak{z}_2) = p$, $\dim(\mathfrak{z}_3) = r$ and $\dim(\mathfrak{z}) = m$. The Lie bracket of $\mathfrak{g}$ is given by

$$[X,Y] = \sum_{i=1}^{m} b_i(X,Y)Z_i,$$

where $\{Z_1,\dots,Z_m\}$ is a basis of $\mathfrak{z}$ passing through the decomposition $\mathfrak{z} = [\mathfrak{l},\mathfrak{l}]\oplus\mathfrak{z}_1\oplus\mathfrak{z}_2\oplus\mathfrak{z}_3$. There a basis $\{L_1,\dots,L_{k-q-s}\}$ of $\mathfrak{l}'$ such that $\{Z_1,\dots,Z_{s+q},L_2,\dots,L_{k-q-s}\}$ is a basis of the subalgebra $\mathfrak{l}_1$ satisfying the second assumption in the proposition. The vector $L_1 \in \mathfrak{z}(\mathfrak{l})$ and there is $X_0 \in \mathfrak{l}_1^{\perp}$ such that the bracket $[X_0,L_1] \notin \mathfrak{l}\oplus\mathfrak{h}$. In particular, there is $p+q+s < i_0 \le m$ such that $b_{i_0}(X_0,L_1) = \alpha \neq 0$. If we complete the vectors $Z_1,\dots,Z_m,L_1,\dots,L_{k-q-s}$ to a basis of $\mathfrak{g}$ passing through the decomposition $\mathfrak{g} = [\mathfrak{l},\mathfrak{l}] \oplus \mathfrak{z}_1 \oplus \mathfrak{z}_2 \oplus \mathfrak{z}_3 \oplus \mathfrak{l}' \oplus \mathfrak{h}' \oplus V$, then the matrix

$$M_1 = \begin{pmatrix} I_s & 0 & 0 \\ 0 & I_q & 0 \\ 0 & 0 & 0 \\ 0 & 0 & C_1 \\ 0 & 0 & D \\ 0 & 0 & 0 \\ 0 & 0 & 0 \end{pmatrix} \in \mathscr{R}(\mathfrak{l},\mathfrak{g},\mathfrak{h}),$$

with

$$D = \begin{pmatrix} 0 & 0 \\ 0 & I_{k-q-s-1} \end{pmatrix} \in M_{k-q-s}(\mathbb{R}),$$

$C_1 = (u_1\ 0) \in M_{r,k-q-s}(\mathbb{R})$, where $u_1 = {}^t(0,\dots,0,x_1,0,\dots,0) \in \mathbb{R}^r$ and x_1 is the i_0 coordinate. Let $C_s = (u_s\ 0) \in M_{s,k-q-s}(\mathbb{R})$ with $u_s = {}^t(a_1,\dots,a_s) \in \mathbb{R}^s$,

$$C_q = (u_q\ 0) \in M_{q,k-q-s}(\mathbb{R}) \quad \text{with } u_q = {}^t(a_{s+1},\dots,a_{s+q}) \in \mathbb{R}^q$$

and

$$C_p = (u_p\ 0) \in M_{p,k-q-s}(\mathbb{R}) \quad \text{with } u_p = {}^t(a_{q+s+1},\dots,a_{q+s+p}) \in \mathbb{R}^p.$$

Let $C_2 = (u_2\ 0) \in M_{m-q-s-p,k-q-s}(\mathbb{R})$, $u_2 = {}^t(a_{q+s+p+1}, \ldots, a_{i_0-1}, x_2, a_{i_0+1}, \ldots, a_m) \in \mathbb{R}^{m-q-s-p}$ such that $x_2 \neq x_1$ and $a_i = \frac{x_2-x_1}{\alpha} b_i(X_0, L_1)$. Then

$$M_2 = \begin{pmatrix} I_s & 0 & C_s \\ 0 & I_q & C_q \\ 0 & 0 & C_p \\ 0 & 0 & C_2 \\ 0 & 0 & D \\ 0 & 0 & 0 \\ 0 & 0 & 0 \end{pmatrix} \in \mathscr{R}(\mathfrak{l}, \mathfrak{g}, \mathfrak{h})$$

and $G \cdot M_1 \neq G \cdot M_2$.

Let $\mathscr{V}_i$ be a neighborhood of M_i for $i = 1, 2$. For $\varepsilon > 0$ small enough, the elements

$$M_{1,\varepsilon} = \begin{pmatrix} I_s & 0 & 0 \\ 0 & I_q & 0 \\ 0 & 0 & 0 \\ 0 & 0 & C_1 \\ 0 & 0 & D_\varepsilon \\ 0 & 0 & 0 \\ 0 & 0 & 0 \end{pmatrix} \in \mathscr{V}_1 \quad \text{and} \quad M_{2,\varepsilon} = \begin{pmatrix} I_s & 0 & C_s \\ 0 & I_q & C_q \\ 0 & 0 & C_p \\ 0 & 0 & C_2 \\ 0 & 0 & D_\varepsilon \\ 0 & 0 & 0 \\ 0 & 0 & 0 \end{pmatrix} \in \mathscr{V}_2,$$

where

$$D_\varepsilon = \begin{pmatrix} \varepsilon & 0 \\ 0 & I_{k-q-s-1} \end{pmatrix} \quad \text{and} \quad \mathrm{Ad}_{\exp \frac{x_2-x_1}{\alpha\varepsilon} X_0} M_{1,\varepsilon} = M_{2,\varepsilon}.$$

This means that $\mathscr{T}(\mathfrak{l}, \mathfrak{g}, \mathfrak{h})$ is not a Hausdorff space. □

4.3 The 3-step case

4.3.1 Some preliminary results

We first prove the following series of lemmas, which will be of use in the section.

Lemma 4.3.1. *Let V be a vector space, E and F two subspaces of V such that $V = E \oplus F$. Then for any $v \in V$,*

$$(v + E) \cap F = P(v),$$

where P is the projection of V on F parallel to E.

Proof. Let $v \in V$ and write $v = v_1 + v_2$ with $v_1 \in E$ and $v_2 \in F$. Then $P(v) = v_2$ and $v + E = v_2 + E$. Let $u \in (v + E) \cap F$, then there exists $w \in E$ such that $u = v_2 + w$ and we have

$$u \in F \Rightarrow v_2 + w \in F \Rightarrow w \in F \Rightarrow w \in E \cap F = \{0\}.$$

Thus, $u = v_2$. □

Lemma 4.3.2. *Let F, K be two finite-dimensional vector spaces and $\mathscr{B} = \{e_1, \ldots, e_m\}$ a basis of F.*

(1) *If $\varphi \in \mathscr{L}(F, K)$ of rank $t > 0$, then there exists $\{e_{j_1}, \ldots, e_{j_t}\}$ a subset of $\mathscr{B}$ such that $\operatorname{Im}\varphi = \mathbb{R}\text{-span}\{\varphi(e_{j_1}), \ldots, \varphi(e_{j_t})\}$.*
(2) *Let $S = \{e_{j_1}, \ldots, e_{j_t}\}$ be a subset of $\mathscr{B}$, then the set*

$$A(S) = \{\varphi \in \mathscr{L}(F, K) \mid \dim \mathbb{R}\text{-span}\{\varphi(e_{j_1}), \ldots, \varphi(e_{j_t})\} = t\}$$

is open in $\mathscr{L}(F, K)$.

Proof. (1) As $\{\varphi(e_1), \ldots, \varphi(e_m)\}$ is a generating family of $\operatorname{Im}(\varphi)$, we can extract from this family a basis of $\operatorname{Im}(\varphi)$.

(2) Let $r = \dim(K)$, fix a basis $\mathscr{B}'$ of K and identify $\mathscr{L}(F, K)$ to the set of matrices $M_{r,m}(\mathbb{R})$ as a topological space. In this context, the set $A(S)$ is identified to

$$A'(S) = \{M \in M_{r,m}(\mathbb{R}) \mid \operatorname{rk}(M') = t\},$$

where $M' \in M_{r,t}(\mathbb{R})$ is the matrix obtained from M by deleting all the columns of index $k \notin \{j_1, \ldots, j_t\}$. Let now

$$J(t, r) = \{(k_1, \ldots, k_t) \in \mathbb{N}^t \mid 1 \le k_1 < \cdots < k_t \le r\}.$$

For $\alpha = (k_1, \ldots, k_t)$ and $M' \in M_{r,t}(\mathbb{R})$, we denote by M'_α the square matrix obtained from M' by deleting all the lines of index $k \notin \{k_1, \ldots, k_t\}$. Then the condition $\operatorname{rk}(M') = t$ is equivalent to

$$\sum_{\alpha \in J(t,r)} [\det(M'_\alpha)]^2 \neq 0,$$

which proves that $A'(S)$ is open and, therefore, $A(S)$ is also open. □

Lemma 4.3.3. *Let F be a finite-dimensional vector space, V a subspace of F and $t = \dim F - \dim V$. For all integer n, let $S_n = \{u_{1,n}, \ldots, u_{t,n}\}$ be a family of linearly independent vectors in F such that:*

(1) *$F = \mathbb{R}\text{-span}(S_n) \oplus V$.*
(2) *For all $1 \le i \le t$, the sequence $(u_{i,n})_n$ converges to some vector u_i.*
(3) *$S = \{u_1, \ldots, u_t\}$ is formed by linearly independent vectors and $F = \mathbb{R}\text{-span}(S) \oplus V$.*

Let P_n denote the projection of F on V parallel to $\mathbb{R}$-span(S_n) and P the projection of F on V parallel to $\mathbb{R}$-span(S). Then the sequence $(P_n)_n$ converges in $\mathscr{L}(V)$ to P.

Proof. Let $m = \dim F$ and $\mathscr{B} = \{e_1, \ldots, e_m\}$ a basis of F such that $\{e_{t+1}, \ldots, e_m\}$ is a basis of V. By the hypothesis (1), the set $\mathscr{B}_n = \{u_{1,n}, \ldots, u_{t,n}, e_{t+1}, \ldots, e_m\}$ is a basis of F for all n and the matrix of P_n in the basis $\mathscr{B}_n$ is

$$Q = \begin{pmatrix} 0_{\mathbb{R}\text{-span}(S_n)} & 0 \\ 0 & \mathrm{Id}_V \end{pmatrix}.$$

If $P_{\mathscr{B}\mathscr{B}_n}$ is the transition base matrix, then the matrix of P_n in $\mathscr{B}$ is

$$Q_n = P_{\mathscr{B}\mathscr{B}_n} Q P^{-1}_{\mathscr{B}\mathscr{B}_n}.$$

Now by (2) and (3), $(P_{\mathscr{B}\mathscr{B}_n})_n$ converges to $P_{\mathscr{B}\mathscr{B}'}$, where $\mathscr{B}' = \{u_1, \ldots, u_t, e_{t+1}, \ldots, e_n\}$. Then $(Q_n)_n$ converges to the matrix $Q' = P_{\mathscr{B}\mathscr{B}'} Q P^{-1}_{\mathscr{B}\mathscr{B}'}$, which is the matrix of P in $\mathscr{B}$. □

Lemma 4.3.4. *Let V, W be two finite-dimensional vector spaces, $\mathscr{B} = \{e_1, \ldots, e_n\}$ a basis of V and $f : V \to W$ a linear map such that $\{f(e_1), \ldots, f(e_k)\}$ is a basis of $\operatorname{Im} f$. Assume that*

$$f(e_{k+j}) = \alpha_{1,j} f(e_1) + \cdots + \alpha_{k,j} f(e_k), \quad 1 \le j \le n-k.$$

Then the family of vectors

$$u_j = e_{k+j} - \alpha_{1,j} e_1 - \cdots - \alpha_{k,j} e_k, \quad 1 \le j \le n-k$$

is a basis of $\ker f$.

Proof. Clearly, the family $\{u_j, 1 \le j \le n-k\}$ is a family of linearly independent vectors and we have $f(u_j) = 0$ for all $1 \le j \le n-k$. As $\dim \ker f = n-k$, the result follows. □

Lemma 4.3.5. *Let V, W be two finite-dimensional vector spaces, $\mathscr{B} = \{e_1, \ldots, e_m\}$ a basis of V and $(f_n : V \to W)_n$ a sequence of linear maps such that:*
1. *$(f_n)_n$ converges to a linear map $f : V \to W$.*
2. *The family $\{f_n(e_1), \ldots, f_n(e_k)\}$ is a basis of $\operatorname{Im} f_n$ for all n.*
3. *The family $\{f(e_1), \ldots, f(e_k)\}$ is a basis of $\operatorname{Im} f$.*

Assume that for all $n \ge 0$ and $1 \le j \le m-k$, we have

$$f_n(e_{k+j}) = \alpha^n_{1,j} f_n(e_1) + \cdots + \alpha^n_{k,j} f_n(e_k)$$

and

$$f(e_{k+j}) = \alpha_{1,j}f(e_1) + \cdots + \alpha_{k,j}f(e_k).$$

Then for all $1 \leq j \leq n-k$ *and* $1 \leq l \leq k$, *the sequence* $(\alpha^n_{l,j})_n$ *converges to* $\alpha_{l,j}$.

Proof. As $(f_n)_n$ converges to f, we have $(f_n(e_{k+j}) - f(e_{k+j}))_n$ converges to zero. Let now $\dim W = r$ and let $u_1, \dots, u_{r-k} \in W$ be such that $\mathcal{B}' = \{f(e_1), \dots, f(e_k), u_1, \dots, u_{r-k}\}$ is a basis of W. As $(f_n(e_i))_n$ converges to $f(e_i)$ for all $1 \leq i \leq k$, there exists $N > 0$ such that for all $n > N$ the family $\{f_n(e_1), \dots, f_n(e_k), u_1, \dots, u_{r-k}\}$ is also a basis of W. Let $S_n = \{f_n(e_2), \dots, f_n(e_k), u_1, \dots, u_{r-k}\}$, $F = \mathbb{R}\text{-span}\{f(e_1)\}$ and q_n the projection of W on F parallel to $\mathbb{R}\text{-span}(S_n)$. Then by Lemma 4.3.3, $(q_n)_n$ converges to the projection q of W on F parallel to $\mathbb{R}\text{-span}(S)$, where $S = \{f(e_2), \dots, f(e_k), u_1, \dots, u_{r-k}\}$ and $(q_n(f_n(e_{k+j})))_n$ converges to $q(f(e_{k+j}))$. Note that $q_n(f_n(e_{k+j})) = q_n(\alpha^n_{1,j}f_n(e_1)) = \alpha^n_{1,j}q_n(f_n(e_1))$ and $q(f(e_{k+j})) = \alpha_{1,j}f(e_1)$. As $(\alpha_{1,j}q_n(f_n(e_1)))_n$ converges to $\alpha_{1,j}f(e_1)$, the sequence $((\alpha^n_{1,j} - \alpha_{1,j})q_n(f_n(e_1)))_n$ converges to zero. But $(q_n(f_n(e_1)))_n$ converges to the nonzero vector $f(e_1)$, then $(\alpha^n_{1,j})_n$ converges to $\alpha_{1,j}$. Using the same argument, we can show that $(\alpha^n_{l,j})_n$ converges to $\alpha_{l,j}$ for all $1 \leq l \leq k$ and $1 \leq j \leq n-k$. □

Lemma 4.3.6. *Let* $(x_n)_n$ *be a sequence in* $\mathbb{R}^q$ *such that:*
(1) *Any subsequence of* $(x_n)_n$ *contains a convergent subsequence.*
(2) *Two convergent subsequences of* $(x_n)_n$ *converge to the same element.*

Then $(x_n)_n$ *is convergent.*

Proof. Suppose that $(x_n)_n$ is not bounded, then for every integer k there exists $(x_{n_k})_{n_k}$ such that $\|x_{n_k}\| > k$. Then obviously $(x_n)_n$ contains a subsequence $(x_{n_k})_{n_k}$ such that $\lim \|x_{n_k}\| = +\infty$, then $(x_{n_k})_{n_k}$ does not have convergent subsequence. Thus, by (1), $(x_n)_n$ is bounded. Let $(x_{n_k})_{n_k}$ be a convergent subsequence of $(x_n)_n$, which converges to y and let $A > 0$ such that $\|x_n - y\| < A$ for all n that is $(x_n)_n$ belongs to the closed ball $B(y, A)$ of center y and radius A. Let U be a neighborhood of y. If $B(y, A) \setminus U$ contains an infinite terms of $(x_n)_n$, then we can find a subsequence, which converges to $y' \neq y$ and then by (2) only finite terms of $(x_n)_n$ are not in U; thus, $(x_n)_n$ converges to y. □

Lemma 4.3.7. *Let* $(f_n : \mathbb{R}^q \to \mathbb{R}^m)_n$ *be a sequence of linear maps and* $(x_n)_n$ *a sequence in* $\mathbb{R}^q$ *such that:*
(1) $(f_n)_n$ *converges to a linear map* $f : \mathbb{R}^q \to \mathbb{R}^m$.
(2) *For all* n, f_n *and* f *are injective.*
(3) *There exists* $x \in \mathbb{R}^q$ *such that* $(f_n(x_n))_n$ *converges to* $f(x)$.

Then $(x_n)_n$ *converges to* x.

Proof. If for all $N > 0$, there exists $n > N$ such that $x_n = 0$ then $(x_n)_n$ contains a subsequence $(x_{n_k})_{n_k}$ such that $x_{n_k} = 0$ for all k and we have $f_{n_k}(x_{n_k}) = 0$ then $(f_n(x_n))_n$

converges to zero. Let $(x_{n_{k_1}})_{n_{k_1}}$ be a subsequence of $(x_n)_n$ such that $x_{n_{k_1}} \neq 0$ for all k_1 and we have to prove that $(x_{n_{k_1}})_{n_{k_1}}$ converges to zero. Note that

$$f_{n_{k_1}}(x_{n_{k_1}}) = \|x_{n_{k_1}}\| f_{n_{k_1}}\left(\frac{x_{n_{k_1}}}{\|x_{n_{k_1}}\|}\right).$$

As the sequence $(\frac{x_{n_{k_1}}}{\|x_{n_{k_1}}\|})_{n_{k_1}}$ is bounded, we can assume that it converges to some $y \neq 0$. Now suppose that $(\|x_{n_{k_1}}\|)_{n_{k_1}}$ does not converge to zero, then up to the choice of a subsequence, we can assume that $(\|x_{n_{k_1}}\|)_{n_{k_1}}$ converges to some $a \in]0, +\infty]$ then $(f_{n_{k_1}}(x_{n_{k_1}}))_{n_{k_1}}$ converges to $af(y) \neq 0$ because $y \neq 0$ and f is injective, which is a contradiction.

Now assume that there exists $N > 0$ such that $x_n \neq 0$ for all $n > N$. In this case, we have to show that $(x_n)_n$ satisfies the conditions (1) and (2) of Lemma 4.3.6. Let $(x_{n_1})_{n_1}$ be a subsequence of $(x_n)_n$, then we can find a subsequence $(x_{n_2})_{n_2}$ of $(x_{n_1})_{n_1}$ such that $(\frac{x_{n_2}}{\|x_{n_2}\|})_{n_2}$ converges to some $y \neq 0$ and using (1), we see that $(f_{n_2}(\frac{x_{n_2}}{\|x_{n_2}\|}))_{n_2}$ converges to $f(y)$. As

$$(f_{n_2}(x_{n_2}))_{n_2} = \left(\|x_{n_2}\| f_{n_2}\left(\frac{x_{n_2}}{\|x_{n_2}\|}\right)\right)_{n_2}$$

converges to $f(x)$, we deduce that $(\|x_{n_2}\|)_{n_2}$ converges to $\frac{\|f(x)\|}{\|f(y)\|}$. In particular, the sequence $(x_{n_2})_{n_2}$ is bounded and contains a convergent subsequence. This shows that any subsequence of $(x_n)_n$ contains a convergent subsequence. Let $(x_{n,1})_{n,1}$ and $(x_{n,2})_{n,2}$ be two convergent subsequences of $(x_n)_n$ such that $(x_{n,1})_{n,1}$ converges to y_1 and $(x_{n,2})_{n,2}$ converges to y_2. Then $(f_{n,1}(x_{n,1}))_{n,1}$ and $(f_{n,2}(x_{n,2}))_{n,2}$ converge to $f(y_1)$ and $f(y_2)$, respectively. By (3), $f(y_1) = f(y_2)$. Then $f(y_1 - y_2) = 0$ and using (2) we deduce that $y_1 = y_2$. Then two convergent subsequences of $(x_n)_n$ converge to the same element. Using Lemma 4.3.6, we conclude that $(x_n)_n$ is convergent. □

4.3.2 On the quotient space $\mathrm{Hom}(\mathfrak{l}, \mathfrak{g})/G$

Describing $\mathrm{Hom}(\mathfrak{l}, \mathfrak{g})$

We assume henceforth that $\mathfrak{g}$ is a 3-step nilpotent Lie algebra and $\mathfrak{l}$ a subalgebra of $\mathfrak{g}$. Let $\mathfrak{g}_0 = [\mathfrak{g}, [\mathfrak{g}, \mathfrak{g}]]$ and $\mathfrak{l}_0 = [\mathfrak{l}, [\mathfrak{l}, \mathfrak{l}]]$. We consider the decompositions

$$\mathfrak{g} = \mathfrak{g}_0 \oplus \mathfrak{g}_1 \oplus \mathfrak{g}_2 \quad \text{and} \quad \mathfrak{l} = \mathfrak{l}_0 \oplus \mathfrak{l}_1 \oplus \mathfrak{l}_2, \tag{4.37}$$

where $\mathfrak{g}_1$ (resp., $\mathfrak{l}_1$) designates a subspace of $\mathfrak{g}$ (of $\mathfrak{l}$, resp.) such that $\mathfrak{g}_0 \oplus \mathfrak{g}_1 = [\mathfrak{g}, \mathfrak{g}]$ (resp., $\mathfrak{l}_0 \oplus \mathfrak{l}_1 = [\mathfrak{l}, \mathfrak{l}]$). $\mathfrak{g}_2$ (resp., $\mathfrak{l}_2$) is a subspace of $\mathfrak{g}$ (of $\mathfrak{l}$, resp.) supplementary to $[\mathfrak{g}, \mathfrak{g}]$ (to $[\mathfrak{l}, \mathfrak{l}]$, resp.) in $\mathfrak{g}$ (in $\mathfrak{l}$, resp.).

Obviously, we can see that $\mathfrak{g}_0$ (resp., $\mathfrak{l}_0$) lies in the center of $\mathfrak{g}$ (of $\mathfrak{l}$, resp.). Any $\varphi \in \mathscr{L}(\mathfrak{l}, \mathfrak{g})$ can be written as

$$\varphi = \begin{pmatrix} A_\varphi & B_\varphi & C_\varphi \\ I_\varphi & D_\varphi & E_\varphi \\ J_\varphi & K_\varphi & F_\varphi \end{pmatrix}, \tag{4.38}$$

where $A_\varphi \in \mathscr{L}(\mathfrak{l}_0, \mathfrak{g}_0)$, $B_\varphi \in \mathscr{L}(\mathfrak{l}_1, \mathfrak{g}_0)$, $C_\varphi \in \mathscr{L}(\mathfrak{l}_2, \mathfrak{g}_0)$, $I_\varphi \in \mathscr{L}(\mathfrak{l}_0, \mathfrak{g}_1)$, $D_\varphi \in \mathscr{L}(\mathfrak{l}_1, \mathfrak{g}_1)$, $E_\varphi \in \mathscr{L}(\mathfrak{l}_2, \mathfrak{g}_1)$, $J_\varphi \in \mathscr{L}(\mathfrak{l}_0, \mathfrak{g}_2)$, $K_\varphi \in \mathscr{L}(\mathfrak{l}_1, \mathfrak{g}_2)$ and $F_\varphi \in \mathscr{L}(\mathfrak{l}_2, \mathfrak{g}_2)$.

For each $\varphi \in \mathscr{L}(\mathfrak{l}, \mathfrak{g})$, we define $\varphi_1 \in \mathscr{L}(\mathfrak{l}, \mathfrak{g})$ by

$$\varphi_1 = \begin{pmatrix} A_\varphi & B_\varphi & 0 \\ 0 & D_\varphi & E_\varphi \\ 0 & 0 & F_\varphi \end{pmatrix}. \tag{4.39}$$

We first remark the following assertion.

Lemma 4.3.8. *An element $\varphi \in \mathscr{L}(\mathfrak{l}, \mathfrak{g})$ is a Lie algebra homomorphism if and only if $I_\varphi = 0$, $J_\varphi = 0$, $K_\varphi = 0$ and $\varphi_1 \in \operatorname{Hom}(\mathfrak{l}, \mathfrak{g})$.*

Proof. We point out first that if $\varphi \in \operatorname{Hom}(\mathfrak{l}, \mathfrak{g})$, then $\varphi(\mathfrak{l}_0) = \varphi([\mathfrak{l}, [\mathfrak{l}, \mathfrak{l}]]) = [\varphi(\mathfrak{l}), [\varphi(\mathfrak{l}), \varphi(\mathfrak{l})]] \subset \mathfrak{g}_0$, in particular $I_\varphi = 0$ and $J_\varphi = 0$. Now $\varphi(\mathfrak{l}_1) \subset \varphi([\mathfrak{l}, \mathfrak{l}]) = [\varphi(\mathfrak{l}), \varphi(\mathfrak{l})] \subset [\mathfrak{g}, \mathfrak{g}]$, so $K_\varphi = 0$ and

$$\varphi = \begin{pmatrix} A_\varphi & B_\varphi & C_\varphi \\ 0 & D_\varphi & E_\varphi \\ 0 & 0 & F_\varphi \end{pmatrix}. \tag{4.40}$$

Let us take $\varphi \in \mathscr{L}(\mathfrak{l}, \mathfrak{g})$ with $I_\varphi = 0$, $J_\varphi = 0$, $K_\varphi = 0$. Then for each $x = x_0 + x_1 + x_2$ and $x' = x'_0 + x'_1 + x'_2 \in \mathfrak{l}$ where $x_i, x'_i \in \mathfrak{l}_i$, $i = 0, 1, 2$, we have the following:

$$\begin{aligned} [\varphi(x), \varphi(x')] &= [A_\varphi(x_0) + B_\varphi(x_1) + D_\varphi(x_1) + C_\varphi(x_2) + E_\varphi(x_2) + F_\varphi(x_2), A_\varphi(x'_0) \\ &\quad + B_\varphi(x'_1) + D_\varphi(x'_1) + C_\varphi(x'_2) + E_\varphi(x'_2) + F_\varphi(x'_2)] \\ &= [D_\varphi(x_1) + E_\varphi(x_2) + F_\varphi(x_2), D_\varphi(x'_1) + E_\varphi(x'_2) + F_\varphi(x'_2)] \\ &= [\varphi_1(x), \varphi_1(x')]. \end{aligned} \tag{4.41}$$

On the other hand,

$$\begin{aligned} \varphi([x, x']) &= \varphi([x_0 + x_1 + x_2, x'_0 + x'_1 + x'_2]) \\ &= \varphi([x_1 + x_2, x'_1 + x'_2]) \\ &= \varphi([x_1, x'_2]) + \varphi([x_2, x'_1]) + \varphi([x_2, x'_2]) \\ &= A_\varphi([x_1, x'_2]) + A_\varphi([x_2, x'_1]) + (A_\varphi + B_\varphi + D_\varphi)([x_2, x'_2]) \\ &= \varphi_1([x, x']). \end{aligned} \tag{4.42}$$

Conversely, let $I_\varphi = 0, J_\varphi = 0, K_\varphi = 0$ and $\varphi_1 \in \mathrm{Hom}(\mathfrak{l}, \mathfrak{g})$, then φ is as in (4.40). Hence,

$$\begin{aligned}\varphi([x,x']) &= \varphi_1([x,x']) \quad \text{by (4.42)}\\ &= [\varphi_1(x), \varphi_1(x')]\\ &= [\varphi(x), \varphi(x')] \quad \text{by (4.41).}\end{aligned}$$

Then for each $\varphi \in \mathscr{L}(\mathfrak{l}, \mathfrak{g})$ with $I_\varphi = 0, J_\varphi = 0, K_\varphi = 0$, we have $\varphi \in \mathrm{Hom}(\mathfrak{l}, \mathfrak{g})$ if and only if $\varphi_1 \in \mathrm{Hom}(\mathfrak{l}, \mathfrak{g})$. □

The *G*-action on Hom($\mathfrak{l}$, $\mathfrak{g}$)

For any $X \in \mathfrak{g}$, the adjoint representation ad_X can be written making use the decomposition (4.37) as

$$\mathrm{ad}_X = \begin{pmatrix} 0 & \Sigma_{1,2}(X) & \Sigma_{1,3}(X)\\ 0 & 0 & \Sigma_{2,3}(X)\\ 0 & 0 & 0\end{pmatrix} \tag{4.43}$$

for some maps $\Sigma_{1,2} : \mathfrak{g} \to \mathscr{L}(\mathfrak{g}_1, \mathfrak{g}_0)$, $\Sigma_{1,3} : \mathfrak{g} \to \mathscr{L}(\mathfrak{g}_2, \mathfrak{g}_0)$ and $\Sigma_{2,3} : \mathfrak{g} \to \mathscr{L}(\mathfrak{g}_2, \mathfrak{g}_1)$.

The adjoint representation $\mathrm{Ad}_{\exp(X)}$ reads therefore

$$\begin{aligned}\mathrm{Ad}_{\exp(X)} &= \mathbb{I}_+ \mathrm{ad}_X + \frac{1}{2}\mathrm{ad}_X^2\\ &= \begin{pmatrix} I_{\mathfrak{g}_0} & \Sigma_{1,2}(X) & \Sigma_{1,3}(X) + \frac{1}{2}\Sigma_{1,2}(X)\Sigma_{2,3}(X)\\ 0 & I_{\mathfrak{g}_1} & \Sigma_{2,3}(X)\\ 0 & 0 & I_{\mathfrak{g}_2}\end{pmatrix}.\end{aligned}$$

Here, $I_{\mathfrak{g}_0}, I_{\mathfrak{g}_1}$ and $I_{\mathfrak{g}_2}$ denote the identity maps of $\mathfrak{g}_0$, $\mathfrak{g}_1$ and $\mathfrak{g}_2$, respectively.

The group G acts on $\mathrm{Hom}(\mathfrak{l}, \mathfrak{g})$ through the following law:

$$\begin{aligned}g \cdot \varphi &= \mathrm{Ad}_g \circ \varphi\\ &= \begin{pmatrix} A_\varphi & B_\varphi + \Sigma_{1,2}(X)D_\varphi & C_\varphi + \Sigma_{1,2}(X)E_\varphi + \Sigma_{1,3}(X)F_\varphi + \frac{1}{2}\Sigma_{1,2}(X)\Sigma_{2,3}(X)F_\varphi\\ 0 & D_\varphi & E_\varphi + \Sigma_{2,3}(X)F_\varphi\\ 0 & 0 & F_\varphi\end{pmatrix},\end{aligned}$$

where $A_\varphi, B_\varphi, C_\varphi, D_\varphi, E_\varphi$ and F_φ are as in formula (4.38), $g = \exp(X)$ and $\Sigma_{1,2}$, $\Sigma_{1,3}$ and $\Sigma_{2,3}$ are as in (4.43). Let now

$$\mathrm{Hom}_1(\mathfrak{l}, \mathfrak{g}) := \{\varphi \in \mathrm{Hom}(\mathfrak{l}, \mathfrak{g}) \mid C_\varphi = 0\}. \tag{4.44}$$

By Lemma 4.3.8, the correspondence $\varphi \mapsto \varphi_1$ gives a map: $\mathrm{Hom}(\mathfrak{l}, \mathfrak{g}) \to \mathrm{Hom}_1(\mathfrak{l}, \mathfrak{g})$. Then G also acts on $\mathrm{Hom}_1(\mathfrak{l}, \mathfrak{g})$ as follows:

$$g * \varphi_1 = \begin{pmatrix} A_\varphi & B_\varphi + \Sigma_{1,2}(X)D_\varphi & 0 \\ 0 & D_\varphi & E_\varphi + \Sigma_{2,3}(X)F_\varphi \\ 0 & 0 & F_\varphi \end{pmatrix}. \tag{4.45}$$

In other words, $g * \varphi_1$ is defined by $(g \cdot \varphi_1)_1$ where $(g \cdot \varphi_1) \in \mathrm{Hom}(\mathfrak{l}, \mathfrak{g})$. One can easily check that $g * \varphi_1$ defines a group action of G on $\mathrm{Hom}_1(\mathfrak{l}, \mathfrak{g})$. Hence, G acts on $\mathrm{Hom}_1(\mathfrak{l}, \mathfrak{g}) \times \mathscr{L}(\mathfrak{l}_2, \mathfrak{g}_0)$ as

$$g \cdot (\varphi_1, C_\varphi) = \left(g * \varphi_1, C_\varphi + \Sigma_{1,2}(X)E_\varphi + \Sigma_{1,3}(X)F_\varphi + \frac{1}{2}\Sigma_{1,2}(X)\Sigma_{2,3}(X)F_\varphi \right). \tag{4.46}$$

We first have the following.

Lemma 4.3.9. *The map*

$$\begin{aligned} \psi : \mathrm{Hom}(\mathfrak{l}, \mathfrak{g}) &\longrightarrow \mathrm{Hom}_1(\mathfrak{l}, \mathfrak{g}) \times \mathscr{L}(\mathfrak{l}_2, \mathfrak{g}_0) \\ \varphi &\longmapsto (\varphi_1, C_\varphi) \end{aligned}$$

is a G-equivariant homeomorphism, where φ_1 is as in (4.39).

Proof. The fact that ψ is a well-defined homeomorphism comes directly from Lemma 4.3.8. Let $g = \exp(X) \in G$ and $\varphi \in \mathrm{Hom}(\mathfrak{l}, \mathfrak{g})$, then

$$\begin{aligned} \psi(g \cdot \varphi) &= \psi(\mathrm{Ad}_g \circ \varphi) \\ &= \left(g * \varphi_1, C_\varphi + \Sigma_{1,2}(X)E_\varphi + \Sigma_{1,3}(X)F_\varphi + \frac{1}{2}\Sigma_{1,2}(X)\Sigma_{2,3}(X)F_\varphi \right) \\ &= g \cdot \psi(\varphi), \end{aligned}$$

which proves the lemma. □

Decomposition of $\mathrm{Hom}_1(\mathfrak{l}, \mathfrak{g})$

Now we consider the linear subspace Δ of $\mathscr{L}(\mathfrak{l}, \mathfrak{g})$ defined by

$$\Delta = \left\{ \varphi \in \mathscr{L}(\mathfrak{l}, \mathfrak{g}) \;\middle|\; \begin{array}{l} A_\varphi = 0, I_\varphi = 0, J_\varphi = 0, D_\varphi = 0, \\ K_\varphi = 0, C_\varphi = 0 \text{ and } F_\varphi = 0 \end{array} \right\} \cong \mathscr{L}(\mathfrak{l}_1, \mathfrak{g}_0) \times \mathscr{L}(\mathfrak{l}_2, \mathfrak{g}_1).$$

For $\varphi_1 \in \mathrm{Hom}_1(\mathfrak{l}, \mathfrak{g})$, we consider the linear map

$$\begin{aligned} l_{\varphi_1} : \mathfrak{g} &\longrightarrow \Delta \\ X &\longmapsto \begin{pmatrix} 0 & \Sigma_{1,2}(X)D_{\varphi_1} & 0 \\ 0 & 0 & \Sigma_{2,3}(X)F_{\varphi_1} \\ 0 & 0 & 0 \end{pmatrix}. \end{aligned}$$

Then from equation (4.45) of the definition of the action of G on $\mathrm{Hom}_1(\mathfrak{l}, \mathfrak{g})$, we obtain immediately the following description of the orbits in $\mathrm{Hom}_1(\mathfrak{l}, \mathfrak{g})$.

Lemma 4.3.10. *The orbit* $G * \varphi_1 = \varphi_0 + (N_{\varphi_1} + \mathrm{Im}(l_{\varphi_1}))$, *where*

$$\varphi_0 = \begin{pmatrix} A_{\varphi_1} & 0 & 0 \\ 0 & D_{\varphi_1} & 0 \\ 0 & 0 & F_{\varphi_1} \end{pmatrix} \quad \text{and} \quad N_{\varphi_1} = \begin{pmatrix} 0 & B_{\varphi_1} & 0 \\ 0 & 0 & E_{\varphi_1} \\ 0 & 0 & 0 \end{pmatrix}.$$

Let $m = \dim \Delta$ and $q = \dim \mathfrak{g}$. For $t = 0, \dots, q$, we define the sets

$$\mathrm{Hom}_1^t(\mathfrak{l}, \mathfrak{g}) := \{\varphi_1 \in \mathrm{Hom}_1(\mathfrak{l}, \mathfrak{g}) \mid \mathrm{rk}(l_{\varphi_1}) = t\}.$$

Then clearly,

$$\mathrm{Hom}_1(\mathfrak{l}, \mathfrak{g}) = \bigcup_{t=0}^{q} \mathrm{Hom}_1^t(\mathfrak{l}, \mathfrak{g}). \tag{4.47}$$

We fix a basis $\{e_1, \dots, e_m\}$ of Δ and let

$$I(m, m-t) = \{(i_1, \dots, i_{m-t}) \in \mathbb{N}^{m-t} \mid 1 \le i_1 < \cdots < i_{m-t} \le m\}.$$

For $\beta = (i_1, \dots, i_{m-t}) \in I(m, m-t)$, we consider the subspace $V_\beta := \bigoplus_{j=1}^{m-t} \mathbb{R}e_{i_j}$ and for any $\varphi_1 \in \mathrm{Hom}_1^t(\mathfrak{l}, \mathfrak{g})$, let $P_{\varphi_1} : \Delta \to \Delta/\mathrm{Im}(l_{\varphi_1})$ and

$$\mathrm{Hom}_{1,\beta}^t(\mathfrak{l}, \mathfrak{g}) := \{\varphi_1 \in \mathrm{Hom}_1^t(\mathfrak{l}, \mathfrak{g}) \mid \det(P_{\varphi_1}(e_{i_1}), \dots, P_{\varphi_1}(e_{i_{m-t}})) \neq 0\}.$$

Then we have the following.

Lemma 4.3.11. *For each* $t = 0, \dots, q = \dim \mathfrak{g}$, *the family* $\{\mathrm{Hom}_{1,\beta}^t(\mathfrak{l}, \mathfrak{g})\}_{\beta \in I(m,m-t)}$ *of subsets in* $\mathrm{Hom}_1^t(\mathfrak{l}, \mathfrak{g})$ *gives an open covering of* $\mathrm{Hom}_1^t(\mathfrak{l}, \mathfrak{g})$.

Proof. We know that for all $\varphi_1 \in \mathrm{Hom}_1^t(\mathfrak{l}, \mathfrak{g})$, the set $\mathrm{Im}(l_{\varphi_1})$ is a linear subspace of Δ of dimension t. There exists therefore $(i_1, \dots, i_{m-t}) \in I(m, m-t)$ such that the family $\{P_{\varphi_1}(e_{i_1}), \dots, P_{\varphi_1}(e_{i_{m-t}})\}$ forms a basis of $\Delta/\mathrm{Im}(l_{\varphi_1})$, and consequently

$$\det(P_{\varphi_1}(e_{i_1}), \dots, P_{\varphi_1}(e_{i_{m-t}})) \neq 0.$$

This shows that $\mathrm{Hom}_1^t(\mathfrak{l},\mathfrak{g}) = \bigcup_{\beta\in I(m,m-t)} \mathrm{Hom}_{1,\beta}^t(\mathfrak{l},\mathfrak{g})$. Now $\det(P_{\varphi_1}(e_{i_1}),\dots,P_{\varphi_1}(e_{i_{m-t}})) \neq 0$ if and only if the family $\{P_{\varphi_1}(e_{i_1}),\dots,P_{\varphi_1}(e_{i_{m-t}})\}$ is a basis of $\Delta/\operatorname{Im} l_{\varphi_1}$, which is equivalent to $\Delta = \operatorname{Im} l_{\varphi_1} \oplus V_\beta$. As $\dim \operatorname{Im} l_{\varphi_1} = t$, we get by [Lemma 4.3.2(1)] that there exists $(j_1,\dots,j_t) \in I(q,t)$ such that the family $\{l_{\varphi_1}(Y_{j_1}),\dots,l_{\varphi_1}(Y_{j_t}),e_{i_1},\dots,e_{i_{m-t}}\}$ forms a basis of Δ, or similarly

$$\sum_{(j_1,\dots,j_t)\in I(k,t)} [\det(l_{\varphi_1}(Y_{j_1}),\dots,l_{\varphi_1}(Y_{j_t}),e_{i_1},\dots,e_{i_{m-t}})]^2 \neq 0.$$

Then

$$\mathrm{Hom}_{1,\beta}^t(\mathfrak{l},\mathfrak{g}) = \left\{\varphi_1 \in \mathrm{Hom}_1^t(\mathfrak{l},\mathfrak{g}) \;\middle|\; \sum_{(j_1,\dots,j_t)\in I(k,t)} [\det(l_{\varphi_1}(Y_{j_1}),\dots,l_{\varphi_1}(Y_{j_t}),e_{i_1},\dots,e_{i_{m-t}})]^2 \neq 0\right\},$$

which is open by continuity of the determinant. □

Proposition 4.3.12. *We have*

$$\mathrm{Hom}_1(\mathfrak{l},\mathfrak{g}) = \bigcup_{t=0}^{q} \bigcup_{\beta\in I(m,m-t)} \mathrm{Hom}_{1,\beta}^t(\mathfrak{l},\mathfrak{g})$$

as a union of G-invariant subsets, where G acts on $\mathrm{Hom}_1(\mathfrak{l},\mathfrak{g})$ *as in* (4.45).

Proof. The decomposition is given by Lemma 4.3.11 and equation (4.47). To see the G-invariance, observe that $D_{\varphi_1} = D_{g*\varphi_1}$ and $F_{\varphi_1} = F_{g*\varphi_1}$, which means that $l_{\varphi_1} = l_{g*\varphi_1}$ and $P_{\varphi_1} = P_{g*\varphi_1}$ for all $\varphi_1 \in \mathrm{Hom}_1(\mathfrak{l},\mathfrak{g})$ and $g \in G$. Then for all $\beta \in I(m,m-t)$ and $0 \leq t \leq q$ the set $\mathrm{Hom}_{1,\beta}^t(\mathfrak{l},\mathfrak{g})$ is G-invariant. □

Let us fix $t = 0,\dots,q$ and $\beta = (i_1,\dots,i_{m-t}) \in I(m,m-t)$. Recall that V_β is a subspace of Δ spanned by $\{e_{i_k}\}_{k=1,\dots,m-t}$. We define the subset $\mathcal{M}_\beta^t(\mathfrak{l},\mathfrak{g})$ of $\mathrm{Hom}_{1,\beta}^t(\mathfrak{l},\mathfrak{g})$ by

$$\mathcal{M}_\beta^t(\mathfrak{l},\mathfrak{g}) = \{\varphi_1 \in \mathrm{Hom}_{1,\beta}^t(\mathfrak{l},\mathfrak{g}) \mid N_{\varphi_1} \in V_\beta\}$$

and we consider the map

$$\begin{aligned} \pi_\beta^t : \mathrm{Hom}_{1,\beta}^t(\mathfrak{l},\mathfrak{g})/G &\longrightarrow \mathcal{M}_\beta^t(\mathfrak{l},\mathfrak{g}) \\ G*\varphi_1 &\longmapsto \varphi_0 + P_{\varphi_1|V_\beta}^{-1}(N_{\varphi_1} + \mathrm{Im}(l_{\varphi_1})). \end{aligned}$$

We next prove the following lemmas.

Lemma 4.3.13. *For each* $\varphi_1 \in \mathrm{Hom}_{1,\beta}^t(\mathfrak{l},\mathfrak{g})$, *the intersection of the G-orbit* $G*\varphi_1$ *and* $\mathcal{M}_\beta^t(\mathfrak{l},\mathfrak{g})$ *in* $\mathrm{Hom}_{1,\beta}^t(\mathfrak{l},\mathfrak{g})$ *is the one set* $\{\varphi_0 + P_{\varphi_1|V_\beta}^{-1}(N_{\varphi_1} + \mathrm{Im}(l_{\varphi_1}))\}$. *In particular, the* *map*

$$\pi_\beta^t : \mathrm{Hom}_{1,\beta}^t(\mathfrak{l},\mathfrak{g})/G \longrightarrow \mathscr{M}_\beta^t(\mathfrak{l},\mathfrak{g}); \quad G * \varphi_1 \longmapsto \varphi_0 + P^{-1}_{\varphi_1|V_\beta}(N_{\varphi_1} + \mathrm{Im}(l_{\varphi_1}))$$

is well-defined.

Proof. Let $\varphi_1 = \varphi_0 + N_{\varphi_1} \in \mathrm{Hom}_{1,\beta}^t(\mathfrak{l},\mathfrak{g})$. Then from Lemma 4.3.10 the orbit $G * \varphi_1 = \varphi_0 + (N_{\varphi_1} + \mathrm{Im}\, l_{\varphi_1})$ and

$$\varphi_0 + P^{-1}_{\varphi_1|V_\beta}(N_{\varphi_1} + \mathrm{Im}\, l_{\varphi_1}) = \varphi_0 + (N_{\varphi_1} + \mathrm{Im}\, l_{\varphi_1}) \cap V_\beta \in \mathscr{M}_\beta^t(\mathfrak{l},\mathfrak{g}).$$

Thus, π_β^t is well-defined. Note now that the intersection of $G * \varphi_1$ and $\mathscr{M}_\beta^t(\mathfrak{l},\mathfrak{g})$ in $\mathrm{Hom}_{1,\beta}^t(\mathfrak{l},\mathfrak{g})$ is not empty as it contains $\pi_\beta^t(G*\varphi_1)$. Let φ_1, φ_1' be two elements in the intersection. Then there exist $v, w \in \mathrm{Im}(l_{\varphi_1})$ such that $\varphi_1 = \varphi_0 + N_{\varphi_1} + v$ and $\varphi_1' = \varphi_0 + N_{\varphi_1} + w$ with $N_{\varphi_1} + v, N_{\varphi_1} + w \in V_\beta$. In particular, $v - w \in V_\beta \cap \mathrm{Im}(l_{\varphi_1}) = \{0\}$, which means that $\varphi_1 = \varphi_1'$. □

Lemma 4.3.14. *The map*

$$\begin{aligned} h : \mathrm{Hom}_{1,\beta}^t(\mathfrak{l},\mathfrak{g}) &\longrightarrow \mathscr{M}_\beta^t(\mathfrak{l},\mathfrak{g}) \\ \varphi_1 &\longmapsto \varphi_0 + P^{-1}_{\varphi_1|V_\beta}(N_{\varphi_1} + \mathrm{Im}\, l_{\varphi_1}) \end{aligned}$$

is continuous.

Proof. To show this lemma, we prove the following fact.

Fact 4.3.15. *The map*

$$\begin{aligned} \mathrm{Hom}_1(\mathfrak{l},\mathfrak{g}) &\longrightarrow \mathscr{L}(\mathfrak{g},\Delta) \\ \varphi_1 &\longmapsto l_{\varphi_1} \end{aligned}$$

is continuous.

Proof. Let $(\varphi_1^{(n)})_n$ be a sequence, which converges to some element φ_1. Then obviously $(D_{\varphi_1^{(n)}})_n$ converges to D_{φ_1} and $(F_{\varphi_1^{(n)}})_n$ converges to F_{φ_1}. Then $(l_{\varphi_1^{(n)}})_n$ converges to l_{φ_1}. □

Now to prove the lemma, let $(\varphi_{1,n})_n$ be a sequence in $\mathrm{Hom}_{1,\beta}^t(\mathfrak{l},\mathfrak{g})$, which converges to an element $\varphi_1 \in \mathrm{Hom}_{1,\beta}^t(\mathfrak{l},\mathfrak{g})$. We have to show that $(h(\varphi_{1,n}))_n$ converges to $h(\varphi_1)$. Note first that for all $\varphi_1 \in \mathrm{Hom}_{1,\beta}^t(\mathfrak{l},\mathfrak{g})$, $h(\varphi_1) = \varphi_0 + (N_{\varphi_1} + \mathrm{Im}\, l_{\varphi_1}) \cap V_\beta$. As $\Delta = \mathrm{Im}\, l_{\varphi_1} \oplus V_\beta$, by Lemma 4.3.1, we see that $h(\varphi_1) = \varphi_0 + q_{\varphi_1}(N_{\varphi_1})$, where q_{φ_1} is the projection on V_β parallel to $\mathrm{Im}\, l_{\varphi_1}$. Let $\{X_1, \dots, X_n\}$ be a basis of $\mathfrak{g}$, using [Lemma 4.3.2(1)], one can find $X_{j_1}, \dots, X_{j_t} \in \{X_1, \dots, X_n\}$ such that

$$\mathrm{Im}\, l_{\varphi_1} = \mathbb{R}\text{-span}\{l_{\varphi_1}(X_{j_1}), \dots, l_{\varphi_1}(X_{j_t})\}.$$

Now by Fact 4.3.15, we get the convergence of the sequence $(l_{\varphi_{1,n}})_n$ to l_{φ_1}. By [Lemma 4.3.2(2)], for $S = \{X_{j_1}, \ldots, X_{j_t}\}$, $A(S) = \{l \in \mathscr{L}(\mathfrak{g}, \Delta) \mid \dim \mathbb{R}\text{-span}\{l(X_{j_1}), \ldots, l(X_{j_t})\} = t\}$ is open in $\mathscr{L}(\mathfrak{g}, \Delta)$. Then there exists $N > 0$ such that for all $n > N$,

$$\operatorname{Im} l_{\varphi_1} = \mathbb{R}\text{-span}\{l_{\varphi_{1,n}}(X_{j_1}), \ldots, l_{\varphi_{1,n}}(X_{j_t})\}.$$

As $(l_{\varphi_{1,n}})_n$ converges to l_{φ_1}, the sequence $(l_{\varphi_{1,n}}(X_{j_k}))_n$ converges to $l_{\varphi_1}(X_{j_k})$ for all $1 \le k \le t$.

By Lemma 4.3.3, let q_n be the projection of Δ on V_β parallel to $\operatorname{Im}(l_{\varphi_{1,n}})$, the sequence $(q_n)_n$ converges to the projection q_{φ_1} of Δ on V_β parallel to $\operatorname{Im} l_{\varphi_1}$. Finally, as $(\varphi_{0,n})_n$ converges to φ_0 and $(N_{\varphi_{1,n}})_n$ converges to N_{φ_1}, we get $(h(\varphi_{1,n}))_n$ converges to $h(\varphi_1)$. □

Lemma 4.3.16. *The map* $\pi^t_\beta : \operatorname{Hom}^t_{1,\beta}(\mathfrak{l}, \mathfrak{g})/G \longrightarrow \mathscr{M}^t_\beta(\mathfrak{l}, \mathfrak{g})$ *defined above is a homeomorphism.*

Proof. To see that π^t_β is surjective, observe that $\mathscr{M}^t_\beta(\mathfrak{l}, \mathfrak{g}) \subset \operatorname{Hom}^t_{1,\beta}(\mathfrak{l}, \mathfrak{g})$ and for $\varphi_1 \in \mathscr{M}^t_\beta(\mathfrak{l}, \mathfrak{g})$ we have $\pi^t_\beta(G * \varphi_1) = \varphi_1$. Let $\varphi_1, \xi_1 \in \operatorname{Hom}^t_{1,\beta}(\mathfrak{l}, \mathfrak{g})$ such that $\pi^t_\beta(G * \varphi_1) = \pi^t_\beta(G * \xi_1)$. Then obviously $\varphi_0 = \xi_0$ and

$$P^{-1}_{\varphi_1|V_\beta}(N_{\varphi_1} + \operatorname{Im} l_{\varphi_1}) = P^{-1}_{\varphi_1|V_\beta}(N_{\varphi_1} + \operatorname{Im} l_{\varphi_1}).$$

As l_{φ_1} depends only on φ_0, we deduce that $l_{\varphi_1} = l_{\xi_1}$, which implies that $P^{-1}_{\varphi_1|V_\beta} = P^{-1}_{\xi_1|V_\beta}$. Hence, $N_{\varphi_1} + \operatorname{Im} l_{\varphi_1} = N_{\xi_1} + \operatorname{Im} l_{\xi_1}$ and in particular $G * \varphi_1 = G * \xi_1$. Thus, π^t_β is injective. Now the following diagram commutes:

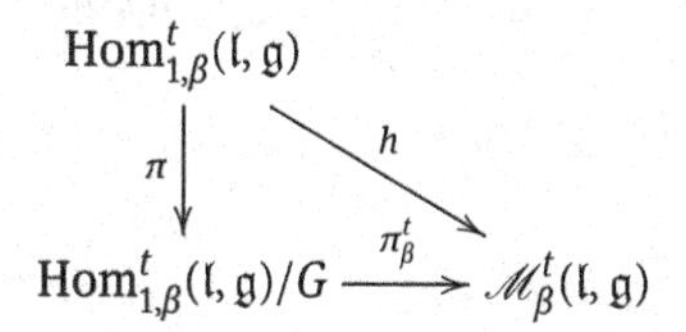

where $h(\varphi_1) = \varphi_0 + P^{-1}_{\varphi_1|V_\beta}(N_{\varphi_1} + \operatorname{Im} l_{\varphi_1})$. Since by Lemma 4.3.14, h is continuous and since π is open then π^t_β is continuous. The quotient canonical map $(\pi^t_\beta)^{-1} = \pi_{|\mathscr{M}^t_\beta(\mathfrak{l},\mathfrak{g})}$ is continuous and then the map π^t_β is a homeomorphism. □

Corollary 4.3.17. *For all* $t = \{0, \ldots, q\}$, *the collection* $\mathrm{S}^t_\beta = (\pi^t_\beta, \operatorname{Hom}^t_{1,\beta}(\mathfrak{l}, \mathfrak{g})/G)_{\beta \in I(m,m-t)}$ *forms a family of local sections of the canonical surjection*

$$\pi^t : \operatorname{Hom}^t_1(\mathfrak{l}, \mathfrak{g}) \to \operatorname{Hom}^t_1(\mathfrak{l}, \mathfrak{g})/G.$$

In particular,

$$\pi^t_\beta(\operatorname{Hom}^t_{1,\beta}(\mathfrak{l}, \mathfrak{g})/G) = \left\{ \varphi_1 \in \operatorname{Hom}_1(\mathfrak{l}, \mathfrak{g}) \;\middle|\; \begin{matrix} \varphi_1 \in \operatorname{Hom}^t_{1,\beta}(\mathfrak{l}, \mathfrak{g}), \\ N_{\varphi_1} \in \mathrm{V}_\beta \end{matrix} \right\}.$$

Decomposition of Hom$(\mathfrak{l}, \mathfrak{g})$
Let $\varphi_1 \in \mathrm{Hom}^t_{1,\beta}(\mathfrak{l}, \mathfrak{g})$ and

$$G_{\varphi_1} = \{g \in G \mid g * \varphi_1 = \varphi_1\}$$

be the isotropy group of φ_1. The group G_{φ_1} acts on $\{\varphi_1\}\times\mathscr{L}(\mathfrak{l}_2, \mathfrak{g}_0)$ through the following law:

$$\exp(X)\cdot(\varphi_1, C) = (\varphi_1, C + \Sigma_{1,2}(X)E_{\varphi_1} + \Sigma_{1,3}(X)F_{\varphi_1}),$$

where $\Sigma_{1,2}$ and $\Sigma_{1,3}$ are as in (4.43). Indeed, for any $g = \exp(X) \in G_{\varphi_1}$, we have $g * \varphi_1 = \varphi_1$, then $\Sigma_{1,2}(X)D_{\varphi_1} = 0$ and $\Sigma_{2,3}(X)F_{\varphi_1} = 0$. We get therefore from (4.46),

$$g\cdot(\varphi_1, C) = (\varphi_1, C + \Sigma_{1,2}(X)E_{\varphi_1} + \Sigma_{1,3}(X)F_{\varphi_1}).$$

Let $\mathfrak{g}_{\pi^t_\beta(G*\varphi_1)} = \log(G_{\pi^t_\beta(G*\varphi_1)})$ and f_{φ_1} be the linear map defined by

$$\begin{aligned} f_{\varphi_1} : \mathfrak{g}_{\pi^t_\beta(G*\varphi_1)} &\longrightarrow \mathscr{L}(\mathfrak{l}_2, \mathfrak{g}_0)\\ X &\longmapsto \Sigma_{1,2}(X)E_{\pi^t_\beta(G*\varphi_1)} + \Sigma_{1,3}(X)F_{\pi^t_\beta(G*\varphi_1)}. \end{aligned}$$

Then the range of f_{φ_1} is a linear subspace of $\mathscr{L}(\mathfrak{l}_2, \mathfrak{g}_0)$ and we can see immediately that we have the following.

Lemma 4.3.18. *For any $g \in G$, we have $f_{\varphi_1} = f_{g*\varphi_1}$. In addition,*

$$G_{\pi^t_\beta(G*\varphi_1)}\cdot(\pi^t_\beta(G * \varphi_1), C) = (\pi^t_\beta(G * \varphi_1), C + \mathrm{Im}(f_{\varphi_1})).$$

Let $m' = \dim(\mathscr{L}(\mathfrak{l}_2, \mathfrak{g}_0))$ and $q' = \dim(\mathfrak{g}_{\pi^t_\beta(G*\varphi_1)})$. For $t' = 0, \dots, q'$, we define the sets

$$\mathrm{Hom}^{t,t'}_{1,\beta}(\mathfrak{l}, \mathfrak{g}) = \{\varphi_1 \in \mathrm{Hom}^t_{1,\beta}(\mathfrak{l}, \mathfrak{g}) \mid \mathrm{rk}(f_{\varphi_1}) = t'\}.$$

Then clearly

$$\mathrm{Hom}^t_{1,\beta}(\mathfrak{l}, \mathfrak{g}) = \bigcup_{t'=0}^{q'} \mathrm{Hom}^{t,t'}_{1,\beta}(\mathfrak{l}, \mathfrak{g}). \tag{4.48}$$

Let us fix a basis $\{e'_1, \dots, e'_{m'}\}$ of $\mathscr{L}(\mathfrak{l}_2, \mathfrak{g}_0)$. For $\beta' = (i'_1, \dots, i'_{m'-t'}) \in I(m', m' - t')$ and $\varphi_1 \in \mathrm{Hom}^{t,t'}_{1,\beta}(\mathfrak{l}, \mathfrak{g})$, we consider the subspace $V_{\beta'} := \bigoplus_{j=1}^{m'-t'} \mathbb{R}e'_{i'_j}$, the quotient map

$$P'_{\varphi_1} : \mathscr{L}(\mathfrak{l}_2, \mathfrak{g}_0) \to \mathscr{L}(\mathfrak{l}_2, \mathfrak{g}_0)/\,\mathrm{Im}(f_{\varphi_1})$$

and the set

$$\mathrm{Hom}^{t,t'}_{1,\beta,\beta'}(\mathfrak{l},\mathfrak{g}) = \{\varphi_1 \in \mathrm{Hom}^{t,t'}_{1,\beta}(\mathfrak{l},\mathfrak{g}) \mid \det(P'_{\varphi_1}(e'_{i'_1}),\dots,P'_{\varphi_1}(e'_{i'_{m'-t'}})) \neq 0\}.$$

Then we get the following.

Lemma 4.3.19. *For each $t = 0,\dots,q = \dim\mathfrak{g}$, $t' = 0,\dots,q' = \dim(\mathfrak{g}_{\pi^t_\beta(G*\varphi_1)})$ and $\beta \in I(m,m-t)$, the family $\{\mathrm{Hom}^{t,t'}_{1,\beta,\beta'}(\mathfrak{l},\mathfrak{g})\}_{\beta'\in I(m',m'-t')}$ of subsets in $\mathrm{Hom}^{t,t'}_{1,\beta}(\mathfrak{l},\mathfrak{g})$ gives an open covering of $\mathrm{Hom}^{t,t'}_{1,\beta}(\mathfrak{l},\mathfrak{g})$.*

Proof. For all $\varphi_1 \in \mathrm{Hom}^{t,t'}_{1,\beta}(\mathfrak{l},\mathfrak{g})$, the set $\mathrm{Im}(f_{\varphi_1})$ is a linear subspace of $\mathscr{L}(\mathfrak{l}_2,\mathfrak{g}_0)$ of dimension t'. There exists therefore $(i'_1,\dots,i'_{m'-t'}) \in I(m',m'-t')$ such that the family $\{P'_{\varphi_1}(e'_{i'_1}),\dots,P'_{\varphi_1}(e'_{i'_{m'-t'}})\}$ forms a basis of $\mathscr{L}(\mathfrak{l}_2,\mathfrak{g}_0)/\mathrm{Im}(f_{\varphi_1})$, and consequently

$$\det(P'_{\varphi_1}(e'_{i'_1}),\dots,P'_{\varphi_1}(e'_{i'_{m'-t'}})) \neq 0.$$

This shows that $\mathrm{Hom}^{t,t'}_{1,\beta}(\mathfrak{l},\mathfrak{g}) = \bigcup_{\beta'\in I(m',m'-t')} \mathrm{Hom}^{t,t'}_{1,\beta,\beta'}(\mathfrak{l},\mathfrak{g})$.

Now to prove that $\mathrm{Hom}^{t,t'}_{1,\beta,\beta'}(\mathfrak{l},\mathfrak{g})$ is open in $\mathrm{Hom}^{t,t'}_{1,\beta}(\mathfrak{l},\mathfrak{g})$, we need the following facts.

Lemma 4.3.20. *Let $\varphi_1 \in \mathrm{Hom}^t_{1,\beta}(\mathfrak{l},\mathfrak{g})$ and assume that $\pi^t_\beta(G*\varphi_1) = \psi_1$. Then $\mathfrak{g}_{\psi_1} = \ker l_{\psi_1}$.*

Proof.

$$\begin{aligned}\mathfrak{g}_{\psi_1} &= \{X \in \mathfrak{g} \mid \exp(X) * \psi_1 = \psi_1\}\\ &= \{X \in \mathfrak{g} \mid \psi_1 + l_{\psi_1}(X) = \psi_1\}\\ &= \{X \in \mathfrak{g} \mid l_{\psi_1}(X) = 0\}\\ &= \ker l_{\psi_1}.\end{aligned}$$

□

Let $\mathscr{B} = \{X_1,\dots,X_q\}$ be a basis of $\mathfrak{g}$, $\varphi_1 \in \mathrm{Hom}^t_{1,\beta}(\mathfrak{l},\mathfrak{g})$ and $\psi_1 = \pi^t_\beta(G*\varphi_1)$. For $\gamma = (j_1,\dots,j_t) \in I(q,t)$ such that $\mathrm{Im}\, l_{\psi_1} = \mathbb{R}\text{-span}\{l_{\psi_1}(X_{j_1}),\dots,l_{\psi_1}(X_{j_t})\}$, we define a linear map $l_{\varphi_1,\gamma}: \mathbb{R}^{q-t} \to \mathfrak{g}$ given by

$$l_{\varphi_1,\gamma}(u_i) = X_{s_i} - \sum_{r=1}^{t} \alpha_{r,s_i} X_{j_r},$$

where $\{u_1,\dots,u_{q-t}\}$ is the canonical basis of $\mathbb{R}^{q-t}$, $\{s_1 < \dots < s_{q-t}\} = \{1,\dots,q\}\backslash\{j_1,\dots,j_t\}$ and

$$l_{\psi_1}(X_{s_i}) = \sum_{r=1}^{t} \alpha_{r,s_i} l_{\psi_1}(X_{j_r}) \quad \forall 1 \leq i \leq q-r.$$

Fact 4.3.21. *We have* $\operatorname{Im} l_{\varphi_1,\gamma} = \mathfrak{g}_{\psi_1}$.

Proof. By Lemma 4.3.4 and Lemma 4.3.20, we have $\operatorname{Im}(l_{\varphi_1,\gamma}) = \ker l_{\psi_1} = \mathfrak{g}_{\psi_1}$. □

Consider now the map $\bar{l}_{\varphi_1,\gamma} : \mathbb{R}^{q-t} \to \mathscr{L}(\mathfrak{l}_2, \mathfrak{g}_0)$ defined by $\bar{l}_{\varphi_1,\gamma} = f_{\varphi_1} \circ l_{\varphi_1,\gamma}$. Then we have the following result.

Lemma 4.3.22. *We have:*
(1) $\bar{l}_{\varphi_1,\gamma}$ *is a linear map.*
(2) $\operatorname{Im}(\bar{l}_{\varphi_1,\gamma}) = \operatorname{Im} f_{\varphi_1}$.

Proof. As $l_{\varphi_1,\gamma}$ and f_{φ_1} are linear maps, the map $\bar{l}_{\varphi_1,\gamma}$ is also linear. Now by Fact 4.3.21,

$$\operatorname{Im}(\bar{l}_{\varphi_1,\gamma}) = f_{\varphi_1}(\operatorname{Im} l_{\varphi_1,\gamma}) = f_{\varphi_1}(\mathfrak{g}_{\psi_1}) = \operatorname{Im}(f_{\varphi_1}).$$ □

Now for $\gamma = (j_1, \dots, j_t) \in I(q,t)$, let

$$\operatorname{Hom}^{t,t'}_{1,\beta,\beta'}(\gamma) = \{\varphi_1 \in \operatorname{Hom}^{t,t'}_{1,\beta,\beta'}(\mathfrak{l}, \mathfrak{g}) \mid \operatorname{rk}(l_{\varphi_1}(X_{j_1}), \dots, l_{\varphi_1}(X_{j_t})) = t\}.$$

Then obviously,

$$\operatorname{Hom}^{t,t'}_{1,\beta,\beta'}(\mathfrak{l}, \mathfrak{g}) = \bigcup_{\gamma \in I(q,t)} \operatorname{Hom}^{t,t'}_{1,\beta,\beta'}(\gamma).$$

To conclude that $\operatorname{Hom}^{t,t'}_{1,\beta,\beta'}(\mathfrak{l}, \mathfrak{g})$ is open in $\operatorname{Hom}^{t,t'}_{1,\beta}(\mathfrak{l}, \mathfrak{g})$, we have to show that for all $\gamma \in I(q,t)$, the set $\operatorname{Hom}^{t,t'}_{1,\beta,\beta'}(\gamma)$ is open in $\operatorname{Hom}^{t,t'}_{1,\beta}(\mathfrak{l}, \mathfrak{g})$. First, note that

$$\operatorname{Hom}^{t,t'}_{1,\beta,\beta'}(\gamma) = \left\{ \varphi_1 \in \operatorname{Hom}^{t,t'}_{1,\beta}(\mathfrak{l}, \mathfrak{g}) \;\middle|\; \begin{array}{l} \operatorname{rk}(l_{\varphi_1}(X_{j_1}), \dots, l_{\varphi_1}(X_{j_t})) = t, \\ \det(P'_{\varphi_1}(e'_{i'_1}), \dots, P'_{\varphi_1}(e'_{i'_{m'-t'}})) \neq 0 \end{array} \right\}.$$

The set

$$A^{t,t'}_{\beta}(\gamma) = \{\varphi_1 \in \operatorname{Hom}^{t,t'}_{1,\beta}(\mathfrak{l}, \mathfrak{g}) \mid \operatorname{rk}(l_{\varphi_1}(X_{j_1}), \dots, l_{\varphi_1}(X_{j_t})) = t\}$$

is open in $\operatorname{Hom}^{t,t'}_{1,\beta}(\mathfrak{l}, \mathfrak{g})$ and

$$\operatorname{Hom}^{t,t'}_{1,\beta,\beta'}(\gamma) = \{\varphi_1 \in A^{t,t'}_{\beta}(\gamma) \mid \det(P'_{\varphi_1}(e'_{i'_1}), \dots, P'_{\varphi_1}(e'_{i'_{m'-t'}})) \neq 0\}.$$

Then to obtain our result, it is sufficient to prove that $\operatorname{Hom}^{t,t'}_{1,\beta,\beta'}(\gamma)$ is open in $A^{t,t'}_{\beta}(\gamma)$. Indeed, the condition $\det(P'_{\varphi_1}(e'_{i'_1}), \dots, P'_{\varphi_1}(e'_{i'_{m'-t'}})) \neq 0$ is equivalent to $\mathscr{L}(\mathfrak{l}_2, \mathfrak{g}_0) = \operatorname{Im} f_{\varphi_1} \oplus$

$V_{\beta'}$. By Lemma 4.3.22, we get

$$\begin{aligned}
&\det(P'_{\varphi_1}(e'_{i'_1}),\dots,P'_{\varphi_1}(e'_{i'_{m'-t'}}))\neq 0\\
&\Leftrightarrow \operatorname{Im}(\bar{l}_{\varphi_1,\gamma})\oplus V_{\beta'}=\mathscr{L}(\mathfrak{l}_2,\mathfrak{g}_0)\\
&\Leftrightarrow \exists\theta\in I(q-t,t'),\quad \theta=(s_1,\dots,s_{t'})/\\
&\quad \det(\bar{l}_{\varphi_1,\gamma}(u_{s_1}),\dots,\bar{l}_{\varphi_1,\gamma}(u_{s_{t'}}),e'_{i'_1},\dots,e'_{i'_{m'-t'}})\neq 0\\
&\Leftrightarrow \sum_{\theta\in I(q-t,t')}[\det(\bar{l}_{\varphi_1,\gamma}(u_{s_1}),\dots,\bar{l}_{\varphi_1,\gamma}(u_{s_{t'}}),e'_{i'_1},\dots,e'_{i'_{m'-t'}})]^2\neq 0.
\end{aligned}$$

Then

$$\begin{aligned}
&\operatorname{Hom}^{t,t'}_{1,\beta,\beta'}(\gamma)\\
&\quad=\left\{\varphi_1\in A^{t,t'}_{\beta}(\gamma)\;\middle|\;\sum_{\theta\in I(q-t,t')}[\det(\bar{l}_{\varphi_1,\gamma}(u_{s_1}),\dots,\bar{l}_{\varphi_1,\gamma}(u_{s_{t'}}),e'_{i'_1},\dots,e'_{i'_{m'-t'}})]^2\neq 0\right\},
\end{aligned}$$

which is clearly an open subset of $A^{t,t'}_{\beta}(\gamma)$. □

As a consequence, we get the following.

Proposition 4.3.23. *We have the following decomposition:*

$$\operatorname{Hom}_1(\mathfrak{l},\mathfrak{g})=\bigcup_{t=0}^{q}\bigcup_{t'=0}^{q'}\bigcup_{\beta\in I(m,m-t)}\bigcup_{\beta'\in I(m',m'-t')}\operatorname{Hom}^{t,t'}_{1,\beta,\beta'}(\mathfrak{l},\mathfrak{g})$$

as a union of G-invariant subsets, where G acts on $\operatorname{Hom}_1(\mathfrak{l},\mathfrak{g})$ *as in* (4.45).

Proof. The decomposition is given by Proposition 4.3.12, Lemma 4.3.19 and equation (4.48). We showed already that for $\beta\in I(m,m-t)$ and $0\leq t\leq q$, the set $\operatorname{Hom}^{t}_{1,\beta}(\mathfrak{l},\mathfrak{g})$ is G-invariant. Likewise, by Lemma 4.3.18 we have $f_{\varphi_1}=f_{g*\varphi_1}$ and then $P'_{\varphi_1}=P'_{g*\varphi_1}$ for all $g\in G$ and $\varphi_1\in\operatorname{Hom}_1(\mathfrak{l},\mathfrak{g})$. As a consequence, for all $\beta\in I(m,m-t)$, $\beta'\in I(m',m'-t')$, $0\leq t\leq q$ and $0\leq t'\leq q'$ the set $\operatorname{Hom}^{t,t'}_{1,\beta,\beta'}(\mathfrak{l},\mathfrak{g})$ is G-invariant. □

Using the map ψ defined as in Lemma 4.3.9, we will identify in the rest of this section the set $\operatorname{Hom}(\mathfrak{l},\mathfrak{g})$ to $\operatorname{Hom}_1(\mathfrak{l},\mathfrak{g})\times\mathscr{L}(\mathfrak{l}_2,\mathfrak{g}_0)$. Let first for $\beta\in I(m,m-t)$, $\beta'\in I(m',m'-t')$, $0\leq t\leq q$ and $0\leq t'\leq q'$,

$$\operatorname{Hom}^{t,t'}_{\beta,\beta'}(\mathfrak{l},\mathfrak{g})=\operatorname{Hom}^{t,t'}_{1,\beta,\beta'}(\mathfrak{l},\mathfrak{g})\times\mathscr{L}(\mathfrak{l}_2,\mathfrak{g}_0).\tag{4.49}$$

Then we have the following.

Proposition 4.3.24. *We have*

$$\mathrm{Hom}(\mathfrak{l}, \mathfrak{g}) = \bigcup_{t=0}^{q} \bigcup_{t'=0}^{q'} \bigcup_{\beta \in I(m,m-t)} \bigcup_{\beta' \in I(m',m'-t')} \mathrm{Hom}_{\beta,\beta'}^{t,t'}(\mathfrak{l}, \mathfrak{g})$$

as a union of G-invariant subsets.

Proof. As in Proposition 4.3.23, for all $\beta \in I(m, m-t)$, $\beta' \in I(m', m'-t')$, $0 \leq t \leq q$ and $0 \leq t' \leq q'$, the set $\mathrm{Hom}_{1,\beta,\beta'}^{t,t'}(\mathfrak{l}, \mathfrak{g})$ is G-invariant where G acts on $\mathrm{Hom}_1(\mathfrak{l}, \mathfrak{g})$ as in (4.45) and then the set $\mathrm{Hom}_{\beta,\beta'}^{t,t'}(\mathfrak{l}, \mathfrak{g})$ is also G-invariant. □

Let now

$$\begin{aligned}
\mathscr{M}_{\beta}^{t,t'}(\mathfrak{l}, \mathfrak{g}) &= \{\varphi_1 \in \mathscr{M}_{\beta}^{t}(\mathfrak{l}, \mathfrak{g}) \mid \mathrm{rk}(f_{\varphi_1}) = t'\} \\
&= \{\varphi_1 \in \mathrm{Hom}_{1,\beta}^{t}(\mathfrak{l}, \mathfrak{g}) \mid N_{\varphi_1} \in V_\beta;\ \mathrm{rk}(f_{\varphi_1}) = t'\} \\
&= \{\varphi_1 \in \mathrm{Hom}_{1,\beta}^{t,t'}(\mathfrak{l}, \mathfrak{g}) \mid N_{\varphi_1} \in V_\beta\},
\end{aligned} \tag{4.50}$$

and

$$\begin{aligned}
\mathscr{M}_{\beta,\beta'}^{t,t'}(\mathfrak{l}, \mathfrak{g}) &= \{\varphi_1 \in \mathscr{M}_{\beta}^{t,t'}(\mathfrak{l}, \mathfrak{g}) \mid \det(P'_{\varphi_1}(e'_{i'_1}), \ldots, P'_{\varphi_1}(e'_{i'_{m'-t'}})) \neq 0\} \\
&= \{\varphi_1 \in \mathrm{Hom}_{1,\beta}^{t,t'}(\mathfrak{l}, \mathfrak{g}) \mid N_{\varphi_1} \in V_\beta,\ \det(P'_{\varphi_1}(e'_{i'_1}), \ldots, P'_{\varphi_1}(e'_{i'_{m'-t'}})) \neq 0\} \\
&\quad \text{(by (4.50))} \\
&= \{\varphi_1 \in \mathrm{Hom}_{1,\beta,\beta'}^{t,t'}(\mathfrak{l}, \mathfrak{g}) \mid N_{\varphi_1} \in V_\beta\} \\
&= \mathrm{Hom}_{1,\beta,\beta'}^{t,t'}(\mathfrak{l}, \mathfrak{g}) \cap \mathscr{M}_{\beta}^{t}(\mathfrak{l}, \mathfrak{g}).
\end{aligned} \tag{4.51}$$

We show next the following lemmas.

Lemma 4.3.25. *Each G-orbit in $\mathrm{Hom}_{\beta,\beta'}^{t,t'}(\mathfrak{l}, \mathfrak{g})$ intersects $\mathscr{M}_{\beta,\beta'}^{t,t'}(\mathfrak{l}, \mathfrak{g}) \times V_{\beta'}$ at exactly one element in $\mathrm{Hom}_{\beta,\beta'}^{t,t'}(\mathfrak{l}, \mathfrak{g})$, if we take an element $g_0 \in G$ such that $g_0 * \varphi_1 \in \mathscr{M}_{\beta}^{t}(\mathfrak{l}, \mathfrak{g})$, then the element in the intersection of $(G \cdot (\varphi_1, C))$ and $\mathscr{M}_{\beta,\beta'}^{t,t'}(\mathfrak{l}, \mathfrak{g}) \times V_{\beta'}$ can be written as*

$$(\pi_{\beta}^{t}(G * \varphi_1), (P'_{\varphi_1|V_{\beta'}})^{-1}(C(g_0) + \mathrm{Im} f_{\varphi_1})),$$

where

$$C(g_0) = C + \Sigma_{1,2}(X_0)E_{\varphi_1} + \Sigma_{1,3}(X_0)F_{\varphi_1} + \frac{1}{2}\Sigma_{1,2}(X_0).\Sigma_{2,3}(X_0)F_{\varphi_1}.$$

Here, $\Sigma_{1,2}$, $\Sigma_{1,3}$ and $\Sigma_{2,3}$ are as in formula (4.43).

Proof. From Lemma 4.3.13, the intersection of the G-orbit $G * \varphi_1$ and $\mathscr{M}_\beta^t(\mathfrak{l},\mathfrak{g})$ in $\mathrm{Hom}_{1,\beta}^t(\mathfrak{l},\mathfrak{g})$ is the one set $\{\varphi_0 + P_{\varphi_1|V_\beta}^{-1}(N_{\varphi_1} + \mathrm{Im}(l_{\varphi_1}))\}$, then there exist $g_0 = \exp(X_0) \in G$ such that

$$g_0 * \varphi_1 = \varphi_0 + P_{\varphi_1|V_\beta}^{-1}(N_{\varphi_1} + \mathrm{Im}(l_{\varphi_1})) = \pi_\beta^t(G * \varphi_1).$$

Now $g_0 \cdot (\varphi_1, C) = (g_0 * \varphi_1, C(g_0))$ and

$$G_{g_0 * \varphi_1} \cdot (g_0 * \varphi_1, C(g_0)) = (g_0 * \varphi_1, C(g_0) + \mathrm{Im} f_{\varphi_1}).$$

Thus,

$$\begin{aligned}(G\cdot(\varphi_1, C)) \cap (\mathscr{M}_{\beta,\beta'}^{t,t'}(\mathfrak{l},\mathfrak{g}) \times V_{\beta'}) &= (G * \varphi_1 \cap \mathscr{M}_{\beta,\beta'}^{t,t'}(\mathfrak{l},\mathfrak{g}), (C(g_0) + \mathrm{Im} f_{\varphi_1}) \cap V_{\beta'}) \\ &= (g_0 * \varphi_1, P'^{-1}_{\varphi_1|V_{\beta'}}(C(g_0) + \mathrm{Im} f_{\varphi_1})).\end{aligned}$$

Let (φ_1, A), (φ_1', A') be in the intersection. Then from Lemma 4.3.13, we have $\varphi_1 = \varphi_1'$. Let now $g = \exp(X)$ and $g' = \exp(X')$ in G such that $g * \varphi_1 = g' * \varphi_1 = \pi_\beta^t(G * \varphi_1)$. Then $g * \varphi_1 = g' * \varphi_1$ is equivalent to $g^{-1}g' \in G_{\varphi_1}$ and there exists $g_1 = \exp(X_1) \in G_{\varphi_1}$ such that $g' = gg_1$. Let $g_2 = \exp(X_2) = gg_1g^{-1}$, then $g_2 \in G_{g*\varphi_1}$ and

$$\begin{aligned}g' \cdot (\varphi_1, C) &= gg_1 \cdot (\varphi_1, C) = gg_1g^{-1}g \cdot (\varphi_1, C) \\ &= g_2 g \cdot (\varphi_1, C) = g_2(g * \varphi_1, C(g)) \\ &= (g * \varphi_1, C(g) + f_{\varphi_1}(X_2)).\end{aligned}$$

Thus, $C(g') - C(g) = f_{\varphi_1}(X_2) \in \mathrm{Im} f_{\varphi_1}$ which is equivalent to

$$C(g) + \mathrm{Im} f_{\varphi_1} = C(g') + \mathrm{Im} f_{\varphi_1},$$

which means that $A = A'$. □

Lemma 4.3.26. *Let $(\varphi_{1,n})_n$ be a sequence of $\mathrm{Hom}_{1,\beta}^t(\mathfrak{l},\mathfrak{g})$, which converges to φ_1, $\pi_\beta^t(G * \varphi_{1,n}) = \psi_{1,n}$ and $\pi_\beta^t(G * \varphi_1) = \psi_1$. Then there exists a convergent sequence $(X_n)_n$ in $\mathfrak{g}$, which converges to some element X such that*

$$\psi_{1,n} = \exp(X_n) * \varphi_{1,n} \quad \textit{and} \quad \psi_1 = \exp(X) * \varphi_1.$$

Proof. Let $\{X_1, \dots, X_i\}$ be a basis of $\mathfrak{g}$ and $\alpha = (i_1, \dots, i_t) \in I(q,t)$ such that $\mathrm{Im}\, l_{\varphi_1} = l_{\varphi_1}(U_\alpha)$ where $U_\alpha = \mathbb{R}\text{-span}\{l_{\varphi_1}(X_{i_1}), \dots, l_{\varphi_1}(X_{i_t})\}$. As $G * \varphi_1 = \varphi_1 + \mathrm{Im}\, l_{\varphi_1}$ and $\pi_\beta^t(G * \varphi_1) \in G * \varphi_1$, there exists $X \in U_\alpha$ such that $\psi_1 = \varphi_1 + l_{\varphi_1}(X)$. By Lemma 4.3.2, there exists $N > 0$ such that for all $n > N$, $\mathrm{Im}\, l_{\varphi_{1,n}} = l_{\varphi_{1,n}}(U_\alpha)$. Thus, there exists $(X_n)_n$ in U_α such that $\psi_{1,n} = \varphi_{1,n} + l_{\varphi_{1,n}}(X_n)$ for all $n > N$. Now from the continuity of the map $\varphi_1 \mapsto \pi_\beta^t(G * \varphi_1)$ and by the convergence of $(\varphi_{1,n})_n$ to φ_1, we deduce that $(\varphi_{1,n} + l_{\varphi_{1,n}}(X_n))_n$ converges to $\{\varphi_1 + l_{\varphi_1}(X)\}$ and then $(l_{\varphi_{1,n}}(X_n))_n$ converges to $l_{\varphi_1}(X)$ (because $(\varphi_{1,n})_n$ converges to φ_1). Let

$l'_{\varphi_{1,n}}$ be the restriction of $l_{\varphi_{1,n}}$ to U_α and l'_{φ_1} the restriction of l_{φ_1} to U_α. Then all the maps $l'_{\varphi_{1,n}}$, $n > N$ are injective and l'_{φ_1} is also injective. The sequence $(l'_{\varphi_{1,n}}(X_n))_n$ converges to $l'_{\varphi_1}(X)$, and obviously $(l'_{\varphi_{1,n}})_n$ converges to l'_{φ_1}. Using Lemma 4.3.7, we conclude that $(X_n)_n$ converges to X. □

Lemma 4.3.27. *Let $(\varphi_{1,n})_n$ be a sequence in $\mathrm{Hom}^{t,t'}_{1,\beta,\beta'}(\mathfrak{l},\mathfrak{g})$, which converges to $\varphi_1 \in \mathrm{Hom}^{t,t'}_{1,\beta,\beta'}(\mathfrak{l},\mathfrak{g})$. Then:*

(1) *There exists $\gamma = (j_1,\dots,j_t) \in I(q,t)$ such that*

$$\mathrm{Im}\, l_{\psi_1} = \mathbb{R}\text{-span}\{l_{\psi_1}(X_{j_1}),\dots,l_{\psi_1}(X_{j_t})\},\quad \psi_1 = \pi^t_\beta(G * \varphi_1)$$

and

$$\mathrm{Im}\, l_{\psi_{1,n}} = \mathbb{R}\text{-span}\{l_{\psi_{1,n}}(X_{j_1}),\dots,l_{\psi_{1,n}}(X_{j_t})\},\quad \psi_{1,n} = \pi^t_\beta(G * \varphi_{1,n}).$$

(2) *The sequence $(\bar{l}_{\varphi_{1,n},\gamma})_n$ converges to $\bar{l}_{\varphi_1,\gamma}$.*

Proof. As the quotient map $\varphi_1 \mapsto G * \varphi_1$ and π^t_β are continuous, the sequence $\{\psi_{1,n}\}$ converges to ψ_1. Then $(l_{\psi_{1,n}})_n$ converges to l_{ψ_1}. Let $\gamma = (j_1,\dots,j_t) \in I(q,t)$ be such that $\mathrm{Im}\, l_{\psi_1} = \mathbb{R}\text{-span}\{l_{\psi_1}(X_{j_1}),\dots,l_{\psi_1}(X_{j_t})\}$. By Lemma 4.3.2, we can assume that

$$\mathrm{Im}\, l_{\psi_{1,n}} = \mathbb{R}\text{-span}\{l_{\psi_{1,n}}(X_{j_1}),\dots,l_{\psi_{1,n}}(X_{j_t})\}.$$

To prove the second result, let $\gamma = (j_1,\dots,j_t) \in I(q,t)$ satisfying (1) and for $\{s_1,\dots,s_{q-t}\} = \{1,\dots,q\} \setminus \{j_1,\dots,j_t\}$, $l_{\psi_{1,n}}(X_{s_i}) = \sum_{r=1}^t \alpha^n_{r,s_i} l_{\psi_{1,n}}(X_{j_r})$ and $l_{\psi_1}(X_{s_i}) = \sum_{r=1}^t \alpha_{r,s_i} l_{\psi_1}(X_{j_r})$. Let $v_{n,s_i} = X_{s_i} - \sum_{r=1}^t \alpha^n_{r,s_i} X_{j_r}$ and $v_{s_i} = X_{s_i} - \sum_{r=1}^t \alpha_{r,s_i} X_{j_r}$. Then by Lemma 4.3.5, $(v_{n,s_i})_n$ converges to v_{s_i}. This shows that $(l_{\varphi_{1,n},\gamma})_n$ converges to $l_{\varphi_1,\gamma}$. Now, for all $1 \le i \le q-t$, we have

$$\bar{l}_{\varphi_{1,n},\gamma}(u_i) = \Sigma_{1,2}(l_{\varphi_{1,n},\gamma}(u_i))E_{\psi_{1,n}} + \Sigma_{1,3}(l_{\varphi_{1,n},\gamma}(u_i))F_{\psi_{1,n}}.$$

As the matrix multiplication is continuous and the maps $\varphi_1 \mapsto E_{\psi_1}$ and $\varphi_1 \mapsto F_{\psi_1}$ are continuous, we deduce that $(\bar{l}_{\varphi_{1,n},\gamma}(u_i))_n$ converges to $\bar{l}_{\varphi_1,\gamma}(u_i)$. Thus, $(\bar{l}_{\varphi_{1,n},\gamma})_n$ converges to $\bar{l}_{\varphi_1,\gamma}$. □

Let $q_{\varphi_1,\gamma}$ be the projection of $\mathscr{L}(\mathfrak{l}_2,\mathfrak{g}_0)$ on V_β parallel to $\mathrm{Im}(\bar{l}_{\varphi_1,\gamma})$. Then we have the following.

Lemma 4.3.28. *We have*

$$h'(\varphi_1, C) = (\pi^t_\beta(G * \varphi_1), q_{\varphi_1,\gamma}(C(\mathfrak{g}_0))),$$

where $C(\mathfrak{g}_0)$ is given in Lemma 4.3.25.

Proof. By [Lemma 4.3.22(2)], we have $\mathrm{Im}(\bar{l}_{\varphi_1,\gamma}) = \mathrm{Im}(f_{\varphi_1})$. Then

$$\begin{aligned}P'^{-1}_{\varphi_1|V_{\beta'}}(C(g_0) + \mathrm{Im}\, f_{\varphi_1}) &= (C(g_0) + \mathrm{Im}(f_{\varphi_1})) \cap V_{\beta'} \\ &= (C(g_0) + \mathrm{Im}(\bar{l}_{\varphi_1,\gamma})) \cap V_{\beta'} \\ &= q_{\varphi_1,\gamma}(C(g_0)) \quad \text{by Lemma 4.3.1.}\end{aligned}$$

□

Lemma 4.3.29. *The map*

$$\begin{aligned}h' : \mathrm{Hom}^{t,t'}_{\beta,\beta'}(\mathfrak{l}, \mathfrak{g}) &\to \mathscr{M}^{t,t'}_{\beta,\beta'}(\mathfrak{l}, \mathfrak{g}) \times V_{\beta'} \\ (\varphi_1, C) &\mapsto (\pi^t_\beta(G * \varphi_1), P'^{-1}_{\varphi_1|V_{\beta'}}(C(g_0) + \mathrm{Im}\, f_{\varphi_1}))\end{aligned}$$

is continuous.

Proof. Let $(\varphi_{1,n}, C_n)_n$ be a sequence in $\mathrm{Hom}^{t,t'}_{\beta,\beta'}(\mathfrak{l}, \mathfrak{g})$, which converges to an element $(\varphi_1, C) \in \mathrm{Hom}^{t,t'}_{\beta,\beta'}(\mathfrak{l}, \mathfrak{g})$. To see that h' is continuous, we have to show that $(h'((\varphi_{1,n}, C_n)))_n$ converges to $h'((\varphi_1, C))$. As $(\varphi_{1,n})_n$ converges to φ_1, by Lemma 4.3.27 there exists $\gamma \in I(q,t)$ such that the sequence $(\bar{l}_{\varphi_{1,n},\gamma})_n$ converges to $\bar{l}_{\varphi_1,\gamma}$. As $\mathscr{L}(\mathfrak{l}_2, \mathfrak{g}_0) = \mathrm{Im}(l_{\varphi_{1,n},\gamma}) \oplus V_{\beta'}$, $\mathscr{L}(\mathfrak{l}_2, \mathfrak{g}_0) = \mathrm{Im}(l_{\varphi_1,\gamma}) \oplus V_{\beta'}$,

$$\mathrm{Im}(\bar{l}_{\varphi_1,\gamma}) = \mathbb{R}\text{-span}\{\bar{l}_{\varphi_1,\gamma}(X_{i_1}), \dots, \bar{l}_{\varphi_1,\gamma}(X_{i_t})\}$$

and

$$\mathrm{Im}(\bar{l}_{\varphi_{1,n},\gamma}) = \mathbb{R}\text{-span}\{\bar{l}_{\varphi_{1,n},\gamma}(X_{i_1}), \dots, \bar{l}_{\varphi_{1,n},\gamma}(X_{i_t})\},$$

where $\gamma = (i_1, \dots, i_t)$. Then by Lemma 4.3.3 the sequence $(q_{\varphi_{1,n},\gamma})_n$ converges to $q_{\varphi_1,\gamma}$. Now $(C_n)_n$ converges to C and by Lemma 4.3.26 there exists a sequence $(X_n)_n$ in $\mathfrak{g}$ such that $(\exp(X_n))_n$ converges to $\exp(X)$. Now by Lemma 4.3.28,

$$h'(\varphi_{1,n}, C_n) = (\pi^t_\beta(G * \varphi_{1,n}), q_{\varphi_{1,n},\gamma}(C_n(\exp(X_n))))$$

and it is clear that $(C_n(\exp(X_n)))_n$ converges to $C(\exp(X))$. Then $(h'((\varphi_{1,n}, C_n)))_n$ converges to $h'((\varphi_1, C))$. □

By Lemma 4.3.25 above, for each G-orbit $\mathcal{O}$ in $\mathrm{Hom}^{t,t'}_{\beta,\beta'}(\mathfrak{l}, \mathfrak{g})$, there uniquely exists an element $A_{\mathcal{O}}$ in $\mathcal{O} \cap (\mathscr{M}^{t,t'}_{\beta,\beta'}(\mathfrak{l}, \mathfrak{g}) \times V_{\beta'})$. We define the map

$$\begin{aligned}\varepsilon^{t,t'}_{\beta,\beta'} : \mathrm{Hom}^{t,t'}_{\beta,\beta'}(\mathfrak{l}, \mathfrak{g})/G &\to \mathscr{M}^{t,t'}_{\beta,\beta'}(\mathfrak{l}, \mathfrak{g}) \times V_{\beta'} \\ \mathcal{O} &\mapsto A_{\mathcal{O}}.\end{aligned}$$

Then $\mathscr{M}^{t,t'}_{\beta,\beta'}(\mathfrak{l}, \mathfrak{g}) \times V_{\beta'}$ is a fundamental domain of the G-action on $\mathrm{Hom}^{t,t'}_{\beta,\beta'}(\mathfrak{l}, \mathfrak{g})$ in the sense below.

Lemma 4.3.30. *The map* $\varepsilon_{\beta,\beta'}^{t,t'} : \mathrm{Hom}_{\beta,\beta'}^{t,t'}(\mathfrak{l},\mathfrak{g})/G \to \mathscr{M}_{\beta,\beta'}^{t,t'}(\mathfrak{l},\mathfrak{g}) \times V_{\beta'}$ *defined above is a homeomorphism.*

Proof. Let $(\varphi_1, C) \neq (\varphi_1', C') \in \mathrm{Hom}_{\beta,\beta'}^{t,t'}(\mathfrak{l},\mathfrak{g})$ such that $G\cdot(\varphi_1, C) = G\cdot(\varphi_1', C')$. Then there exist g_0 and g_0' in G such that

$$\varepsilon_{\beta,\beta'}^{t,t'}(G\cdot(\varphi_1,C)) = (g_0 * \varphi_1, P'^{-1}_{\varphi_1|V_{\beta'}}(C(g_0) + \mathrm{Im}(f_{\varphi_1})))$$

and

$$\varepsilon_{\beta,\beta'}^{t,t'}(G\cdot(\varphi_1',C')) = (g_0' * \varphi_1', P'^{-1}_{\varphi_1'|V_{\beta'}}(C'(g_0') + \mathrm{Im}(f_{\varphi_1'}))).$$

Since $G\cdot(\varphi_1, C) = G\cdot(\varphi_1', C')$, there exists $g \in G$ such that $(\varphi_1, C) = g\cdot(\varphi_1', C') = (g * \varphi_1', C'(g))$. We thus get the following:

$$\begin{aligned} G * \varphi_1 = G * (g * \varphi_1') = G * \varphi_1' &\Longleftrightarrow \pi_\beta^t(G * \varphi_1) = \pi_\beta^t(G * \varphi_1') \\ &\Longleftrightarrow g_0 * \varphi_1 = g_0' * \varphi_1'. \end{aligned}$$

This means in particular that $f_{\varphi_1} = f_{\varphi_1'}$. Besides, there exists $g_1' = \exp(X_1') \in G_{g_0' * \varphi_1'}$ such that $g_1'g_0'\cdot(\varphi_1', C') = g_0\cdot(\varphi_1, C)$. This entails that

$$(g_0 * \varphi_1, C(g_0)) = (g_0' * \varphi_1', C'(g_0') + \Sigma_{1,2}(X_1')E_{g_0' * \varphi_1'} + \Sigma_{1,3}(X_1')F_{g_0' * \varphi_1'}),$$

which is equivalent to

$$C(g_0) + \mathrm{Im}\, f_{\varphi_1} = C'(g_0') + f_{\varphi_1'}(X_1') + \mathrm{Im}\, f_{\varphi_1} = C'(g_0') + \mathrm{Im}\, f_{\varphi_1'}.$$

Thus, $\varepsilon_{\beta,\beta'}^{t,t'}(G\cdot(\varphi_1, C)) = \varepsilon_{\beta,\beta'}^{t,t'}(G\cdot(\varphi_1', C'))$ and $\varepsilon_{\beta,\beta'}^{t,t'}$ is a well-defined map. We now prove that the map $\varepsilon_{\beta,\beta'}^{t,t'}$ is a homeomorphism. In fact, we first show that $\varepsilon_{\beta,\beta'}^{t,t'}$ is a bijection. Let (φ_1, C) and (φ_1', C') be in $\mathrm{Hom}_{\beta,\beta'}^{t,t'}(\mathfrak{l},\mathfrak{g})$ such that

$$\varepsilon_{\beta,\beta'}^{t,t'}(G\cdot(\varphi_1, C)) = \varepsilon_{\beta,\beta'}^{t,t'}(G\cdot(\varphi_1', C')).$$

Then there exist $g_0, g_0' \in G$ such that

$$(g_0 * \varphi_1, P'^{-1}_{\varphi_1|V_{\beta'}}(C(g_0) + \mathrm{Im}\, f_{\varphi_1})) = (g_0' * \varphi_1', P'^{-1}_{\varphi_1'|V_{\beta'}}(C'(g_0') + \mathrm{Im}\, f_{\varphi_1'})).$$

This implies that $g_0 * \varphi_1 = g_0' * \varphi_1'$, which means that $f_{\varphi_1} = f_{\varphi_1'}$ and $P'_{\varphi_1} = P'_{\varphi_1'}$. Then from the equality

$$P'^{-1}_{\varphi_1|V_{\beta'}}(C(g_0) + \mathrm{Im}\, f_{\varphi_1}) = P'^{-1}_{\varphi_1'|V_{\beta'}}(C'(g_0') + \mathrm{Im}\, f_{\varphi_1'}),$$

we get $C(g_0) + \mathrm{Im} f_{\varphi_1} = C'(g'_0) + \mathrm{Im} f_{\varphi'_1}$. Then $G * \varphi_1 = G * \varphi'_1$ and

$$G_{g_0 * \varphi_1} \cdot (g_0 * \varphi_1, C(g_0)) = G_{g'_0 * \varphi'_1} \cdot (g'_0 * \varphi'_1, C'(g'_0)),$$

which is equivalent to $G \cdot (\varphi_1, C) = G \cdot (\varphi'_1, C')$ and the map $\varepsilon^{t,t'}_{\beta,\beta'}$ is injective. Let now $(\psi_1, C_1) \in \mathscr{M}^{t,t'}_{\beta,\beta'}(\mathfrak{l}, \mathfrak{g}) \times V_{\beta'}$. Since the map π^t_β is a homeomorphism, there exist $\varphi_1 \in \mathrm{Hom}^{t,t'}_{1,\beta,\beta'}(\mathfrak{l}, \mathfrak{g})$ and $g' \in G$ such that

$$(G * \varphi_1) \cap \mathscr{M}^{t,t'}_{\beta,\beta'}(\mathfrak{l}, \mathfrak{g}) = \psi_1 = g' * \varphi_1.$$

Hence, there exists $C \in \mathscr{L}(\mathfrak{l}_2, \mathfrak{g}_0)$, such that $g' \cdot (\varphi_1, C) = (g' * \varphi_1, C(g'))$,

$$C_1 = P'^{-1}_{\psi_1|V_{\beta'}}(C(g') + \mathrm{Im} f_{\psi_1})$$

and $\varepsilon^{t,t'}_{\beta,\beta'}(G \cdot (\varphi_1, C)) = (\psi_1, C_1)$ and then the map $\varepsilon^{t,t'}_{\beta,\beta'}$ is surjective. Now the diagram below commutes

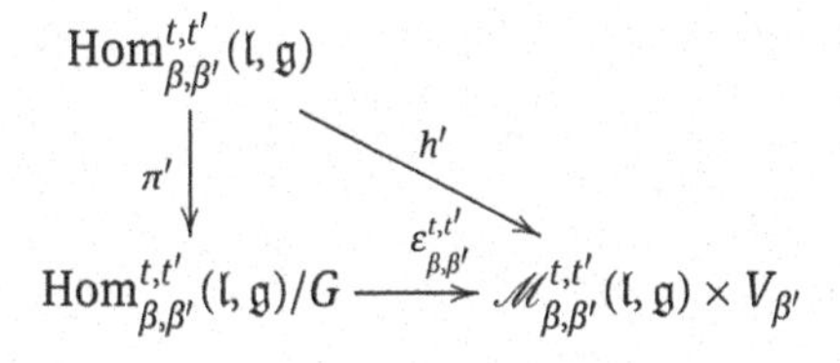

where $h'((\varphi_1, C)) = (\pi^t_\beta(G * \varphi_1), P'^{-1}_{\varphi_1|V_{\beta'}}(C(g_0) + \mathrm{Im}(f_{\varphi_1}))$. Since by Lemma 4.3.29, h' is continuous and since π' is open then $\varepsilon^{t,t'}_{\beta,\beta'}$ is continuous. Now the quotient canonical map

$$(\varepsilon^{t,t'}_{\beta,\beta'})^{-1} = \pi'_{|\mathscr{M}^{t,t'}_{\beta,\beta'}(\mathfrak{l},\mathfrak{g}) \times V_{\beta'}}$$

is continuous and then the map $\varepsilon^{t,t'}_{\beta,\beta'}$ is a homeomorphism. □

As an immediate consequence, we get the following.

Proposition 4.3.31. *We have the following:*

$$\mathrm{Hom}(\mathfrak{l}, \mathfrak{g})/G = \bigcup_{\substack{0 \le t \le q \\ 0 \le t' \le q'}} \bigcup_{\substack{\beta \in I(m, m-t) \\ \beta' \in I(m', m'-t')}} \mathrm{Hom}^{t,t'}_{\beta,\beta'}(\mathfrak{l}, \mathfrak{g})/G, \tag{4.52}$$

where for all $t \in \{0, \ldots, q\}$, $t' \in \{0, \ldots, q'\}$, $\beta \in I(m, m-t)$ and $\beta' \in I(m', m'-t')$, $\mathrm{Hom}^{t,t'}_{\beta,\beta'}(\mathfrak{l}, \mathfrak{g})$ is defined as in formula (4.49) and the set $\mathrm{Hom}^{t,t'}_{\beta,\beta'}(\mathfrak{l}, \mathfrak{g})/G$ is homeomorphic to $\mathscr{M}^{t,t'}_{\beta,\beta'}(\mathfrak{l}, \mathfrak{g}) \times V_{\beta'}$.

Let now

$$\mathrm{Hom}_1^{t,t'}(\mathfrak{l},\mathfrak{g}) = \bigcup_{\substack{\beta\in I(m,m-t)\\ \beta'\in I(m',m'-t')}} \mathrm{Hom}_{1,\beta,\beta'}^{t,t'}(\mathfrak{l},\mathfrak{g}) \tag{4.53}$$

and

$$\mathrm{Hom}^{t,t'}(\mathfrak{l},\mathfrak{g}) = \mathrm{Hom}_1^{t,t'}(\mathfrak{l},\mathfrak{g}) \times \mathscr{L}(\mathfrak{l}_2,\mathfrak{g}_2). \tag{4.54}$$

We then show the following.

Lemma 4.3.32. *Refer to Lemma (4.53) and Corollary (4.54). Then the collection*

$$\mathrm{S}_{\beta,\beta'}^{t,t'} = (\varepsilon_{\beta,\beta'}^{t,t'}, \mathrm{Hom}_{\beta,\beta'}^{t,t'}(\mathfrak{l},\mathfrak{g})/G)_{\substack{\beta\in I(m,m-t)\\ \beta'\in I(m',m'-t')}} \quad (t = 0,\ldots,q \text{ and } t' = 0,\ldots,q')$$

constitutes a family of local sections of the canonical surjection

$$\pi^{t,t'} : \mathrm{Hom}^{t,t'}(\mathfrak{l},\mathfrak{g}) \to \mathrm{Hom}^{t,t'}(\mathfrak{l},\mathfrak{g})/G.$$

Proof. We have to show that $\pi^{t,t'} \circ \varepsilon_{\beta,\beta'}^{t,t'} = \mathrm{Id}_{\mathrm{Hom}_{\beta,\beta'}^{t,t'}(\mathfrak{l},\mathfrak{g})/G}$ for all t,t',β,β'. Let $\varphi = (\varphi_1, C_\varphi) \in \mathrm{Hom}(\mathfrak{l},\mathfrak{g})$ be such that the orbit $G\cdot(\varphi_1, C_\varphi) \in \mathrm{Hom}_{\beta,\beta'}^{t,t'}(\mathfrak{l},\mathfrak{g})/G$. Then

$$\pi^{t,t'} \circ \varepsilon_{\beta,\beta'}^{t,t'}(G\cdot(\varphi_1, C_\varphi)) = \pi^{t,t'}(g_0 * \varphi_1, P'^{-1}_{\varphi_1|V_{\beta'}}(C_\varphi(g_0) + \mathrm{Im}(f_{\varphi_1})).$$

There exists $g_1 \in G_{g_0*\varphi_1}$ such that

$$g_1 g_0 \cdot (\varphi_1, C_\varphi) = (g_0 * \varphi_1, (C_\varphi(g_0) + \mathrm{Im}\, f_{\varphi_1}) \cap V_{\beta'}).$$

Thus,

$$\begin{aligned}\pi^{t,t'}(g_0 * \varphi_1, P'^{-1}_{\varphi_1|V_{\beta'}}(C_\varphi(g_0) + \mathrm{Im}(f_{\varphi_1})) &= \pi^{t,t'}(g_1 g_0 \cdot (\varphi_1, C_\varphi)) \\ &= G\cdot(g_1 g_0 \cdot (\varphi_1, C_\varphi)) \\ &= G\cdot(\varphi_1, C_\varphi).\end{aligned}$$

In particular,

$$\varepsilon_{\beta,\beta'}^{t,t'}(\mathrm{Hom}_{\beta,\beta'}^{t,t'}(\mathfrak{l},\mathfrak{g})/G) = \left\{\varphi \in \mathrm{Hom}(\mathfrak{l},\mathfrak{g}) \;\middle|\; \begin{array}{l}\varphi_1 \in \mathscr{M}_{\beta,\beta'}^{t,t'}(\mathfrak{l},\mathfrak{g}), \\ C_\varphi \in V_{\beta'}\end{array}\right\}. \qquad \square$$

Write

$$\mathrm{Hom}^{t,t'}(\mathfrak{l},\mathfrak{g}) = \bigcup_{\substack{\beta\in I(m,m-t)\\ \beta'\in I(m',m'-t')}} \mathrm{Hom}_{\beta,\beta'}^{t,t'}(\mathfrak{l},\mathfrak{g}).$$

Then from Proposition 4.3.31, the set $\mathrm{Hom}^{t,t'}(\mathfrak{l},\mathfrak{g})$ is a G-invariant subset. More precisely all of the subsets of the union are G-invariant and open in $\mathrm{Hom}^{t,t'}(\mathfrak{l},\mathfrak{g})$. Our main result is the following.

Theorem 4.3.33. *The writing*

$$\mathrm{Hom}(\mathfrak{l},\mathfrak{g})/G = \bigcup_{\substack{0\le t\le q\\ 0\le t'\le q'}} \mathrm{Hom}^{t,t'}(\mathfrak{l},\mathfrak{g})/G$$

is a decomposition of $\mathrm{Hom}(\mathfrak{l},\mathfrak{g})/G$ *as a union of Hausdorff subspaces. The sets* $\mathrm{Hom}^{t,t'}(\mathfrak{l},\mathfrak{g})/G$ *may fail to be open in* $\mathrm{Hom}(\mathfrak{l},\mathfrak{g})/G$.

To prove this result, we need the following lemma.

Lemma 4.3.34. *Let* $\varphi = (\varphi_1, C)$ *and* $\xi = (\xi_1, C')$ *be two elements in* $\mathrm{Hom}^{t,t'}(\mathfrak{l},\mathfrak{g})$. *If* $[\varphi_1]$ *and* $[\xi_1]$ *are separated in* $\mathrm{Hom}_1^{t,t'}(\mathfrak{l},\mathfrak{g})/G$, *then so are* $[\varphi]$ *and* $[\xi]$ *in* $\mathrm{Hom}^{t,t'}(\mathfrak{l},\mathfrak{g})/G$.

Proof. We consider the following diagram:

$$\begin{array}{ccc} \mathrm{Hom}^{t,t'}(\mathfrak{l},\mathfrak{g}) & \xrightarrow{P_1} & \mathrm{Hom}_1^{t,t'}(\mathfrak{l},\mathfrak{g}) \\ \downarrow{\scriptstyle \pi_1} & & \downarrow{\scriptstyle \pi_2} \\ \mathrm{Hom}^{t,t'}(\mathfrak{l},\mathfrak{g})/G & \xrightarrow[\widetilde{P_1}]{} & \mathrm{Hom}_1^{t,t'}(\mathfrak{l},\mathfrak{g})/G \end{array}$$

where π_1, π_2 are the quotient maps, $P_1(\varphi_1, C) = \varphi_1$ and $\widetilde{P_1}(G\cdot(\varphi_1, C)) = G * \varphi_1$. Then obviously this diagram commutes and the maps P_1 and $\widetilde{P_1}$ are continuous. Assume that $[P_1(\varphi)]$ and $[P_1(\xi)]$ are separated, then there exist a neighborhood U_1 of $\pi_2\circ P_1(\varphi)$ and U_2 of $\pi_2\circ P_1(\xi)$ such that $U_1\cap U_2 = \varnothing$. Now $\pi_2\circ P_1(\varphi) = \widetilde{P}_1\circ\pi_1(\varphi)\in U_1$ and $\pi_2\circ P_1(\xi) = \widetilde{P}_1\circ\pi_1(\xi)\in U_2$. Thus, $\pi_1(\varphi)\in\widetilde{P}_1^{-1}(U_1)$, $\pi_1(\xi)\in\widetilde{P}_1^{-1}(U_2)$ and we have $\widetilde{P}_1^{-1}(U_1)\cap\widetilde{P}_1^{-1}(U_2) = \varnothing$. □

Proof of Theorem 4.3.33. Let $\varphi = (\varphi_1, C)$ and $\xi = (\xi_1, C')$ be two elements in $\mathrm{Hom}^{t,t'}(\mathfrak{l},\mathfrak{g})$ and assume that $[\varphi]$ and $[\xi]$ are not separated. From Lemma 4.3.34, $[\varphi_1]$ and $[\xi_1]$ are not separated in $\mathrm{Hom}_1^{t,t'}(\mathfrak{l},\mathfrak{g})/G$. Then there exist $(\varphi_{1,n})_n\subset\mathrm{Hom}_1^{t}(\mathfrak{l},\mathfrak{g})$ and $g_n = \exp(X_n)\in G$ such that $\varphi_{1,n}$ converges to φ_1 and $g_n * \varphi_{1,n}$ converges to ξ_1 in $\mathrm{Hom}_1^{t}(\mathfrak{l},\mathfrak{g})$. This means $A_{\varphi_1} = A_{\xi_1}$, $D_{\varphi_1} = D_{\xi_1}$ and $F_{\varphi_1} = F_{\xi_1}$. In particular, $l_{\varphi_1} = l_{\xi_1}$ and $P_{\varphi_1} = P_{\xi_1}$. Thus $[\varphi_1]$ and $[\xi_1]$ belong to the open set $\mathrm{Hom}_{1,\beta}^{t,t'}(\mathfrak{l},\mathfrak{g})/G$ for some $\beta\in I(m, m-t)$. Now $\mathrm{Hom}_{1,\beta}^{t,t'}(\mathfrak{l},\mathfrak{g})/G$ is a Hausdorff space as it is included in $\mathrm{Hom}_{1,\beta}^{t}(\mathfrak{l},\mathfrak{g})/G$, which is a Hausdorff space as being homeomorphic to $\mathscr{M}_\beta^t(\mathfrak{l},\mathfrak{g})$ by Lemma 4.3.16. Then $[\varphi_1] = [\xi_1]$ and this implies that $f_{\varphi_1} = f_{\xi_1}$ and $P'_{\varphi_1} = P'_{\xi_1}$. Finally, $[\varphi]$ and $[\xi]$ belong to the open set

$\mathrm{Hom}^{t,t'}_{\beta,\beta'}(\mathfrak{l},\mathfrak{g})/G$ for some $\beta \in I(m, m-t)$ and $\beta' \in I(m', m'-t')$. Now from Proposition 4.3.31, $\mathrm{Hom}^{t,t'}_{\beta,\beta'}(\mathfrak{l},\mathfrak{g})/G$ is homeomorphic to the Hausdorff space $\mathscr{M}^{t,t'}_{\beta,\beta'}(\mathfrak{l},\mathfrak{g}) \times V_{\beta'}$, then $\mathrm{Hom}^{t,t'}_{\beta,\beta'}(\mathfrak{l},\mathfrak{g})/G$ is a Hausdorff space and $[\varphi] = [\xi]$. □

4.3.3 Description of the parameter and the deformation spaces

We use the same setting and notation. Let us take a subalgebra $\mathfrak{h}$ of $\mathfrak{g}$ and consider the decompositions

$$\mathfrak{g} = (\mathfrak{g}_0 \cap \mathfrak{h}) \oplus \mathfrak{g}_0' \oplus \mathfrak{h}' \oplus V \oplus \mathfrak{h}'' \oplus W \quad \text{and} \quad \mathfrak{l} = \mathfrak{l}_0 \oplus \mathfrak{l}_1 \oplus \mathfrak{l}_2,$$

where $\mathfrak{g}_0'$, $\mathfrak{h}'$ and $\mathfrak{h}''$ designate some subspaces of $\mathfrak{g}$ such that $\mathfrak{g}_0 = (\mathfrak{g}_0\cap\mathfrak{h})\oplus\mathfrak{g}_0'$, $[\mathfrak{g},\mathfrak{g}]\cap\mathfrak{h} = (\mathfrak{g}_0 \cap \mathfrak{h}) \oplus \mathfrak{h}'$, $\mathfrak{h} = [\mathfrak{g},\mathfrak{g}] \cap \mathfrak{h} \oplus \mathfrak{h}''$, V is a linear subspace supplementary to $(\mathfrak{g}_0 \cap \mathfrak{h}) \oplus \mathfrak{g}_0' \oplus \mathfrak{h}'$ in $[\mathfrak{g},\mathfrak{g}]$ and W a linear supplementary subspace to $(\mathfrak{g}_0 \cap \mathfrak{h}) \oplus \mathfrak{g}_0' \oplus \mathfrak{h}' \oplus V \oplus \mathfrak{h}''$ in $\mathfrak{g}$.

Then with respect to these decompositions, the adjoint representation Ad_g, $g = \exp(X) \in G$ can once again be written down as

$$\mathrm{Ad}_g = \begin{pmatrix} I_1 & 0 & \sigma_{1,3}(X) & \delta_{1,4}(X) & \gamma_{1,5}(X) & \omega_{1,6}(X) \\ 0 & I_2 & \sigma_{2,3}(X) & \delta_{2,4}(X) & \gamma_{2,5}(X) & \omega_{2,6}(X) \\ 0 & 0 & I_3 & 0 & \gamma_{3,5}(X) & \omega_{3,6}(X) \\ 0 & 0 & 0 & I_4 & \gamma_{4,5}(X) & \omega_{4,6}(X) \\ 0 & 0 & 0 & 0 & I_5 & 0 \\ 0 & 0 & 0 & 0 & 0 & I_6 \end{pmatrix},$$

where

$$\Sigma_{1,2}(X) = \begin{pmatrix} \sigma_{1,3}(X) & \delta_{1,4}(X) \\ \sigma_{2,3}(X) & \delta_{2,4}(X) \end{pmatrix},$$

$$\Sigma_{1,3}(X) + \frac{1}{2}\Sigma_{1,2}(X)\Sigma_{2,3}(X) = \begin{pmatrix} \gamma_{1,5}(X) & \omega_{1,6}(X) \\ \gamma_{2,5}(X) & \omega_{2,6}(X) \end{pmatrix}$$

and

$$\Sigma_{2,3}(X) = \begin{pmatrix} \gamma_{3,5}(X) & \omega_{3,6}(X) \\ \gamma_{4,5}(X) & \omega_{4,6}(X) \end{pmatrix}$$

with $\sigma_{1,3}(X) \in \mathscr{L}(\mathfrak{h}', \mathfrak{g}_0 \cap \mathfrak{h})$, $\delta_{1,4}(X) \in \mathscr{L}(V, \mathfrak{g}_0 \cap \mathfrak{h})$, $\sigma_{2,3}(X) \in \mathscr{L}(\mathfrak{h}', \mathfrak{g}_0')$, $\delta_{2,4}(X) \in \mathscr{L}(V, \mathfrak{g}_0')$, $\gamma_{1,5}(X) \in \mathscr{L}(\mathfrak{h}'', \mathfrak{g}_0\cap\mathfrak{h})$, $\omega_{1,6}(X) \in \mathscr{L}(W, \mathfrak{g}_0\cap\mathfrak{h})$, $\gamma_{2,5}(X) \in \mathscr{L}(\mathfrak{h}'', \mathfrak{g}_0')$, $\omega_{2,6}(X) \in \mathscr{L}(W, \mathfrak{g}_0')$, $\gamma_{3,5}(X) \in \mathscr{L}(\mathfrak{h}'', \mathfrak{h}')$, $\omega_{3,6}(X) \in \mathscr{L}(W, \mathfrak{h}')$, $\gamma_{4,5}(X) \in \mathscr{L}(\mathfrak{h}'', V)$ and $\omega_{4,6}(X) \in \mathscr{L}(W, V)$. Here, I_1, I_2, I_3, I_4, I_5 and I_6 designate the identity maps of $\mathfrak{g}_0\cap\mathfrak{h}$, $\mathfrak{g}_0'$, $\mathfrak{h}'$, V, $\mathfrak{h}''$

and W, respectively. This leads to the fact that any element of $\operatorname{Hom}(\mathfrak{l}, \mathfrak{g})$ can be written accordingly, as a matrix

$$\varphi := \varphi(A, B, C) = \begin{pmatrix} A_1 & B_1 & C_1 \\ A_2 & B_2 & C_2 \\ 0 & B_3 & C_3 \\ 0 & B_4 & C_4 \\ 0 & 0 & C_5 \\ 0 & 0 & C_6 \end{pmatrix},$$

where

$$A = \begin{pmatrix} A_1 \\ A_2 \end{pmatrix}, \quad B = \begin{pmatrix} B_1 \\ B_2 \\ B_3 \\ B_4 \end{pmatrix} \quad \text{and} \quad C = \begin{pmatrix} C_1 \\ C_2 \\ C_3 \\ C_4 \\ C_5 \\ C_6 \end{pmatrix}.$$

Here, $A_1 \in \mathscr{L}(\mathfrak{l}_0, \mathfrak{g}_0 \cap \mathfrak{h})$, $A_2 \in \mathscr{L}(\mathfrak{l}_0, \mathfrak{g}_0')$, $B_1 \in \mathscr{L}(\mathfrak{l}_1, \mathfrak{g}_0 \cap \mathfrak{h})$, $B_2 \in \mathscr{L}(\mathfrak{l}_1, \mathfrak{g}_0')$, $B_3 \in \mathscr{L}(\mathfrak{l}_1, \mathfrak{h}')$, $B_4 \in \mathscr{L}(\mathfrak{l}_1, V)$, $C_1 \in \mathscr{L}(\mathfrak{l}_2, \mathfrak{g}_0 \cap \mathfrak{h})$, $C_2 \in \mathscr{L}(\mathfrak{l}_2, \mathfrak{g}_0')$, $C_3 \in \mathscr{L}(\mathfrak{l}_2, \mathfrak{h}')$, $C_4 \in \mathscr{L}(\mathfrak{l}_2, V)$, $C_5 \in \mathscr{L}(\mathfrak{l}_2, \mathfrak{h}'')$ and $C_6 \in \mathscr{L}(\mathfrak{l}_2, W)$.

We can now state our first result.

Theorem 4.3.35. *Let G be a 3-step nilpotent Lie group, H a connected subgroup of G and Γ a discontinuous group for G/H. The syndetic hull of Γ in G and its Lie algebra are denoted by L and $\mathfrak{l}$, respectively. Then the parameter space $\mathscr{R}(\mathfrak{l}, \mathfrak{g}, \mathfrak{h})$ writes as a disjoint union $\mathscr{R}_1 \sqcup \mathscr{R}_2$, where $\mathscr{R}_1$ is the open set defined by*

$$\mathscr{R}_1 := \left\{ \varphi(A, B, C) \in \operatorname{Hom}(\mathfrak{l}, \mathfrak{g}) \;\middle|\; \begin{array}{l} \operatorname{rk}(C_6) = \dim(\mathfrak{l}_2), \\ \operatorname{rk}(B_4) = \dim(\mathfrak{l}_1) \\ \textit{and } \operatorname{rk}(A_2) = \dim(\mathfrak{l}_0) \end{array} \right\},$$

$$\mathscr{R}_2 := \left\{ \varphi(A, B, C) \in \operatorname{Hom}(\mathfrak{l}, \mathfrak{g}) \;\middle|\; \begin{array}{l} \operatorname{rk}(B_4) + \operatorname{rk}(C_6) < \dim(\mathfrak{l}_1 \oplus \mathfrak{l}_2) \textit{ and} \\ \operatorname{rk}(M_{\varphi,X}) = \dim(\mathfrak{l}) \textit{ for all } X \in \mathfrak{g} \end{array} \right\}$$

which may fail to be open. Here,

$$M_{\varphi,X} = \begin{pmatrix} A_2 & B_2 + \sigma_{2,3}(X)B_3 + \delta_{2,4}(X)B_4 & \begin{array}{c} C_2 + \sigma_{2,3}(X)C_3 + \delta_{2,4}(X)C_4 \\ +\gamma_{2,5}(X)C_5 + \omega_{2,6}(X)C_6 \end{array} \\ 0 & B_4 & C_4 + \gamma_{4,5}(X)C_5 + \omega_{4,6}(X)C_6 \\ 0 & 0 & C_6 \end{pmatrix}.$$

Proof. As the pair (G, H) has the Lipsman property (cf. Definition 2.1.10), Theorem 2.2.2 enables us to state that

$$\mathscr{R}(\mathfrak{l}, \mathfrak{g}, \mathfrak{h}) = \left\{ \varphi \in \operatorname{Hom}(\mathfrak{l}, \mathfrak{g}) \;\middle|\; \begin{array}{l} \dim \varphi(\mathfrak{l}) = \dim(\mathfrak{l}), \\ \operatorname{Ad}_g \circ \varphi(\mathfrak{l}) \cap \mathfrak{h} = \{0\} \text{ for all } g = \exp(X) \in G \end{array} \right\}. \tag{4.55}$$

Now

$$\operatorname{Ad}_g \circ \varphi = \begin{pmatrix} A_1 & B_1 + \sigma_{1,3}(X)B_3 + \delta_{1,4}(X)B_4 & C_1 + \sigma_{1,3}(X)C_3 + \delta_{1,4}(X)C_4 \\ & & +\gamma_{1,5}(X)C_5 + \omega_{1,6}(X)C_6 \\ A_2 & B_2 + \sigma_{2,3}(X)B_3 + \delta_{2,4}(X)B_4 & C_2 + \sigma_{2,3}(x)C_3 + \delta_{2,4}(X)C_4 \\ & & +\gamma_{2,5}(X)C_5 + \omega_{2,6}(X)C_6 \\ 0 & B_3 & C_3 + \gamma_{3,5}(X)C_5 + \omega_{3,6}(X)C_6 \\ 0 & B_4 & C_4 + \gamma_{4,5}(X)C_5 + \omega_{4,6}(X)C_6 \\ 0 & 0 & C_5 \\ 0 & 0 & C_6 \end{pmatrix},$$

which means that the condition $\operatorname{Ad}_g \circ \varphi(\mathfrak{l}) \cap \mathfrak{h} = \{0\}$ is equivalent to the fact that $\operatorname{rk}(M_{\varphi,X}) = \dim(\mathfrak{l})$, which is in turn equivalent to

$$\operatorname{rk}(C_6) = \dim(\mathfrak{l}_2), \quad \operatorname{rk}(B_4) = \dim(\mathfrak{l}_1) \quad \text{and} \quad \operatorname{rk}(A_2) = \dim(\mathfrak{l}_0),$$

or

$$\begin{cases} \operatorname{rk}(C_6) < \dim(\mathfrak{l}_2), & \operatorname{rk}(B_4) = \dim(\mathfrak{l}_1) \quad \text{and} \quad \operatorname{rk}(M_{\varphi,X}) = \dim(\mathfrak{l}) \quad \text{or} \\ \operatorname{rk}(B_4) < \dim(\mathfrak{l}_1), & \operatorname{rk}(C_6) = \dim(\mathfrak{l}_2) \quad \text{and} \quad \operatorname{rk}(M_{\varphi,X}) = \dim(\mathfrak{l}) \quad \text{or} \\ \operatorname{rk}(C_6) < \dim(\mathfrak{l}_2), & \operatorname{rk}(B_4) < \dim(\mathfrak{l}_1) \quad \text{and} \quad \operatorname{rk}(M_{\varphi,X}) = \dim(\mathfrak{l}). \end{cases}$$

The latter three cases are equivalent to say that

$$\operatorname{rk}(B_4) + \operatorname{rk}(C_6) < \dim(\mathfrak{l}_1 \oplus \mathfrak{l}_2).$$ □

We are now ready to present our main result in this section.

Theorem 4.3.36. *Let $\mathfrak{g}$, $\mathfrak{h}$ and $\mathfrak{l}$ be as before. The deformation space reads*

$$\mathscr{T}(\mathfrak{l}, \mathfrak{g}, \mathfrak{h}) = \bigcup_{i=1}^{2} \bigcup_{\substack{0 \le t \le q \\ 0 \le t' \le q'}} \bigcup_{\substack{\beta \in I(m,m-t) \\ \beta' \in I(m',m'-t')}} \mathscr{T}_{t,t',\beta,\beta',i}(\mathfrak{l}, \mathfrak{g}, \mathfrak{h}),$$

where for $\beta = (i_1, \ldots, i_{m-t})$ *and* $\beta' = (i'_1, \ldots, i'_{m'-t'})$ *the set* $\mathcal{T}_{t,t',\beta,\beta',1}$ *is homeomorphic to the semialgebraic subset in* $\mathcal{L}(\mathfrak{l}, \mathfrak{g})$,

$$\mathcal{T}_{t,t',\beta,\beta',1} \simeq \left\{ \varphi(A, B, C) \in \operatorname{Hom}(\mathfrak{l}, \mathfrak{g}) \;\middle|\; \begin{array}{l} \varphi_1 \in \mathcal{M}^{t,t'}_{\beta,\beta'}(\mathfrak{l}, \mathfrak{g}), \\ \begin{pmatrix} C_1 \\ C_2 \end{pmatrix} \in V_{\beta'} \text{ and} \\ \operatorname{rk}\begin{pmatrix} A_2 & 0 & 0 \\ 0 & B_4 & 0 \\ 0 & 0 & C_6 \end{pmatrix} = \dim(\mathfrak{l}) \end{array} \right\}$$

and $\mathcal{T}_{t,t',\beta,\beta',2}$ *is homeomorphic to*

$$\mathcal{T}_{t,t',\beta,\beta',2} \simeq \left\{ \varphi(A, B, C) \in \operatorname{Hom}(\mathfrak{l}, \mathfrak{g}) \;\middle|\; \begin{array}{l} \varphi_1 \in \mathcal{M}^{t,t'}_{\beta,\beta'}(\mathfrak{l}, \mathfrak{g}), \\ \begin{pmatrix} C_1 \\ C_2 \end{pmatrix} \in V_{\beta'}, \\ \operatorname{rk}(B_4) + \operatorname{rk}(C_6) < \dim(\mathfrak{l}_1 \oplus \mathfrak{l}_2) \\ \text{and } \operatorname{rk}(M_{\varphi,X}) = \dim(\mathfrak{l}) \text{ for all } X \in \mathfrak{g} \end{array} \right\}.$$

Proof. Recall first that

$$\mathcal{R}(\mathfrak{l}, \mathfrak{g}, \mathfrak{h}) = \bigcup_{i=1}^{2} \bigcup_{\substack{0 \le t \le q \\ 0 \le t' \le q'}} \bigcup_{\substack{\beta \in I(m,m-t) \\ \beta' \in I(m',m'-t')}} \operatorname{Hom}^{t,t'}_{\beta,\beta'}(\mathfrak{l}, \mathfrak{g}) \cap \mathcal{R}_i.$$

On the other hand, $\operatorname{Hom}^{t,t'}_{\beta,\beta'}(\mathfrak{l}, \mathfrak{g}) \cap \mathcal{R}_i$ is a G-invariant set as in formula (4.52). Hence,

$$\mathcal{T}(\mathfrak{l}, \mathfrak{g}, \mathfrak{h}) = \bigcup_{i=1}^{2} \bigcup_{\substack{0 \le t \le q \\ 0 \le t' \le q'}} \bigcup_{\substack{\beta \in I(m,m-t) \\ \beta' \in I(m',m'-t')}} (\operatorname{Hom}^{t,t'}_{\beta,\beta'}(\mathfrak{l}, \mathfrak{g}) \cap \mathcal{R}_i)/G,$$

and the result follows from Theorem 4.3.35. Now to see the semialgebraicness of $\mathcal{T}_{t,t',\beta,\beta',1}$, we state first the following claim, which explains the semialgebraicness of $\operatorname{Hom}(\mathfrak{l}, \mathfrak{g})$ in $\mathcal{L}(\mathfrak{l}, \mathfrak{g})$.

Lemma 4.3.37. *The set* $\operatorname{Hom}(\mathfrak{l}, \mathfrak{g})$ *is algebraic in* $\mathcal{L}(\mathfrak{l}, \mathfrak{g})$.

Proof. Let $\{Y_1, \ldots, Y_k\}$ be a basis of $\mathfrak{l}$ and $\{X_1, \ldots, X_q\}$ a basis of $\mathfrak{g}$. Assume that the Lie brackets of $\mathfrak{l}$ are given by $[Y_i, Y_j] = \sum_{u=1}^{k} c^u_{ij} Y_u$ for all $1 \le i, j \le k$ and the Lie brackets of $\mathfrak{g}$ are given by $[X_s, X_{s'}] = \sum_{v=1}^{q} d^v_{ss'} X_v$ for all $1 \le s, s' \le q$. Let now $\varphi \in \mathcal{L}(\mathfrak{l}, \mathfrak{g})$ and assume that $\varphi(Y_i) = \sum_{s=1}^{q} a_{si}X_s$ for all $1 \leq i \leq k$. Now

$$\begin{aligned}\mathrm{Hom}(\mathfrak{l},\mathfrak{g}) &= \{\varphi \in \mathscr{L}(\mathfrak{l},\mathfrak{g}) \mid \varphi([Y,T]) = [\varphi(Y),\varphi(T)],\ \forall Y,T \in \mathfrak{l}\}\\ &= \{\varphi \in \mathscr{L}(\mathfrak{l},\mathfrak{g}) \mid \varphi([Y_i,Y_j]) = [\varphi(Y_i),\varphi(Y_j)],\ \forall 1 \leq i,j \leq k\}.\end{aligned}$$

As

$$\varphi([Y_i,Y_j]) = \sum_{u=1}^{k} c_{i,j}^{u}\varphi(Y_u) = \sum_{u=1}^{k}\sum_{v=1}^{q} c_{i,j}^{u}a_{vu}X_v = \sum_{v=1}^{q}\left(\sum_{u=1}^{k} c_{i,j}^{u}a_{vu}\right)X_v,$$

and

$$\begin{aligned}[\varphi(Y_i),\varphi(Y_j)] &= \left[\sum_{s=1}^{q} a_{si}X_s, \sum_{s'=1}^{q} a_{s'j}X_{s'}\right] = \sum_{s,s'=1}^{q} a_{si}a_{s'j}[X_s,X_{s'}]\\ &= \sum_{s,s'=1}^{q}\sum_{v=1}^{q} a_{si}a_{s'j}d_{ss'}^{v}X_v = \sum_{v=1}^{q}\left(\sum_{s,s'=1}^{q} a_{si}a_{s'j}d_{ss'}^{v}\right)X_v,\end{aligned}$$

we get

$$[\varphi(Y_i),\varphi(Y_j)] = \varphi([Y_i,Y_j]) \Leftrightarrow \sum_{u=1}^{k} c_{i,j}^{u}a_{vu} = \sum_{s,s'=1}^{q} a_{si}a_{s'j}d_{ss'}^{v}$$

for any $v = 1,\dots,q$. Hence, if we identify $\mathscr{L}(\mathfrak{l},\mathfrak{g})$ to $M_{q,k}(\mathbb{R})$ via the map $\varphi \mapsto M_\varphi = (\varphi(Y_1)|\cdots|\varphi(Y_k))$, then

$$\mathrm{Hom}(\mathfrak{l},\mathfrak{g}) = \left\{(a_{si})_{\substack{1\leq s\leq q\\ 1\leq i\leq k}} \;\middle|\; \sum_{u=1}^{k} c_{i,j}^{u}a_{vu} = \sum_{s,s'=1}^{q} a_{si}a_{s'j}d_{ss'}^{v},\ 1 \leq i,j \leq k\right\},$$

which is an algebraic set. □

Now we see that the conditions

$$\begin{pmatrix} C_1\\ C_2\end{pmatrix} \in V_{\beta'} \quad \text{and} \quad \mathrm{rk}\begin{pmatrix} A_2 & 0 & 0\\ 0 & B_4 & 0\\ 0 & 0 & C_6\end{pmatrix} = \dim(\mathfrak{l})$$

are semialgebraic conditions and as $\mathrm{Hom}_{1,\beta,\beta'}^{t,t'}(\mathfrak{l},\mathfrak{g})$ is semialgebraic and the condition $N_{\varphi_1} \in V_\beta$ is an algebraic condition. Then by (4.51), we conclude that $\mathscr{M}_{\beta,\beta'}^{t,t'}(\mathfrak{l},\mathfrak{g})$ is semialgebraic. □

4.3.4 Hausdorffness of the deformation space

This section aims to study the Hausdorfness of the deformation space $\mathscr{T}(\mathfrak{l}, \mathfrak{g}, \mathfrak{h})$ in the setting where $\mathfrak{g}$ is 3-step nilpotent. Let

$$\begin{array}{rcl} p : \mathscr{R}(\mathfrak{l}, \mathfrak{g}, \mathfrak{h}) & \longrightarrow & \mathrm{Hom}_1(\mathfrak{l}, \mathfrak{g}) \\ (\varphi_1, C) & \longmapsto & \varphi_1, \end{array}$$

where $\mathrm{Hom}_1(\mathfrak{l}, \mathfrak{g})$ is as in (4.44), φ_1 is as in (4.39) and

$$(\varphi_1, C) \in \mathscr{R}(\mathfrak{l}, \mathfrak{g}, \mathfrak{h}) \subset \mathrm{Hom}_1(\mathfrak{l}, \mathfrak{g}) \times \mathscr{L}(\mathfrak{l}_2, \mathfrak{g}_0).$$

Then p is a G-equivariant map and we can state the following.

Theorem 4.3.38. *Let $G = \exp \mathfrak{g}$ be a 3-step nilpotent Lie group, $H = \exp \mathfrak{h}$ a closed, connected subgroup of G, Γ a discontinuous group for the homogeneous space G/H and $L = \exp \mathfrak{l}$ its syndetic hull. If the dimensions of G-orbits in $\mathscr{R}(\mathfrak{l}, \mathfrak{g}, \mathfrak{h})$ and those in $p(\mathscr{R}(\mathfrak{l}, \mathfrak{g}, \mathfrak{h}))$ are constant, respectively, then $\mathscr{T}(\mathfrak{l}, \mathfrak{g}, \mathfrak{h})$ is a Hausdorff space.*

Proof. In such a situation, there is $t \in \{0, \ldots, q\}$ and $t' \in \{0, \ldots, q'\}$ such that $\mathscr{R}(\mathfrak{l}, \mathfrak{g}, \mathfrak{h}) \subset \mathrm{Hom}^{t,t'}(\mathfrak{l}, \mathfrak{g})$. Indeed, let $\varphi = (\varphi_1, C_\varphi) \in \mathscr{R}(\mathfrak{l}, \mathfrak{g}, \mathfrak{h})$ and assume that $\mathrm{rk}\, l_{\varphi_1} = t$ and $\mathrm{rk} f_{\varphi_1} = t'$. As $G * \varphi_1 = \varphi_0 + \mathrm{Im}(l_{\varphi_1})$, we have $\dim G * \varphi_1 = t$ and

$$\begin{aligned} \dim G \cdot \varphi &= \dim G \cdot (\varphi_1, C_\varphi) \\ &= \dim G * \varphi_1 + \dim(\pi_\beta^t(G * \varphi_1), C_\varphi + \mathrm{Im} f_{\varphi_1}) \\ &= \dim(\varphi_0 + (N_{\varphi_1} + \mathrm{Im}\, l_{\varphi_1})) + \mathrm{rk} f_{\varphi_1} \\ &= \mathrm{rk}\, l_{\varphi_1} + \mathrm{rk} f_{\varphi_1} = t + t'. \end{aligned}$$

Since the dimensions of G-orbits of $\mathscr{R}(\mathfrak{l}, \mathfrak{g}, \mathfrak{h})$ and of $p(\mathscr{R}(\mathfrak{l}, \mathfrak{g}, \mathfrak{h}))$ are constant, then so are t and t'. The deformation space is therefore contained in $\mathrm{Hom}^{t,t'}(\mathfrak{l}, \mathfrak{g})/G$, which is a Hausdorff space by Theorem 4.3.33. □

4.3.5 Illustrating examples

For the convenience of the readers, we develop the following series of examples for which the hypotheses of Theorem 4.3.38 are met and then the corresponding deformation space turns out to be a Hausdorff space. Explicit computations are also developed in the paper [19]. Let $\mathfrak{g} = \mathbb{R}\text{-span}\{X_0, X_1, X_2, X_3\}$ be the (3-step nilpotent) threadlike Lie algebra, whose pairwise brackets equal zero, except the following:

$$[X_0, X_i] = X_{i+1}, \quad i = 1, 2.$$

The center of $\mathfrak{g}$ is the space $\mathbb{R}\text{-span}\{X_3\}$, $\mathfrak{g}_1 = \mathbb{R}\text{-span}\{X_2\}$, $\mathfrak{g}_2 = \mathbb{R}\text{-span}\{X_0, X_1\}$ and for

$$X = x_0X_0 + x_1X_1 + x_2X_2 + x_3X_3 \in \mathfrak{g},$$

we have

$$\mathrm{ad}_X(X_0) = -x_1X_2 - x_2X_3, \quad \mathrm{ad}_X(X_1) = x_0X_2 \quad \text{and} \quad \mathrm{ad}_X(X_2) = x_0X_3.$$

On the other hand, through the basis $\mathscr{B} = \{X_3, X_2, X_1, X_0\}$, the matrices of the endomorphisms ad_X and ad_X^2 are written as

$$\mathrm{ad}_X = \begin{pmatrix} 0 & x_0 & 0 & -x_2 \\ 0 & 0 & x_0 & -x_1 \\ 0 & 0 & 0 & 0 \\ 0 & 0 & 0 & 0 \end{pmatrix}, \quad \mathrm{ad}_X^2 = \begin{pmatrix} 0 & 0 & x_0^2 & -x_0x_1 \\ 0 & 0 & 0 & 0 \\ 0 & 0 & 0 & 0 \\ 0 & 0 & 0 & 0 \end{pmatrix}.$$

Hence, the matrix of the adjoint representation $\mathrm{Ad}_{\exp(X)}$ can be expressed as

$$\mathrm{Ad}_{\exp(X)} = \begin{pmatrix} 1 & x_0 & \frac{1}{2}x_0^2 & -x_2 - \frac{1}{2}x_0x_1 \\ 0 & 1 & x_0 & -x_1 \\ 0 & 0 & 1 & 0 \\ 0 & 0 & 0 & 1 \end{pmatrix},$$

and finally

$$\mathrm{Ad}_{\exp(X)} \circ \begin{pmatrix} a \\ b \\ c \\ d \end{pmatrix} = \begin{pmatrix} a + bx_0 + \frac{1}{2}x_0^2c - d(x_2 + \frac{1}{2}x_0x_1) \\ b + cx_0 - dx_1 \\ c \\ d \end{pmatrix}, \tag{4.56}$$

where the vector ${}^t(a\,b\,c\,d)$ represents a vector of $\mathfrak{g}$ through the basis $\mathscr{B}$.

Example 4.3.39. Let $\mathfrak{h} = \mathbb{R}\text{-span}\{X_1, X_2, X_3\}$ and $\mathfrak{l} = \mathbb{R}\text{-span}\{X_0\}$. Then if G, H designate the Lie groups associated to $\mathfrak{g}$ and $\mathfrak{h}$, respectively, and $\Gamma = \exp(\mathbb{Z}X_0)$, then obviously the resulting Clifford–Klein form $\Gamma\backslash G/H$ turns out to be compact. Clearly, $\Gamma \simeq \mathbb{Z}$, $G/H \simeq \mathbb{R}$ and $\Gamma\backslash G/H \simeq S^1$. As such, it is straightforward that through the basis $\mathscr{B}$, any $\varphi \in \mathrm{Hom}(\mathfrak{l}, \mathfrak{g})$ is given by

$$\varphi : \mathfrak{l} \to \mathfrak{g}; \quad \lambda X_0 \mapsto \lambda(dX_0 + cX_1 + bX_2 + aX_3)$$

and

$$\varphi = \begin{pmatrix} a \\ b \\ c \\ d \end{pmatrix} \in \mathscr{R}(\mathfrak{l}, \mathfrak{g}, \mathfrak{h}) \Longleftrightarrow d \neq 0. \tag{4.57}$$

Indeed, from the set equality (4.55),

$$\begin{aligned} \mathscr{R}(\mathfrak{l}, \mathfrak{g}, \mathfrak{h}) &= \left\{ \varphi = \begin{pmatrix} a \\ b \\ c \\ d \end{pmatrix} \in \operatorname{Hom}(\mathfrak{l}, \mathfrak{g}) \,\middle|\, \operatorname{Ad}_{\exp(X)} \circ \varphi(\mathfrak{l}) \oplus \mathfrak{h} = \mathfrak{g} \right\} \\ &= \left\{ \begin{pmatrix} a \\ b \\ c \\ d \end{pmatrix} \in \operatorname{Hom}(\mathfrak{l}, \mathfrak{g}) \,\middle|\, d \neq 0 \right\}. \end{aligned}$$

On the other hand, and according to our construction, for φ as in equation (4.57),

$$\varphi_1 = \begin{pmatrix} 0 \\ b \\ c \\ d \end{pmatrix}.$$

By equation (4.56), we get

$$\operatorname{Ad}_{\exp(X)} \circ \varphi - \varphi = \begin{pmatrix} bx_0 + \frac{1}{2}x_0^2 c - d(x_2 + \frac{1}{2}x_0 x_1) \\ cx_0 - dx_1 \\ 0 \\ 0 \end{pmatrix}$$

and

$$\operatorname{Ad}_{\exp(X)} * \varphi_1 - \varphi_1 = \begin{pmatrix} 0 \\ cx_0 - dx_1 \\ 0 \\ 0 \end{pmatrix},$$

which means that

$$\begin{aligned} G_\varphi &= \left\{ \exp(x_0 X_0 + x_1 X_1 + x_2 X_2 + x_3 X_3) \in G \,\middle|\, x_1 = \frac{c}{d} x_0,\ x_2 = \frac{b}{d} x_0 \right\} \quad \text{and} \\ G_{\varphi_1} &= \left\{ \exp(x_0 X_0 + x_1 X_1 + x_2 X_2 + x_3 X_3) \in G \,\middle|\, x_1 = \frac{c}{d} x_0 \right\}. \end{aligned}$$

Finally, $\dim G\cdot\varphi = 2$ and $\dim G * \varphi_1 = 1$ for any $\varphi \in \mathscr{R}(\mathfrak{l},\mathfrak{g},\mathfrak{h})$. Then by Theorem 4.3.38, the deformation space $\mathscr{T}(\mathfrak{l},\mathfrak{g},\mathfrak{h})$ is a Hausdorff space.

Example 4.3.40. Let now $\mathfrak{h} = \mathbb{R}\text{-span}\{X_0\}$ and $\mathfrak{l} = \mathbb{R}\text{-span}\{X_1, X_2, X_3\}$. Then again the resulting Clifford–Klein form $\Gamma\backslash G/H$ is compact. Clearly, $\Gamma \simeq \mathbb{Z}^3$, $G/H \simeq \mathbb{R}^3$ and $\Gamma\backslash G/H$ is homeomorphic to the 3-dimensional torus. As $\varphi \in \mathscr{R}(\mathfrak{l},\mathfrak{g},\mathfrak{h})$ if and only if $\varphi(\mathfrak{l}) = \mathfrak{l}$, φ takes the following form:

$$\varphi = \begin{pmatrix} a_1 & a_2 & a_3 \\ b_1 & b_2 & b_3 \\ c_1 & c_2 & c_3 \\ 0 & 0 & 0 \end{pmatrix}.$$

Hence,

$$\begin{aligned}\operatorname{Ad}_{\exp(X)} \circ \varphi &= \begin{pmatrix} a_1 + b_1x_0 + \frac{1}{2}x_0^2c_1 & a_2 + b_2x_0 + \frac{1}{2}x_0^2c_2 & a_3 + b_3x_0 + \frac{1}{2}x_0^2c_3 \\ b_1 + x_0c_1 & b_2 + x_0c_2 & b_3 + x_0c_3 \\ c_1 & c_2 & c_3 \\ 0 & 0 & 0 \end{pmatrix} \\ &= \varphi + x_0 \begin{pmatrix} b_1 & b_2 & b_3 \\ c_1 & c_2 & c_3 \\ 0 & 0 & 0 \\ 0 & 0 & 0 \end{pmatrix} + \frac{1}{2}x_0^2 \begin{pmatrix} c_1 & c_2 & c_3 \\ 0 & 0 & 0 \\ 0 & 0 & 0 \\ 0 & 0 & 0 \end{pmatrix},\end{aligned}$$

and then $\dim G\cdot\varphi = 1$. Besides,

$$\varphi_1 = \begin{pmatrix} 0 & 0 & 0 \\ b_1 & b_2 & b_3 \\ c_1 & c_2 & c_3 \\ 0 & 0 & 0 \end{pmatrix},$$

$$\operatorname{Ad}_{\exp(X)} * \varphi_1 = \varphi_1 + x_0 \begin{pmatrix} 0 & 0 & 0 \\ c_1 & c_2 & c_3 \\ 0 & 0 & 0 \\ 0 & 0 & 0 \end{pmatrix},$$

and likewise $\dim G * \varphi_1 = 1$ for any $\varphi \in \mathscr{R}(\mathfrak{l},\mathfrak{g},\mathfrak{h})$. Then by Theorem 4.3.38, the deformation space is a Hausdorff space.

Example 4.3.41. The following example treats a noncompact Clifford–Klein form case. Let $\mathfrak{h} = \mathbb{R}\text{-span}\{X_3\}$ and $\mathfrak{l} = \mathbb{R}\text{-span}\{X_1, X_2\}$. Clearly, $\Gamma \simeq \mathbb{Z}^2$ and $G/H \simeq \mathbb{R}^3$. Therefore, $\Gamma\backslash G/H$ is not compact. We first prove the following.

Claim 4.3.42. *For any $\varphi \in \mathscr{R}(\mathfrak{l},\mathfrak{g},\mathfrak{h})$, we have $\varphi(\mathfrak{l}) \subset \mathbb{R}\text{-span}\{X_1, X_2, X_3\}$.*

Proof. If not, there exists $v = X_0 + u \in \varphi(\mathfrak{l})$ for some $u \in \mathbb{R}\text{-span}\{X_1, X_2, X_3\}$. Since $\dim(\varphi(\mathfrak{l})) = 2$, there exists $w \neq 0$ such that $w \in \varphi(\mathfrak{l}) \cap \mathbb{R}\text{-span}\{X_1, X_2, X_3\}$. Thus, $\mathbb{R}\text{-span}\{[v,[v,w]],[v,w],w\} \cap \mathfrak{h} \neq \{0\}$, which leads to a contradiction as $\varphi(\mathfrak{l}) \cap \mathfrak{h} = \{0\}$. □

Now, any $\varphi \in \mathscr{R}(\mathfrak{l}, \mathfrak{g}, \mathfrak{h})$ reads

$$\varphi = \begin{pmatrix} a_1 & a_2 \\ b_1 & b_2 \\ c_1 & c_2 \\ 0 & 0 \end{pmatrix} \tag{4.58}$$

and we have the following.

Claim 4.3.43. *Let $\varphi \in \mathscr{R}(\mathfrak{l}, \mathfrak{g}, \mathfrak{h})$ be as in equation* (4.58), *then* $(c_1, c_2) \neq (0,0)$.

Proof. From Claim 4.3.42, any $\varphi \in \mathscr{R}(\mathfrak{l}, \mathfrak{g}, \mathfrak{h})$ reads

$$\varphi = \begin{pmatrix} a_1 & a_2 \\ b_1 & b_2 \\ c_1 & c_2 \\ 0 & 0 \end{pmatrix}.$$

Now, from the set equality (4.55),

$$\begin{aligned} \varphi \in \mathscr{R}(\mathfrak{l}, \mathfrak{g}, \mathfrak{h}) &\Leftrightarrow \dim \mathfrak{h} + \dim \varphi(\mathfrak{l}) = 3, \quad \varphi \in \mathscr{R}(\mathfrak{l}, \mathfrak{g}, \mathfrak{h}) \\ &\Leftrightarrow \det \begin{pmatrix} b_1 & b_2 \\ c_1 & c_2 \end{pmatrix} \neq 0. \end{aligned}$$

□

Now for $\varphi \in \mathscr{R}(\mathfrak{l}, \mathfrak{g}, \mathfrak{h})$ be as in equation (4.58),

$$\mathrm{Ad}_{\exp(X)} \circ \varphi = \varphi + x_0 \begin{pmatrix} b_1 & b_2 \\ c_1 & c_2 \\ 0 & 0 \\ 0 & 0 \end{pmatrix} + \frac{1}{2} x_0^2 \begin{pmatrix} c_1 & c_2 \\ 0 & 0 \\ 0 & 0 \\ 0 & 0 \end{pmatrix}$$

and then $\dim G \cdot \varphi = 1$. It is also obviously the case for

$$\varphi_1 = \begin{pmatrix} 0 & 0 \\ b_1 & b_2 \\ c_1 & c_2 \\ 0 & 0 \end{pmatrix}$$

as

$$\mathrm{Ad}_{\exp(X)} * \varphi_1 = \varphi_1 + x_0 \begin{pmatrix} 0 & 0 \\ c_1 & c_2 \\ 0 & 0 \\ 0 & 0 \end{pmatrix}.$$

Then by Theorem 4.3.38, the deformation space is a Hausdorff space.

4.4 Deformation space of threadlike nilmanifolds

Throughout the section, we note $G = \exp(\mathfrak{g})$ a threadlike nilpotent Lie group as defined in Subsection 1.1.3. We fix a basis $\mathscr{B} = \{X, Y_1, \ldots, Y_n\}$ of $\mathfrak{g}$ with nontrivial Lie brackets defined in (1.7). Recall the subspace $\mathfrak{g}_0 = \mathbb{R}\text{-span}\{Y_1, \ldots, Y_n\}$, which is an Abelian ideal of $\mathfrak{g}$ of codimension one and let $G_0 = \exp(\mathfrak{g}_0)$. The center $\mathfrak{z}(\mathfrak{g})$ of $\mathfrak{g}$ is however one- dimensional and it is the space $\mathbb{R}\text{-span}\{Y_n\}$. It turns out therefore that any threadlike Lie group belongs to the family of nilpotent Lie groups referred to as to be special, which admit one-codimensional normal Abelian subgroup, hence of the form $\mathbb{R} \ltimes \mathbb{R}^n$. The study of the deformation space of the action of discontinuous groups for special homogeneous spaces, especially its explicit determination seems to be subtle and problematic for some structural and technical reasons. However, we know already how to characterize the proper action of discontinuous groups on homogeneous spaces in this setup (cf. Chapter 2).

4.4.1 Description of Hom($\mathfrak{l}$, $\mathfrak{g}$)

Let $\Gamma \simeq \mathbb{Z}^k$ be a discrete finitely generated subgroup of G and $L = \exp \mathfrak{l}$ its syndetic hull. Our main result in this section consists in giving an explicit description of $\mathrm{Hom}(\mathfrak{l}, \mathfrak{g})$, the set of all algebras homomorphisms from $\mathfrak{l}$ to $\mathfrak{g}$. Toward such a purpose, recall first the result of Lemma 1.1.13 concerning the description of the structure of Lie subalgebras of threadlike algebras and we consider the sets:

$$H_{0,k} = \left\{ \begin{pmatrix} {}^t\vec{0} \\ N \end{pmatrix} \in M_{n+1,k}(\mathbb{R}) : N \in M_{n,k}(\mathbb{R}) \right\} \simeq M_{n,k}(\mathbb{R}), \tag{4.59}$$

and for any $j \in \{1, \ldots, k\}$:

$$\begin{aligned} H_{j,k} = \Bigg\{ & \begin{pmatrix} \lambda_1 T \cdots \lambda_{j-1} T & T & \lambda_{j+1} T \cdots \lambda_k T \\ z_1 \;\cdots\; z_{j-1} & z_j & z_{j+1} \;\cdots\; z_k \end{pmatrix} \in M_{n+1,k}(\mathbb{R}) : {}^tT \in \mathbb{R}^\times \times \mathbb{R}^{n-1}, \\ & (z_1, \ldots, z_k) \in \mathbb{R}^k, (\lambda_1, \ldots, \check{\lambda}_j, \ldots, \lambda_k) \in \mathbb{R}^{k-1} \Bigg\} \\ \simeq\ & \mathbb{R}^\times \times \mathbb{R}^{n-1} \times \mathbb{R}^{k-1} \times \mathbb{R}^k. \end{aligned} \tag{4.60}$$

The following upshot accurately describes the structure of the set $\mathrm{Hom}(\mathfrak{l},\mathfrak{g})$, which is one of the main means to study the parameter and the deformation spaces.

Theorem 4.4.1. *With the same notation and hypotheses, we have*

$$\mathrm{Hom}(\mathfrak{l},\mathfrak{g}) = \bigcup_{j=0}^{k} H_{j,k}.$$

Proof. We identify any element $T = xX + \sum_{i=1}^{n} y_i Y_i \in \mathfrak{g}$ by the column vector ${}^t(x, y_1, \dots, y_n)$. We define now the subset $M'_{n+1,k}(\mathbb{R}) = \{(T_1,\dots,T_k) \in M_{n+1,k}(\mathbb{R}) : [T_s, T_r] = 0,\ 1 \le r,s \le k\}$. Then having fixed a basis $\mathscr{B}_{\mathfrak{l}}$ of $\mathfrak{l}$, it appears clear that the map

$$\Psi : \mathrm{Hom}(\mathfrak{l},\mathfrak{g}) \longrightarrow M'_{n+1,k}(\mathbb{R}), \tag{4.61}$$

which associates to any element of $\mathrm{Hom}(\mathfrak{l},\mathfrak{g})$ its matrix written through the bases $\mathscr{B}_{\mathfrak{l}}$ and $\mathscr{B}$ is a homeomorphism. Let now

$$M = \begin{pmatrix} x_1 & \cdots & x_k \\ y_{11} & \cdots & y_{1k} \\ \vdots & & \vdots \\ y_{n1} & \cdots & y_{kn} \end{pmatrix} \in M'_{n+1,k}(\mathbb{R}),$$

then for any $1 \le s,r \le k$, $[T_s, T_r] = 0$ where $T_s = x_s X + \sum_{i=1}^{n} y_{is} Y_i$ and $T_r = x_r X + \sum_{i=1}^{n} y_{ir} Y_i$. This gives rise to the following equation:

$$x_s y_{ir} - x_r y_{is} = 0 \quad \text{for all } 1 \le r,s \le k \text{ and all } 1 \le i \le n-1. \tag{4.62}$$

In order to find the solutions of (4.62), we shall discuss the two following dichotomous cases. Assume in a first time that the first line of M is zero, that is, $x_j = 0$, $1 \le j \le k$. In this case, M satisfies (4.62) and obviously belongs to $M'_{n+1,k}(\mathbb{R})$. Suppose now that the first line of M is not zero. There exists then $j \in \{1,\dots,k\}$ satisfying $x_j \ne 0$. So, $M \in M'_{n+1,k}(\mathbb{R})$ if and only if there exists $\Lambda_j = (\lambda_1,\dots,\lambda_{j-1},\lambda_{j+1},\dots,\lambda_k) \in \mathbb{R}^{k-1}$ such that for $s \ne j$, we have $T'_s = \lambda_s T'_j$, where $T'_s = {}^t(x_s, y_{1s},\dots,y_{n-1s})$. According to this discussion, we get that $M'_{n+1,k}(\mathbb{R}) = \bigcup_{j=0}^{k} H_{j,k}$ where $H_{j,k}$, $0 = 1,\dots,k$ are as determined by equations (4.59) and (4.60). This achieves the proof of the theorem. □

Note that the set of all injective algebras homomorphisms from $\mathfrak{l}$ to $\mathfrak{g}$ denoted by $\mathrm{Hom}^0(\mathfrak{l},\mathfrak{g})$, rather than $\mathrm{Hom}(\mathfrak{l},\mathfrak{g})$ itself will be of interest in the next section, merely because it is involved in deformations. The following result accurately determines the stratification of such a set.

Proposition 4.4.2. *Let $k \in \{1,\dots,n+1\}$, there exists a finite set I_k such that $0 \in I_k$ and $\mathrm{Hom}^0(\mathfrak{l},\mathfrak{g}) = \coprod_{j\in I_k} K_{j,k}$, where:*

(i) *If $k > 2$, then $I_k = \{0\}$ and for any $k \in \{1, \dots, n\}$ we have*

$$K_{0,k} = \left\{ \begin{pmatrix} {}^t\vec{0} \\ N \end{pmatrix} : N \in M^0_{n,k}(\mathbb{R}) \right\} \simeq M^0_{n,k}(\mathbb{R}). \tag{4.63}$$

Here, $M^0_{n,m}(\mathbb{R})$ denotes the set of all matrix of n rows, m columns and of maximal rank.

(ii) *If $k = 2$, then $I_2 = \{0, 1, 2, 3\}$ and*

$$K_{1,2} = \left\{ \begin{pmatrix} x & 0 \\ \vec{y} & \vec{0} \\ z_1 & z_2 \end{pmatrix} \in M_{n+1,2}(\mathbb{R}) : \, xz_2 \neq 0 \right\},$$

$$K_{2,2} = \left\{ \begin{pmatrix} 0 & x \\ \vec{0} & \vec{y} \\ z_1 & z_2 \end{pmatrix} \in M_{n+1,2}(\mathbb{R}) : \, xz_1 \neq 0 \right\},$$

$$K_{3,2} = \left\{ \begin{pmatrix} x & \lambda x \\ \vec{y} & \lambda\vec{y} \\ z_1 & z_2 \end{pmatrix} \in M_{n+1,2}(\mathbb{R}) : \, \lambda x \neq 0, \, \lambda z_1 - z_2 \neq 0 \right\}.$$

(iii) *If $k = 1$, then $I_1 = \{0, 1\}$ and*

$$K_{1,1} = H_{1,1} = \left\{ \begin{pmatrix} x \\ \vec{y} \end{pmatrix} : x \in \mathbb{R}^\times, \vec{y} \in \mathbb{R}^n \right\} \simeq \mathbb{R}^\times \times \mathbb{R}^n.$$

Proof. Let $k \in \{1, \dots, n+1\}$, $j \in \{1, \dots, k\}$ and $M \in H_{j,k}$. It is not hard to check that $\operatorname{rank}(M) \leq 2$. In addition, M is of maximal rank if and only if $\operatorname{rank}(M) = k$. So, it appears clear that if $k > 2$, we have $\operatorname{Hom}^0(\mathfrak{l}, \mathfrak{g}) \cap H_{j,k} = \emptyset$ for all $j = 1, \dots, k$. This proves the first assertion of the proposition. Suppose now that $k = 2$ and choose in a first time

$$M = \begin{pmatrix} x & \lambda x \\ \vec{y} & \lambda\vec{y} \\ z_1 & z_2 \end{pmatrix} \in H_{1,2},$$

where $x \in \mathbb{R}^\times$, ${}^t\vec{y} \in \mathbb{R}^{n-1}$, $\lambda \in \mathbb{R}$ and $(z_1, z_2) \in \mathbb{R}^2$. The condition M is of maximal rank is equivalent to $\lambda z_1 - z_2 \neq 0$. If furthermore, we choose

$$M = \begin{pmatrix} \lambda x & x \\ \lambda\vec{y} & \vec{y} \\ z_1 & z_2 \end{pmatrix} \in H_{2,2}$$

for some $x \in \mathbb{R}^\times$, ${}^t\vec{y} \in \mathbb{R}^{n-1}$ and $(z_1, z_2) \in \mathbb{R}^2$, we get M is of maximal rank if and only if $\lambda z_2 - z_1 \neq 0$. Therefore,

$$\mathrm{Hom}^0(\mathfrak{l}, \mathfrak{g}) \cap H_{1,2} = \left\{ M = \begin{pmatrix} x & \lambda x \\ \vec{y} & \lambda \vec{y} \\ z_1 & z_2 \end{pmatrix} \in H_{1,2} : \lambda z_1 - z_2 \neq 0 \right\}$$

and

$$\mathrm{Hom}^0(\mathfrak{l}, \mathfrak{g}) \cap H_{2,2} = \left\{ M = \begin{pmatrix} \lambda x & x \\ \lambda \vec{y} & \vec{y} \\ z_1 & z_2 \end{pmatrix} \in H_{2,2} : \lambda z_2 - z_1 \neq 0 \right\}.$$

It is then easy to see that $\mathrm{Hom}^0(\mathfrak{l}, \mathfrak{g}) \cap (H_{1,2} \cup H_{2,2})$ is equal to the disjoint union of the sets $K_{j,2}, j = 1, 2, 3$ defined above. So we end up with the following decomposition $\mathrm{Hom}^0(\mathfrak{l}, \mathfrak{g}) = \coprod_{j=0}^{3} K_{j,2}$. Finally, if $k = 1$ then any homomorphism in $H_{1,1}$ is injective, therefore, $K_{1,1} = H_{1,1}$ and then $\mathrm{Hom}^0(\mathfrak{l}, \mathfrak{g}) = K_{1,1} \coprod K_{0,1}$. □

Remark 4.4.3. It is worth noting at this step that the fact that $\mathrm{Hom}^0(\mathfrak{l}, \mathfrak{g}) \cap H_{j,k} = \emptyset$ for $k > 2$ is justified by the fact that any discrete subgroup $\Gamma \not\subset G_0$ must be of rank ≤ 2 as shows the following proposition.

Proposition 4.4.4. *Let Γ be an Abelian discrete subgroup of G such that $\Gamma \not\subset G_0$. Then* $\mathrm{rank}(\Gamma) \in \{1, 2\}$.

Proof. Let L be the syndetic hull of Γ and $\mathfrak{l}$ its corresponding Lie algebra. It is clear that L is Abelian and $L \not\subset G_0$. So, we can suppose that $X \in \mathfrak{l}$. Let now $T = \sum_{i=1}^{n} \alpha_i Y_i \in \mathfrak{l}$, for some $\alpha_i \in \mathbb{R}$, then $[X, T] = 0$. This gives $T = \alpha_n Y_n$ and then $\mathfrak{l} \subset \mathbb{R}\text{-span}\{X, Y_n\}$. We conclude that $\mathrm{rank}(\Gamma) = \dim L \leq 2$. □

4.4.2 Description of the parameter space $\mathscr{R}(\Gamma, G, H)$

The most important problem in the study of the deformation space of discontinuous groups is the description of the parameter set $\mathscr{R}(\Gamma, G, H)$ given as in equation (3.1). Let G act on $M'_{n+1,k}(\mathbb{R})$ by

$$g \cdot M = \mathrm{Ad}_{g^{-1}} \cdot M, \quad M \in M'_{n+1,k}(\mathbb{R}), \quad g \in G.$$

Here, we view $\mathrm{Ad}_{g^{-1}}$ as a real valued matrix for any $g \in G$. Taking into account the action of G on $\mathrm{Hom}(\mathfrak{l}, \mathfrak{g})$ defined in (3.3), the following lemma is immediate.

Lemma 4.4.5. *The map Ψ defined in (4.61) is G-equivariant. That is, for any $\psi \in \mathrm{Hom}(\mathfrak{l}, \mathfrak{g})$ and $g \in G$, we have $\Psi(g \cdot \psi) = g \cdot \Psi(\psi)$.*

The following result is immediate.

Lemma 4.4.6. *Let G be a threadlike nilpotent Lie group, $H = \exp \mathfrak{h}$ be a closed connected subgroup of G and let Γ be an Abelian discrete subgroup of G. In light of Theorem 3.2.4, the set $\mathscr{R}(\Gamma, G, H)$ is homeomorphic to*

$$\{M \in M'_{n+1,k}(\mathbb{R}) : \operatorname{rank}(M ⋒ g \cdot M_{\mathfrak{h},\mathscr{B}}) = k + p \text{ for any } g \in G\}, \tag{4.64}$$

where $\mathscr{B} = \{X, Y_1, \dots, Y_n\}$ and the symbol ⋒ merely means the concatenation of the matrices written through $\mathscr{B}$.

Proof. If $M \in \mathscr{R}(\Gamma, G, H)$, then $M \in M^0_{n+1,k}(\mathbb{R})$, which gives that $\operatorname{rank}(M) = k$. Now using Theorem 2.3.2, the proper action of L on G/H is equivalent to the fact that $\mathfrak{l} \cap \operatorname{Ad}_g \mathfrak{h} = \{0\}$ for any $g \in G$, which means that $\operatorname{rank}(M ⋒ g \cdot M_{\mathfrak{h},\mathscr{B}}) = k + p$. Note finally that the condition $\operatorname{rank}(M) = k$ is irrelevant at this stage, as was to be shown. □

Proposition 4.4.7. *Let G be a threadlike Lie group and H a connected Lie subgroup of G. Then $\mathscr{R}(\Gamma, G, H) = \coprod_{j \in I_k} R_{j,k}$ where $R_{j,k} = \mathscr{R}(\Gamma, G, H) \cap K_{j,k}$. More precisely, one has:*
(i) *If $k > 2$, then $\mathscr{R}(\Gamma, G, H) = R_{0,k}$.*
(ii) *If $k = 2$, then $\mathscr{R}(\Gamma, G, H) = \coprod_{j=0}^{3} R_{j,2}$.*
(iii) *If $k = 1$, then $\mathscr{R}(\Gamma, G, H) = R_{0,1} \coprod R_{1,1}$.*

Proof. This result stems immediately from Proposition 4.4.2, which describes the structure of $\operatorname{Hom}^0(\mathfrak{l}, \mathfrak{g})$. □

We now determine the parameter space according to the values of $k = \operatorname{rank}(\Gamma)$. Toward that purpose, let q denote the codimension of $\mathfrak{h}$ in $\mathfrak{g}$ and introduce the matrix $A(t), t \in \mathbb{R}$ of $(\operatorname{Ad}_{\exp tX})_{|\mathfrak{g}_0}$ written with respect to the strong Malcev basis $\mathscr{B}_0 = \{Y_1, \dots, Y_n\}$ of $\mathfrak{g}_0$. So, a routine computation shows that

$$A(t) = \begin{pmatrix} 1 & 0 & \cdots & \cdots & 0 \\ t & 1 & \ddots & & \vdots \\ \frac{t^2}{2} & t & 1 & \ddots & \vdots \\ \vdots & \ddots & \ddots & \ddots & 0 \\ \frac{t^{n-1}}{(n-1)!} & \cdots & \frac{t^2}{2} & t & 1 \end{pmatrix}.$$

The following proposition deals with the description of $R_{0,k}$, which coincides with the parameter space in the case where $2 < k \leq q$.

Proposition 4.4.8 (The parameter space for $k > 2$). *We keep the same hypotheses and notation as before. Then:*

(i) *If* $\mathfrak{h} \not\subset \mathfrak{g}_0$, *then*

$$R_{0,k} = \left\{ \begin{pmatrix} {}^t\vec{0} \\ N_1 \\ N_2 \end{pmatrix} \in H_{0,k} : N_1 \in M^0_{q,k}(\mathbb{R}), N_2 \in M_{n-q,k}(\mathbb{R}) \right\}$$
$$\simeq M^0_{q,k}(\mathbb{R}) \times M_{n-q,k}(\mathbb{R}). \tag{4.65}$$

(ii) *If* $\mathfrak{h} \subset \mathfrak{g}_0$, *then*

$$R_{0,k} = \left\{ \begin{pmatrix} {}^t\vec{0} \\ N \end{pmatrix} \in H_{0,k} : \operatorname{rank}(A(t)N \Cap M_{\mathfrak{h},\mathscr{B}_0}) = k + p \textit{ for any } t \in \mathbb{R} \right\}.$$

Proof. Thanks to Lemma 4.4.7, $\mathscr{R}(\Gamma, G, H) = \emptyset$ for $k > q$. We can from now on suppose that $k \leq q$. Suppose in first time that $\mathfrak{h} \not\subset \mathfrak{g}_0$. We note $\mathfrak{h}_0 = \mathfrak{h} \cap \mathfrak{g}_0$, which is an ideal of $\mathfrak{g}$. Let

$$M = \begin{pmatrix} {}^t\vec{0} \\ N \end{pmatrix} \in H_{0,k}.$$

So, equation (4.64) is equivalent to $\operatorname{rank}(M \Cap M_{\mathfrak{h}_0,\mathscr{B}}) = k + p - 1$, which is in turn equivalent to the fact that

$$\operatorname{rank}(N \Cap M_{\mathfrak{h}_0,\mathscr{B}_0}) = k + p - 1. \tag{4.66}$$

According to our choice of the strong Malcev basis $\mathscr{B}$, we get that

$$M_{\mathfrak{h}_0,\mathscr{B}_0} = \begin{pmatrix} (0) \\ I_{p-1} \end{pmatrix} \in M_{n,p-1}(\mathbb{R}),$$

where I_{p-1} designates the identity matrix of $M_{p-1}(\mathbb{R})$. We now write

$$M = \begin{pmatrix} {}^t\vec{0} \\ N_1 \\ N_2 \end{pmatrix},$$

where $N_1 \in M^0_{q,k}(\mathbb{R})$ and $N_2 \in M_{n-q,k}(\mathbb{R})$, we get that equation (4.66) is equivalent to the fact $\operatorname{rank}(N_1) = k$. We hence end up with

$$R_{0,k} = \mathscr{R}(\Gamma, G, H) \cap H_{0,k} = \left\{ \begin{pmatrix} {}^t\vec{0} \\ N_1 \\ N_2 \end{pmatrix} : N_1 \in M^0_{q,k}(\mathbb{R}), N_2 \in M_{n-q,k}(\mathbb{R}) \right\}.$$

We now treat the case where $\mathfrak{h} \subset \mathfrak{g}_0$. So, it is not hard to see that $\mathrm{Ad}_G \mathfrak{h} = \bigcup_{t\in\mathbb{R}} \mathrm{Ad}_{\exp tX} \mathfrak{h}$. We get therefore that

$$\begin{aligned} M \in \mathscr{R}(\Gamma, G, H) &\Leftrightarrow \mathrm{rank}(M ⋒ \exp tX \cdot M_{\mathfrak{h},\mathscr{B}}) = k + p \quad \text{for all } t \in \mathbb{R} \\ &\Leftrightarrow \mathrm{rank}(\exp tX \cdot N ⋒ M_{\mathfrak{h}\mathscr{B}_0}) = k + p \quad \text{for all } t \in \mathbb{R} \\ &\Leftrightarrow \mathrm{rank}(A(t)N ⋒ M_{\mathfrak{h},\mathscr{B}_0}) = k + p \quad \text{for all } t \in \mathbb{R}, \end{aligned}$$

which completes the proof of the proposition. □

We assume henceforth that $\mathrm{rank}(\Gamma) \in \{1, 2\}$. We will be dealing with these subsequent cases separately. The following upshot exhibits an accurate description of the parameter space when $k = 2$.

Proposition 4.4.9 (The parameter space for $k = 2$). *Assume that $k = 2$. Then $R_{0,2}$ being described in Proposition 4.4.8, we have:*

(i) *If $\mathfrak{h} \not\subset \mathfrak{g}_0$, then:*

i$_1$. *If $q = n$, then*

$$R_{1,2} = \left\{ \begin{pmatrix} x & 0 \\ y_1 & 0 \\ \vec{y} & \vec{0} \\ z_1 & z_2 \end{pmatrix} M_{n+1,2}(\mathbb{R}) : y_1 x z_2 \neq 0 \right\},$$

$$R_{2,2} = \left\{ \begin{pmatrix} 0 & x \\ 0 & y_1 \\ \vec{0} & \vec{y} \\ z_1 & z_2 \end{pmatrix} \in M_{n+1,2}(\mathbb{R}) : y_1 x z_1 \neq 0 \right\}$$

and

$$R_{3,2} = \left\{ \begin{pmatrix} x & \lambda x \\ y_1 & \lambda y_1 \\ \vec{y} & \lambda \vec{y} \\ z_1 & z_2 \end{pmatrix} \in M_{n+1,2}(\mathbb{R}) : \lambda x(\lambda z_1 - z_2) y_1 \neq 0 \right\}.$$

i$_2$. *If $q < n$, then $R_{j,2} = \emptyset$, $j = 1, 2, 3$.*

(ii) *If $\mathfrak{h} \subset \mathfrak{g}_0$, then $R_{1,2} = R_{2,2} = R_{3,2} = \emptyset$ if $Y_n \in \mathfrak{h}$. Otherwise, $R_{i,2} = K_{i,2}$, $i = 1, 2, 3$.*

Proof. Let

$$M = \begin{pmatrix} x & \lambda x \\ \vec{y} & \lambda \vec{y} \\ z_1 & z_2 \end{pmatrix} \in \mathrm{Hom}^0(\mathfrak{l}, \mathfrak{g}),$$

where $x \in \mathbb{R}^\times$, ${}^t\vec{y} = (y_1, \dots, y_{n-1}) \in \mathbb{R}^{n-1}$ and $(z_1, z_2) \in \mathbb{R}^2$ such that $\lambda z_1 - z_2 \in \mathbb{R}^\times$. We tackle first the case where $\mathfrak{h} \not\subset \mathfrak{g}_0$. In the case where $\mathfrak{h} = \mathbb{R}X$, a simple computation shows that $\mathrm{Ad}_G \mathfrak{h} = \mathbb{R}X + [X, \mathfrak{g}_0]$. Hence, the assertion $\mathrm{rank}(M ⋒ g \cdot M_{\mathfrak{h},\mathscr{B}}) = 3$ for any $g \in G$, is equivalent to $\mathrm{rank}(M ⋒ {}^t(1, 0, \alpha_2, \dots, \alpha_n)) = 3$, for all $\alpha_2, \dots, \alpha_n \in \mathbb{R}$, which is in turn equivalent to $y_1 \in \mathbb{R}^\times$. Therefore, $M \in R_{1,2} \cup R_{3,2}$ if and only if $y_1 \in \mathbb{R}^\times$. Similar computations show that for

$$M = \begin{pmatrix} 0 & x \\ \vec{0} & \vec{y} \\ z_1 & z_2 \end{pmatrix} \in K_{2,2},$$

one gets that $M \in R_{2,2}$ if and only if $y_1 \in \mathbb{R}^\times$. Suppose now that $\mathbb{R}X \subsetneq \mathfrak{h}$, that is $q < n$, we have that the vector $Y_n = {}^t(0, \dots, 0, 1) \in \mathfrak{h}$ and it is a linear combination of the columns of M. So $\mathrm{rank}(M ⋒ M_{\mathfrak{h},\mathscr{B}}) < p + 2$ and then $\mathscr{R}(\Gamma, G, H) \cap K_{j,2} = R_{j,2} = \emptyset$, $j = 1, 2, 3$.

Let finally $\mathfrak{h} \subset \mathfrak{g}_0$. If $Y_n \in \mathfrak{h}$, then $\mathrm{rank}(M ⋒ M_{\mathfrak{h},\mathscr{B}}) < p + 2$, which gives $\mathscr{R}(\Gamma, G, H) \cap K_{j,2} = \emptyset$ $(j = 1, 2, 3)$. Otherwise, $M \in \mathscr{R}(\Gamma, G, H)$ is equivalent to $\mathrm{rank}(M ⋒ \exp tX \cdot M_{\mathfrak{h},\mathscr{B}}) = 2+p$ for all $t \in \mathbb{R}$, that is, $\mathrm{rank}(M) = 2$ and then $M \in \mathrm{Hom}^0(\mathfrak{l}, \mathfrak{g})$. Thus, we have $R_{j,2} = K_{j,2}$, $j = 1, 2, 3$, which completes the proof in this case. □

Similar arguments are used to prove the following.

Proposition 4.4.10 (The parameter space for $k = 1$). *Assume that $k = 1$. The layer $R_{0,1}$ being described in Proposition 4.4.8, we get:*
(i) *If $\mathfrak{h} \not\subset \mathfrak{g}_0$, then*

$$R_{1,1} = \left\{ \begin{pmatrix} x \\ y_1 \\ \vec{y} \end{pmatrix} : x \in \mathbb{R}^\times, y_1 \in \mathbb{R}^\times, \vec{y} \in \mathbb{R}^{n-1} \right\} \simeq (\mathbb{R}^\times)^2 \times \mathbb{R}^{n-1}.$$

(ii) *If $\mathfrak{h} \subset \mathfrak{g}_0$, then $R_{1,1} = H_{1,1} = K_{1,1}$.*

4.4.3 Description of the deformation space $\mathscr{T}(\Gamma, G, H)$

This section aims to describe the deformation space of the action of an Abelian discrete subgroup $\Gamma \subset G$ on a threadlike homogeneous space G/H. This description strongly relies on the comprehensive details about the parameter space provided in the previous section.

Now we can state the following.

Proposition 4.4.11. *We keep the same hypotheses and notation. The disjoint components $R_{j,k}$ involved through the description of the parameter space $\mathscr{R}(\Gamma, G, H)$ are G-invariant. More precisely, we have the following:*
(i) *If $k > 2$, then $\mathscr{T}(\Gamma, G, H) = R_{0,k}/G$.*

(ii) *If* $k = 2$, *then* $\mathscr{T}(\Gamma, G, H) = \coprod_{j=0}^{3}(R_{j,2}/G)$.
(iii) *If* $k = 1$, *then* $\mathscr{T}(\Gamma, G, H) = (R_{0,1}/G)\coprod(R_{1,1}/G)$.

Proof. Let us first prove that the set $R_{0,k}$ is G-stable. It is clear that the G-action on $R_{0,k}$ is reduced to the action of $\exp \mathbb{R}X$. Let

$$M = \begin{pmatrix} {}^t\vec{0} \\ N \end{pmatrix} \in R_{0,k}$$

and $t \in \mathbb{R}$, then

$$\exp tX \cdot M = \begin{pmatrix} {}^t\vec{0} \\ A(t)N \end{pmatrix}.$$

From the G-invariance of $\mathscr{R}(\Gamma, G, H)$, we get that $\exp tX \cdot M \in \mathscr{R}(\Gamma, G, H)$, so we are done in this case.

Suppose now that $k = 2$. Let

$$M = \begin{pmatrix} x & \lambda x \\ y_1 & \lambda y_1 \\ \vec{y} & \lambda \vec{y} \\ z_1 & z_2 \end{pmatrix} \in R_{3,2},$$

where ${}^t\vec{y} = (y_2, \ldots, y_{n-1}) \in \mathbb{R}^{n-2}$ and let $g = \exp(tX + a_1Y_1 + \cdots + a_nY_n) \in G$ for some $t, a_1, \ldots, a_n \in \mathbb{R}$, then a routine computation shows that

$$g \cdot M = \begin{pmatrix} x & \lambda x \\ y_1 & \lambda y_1 \\ \vec{y'} & \lambda \vec{y'} \\ z_1' & z_2' \end{pmatrix},$$

where ${}^t\vec{y'} = (y_2', \ldots, y_{n-1}')$ is such that

$$y_i' = y_i + \sum_{j=1}^{i-1} \frac{t^{j-1}}{j!}(ty_{i-j} - xa_{i-j}), \quad i = 2, \ldots, n-1 \tag{4.67}$$

and $z_i' = z_i + \sum_{j=1}^{n-1} \frac{t^{j-1}}{j!}(ty_{n-j} - xa_{n-j})$, $i = 1, 2$. As $\lambda z_1' - z_2' = \lambda z_1 - z_2$, we get that $g \cdot M \in R_{3,2}$ as was to be shown. We opt for the same arguments to show that $R_{1,2}$ and $R_{2,2}$ are G-invariant as well. For $k = 1$, $M = {}^t(x, y_1, \ldots, y_{n-1}) \in R_{1,1}$ and $g \in G$, we have

$$g \cdot M = \begin{pmatrix} x \\ y_1 \\ \vec{y'} \end{pmatrix},$$

where ${}^t\overrightarrow{y'} = (y'_2, \dots, y'_{n-1})$ is given as in equation (4.67). So, the same arguments allow us to conclude. □

We are now ready to give an explicit description of the deformation space $\mathscr{T}(\Gamma, G, H)$. Toward this purpose, we can divide the task into three parts as in the previous section. More precisely, we shall define a cross-section of $R_{j,k}/G$ denoted by $\mathscr{T}_{j,k}$ for any $j \in I_k$ and $k \in \{1, \dots, n\}$. Recall that if $q < k$ and $k \geq 2$, then we got $\mathscr{R}(\Gamma, G, H) = \emptyset$. We suppose then that $k \leq q$. Let $m, n \in \mathbb{N}$ and denote for any $1 \leq r \leq n$ and any $1 \leq s \leq m$ by $M_{n,m}(r, s, \mathbb{R})$ the subset of $M_{n,m}(\mathbb{R})$ defined by

$$M_{n,m}(r,s,\mathbb{R}) = \left\{ \begin{matrix} \\ \\ \\ (r) \\ \\ \\ \\ \end{matrix} \begin{pmatrix} 0 & \cdots & 0 & \overset{(s)}{0} & 0 & \cdots & 0 \\ \vdots & & \vdots & \vdots & \vdots & & \vdots \\ 0 & \cdots & 0 & 0 & 0 & \cdots & 0 \\ 0 & \cdots & 0 & x_{rs} & * & \cdots & * \\ * & \cdots & * & * & * & \cdots & * \\ \vdots & & \vdots & \vdots & \vdots & & \vdots \\ * & \cdots & * & * & * & \cdots & * \end{pmatrix} \in M_{n,m}(\mathbb{R}) : x_{rs} \in \mathbb{R}^\times \right\},$$

and

$$M'_{n,m}(r,s,\mathbb{R}) = \{M \in M_{n,m}(r,s,\mathbb{R}) : x_{(r+1)s} = 0\}.$$

We now consider the set $R_{0,k}(r,s) = R_{0,k} \cap M_{n+1,k}(r,s,\mathbb{R})$. For any $k \in \{1, \dots, n\}$, let J_k designate the set of all $(r,s) \in \{1 \leq r \leq n+1\} \times \{1 \leq s \leq k\}$ for which $R_{0,k}(r,s) \neq \emptyset$. Then $(1,s) \notin J_k$ and

$$R_{0,k} = \coprod_{(r,s)\in J_k} R_{0,k}(r,s)$$

whenever $k > 2$. Moreover, it is not hard to see that $R_{0,k}(r,s)$ is G-invariant. Then

$$R_{0,k}/G = \coprod_{(r,s)\in J_k} (R_{0,k}(r,s)/G).$$

For $(r,s) \in J_k$, let $\mathscr{T}_{0,k}(r,s) = R_{0,k}(r,s) \cap M'_{n+1,k}(r,s,\mathbb{R}) = R_{0,k} \cap M'_{n+1,k}(r,s,\mathbb{R})$. We have the following.

Proposition 4.4.12. *We keep all our notation as above. Then we have:*
(i) *$\mathscr{T}_{0,k}(r,s)$ is homeomorphic to $R_{0,k}(r,s)/G$ for any $(r,s) \in J_k$.*
(ii) *$\mathscr{T}_{0,k} = \coprod_{(r,s)\in J_k} \mathscr{T}_{0,k}(r,s)$.*

If in particular $k > 2$, then $\mathscr{T}(\Gamma, G, H) \simeq \mathscr{T}_{0,k}$.

Proof. Let $(r,s) \in J_k$. We show that $\mathscr{T}_{0,k}(r,s)$ is a cross-section of all adjoint orbits of $R_{0,k}(r,s)$. It is clear that the G-action on $R_{0,k}(r,s)$ is reduced to the action of $\exp \mathbb{R}X$. More precisely, let

$$M = \begin{pmatrix} {}^t\vec{0} \\ N \end{pmatrix} \in R_{0,k}(r,s),$$

then

$$G \cdot M = [M] = \left\{ \begin{pmatrix} {}^t\vec{0} \\ A(t)N \end{pmatrix} : t \in \mathbb{R} \right\}.$$

Noting $N = \{(a_{i,j}), 1 \leq i \leq n,\ 1 \leq j \leq k\}$, we get $a_{r-1,s} \neq 0$. Let

$$t_M = \begin{cases} -\frac{a_{r,s}}{a_{r-1,s}} & \text{if } r < n+1, \\ 0 & \text{if } r = n+1. \end{cases}$$

We can then show that

$$\left\{ \begin{pmatrix} {}^t\vec{0} \\ A(t_M)N \end{pmatrix} \right\} = G \cdot M \cap \mathscr{T}_{0,k}(r,s). \tag{4.68}$$

Remark first that if $r = n+1$ then $k = 1$ and $G \cdot M = M$, so (4.68) holds. Suppose now that $r \leq n$. It is then clear that

$$\begin{pmatrix} {}^t\vec{0} \\ A(t_M)N \end{pmatrix} \in G \cdot M \cap \mathscr{T}_{0,k}(r,s)$$

using the G-invariance of the layer $R_{0,k}(r,s)$. Conversely, if

$$\begin{pmatrix} {}^t\vec{0} \\ A(t)N \end{pmatrix} \in \mathscr{T}_{0,k}(r,s),$$

then by an easy computation, we can see that $a_{r,s} + t a_{r-1,s} = 0$, which gives that $t = t_M$. The next step consists in showing that the map

$$\begin{array}{rcl} (\Phi_{0,k})_{(r,s)} : R_{0,k}(r,s)/G & \to & \mathscr{T}_{0,k}(r,s) \\ {[M]} & \mapsto & \begin{pmatrix} {}^t\vec{0} \\ A(t_M)N \end{pmatrix} \end{array}$$

is bijective. First of all, it is clear that $(\Phi_{0,k})_{(r,s)}$ is well-defined. In fact, let $M_1, M_2 \in R_{0,k}(r,s)$ such that $[M_1] = [M_2]$. Then $(\Phi_{0,k})_{(r,s)}([M_2]) = G \cdot M_2 \cap \mathscr{T}_{0,k}(r,s) = G \cdot M_1 \cap$

$\mathscr{T}_{0,k}(r,s) = (\Phi_{0,k})_{(r,s)}([M_1])$. For

$$M_1 = \begin{pmatrix} {}^t\vec{0} \\ N_1 \end{pmatrix} \quad \text{and} \quad M_2 = \begin{pmatrix} {}^t\vec{0} \\ N_2 \end{pmatrix} \in R_{0,k}(r,s)$$

such that $(\Phi_{0,k})_{(r,s)}([M_1]) = (\Phi_{0,k})_{(r,s)}([M_2])$, we have

$$\begin{pmatrix} {}^t\vec{0} \\ A(t_{M_1})N_1 \end{pmatrix} = \begin{pmatrix} {}^t\vec{0} \\ A(t_{M_2})N_2 \end{pmatrix} \in G \cdot M_1 \cap G \cdot M_2.$$

It follows therefore that $[M_1] = [M_2]$, which leads to the injectivity of $(\Phi_{0,k})_{(r,s)}$. Now, to see that $(\Phi_{0,k})_{(r,s)}$ is surjective, it is sufficient to verify that for all $M \in \mathscr{T}_{0,k}(r,s)$, we have $(\Phi_{0,k})_{(r,s)}([M]) = M$ as $G \cdot M \cap \mathscr{T}_{0,k}(r,s) = \{M\}$. To achieve the proof, we prove that $(\Phi_{0,k})_{(r,s)}$ is bicontinuous. Let $(\pi_{0,k})_{(r,s)}$ be the canonical surjection $(\pi_{0,k})_{(r,s)} : R_{0,k}(r,s) \to R_{0,k}(r,s)/G$. Thus, we can easily see the continuity of $(\widetilde{\Phi}_{0,k})_{(r,s)} = (\Phi_{0,k})_{(r,s)} \circ (\pi_{0,k})_{(r,s)}$, which is equivalent to the continuity of $(\Phi_{0,k})_{(r,s)}$. Finally, it is clear that $((\Phi_{0,k})_{(r,s)})^{-1} = ((\pi_{0,k})_{(r,s)})_{|\mathscr{T}_{0,k}(r,s)}$, so the bicontinuity follows. □

Proposition 4.4.13. *Assume that $k = 2$, then $\mathscr{T}_{0,2}$ is described in Proposition 4.4.12 and $R_{j,2}/G$ is homeomorphic to $\mathscr{T}_{j,2}$ for $j = 1,2,3$ given as follows:*
(i) *If $\mathfrak{h} \not\subset \mathfrak{g}_0$, then we have the following subcases:*
  (i$_1$) *if $q < n$, then $\mathscr{T}_{j,2} = \emptyset$, $j = 1,2,3$.*
  (i$_2$) *if $q = n$, then*

$$\mathscr{T}_{1,2} = \left\{ \begin{pmatrix} x & 0 \\ y & 0 \\ \vec{0} & \vec{0} \\ 0 & \beta \end{pmatrix} \in M_{n+1,2}(\mathbb{R}) : x \in \mathbb{R}^\times, \beta \in \mathbb{R}^\times, y \in \mathbb{R}^\times \right\},$$

$$\mathscr{T}_{2,2} = \left\{ \begin{pmatrix} 0 & x \\ 0 & y \\ \vec{0} & \vec{0} \\ \beta & 0 \end{pmatrix} \in M_{n+1,2}(\mathbb{R}) : x \in \mathbb{R}^\times, y \in \mathbb{R}, \beta \in \mathbb{R}^\times \right\}$$

and

$$\mathscr{T}_{3,2} = \left\{ \begin{pmatrix} x & \lambda x \\ y & \lambda y \\ \vec{0} & \vec{0} \\ 0 & \beta \end{pmatrix} \in M_{n+1,2}(\mathbb{R}) : x \in \mathbb{R}^\times, y \in \mathbb{R}, \beta \in \mathbb{R}^\times, \lambda \in \mathbb{R}^\times \right\}. \tag{4.69}$$

(ii) *If $\mathfrak{h} \subset \mathfrak{g}_0$, then:*
  (ii$_1$) *if $Y_n \in \mathfrak{h}$, then $\mathscr{T}_{j,2} = \emptyset$, $j = 1,2,3$.*

(ii$_2$) *If* $Y_n \notin \mathfrak{h}$, *then* $\mathscr{T}_{j,2}, j = 1,2,3$ *are given by*

$$\mathscr{T}_{1,2} = \left\{ \begin{pmatrix} x & 0 \\ y & 0 \\ \vec{0} & \vec{0} \\ 0 & \beta \end{pmatrix} \in M_{n+1,2}(\mathbb{R}) : x \in \mathbb{R}^{\times}, y \in \mathbb{R}, \beta \in \mathbb{R}^{\times} \right\},$$

$$\mathscr{T}_{2,2} = \left\{ \begin{pmatrix} 0 & x \\ 0 & y \\ \vec{0} & \vec{0} \\ \beta & 0 \end{pmatrix} \in M_{n+1,2}(\mathbb{R}) : x \in \mathbb{R}^{\times}, \beta \in \mathbb{R}^{\times} \right\}$$

and

$$\mathscr{T}_{3,2} = \left\{ \begin{pmatrix} x & \lambda x \\ y & \lambda y \\ \vec{0} & \vec{0} \\ 0 & \beta \end{pmatrix} \in M_{n+1,2}(\mathbb{R}) : x \in \mathbb{R}^{\times}, \beta \in \mathbb{R}^{\times}, \lambda \in \mathbb{R}^{\times} \right\}.$$

Proof. It is clear that whenever $R_{j,2} = \emptyset$, $j = 1,2,3$ we have $\mathscr{T}_{j,2} = \emptyset$, $j = 1,2,3$ and therefore $\mathscr{T}(\Gamma, G, H) \simeq \mathscr{T}_{0,2}$. So, we only have to treat the case when $\mathfrak{h} \not\subset \mathfrak{g}_0$ for $q = n$ and the case when $Y_n \notin \mathfrak{h}$ and $\mathfrak{h} \subset \mathfrak{g}_0$. Let in a first time

$$M = \begin{pmatrix} x & \lambda x \\ y_1 & \lambda y_1 \\ \vec{y} & \lambda \vec{y} \\ z & \lambda z + \beta \end{pmatrix} \in R_{3,2}.$$

We get that

$$G \cdot M = \left\{ \begin{pmatrix} x & \lambda x \\ y_1 & \lambda y_1 \\ \vec{a} & \lambda \vec{a} \\ b & \lambda b + \beta \end{pmatrix} : b \in \mathbb{R}, {}^t\vec{a} \in \mathbb{R}^{n-2} \right\} \tag{4.70}$$

and then

$$G \cdot M \cap \mathscr{T}_{3,2} = \left\{ \begin{pmatrix} x & \lambda x \\ y_1 & \lambda y_1 \\ \vec{0} & \vec{0} \\ 0 & \beta \end{pmatrix} \right\}.$$

The other cases follow. This achieves the proof of the proposition. □

The same analysis gives us the following.

Proposition 4.4.14. *Assume that $k = 1$. The layer $\mathscr{T}_{0,1}$ being described in Proposition 4.4.12, we have that $R_{1,1}/G$ is homeomorphic to $\mathscr{T}_{1,1}$, where:*

(i) *If $\mathfrak{h} \not\subset \mathfrak{g}_0$, then*

$$\mathscr{T}_{1,1} = \left\{ \begin{pmatrix} x \\ y_1 \\ \vec{0} \end{pmatrix} \in M_{n+1,1}(\mathbb{R}) : x \in \mathbb{R}^{\times}, y_1 \in \mathbb{R}^{\times} \right\}.$$

(ii) *If $\mathfrak{h} \subset \mathfrak{g}_0$, then*

$$\mathscr{T}_{1,1} = \left\{ \begin{pmatrix} x \\ y_1 \\ \vec{0} \end{pmatrix} \in M_{n+1,1}(\mathbb{R}) : x \in \mathbb{R}^{\times}, y_1 \in \mathbb{R} \right\}.$$

Summarizing Sections 4.4.2 and 4.4.3, we get the following.

Theorem 4.4.15. *Let G be a threadlike nilpotent Lie group, H a closed, connected subgroup of G and $\Gamma \simeq \mathbb{Z}^k$ a discrete subgroup of G. Then the deformation space $\mathscr{T}(\Gamma, G, H)$ is described as follows:*

$$\mathscr{T}(\Gamma, G, H) = \coprod_{(r,s)\in J_k} \mathscr{T}_{0,k}^{r,s}(\Gamma, G, H) \coprod_{j\in I_k\setminus\{0\}} \mathscr{T}_{j,k}(\Gamma, G, H),$$

where $\mathscr{T}_{0,k}^{r,s}(\Gamma, G, H)$ is homeomorphic to $\mathscr{T}_{0,k}(r,s)$ for any $(r,s) \in J_k$ and $\mathscr{T}_{j,k}(\Gamma, G, H)$ to $\mathscr{T}_{j,k}$ for any $j \in I_k \setminus \{0\}$.

4.4.4 Case of non-Abelian discontinuous groups

A first result on algebra homomorphisms

Let G be a threadlike Lie group, Γ a discrete non-Abelian subgroup of G and $L = \exp \mathfrak{l}$ be its syndetic hull. The set of all injective homomorphisms from $\mathfrak{l}$ to $\mathfrak{g}$ denoted by $\mathrm{Hom}^0(\mathfrak{l}, \mathfrak{g})$, which rather than $\mathrm{Hom}(\mathfrak{l}, \mathfrak{g})$ itself will be of interest in the next section, merely because it is involved in deformations and is viewed as a starting means to study the parameter and the deformation spaces. Our main result in this section consists in giving an explicit description of its structure. In general, this set fails in most of the cases to be equipped with a smooth manifold structure. Clearly, $\mathrm{Hom}(\mathfrak{l}, \mathfrak{g})$ is homeomorphically identified to a set $\mathscr{U}$ of $\mathscr{M}_{n+1,k}(\mathbb{R})$ and $\mathrm{Hom}^0(\mathfrak{l}, \mathfrak{g})$ of all injective homomorphisms to the subset $\mathscr{U}^0$ of $\mathscr{U}$ consisting of the totality of matrices in $\mathscr{U}$ of maximal rank. Obviously, the set $\mathscr{U}$ is closed and algebraic in $\mathscr{M}_{n+1,k}(\mathbb{R})$ and $\mathscr{U}^0$ is semialgebraic and open in $\mathscr{U}$. Our first result is the following.

Theorem 4.4.16. *Let $\mathfrak{g}$ be a threadlike Lie algebra and $\mathfrak{l}$ a k-dimensional non-Abelian subalgebra of $\mathfrak{g}$. Then $\mathrm{Hom}^0(\mathfrak{l},\mathfrak{g})$ is endowed with a smooth manifold structure of dimension $n+4$ for $k=3$ and of dimension $n+k$ whenever $k>3$. For $k=3$, $\mathrm{Hom}^0(\mathfrak{l},\mathfrak{g})$ is a disjoint union of an open dense smooth manifold and a closed smooth manifold of dimension $n+3$.*

Proof. We fix a basis $\mathscr{B}=\{X,Y_1,\dots,Y_n\}$ of $\mathfrak{g}$ with nontrivial Lie brackets defined in (1.1.13) such that $\mathfrak{l}=\mathrm{span}\{X,Y_p,\dots,Y_n\}$, where $p=n-k+2$. We identify any element $T=tX+\sum_{i=1}^n t_iY_i\in\mathfrak{g}$ to the column vector ${}^t(t,t_1,\dots,t_n)$. Let

$$\mathscr{U}=\left\{\lfloor T_1,\dots,T_k\rfloor\in M_{n+1,k}(\mathbb{R})\;\middle|\;\begin{array}{l}[T_1,T_j]=T_{j+1},2\le j\le k-1,\\ [T_r,T_s]=0,2\le r,s\le k,\\ [T_1,T_k]=0\end{array}\right\},$$

where the symbol $\lfloor T_1,\dots,T_k\rfloor$ merely designates the matrix constituted by means of the columns $T_1,\dots,T_k$. We first show that $\mathrm{Hom}(\mathfrak{l},\mathfrak{g})$ is homeomorphic to $\mathscr{U}$. Any $\varphi\in\mathrm{Hom}(\mathfrak{l},\mathfrak{g})$ is determined by $\varphi(X)$ and $\varphi(Y_j)$ for $j=p,\dots,n$, where

$$\varphi(X)=xX+\sum_{i=1}^n y_iY_i\quad\text{and}\quad\varphi(Y_j)=x_jX+\sum_{i=1}^n y_{ij}Y_i,$$

for some $x,x_j,y_i,y_{ij}\in\mathbb{R}$. Let $\Psi:\mathrm{Hom}(\mathfrak{l},\mathfrak{g})\to M_{n+1,k}(\mathbb{R})$ be the injection map defined by

$$\Psi(\varphi)=\lfloor\varphi(X),\varphi(Y_p),\dots,\varphi(Y_n)\rfloor.\tag{4.71}$$

Now $\varphi\in\mathrm{Hom}(\mathfrak{l},\mathfrak{g})$ satisfies

$$\varphi([X,Y_j])=[\varphi(X),\varphi(Y_j)]\quad\text{and}\quad[\varphi(Y_i),\varphi(Y_j)]=0,\quad i,j=p,\dots,n.\tag{4.72}$$

This entails that $\Psi(\mathrm{Hom}(\mathfrak{l},\mathfrak{g}))\subset\mathscr{U}$. Let conversely $\lfloor T_1,\dots,T_k\rfloor\in\mathscr{U}$, we can define an algebras homomorphism: $\varphi:\mathfrak{l}\to\mathfrak{g}$ satisfying

$$\varphi(X)=T_1\quad\text{and}\quad\varphi(Y_{n-k+j})=T_j,\quad j=2,\dots,k$$

in such a way that $\Psi(\varphi)=\lfloor T_1,\dots,T_k\rfloor$. Thus, $\Psi(\mathrm{Hom}(\mathfrak{l},\mathfrak{g}))=\mathscr{U}$. By identifying $\mathfrak{g}^k=\mathfrak{g}\times\cdots\times\mathfrak{g}$ to the space $M_{n+1,k}(\mathbb{R})$, we can easily see the bicontinuity of Ψ. We now identify any homomorphism $\varphi\in\mathrm{Hom}(\mathfrak{l},\mathfrak{g})$ through its corresponding matrix $\Psi(\varphi)\in\mathscr{U}$ subject of deal from now on. Let $M=\lfloor U,V_p,\dots,V_n\rfloor\in M_{n+1,k}(\mathbb{R})$,

where

$$U = uX + \sum_{i=1}^{n} u_i Y_i \quad \text{and} \quad V_j = v_j X + \sum_{i=1}^{n} v_{i,j} Y_i, \quad p \le j \le n.$$

From equations (4.72), $M \in \mathscr{U}$ if and only if

$$\begin{cases} [U, V_j] = V_{j+1} & \text{for any } p \le j \le n-1, \\ [U, V_n] = 0, \\ [V_i, V_j] = 0 & \text{for any } p \le i, j \le n. \end{cases}$$

This gives rise to the following equations:

$$\begin{cases} v_{j+1} = v_{1,j+1} = 0, \quad p \le j \le n-1, & (1) \\ uv_{i,j} - u_i v_j = v_{i+1,j+1}, \quad 1 \le i \le n-1, p \le j \le n-1, & (2) \\ uv_{i,n} - u_i v_n = 0, \quad 1 \le i \le n-1, & (3) \\ v_s v_{i,r} - v_r v_{i,s} = 0, \quad p \le r, s \le n, 1 \le i \le n-1. & (4) \end{cases} \tag{4.73}$$

We now focus on the system (4.73). Thanks to equation (1), equation (2) gives that $v_{i,p+j} = 0$ for all $i = 2, \ldots, n-p$ and $j = 2, \ldots, n-p$ satisfying $i \le j$. M thus shapes as

$$M = \begin{pmatrix} u & v_p & 0 & 0 & \cdots & 0 \\ u_1 & v_{1,p} & 0 & 0 & \cdots & 0 \\ u_2 & v_{2,p} & v_{2,p+1} & 0 & \ddots & 0 \\ \vdots & \vdots & \vdots & \ddots & \ddots & \vdots \\ u_{n-p} & v_{n-p,p} & v_{n-p,p+1} & v_{n-p,p+2} & \ddots & 0 \\ u_{n-p+1} & v_{n-p+1,p} & v_{n-p+1,p+1} & v_{n-p+1,p+2} & \cdots & v_{n-p+1,n} \\ \vdots & \vdots & \vdots & \vdots & & \vdots \\ u_n & v_{n,p} & v_{n,p+1} & v_{n,p+2} & \cdots & v_{n,n} \end{pmatrix}.$$

We are conclusively led to the following discussions:

Case 1: If $u \neq 0$, then by equation (3), we have

$$v_{i,n} = 0 \quad \text{for all } 1 \le i \le n-1, \tag{4.74}$$

as $v_n = 0$. Assume for a while that $v_p = 0$. Then equation (2) gives

$$M = \begin{pmatrix} u & v_p & 0 & \cdots & 0 \\ u_1 & v_{1,p} & 0 & \cdots & 0 \\ u_2 & v_{2,p} & uv_{1,p} & & 0 \\ \vdots & \vdots & \vdots & \ddots & \vdots \\ u_{n-p} & v_{n-p,p} & uv_{n-p-1,p} & \ddots & 0 \\ u_{n-p+1} & v_{n-p+1,p} & uv_{n-p,p} & \cdots & u^{n-p}v_{n-p,n-1} \\ \vdots & \vdots & \vdots & & \vdots \\ u_n & v_{n,p} & uv_{n-1,p} & \cdots & u^{n-p}v_{n-1,n-1} \end{pmatrix}.$$

By (4.74), we have $v_{i,p} = 0$ for all $i = 1, \dots, p-1$. This shows that M is of the form

$$M_0(U,V) := \begin{pmatrix} u & 0 & 0 & \cdots & 0 \\ u_1 & 0 & 0 & \cdots & 0 \\ \vdots & \vdots & \vdots & & \vdots \\ u_{p-1} & 0 & 0 & \cdots & 0 \\ u_p & v_{p,p} & 0 & \cdots & 0 \\ u_{p+1} & v_{p+1,p} & uv_{p,p} & \ddots & \vdots \\ \vdots & \vdots & \vdots & \ddots & 0 \\ u_n & v_{n,p} & uv_{n-1,p} & \cdots & u^{n-p}v_{p,p} \end{pmatrix}.$$

Set

$$H_{0,k} = \{M_0(U,V) \in M_{n+1,k}(\mathbb{R}) : u \in \mathbb{R}^\times\}. \tag{4.75}$$

Let now $v_p \neq 0$. Using equations (2), (4) and (4.74), there exists $\lambda \in \mathbb{R}^\times$ such that $V_p = {}^t(\lambda u, \lambda u_1, \dots, \lambda u_{n-2}, v_{n-1,p}, v_{n,p})$. So, $V_{p+1} = {}^t(0, \dots, 0, u(v_{n-1,p} - \lambda u_{n-1}))$ and $V_i = \vec{0}$ for all $i = p+2, \dots, n$. We get finally that M belongs to the set defined as

$$H_{1,k} = \{M_1(U,V,\lambda) \in M_{n+1,k}(\mathbb{R}) : \lambda u \in \mathbb{R}^\times\}, \tag{4.76}$$

where

$$M_1(U,V,\lambda) := \begin{pmatrix} u & \lambda u & 0 & 0 & \cdots & 0 \\ u_1 & \lambda u_1 & 0 & 0 & & 0 \\ \vdots & \vdots & \vdots & & & \\ u_{n-2} & \lambda u_{n-2} & 0 & \vdots & & \vdots \\ u_{n-1} & v_{n-1} & 0 & & & \\ u_n & v_n & u(v_{n-1} - \lambda u_{n-1}) & 0 & \cdots & 0 \end{pmatrix}.$$

Case 2: If $u = 0$, then $V_{p+1} = {}^t(0,0,-u_1v_p,\dots,-u_{n-1}v_p)$ and $V_i = \vec{0}$ for all $i = p+2,\dots,n$ by equation (2). When $v_p \neq 0$, we have by equation (4) that $u_j = 0$ for all $j = 1,\dots,n-2$ that is, $U = {}^t(0,\dots,0,u_{n-1},u_n)$ and then M reads

$$M_2(U,V) := \begin{pmatrix} 0 & v & 0 & 0 & \cdots & 0 \\ 0 & v_1 & 0 & 0 & \cdots & 0 \\ \vdots & \vdots & \vdots & \vdots & & \vdots \\ 0 & v_{n-2} & 0 & \vdots & & \vdots \\ u_{n-1} & v_{n-1} & 0 & & & \\ u_n & v_n & -vu_{n-1} & 0 & \cdots & 0 \end{pmatrix}.$$

Set again

$$H_{2,k} = \{M_2(U,V) \in M_{n+1,k}(\mathbb{R}) : v \in \mathbb{R}^\times\}. \tag{4.77}$$

Otherwise,

$$M \in H_{3,k} = \{\lfloor U,V,\vec{0},\dots,\vec{0}\rfloor \in M_{n+1,k}(\mathbb{R}) : {}^tU \text{ and } {}^tV \in \{0\}\times\mathbb{R}^n\}. \tag{4.78}$$

To determine a stratification of $\mathrm{Hom}^0(\mathfrak{l},\mathfrak{g})$, we note for all $k \in \{3,\dots,n+1\}$, $j \in \{0,\dots,k\}$ and $K_{j,k} = \mathrm{Hom}^0(\mathfrak{l},\mathfrak{g}) \cap H_{j,k}$. So, it is not hard to check that for all $M \in H_{j,k}$, we have $\mathrm{rank}(M) \leq 3$ for $j \in \{1,2\}$ and that $\mathrm{Hom}^0(\mathfrak{l},\mathfrak{g}) \cap H_{3,k} = \emptyset$ as $\mathfrak{l}$ is not Abelian. In addition, M is of maximal rank if and only if $\mathrm{rank}(M) = k$. So, $\mathrm{Hom}^0(\mathfrak{l},\mathfrak{g}) \cap H_{j,k} = \emptyset$ for $j \in \{1,2,3\}$ whenever $k > 3$. With the above in mind, the following result is immediate.

Proposition 4.4.17. *Let $\mathfrak{g}$ be a threadlike Lie algebra and $\mathfrak{l}$ a k-dimensional non-Abelian subalgebra of $\mathfrak{g}$. We have the following:*
(1) *If $k > 3$, then $\mathrm{Hom}^0(\mathfrak{l},\mathfrak{g})$ is homeomorphic to $K_{0,k} = \{M \in H_{0,k} : v_{p,p} \neq 0\}$.*
(2) *If $k = 3$, then $\mathrm{Hom}^0(\mathfrak{l},\mathfrak{g}) = \widetilde{K_{1,3}} \coprod K_{2,3}$, where*

$$K_{1,3} = \{M \in H_{1,3} : v_{n-1} - \lambda u_{n-1} \in \mathbb{R}^\times\}, \tag{4.79}$$
$$\widetilde{K_{1,3}} = K_{1,3} \coprod K_{0,3} \tag{4.80}$$

and

$$K_{2,3} = \{M \in H_{2,3} : u_{n-1} \in \mathbb{R}^\times\}. \tag{4.81}$$

We now look at the maps $\phi_0 : \mathbb{R}^{n+1}\times\mathbb{R}^{k-1} \ni (U,V) \mapsto M_0(U,V)$, $\phi_1 : \mathbb{R}^{n+1}\times\mathbb{R}^2\times\mathbb{R}^\times \ni (U,V,\lambda) \mapsto M_1(U,V,\lambda)$ and $\phi_2 : \mathbb{R}^2 \times \mathbb{R}^{n+1} \ni (U,V) \mapsto M_2(U,V)$, which are clearly C^∞ embeddings on their closed images regarded as subsets of $M_{n+1,k}(\mathbb{R})$. This shows that $K_{0,k}$, $K_{1,3}$ and $K_{2,3}$ and also $\widetilde{K_{1,3}}$ are endowed with smooth manifold structures with the mentioned dimensions. Hence, Lemma 4.1.30 allows to close the proof. Let $M_2(U,V) \in K_{2,3}$. Then clearly it is the limit of the sequence

$$M^s(U,V) = \begin{pmatrix} \frac{1}{s} & v & 0 \\ \frac{v_1}{sv} & v_1 & 0 \\ \vdots & \vdots & \vdots \\ \frac{v_{n-2}}{sv} & v_{n-2} & 0 \\ u_{n-1} & v_{n-1} & 0 \\ u_n & v_n & \frac{1}{s}v_{n-1} - vu_{n-1} \end{pmatrix},$$

which shows that $\widetilde{K_{1,3}}$ is dense in $\mathrm{Hom}^0(\mathfrak{l},\mathfrak{g})$. On the other hand, it is not hard to check in this case that

$$\mathrm{Hom}^0(\mathfrak{l},\mathfrak{g}) = \left\{ M(U,V) \in M_{n+1,3}(\mathbb{R}) \,\middle|\, \begin{array}{l} uv_{n-1} - vu_{n-1} \neq 0, \\ uv_i - vu_i = 0, 1 \leq i \leq n-2 \end{array} \right\},$$

where keeping the same notation as above

$$M(U,V) = \begin{pmatrix} u & v & 0 \\ u_1 & v_1 & 0 \\ \vdots & \vdots & \vdots \\ u_{n-1} & v_{n-1} & 0 \\ u_n & v_n & uv_{n-1} - vu_{n-1} \end{pmatrix}.$$

Let $\mathscr{V} = \{M(U,V) \in M_{n+1,3}(\mathbb{R}) : uv_{n-1} - vu_{n-1} \neq 0\}$, which is a smooth manifold. It is clear that $\mathrm{Hom}^0(\mathfrak{l},\mathfrak{g}) = f^{-1}(\{0\})$, where $f : \mathscr{V} \longrightarrow \mathbb{R}^{n-2}$, $M(U,V) \longmapsto (uv_1 - vu_1, \ldots, uv_{n-2} - vu_{n-2})$. Clearly, the zero point of $\mathbb{R}^{n-2}$ is a regular value of the C^∞– function f, which shows that $\mathrm{Hom}^0(\mathfrak{l},\mathfrak{g})$ is a manifold. This achieves the proof of the theorem. □

We get right away the following description of the set $\mathrm{Hom}(\mathfrak{l},\mathfrak{g})$. We have the following.

Corollary 4.4.18. *Let $\mathfrak{g}$ be a threadlike Lie algebra and $\mathfrak{l}$ a k-dimensional non-Abelian subalgebra of $\mathfrak{g}$. Then $\mathrm{Hom}(\mathfrak{l},\mathfrak{g})$ is homeomorphic to the disjoint union $\coprod_{j=0}^{3} H_{j,k}$.*

The following result is a direct consequence from above.

Corollary 4.4.19. *The Lie group $\mathrm{Aut}(\mathfrak{g})$ of automorphisms of the n-step threadlike Lie algebra $\mathfrak{g}$ is of dimension $2n+1$ whenever $n > 2$ and of dimension 6 for $n = 2$.*

Remark 4.4.20. The last result is a direct consequence of Theorem 4.4.16. For $n = 2$, $\mathfrak{g}$ is the three-dimensional Heisenberg Lie algebra. It is then clear that $\widetilde{K_{1,3}} \coprod K_{2,3}$ is nothing but the set

$$\left\{\begin{pmatrix} u_1 & v_1 & 0 \\ u_2 & v_2 & 0 \\ u_3 & v_3 & u_1v_2 - u_2v_1 \end{pmatrix} \in M_3(\mathbb{R}) : u_1v_2 - u_2v_1 \neq 0\right\},$$

and the result follows.

An insight on *G*-orbits

As above, the group G acts on $\mathrm{Hom}(\mathfrak{l}, \mathfrak{g})$ through the law $g \star \psi = \mathrm{Ad}_g \circ \psi$ as in equation (3.3). Let

$$M = \begin{pmatrix} u & \vec{0} \\ U_0 & N \end{pmatrix} \in K_{0,k}, \tag{4.82}$$

where ${}^tU_0 = (u_1, \dots, u_n)$, $N = \lfloor W_p, \dots, W_n \rfloor$ with ${}^tW_p = (0, \dots, 0, v_p, \dots, v_n) \in \mathbb{R}^n$, ${}^tW_{p+j} = (0, \dots, 0, u^j v_p, \dots, u^j v_{n-j}), j = 1, \dots, n-p$, and $g = \exp(xX + y_1Y_1 + \dots + y_nY_n) \in G$ with $x, y_1, \dots, y_n \in \mathbb{R}$. We have

$$g \star M = \begin{pmatrix} u & \vec{0} \\ U_0' & A(x)N \end{pmatrix}, \tag{4.83}$$

where

$$A(x) = \begin{pmatrix} 1 & 0 & \cdots & \cdots & 0 \\ x & 1 & \ddots & & \vdots \\ \frac{x^2}{2} & x & 1 & \ddots & \vdots \\ \vdots & \ddots & \ddots & \ddots & 0 \\ \frac{x^{n-1}}{(n-1)!} & \cdots & \frac{x^2}{2} & x & 1 \end{pmatrix}$$

denotes the matrix of $(\mathrm{Ad}_{\exp xX})_{|\mathfrak{g}_0}$ written through the basis $\mathscr{B}_0 = \{Y_1, \dots, Y_n\}$ and U_0' is the column vector associated to $g \star U - uX$ for $U = {}^t(u, U_0)$. Let on the other hand $k = 3$ and $M = M_1(U, V, \lambda) \in \widetilde{K_{1,3}}$. A routine computation shows that for g as above,

$$g \star M = \begin{pmatrix} u & \lambda u & 0 \\ u_1 & \lambda u_1 & 0 \\ u_2' & \lambda u_2' & 0 \\ \vdots & \vdots & \vdots \\ u_{n-2}' & \lambda u_{n-2}' & 0 \\ u_{n-1}' & v_{n-1}' & 0 \\ u_n' & v_n' & u(v_{n-1} - \lambda u_{n-1}) \end{pmatrix}, \tag{4.84}$$

where

$$u'_i = u_i + \sum_{j=1}^{i-1} \frac{x^{j-1}}{j!}(xu_{i-j} - uy_{i-j}), \quad i = 2, \ldots, n, \tag{4.85}$$

$$v'_{n-1} = v_{n-1} + \lambda \sum_{j=1}^{n-2} \frac{x^{j-1}}{j!}(xu_{n-j-1} - uy_{n-j-1}) \tag{4.86}$$

and

$$v'_n = v_n + (xv_{n-1} - \lambda uy_{n-1}) + \lambda \sum_{j=2}^{n-2} \frac{x^{j-1}}{j!}(xu_{n-j-1} - uy_{n-j-1}). \tag{4.87}$$

When submitted to the layer $K_{2,3}$, the action of G is likewise described as in equation (4.84) for $\lambda = 0$ and after having substituted the first two columns. The following upshot is a direct consequence from above.

Corollary 4.4.21. *Let* $\mathfrak{g}$ *be a threadlike Lie algebra, and* $\mathfrak{l}$ *a subalgebra of* $\mathfrak{g}$. *Then* $G = \exp(\mathfrak{g})$ *acts on* $\mathrm{Hom}^0(\mathfrak{l}, \mathfrak{g})$ *with constant dimension orbits if and only if* $\dim \mathfrak{l} > 2$.

Proof. Let $\mathfrak{l}$ be non-Abelian. Then the G-orbits in $K_{0,k}$ are uniformly n-dimensional. For $k = 3$, the orbits in $K_{1,3}$ and in $K_{2,3}$ are also n-dimensional as in equation (4.84). Suppose now that $\mathfrak{l}$ is Abelian, then the G-orbits in $\mathrm{Hom}^0(\mathfrak{l}, \mathfrak{g})$ are uniformly one-dimensional whenever $\dim \mathfrak{l} > 2$. For $k = 1, 2$, we have $\{\dim(G \star M) : M \in \mathrm{Hom}^0(\mathfrak{l}, \mathfrak{g})\} = \{1, n-1\}$. □

We are now in measure to prove the following result.

Theorem 4.4.22. *Let G be a threadlike Lie group and Γ a non-Abelian discrete subgroup of G of rank k. Then:*
1. *For any $H \in \mathfrak{M}_{gp}(\Gamma : G)$, the deformation space $\mathscr{T}(\Gamma, G, H)$ is a Hausdorff space.*
2. *For $k > 3$, $\mathscr{T}(\Gamma, G, H)$ is endowed with a smooth manifold structure.*
3. *For $k = 3$, $\mathscr{T}(\Gamma, G, H)$ is a disjoint union of an open dense smooth manifold and a closed, smooth manifold.*

Proof. The proof will be divided into several steps. With the above in mind, we first prove the following.

Lemma 4.4.23. *The disjoint components $R_{j,k}, j = 0, 1, 2$ are G-invariant.*

Proof. Let $M \in R_{0,k}$ be as defined in (4.82) and $g = \exp(xX + y_1Y_1 + \cdots + y_nY_n) \in G$, where $x, y_1, \ldots, y_n \in \mathbb{R}$. By equation (4.83), $g \star M \in R_{0,k}$. Suppose now that $k = 3$. Let $M = M_1(U, V, \lambda) \in R_{1,3}$ and $g \in G$, then $g \star M = M_1(U', V', \lambda)$ by equation (4.84), where $u'_i, i = 2, \ldots, n$, v'_{n-1} and v'_n are as in equations (4.85), (4.86) and (4.87), respectively. As $(v'_{n-1} - \lambda u'_{n-1}) = (v_{n-1} - \lambda u_{n-1})$, we get that $g \star M \in R_{1,3}$. We opt for the same arguments to show that $R_{2,3}$ is G-invariant. □

We now prove the following.

Lemma 4.4.24. *We keep all our notation as above, the set $R_{0,k}/G$ is homeomorphic to the set $T_{0,k}$ given by*

$$T_{0,k} = \left\{ M_0(U,V) \in R_{0,k} \;\middle|\; \begin{array}{l} U = {}^t(u, u_1, 0, \dots, 0), \\ V = {}^t(0, \dots, 0, v_p, 0, v_{p+2}, \dots, v_n) \end{array} \right\}. \tag{4.88}$$

If $k = 3$, then $R_{j,3}/G$ is homeomorphic to $T_{j,3}$ for $j = 1, 2$ given as follows:

(i) *If $\mathfrak{h} \not\subset \mathfrak{g}_0$, then $\mathfrak{h} = \operatorname{span}\{X + h_1 Y_1 + \cdots + h_n Y_n\}$ for some $h_1, \dots, h_n \in \mathbb{R}$ and*

$$T_{1,3} = \left\{ M_1(U,V,\lambda) \in R_{1,3} \;\middle|\; \begin{array}{l} U = {}^t(u, u_1, 0, \dots, 0), \\ V = {}^t(0, \dots, 0, v_{n-1}, 0) \end{array} \right\} \tag{4.89}$$

and

$$T_{2,3} = \left\{ M_2(U,V) \in R_{2,3} \;\middle|\; \begin{array}{l} U = {}^t(0, \dots, 0, u_{n-1}, 0), \\ V = {}^t(v, v_1, 0, \dots, 0) \end{array} \right\}. \tag{4.90}$$

(ii) *If $\mathfrak{h} \subset \mathfrak{g}_0$, then $\mathscr{I}_{\mathscr{B}}^{\mathfrak{h}} \subset \{1, \dots, n-2\}$ and $T_{1,3}$ and $T_{2,3}$ are defined respectively as in (4.89) and (4.90).*

Proof. We show that $T_{0,k}$ is a cross-section of all adjoint orbits of $R_{0,k}$. Let $M \in R_{0,k}$ as in (4.82), then by (4.83) we have

$$G \star M = [M] = \left\{ \begin{pmatrix} u & \vec{0} \\ A(x)U_0 - uB(x)Y & A(x)N \end{pmatrix} : {}^tY \in \mathbb{R}^{n-1},\ x \in \mathbb{R} \right\}, \tag{4.91}$$

where

$$B(x) = \begin{pmatrix} 0 & 0 & \cdots & \cdots & 0 \\ 1 & 0 & \cdots & \cdots & 0 \\ \frac{x}{2} & 1 & \ddots & & \vdots \\ \frac{x^2}{3!} & \frac{x}{2} & 1 & \ddots & \vdots \\ \vdots & \ddots & \ddots & \ddots & 0 \\ \frac{x^{n-2}}{(n-1)!} & \cdots & \frac{x^2}{3!} & \frac{x}{2} & 1 \end{pmatrix} \in M_{n,n-1}(\mathbb{R}).$$

Noting

$$A(x) = \begin{pmatrix} a \\ A'(x) \end{pmatrix}$$

with $a = (1, 0, \dots, 0)$ and

$$B(x) = \begin{pmatrix} 0 \\ B'(x) \end{pmatrix},$$

we pose $x_M = -\frac{v_{p+1}}{v_p}$ and Y_M the unique solution of the linear equation

$$uB'(x_M)\, Y = A'(x_M)\, U_0.$$

We then see that

$$G \star M \cap T_{0,k} = \left\{ \begin{pmatrix} u & 0 \\ A(x_M)U_0 - uB(x_M)Y_M & A(x_M)N \end{pmatrix} \right\} = \{X_M\}. \tag{4.92}$$

We now prove that the map:

$$\begin{array}{rcl} \Phi_{0,k} : R_{0,k}/G & \to & T_{0,k} \\ {[M]} & \mapsto & X_M \end{array} \tag{4.93}$$

is a homeomorphism. First of all, it is clear that $\Phi_{0,k}$ is well-defined. The injectivity $\Phi_{0,k}$ is obvious. To see that $\Phi_{0,k}$ is surjective, it is sufficient to verify that for all $M \in T_{0,k}$, we have $\Phi_{0,k}([M]) = M$ as $G \star M \cap T_{0,k} = \{M\}$. Let $\pi_{0,k}$ be the canonical surjection $\pi_{0,k} : R_{0,k} \to R_{0,k}/G$. Thus, we can easily see the continuity of $\Phi_{0,k} \circ \pi_{0,k}$, which is equivalent to the continuity of $\Phi_{0,k}$. Finally, it is clear that $\Phi_{0,k}^{-1} = (\pi_{0,k})_{|T_{0,k}}$, so the bicontinuity follows.

Assume now that $k = 3$. Let $M = M_1(U, V, \lambda) \in R_{1,3}$ where ${}^tU = (u, u_1, \dots, u_n)$ and $V = {}^t(0, \dots, 0, v_{n-1}, v_n)$. We get that

$$G \star M = \left\{ M_1(U', V', \lambda) \in R_{1,3} \;\middle|\; \begin{array}{l} U' = {}^t(u, u_1, u_2' \dots, u_n') \in \mathbb{R}^{n+1}, \\ V' = {}^t(0, \dots, 0, v_{n-1}', v_n') \in \mathbb{R}^{n+1}, \\ v_{n-1}' - \lambda u_{n-1}' = v_{n-1} - \lambda u_{n-1} \end{array} \right\}. \tag{4.94}$$

There exists then $v_{n-1}^0 \in \mathbb{R}^\times$ such that $G \star M \cap T_{1,3} = \{M_1(U^0, V^0, \lambda)\} \subset R_{1,3}$ where $U^0 = {}^t(u, u_1, 0, \dots, 0)$, $V^0 = {}^t(0, \dots, 0, v_{n-1}^0, 0)$ and the map:

$$\begin{array}{rcl} \Phi_{1,3} : R_{1,3}/G & \to & T_{1,3} \\ {[M]} & \mapsto & M_1(U^0, V^0, \lambda) \end{array} \tag{4.95}$$

is a homeomorphism. The other cases follow. This achieves the proof of the lemma. □

We finally prove that the deformation space is a Hausdorff space. When $k > 3$,

$$\mathscr{T}(\Gamma, G, H) = R_{0,k}/G$$

and the result is immediate by Lemma 4.4.24. Otherwise, $\mathscr{T}(\Gamma, G, H) = \coprod_{j=0}^{2}(R_{j,3}/G)$. Let $[M_1]$ and $[M_2]$ be elements of $\mathscr{T}(\Gamma, G, H)$ such that $[M_1] \neq [M_2]$. With the above in mind, we can choose the matrices M_j of the following form:

$$M_j = \begin{pmatrix} u_j & v_j & 0 \\ u_{1,j} & v_{1,j} & 0 \\ \vec{0} & \vec{0} & \vec{0} \\ \alpha_j & \alpha_j' & 0 \\ 0 & 0 & \beta_j \end{pmatrix}, \quad j = 1, 2.$$

Now $[M_1] \neq [M_2]$ is equivalent to

$$\left(\beta_1, \begin{pmatrix} u_1 & v_1 \\ u_{1,1} & v_{1,1} \end{pmatrix}\right) \neq \left(\beta_2, \begin{pmatrix} u_2 & v_2 \\ u_{1,2} & v_{1,2} \end{pmatrix}\right)$$

which enough to conclude. As for the smooth structure of the deformation space, the result is immediate whenever $k > 3$. Suppose now that $k = 3$. Lemma 4.4.24 gives that $R_{2,3}/G$ is a smooth manifold. Moreover, as $R_{2,3}$ is closed in $\mathscr{R}(\mathfrak{l}, \mathfrak{g}, \mathfrak{h})$ and G-invariant, we see that $\pi(R_{2,3})$ is closed in $\mathscr{T}(\Gamma, G, H)$ where $\pi : \mathscr{R}(\Gamma, G, H) \to \mathscr{T}(\Gamma, G, H)$ is the canonical surjection. Now we denote by $\widetilde{T_{1,3}} = T_{0,3} \coprod T_{1,3}$. Using the same arguments of Lemma 4.4.24, we can prove that the map

$$\begin{aligned} \widetilde{\Phi_{1,3}} : \widetilde{R_{1,3}}/G &\to \widetilde{T_{1,3}} \\ [M] &\mapsto G \star M \cap \widetilde{T_{1,3}} \end{aligned}$$

is a homeomorphism. We get therefore that $\widetilde{R_{1,3}}/G$ is a smooth manifold, and which is clearly by Theorem 4.4.16, a dense subset of $\mathscr{T}(\Gamma, G, H)$. □

5 Local and strong local rigidity

Let G be a Lie group and $\Gamma \subset G$ a finitely generated discrete subgroup. Let $\mathrm{Hom}(\Gamma, G)$ denote as earlier the set of deformation parameters of Γ in G, that is, the space of all homomorphisms $\Gamma \to G$ endowed with the topology of pointwise convergence. The subgroup Γ is said to be *locally rigid* if there is a neighborhood Ω of the inclusion map $\rho_0 : \Gamma \to G$ in $\mathrm{Hom}(\Gamma, G)$ such that any $\rho \in \Omega$ is conjugate to ρ_0 under the action of G (such ρ is called a *trivial deformation*).

The study of local rigidity plays a crucial role in the deformation theory of discontinuous groups. For instance, in the case where G is a compact linear simple Lie group and H its maximal compact subgroup, there exists an equivalence between the existence of a uniform lattice $\varphi : \Gamma \to G$ such that $\varphi \in \mathscr{R}(\Gamma, G, H)$ is not locally rigid and the fact that G is locally isomorphic to $SL_2(\mathbb{R})$.

On the hand, for an irreducible Riemannian symmetric space G/H of dimension ≥ 3 with a compact subgroup H and Γ a uniform lattice of G/H, there does not exist any essential deformation of Γ. This result, proved by Selberg and Weil (cf. [123]), claims that the deformation space is discrete in this context and can be regarded as the original model for various kinds of rigidity theorems in Riemannian geometry. Besides, the local rigidity does not hold in general in the non-Riemannian case as remarked first by T. Kobayashi. So once again referring to Chapter 4, it does make sense to determine explicitly the deformation space of a discontinuous action as it provides comprehensive information of its local structure, namely the Hausdorfness and the local rigidity.

Throughout the chapter, we pay attention to the local rigidity property using some previous results. We substantiate the local rigidity conjecture in the nilpotent setting, which states that the local rigidity fails to hold for any nontrivial discontinuous group of a nilpotent homogeneous space. We further extend our study to many exponential and solvable settings. In this context, deformations of discontinuous groups means purely deformations of group homomorphisms unlike the semisimple setting, where the G-action on G/H is not always effective, and thus the space of group theoretic deformations (formal deformations) could be larger than geometric deformation spaces. We determine the corresponding deformation space and also its quotient modulo uneffective parts when $\mathrm{rank}\,\Gamma = 1$. Unlike the context of the exponential solvable case, we prove the existence of formal colored discontinuous groups. That is, the parameter space admits a mixture of locally rigid and formally nonrigid deformations.

We are also concerned with an analogue of the so-called Selberg–Weil–Kobayashi local rigidity theorem in the context of a real exponential group G and H a maximal subgroup of G, where the local rigidity property is shown to hold if and only if the group G is isomorphic to the group $\mathrm{Aff}(\mathbb{R})$ of affine transformations of the real line. One substantial ingredient, which has made such an achievement possible, is that any Abelian discrete subgroup of G admits a syndetic hull in G, a unique connected analytic subgroup of G containing it cocompactly (cf. Chapter 1).

https://doi.org/10.1515/9783110765304-005

5.1 The local rigidity conjecture

5.1.1 The concept of (strong) local rigidity

A. Weil [130] introduced the notion of local rigidity of homomorphisms in the case where the subgroup H is compact. T. Kobayashi [97] generalized it in the case where H is not compact. For the noncompact setting, the local rigidity does not hold in the general non-Riemannian case and has been studied in [31, 34, 93, 95, 98]. For comprehensible information, we further refer the readers to [57, 73, 77, 92, 92, 94, 97] and some references therein. For $\varphi \in \mathscr{R}(\Gamma, G, H)$, the discontinuous group $\varphi(\Gamma)$ for the homogeneous space G/H is said to be *locally rigid* (resp., *strongly locally rigid*) (cf. [97]) as a discontinuous group of G/H if the orbit of φ through the inner conjugation is open in $\mathscr{R}(\Gamma, G, H)$ (resp., in $\mathrm{Hom}(\Gamma, G)$). This means equivalently that any point sufficiently close to φ should be conjugate to φ under an inner automorphism of G. So, the homomorphisms, which are locally rigid are those which correspond to those which are isolated points in the deformation space $\mathscr{T}(\Gamma, G, H)$. When every point in $\mathscr{R}(\Gamma, G, H)$ is locally rigid, the deformation space turns out to be discrete and then we say that the Clifford–Klein form $\Gamma\backslash G/H$ cannot deform continuously through the deformation of Γ in G. If a given $\varphi \in \mathscr{R}(\Gamma, G, H)$ is not locally rigid, we say that it admits a *continuous deformation* and that the related Clifford–Klein form is continuously deformable.

5.1.2 The nilpotent setting

We first restrict to the setting of nilpotent Lie groups. We substantiate the following local rigidity conjecture for nilpotent Lie groups.

Conjecture 5.1.1 (A. Baklouti, cf. [11]). *Let G be a connected simply connected nilpotent Lie group, H a connected subgroup of G and Γ a nontrivial discontinuous group for G/H. Then the local rigidity fails to hold.*

Note that we have a first element of an answer in the case of Abelian discontinuous groups. We have the following.

Proposition 5.1.2. *Conjecture* 5.1.1 *holds in the case of Abelian discontinuous groups.*

Proof. Let $\varphi \in \mathscr{R}(\mathfrak{l}, \mathfrak{g}, \mathfrak{h})$ and $L_\varphi = \mathbb{R}_{>0} \cdot \varphi$. Clearly, $L_\varphi \subset \mathscr{R}(\mathfrak{l}, \mathfrak{g}, \mathfrak{h})$ and $G \cdot \psi \cap L_\varphi = \{\psi\}$ for any $\psi \in L_\varphi$, as G acts unipotently on $\mathrm{Hom}(\mathfrak{l}, \mathfrak{g})$. The projection

$$\pi : L_\varphi \to \mathbb{R}_{>0} \cdot [G \cdot \varphi] \subset \mathscr{T}(\mathfrak{l}, \mathfrak{g}, \mathfrak{h})$$

is bijective, continuous and open and, therefore, bicontinuous. This means that $G \cdot \varphi$ cannot be open in $\mathbb{R}_{>0} \cdot [G \cdot \varphi]$, which achieves the proof. □

Evidently, this proof is not adaptable to the case where the discontinuous group is no longer Abelian. We shall provide later on (cf. Remark 5.4.28) a counterexample showing its failure in the setting of general exponential Lie groups.

5.1.3 Case of 2-step nilpotent Lie groups

As a direct consequence from Theorem 4.2.6, we prove that the rigidity property fails to hold. We will prove the following.

Theorem 5.1.3. *Let G be a connected and simply connected 2-step nilpotent Lie group. Then Conjecture* 5.1.1 *holds.*

Proof. Note first that $\mathbb{R}_+^\times$ acts on $\mathscr{T}_{t,\beta,i}$, $i = 1, 2$, by left multiplication

$$\begin{aligned} \mathbb{R}_+^\times \times \mathscr{T}_{t,\beta,i} &\longrightarrow \mathscr{T}_{t,\beta,i} \\ (\lambda, \varphi(A,B)) &\longmapsto \lambda \cdot \varphi = \varphi(\lambda^2 A, \lambda B). \end{aligned}$$

Then this action is well-defined. Indeed, we define an $\mathbb{R}_+^\star$-action ρ on $\mathfrak{l} = [\mathfrak{l}, \mathfrak{l}] \oplus \mathfrak{l}'$ by

$$\rho(\lambda)(X) := \begin{cases} \lambda^2 X & \text{for } X \in [\mathfrak{l}, \mathfrak{l}], \\ \lambda X & \text{for } X \in \mathfrak{l}'. \end{cases}$$

Since $\mathfrak{l}$ is 2-step nilpotent, this action preserves the Lie algebra structure of $\mathfrak{l}$. Therefore, this induces an $\mathbb{R}_+^\star$-action on $\mathscr{L}(\mathfrak{l}, \mathfrak{g})$. Its restriction is nothing but the $\mathbb{R}_+^\star$-action on $\mathscr{T}_{t,\beta,i}$. Suppose that there is a local rigid homomorphism $\varphi(A,B) \in \mathscr{R}(\mathfrak{l}, \mathfrak{g}, \mathfrak{h})$, then its class $[\varphi(A,B)]$ is an open point in $\mathscr{T}(\mathfrak{l}, \mathfrak{g}, \mathfrak{h})$ and there exists β and t such that $[\varphi(A,B)] \in \mathscr{T}_{t,\beta,i}$ for some $i \in \{1,2\}$. It follows that the image of $[\varphi(A,B)]$ denoted also by $[\varphi(A,B)]$, in the semialgebraic set $\mathscr{T}_{t,\beta,i}$ is an isolated point. Therefore, any continuous action of $\mathbb{R}_+^\times$ on $\mathscr{T}_{t,\beta,i}$ fixes $[\varphi(A,B)]$. This means that for any $\lambda \in \mathbb{R}_+^\times$ there is $g(\lambda) \in G$ such that

$$\lambda \cdot \varphi(A,B) = g(\lambda) \cdot \varphi(A,B) = \mathrm{Ad}_{g(\lambda)} \varphi(A,B),$$

for some $g(\lambda) \in G$ where $\mathrm{Ad}_{g(\lambda)}$ is defined as in (4.33). This is equivalent to

$$\lambda \cdot \varphi(A,B) = \begin{pmatrix} A_1 & B_1 + \sigma_1(g(\lambda))B_3 + \delta_1(g(\lambda))B_4 \\ A_2 & B_2 + \sigma_2(g(\lambda))B_3 + \delta_2(g(\lambda))B_4 \\ 0 & B_3 \\ 0 & B_4 \end{pmatrix}$$

for any $\lambda \in \mathbb{R}_+^\times$, and this implies that

$$\begin{pmatrix} B_3 \\ B_4 \end{pmatrix} = 0, \quad \begin{pmatrix} A_1 \\ A_2 \end{pmatrix} = 0 \quad \text{and} \quad \begin{pmatrix} B_1 \\ B_2 \end{pmatrix} = 0,$$

which is a contradiction with the injectivity of $\varphi(A,B)$. □

Remark 5.1.4. Recall the context of Subsection 4.1.9. Let Γ be a nontrivial discontinuous group for the homogenous space P/Δ, where $P = \exp(\mathfrak{p})$ and $\mathfrak{p} = \mathfrak{h}_{2n+1} \times \mathfrak{h}_{2n+1}$. Then the following shows an alternative argument of the proof of local rigidity Conjecture 5.1.1 in this case. Indeed, we consider the action of $\mathbb{R}_+^*$ on $R(\mathfrak{l}, \mathfrak{p}, \mathfrak{D})$ defined by

$$\begin{aligned} \mathbb{R}_+^* \times R(\mathfrak{l}, \mathfrak{p}, \mathfrak{D}) &\longrightarrow R(\mathfrak{l}, \mathfrak{p}, \mathfrak{D}) \\ (t, M = M(x, y, A, B)) &\longmapsto t \star M := M(t^2x, t^2y, tA, tB). \end{aligned}$$

The group $\mathbb{R}_+^*$ acts on $\mathscr{T}(\mathfrak{l}, \mathfrak{p}, \mathfrak{D})$ by the following:

$$t \cdot [M] = [t \star M], \quad M \in R(\mathfrak{l}, \mathfrak{p}, \mathfrak{D}), \quad t \in \mathbb{R}_+^*.$$

According to (4.24),

$$\mathrm{Ad}_{\exp X} \cdot M = M(x + (u - v)A + uB, y + vA, A, B) := \mathrm{Ad}_{(u,v)} \cdot M, \quad u, v \in \mathbb{R}^{2n}.$$

Then

$$t \star \mathrm{Ad}_{(u,v)} \cdot M = M(t^2x + t^2(u - v)A + t^2uB, t^2y + t^2vA, t^2A, t^2B) = \mathrm{Ad}_{(tu,tv)} \cdot (t \star M).$$

That is, the $\mathbb{R}_+^*$-action on $\mathscr{T}(\mathfrak{l}, \mathfrak{p}, \mathfrak{D})$ is well-defined. Moreover, for $s, t \in \mathbb{R}_+^*$, $t \cdot [M] = s \cdot [M]$ if and only if $t \star M = G \cdot (s \star M)$. Hence $s = t$ by formula (4.24). This shows that $[M]$ lies in a one-dimensional curve and, therefore, it cannot be an open point inside $\mathscr{T}(\mathfrak{l}, \mathfrak{p}, \mathfrak{D})$.

5.1.4 The threadlike case

As a direct consequence from Theorem 4.4.15, we get the following result concerning the property of local rigidity in the threadlike case. We have the following.

Theorem 5.1.5. *Conjecture* 5.1.1 *holds for threadlike nilpotent Lie groups.*

5.2 Local rigidity for exponential Lie groups

We now focus on the exponential setting. We first treat some general cases where the statement of Conjecture 5.1.1 holds. Let us start with the setting where $H \subset G$ a normal connected subgroup of an exponential solvable Lie group G, and $\Gamma \simeq \mathbb{Z}^k$. A direct consequence from Theorem 3.3.1 is the following important fact concerning the topological features of Clifford–Klein forms.

Theorem 5.2.1. *Let G be an exponential solvable Lie group, $H \subset G$ a normal connected subgroup and $\Gamma \simeq \mathbb{Z}^k$ a discrete subgroup of G. Then $\mathscr{R}(\Gamma, G, H)$ is an open set in $\mathrm{Hom}(\Gamma, G)$. If in addition H contains $[G, G]$, then every Clifford–Klein form $\Gamma\backslash G/H$ is continuously deformable. Actually, the local rigidity property fails to hold.*

Proof. As G is an exponential solvable Lie group, the discrete subgroup $\Gamma \simeq \mathbb{Z}^k$ of G admits a syndetic hull $L = \exp(\mathfrak{l})$. Then $\mathscr{R}(\Gamma, G, H)$ is homeomorphic to the set

$$\left\{\psi \in \mathrm{Hom}(\mathfrak{l}, \mathfrak{g}) \;\middle|\; \begin{array}{l} \dim \psi(\mathfrak{l}) = \dim \mathfrak{l} \text{ and} \\ \psi(\mathfrak{l}) \cap \mathfrak{h} = \{0\} \end{array}\right\}$$

as provided by Proposition 3.3.7. It follows therefore that $\mathscr{R}(\Gamma, G, H)$ is homeomorphic to the set

$$\{\psi \in \mathrm{Hom}(\mathfrak{l}, \mathfrak{g}) : \psi \text{ is injective and } \psi(\mathfrak{l}) \cap \mathfrak{h} = \{0\}\}$$

as $\mathfrak{h}$ is an ideal of $\mathfrak{g}$. Let $\mathscr{R}'(\Gamma, G, H) = \{\psi \in \mathrm{Hom}(\mathfrak{l}, \mathfrak{g}) : \psi \text{ is injective}\}$, then clearly $R'(\Gamma, G, H)$ is an open set of $\mathrm{Hom}(\mathfrak{l}, \mathfrak{g})$ and that $\mathscr{R}(\Gamma, G, H) \subset R'(\Gamma, G, H)$. It is then sufficient to see that $\mathscr{R}(\Gamma, G, H)$ is an open subset of $\mathscr{R}'(\Gamma, G, H)$. Let $(X_1, \ldots, X_s)$ be the basis of $\mathfrak{h}$ as indicated above and pick a basis $(Y_1, \ldots, Y_k)$ of $\mathfrak{l}$. Then $(\psi(Y_1), \ldots, \psi(Y_k))$ is a basis of $\psi(\mathfrak{l})$, which is a complementary basis of $\mathfrak{h}$ inside the subalgebra $\mathfrak{h} \oplus \psi(\mathfrak{l})$. It follows therefore that the rank of the matrix

$$\begin{pmatrix} I_s & \psi(Y_1), \ldots, \psi(Y_k) \\ (0) & (*) \end{pmatrix}$$

is exactly $k + s$ for every $\psi \in \mathscr{R}(\Gamma, G, H)$. Such a matter shows that $\mathscr{R}(\Gamma, G, H)$ appears to be the complementary set of the zeros of a polynomial function. This achieves the proof of the first part of the theorem. As for the local rigidity, referring back to Subsection 3.3.1 let $[\psi] \in \mathscr{T}(\Gamma, G, H)$, which is open; then $[\psi]$ is open in $\mathscr{T}_\alpha$ for some $\alpha \in I_s(n, k)$. It follows that $[\psi]$ is open in $GL_k(\mathbb{R}) \times \eta_\alpha(\mathscr{V}_\alpha/G)$. Now the projection on the first component is an open map, which means that the image of $[\psi]$ by the first projection is open and consists of one single point in $GL_k(\mathbb{R})$, which is absurd. This achieves the proof of the theorem. □

Remark 5.2.2. The first statement of Theorem 5.2.1 can be obtained differently. Indeed, if we consider the set equality given by equation (3.12), it comes out that the set $R_k(\mathfrak{g}, \mathfrak{h})$ appears to be an open subset of $\mathrm{Hom}(\mathfrak{l}, \mathfrak{g})$, which is identified to the space $\mathscr{V}$. As $\mathscr{R}(\Gamma, G, H)$ is homeomorphic to $R_k(\mathfrak{g}, \mathfrak{h})$ as provided by Lemma 3.3.9, we are done. It is somehow noteworthy to point out that the arguments given to prove the first statement of last theorem could run once we consider the case where G is nilpotent and replace the hypothesis H normal by the assumption that the Clifford–Klein form in question is compact.

5.2.1 A local rigidity theorem where $[L, L] = [G, G]$

Back now to our settings and notation of Section 3.4.2, in particular G is completely solvable. Consider first the natural continuous action of $\mathrm{Aut}(\mathfrak{l})$ on $\mathrm{Hom}(\mathfrak{l}, \mathfrak{g})$, which respects $\mathscr{R}(\Gamma, G, H)$. Then we have the following.

Theorem 5.2.3. *Let G be a completely solvable Lie group, H a connected subgroup of G and Γ a discontinuous group for G/H such that $[L, L] = [G, G]$. Then $\mathscr{R}(\Gamma, G, H)$ is an open set in $\mathrm{Hom}(\Gamma, G)$ and semialgebraic. Moreover, for $\varphi \in \mathscr{R}(\Gamma, G, H)$ the following assertions are equivalent:*
(ı) *φ is strongly locally rigid.*
(ıı) *φ is locally rigid.*
(ııı) *The orbit $\varphi\,\mathrm{Aut}(\mathfrak{l})$ is open in $\mathrm{Hom}(\mathfrak{l}, \mathfrak{g})$ and*

$$\dim \mathrm{Aut}(\mathfrak{l}) + \dim \varphi(\mathfrak{l})^{\perp} = \dim \mathfrak{g}, \tag{5.1}$$

where $\varphi(\mathfrak{l})^{\perp} = \{Y \in \mathfrak{g},\ [X, Y] = 0 \text{ for all } X \in \varphi(\mathfrak{l})\}$.

Proof. Recall that $\mathscr{R}(\Gamma, G, H)$ is open by Proposition 3.4.6, then clearly (ı) and (ıı) are equivalent. Let $\varphi \in \mathscr{R}(\Gamma, G, H)$, M be the corresponding matrix to φ and $\xi'_{\alpha\beta}(M) = (A, x, W)$. Suppose that $G \cdot M$ is open. We have

$$G \cdot M = c_{A,x,W,\alpha}(G) \cdot M \quad \text{and} \quad \mathrm{Aut}(\mathfrak{l}) \cdot M = \dot{\bigcup_{a \in \mathrm{Aut}(\mathfrak{l})}} a c_{A,x,W,\alpha}(G) \cdot M. \tag{5.2}$$

Then $\mathrm{Aut}(\mathfrak{l})M$ is a union of open subsets. The $\mathrm{Aut}(\mathfrak{l})$-orbit of M is identified via $\xi_{\alpha\beta}$ with the set $\mathrm{Aut}(\mathfrak{l}) \times \{(x, W)\}$. The restriction of the canonical surjection from $\mathscr{R}(\Gamma, G, H) \to \mathscr{T}(\Gamma, G, H)$ to the $\mathrm{Aut}(\mathfrak{l})$-orbit of M is a continuous map, and from (3.26) its image is homeomorphic to the homogeneous space $\mathrm{Aut}(\mathfrak{l})/c_{x,W,\alpha}(G)$. Now the strong local rigidity of φ implies that the image of φ in the homogeneous Hausdorff space $\mathrm{Aut}(\mathfrak{l})/c_{x,W,\alpha}(G)$ is open and closed. This means that this point is a connected component of $\mathrm{Aut}(\mathfrak{l})/c_{x,W,\alpha}(G)$, in particular $\dim(\mathrm{Aut}(\mathfrak{l})/c_{x,W,\alpha}(G)) = 0$. We now prove the following.

Lemma 5.2.4. $\dim c_{x,W,\alpha}(G) = \dim(\mathfrak{g}) - \dim W^{\perp}$.

Proof. By the definition (3.21) of $c_{W,\alpha}$, we have

$$\begin{aligned}
\ker(c_{W,\alpha}) &= \{g \in G,\ \mathrm{Ad}_g\, \eta_\alpha^{-1}(W) = \eta_\alpha^{-1}(W)\} \\
&= \{\exp(X),\ X \in \mathfrak{g} \text{ and } \mathrm{Ad}_{\exp(X)}(Y) = Y, \text{ for any } Y \in W\} \\
&= \{\exp(X),\ [X, Y] = 0, \text{ for any } Y \in W\} = \exp(W^{\perp}).
\end{aligned}$$

□

Using Lemma 5.2.4, we get $\dim \mathrm{Aut}(\mathfrak{l}) + \dim \varphi(\mathfrak{l})^{\perp} = \dim \mathfrak{g}$. Suppose now that $(\imath\imath\imath)$ holds. By the connectedness of G, we see that $c_{x,W,\alpha}(G)$ is the connected component of the identity of $\mathrm{Aut}(\mathfrak{l})$ and $c_{x,W,\alpha}(G) = c_{A,x,W,\alpha}(G)$. To prove that $G \cdot M$ is open, it is sufficient using (5.2) to see that $c_{A,x,W,\alpha}(G) \cdot M$ is open. Let $y \in c_{x,W,\alpha}(G) \cdot M$ and pick an open neighborhood V_0 of the identity in $\mathrm{Aut}(\mathfrak{l})$ included in $c_{x,W,\alpha}(G)$. Il follows therefore that $V_0 \cdot y \subset c_{x,W,\alpha}(G) \cdot M$ and it is enough to show that $V_0 \cdot y$ is open in $\mathrm{Hom}(\mathfrak{l}, \mathfrak{g})$. Recall that the map $\mathrm{Aut}(\mathfrak{l}) \to \mathrm{Aut}(\mathfrak{l}) \cdot M$, $a \mapsto Ma^{-1}$ is a homeomorphism. Then $V_0 \cdot y$ is open in $\mathrm{Aut}(\mathfrak{l}) \cdot M$, and thus in $\mathrm{Hom}(\mathfrak{l}, \mathfrak{g})$. □

The following important consequence is therefore immediate.

Corollary 5.2.5. *We keep the same hypotheses and notation. If the group* $\mathrm{Aut}(\mathfrak{l})$ *is not solvable, then the local rigidity fails to hold.*

5.3 Selberg–Weil–Kobayashi local rigidity theorem

In the case of semisimple Lie groups, local rigidity was first proved by Selberg [123] for uniform lattices in the case $G = SL_n(\mathbb{R})$, $n \geq 3$, and by Calabi [51] for uniform lattices in the case $G = PO(n, 1) = \mathrm{Iso}(\mathbb{H}^n)$, $n \geq 3$. Then Weil [129] generalized these results to any uniform irreducible lattice in any G, assuming that G is not locally isomorphic to $SL_2(\mathbb{R})$ in which case lattices have nontrivial deformations. More generally, these results may be formulated (cf. [130]).

Theorem 5.3.1 (Local rigidity theorem—Selberg and Weil). *Let G be a noncompact linear simple Lie group and H its maximal compact subgroup, then the following assertions on G are equivalent:*
(1) *There exists a uniform lattice $\varphi : \Gamma \to G$ such that $\varphi \in \mathscr{R}(\Gamma, G, H)$ admits continuous deformations.*
(2) *G is locally isomorphic to $SL_2(\mathbb{R})$.*

The following theorem (cf. [94]) produces some irreducible non-Riemannian symmetric spaces of arbitrary high dimension endowed with a uniform lattice for which the local rigidity theorem does not hold.

Theorem 5.3.2. *We keep the same assumptions as in Theorem* 5.3.1 *and let* $(G', H') := (G \times G, \Delta_G)$, *where Δ_G denotes the diagonal group. Then the following are equivalent:*
(1) *There exists a uniform lattice $\varphi : \Gamma \to G$ such that $\varphi \times 1 \in \mathscr{R}(\Gamma, G', H')$ admits continuous deformations.*
(2) *G is locally isomorphic to $SO(n, 1)$ or $SU(n, 1)$.*
(3) *G does not have Kazhdan's property* (T).

Note that Theorem 5.3.1 was formulated so that these two rigidity theorems can be compared. As such, the result of Theorem 5.3.2 produces some irreducible non-

Riemannian symmetric spaces of arbitrary high dimension endowed with a uniform lattice for which local rigidity does not hold. For the Riemannian case, this is very rare as mentioned earlier.

The aim of the present subsection is to derive an analogue to such results in the setting of exponential solvable Lie groups. More details could be found in the references [1, 14, 33]. Let us start by studying the case of maximal homogeneous spaces.

5.3.1 Maximal exponential homogeneous spaces

Our intention in this section is to state some preliminary results with regard to the parameter and the deformation spaces of discontinuous groups acting on maximal homogenous spaces for which the underlying group in question is exponential. As in Theorem 1.1.15 and Remark 1.1.17, maximal subgroups are of codimension one or two and we have the following.

Theorem 5.3.3. *Let $\mathfrak{g}$ be a exponential Lie algebra and $\mathfrak{h}$ a maximal subalgebra of $\mathfrak{g}$, which is not an ideal. Then $\mathfrak{h}$ is of codimension one or two in $\mathfrak{g}$ and we have the following:*

(1) *If $\mathfrak{h}$ is of codimension one, then there exist a codimension one ideal $\mathfrak{g}_0$ of $\mathfrak{h}$, which is a codimension two ideal in $\mathfrak{g}$, and two elements A, X in $\mathfrak{g}$ such that*

$$\mathfrak{g} = \mathfrak{h} \oplus \mathbb{R}X, \quad \mathfrak{h} = \mathfrak{g}_0 \oplus \mathbb{R}A$$

and

$$[A, X] = X \bmod \mathfrak{g}_0. \tag{5.3}$$

(2) *If $\mathfrak{h}$ is of codimension two, then there exist a codimension one ideal $\mathfrak{g}_0$ of $\mathfrak{h}$, which is an ideal of $\mathfrak{g}$ of codimension 3, as well as three elements A, X, Y in $\mathfrak{g}$ and a nonzero real number α such that*

$$\mathfrak{g} = \mathfrak{h} \oplus \mathbb{R}X \oplus \mathbb{R}Y, \quad \mathfrak{h} = \mathfrak{g}_0 \oplus \mathbb{R}A,$$
$$[A, X + iY] = (\alpha + i)(X + iY) \bmod \mathfrak{g}_0 \tag{5.4}$$

and

$$[X, Y] = 0 \bmod \mathfrak{g}_0. \tag{5.5}$$

A *maximal basis of $\mathfrak{g}$ adapted to $\mathfrak{h}$* is any basis $\mathscr{B}$ constituted of a basis of $\mathfrak{g}_0$ together with the vectors $\{A, X\}$ (resp., $\{A, X, Y\}$) organized according to the order above. Making use of Theorem 2.4.4 dealing with the proper action of solvable maximal homogeneous spaces, the following turns out to be a direct consequence.

Corollary 5.3.4. *Let G be an exponential Lie group, H a maximal subgroup of G and Γ a discontinuous group for the homogenous space G/H. Then Γ is isomorphic to $\mathbb{Z}$ or $\mathbb{Z}^2$.*

Proof. As G is a exponential solvable, Γ admits a unique syndetic hull $L = \exp(\mathfrak{l})$, which turns out to be Abelian as Γ does. Now thanks to Theorem 2.4.4, there is equivalence of the proper and free actions of L on G/H. If $\mathfrak{h}$ denotes the Lie algebra associated to H, we get right away that $\mathfrak{l} \cap \mathfrak{h} = \{0\}$ and by the maximality of $\mathfrak{h}$, Γ is of rank $\dim \mathfrak{l} \leq 2$. □

The parameter space, revisited

Let $H = \exp \mathfrak{h}$ be a connected subgroup of an exponential Lie group $G = \exp \mathfrak{g}$, Γ a discontinuous group for the homogenous space G/H and $L = \exp \mathfrak{l}$ its syndetic hull. We designate by $\mathrm{Hom}(\mathfrak{l}, \mathfrak{g})$ the set of all algebra homomorphisms from $\mathfrak{l}$ to $\mathfrak{g}$. Let $\mathscr{B}$ be a maximal basis of $\mathfrak{g}$ adapted to $\mathfrak{h}$. We identify $\mathfrak{g}$, $\mathfrak{l}$ and the space of the linear maps $\mathscr{L}(\mathfrak{l}, \mathfrak{g})$ with $\mathbb{R}^n$, $\mathbb{R}^k$ and the space of matrices $\mathscr{M}_{n,k}(\mathbb{R})$, respectively. Having fixed a basis $\mathscr{B}_{\mathfrak{l}}$ of $\mathfrak{l}$, it appears clear that the map

$$\Psi : \mathrm{Hom}(\mathfrak{l}, \mathfrak{g}) \longrightarrow \mathscr{M}_{n,k}(\mathbb{R}), \tag{5.6}$$

which associates to any element of $\mathrm{Hom}(\mathfrak{l}, \mathfrak{g})$ its matrix written in the bases $\mathscr{B}_{\mathfrak{l}}$ and the basis $\mathscr{B}$ of $\mathfrak{g}$ is a homeomorphism on its range. Throughout the whole text, the set $\mathrm{Hom}(\mathfrak{l}, \mathfrak{g})$ is therefore homeomorphically identified to a set $\mathscr{U}$ of $\mathscr{M}_{n,k}(\mathbb{R})$ and $\mathrm{Hom}^{\circ}(\mathfrak{l}, \mathfrak{g})$ of all injective homomorphisms to the subset $\mathscr{U}^{\circ}$ of $\mathscr{U}$ consisting of all matrices in $\mathscr{U}$ of maximal rank. Obviously, the set $\mathscr{U}$ is closed and algebraic in $\mathscr{M}_{n,k}(\mathbb{R})$. In addition, the set $\mathscr{U}^{\circ}$ is semialgebraic and open in $\mathscr{U}$. Let G act on $\mathscr{M}_{n,k}(\mathbb{R})$ by

$$g \cdot M = \mathrm{Ad}_g \times M, \quad M \in \mathscr{M}_{n,k}(\mathbb{R}) \quad \text{and} \quad g \in G, \tag{5.7}$$

where Ad_g is viewed as a real valued matrix for any $g \in G$. In the setting where $\mathfrak{g} = \mathbb{R}X \oplus \mathfrak{h}$ with $\mathfrak{h} = \mathbb{R}A \oplus \mathfrak{g}_0$ and $[A, X] = X \bmod(\mathfrak{g}_0)$, for $g = \exp aA \cdot \exp xX \cdot g_0 \in G$ where a and x are in $\mathbb{R}$ and $g_0 \in G_0$, a routine computation shows that

$$g \cdot \begin{pmatrix} a_0 \\ x_0 \\ \vec{z} \end{pmatrix} = \begin{pmatrix} a_0 \\ (x_0 - xa_0)e^a \\ \vec{z'}(g) \end{pmatrix}, \tag{5.8}$$

where $\vec{z'}(g) \in \mathbb{R}^{n-2}$ is an analytic expression depending upon g (which means that $T \mapsto \vec{z'}(\exp T)$ is an analytic function on the vector space $\mathfrak{g}$). In the case where $\mathfrak{g} = \mathbb{R}X \oplus \mathbb{R}Y \oplus \mathfrak{h}$ with $\mathfrak{h} = \mathbb{R}A \oplus \mathfrak{g}_0$, $[A, X + iY] = (\alpha + i)(X + iY) \bmod(\mathfrak{g}_0)$ and $[X, Y] =$

$0 \bmod(\mathfrak{g}_0)$. For $g = \exp aA \cdot \exp xX \cdot \exp yY \cdot g_0$, where a, x and $y \in \mathbb{R}$ and $g_0 \in G$,

$$g \cdot \begin{pmatrix} a_1 \\ x_1 \\ y_1 \\ \vec{z_1} \end{pmatrix} = \begin{pmatrix} a_1 \\ e^{\alpha a}((x_1 - a_1(\alpha x + y)) \cos a + (y_1 + a_1(x - \alpha y)) \sin a) \\ e^{\alpha a}((y_1 + a_1(x - \alpha y)) \cos a - (x_1 - a_1(\alpha x + y)) \sin a) \\ \vec{z_1'}(g) \end{pmatrix}, \tag{5.9}$$

where also $\vec{z'}(g) \in \mathbb{R}^{n-3}$ is an analytic expression depending upon g. When Γ is of rank two, then

$$g \cdot \begin{pmatrix} a_1 & a_2 \\ x_1 & x_2 \\ y_1 & y_2 \\ \vec{z_1} & \vec{z_2} \end{pmatrix} = \begin{pmatrix} a_1 & a_2 \\ x_1' & x_2' \\ y_1' & y_2' \\ \vec{z_1'}(g) & \vec{z_2'}(g) \end{pmatrix}, \tag{5.10}$$

where for $i = 1, 2$, $\vec{z_i'}(g) \in \mathbb{R}^{n-3}$ depends analytically on $g \in G$,

$$x_i' = e^{\alpha a}((x_i - a_i(\alpha x + y)) \cos a + (y_i + a_i(x - \alpha y)) \sin a)$$

and

$$y_i' = e^{\alpha a}((y_i + a_i(x - \alpha y)) \cos a - (x_i - a_i(\alpha x + y)) \sin a).$$

Taking into account the action of G on $\mathrm{Hom}(\mathfrak{l}, \mathfrak{g})$ defined in (3.3), the following lemma is immediate.

Lemma 5.3.5. *The map Ψ defined in (5.6) is G-equivariant with regard to the actions (3.3) and (5.7). That is, for any $\psi \in \mathrm{Hom}(\mathfrak{l}, \mathfrak{g})$ and $g \in G$, we have $\Psi(g \cdot \psi) = g \cdot \Psi(\psi)$.*

Assume now that $\mathfrak{l}$ is Abelian. Thus, $\psi \in \mathscr{L}(\mathfrak{l}, \mathfrak{g})$ is an algebra homomorphism if and only if ψ is a linear map satisfying $[\psi(X), \psi(Y)] = 0$ for all X and Y in $\mathfrak{l}$, which gives rise to the following expression:

$$\mathrm{Hom}(\mathfrak{l}, \mathfrak{g}) = \{\psi \in \mathscr{L}(\mathfrak{l}, \mathfrak{g}),\ [\psi(X), \psi(Y)] = 0 \text{ for all } X \text{ and } Y \text{ in } \mathfrak{l}\}.$$

With the above in mind, for $k = 2$, the set $\mathscr{U}$ (resp., $\mathscr{U}^\circ$) is homeomorphic to $\{(M_1, M_2) \in \mathscr{M}_{n,2}(\mathbb{R}),\ [M_1, M_2] = 0\}$ (resp., to the same set of matrices of maximal rank), with the convention that the Lie bracket $[M_1, M_2]$ represents the corresponding vectors of the Lie algebra defining the associated homomorphism. When $k = 1$, $\mathrm{Hom}(\mathfrak{l}, \mathfrak{g}) = \mathscr{M}_{n,1}(\mathbb{R})$.

We now prove the following propositions toward the determination of the parameter space $\mathscr{R}(\mathfrak{l}, \mathfrak{g}, \mathfrak{h})$, which are the key points of many uses later. The first deals with the case where $k = 1$.

Proposition 5.3.6. *Let G be an exponential Lie group of dimension n, H a nonnormal connected maximal subgroup of G and Γ a discontinuous subgroup for G/H of rank one. Then $\mathscr{R}(\mathfrak{l}, \mathfrak{g}, \mathfrak{h})$ is a smooth manifold and described more precisely as follows:*

(i) *If* $\operatorname{codim}(\mathfrak{h}) = 1$, $\mathscr{R}(\mathfrak{l}, \mathfrak{g}, \mathfrak{h})$ *is homeomorphic to*

$$\mathscr{R}_{1,1}^{n} = \{{}^{t}(0\; x\; \vec{z}) \in \mathscr{M}_{n,1}(\mathbb{R}),\ x \in \mathbb{R}^{\times}\} \simeq \mathbb{R}^{\times} \times \mathbb{R}^{n-2}. \tag{5.11}$$

(ii) *If* $\operatorname{codim}(\mathfrak{h}) = 2$, *then* $\mathscr{R}(\mathfrak{l}, \mathfrak{g}, \mathfrak{h})$ *is homeomorphic to*

$$\mathscr{R}_{1,2}^{n} = \{{}^{t}(0\; x\; y\; \vec{z}) \in \mathscr{M}_{n,1}(\mathbb{R}),\ x^2 + y^2 \neq 0\} \simeq \mathbb{R}^2 \setminus \{(0,0)\} \times \mathbb{R}^{n-3}. \tag{5.12}$$

In particular, $\mathscr{R}_{1,2}^{3} \simeq \mathbb{R}^2 \setminus \{(0,0)\}$.

Proof. Let us assume first that $\mathfrak{h}$ is of codimension one and let $M = {}^{t}(a_0\; x_0\; \vec{z}) \in \mathscr{M}_{n,1}(\mathbb{R})$. From (5.8), $M \in \mathscr{R}(\mathfrak{l}, \mathfrak{g}, \mathfrak{h})$ if and only if

$$\begin{pmatrix} a_0 \\ (x_0 - xa_0)e^{a} \\ \vec{z'}(g) \end{pmatrix} \notin \mathfrak{h}$$

for all reals x and a. This means $x_0 - xa_0 \neq 0$ for all $x \in \mathbb{R}$, which is in turn equivalent to $x_0 \in \mathbb{R}^{\times}$ and $a_0 = 0$, as was to be shown. Assume now that $\mathfrak{h}$ is of codimension two. Then, for $M = {}^{t}(a_1\; x_1\; y_1 \vec{z_1}) \in \mathscr{M}_{n,1}(\mathbb{R})$, we have by similar arguments:

$$\begin{aligned} M \in \mathscr{R}(\mathfrak{l}, \mathfrak{g}, \mathfrak{h}) &\Leftrightarrow (x_1 - a_1(\alpha x + y))^2 + (y_1 + a_1(x - \alpha y))^2 \neq 0 \quad \text{for all } x, y \in \mathbb{R} \\ &\Leftrightarrow x_1^2 + y_1^2 \neq 0 \quad \text{and} \quad a_1 = 0, \end{aligned}$$

which achieves the proof in this case. □

We now move to the case $k = 2$. The following proposition exhibits a description of the parameter space in this case.

Proposition 5.3.7. *Let G be an exponential Lie group of dimension n, H a nonnormal connected maximal subgroup of G and Γ a discontinuous group for G/H of rank two. Then:*

(i) *If* $\operatorname{codim}(\mathfrak{h}) = 1$, *then* $\mathscr{R}(\mathfrak{l}, \mathfrak{g}, \mathfrak{h}) = \emptyset$.

(ii) *If* $\operatorname{codim}(\mathfrak{h}) = 2$, *then* $\mathscr{R}(\mathfrak{l}, \mathfrak{g}, \mathfrak{h})$ *is homeomorphic to*

$$\mathscr{R}_{2,2}^{n} = \left\{ \begin{pmatrix} 0 & 0 \\ x_1 & x_2 \\ y_1 & y_2 \\ \vec{z_1} & \vec{z_2} \end{pmatrix} \in \mathscr{U},\ x_1y_2 - x_2y_1 \neq 0 \right\}. \tag{5.13}$$

In particular, $\mathscr{R}_{2,2}^{3} = GL_2(\mathbb{R})$.

Proof. When $\mathfrak{h}$ is of codimension one, then $\mathfrak{h} \cap \varphi(\mathfrak{l})$ is not trivial for any $\varphi \in \mathscr{R}(\mathfrak{l}, \mathfrak{g}, \mathfrak{h})$, which is a contradiction as $\mathfrak{l}$ is of dimension two. Hence $\mathscr{R}(\mathfrak{l}, \mathfrak{g}, \mathfrak{h}) = \emptyset$. Suppose now that $\mathfrak{h}$ is of codimension two. Then any

$$M = \begin{pmatrix} a_1 & a_2 \\ x_1 & x_2 \\ y_1 & y_2 \\ \vec{z_1} & \vec{z_2} \end{pmatrix} \in \mathscr{U}$$

fulfills the quadratic equation $a_1 x_2 - a_2 x_1 = a_1 y_2 - a_2 y_1 = 0$. Then $M \in \mathscr{R}(\mathfrak{l}, \mathfrak{g}, \mathfrak{h})$ if and only if the matrix

$$\begin{pmatrix} x_1' & x_2' \\ y_1' & y_2' \end{pmatrix}$$

is regular, where for $j \in \{1, 2\}$,

$$z_j' = x_j' + iy_j' = e^{(\alpha - i)a}\big[(x_j - a_j(\alpha x + y)) + i(y_j + a_j(x - \alpha y))\big],$$

as in equation (5.10). This gives in turn that $a_1 = a_2 = 0$ and $x_1 y_2 - x_2 y_2 \neq 0$, as was to be shown. □

5.3.2 Local rigidity for small-dimensional exponential Lie groups

This subsection aims to study the local rigidity proprieties for which the underlying group in question is assumed to be exponential and of low dimension. We have the following.

Lemma 5.3.8.

(1) *Let $\mathfrak{g}$ be the Lie algebra of the group* $\mathrm{Aff}(\mathbb{R})$ *(as in Example* 1.1.4) *and Γ a discontinuous group for* $\exp(\mathfrak{g})/\exp(\mathfrak{h})$. *Then the local rigidity property holds.*

(2) *Let $G = G_\alpha(\alpha \in \mathbb{R}^\times)$ be the Lie group with the Lie algebra $\mathfrak{g}_\alpha$ spanned by A, X, Y with $[A, X + iY] = (\alpha + i)(X + iY)$, $\mathfrak{h} = \mathbb{R}A$ and Γ a nontrivial discontinuous group for* $\exp(\mathfrak{g})/\exp(\mathfrak{h})$. *Then the local rigidity property fails to hold.*

Proof. (1) Assume that Γ is not trivial. Thanks to equation (5.8), it is clear that any G-orbit in $\mathscr{R}^2_{1,1} \simeq \mathbb{R}^\times$ is either homeomorphic to $(0, +\infty)$ or to $(-\infty, 0)$, which is enough to conclude.

(2) Now $\mathfrak{h}$ is of codimension two. We argue separately according to the dimension of $\mathfrak{l}$. Suppose first that $\mathfrak{l}$ is of dimension two. For $M \in \mathscr{R}^3_{2,2} \simeq GL_2(\mathbb{R})$ as in (5.13), equa-

tion (5.9) shows that the orbit of M through the action of G reads:

$$G \cdot M = \left\{ \begin{pmatrix} 0 & 0 \\ e^{\alpha a}(x_1 \cos a + y_1 \sin a) & e^{\alpha a}(x_2 \cos a + y_2 \sin a) \\ e^{\alpha a}(-x_1 \sin a + y_1 \cos a) & e^{\alpha a}(-x_2 \sin a + y_2 \cos a) \end{pmatrix}, \ a \in \mathbb{R} \right\},$$

which is not an open set in $\mathscr{R}^3_{2,2}$. Suppose now that $\mathfrak{l}$ is of dimension one. It is clear thanks to equation (5.10) that the G-orbit of $M \in \mathscr{R}^3_{1,2}$ as in (5.12),

$$G \cdot M = \{{}^t(0 \quad e^{\alpha a}(x \cos a + y \sin a) \quad e^{\alpha a}(-x \sin a + y \cos a)), \ a \in \mathbb{R}\}.$$

So, the local rigidity property fails to hold globally on $\mathscr{R}(\mathfrak{l}, \mathfrak{g}, \mathfrak{h})$ in this case. □

Now, our so-called analogue of Selberg–Weil–Kobayashi rigidity theorem in the context of maximal exponential homogeneous spaces is stated as follows.

Theorem 5.3.9 (An analogue of Selberg–Weil–Kobayashi rigidity theorem). *Let G be an exponential Lie group, H a nonnormal connected maximal subgroup of G and Γ a discontinuous group for G/H. Then the following assertions are equivalent:*

(i) *G is isomorphic to the group* Aff($\mathbb{R}$).

(ii) *Every homomorphism in $\mathscr{R}(\Gamma, G, H)$ is locally rigid.*

(iii) *Some homomorphism in $\mathscr{R}(\Gamma, G, H)$ is locally rigid.*

The proof of Theorem 5.3.9 will be divided into different steps. First, Table 5.1 establishes the parallelism, which can be built to compare the two settings.

Table 5.1: Summarizing Selberg–Weil–Kobayashi rigidity results.

G is a noncompact simple Lie group	G is an exponential Lie group
$H \subset G$ is a maximal (compact) or H is the diagonal Δ_G of $G \times G$	$H \subset G$ is a maximal (nonnormal)
The Clifford–Klein form $\Gamma \backslash G/H$ is *compact*	$\Gamma \backslash G/H$ is *arbitrary*
no continuous deformations	no continuous deformations
G is *not* locally isomorphic to $SL_2(\mathbb{R})$ for a Riemannian space G/H or G is *not* locally isomorphic to $SO(n,1)$ or $SU(n,1)$ for $(G \times G)/\Delta_G$	G is isomorphic to Aff($\mathbb{R}$)

5.3.3 Passing through the quotients

We keep all our assumptions and notation. $G = \exp \mathfrak{g}$ be an exponential Lie group, $H = \exp \mathfrak{h}$ a nonnormal connected maximal subgroup of G, Γ be a nontrivial discontinuous subgroup of G and $L = \exp \mathfrak{l}$ its syndetic hull. According to Lemma 5.3.11, $\mathfrak{l}$ is an Abelian subalgebra of $\mathfrak{g}$ of dimension one or two. We denote by $\mathfrak{g}_0$ the ideal defined as in Theorem 5.3.3 and G_0 its corresponding Lie group. Let $K = \exp \mathfrak{k} \subset H$ be a normal subgroup of G. We consider the quotient group $\overline{G} = G/K$, $\pi : G \to \overline{G}$ the canonical projection homomorphism and $d_e\pi : \mathfrak{g} \to \overline{\mathfrak{g}}$, $X \mapsto \overline{X}$ its derivative. Likewise, we denote $\overline{H} = \pi(H) := \exp(\overline{\mathfrak{h}})$ and $\overline{L} = \pi(L) := \exp(\overline{\mathfrak{l}})$. Since L is the syndetic hull of Γ, then L is a subgroup of G acting properly on G/H and then freely. So, $L \cap H$ is trivial and the homomorphism $\tilde{\pi} = \pi_{|L} : L \to \overline{L}$ appears to be a groups isomorphism. Thus, the Lie algebra $\overline{\mathfrak{l}} = \log \overline{L}$ is isomorphic to $\mathfrak{l}$, and clearly $\overline{L}$ is isomorphic to L.

Our intention now is to investigate the connection between the parameter spaces $\mathscr{R}(\mathfrak{l}, \mathfrak{g}, \mathfrak{h})$ and $\mathscr{R}(\overline{\mathfrak{l}}, \overline{\mathfrak{g}}, \overline{\mathfrak{h}})$. It consists in fact in constructing an open surjective map between these spaces. We recall the notation $\operatorname{Hom}(\mathfrak{l}, \mathfrak{g})$ (resp., $\operatorname{Hom}(\overline{\mathfrak{l}}, \overline{\mathfrak{g}})$) assigned to the set of all algebra homomorphisms from $\mathfrak{l}$ to $\mathfrak{g}$ (resp., from $\overline{\mathfrak{l}}$ to $\overline{\mathfrak{g}}$). The group G (resp., $\overline{G}$) acts on $\operatorname{Hom}(\mathfrak{l}, \mathfrak{g})$ (resp., $\operatorname{Hom}(\overline{\mathfrak{l}}, \overline{\mathfrak{g}})$) as in (3.3). We consider the map

$$\begin{array}{rcl} \xi : \operatorname{Hom}(\mathfrak{l}, \mathfrak{g}) & \to & \operatorname{Hom}(\overline{\mathfrak{l}}, \overline{\mathfrak{g}}) \\ \varphi & \mapsto & \overline{\varphi} = d_e\pi \circ \varphi \circ d_e\tilde{\pi}^{-1}. \end{array}$$

Lemma 5.3.10. *The group G acts on $\xi(\operatorname{Hom}(\mathfrak{l}, \mathfrak{g}))$ by the law*

$$g \cdot \xi(\varphi) = \overline{g} \cdot \xi(\varphi) \tag{5.14}$$

making of ξ a G-equivariant map on its range.

Proof. Clearly, the law (5.14) above defines a G-action on $\xi(\operatorname{Hom}(\mathfrak{l}, \mathfrak{g}))$. Let, on the other hand, $g \in G$ and $\varphi \in \operatorname{Hom}(\mathfrak{l}, \mathfrak{g})$. Then, for $X \in \mathfrak{l}$, we have

$$\begin{aligned} \xi(g \cdot \varphi)(\overline{X}) &= \xi(\operatorname{Ad}_g \varphi)(\overline{X}) \\ &= \overline{\operatorname{Ad}_g \varphi(X)} \\ &= \operatorname{Ad}_g \varphi(X) + \mathfrak{k} \\ &= \operatorname{Ad}_g(\varphi(X) + \mathfrak{k}) \\ &= \operatorname{Ad}_g \overline{\varphi(X)} \\ &= \operatorname{Ad}_g \operatorname{Ad}_K \overline{\varphi(X)} \\ &= \operatorname{Ad}_{gK} \overline{\varphi(X)} \\ &= \overline{g} \cdot \xi(\varphi)(\overline{X}), \end{aligned}$$

which shows that ξ is G-equivariant. □

Weakening the assumptions of Theorem 2.4.6, we have the following.

Lemma 5.3.11. *Let G be an exponential Lie group and H be a maximal subgroup of G. Then any subgroup Γ of G acting on G/H freely is Abelian.*

Proof. Assume the same notation as in Subsection 5.3.3. One can easily see that $\overline{\Gamma}$ acts freely on $\overline{G}/\overline{H}$. In fact, let $\overline{h} \in \overline{H}$, $\overline{\gamma} \in \overline{\Gamma}$ and $\overline{t} \in \overline{G}$ such that $\overline{h} = \overline{t}\overline{\gamma}\overline{t}^{-1}$. Then $h = tyt^{-1}t_0$ for some $t_0 \in G_0$, which implies that $ht_0^{-1} = tyt^{-1} \in H \cap t\Gamma t^{-1}$. Finally, $h = t_0$ as Γ acts freely on G/H and, therefore, $\overline{h} = \overline{e}$. On the other hand, provided that $\Gamma \cap G_0$ is trivial, the homomorphism $\tilde{\pi} = \pi_{|\Gamma} : \Gamma \to \overline{\Gamma}$ appears to be a group isomorphism. To prove that Γ is an Abelian subgroup of G, it is sufficient to prove that $\overline{\Gamma}$ is an Abelian subgroup of $\overline{G}$.

When H is normal, it is of codimension one, and by taking $G_0 = H$, the result follows as $\overline{G}$ is of dimension one. Assume henceforth that H is not normal and denote by $\overline{\mathfrak{g}}$ the Lie algebra associated to $\overline{G}$. We also denote by $\mathfrak{g}_0$ the ideal defined as in Theorem 5.3.3 and G_0 its corresponding Lie group. Suppose in a first time that $\mathfrak{h}$ is of dimension one. It is clear that $\overline{\mathfrak{g}}$ is the Lie algebra of two dimension two spanned by the vectors A, X satisfying the bracket relation $[A, X] = X$ and that $\overline{\mathfrak{h}} = \mathbb{R}A$. We identify $\overline{G}$ to $\mathbb{R}^2$ with the multiplication law:

$$(a, x) \cdot (a', x') = (a + a', x' + xe^{-a'}).$$

We note $\overline{\mathfrak{g}}_1 = \mathbb{R}X$ and $\overline{G_1}$ its Lie group. Suppose that $\overline{\Gamma} \not\subset \overline{G_1}$. Then there exist $a \in \mathbb{R}^\times$ and $x \in \mathbb{R}^\times$ such that $\overline{g} = (a, x) = \exp aA \cdot \exp xX$ belongs to $\overline{\Gamma}$. But this gives

$$\exp\left(\frac{x}{e^{-a} - 1}\right)X \cdot \overline{g} \cdot \exp\left(\frac{-x}{e^{-a} - 1}\right)X \in H,$$

which contradicts the fact that $\overline{\Gamma}$ acts freely on $\overline{G}/\overline{H}$. This completes the proof in this case. Suppose now that $\mathfrak{h}$ is of dimension two. Then $\overline{\mathfrak{g}}$ is the Lie algebra spanned by A, X, Y with $[A, X + iY] = (\alpha + i)(X + iY)$, $\alpha \in \mathbb{R}^\times$ and $\overline{\mathfrak{h}} = \mathbb{R}A$. We identify $\overline{G}$ to $\mathbb{R}^3$ with the multiplication law:

$$(a, x, y) \cdot (a', x', y') = (a + a', x' + e^{-\alpha a'}(x \cos a' - y \sin a'), y' + e^{-\alpha a'}(y \cos a' + x \sin a')).$$

We note $\overline{\mathfrak{g}}_2 = \mathbb{R}X \oplus \mathbb{R}Y$ and $\overline{G_2}$ its Lie group. If $\overline{\Gamma} \not\subset \overline{G_2}$, then there exist $a \in \mathbb{R}^\times$ and $(x, y) \in \mathbb{R}^2 \setminus \{(0, 0)\}$ such that $\overline{g} = (a, x, y) = \exp aA \cdot \exp xX \cdot \exp yY$ is an element of $\overline{\Gamma}$. It comes by a simple computation that

$$\exp x_0 X \cdot \exp y_0 Y \cdot \overline{g} \cdot \exp(-x_0 X) \cdot \exp(-y_0 Y) \in H,$$

where

$$x_0 = \frac{-e^{a\alpha}((\cos a - e^{a\alpha})x + y \sin a)}{(\cos a - e^{a\alpha})^2 + \sin^2 a}$$

and

$$y_0 = \frac{e^{a\alpha}(x\sin a - (\cos a - e^{a\alpha})y)}{(\cos a - e^{a\alpha})^2 + \sin^2 a}.$$

This is impossible provided that $\overline{\Gamma}$ acts freely on the homogeneous space $\overline{G}/\overline{H}$. So, $\overline{\Gamma} \subset \overline{G_2}$ and $\overline{\Gamma}$ is an Abelian subgroup of $\overline{G}$. This completes the proof of the lemma. □

5.3.4 Proof of Theorem 5.3.9

Thanks to Lemma 5.3.8, we only need to show the implication $(iii) \to (i)$. We shall assume in what follows that the Lie group G is not isomorphic to $\mathrm{Aff}(\mathbb{R})$ and show that there is no locally rigid homomorphisms inside the parameter space. We will separately tackle the following cases.

Case where rank(Γ) = 2

We will take $\mathfrak{k} = \mathfrak{g}_0$ in this case and the symbol “bar” designates along this subsection the related quotient objects. The following upshot is the starting means to study the connection between the parameter spaces $\mathscr{R}(\mathfrak{l}, \mathfrak{g}, \mathfrak{h})$ and $\mathscr{R}(\bar{\mathfrak{l}}, \bar{\mathfrak{g}}, \bar{\mathfrak{h}})$.

Lemma 5.3.12. *We keep the same assumptions and notation as before and let Γ be of rank two. The restriction $\xi' := \xi_{|\mathscr{R}(\mathfrak{l},\mathfrak{g},\mathfrak{h})}$ of ξ to $\mathscr{R}(\mathfrak{l}, \mathfrak{g}, \mathfrak{h})$ is open and onto $\mathscr{R}(\bar{\mathfrak{l}}, \bar{\mathfrak{g}}, \bar{\mathfrak{h}})$, and thus G-equivariant.*

Proof. Consider first the natural continuous action of $\mathrm{Aut}(\mathfrak{l})$ on $\mathrm{Hom}(\mathfrak{l}, \mathfrak{g})$, which respects $\mathscr{R}(\lambda, \gamma, \mathfrak{h})$. This action induces a subsequent action of $GL_2(\mathbb{R})$ on the set $\mathscr{R}^n_{2,2}$ as defined in (5.13). More precisely, for ${}^t(0\ N_0\ \vec{z}) \in \mathscr{R}^n_{2,2}$ with $N_0 \in GL_2(\mathbb{R})$ and $\vec{z} \in \mathscr{M}_{n-3,2}(\mathbb{R})$ and $N \in GL_2(\mathbb{R})$, a matrix-like expression of this action can be described as follows:

$$MN^{-1} = \begin{pmatrix} \vec{0} \\ N_0 N^{-1} \\ \vec{z} N^{-1} \end{pmatrix}.$$

It is clear that $\bar{\mathfrak{g}} = \mathfrak{g}_\alpha$ for some $\alpha \in \mathbb{R}^\times$ and $\bar{\mathfrak{h}} = \mathbb{R}A$, where $\mathfrak{g}_\alpha$ is as in Lemma 5.3.8. According to equation (5.13), we get $\xi(\mathscr{R}^n_{2,2}) \subset \mathscr{R}^3_{2,2}$. Conversely, let $M = {}^t(0\ N) \in \mathscr{R}^3_{2,2}$ as in (5.13) with $N \in GL_2(\mathbb{R})$. Let $M_0 = {}^t(0\ N_0\ \vec{z_0}) \in \mathscr{R}^n_{2,2}$. Then $P = M_0 N_0^{-1} N = {}^t(0\ N\ \vec{z}) \in \mathscr{R}^n_{2,2}$ satisfies $\xi(P) = M$. This shows that ξ' is onto and, therefore, G-equivariant by Lemma 5.3.10. Let $\mathscr{A} := \{{}^t(0\ I_2\ \vec{z_0}) \in \mathscr{M}_{n,2}(\mathbb{R})\} \cap \mathscr{U}$ and

$$\begin{aligned} \psi &: \mathscr{R}^n_{2,2} \to GL_2(\mathbb{R}) \times \mathscr{A}, \\ M &= {}^t(0\ N\ \vec{z}) \mapsto (N, MN^{-1}) = (N, {}^t(0\ I_2\ \vec{z}N^{-1})). \end{aligned}$$

Clearly, ψ is well-defined and injective. On the other hand, let $(N, M') \in GL_2(\mathbb{R}) \times \mathscr{A}$, then $M = M'N \in \mathscr{R}^n_{2,2}$ and $\psi(M) = (N, M')$, which shows that ψ is onto. Finally, ψ is bijective and clearly a homeomorphism. On the other hand, it is immediate to see that $\xi = \iota \circ p_1 \circ \psi$, where $\iota : GL_2(\mathbb{R}) \to \mathscr{R}^3_{2,2}$, $N \mapsto {}^t(0\ N)$ and $p_1 : GL_2(\mathbb{R}) \times \mathscr{A} \to GL_2(\mathbb{R})$ the natural projection map. As ι is a homeomorphism and p_1 is open, ξ is an open map. □

Let now Ω be a G-orbit in $\mathscr{R}(\mathfrak{l}, \mathfrak{g}, \mathfrak{h})$. Then $\xi(\Omega)$ is a G-orbit in $\mathscr{R}(\bar{\mathfrak{l}}, \bar{\mathfrak{g}}, \bar{\mathfrak{h}})$, which is not open by Lemma 5.3.8, nor is Ω by Lemma 5.3.12 as was to be shown. This achieves the proof in this case.

Case where rank(Γ) = 1

We fix as above $\mathfrak{k}$ an ideal of $\mathfrak{g}$ included in $\mathfrak{g}_0$. The restriction $\xi' := \xi_{|\mathscr{R}(\mathfrak{l},\mathfrak{g},\mathfrak{h})}$ of ξ to $\mathscr{R}(\mathfrak{l}, \mathfrak{g}, \mathfrak{h})$ is also open and onto $\mathscr{R}(\bar{\mathfrak{l}}, \bar{\mathfrak{g}}, \bar{\mathfrak{h}})$ as being a natural projection. It is immediate that in case where $\mathfrak{h}$ is of codimension two. Lemma 5.3.8 gives us that the local rigidity fails at the level of $\mathscr{R}(\bar{\mathfrak{l}}, \bar{\mathfrak{g}}, \bar{\mathfrak{h}})$, upon the consideration of the case $\mathfrak{k} = \mathfrak{g}_0$, which is enough to conclude.

Suppose now that $\mathfrak{h}$ is of codimension one. In this case, we shall state the local rigidity property separately according to the dimension of the Lie algebra $\mathfrak{g}_0$. In the case where $\dim \mathfrak{g}_0 = 1$, $\mathfrak{g}$ is the Lie algebra spanned by the vectors A, X, Y such that $Y \in \mathfrak{g}_0$ and satisfying the brackets relations

$$[A, X] = X + \alpha Y, \quad [A, Y] = \beta Y$$

for some $\alpha, \beta \in \mathbb{R}$. Thanks to Jacobi's identity, one easily gets $[X, Y] = 0$. As $\mathscr{R}(\mathfrak{l}, \mathfrak{g}, \mathfrak{h}) \simeq \mathbb{R}^\times X + \mathbb{R}Y$ by Proposition 5.3.6. Then the one parameter subgroups $\exp \mathbb{R}X$ and $\exp \mathbb{R}Y$ clearly belong to the stabilizer of any element of $\mathscr{R}(\mathfrak{l}, \mathfrak{g}, \mathfrak{h})$, which in turn means that the orbit is of dimension one. So we are done in this case.

We now pay attention to the case where $\dim \mathfrak{g}_0 = 2$. Then $\mathfrak{h}$ is the Lie algebra of dimension three, which admits a basis $\mathscr{B} = \{A, Y, T\}$ with $T, Y \in \mathfrak{g}_0$, $\mathfrak{g} = \mathbb{R}X \oplus \mathfrak{h}$ and $[A, X] = X \bmod \mathfrak{g}_0$. It comes out on the other hand that the parameter space $\mathscr{R}(\mathfrak{l}, \mathfrak{g}, \mathfrak{h})$ is a smooth manifold of dimension three and is homeomorphic to $\mathbb{R}^\times X + \mathbb{R}Y + \mathbb{R}T$ by Proposition 5.3.6.

Assume first that the nilpotent Lie algebra $\mathfrak{g}' = [\mathfrak{g}, \mathfrak{g}]$ is of dimension three. Then $\mathfrak{g}'$ is either Abelian or isomorphic to the three-dimensional Heisenberg algebra. This means that for all $v \in \mathfrak{g}'$, $\dim G' \cdot v \leq 1$ where $G' = \exp(\mathfrak{g}')$. But $G = \exp(\mathbb{R}A)G'$ and $\mathscr{R}(\mathfrak{l}, \mathfrak{g}, \mathfrak{h}) \subset \mathfrak{g}'$, which is enough to conclude.

Assume now that $\dim \mathfrak{g}' \leq 2$. If the vectors $[Y, T]$, $[X, T]$, $[X, Y]$, $[A, Y]$ and $[A, T]$ are trivial, then obviously $\dim G \cdot X_0 \leq 2$ for any $X_0 \in \mathscr{R}(\mathfrak{l}, \mathfrak{g}, \mathfrak{h})$. Otherwise, we can and do assume that these vectors generate a one-dimensional subalgebra of $\mathfrak{g}_0$ and $\mathbb{R}u_0$. For $X_0 \in \mathscr{R}(\mathfrak{l}, \mathfrak{g}, \mathfrak{h})$, there exists a linear form $f_{X_0} \in \mathfrak{g}^*$, such that for any for $Z =$

$aA + xX + yY + tT \in \mathfrak{g}$, we have

$$[Z, X_0] = a[A, X_0] + f_{X_0}(Z)u_0.$$

The Lie algebra of the stabilizer $G(X_0)$ reads

$$\mathfrak{g}(X_0) = \{Z \in \mathfrak{g}; a = 0 \text{ and } f_{X_0}(Z) = 0\}.$$

Then $\dim \mathfrak{g}(X_0) \geq 2$, or equivalently $\dim G \cdot X_0 \leq 2$, which achieves the proof in this case.

We finally tackle the case where $\dim \mathfrak{g}_0 > 2$. There exists therefore an ideal $\mathfrak{k} \subset \mathfrak{g}_0$ of $\mathfrak{g}$ such that $\dim \mathfrak{g}_0/\mathfrak{k} = 1$ or 2. Passing to the quotient with $\mathfrak{k}$, we are conclusively led to one of the situations above where the local rigidity fails to hold. To complete the proof, we point out that the restriction map is also open and onto in this case. □

Remark 5.3.13. When we remove the assumption on the maximal subgroup H to be nonnormal in G, we get the following.

Corollary 5.3.14. *Let G be an exponential Lie group, H a connected maximal subgroup of G and Γ a discontinuous group for G/H. If the parameter space $\mathscr{R}(\Gamma, G, H)$ admits a locally rigid homomorphism, then G is isomorphic to the group* $\mathrm{Aff}(\mathbb{R})$.

Proof. If H is normal in G, then Γ stands to be Abelian as in the proof of Lemma 5.3.11. Looking at its syndetic hull $\mathfrak{l}$, the parameter space $\mathscr{R}(\mathfrak{l}, \mathfrak{g}, \mathfrak{h})$ turns out to be semialgebraic and open in $\mathrm{Hom}(\mathfrak{l}, \mathfrak{g}) \simeq \mathfrak{g}$. No G-orbit in $\mathscr{R}(\mathfrak{l}, \mathfrak{g}, \mathfrak{h})$ is therefore open by a reason of dimensions. This means that H is not normal in G and the result follows by Theorem 5.3.9. □

5.4 Criteria for local rigidity

5.4.1 Necessary condition for local rigidity using the automorphism group $\mathrm{Aut}(\mathfrak{l})$

Let $G_{\varphi(\mathfrak{l})}$ be the stabilizer in G of $\varphi(\mathfrak{l})$ and ρ_φ the homomorphism defined as

$$\rho_\varphi : G_{\varphi(\mathfrak{l})} \longrightarrow \mathrm{Aut}(\mathfrak{l}), \quad g \longmapsto \psi \circ \mathrm{Ad}_g \circ \varphi, \tag{5.15}$$

where ψ is the inverse of the isomorphism $\mathfrak{l} \overset{\varphi}{\to} \varphi(\mathfrak{l})$.

Let $\mathrm{Aut}^\circ(\mathfrak{l})$ be the connected component of the identity of $\mathrm{Aut}(\mathfrak{l})$. Our first result on rigidity is the following necessary condition for a locally rigid homomorphism.

Theorem 5.4.1. *Let G be an exponential Lie group, H a connected Lie subgroup of G and Γ a discontinuous subgroup for G/H. Let $\mathfrak{l}$ be the Lie algebra of the syndetic hull of Γ. If $\varphi \in \mathscr{R}(\mathfrak{l}, \mathfrak{g}, \mathfrak{h})$ is locally rigid, then* $\mathrm{Aut}^\circ(\mathfrak{l})$ *is contained in the range of the homomorphism ρ_φ defined in* (5.15).

Proof. Note first that as a direct aftermath of Theorem 3.2.1, the group $\mathrm{Aut}(\mathfrak{l})$ acts continuously on the parameter space via the law

$$a \cdot \varphi = \varphi \circ a^{-1}, \quad a \in \mathrm{Aut}(\mathfrak{l}).$$

This action commutes with the action of G, thus $\mathrm{Aut}(\mathfrak{l})$ acts on the deformation space and we can state the following result.

Lemma 5.4.2. *The action of* $\mathrm{Aut}(\mathfrak{l})$ *on* $\mathscr{T}(\mathfrak{l}, \mathfrak{g}, \mathfrak{h})$, *given by* $a \cdot [\varphi] = [\varphi \circ a^{-1}]$, *is well-defined and continuous. In particular, for any* $[\varphi] \in \mathscr{T}(\mathfrak{l}, \mathfrak{g}, \mathfrak{h})$, *the map* $i_{[\varphi]} : \mathrm{Aut}(\mathfrak{l}) \to \mathscr{O}_{[\varphi]}$, $a \mapsto [\varphi \circ a^{-1}]$ *is continuous, where* $\mathscr{O}_{[\varphi]}$ *is the orbit of* $[\varphi]$ *under the action of* $\mathrm{Aut}(\mathfrak{l})$.

Proof. The fact that the action of $\mathrm{Aut}(\mathfrak{l})$ is well-defined is trivial. We now prove that the action is continuous. Indeed, remark first that the action of $\mathrm{Aut}(\mathfrak{l})$ on $\mathscr{R}(\mathfrak{l}, \mathfrak{g}, \mathfrak{h})$ is continuous. On the other hand, the quotient map $\mathscr{R}(\mathfrak{l}, \mathfrak{g}, \mathfrak{h}) \to \mathscr{T}(\mathfrak{l}, \mathfrak{g}, \mathfrak{h})$ is an open continuous map. As the following diagram

$$\begin{array}{ccc}
\mathrm{Aut}(\mathfrak{l}) \times \mathscr{R}(\mathfrak{l}, \mathfrak{g}, \mathfrak{h}) & \xrightarrow{a} & \mathscr{R}(\mathfrak{l}, \mathfrak{g}, \mathfrak{h}) \\
\pi \downarrow & & \downarrow \pi \\
\mathrm{Aut}(\mathfrak{l}) \times \mathscr{T}(\mathfrak{l}, \mathfrak{g}, \mathfrak{h}) & \xrightarrow{a} & \mathscr{T}(\mathfrak{l}, \mathfrak{g}, \mathfrak{h})
\end{array}$$

commutes, we are done. □

Back now to the proof of the theorem. Let φ be locally rigid homomorphism then the class $[\varphi]$ is an open point in $\mathscr{T}(\mathfrak{l}, \mathfrak{g}, \mathfrak{h})$. As a direct matter of the continuity of the action of $\mathrm{Aut}(\mathfrak{l})$ on $\mathscr{T}(\mathfrak{l}, \mathfrak{g}, \mathfrak{h})$, we observe that the orbit $\mathscr{O}_{[\varphi]}$ is a discrete topological space. By the continuity of the map $i_{[\varphi]}$, we conclude that $\mathrm{Aut}^{\circ}(\mathfrak{l})$ fixes $[\varphi]$. Let $a \in \mathrm{Aut}^{\circ}(\mathfrak{l})$ then $[\varphi \circ a^{-1}] = [\varphi]$, which is equivalent to the existence of $g \in G$ such that $\mathrm{Ad}_g \circ \varphi = \varphi \circ a^{-1}$, in particular, $g \in G_{\varphi(\mathfrak{l})}$ and $a^{-1} = \psi \circ \mathrm{Ad}_g \circ \varphi \in \mathrm{Im}(\rho_\varphi)$. □

We now introduce the following definitions.

Definition 5.4.3. Let $\mathscr{D}(\mathfrak{l})$ be the derivations algebra of $\mathfrak{l}$. It is the Lie algebra of $\mathrm{Aut}(\mathfrak{l})$. A solvable (resp., a completely solvable, a nilpotent) Lie algebra for which $\mathscr{D}(\mathfrak{l})$ is solvable (resp., a completely solvable, a nilpotent) Lie algebra, is said to be *characteristically* solvable (resp., completely solvable, nilpotent) Lie algebra.

As a direct consequence of Theorem 5.4.1, we get the following.

Theorem 5.4.4. *The local rigidity fails to hold, in the context of the three following situations:*

(1) $\mathfrak{l}$ *is not characteristically solvable Lie algebra.*
(2) $\mathfrak{g}$ *is completely solvable and* $\mathfrak{l}$ *is not characteristically completely solvable Lie algebra.*
(3) $\mathfrak{g}$ *is nilpotent and* $\mathfrak{l}$ *is not characteristically nilpotent Lie algebra.*

Proof. For any $\varphi \in \mathscr{R}(\mathfrak{l}, \mathfrak{g}, \mathfrak{h})$, the subgroup $G_{\varphi(\mathfrak{l})}$ is a solvable Lie group and $\mathrm{Im}(\rho_\varphi)$ is a quotient of $G_{\varphi(\mathfrak{l})}$, then any subgroup of $\mathrm{Im}(\rho_\varphi)$ is also solvable. In particular, if there is a locally rigid homomorphism φ say, by Theorem 5.4.1, we conclude that $\mathrm{Aut}^\circ(\mathfrak{l})$ is a subgroup of $\mathrm{Im}(\rho_\varphi)$, and thus solvable. Now if G is completely solvable, the subgroups $G_{\varphi(\mathfrak{l})}$ and $\mathrm{Ker}(\rho_\varphi)$ are connected closed subgroups of G and the quotient group $G_{\varphi(\mathfrak{l})}/\ker(\rho_\varphi) \cong \mathrm{Im}(\rho_\varphi)$ is a completely solvable Lie group. If φ is locally rigid, then the same argument induces that $\mathrm{Aut}^\circ(\mathfrak{l})$ is completely solvable. When $\mathfrak{g}$ is nilpotent, the same arguments work out. □

5.4.2 Case of graded Lie subalgebras

We will derive some direct consequences from Theorem 5.4.4, showing that the rigidity property fails to hold. This gives therefore a positive answer to Conjecture 5.1.1. Our first result is a direct consequence of the Theorem 5.4.4.

Corollary 5.4.5. *Let $\mathfrak{l} = \mathfrak{g}$ or $\mathfrak{g}$ is nilpotent of dimension less or than equal to* 7, *then Conjecture* 5.1.1 *holds.*

Proof. If $\mathfrak{l} = \mathfrak{g}$, then the deformation space is homeomorphic to $\mathrm{Aut}(\mathfrak{g})/\mathrm{Ad}_G$, which is smooth manifold of dimension greater than one. The deformation space contains therefore no open points. If $\dim(\mathfrak{g}) \leq 7$ and $\mathfrak{l} \neq \mathfrak{g}$, then $\dim(\mathfrak{l}) \leq 6$ and it is well known that there is no characteristically nilpotent Lie algebra $\mathfrak{l}$ for which $\dim(\mathfrak{l}) < 7$ (cf. [109]). □

Definition 5.4.6. Recall that a graded Lie algebra is any Lie algebra $\mathfrak{l}$ with a decomposition

$$\mathfrak{l} = \bigoplus_{d \in \mathbb{Z}} \mathfrak{l}_d,$$

where $\mathfrak{l}_d, d \in \mathbb{Z}$ are some subspaces of $\mathfrak{l}$, such that

$$[\mathfrak{l}_i, \mathfrak{l}_j] \subset \mathfrak{l}_{i+j}, \quad i, j \in \mathbb{Z}.$$

For any $t \in \mathbb{R}_+^*$, the dilation λ_t defined on $\mathfrak{l}$ by $\lambda_t(v) = t^d v$ for all $v \in \mathfrak{l}_d$ and extended on $\mathfrak{l}$ by linearity, is a nonnilpotent automorphism of $\mathfrak{l}$ and belongs to $\mathrm{Aut}^\circ(\mathfrak{l})$. Then $\mathfrak{l}$ is not characteristically nilpotent.

Corollary 5.4.7. *Assume that $\mathfrak{g}$ is nilpotent and $\mathfrak{l}$ is a graded Lie algebra, then Conjecture* 5.1.1 *is positively answered.*

Remark 5.4.8.

(1) Note that if $\mathfrak{g}$ is a 2-step or a threadlike Lie algebra, then any subalgebra of $\mathfrak{g}$ is either Abelian or a 2-step or a threadlike Lie algebra and, therefore, a graded Lie

algebra. It comes out that the nonlocal rigidity theorems obtained above, are direct consequences of the last corollary.

(2) The necessary condition for the rigidity in Theorem 5.4.4 involves only the structure of $\mathfrak{g}$ and $\mathfrak{l}$.
(3) The following examples reveal that the additional assumption on $\mathfrak{l}$ to be characteristically nilpotent might be irrelevant.

Example 5.4.9. Let $\mathfrak{l}$ be a characteristically nilpotent Lie algebra and $G = \mathbb{R}^n \times L$. Then $[\mathfrak{g}, \mathfrak{g}] = [\mathfrak{l}, \mathfrak{l}]$. In this case, we know already that a homomorphism $\varphi \in \mathscr{R}(\mathfrak{l}, \mathfrak{g}, \mathfrak{h})$ can be locally rigid only if

$$\dim \varphi(\mathfrak{l})^{\perp} + \dim(\mathrm{Aut}(\mathfrak{l})) = \dim(\mathfrak{g}).$$

But

$$\dim \varphi(\mathfrak{l})^{\perp} \geq \dim(\mathfrak{z}(\varphi(\mathfrak{l})) + n \geq n + 1,$$

and

$$\dim(\mathrm{Aut}(\mathfrak{l})) \geq \dim(\mathfrak{l}).$$

So, Conjecture 5.1.1 has also a positive answer in this setup.

Example 5.4.10. Suppose that $\mathfrak{g} = \mathfrak{h} \oplus \mathfrak{l}$ as a sum of two ideals of $\mathfrak{g}$. Then it is easy to see that the parameter space is identified to the product $\mathrm{Hom}(\mathfrak{l}, \mathfrak{h}) \times \mathrm{Aut}(\mathfrak{l})$ and the deformation space can be written as

$$\mathscr{T}(\mathfrak{l}, \mathfrak{g}, \mathfrak{h}) = \mathrm{Hom}(\mathfrak{l}, \mathfrak{h}) / \mathrm{Ad}_H \times \mathrm{Aut}(\mathfrak{l}) / \mathrm{Ad}_L .$$

Then if $\mathscr{T}(\mathfrak{l}, \mathfrak{g}, \mathfrak{h})$ contains an open point, its image by the second projection must be open, which is impossible. So, a positive answer to question 5.1.1 holds also in this context. Remark finally that the condition on $\mathfrak{l}$ to be characteristically nilpotent is not useful in this case.

Let us derive first some immediate consequences from Theorem 5.4.4. Remark that if $\mathfrak{l}$ is a Heisenberg algebra, then the automorphisms group of $\mathfrak{l}$ contains a symplectic subgroup, which is simple. Then we get the following result.

Corollary 5.4.11. *If $\mathfrak{g}$ is completely solvable and $\mathfrak{l}$ is a Heisenberg algebra or more generally $\mathfrak{l}$ is a direct sum of two ideals one of them is a Heisenberg algebra, then the local rigidity fails to hold.*

Proof. Indeed, let us write $\mathfrak{l} = \mathfrak{k} \oplus \mathfrak{a}$, where $\mathfrak{k}$ is a Heisenberg algebra and $\mathfrak{a}$ is an ideal of $\mathfrak{g}$. Then $\mathrm{Aut}(\mathfrak{k})$ is a subgroup of $\mathrm{Aut}(\mathfrak{l})$, which therefore contains a symplectic group that is simple. □

We now tackle the general situation of exponential Lie groups when Γ is Abelian. In this case, we can already note the following.

Corollary 5.4.12. *If $\mathfrak{g}$ is exponential, $\mathfrak{l}$ is Abelian and* $\dim(\mathfrak{l}) \geq 2$, *then the local rigidity fails to hold.*

Proof. If $\mathfrak{l}$ is Abelian and $\dim(\mathfrak{l}) \geq 2$, then $\mathrm{Aut}(\mathfrak{l}) = GL_k(\mathbb{R})$, which is a reductive group. □

5.4.3 Abelian discontinuous groups

We now prove a main result, which deals with the local rigidity property in the case where Γ is Abelian. We have the following.

Theorem 5.4.13. *Let G be an exponential Lie group, H a connected Lie subgroup of G and Γ a Abelian discontinuous subgroup for G/H. Then $\mathscr{R}(\Gamma, G, H)$ admits a locally rigid homomorphism if and only if G is isomorphic to* $\mathrm{Aff}(\mathbb{R})$ *and H is maximal and nonnormal in G.*

Proof. We proceed through different steps. When the rank of Γ is larger than 1, then Corollary 5.4.12 gives us the answer. We then assume in the rest of the proof that Γ is of rank one. In this case, there exists $X_0 \in \mathfrak{g}$ such that $\Gamma = \exp(\mathbb{Z}X_0)$ and $\mathfrak{l} = \mathbb{R}X_0$. The set $\mathrm{Hom}(\mathfrak{l}, \mathfrak{g})$ can therefore be identified to $\mathfrak{g}$ via the canonical map $\varphi \mapsto \varphi(X_0)$. Assume in a first time that H is contained in a connected normal subgroup of G. We will be proving some intermediary results, and the first is the following.

Lemma 5.4.14. *If H is a normal subgroup of G, then any homomorphism in $\mathscr{R}(\Gamma, G, H)$ is continuously deformable.*

Proof. If H is normal, then the parameter space reads

$$\mathscr{R}(\Gamma, G, H) = \{\varphi \in \mathfrak{g},\ \varphi \notin \mathfrak{h}\},$$

which is an open Zariski dense subset of $\mathfrak{g}$. Let $\varphi \in \mathscr{R}(\Gamma, G, H)$ be a locally rigid homomorphism, then the orbit $\mathrm{Ad}_G \cdot \varphi$ is open in $\mathscr{R}(\Gamma, G, H)$ and, therefore, in $\mathfrak{g}$. This means that the stabilizer of φ in $\mathfrak{g}$ is trivial, which is impossible because it contains the one-dimensional subgroup $\exp(\varphi(\mathfrak{l}))$. □

Assume now that a normal connected subgroup contains H, then we have the following.

Lemma 5.4.15. *If H is contained in a normal connected Lie subgroup of G, then the local rigidity fails to hold.*

Proof. Let H_1 be a connected normal subgroup containing H and let φ be a locally rigid homomorphism in $\mathscr{R}(\Gamma, G, H)$. As $\mathscr{R}(\Gamma, G, H_1) \subset \mathscr{R}(\Gamma, G, H)$ and $\mathscr{R}(\Gamma, G, H_1)$ is dense in

$\mathfrak{g}$, then $\mathscr{R}(\Gamma, G, H_1)$ turns out to be a dense subset of $\mathscr{R}(\Gamma, G, H)$. There exists therefore a sequence $(\varphi_n)_n$ in $\mathscr{R}(\Gamma, G, H_1)$, which converges to φ. As the orbit of φ is open in $\mathscr{R}(\Gamma, G, H)$, the element φ_n belongs to the orbit of φ for n sufficiently large. It comes out that φ_n is also locally rigid. This means that the orbit $\mathrm{Ad}_G\, \varphi_n$ is open in $\mathscr{R}(\Gamma, G, H)$ and contained in $\mathscr{R}(\Gamma, G, H_1)$. Then $\mathrm{Ad}_G\, \varphi_n$ is open in $\mathscr{R}(\Gamma, G, H_1)$, which leads to a contradiction thanks to Lemma 5.4.14 as H_1 is normal in G. □

As an immediate consequence, we have the following.

Lemma 5.4.16. *Suppose that $\varphi \in \mathscr{R}(\Gamma, G, H)$ is locally rigid, then $\varphi(\mathfrak{l}) \subset [\mathfrak{g}, \mathfrak{g}]$.*

Proof. Consider the Lie subgroup $G_{\varphi(\mathfrak{l})}$ and denote by $\mathfrak{g}_\varphi$ its Lie algebra. From Proposition 5.4.1, for all $X \in \mathfrak{g}_\varphi$ and Y in $\varphi(\mathfrak{l})$ we have

$$\mathrm{Ad}_{\exp(tX)}(Y) = f(t)Y,$$

for some analytic function f defined on $\mathbb{R}$. Then we get

$$[X, Y] = \frac{d}{dt} \mathrm{Ad}_{\exp(tX)}(Y)_{|t=0} = f'(0)Y.$$

To get that $Y \in [\mathfrak{g}, \mathfrak{g}]$, it is sufficient to see that there exists $X \in \mathfrak{g}_\varphi$ such that $[X, Y] = \alpha Y$ for some nonzero $\alpha \in \mathbb{R}$. Indeed, if for all X in $\mathfrak{g}_\varphi$ the bracket $[X, Y] = 0$ then the kernel of the homomorphism ρ_φ contains $G^\circ_{\varphi(\mathfrak{l})}$, the connected component of the identity of $G_{\varphi(\mathfrak{l})}$. This means that the quotient group $G_{\varphi(\mathfrak{l})}/G^\circ_{\varphi(\mathfrak{l})}$ is a discrete group and then $\mathrm{Aut}^\circ(\mathfrak{l})$ is trivial. But this is impossible as $\mathrm{Aut}^\circ(\mathfrak{l}) = \mathbb{R}^\times_+ = \{x \in \mathbb{R} : x > 0\}$. □

Assume henceforth that H is not contained in any normal connected subgroup of G. Let H_1 be a maximal connected nonnormal subgroup containing H and denote by $\mathfrak{h}_1$ its corresponding Lie subalgebra. Then by Theorem 5.3.3, we have the following:

(1) H_1 is of codimension 1 in G, and there exist A and $X \in \mathfrak{g}$ such that

$$\mathfrak{g} = \mathbb{R}X \oplus \mathbb{R}A \oplus \mathfrak{g}_0 \quad \text{and} \quad \mathfrak{h}_1 = \mathbb{R}A \oplus \mathfrak{g}_0,$$

where $\mathfrak{g}_0$ is an ideal of $\mathfrak{g}$ and $[A, X] = X \,\mathrm{mod}(\mathfrak{g}_0)$.

(2) H_1 is of codimension 2 in G, and there exist $A, X, Y \in \mathfrak{g}$ such that

$$\mathfrak{g} = \mathbb{R}X \oplus \mathbb{R}Y \oplus \mathbb{R}A \oplus \mathfrak{g}_0 \quad \text{and} \quad \mathfrak{h}_1 = \mathfrak{g}_0 \oplus \mathbb{R}A,$$

where $\mathfrak{g}_0$ is an ideal of $\mathfrak{g}$ and $[A, X + iY] = (\alpha + i)(X + iY)\,\mathrm{mod}(\mathfrak{g}_0)$ for some nonzero real number α.

The following proposition stems directly from Proposition 5.3.6.

Proposition 5.4.17. *Let H_1 be a connected maximal nonnormal Lie subgroup of G. Then we have:*

(1) *If the codimension of* H_1 *is equal to* 1, *then* $\mathscr{R}(\Gamma, G, H_1) = \mathfrak{g}_0 \oplus \mathbb{R}^* X$.
(2) *If the codimension of* H_1 *is equal to* 2, *then* $\mathscr{R}(\Gamma, G, H_1) = \mathfrak{g}_0 \oplus (\mathbb{R}X \oplus \mathbb{R}Y \setminus \{(0,0)\})$.

Let V denote the subalgebra $\mathbb{R}X \oplus \mathfrak{g}_0$ of $\mathfrak{g}$ if H_1 is of codimension 1 and $\mathbb{R}X \oplus \mathbb{R}Y \oplus \mathfrak{g}_0$ otherwise. Then it is clear that $[\mathfrak{g}, \mathfrak{g}] \subset V$ and $\mathscr{R}(\Gamma, G, H_1)$ is an open dense subset of V. Let now φ be a locally rigid homomorphism. By Lemma 5.4.16, $\varphi \in [\mathfrak{g}, \mathfrak{g}]$. There exists then a sequence $(\varphi_n)_n \subset \mathscr{R}(\Gamma, G, H_1)$, which converges to φ. In particular, for n sufficiently large, φ_n belongs to the orbit $\mathrm{Ad}_G\, \varphi$. It follows that the orbit $\mathrm{Ad}_G\, \varphi$ is open in $\mathscr{R}(\Gamma, G, H_1)$, which means that φ is locally rigid in $\mathscr{R}(\Gamma, G, H_1)$. Using Theorem 5.3.9, we can conclude that G is isomorphic to $\mathrm{Aff}(\mathbb{R})$. This achieves the proof of the theorem. □

As an immediate consequence, we get the following.

Corollary 5.4.18. *Let* G *be an exponential Lie group,* H *a connected Lie subgroup of* G *and* Γ *an Abelian discontinuous group for* G/H. *Then the local rigidity property fails to hold if and only if* $\dim G \neq 2$ *or otherwise* H *is normal in* G.

5.4.4 Removing the assumption on Γ to admit a syndetic hull

The result of this section is also a necessary condition for the local rigidity for a general Lie group, which is not necessarily exponential. Let G be a Lie group, H closed subgroup of G and Γ a discontinuous group for G/H. Let

$$N(H) := \{\sigma \in \mathrm{Aut}(G), \sigma(H) \subset H\}$$

be the stabilizer of H, which is a closed subgroup of $\mathrm{Aut}(G)$ and denote by $N^\circ(H)$ the connected component of the identity element of $N(H)$. The automorphism group $\mathrm{Aut}(G)$ acts naturally on $\mathrm{Hom}(\Gamma, G)$ by composition on the left. For $\varphi \in \mathrm{Hom}(\Gamma, G)$, let

$$S_\varphi := \{\sigma \in \mathrm{Aut}(G), \sigma \circ \varphi = \varphi\}$$

denote the stabilizer of φ. Recall that the group of inner automorphisms $I(G)$ forms a normal subgroup of $\mathrm{Aut}(G)$. Then the set

$$I_\varphi(G) = S_\varphi I(G)$$

is a subgroup of $\mathrm{Aut}(G)$ and we have the following.

Theorem 5.4.19. *Let* G *be a Lie group,* H *a connected Lie subgroup of* G *and* Γ *a discontinuous subgroup for* G/H. *If an infinitesimal deformation* $\varphi \in \mathscr{R}(\Gamma, G, H)$ *is locally rigid, then* $N^\circ(H)$ *is contained in* $I_\varphi(G)$.

Proof. We first need to prove the following lemmas.

Lemma 5.4.20. *Let $\sigma \in N(H)$ and $\varphi \in \mathrm{Hom}(\Gamma, G)$. Then:*
(ı) *$\varphi(\Gamma)$ acts properly on G/H if and only if $\sigma \circ \varphi(\Gamma)$ acts properly on G/H.*
(ıı) *$\varphi(\Gamma)$ acts freely on G/H if and only if $\sigma \circ \varphi(\Gamma)$ acts freely on G/H.*

In particular, the action of the subgroup $N(H)$ on $\mathrm{Hom}(\Gamma, G)$ leaves the parameter space $\mathscr{R}(\Gamma, G, H)$ stable.

Proof. Let S be a compact set in G and $\gamma \in \varphi(\Gamma)$, we have

$$\gamma SH \cap SH \neq \emptyset \quad \text{if and only if} \quad \sigma(\gamma)\sigma(S)H \cap \sigma(S)H \neq \emptyset.$$

With the above in mind, we also get

$$\sigma \circ \varphi(\Gamma)_{\sigma(S)} = \sigma(\varphi(\Gamma)_S),$$

which gives (ı). For the second assumption, it is sufficient to see that for $g \in G$ we have

$$\varphi(\Gamma) \cap gHg^{-1} = \{e\} \quad \text{if and only if} \quad \sigma \circ \varphi(\Gamma) \cap \sigma(g)H\sigma(g)^{-1} = \{e\}.$$ □

The next lemma describes the action of the connected subgroup $N^\circ(H)$ on the deformation space $\mathscr{T}(\Gamma, G, H)$.

Lemma 5.4.21. *The connected subgroup $N^\circ(H)$ acts continuously on $\mathscr{T}(\Gamma, G, H)$ through the law: $\sigma \cdot [\varphi] = [\sigma \circ \varphi]$. In particular, for any $[\varphi] \in \mathscr{T}(\Gamma, G, H)$, the map $j_{[\varphi]} : N^\circ(H) \to \mathscr{F}_{[\varphi]}$, $\sigma \mapsto [\sigma \circ \varphi]$ is continuous, where $\mathscr{F}_{[\varphi]}$ is the orbit of $[\varphi]$ under the action of $N^\circ(H)$.*

Proof. As a direct consequence of the normality of $I(G)$, the action is well-defined. The subgroup $N^\circ(H)$ acts continuously on $\mathscr{R}(\Gamma, G, H)$ and the quotient map $\mathscr{R}(\Gamma, G, H) \to \mathscr{T}(\Gamma, G, H)$ is open and continuous. Then obviously the action of $N^\circ(H)$ on the deformation space $\mathscr{T}(\Gamma, G, H)$ is continuous. □

Let $[\varphi]$ be a point in $\mathscr{T}(\Gamma, G, H)$. From Lemma 5.4.21, the orbit $\mathscr{F}_{[\varphi]}$ is connected. Assume that φ is a locally rigid homomorphism, or equivalently $[\varphi]$ is an open point in $\mathscr{T}(\Gamma, G, H)$. Then the orbit $\mathscr{F}_{[\varphi]} = \{[\varphi]\}$. It follows that for any $\sigma \in N^\circ(H)$ there exists $g \in G$ such that

$$\sigma \circ \varphi = \mathscr{T}_g \circ \varphi,$$

where $\mathscr{T}_g \in I(G)$ is the conjugation map by g. This implies that $\mathscr{T}_{g^{-1}} \circ \sigma \in S_\varphi$ and, therefore, $\sigma \in I_\varphi(G)$. This completes the proof of Theorem 5.4.19. □

5.4.5 Exponential Lie algebras of type $\mathscr{T}$

We now assume that G is an exponential Lie group and H a connected Lie subgroup of G. In this case, the map

$$\begin{array}{rccl} E: & \mathrm{Aut}(G) & \longrightarrow & \mathrm{Aut}(\mathfrak{g}) \\ & \sigma & \longmapsto & \mathrm{Log} \circ \sigma \circ \exp \end{array} \tag{5.16}$$

is a C^∞ isomorphism, with $E(I(G)) = \mathrm{Ad}_G$. Let $\mathfrak{h} = \mathrm{Log}(H)$ be the Lie algebra associated to H and

$$\mathfrak{n}(\mathfrak{h}) = \{\psi \in \mathrm{Aut}(\mathfrak{g}) \text{ and } \psi(\mathfrak{h}) \subset \mathfrak{h}\}$$

the stabilizer of the Lie subalgebra $\mathfrak{h}$ in $\mathrm{Aut}(\mathfrak{g})$, then $E(N(H)) = \mathfrak{n}(\mathfrak{h})$. Similarly, for any $\varphi \in \mathrm{Hom}(\Gamma, G)$ let

$$\mathfrak{s}_\varphi = \{\psi \in \mathrm{Aut}(\mathfrak{g}), \psi(X) = X \text{ for all } X \in \mathrm{Log} \circ \varphi(\Gamma)\},$$

then $E(S_\varphi) = \mathfrak{s}_\varphi$.

Definition 5.4.22. A solvable Lie algebra is called a type $\mathscr{T}$ if for all $X \in \mathfrak{g}$ we have $\mathrm{sp}(\mathrm{ad}_X) \cap \mathbb{R}^\times = \emptyset$, where $\mathrm{sp}(\mathrm{ad}_X)$ denotes the set of all eigenvalues of the endomorphism ad_X. Here, $\mathbb{R}^\times = \{x \in \mathbb{R} : x \neq 0\}$.

Definition 5.4.23. Following Definition 5.4.6, a graded Lie algebra $\mathfrak{g} = \bigoplus_{d\in\mathbb{Z}} \mathfrak{g}_d$ is called positive if $\mathfrak{g}_d = \{0\}$ for any $d < 0$. A Lie subalgebra $\mathfrak{h}$ of a graded Lie algebra $\mathfrak{g}$ is called subgraded algebra if $\mathfrak{h} = \bigoplus_{d\in\mathbb{Z}} (\mathfrak{h} \cap \mathfrak{g}_d)$.

Theorem 5.4.24. *Let $\mathfrak{g}$ be a positive graded exponential Lie algebra of type $\mathscr{T}$, G the Lie group associated to $\mathfrak{g}$ and $H = \exp(\mathfrak{h})$ a connected subgroup of G. If the subalgebra $\mathfrak{g}_0$ is Abelian and $\mathfrak{h}$ is a subgraded subalgebra of $\mathfrak{g}$, then for any discontinuous group for G/H, the local rigidity property fails to hold.*

Proof. By Lie's theorem, there exists a basis of the complexified Lie algebra $\mathfrak{g}_\mathbb{C} = \mathbb{C} \otimes_\mathbb{R} \mathfrak{g}$ and some linear forms $\alpha_1, \dots, \alpha_n$ on $\mathfrak{g}_\mathbb{C}$ such that the matrix of ad_X written in this basis is upper triangular with $\alpha_1(X), \dots, \alpha_n(X)$ on the diagonal. For any $X \in \mathfrak{g}$, let us write

$$\alpha_j(X) = \alpha_{j0}(X) + i\alpha_{j1}(X), \quad 1 \leq j \leq n.$$

Then clearly $\alpha_{j0}(X)$ and $\alpha_{j1}(X)$ are linear forms on $\mathfrak{g}$, and we have the following facts.

Lemma 5.4.25. *Assume that $\mathfrak{g}$ is exponential of type $\mathscr{T}$. Then for all $1 \leq j \leq n$, there exists a real number $a_j \neq 0$ such that $\alpha_j = a_j\alpha_{j1} + i\alpha_{j1}$.*

Proof. As $\mathfrak{g}$ is exponential, any root of $\mathfrak{g}$ is of the form $(1 + i\alpha)\lambda$, where α denotes a real number and λ a linear form on $\mathfrak{g}$. Now the fact that $\mathfrak{g}$ is of type $\mathscr{T}$ allows us to conclude.

Indeed, let $\alpha_j = (1 + i\alpha)\lambda_j$ for some $\alpha_j \in \mathfrak{g}^*$ If λ_j is trivial, we are done. Otherwise, $\alpha \neq 0$ and $\alpha_j = \frac{1}{\alpha}\alpha_{j1} + i\alpha_{j1}$. □

For $X \in \mathfrak{g}$, let $\mathrm{sp}(\mathrm{ad}_X) = \{\alpha_1(X), \dots, \alpha_n(X)\}$. Then

$$\mathrm{sp}(\mathrm{Ad}_{\exp(X)}) = \{e^{\alpha_1(X)}, \dots, e^{\alpha_n(X)}\}.$$

We denote by $\mathrm{sp}(\mathrm{Ad}_G)$ the subset of $\mathbb{C}$, obtained as the union of all the spectrum $\mathrm{sp}(\mathrm{Ad}_g)$, $g \in G$.

Lemma 5.4.26. *If $\mathfrak{g}$ is an exponential Lie algebra of type $\mathscr{T}$, then the set $\mathrm{sp}(\mathrm{Ad}_G) \cap \mathbb{R}$ is a countable set.*

Proof. Suppose that $e^{\alpha_j(X)} \in \mathbb{R}$. Then $\alpha_{j1}(X) \in 2\pi\mathbb{Z}$ and $e^{\alpha_j(X)} \in e^{2a_j\pi\mathbb{Z}}$. In particular,

$$\mathrm{sp}(\mathrm{Ad}_G) \subset \bigcup_{j=1}^{n} e^{2a_j\pi\mathbb{Z}}.$$ □

As $\mathfrak{g}$ is a positive graded algebra, any element $X \in \mathfrak{g}$ can be uniquely written as

$$X = \sum_{d \geq 0} X_d, \quad X_d \in \mathfrak{g}_d.$$

We have the following.

Lemma 5.4.27. *Let $X, Y \in \mathfrak{g}$ and $d_0 = \min\{d \geq 0,\ Y_d \neq 0\}$. Then we have*

$$\mathrm{Ad}_{\exp(X)}(Y) = \mathrm{Ad}_{\exp(X_0)}(Y_{d_0}) \bmod(\mathfrak{g}_{\geq d_0+1}),$$

where for $k \geq 0$, $\mathfrak{g}_{\geq k} = \bigoplus_{d \geq k} \mathfrak{g}_d$.

Proof. The subspace $\mathfrak{g}_{\geq k}$ is an ideal of $\mathfrak{g}$ for all k. Then

$$\mathrm{Ad}_{\exp(X)}(Y) = \mathrm{Ad}_{\exp(X)}(Y_{d_0}) \bmod(\mathfrak{g}_{\geq d_0+1}).$$

For all $n \geq 0$ and $X = X_0 + X'$ with $X' \in \mathfrak{g}_{\geq 1}$, we have

$$(\mathrm{ad}_X)^n(Y_{d_0}) = (\mathrm{ad}_{X_0})^n(Y_{d_0}) \bmod(\mathfrak{g}_{\geq d_0+1}).$$

Then the result follows from the equality $\mathrm{Ad}_{\exp(X)} = \exp(\mathrm{ad}_X)$. □

We now come back to the proof of Theorem 5.4.24. By hypothesis, for any $t > 0$ the dilation λ_t defined as previously by

$$\lambda_t(X) = t^d X \quad \text{for all } X \in \mathfrak{g}_d$$

is an automorphism of $\mathfrak{g}$ and $\lambda_t(\mathfrak{h}) = \mathfrak{h}$. More precisely λ_t is an element of the connected component $\mathfrak{n}^0(\mathfrak{h})$ of the identity of $\mathfrak{n}(\mathfrak{h})$. Let Γ be a discontinuous group for G/H and

$\varphi : \Gamma \to G$ a locally rigid homomorphism. By Theorem 5.4.19, $N^\circ(H)$ is included in $I_\varphi(G)$. Using the isomorphism E given in equation (5.16), we can see that $\mathfrak{n}^\circ(\mathfrak{h})$ is a subgroup of $E(I_\varphi(G)) = \mathrm{Ad}_G\,\mathfrak{s}_\varphi$. In particular, for all $t > 0$, there exist $X \in \mathfrak{g}$ and $\sigma \in \mathfrak{s}_\varphi$ such that

$$\lambda_t = \mathrm{Ad}_{\exp(X)} \circ \sigma.$$

By Theorem 5.4.13, we can and do assume that the subgroup Γ is not Abelian, which means that $\varphi(\Gamma) \cap [G, G] \neq \{e\}$. There exists therefore $Y \in [\mathfrak{g}, \mathfrak{g}]$ such that $Y \in \mathrm{Log}(\varphi(\Gamma))$. In this case, we obtain

$$\sigma(Y) = Y \quad \text{and} \quad \lambda_t(Y) = \mathrm{Ad}_{\exp(X)}(Y).$$

Then by Lemma 5.4.27, we can write

$$t^{d_0} Y_{d_0} = \mathrm{Ad}_{\exp(X_0)}(Y_{d_0}) \bmod (\mathfrak{g}_{\geq d_0+1}),$$

which gives the equality $t^{d_0} Y_{d_0} = \mathrm{Ad}_{\exp(X_0)}(Y_{d_0})$ and then $\mathbb{R}_+^* \subset \mathrm{sp}(\mathrm{Ad}_G)$. By Lemma 5.4.26, $\mathfrak{g}$ is not of type $\mathscr{T}$. This completes the proof of the theorem. □

Remark 5.4.28.

(1) The first example in Lemma 5.3.8 reveals that the rigidity holds everywhere despite the fact that Γ is Abelian. This shows that the result of Proposition 6.3.1 also fails in the context of exponential Lie groups.

(2) For semisimple case G/H, if we drop the assumption that $\Gamma \backslash G/H$ is compact in Theorem 5.3.1, then the feature changes. Namely, there are more examples where Γ is not locally rigid. For instance, we see that this is the case if $\mathrm{rank}_{\mathbb{R}} H > 1$ by the criterion of [94].

5.5 Local rigidity in the solvable case

The present section deals with the general context when G is a connected solvable Lie group and H a maximal nonnormal subgroup of G. We discuss an analogue of the Selberg–Weil–Kobayashi local rigidity theorem (cf. Theorem 5.3.9) in this setting. In contrast to the semisimple case, the G-action on G/H is not always effective, and thus the space of group theoretic deformations $\mathscr{T}(\Gamma, G, G/H)$ could be larger than geometric deformation spaces. We determine $\mathscr{T}(\Gamma, G, G/H)$ and also its quotient modulo uneffective parts when rank $\Gamma = 1$. Unlike the context of exponential solvable case, we prove the existence of formal colored discontinuous groups. That is, the parameter space admits a mixture of locally rigid and formally nonrigid deformations.

Therefore, our deformations of discontinuous groups may not give rise to deformations of actions of discontinuous groups. It should be noted that many of deformations studied here are such that nongeometric deformations of discontinuous groups, and thus we say them to be formal deformation. As in Chapter 2, we saw that any discontinuous group for the homogeneous space G/H appears to be Abelian and of rank ≤ 2 (cf. Theorem 2.4.6), but fails in general to admit a syndic hull. Consequently, the problems faced toward the study of the rigidity question are considerable.

Unlike the setting of exponential Lie groups, a new phenomenon shows up. That is, the parameter space $\mathscr{R}(\Gamma, G, H)$ may admit a mixture of a locally rigid and formally nonrigid homomorphisms (cf. Theorem 5.5.2), in which case the discrete subgroup are said to be a formally *colored* discontinuous group for G/H. We are then submitted to study three dichotomous situations. The first one consists in proving that the local rigidity holds only when G is isomorphic to the group Aff($\mathbb{R}$). The second says that no local rigidity phenomenon can occur. Our technique to tackle these cases consists in reducing the study of the problem to some lower-dimensional exponential Lie groups via an open descending map, where explicit computations are shown to be efficiently carried out. In the third situation, we explicitly build up an infinite family of solvable connected Lie groups G_n (where n is a positive integer) with a maximal nonnormal subgroup H; each of them admits a formally colored discontinuous group for G_n/H.

In the case where Γ is of rank two, we give an explicit description of the parameter and provide some examples. We deeply believe that the rigidity property fails in this case, as the studied examples reveal. In addition, we conclusively provide an explicit description of the deformation space in the case where G acts effectively on G/H (cf. Proposition 5.5.13).

5.5.1 The notion of colored discrete subgroups

Let G be a locally compact group and Γ a discrete subgroup of G. Recall first the set $\pitchfork_{gp}(\Gamma : G)$ as in Definition 2.1.1, consisting of all closed connected subgroups H for which $SHS^{-1} \cap \Gamma$ is compact for any compact set S in G.

Definition 5.5.1. Let Γ be a discrete subgroup of G.

(1) The subgroup Γ is said to be a colored subgroup of G if there exists $H \in \pitchfork_{gp}(\Gamma, G)$ such that $\mathscr{R}(\Gamma, G, H)$ contains at least a mixture of a locally rigid and a nonlocally rigid homomorphisms.

(2) For a given connected subgroup H of G, a colored discontinuous group for G/H is abbreviated to be a H-colored discrete subgroup of G.

(3) We shall adapt the terminology of formally colored (and formally H-colored) discrete subgroups.

5.5.2 The rank-one solvable case

The parameter space

As a straightforward consequence of Theorem 2.4.12 and Proposition 2.4.13, we get the following simpler description of the parameter space:

$$\mathscr{R}(\Gamma, G, H) = \{\varphi \in \operatorname{Hom}(\Gamma, G) \mid \varphi(\Gamma) \text{ is nontrivial and acts freely on } G/H\}.$$

This therefore says that the parameter space is explicitly given as follows:

In case 1,

$$\mathscr{R}(\Gamma, G, H) = \left\{ \begin{pmatrix} 0 \\ x \\ g_0 \end{pmatrix} ; x \in \mathbb{R}^{\times} \text{ and } g_0 \in G_0 \right\}. \tag{5.17}$$

In case 2 and $\alpha \neq 0$,

$$\mathscr{R}(\Gamma, G, H) = \left\{ \begin{pmatrix} 0 \\ v \\ g_0 \end{pmatrix} \in G; v \in \mathbb{R}^2 \setminus \{(0,0)\} \text{ and } g_0 \in G_0 \right\}. \tag{5.18}$$

In case 2 and if $\alpha = 0$,

$$\mathscr{R}(\Gamma, G, H) = \left\{ \begin{pmatrix} 2k\pi \\ v \\ g_0 \end{pmatrix} \in G; k \in \mathbb{Z},\ v \in \mathbb{R}^2 \setminus \{(0,0)\} \text{ and } g_0 \in G_0 \right\}. \tag{5.19}$$

In case 3,

$$\mathscr{R}(\Gamma, G, H) = \left\{ \begin{pmatrix} 0 \\ 2k\pi \\ v \\ g_0 \end{pmatrix} \in G \;\middle|\; \begin{array}{l} k \in \mathbb{Z}, \\ v \in \mathbb{R}^2 \setminus \{(0,0)\} \text{ and } g_0 \in G_0 \end{array} \right\}. \tag{5.20}$$

We now treat the question of local rigidity of discontinuous action on a maximal homogeneous space for the rank-one case. We will proceed through case by case attack according to Theorem 1.1.15.

We now prove the following.

Theorem 5.5.2. *Assume the context of Theorem* 1.1.15, *then:*
(1) *In case* 1, *the local rigidity holds if and only if G is isomorphic to* Aff($\mathbb{R}$).
(2) *In case* 2, *the local rigidity fails to hold.*

In particular, G admits no formally colored discrete subgroups in these two cases.

(3) *In case 3 and if* $\dim G = 4$, *then the local rigidity holds. Otherwise, for any integer* $n > 4$, *there exists a solvable nonexponential connected and simply connected Lie group* G_n *of dimension n, which admits at least a formally H-colored discrete subgroup, where H stands for a maximal nonnormal subgroup of* G_n.

Proof. (1) If $\mathfrak{g}_0$ is trivial, then G is the group $\mathrm{Aff}(\mathbb{R})$ for which the Lie algebra is given by $\mathfrak{g} = \mathbb{R}\text{-span}\{X, Y\}$ such that $[X, Y] = Y$ with $\mathfrak{h}$ is isomorphic to $\mathbb{R}X$. Let Γ for any discontinuous group for $\exp(\mathfrak{g})/\exp(\mathfrak{h})$, and the local rigidity property holds. Indeed, if Γ is nontrivial, it is isomorphic to $\exp(\mathbb{Z}Y)$. The corresponding parameter space is then homeomorphic to $\mathbb{R}^{\times}Y$. For $\varphi = aY \in \mathscr{R}(\Gamma, G, H)$ with $a \in \mathbb{R}^{\times}$, we have

$$G \cdot \varphi = \{ae^{b}Y,\ b \in \mathbb{R}\}.$$

This means that $\mathscr{R}(\Gamma, G, H)$ only admits two open orbits. In other words, the parameter space reads

$$\mathscr{R}(\Gamma, G, H) = \left\{\begin{pmatrix} 0 \\ y \end{pmatrix}, y \in \mathbb{R}^{\times}\right\}$$

as in Theorem 5.5.2 and the deformation space $\mathscr{T}(\Gamma, G, H)$ turns out to be a union of two open orbits. Now, if $\mathfrak{g}_0$ is not trivial then $\dim(\mathfrak{g}) \geq 3$, then there is a ideal $\mathfrak{g}_1$, which is included in $\mathfrak{h}$ such that $\operatorname{codim} \mathfrak{g}_1 = 3$ or 4 in $\mathfrak{g}$. Let $\overline{G} = G/G_1$, $\overline{H} = H/G_1$ and $\overline{\Gamma} = \Gamma G_1/G_1$, then the canonical surjection

$$\pi : \mathscr{R}(\Gamma, G, H) \longrightarrow \mathscr{R}(\overline{\Gamma}, \overline{G}, \overline{H})$$

is an open map. If $G \cdot \varphi$ is an open orbit in $\mathscr{R}(\Gamma, G, H)$, then $\pi(G \cdot \varphi) = \overline{G} \cdot \overline{\varphi}$ is an open orbit in $\mathscr{R}(\overline{\Gamma}, \overline{G}, \overline{H})$ and the result follows from the following lemma.

Lemma 5.5.3. *If* $\dim(G) = 3$ *or* 4, *then the parameter space admits no open orbits.*

Proof. If G is exponential, then the result follows from Theorem 5.3.9. Assume therefore that G is solvable and nonexponential. By Theorem 1.1.15, $\mathfrak{g}$ is isomorphic to one of the following algebras:

($\imath$) $\mathfrak{g} = \mathbb{R}\text{-span}\{A, X, Y\}$ where $[A, X] = \alpha X - Y$ and $[A, Y] = X - \alpha Y$.

($\imath\imath$) $\mathfrak{g} = \mathbb{R}\text{-span}\{A, B, X, Y\}$ such that $[A, X + iY] = X + iY$ and $[B, X + iY] = i(X + iY)$.

($\imath\imath\imath$) $\mathfrak{g} = \mathbb{R}\text{-span}\{A, B, X, Y\}$ such that $[B, X] = Y$, $[A, B] = \alpha B - X$, $[A, X] = B + \alpha X$ and $[A, Y] = 2\alpha Y$.

($\imath$v) $\mathfrak{g} = \mathbb{R}\text{-span}\{A, X, Y, Z\}$ where $[A, X] = \alpha X - Y$ and $[A, Y] = X - \alpha Y$.

(v) $\mathfrak{g} = \mathbb{R}\text{-span}\{A, B, X, Y\}$ such that $[A, B] = \alpha B$, $[A, X] = \beta X - Y$ and $[A, Y] = X + \beta Y$.

The four first algebras do not admit any maximal subalgebras of codimension one for which, the associated Lie subgroup is not normal. We only have to treat the fifth case. In this case, the unique ideal of $\mathfrak{g}$ of codimension 2 is $\mathfrak{g}_0 = \mathbb{R}X \oplus \mathbb{R}Y$. It follows that $\mathfrak{h}$ is generated by the basis $\{X, Y, A + \lambda B\}$ for some $\lambda \in \mathbb{R}$, and finally $\mathfrak{h}$ is isomorphic to

$\mathbb{R}$-span$\{X, Y, A'\}$ such that for some $B' \in \mathfrak{g} \setminus \mathfrak{h}$, we have $[A', B'] = \alpha B'$, $[A', X] = \beta X - Y$ and $[A', Y] = X + \beta Y$. From Theorem 5.5.2, the parameter space reads

$$\mathscr{R}(\Gamma, G, H) = \left\{ \begin{pmatrix} 0 \\ b' \\ v' \end{pmatrix}, b' \in \mathbb{R}^{\times} \text{ and } v' \in \mathbb{R}^2 \right\} \simeq \mathbb{R}^{\times} \times \mathbb{R}^2.$$

Let $\varphi = {}^t(0, b', v') \in \mathscr{R}(\Gamma, G, H)$ and $g = (a, b, v) \in G$,

$$g\varphi g^{-1} = {}^t(0, b' e^{\alpha a}, e^{\beta a} r(a) v').$$

Clearly, $\dim G \cdot \varphi = 1 < \dim \mathscr{R}(\Gamma, G, H)$. Then the property of rigidity fails to hold. This ends the proof of the lemma. □

(2) If $\alpha \neq 0$, from Theorem 5.5.2,

$$\mathscr{R}(\Gamma, G, H) \simeq \mathbb{R}^2 \setminus \{(0,0)\} \times G_0.$$

Let $\varphi = {}^t(0, v, g_0) \in \mathscr{R}(\Gamma, G, H)$ as in equation (5.18) and $g = (a_0, v_0, g_0') \in G$, then

$$g\varphi g^{-1} = {}^t(0, e^{\alpha a_0} r(a_0) v) \operatorname{mod}(G_0). \tag{5.21}$$

Then the orbit of φ is not open in $\mathscr{R}(\Gamma, G, H)$. Now, if $\alpha = 0$,

$$\mathscr{R}(\Gamma, G, H) \simeq 2\pi\mathbb{Z} \times \mathbb{R}^2 \setminus \{(0,0)\} \times G_0.$$

Let $\varphi = {}^t(2k\pi, v, g_0) \in \mathscr{R}(\Gamma, G, H)$ as in equation (5.19) and $g = (a_0, v_0, g_0') \in G$, then

$$g\varphi g^{-1} = {}^t(2k\pi, r(a_0) v) \operatorname{mod}(G_0), \tag{5.22}$$

which gives that $G\cdot\varphi$ is not open in $\mathscr{R}(\Gamma, G, H)$. This achieves the proof of Theorem 5.5.2.

(3) We now look at case 3. If $\dim G = 4$, then $\mathfrak{g}_0$ is trivial, and from equation (5.20) the parameter space reads

$$\mathscr{R}(\Gamma, G, H) = \left\{ \begin{pmatrix} 0 \\ 2k\pi \\ v \end{pmatrix} \in G \;\middle|\; \begin{array}{l} k \in \mathbb{Z}, \\ v \in \mathbb{R}^2 \setminus \{(0,0)\} \end{array} \right\}.$$

Let $\varphi = {}^t(0, 2k\pi, v) \in \mathscr{R}(\Gamma, G, H)$ and $g = (a', b', v') \in G$. Then

$$g\varphi g^{-1} = {}^t(0, 2k\pi, e^{a'} r(b') v),$$

which means that $G \cdot \varphi = {}^t(0, 2k\pi, \mathbb{R}^2 \setminus \{(0,0)\})$, and consequently $G \cdot \varphi$ is an open orbit of $\mathscr{R}(\Gamma, G, H)$.

We consider next the following infinite family of solvable Lie algebras $\mathfrak{g}_n$:= Lie(G_n), where

$$\mathfrak{g}_n = \mathbb{R}\text{-span}\{A, B, X, Y, C_1, \dots, C_n\}$$

such that

$$[A, X + iY] = X + iY, \quad [B, X + iY] = i(X + iY) \quad \text{and} \quad [A, C_j] = [B, C_j] = C_j$$

for all $1 \leq j \leq n$ and $\mathfrak{h} = \mathbb{R}\text{-span}\{A, B, C_1, \dots, C_n\}$. Let $g = (a, b, v, c_1, \dots, c_n)$ and $g' = (a', b', v', c_1', \dots, c_n') \in G$, then the multiplication group is given by

$$gg' = (a + a', b + b', e^{-a'} r(-b')v + v', c_1 e^{-a'-b'} + c_1', \dots, c_n e^{-a'-b'} + c_n').$$

As before, a routine computation shows that

$$\mathscr{R}(\Gamma, G_n, H) = \left\{ \begin{pmatrix} 0 \\ 2k\pi \\ v' \\ c_1' \\ \vdots \\ c_n' \end{pmatrix} \middle| \begin{array}{l} k \in \mathbb{Z}, \\ v' \in \mathbb{R}^2 \setminus \{(0,0)\}, \\ c_j' \in \mathbb{R} \text{ for } 1 \leq j \leq n \end{array} \right\} \simeq \mathbb{Z} \times \mathbb{R}^2 \setminus \{(0,0)\} \times \mathbb{R}^n.$$

Let $\varphi = {}^t(0, 2k\pi, v', c_1', \dots, c_n') \in \mathscr{R}(\Gamma, G_n, H)$ and $g = (a, b, v, c_1, \dots, c_n) \in G_n$. Then

$$g\varphi g^{-1} = {}^t(0, 2k\pi, e^a r(b)v', e^{a+b}(c_1(e^{-2k\pi} - 1) + c_1'), \dots, e^{a+b}(c_n(e^{-2k\pi} - 1) + c_n')).$$

If $k = 0$, then $g\varphi g^{-1} = {}^t(0, 0, e^a r(b)v', e^{a+b} c_1', \dots, e^{a+b} c_n')$. We have

$$\exp(\mathbb{R}X) \cdot \exp(\mathbb{R}Y) \cdot \exp(\mathbb{R}C_1) \cdots \exp(\mathbb{R}C_n) = \text{Stab}_\varphi := \{t \in G_n : t \cdot \varphi = \varphi\},$$

then $\dim G_n \cdot \varphi = 2 < \dim \mathscr{R}(\Gamma, G_n, H)$, and finally the local rigidity fails to hold. Now, if $k \neq 0$,

$$G_n \cdot \varphi = {}^t(0, 2k\pi, \mathbb{R}^2 \setminus \{(0,0)\}\mathbb{R}^n),$$

which is an open orbit sitting inside $\mathscr{R}(\Gamma, G_n, H)$. This achieves the proof of Theorem 5.5.2. □

Remark 5.5.4. Up to this point, the Selberg–Weil–Kobayashi local rigidity theorem for the rank-one case of solvable Lie groups can be schematized, with respect to the simple noncompact linear case, as in Table 5.2

Table 5.2: Summarizing local rigidity results.

(A): G is a noncompact simple linear Lie group, H a maximal compact subgroup and Γ is a uniform lattice	$\mathscr{R}(\Gamma, G, H)$ admits a continuous deformation $\Longleftrightarrow$ G is locally isomorphic to $SL_2(\mathbb{R})$		
(B): $G' = G \times G$, $H' = \Delta_{G'}$ and Γ is a uniform lattice of G (G as in (A))	$\mathscr{R}(\Gamma, G', H')$ admits a continuous deformation $\Longleftrightarrow$ G is locally isomorphic to $SO(n, 1)$ or $SU(n, 1)$		
(C): G is exponential solvable, H a maximal nonnormal subgroup of G and Γ a discontinuous group for G/H	$\mathscr{R}(\Gamma, G, H)$ admits a locally rigid deformation $\Longleftrightarrow$ G is locally isomorphic to $\mathrm{Aff}(\mathbb{R})$		
	Case 1	Case 2	Case 3
(D): G is solvable connected and simply connected Lie group, H a maximal nonnormal subgroup of G and Γ a discontinuous group for G/H of rank one	Local rigidity $\Longleftrightarrow$ $G \simeq \mathrm{Aff}(\mathbb{R})$	The local rigidity fails	if $\dim G \leq 4$, then the local rigidity holds. For any $n > 4$, there exists a Lie group G_n as in (D), which admits at least a formally H-colored discontinuous group

5.5.3 The setting where the action of G on G/H is effective

Let G_0 denote the kernel of the action of G on G/H. It is the maximal normal subgroup of G included in H, $G_0 = \bigcap_{g\in G} gHg^{-1}$. Let us opt for the notation $\mathscr{R}(\Gamma, G, X)$ and $\mathscr{T}(\Gamma, G, X)$ (instead of $\mathscr{R}(\Gamma, G, H)$ and $\mathscr{T}(\Gamma, G, H)$), in the setting where the action of G on X is effective. In such a case, G_0 is trivial and then obviously $\dim G \leq 4$. As an immediate consequence from Theorem 5.5.2, the following is then immediate.

Corollary 5.5.5. *Let us keep the same hypotheses and notation. For a discontinuous group $\Gamma \simeq \mathbb{Z}$ for $X = G/H$, the following holds:*
1. *If (G, H) is as in case 1, $\mathscr{T}(\mathbb{Z}, G, X)$ consists of two isolated points (i. e., the topology is discrete).*
2. *If (G, H) is as in case 2, $\mathscr{T}(\mathbb{Z}, G, X)$ has no isolated points.*
3. *If (G, H) is as in case 3, $\mathscr{T}(\mathbb{Z}, G, X) \simeq \mathbb{Z}$ consists only of isolated points (i. e., the topology is discrete).*

To seek the kind of perturbations of Γ in G with no contribution to the action on X, one can consider the following equivalence relation on $\mathrm{Hom}(\Gamma, G)$:

$$\varphi_1 \sim_{G_0} \varphi_2, \quad \varphi_1, \varphi_2 \in \mathrm{Hom}(\Gamma, G) \tag{5.23}$$

if there exists $\overline{g} \in \overline{G}$ such that $\pi \circ \varphi_1 = (\pi \circ \varphi_2)^{\overline{g}}$ in $\mathrm{Hom}(\overline{\Gamma}, G)$.

We then get the following.

Corollary 5.5.6. *Let us keep the same hypotheses and notation. For a discontinuous group $\Gamma \simeq \mathbb{Z}$ for G/H, the following holds:*

(1) *If (G, H) is as in case 1, $\mathscr{R}(\mathbb{Z}, G, H)/\sim_{G_0}$ consists of isolated points (i. e., the topology is discrete).*
(2) *If (G, H) is as in case 2, $\mathscr{R}(\mathbb{Z}, G, H)/\sim_{G_0}$ has no isolated points.*
(3) *If (G, H) is as in case 3, $\mathscr{R}(\mathbb{Z}, G, H)/\sim_{G_0}$ consists only of isolated points (i. e., the topology is discrete).*

Proof. Let $\Gamma \cong \mathbb{Z}$ be a discontinuous group for G/H. In this situation, the spaces G/H and $\overline{G}/\overline{H}$ are diffeomorphic, where as before $\overline{G} = G/G_0$ and $\overline{H} = H/G_0$. As in Lemma 2.4.7, the action of $\overline{\Gamma}$ on $\overline{G}/\overline{H}$ is discontinuous and obviously the projection

$$\pi : \mathscr{R}(\mathbb{Z}, G, H) \longrightarrow \mathscr{R}(\mathbb{Z}, \overline{G}, \overline{H})$$

is open. As $\mathrm{Hom}(\mathbb{Z}, G)$ is identified to G as a set, the group G_0 acts by left multiplication on $\mathscr{R}(\mathbb{Z}, G, H)$ by

$$(g_0 \cdot \varphi)(\gamma) = g_0\varphi(\gamma), \quad \forall g_0 \in G, \gamma \in \Gamma.$$

Then the projection map π factors through the quotient map

$$p : \mathscr{R}(\mathbb{Z}, G, H) \longrightarrow \mathscr{R}(\mathbb{Z}, G, H)/G_0,$$

which also an open map. Then the induced map

$$i : \mathscr{R}(\mathbb{Z}, G, H)/G_0 \longrightarrow \mathscr{R}(\mathbb{Z}, \overline{G}, \overline{H})$$

defined by $\pi = i \circ p$ is a homeomorphism. The quotient group $\overline{G}$ acts also on $\mathscr{R}(\mathbb{Z}, G, H)/G_0$ by $\overline{g} \cdot p(\varphi) = p(g \cdot \varphi)$ and the double cosets space $G_0 \backslash \mathscr{R}(\mathbb{Z}, G, H)/\overline{G}$ is identified to the set of the equivalence classes $\mathscr{R}(\mathbb{Z}, G, H)/\sim_{G_0}$. Now

$$\overline{g} \cdot i(p(\varphi)) = \overline{g} \cdot \pi(\varphi) = \pi(g \cdot \varphi) = i \circ p(g \cdot \varphi) = i(\overline{g}\dot{p}(\varphi)).$$

Thus, i induces a homeomorphism

$$\overline{i} : \mathscr{R}(\mathbb{Z}, G, H)/\sim_{G_0} \longrightarrow \mathscr{T}(\mathbb{Z}, \overline{G}, \overline{H}).$$

But the spaces $\mathscr{T}(\mathbb{Z}, \overline{G}, \overline{H})$ and $\mathscr{R}(\mathbb{Z}, \overline{G}, X)$ are homeomorphic, then Corollary 5.5.5 allows to conclude. □

5.5.4 The rank-two case

This section aims to put the emphasis on the case where Γ is of rank 2. Unlike the context of exponential solvable Lie groups, the authors do not know so far how to deal with this case. However, we can get an accurate description of the parameter space $\mathscr{R}(\mathbb{Z}^2, G, H)$. We first prove the following.

Proposition 5.5.7. *Let G be a connected simply connected solvable Lie group, H a non-normal maximal subgroup of G and Γ a subgroup of G of rank 2. Then:*

(1) *In case 2, Γ is a discrete subgroup acting properly on G/H if and only if Γ is generated by two elements $\gamma_i = (a_i, v_i, g_i)$, $i = 1, 2$ satisfying:*
   (i) $a_i \in 2\pi\mathbb{Z}$ *and* $\det(v_1, v_2) \neq 0$ *if* $\alpha = 0$.
   (ii) $a_i = 0$ *and* $\det(v_1, v_2) \neq 0$ *if* $\alpha \neq 0$.

(2) *In case 3, Γ is a discrete subgroup acting properly on G/H if and only if Γ is generated by $\gamma_i = (a_i, bi, v_i, g_i)$, $i = 1, 2$ such that $a_i = 0$, $b_i \in 2\pi\mathbb{Z}$ and* $\det(v_1, v_2) \neq 0$.

As proper action is also free in our context, by Lemmas 2.4.10 and 2.4.11 the subgroup Γ acts properly on G/H only if the two elements γ_1, γ_2 satisfy

$$\gamma_i = \begin{cases} (0, v_i, g_i) & \text{if } \alpha \neq 0 \text{ in case 2,} \\ (2k_i\pi, v_i, g_i) & \text{if } \alpha = 0 \text{ in case 2,} \\ (0, 2k_i\pi, v_i, g_i) & \text{in case 3.} \end{cases} \tag{5.24}$$

Then the proposition is a direct consequence of the following result.

Lemma 5.5.8. *Let Γ be a subgroup of G generated by two elements γ_1, γ_2 satisfying* (5.24). *Then the action of Γ on G/H is proper if and only if* $\det(v_1, v_2) \neq 0$.

Proof. As before, the nonproperness of the action of Γ on G/H is equivalent to the existence of a convergent sequence $(s_n)_n$ in G and a sequence $(\gamma_n)_n \in \Gamma$ with $\{\gamma_n, n \in \mathbb{N}\}$ is noncompact such that $(\gamma_n s_n)_n$ converges modulo H. For $\gamma_n = \gamma_1^{\lambda_n}\gamma_2^{\mu_n}$ and

$$s_n = \begin{cases} (t_n, w_n, g_n) & \text{in case 2,} \\ (t_n, f_n, w_n, g_n) & \text{in case 3,} \end{cases}$$

we obtain

$$\gamma_n s_n = \begin{cases} (0, \lambda_n v_1 + \mu_n v_2 + e^{\alpha t_n} r(t_n) w_n, 0) \bmod(H) & \text{in case 2,} \\ (0, 0, \lambda_n v_1 + \mu_n v_2 + e^{t_n} r(f_n) w_n, 0) \bmod(H) & \text{in case 3.} \end{cases}$$

Now $(t_n)_n$, $(f_n)_n$ and $(w_n)_n$ are convergent, then the nonproperness of the action is equivalent to the existence of a sequence $(\lambda_n, \mu_n) \subset \mathbb{Z}^2$ such that $(\lambda_n v_1 + \mu_n v_2)_n$ converges, which is also equivalent to the condition $\det(v_1, v_2) = 0$. ☐

Remark 5.5.9. In the rank-two case, there no equivalence between free action and proper action. In fact, in Proposition 5.5.7, if we replace the word properly by freely, and the statement $\det(v_1, v_2) \neq 0$ by the fact that v_1, v_2 are $\mathbb{Q}$-linearly independent, we obtain a characterization of the free action.

Now as a direct consequence of Proposition 5.5.7 and Proposition 2.4.13, we get the following.

Theorem 5.5.10 (The rank-two case). *Let G be a connected simply connected solvable Lie group, H a connected nonnormal maximal subgroup of G and $\Gamma \simeq \mathbb{Z}^2$ a discontinuous group for G/H of rank two. Then the parameter is given in the following cases by:*
(i) *In case 2, we have*

$$\mathscr{R}(\mathbb{Z}^2, G, H) = \left\{ \begin{pmatrix} 0 & 0 \\ v_1 & v_2 \\ g_1 & g_2 \end{pmatrix} \in \mathrm{Hom}(\Gamma, G) \;\middle|\; \begin{array}{l} \det(v_1, v_2) \neq 0, \\ g_1, g_2 \in G_0 \end{array} \right\}, \quad \text{if } \alpha \neq 0$$

and

$$\mathscr{R}(\mathbb{Z}^2, G, H) = \left\{ \begin{pmatrix} 2k_1\pi & 2k_2\pi \\ v_1 & v_2 \\ g_1 & g_2 \end{pmatrix} \in \mathrm{Hom}(\Gamma, G) \;\middle|\; \begin{array}{l} \det(v_1, v_2) \neq 0, \\ g_1, g_2 \in G_0 \end{array} \right\}, \quad \text{if } \alpha = 0.$$

(ii) *In case 3, we have*

$$\mathscr{R}(\mathbb{Z}^2, G, H) = \left\{ \begin{pmatrix} 0 & 0 \\ 2k_1\pi & 2k_2\pi \\ v_1 & v_2 \\ g_1 & g_2 \end{pmatrix} \in \mathrm{Hom}(\Gamma, G) \;\middle|\; \begin{array}{l} k_1, k_2 \in \mathbb{Z}, \\ \det(v_1, v_2) \neq 0, \\ g_1, g_2 \in G_0 \end{array} \right\}.$$

Finally, the following examples show that the local rigidity property fails to hold in the rank-two case.

Example 5.5.11. Let $\mathfrak{g} = \mathbb{R}\text{-span}\{A, X, Y\}$ be such that $[A, X + iY] = (\alpha + i)(X + iY)$ for some $\alpha \in \mathbb{R}$. Let $g = (a, v)$ and $g' = (a', v') \in G$, then by equation (2.19), we have

$$gg' = (a + a', e^{-\alpha a'} r(-a')v + v').$$

Let $\mathfrak{h} = \mathbb{R}A \oplus \mathbb{R}B$ and $\Gamma = \langle \exp X, \exp Y \rangle$ the discrete subgroup generated by $\exp X$ and $\exp Y$. If $\alpha \neq 0$, then G is exponential solvable and Theorem 5.5.2 shows that the local rigidity fails to hold.

Now, if $\alpha = 0$, which means that G is no longer exponential solvable. Then from Theorem 5.5.10,

$$\mathscr{R}(\Gamma, G, H) = \left\{ \begin{pmatrix} 2k\pi & 2k'\pi \\ v & v' \end{pmatrix} \in \operatorname{Hom}(\Gamma, G) \;\middle|\; \begin{array}{l} k, k' \in \mathbb{Z}, \\ \det(v, v') \neq 0 \end{array} \right\}.$$

Let

$$\varphi = \begin{pmatrix} 2k\pi & 2k'\pi \\ v & v' \end{pmatrix} \in \mathscr{R}(\Gamma, G, H)$$

and $g = (a_0, v_0) \in G$. Then

$$g\varphi g^{-1} = \begin{pmatrix} 2k\pi & 2k'\pi \\ e^{\alpha a_0} r(a_0) v & e^{\alpha a_0} r(a_0) v' \end{pmatrix},$$

which means that $G \cdot \varphi$ is not an open orbit and the local rigidity fails to hold.

Example 5.5.12. $\mathfrak{g} = \mathbb{R}\text{-span}\{A, B, X, Y\}$ such that $[A, X + iY] = X + iY$ and $[B, X + iY] = i(X + iY)$. Let $g = (a, b, v)$ and $g' = (a', b', v') \in G$. Then by equation (2.18), we have

$$gg' = (a + a', b + b', e^{-a'} r(-b')v + v').$$

Let as earlier $\mathfrak{h} = \mathbb{R}A \oplus \mathbb{R}B$ and $\Gamma = \langle \exp X, \exp Y \rangle$. Then as in Theorem 5.5.10,

$$\mathscr{R}(\Gamma, G, H) = \left\{ \begin{pmatrix} 0 & 0 \\ 2k\pi & 2k'\pi \\ v & v' \end{pmatrix} \in \operatorname{Hom}(\Gamma, G) \;\middle|\; \begin{array}{l} k, k' \in \mathbb{Z}, \\ \det(v, v') \neq 0 \end{array} \right\}.$$

Let

$$\varphi = \begin{pmatrix} 0 & 0 \\ 2k\pi & 2k'\pi \\ v & v' \end{pmatrix} \in \mathscr{R}(\Gamma, G, H)$$

and $g = (a_0, b_0, v_0) \in G$. Then

$$g\varphi g^{-1} = \begin{pmatrix} 0 & 0 \\ 2k\pi & 2k'\pi \\ e^{a_0} r(b_0) v & e^{a_0} r(b_0) v' \end{pmatrix},$$

which shows that $G \cdot \varphi$ is not an open orbit. So, the local rigidity fails to hold.

Theorem 5.5.2 and Examples 5.5.11 and 5.5.12 above allow us to write down explicitly the deformation space $\mathscr{T}(\Gamma, G, X)$ in the setting where G acts effectively on X. Indeed, the following holds.

Proposition 5.5.13. *Assume that G acts effectively on X and Γ a discontinuous group for G/H, then:*

(1) *If* $\dim(G) = 2$, *then* $\mathscr{T}(\Gamma, G, X)$ *consists of two isolated points.*
(2) *If* $\dim(G) = 3$, *then we are in case 2. In this case, we have:*
   (a) *If* Γ *is of rank one, then* $\mathscr{T}(\Gamma, G, X)$ *is homeomorphic to:*
      (i) $\mathbb{R}_{+}^{\times} \times \mathbb{Z}$, *if* $\alpha = 0$.
      (ii) S^1, *if* $\alpha \neq 0$.
   (b) *If the rank of* Γ *is two, then* $\mathscr{T}(\Gamma, G, X)$ *is homeomorphic to:*
      (i) $\mathbb{Z}^2 \times GL_2(\mathbb{R})/SO_2$, *if* $\alpha = 0$.
      (ii) $GL_2(\mathbb{R})/\mathbb{R}$, *if* $\alpha \neq 0$, *where* $\mathbb{R}$ *is embedded as a subgroup via* $t \mapsto e^{\alpha t} r(t)$.
(3) *If* $\dim(G) = 4$, *then for the rank-one case* $\mathscr{T}(\Gamma, G, X)$ *is homeomorphic to* $\mathbb{Z}$ *for the rank-one case and to* $\mathbb{Z}^2 \times GL_2(\mathbb{R})/P$ *for the rank-two case, where P is the two-dimensional connected Lie subgroup of* $GL_2(\mathbb{R})$ *image of* $\mathbb{R}^2$ *by the homomorphism* $(a, b) \mapsto e^a r(b)$.

Proof. (1) comes directly from Theorem 5.5.2. For (2), if Γ is of rank one, then equations (5.18) and (5.19) give us the result provided the expressions of the corresponding actions as in equations (5.21) and (5.22). If Γ is of rank two or $\dim(G) = 4$, then the result comes from Examples 5.5.11 and 5.5.12 and Corollary 5.5.5. □

5.6 The case of Diamond groups

The *diamond algebra* $\mathfrak{g}$ is defined as the direct sum of the Heisenberg Lie algebra $\mathfrak{h}_{2n+1}$ as defined in Subsection 1.1.2 and an n-dimensional Abelian Lie algebra $\mathfrak{a} = \bigoplus_{l=1}^{n} \mathbb{R}A_i$ with the additional nontrivial brackets

$$[A_l, X_l] = Y_l \quad \text{and} \quad [A_l, Y_l] = -X_l, \quad \text{for } l = 1, \ldots, n. \tag{5.25}$$

A Lie algebra $\mathfrak{l}$ is called a graded Lie algebra if there is a decomposition

$$\mathfrak{l} = \bigoplus_{d \in \mathbb{Z}} \mathfrak{l}_d,$$

where $\mathfrak{l}_d, d \in \mathbb{Z}$ are a subspaces of $\mathfrak{l}$ such that

$$[\mathfrak{l}_i, \mathfrak{l}_j] \subset \mathfrak{l}_{i+j}, \quad i, j \in \mathbb{Z}.$$

From the brackets relations (5.25) and (1.5), the diamond algebra $\mathfrak{g}$ is a graded Lie algebra with the decomposition

$$\mathfrak{g} = \mathfrak{g}_0 \oplus \mathfrak{g}_1 \oplus \mathfrak{g}_2,$$

where

$$\mathfrak{g}_0 = \mathfrak{a}, \quad \mathfrak{g}_1 = \bigoplus_{i=1}^{n} (\mathbb{R}X_i \oplus \mathbb{R}Y_i),$$

and $\mathfrak{g}_2 = \mathbb{R}Z$ denotes the center of $\mathfrak{g}$. As a consequence, for all $t \in \mathbb{R}_+^* := \{t \in \mathbb{R}, t > 0\}$, the dilation $\mu_t : \mathfrak{g} \to \mathfrak{g}$ defined for $v \in \mathfrak{g}_d$ by $\mu_t(v) = t^d v$ is an automorphism of $\mathfrak{g}$.

Definition 5.6.1. A subalgebra $\mathfrak{h}$ of a graded Lie algebra $\mathfrak{l}$ is called *subgraded* subalgebra if

$$\mathfrak{h} = \bigoplus_{d \in \mathbb{Z}} (\mathfrak{h} \cap \mathfrak{l}_d).$$

The subgraded subalgebras of $\mathfrak{g}$ are the subalgebras of $\mathfrak{g}$, which are stable by the one parameter family of dilations μ_t, $t \in \mathbb{R}_+^*$.

Compatible subalgebras

A subalgebra $\mathfrak{h}$ of $\mathfrak{g}$ is said to be compatible with a given Levi decomposition if it is a direct sum $\mathfrak{u} \oplus \mathfrak{h}_0$, where $\mathfrak{u} = \mathfrak{a} \cap \mathfrak{h}$ and $\mathfrak{h}_0 = \mathfrak{h}_{2n+1} \cap \mathfrak{h}$. Any subgraded subalgebra is compatible with a given Levi decomposition, and conversely a subalgebra, which is compatible with a given Levi decomposition and contains the center of $\mathfrak{g}$ is subgraded subalgebra. In general, a subalgebra, which is compatible with a given Levi decomposition, is not subgraded. The subalgebra generated by $X_1 + Z$ is compatible with a given Levi decomposition but not subgraded subalgebra of $\mathfrak{g}$. Nevertheless, we have the following statement.

Proposition 5.6.2. *Let $\mathfrak{h}$ be a subalgebra of $\mathfrak{g}$, which is compatible with a given Levi-decomposition. Then there exists $g \in G$ such that $\mathrm{Ad}_g(\mathfrak{h})$ is a subgraded subalgebra.*

Assume for the rest of this paragraph that $\mathfrak{h} = \mathfrak{u} \oplus \mathfrak{h}_0$, with $\mathfrak{u}$ and $\mathfrak{h}_0$ as before. To prove Proposition 5.6.2, we need some preliminary remarks. Let $\underline{A}$ be an element of $\mathfrak{u}$. We define the *support of* $\underline{A}$ as the unique subset $I_{\underline{A}} \subseteq \{1, \dots, n\}$ such that

$$\underline{A} = \sum_{i \in I_{\underline{A}}} \alpha_i A_i \quad \text{with } \alpha_i \neq 0, \ \forall\, i \in I_{\underline{A}}. \tag{5.26}$$

We define also the *support of* $\mathfrak{h}$ as the set $I := \bigcup_{\underline{A} \in \mathfrak{u}} I_{\underline{A}}$. Obviously, there exists $\underline{A} \in \mathfrak{u}$ such that $I = I_{\underline{A}}$; such a vector is called *generic element of* $\mathfrak{u}$. Denote by $\mathfrak{U}_I$ the linear subspace of $\mathfrak{g}$ generated by X_i, Y_i, $i \in I$ and let $\mathfrak{h}_I = \mathbb{R}Z \oplus \mathfrak{U}_I$. Note that $\mathfrak{h}_{2n+1} = \mathfrak{U}_I \oplus \mathfrak{h}_{I^c}$ and $[\mathfrak{h}_I, \mathfrak{h}_{I^c}] = [\mathfrak{u}, \mathfrak{h}_{I^c}] = \{0\}$.

Lemma 5.6.3. *Let I be the support of $\mathfrak{h}$. Then*

$$\mathfrak{h} = \mathfrak{u} \oplus (\mathfrak{h}_0 \cap \mathfrak{U}_I) \oplus (\mathfrak{h}_0 \cap \mathfrak{h}_{I^C}),$$

as a direct sum of linear subspaces.

Proof. Let $v \in \mathfrak{h}_0$ such that $v = v_I + v_{I^C}$, where $v_I \in \mathfrak{U}_I$ and $v_{I^C} \in \mathfrak{h}_{I^C}$. To conclude, we have to prove that $v_I \in \mathfrak{h}$. Let $\underline{A}$ be a generic element of $\mathfrak{u}$. Then we have $\mathrm{ad}(\underline{A})^m v = \mathrm{ad}(\underline{A})^m v_I$ and $\mathrm{ad}(\underline{A})^m v_I \in \mathfrak{h}$ for all $m > 0$, where $\mathrm{ad}(\underline{A})^m v_I = [\underline{A}, \mathrm{ad}(\underline{A})^{m-1} v_I]$. Let $\underline{A}$ be written as in (5.26) and

$$v_I = \sum_{i \in I} v_i, \quad \text{with } v_i \in \mathfrak{U}_{\{i\}}.$$

Consider the set of strictly positive integers $\{\beta_1, < \cdots, < \beta_q\} := \{|\alpha_i|, i \in I\}$. Then from the expressions

$$\underline{A} = \sum_{i=1}^{q} \beta_i \sum_{|\alpha_j| = \beta_i} \frac{\alpha_j}{|\alpha_j|} A_j \quad \text{and} \quad v_I = \sum_{i=1}^{q} \sum_{|\alpha_j| = \beta_i} v_j,$$

we deduce that

$$\mathrm{ad}(\underline{A})^m v_I = \sum_{i=1}^{q} \beta_i^m \sum_{|\alpha_j| = \beta_i} \left(\frac{\alpha_j}{|\alpha_j|} \right)^m \mathrm{ad}(A_j)^m v_j.$$

Now observe that $\mathrm{ad}(A_j)^4 v_j = v_j$. Then

$$\mathrm{ad}(\underline{A})^{4m} v_I = \sum_{i=1}^{q} \beta_i^{4m} \sum_{|\alpha_j| = \beta_i} v_j.$$

For $u_i = \sum_{|\alpha_j| = \beta_i} v_j$, the sequence $w_m = \sum_{i=1}^{q} \beta_i^{4m} u_i = \mathrm{ad}(\underline{A})^{4m} v_I$ is contained in the subspace L generated by the free family $\{u_1, \dots, u_q\}$. The matrix associated to the coordinates of $w_1, \dots, w_q$ via the basis $\{u_1, \dots, u_q\}$ is the Vandermonde matrix

$$\begin{pmatrix} \beta_1^4 & \cdots & \beta_1^{4q} \\ \vdots & & \vdots \\ \beta_q^4 & \cdots & \beta_q^{4q} \end{pmatrix},$$

which is nonsingular, because $\beta_i^4 \neq \beta_j^4$ for all $i \neq j$. This means that the vectors $w_1, \dots, w_q$ form a basis of L. As $w_m \in \mathfrak{h}$ for all $m > 0$, the subspace $L \subset \mathfrak{h}$, in particular, the vector $v_I = \sum_{i=1}^{q} u_i \in \mathfrak{h}$. □

Recall now the following result, object of Proposition 1.1.12, which asserts the following. Let $\mathfrak{h}$ be a Lie subalgebra of $\mathfrak{h}_{2n+1}$, such that $\mathfrak{z}(\mathfrak{h}_{2n+1}) \not\subset \mathfrak{h}$. Then $\dim \mathfrak{h} \leq n$ and there exists a basis $\mathscr{B}_{\mathfrak{h}} = \{Z, X_1' \ldots, X_n', Y_1', \ldots, Y_n'\}$ of $\mathfrak{h}_{2n+1}$ with the Lie commutation relations

$$[X_i', Y_j'] = \delta_{i,j} Z, \quad i, j = 1, \ldots, n, \tag{5.27}$$

such that $\mathfrak{h}$ is generated by $X_1', \ldots, X_p'$, where $p = \dim \mathfrak{h}$. The symbol $\delta_{i,j}$ designates here the Kronecker index.

Proof of Proposition 5.6.2. If $\mathfrak{h}$ contains the center, then obviously $\mathfrak{h}$ is subgraded. Assume that $\mathfrak{h}$ does not contain the center. Using Lemma 5.6.3, where we replace $\mathfrak{h}_{2n+1}$ by $\mathfrak{h}_{I^c}$ and $\mathfrak{h}$ by $\mathfrak{h} \cap \mathfrak{h}_{I^c}$, we deduce the existence of a basis $\{Z, X_1' \ldots, X_m', Y_1', \ldots, Y_m'\}$ of $\mathfrak{h}_{I^c}$ satisfying the relation (5.27) and such that $X_1' \ldots, X_p'$ is a basis of $\mathfrak{h} \cap \mathfrak{h}_{I^c}$, where $p = \dim(\mathfrak{h} \cap \mathfrak{h}_{I^c})$ and $2m+1 = \dim \mathfrak{h}_{I^c}$. For $X_i' = X_i'' + \alpha_i Z$, with $X_i'' \in \mathfrak{g}_1$, let $X = \sum_{i=1}^{p} \alpha_i Y_i'$. Then a direct calculation shows that $\mathrm{Ad}_{\exp(X)}(\mathfrak{h} \cap \mathfrak{h}_{I^c})$ is the linear span of $X_1'', \ldots, X_p''$, in particular, $\mathrm{Ad}_{\exp(X)}(\mathfrak{h} \cap \mathfrak{h}_{I^c}) \subset \mathfrak{g}_1$. Observe that for all $X \in \mathfrak{h}_{I^c}$, we have $[X, \mathfrak{u}] = 0$ and $[X, \mathfrak{U}_I] = 0$. Thus, $\mathrm{Ad}_{\exp(X)}(\mathfrak{u}) = \mathfrak{u}$ and $\mathrm{Ad}_{\exp(X)}(\mathfrak{U}_I \cap \mathfrak{h}) = \mathfrak{U}_I \cap \mathfrak{h}$. Using Lemma 5.6.3, we conclude that

$$\mathrm{Ad}_{\exp(X)}(\mathfrak{h}) = \mathfrak{u} \oplus (\mathfrak{U}_I \cap \mathfrak{h}) \oplus \mathrm{Ad}_{\exp(X)}(\mathfrak{h} \cap \mathfrak{h}_{I^c}),$$

which is graded since $\mathfrak{U}_I \cap \mathfrak{h} \oplus \mathrm{Ad}_{\exp(X)}(\mathfrak{h} \cap \mathfrak{h}_{I^c}) \subseteq \mathfrak{g}_1$. □

5.6.1 Dilation invariant subgroups

The Heisenberg group H_{2n+1} is the connected simply connected Lie group of dimension $2n+1$ associated to $\mathfrak{h}_{2n+1}$. As $\mathfrak{h}_{2n+1}$ is nilpotent, H_{2n+1} is an exponential Lie group. We identify H_{2n+1} with the affine space $\mathbb{R}^{2n+1} = (\mathbb{R}^2)^n \times \mathbb{R}$. Any element x of H_{2n+1} can be written as

$$x = (x_1, \ldots, x_n, z), \quad \text{where } x_i = \begin{pmatrix} \alpha_i \\ \beta_i \end{pmatrix} \in \mathbb{R}^2.$$

Thanks to the Campbell–Backer–Hausdorff formula for 2-step nilpotent Lie groups,

$$C(X, Y) = X + Y + \frac{1}{2}[X, Y],$$

the multiplication is obtained by the formula

$$xx' = \left(x_1 + x_1', \ldots, x_n + x_n', z + z' + \frac{1}{2} \sum_{l=1}^{n} b(x_l, x_l') \right),$$

where b is the standard nondegenerate skew-symmetric bilinear form on $\mathbb{R}^2$, given by $b(x_l, x_l') = \det(x_l, x_l')$.

As $[\mathfrak{g},\mathfrak{g}] = \mathfrak{h}_{2n+1}$, the family $\{A_1,\dots,A_n\}$ is a coexponential basis to $\mathfrak{h}_{2n+1}$ in $\mathfrak{g}$. The *diamond group* is identified to the product space $\mathbb{R}^n \times H_{2n+1}$ and the map

$$\mathfrak{a} \times \mathfrak{h}_{2n+1} \to G, \quad (A,X) \mapsto \exp(A)\exp(X)$$

is a diffeomorphism.

The Abelian group $\exp(\mathfrak{a}) = \mathbb{R}^n$ acts on $\mathfrak{h}_{2n+1}$ via the adjoint representation

$$\mathrm{Ad}_{\exp(t_iA_i)}(\alpha_j X_j + \beta_j Y_j) = \begin{cases} \alpha_j X_j + \beta_j Y_j & \text{if } i \neq j, \\ \alpha_j' X_j + \beta_j' Y_j & \text{if } i = j, \end{cases}$$

where $\binom{\alpha_j'}{\beta_j'} = r(t_j)\binom{\alpha_j}{\beta_j}$ and $r(t_i)$ is the rotation transformation

$$r(t_i) = \begin{pmatrix} \cos(t_i) & -\sin(t_i) \\ \sin(t_i) & \cos(t_i) \end{pmatrix},$$

through the basis $\{X_i, Y_i\}$. Thus, the diamond group is the semidirect product $G = \mathbb{R}^n \ltimes H_{2n+1}$, where $\mathbb{R}^n$ acts on H_{2n+1} as follows:

$$(t_1,\dots,t_n)\cdot x = (r(t_1)x_1,\dots,r(t_n)x_n,z).$$

More precisely, let us consider the projections:

$$\xi_l : G \to \mathbb{R}^2, \quad \tau_l : G \to \mathbb{R} \quad \text{and} \quad \zeta : G \to \mathbb{R}$$

defined by $\xi_l(g) = x_l$, $\tau_l(g) = t_l$ for $l = 1,\dots,n$ and $\zeta(g) = z$, where

$$g = (t_1,\dots,t_n,x_1,\dots,x_n,z).$$

The product of two elements $g, g' \in G$ is determined by the formulas

$$\xi_l(gg') = r_l(g')\xi_l(g) + \xi_l(g'), \quad \tau_l(gg') = \tau_l(g) + \tau_l(g'), \tag{5.28}$$

and

$$\zeta(gg') = \zeta(g) + \zeta(g') + \frac{1}{2}\sum_{l=1}^{n} b(r_l(g')\xi_l(g), \xi_l(g')), \tag{5.29}$$

where $r_l(g) = r(\tau_l(g))^{-1}$. As a consequence, the action of G on itself by conjugation is given by

$$\delta_l(gg'g^{-1}) = \delta_l(g'), \quad \xi_l(gg'g^{-1}) = r_l(g)^{-1}((r_l(g') - I)\xi_l(g) + \xi_l(g')), \tag{5.30}$$

$$\begin{aligned}\zeta(gg'g^{-1}) = \zeta(g') &+ \frac{1}{2}\sum_{l=1}^{n} b(r_l(g')\xi_l(g), \xi_l(g')) \\ &- \frac{1}{2}\sum_{l=1}^{n} b(r_l(g')\xi_l(g), \xi_l(g)) + b(\xi_l(g'), \xi_l(g)).\end{aligned} \tag{5.31}$$

Compatible subgroups

Let H_0 be a subgroup of H_{2n+1} and U a subgroup of $\exp(\mathfrak{a})$, such that U normalizes H_0. Then the semidirect product $U \ltimes H_0$ is a well-defined subgroup of G.

Definition 5.6.4 (cf. [89]). A closed subgroup H of G is said to be compatible with a given Levi decomposition if it is a semidirect product $U \ltimes H_0$, where $U = \exp(\mathfrak{a}) \cap H$ and $H_0 = H \cap H_{2n+1}$.

Proposition 5.6.5. *Let $H = U \ltimes H_0$ be a subgroup of G compatible with a given Levi decomposition and $L(H_0)$ the syndetic hull of H_0. Then U normalizes $L(H_0)$ and the subgroup $U \ltimes L(H_0)$ contains $U \ltimes H_0$ cocompactly. In particular, if U is connected and $U \ltimes L(H_0)$ is a syndetic hull of H.*

Note that since a compatible subgroup with a given Levi decomposition is closed, H_0 is closed subgroup of H_{2n+1}. Thus, the syndetic hull of H_0 exists and it is unique. Similarly, any closed subgroup of the Abelian group $\exp(\mathfrak{a})$ has a unique syndetic hull. More precisely, we can state the following.

Lemma 5.6.6. *Let K be a closed subgroup of a completely solvable Lie group. Then the syndetic hull of K is the connected Lie subgroup K' generated by the one parameter subgroups $\{\exp(tX) \mid t \in \mathbb{R}\}$, for all $\exp(X) \in K$.*

Proof. As G is a completely solvable, K admits a unique syndetic hull $L(K)$; cf. Lemma 1.4.2. Clearly, K' is connected subgroup containing K and $K' \subset L(K)$. The homogeneous space $L(K)/K'$ is homeomorphic to $\mathbb{R}^d$, where $d = \dim L(K) - \dim K'$ and the map

$$L(K)/K \to L(K)/K', \quad xK \mapsto xK'$$

is a surjective and continuous. As $L(K)/K$ is compact, $d = 0$ and $L(K) = K'$. □

Proof of Proposition 5.6.5. Using Lemma 5.6.6, it is sufficient to show that $U \ltimes H_0'$ is a subgroup of G. Or equivalently, U normalizes H_0'. Let $\exp(tX)$ be a generator of H_0' and $\exp(A)$ an element of U, then

$$\exp(A)\exp(tX)\exp(-A) = \exp(t\,\mathrm{Ad}_{\exp(A)}(X)) \quad \text{and} \quad \mathrm{Ad}_{\exp(A)}(X) \in H_0.$$

Thus, the one parameter subgroup $\exp(t\,\mathrm{Ad}_{\exp(A)}(X))$ is also a generator of H_0'.

For the second statement, let S be a compact set in $L(H_0)$ such that $L(H_0) = H_0 S$. Then $U \ltimes L(H_0) = U \ltimes H_0 L = (U \ltimes H_0)(0, S)$ and $(0, S)$ is a compact set in $U \ltimes L(H_0)$. □

Lemma 5.6.7. *Let H be a subgroup of G, which is compatible with a given Levi decomposition. Assume that H_{2n+1} is the syndetic hull of H_0. If $L(U)$ is the syndetic hull of U, then the connected subgroup $L(U) \ltimes H_{2n+1}$ is a syndetic hull of H.*

Proof. If $L(U) = SU$ for a compact set S in $\exp(\mathfrak{a})$ and $H_{2n+1} = H_0 S'$ for S' a compact set of H_{2n+1}. Then

$$\begin{aligned} L(U) \ltimes H_{2n+1} &= (S,0)(U \ltimes H_0)(0,S') \\ &= (S,0)(0,S'^{-1})(U \ltimes H_0) \\ &= (S,S'^{-1})H. \end{aligned}$$ □

Subgraded and dilation-invariant subgroups

Recall that, if L is a connected simply connected Lie groups, then any automorphism μ of the Lie algebra $\mathfrak{l}$ of L lifts to an automorphism $\tilde{\mu}$ of G, by the formula

$$\tilde{\mu}(\exp(X)) = \exp(\mu(X)), \quad \text{for all } X \in \mathfrak{l}.$$

In particular, if $\mathfrak{l}$ is graded Lie algebra, for any $t \in \mathbb{R}_+^*$ the dilation automorphism μ_t induces an automorphism of L. This leads us to define an action of $\mathbb{R}_+^*$ on L.

Definition 5.6.8. Assume that L is a connected simply connected Lie group with a graded Lie algebra $\mathfrak{l}$. A closed subgroup H of L is said to be:
(1) dilation-invariant, if H is stable under the action of $\mathbb{R}_+^*$ on L.
(2) subgraded, if H is connected and its Lie subalgebra is a subgraded subalgebra of $\mathfrak{l}$.

If H is subgraded, then obviously H is a dilation-invariant subgroup. Conversely, a dilation-invariant subgroup is not necessarily connected; and a closed subgroup with a subgraded Lie subalgebra may or not be dilation-invariant. As an example, in our situation, for $g \in G$ the dilation action is given by

$$\tau_l(\tilde{\mu}_t(g)) = \tau_l(g), \quad \xi_l(\tilde{\mu}_t(g)) = t\xi_l(g) \quad \text{and} \quad \zeta(\tilde{\mu}_t(g)) = t^2\zeta(g). \tag{5.32}$$

The center of G,

$$Z(G) = \{g \in G, \xi_l(g) = 0 \text{ and } \tau_l(g) \in 2\pi\mathbb{Z} \text{ for all } l = 1, \dots, n\}$$

and the subgroup $H = \exp(\mathbb{Z}X)\exp(\mathbb{R}Z)$ for $X \in \mathfrak{g}_1$ are both nonconnected and have a same Lie subalgebra the center $\mathfrak{g}_2$, which is subgraded. The center $Z(G)$ is dilation-invariant but H is not. The following lemma is a characterization of dilation-invariant subgroups in G.

Lemma 5.6.9. *A closed subgroup H of G is a dilation-invariant subgroup if and only if H is compatible with a given Levi decomposition such that $H \cap H_{2n+1}$ is a subgraded subgroup.*

Proof. Suppose that H is dilation-invariant and let U and H_0 be as in Definition 5.6.4. As H_{2n+1} is a dilation-invariant subgroup, then so is H_0. Clearly, U normalizes H_0 and $U \ltimes H_0$ is a closed subgroup of H. Let $h = \exp(A)\exp(X) \in H$. As H is closed

$$\lim_{t\to 0} \tilde{\mu}_t(\exp(A)\exp(X)) = \lim_{t\to 0} \exp(A)\exp(\mu_t(X)) = \exp(A) \in U,$$

thus $\exp(X) \in H_0$ and $H = U \ltimes H_0$. As $\lim_{t\to 0} \exp(\mu_t(X)) = e$ for all $\exp(X)$ in H_0, the dilation invariance of H_0 leads to conclude that H_0 is connected and the Lie subalgebra of H_0 is $\mathfrak{h}_0 = \{X, \exp(X) \in H_0\}$. Let $X \in \mathfrak{h}_0$, and write $X = X' + X''$ with $X' \in \mathfrak{g}_1$ and $X'' \in \mathfrak{g}_2$. For $t \neq 0$, we have $X - \frac{1}{t}\mu_t(X) = tX'' \in \mathfrak{h}_0$. Then $X', X'' \in \mathfrak{h}_0$ and $\mathfrak{h}_0 = \mathfrak{h}_0 \cap \mathfrak{g}_1 \oplus \mathfrak{h}_0 \cap \mathfrak{g}_2$.

Conversely, if H_0 is subgraded then H_0 is dilation-invariant and we have $\tilde{\mu}_t(U \ltimes H_0) = U \ltimes \tilde{\mu}_t(H_0) = U \ltimes H_0$. □

5.6.2 Strong local rigidity results

The main results in this section are as follows.

Theorem 5.6.10. *Let G be the diamond group and Γ a nontrivial finitely generated subgroup of G (not necessarily discrete). Then there is no open G-orbit in* $\mathrm{Hom}(\Gamma, G)$. *In particular, if Γ is a discontinuous group for a homogeneous space G/H, then the strong local rigidity property fails to hold.*

Theorem 5.6.11. *Assume that H is a dilation-invariant subgroup of G and Γ is a nontrivial discontinuous group for G/H. Then for every $\varphi \in \mathscr{R}(\Gamma, G, H)$, local rigidity property fails to hold.*

Corollary 5.6.12. *Let $H = U \ltimes H_0$ be a subgroup of G compatible with a given Levi decomposition. Assume that one of the following statements is satisfied:*

(i) *The subgroup $H_0 \cap Z(H_{2n+1})$ is nontrivial.*
(ii) *The subgroup U is connected, which is the case when H is connected.*

Then for any nontrivial discontinuous group for G/H, local rigidity property fails to hold.

To prove these results, we need some preliminary results. The following observation is an important tool. The action of $\mathbb{R}_+^*$ on G defined by (5.32) induces a natural action of $\mathbb{R}_+^*$ on $\mathrm{Hom}(\Gamma, G)$ by the formula $t \cdot \varphi = \tilde{\mu}_t \circ \varphi$.

Lemma 5.6.13. *The map $\mathbb{R}_+^* \times \mathrm{Hom}(\Gamma, G)/G \to \mathrm{Hom}(\Gamma, G)/G$, $(t, [\varphi]) \mapsto [t \cdot \varphi]$ gives a well-defined continuous action of $\mathbb{R}_+^*$ on $\mathrm{Hom}(\Gamma, G)/G$. In particular, for all $\varphi \in \mathrm{Hom}(\Gamma, G)$, the map $\mathbb{R}_+^* \to \mathrm{Hom}(\Gamma, G)/G$, $t \mapsto t \cdot [\varphi]$ is continuous.*

Proof. The fact that the map is a well-defined action is derived from the equality

$$t \cdot (g \cdot \varphi) = (t \cdot g) \cdot (t \cdot \varphi).$$

Since the quotient map $\operatorname{Hom}(\Gamma, G) \to \operatorname{Hom}(\Gamma, G)/G$ is open and the action of $\mathbb{R}_+^*$ on $\operatorname{Hom}(\Gamma, G)$ is continuous, the action of $\mathbb{R}_+^*$ on $\operatorname{Hom}(\Gamma, G)/G$ is also continuous. □

The aim now is to prove Theorem 5.6.10. Let $\gamma_1, \ldots, \gamma_k$ be the generators of Γ and identify $\operatorname{Hom}(\Gamma, G)$ to a subspace of G^k via the injection $\varphi \mapsto (\varphi(\gamma_1), \ldots, \varphi(\gamma_k))$. To simplify the notation, we write φ_i instead of $\varphi(\gamma_i)$. Let $\operatorname{Fix}(\Gamma, G)$ be the set of fixed points in $\operatorname{Hom}(\Gamma, G)/G$ under the action of $\mathbb{R}_+^*$. Then as a direct consequence of Lemma 5.6.13, we obtain the following.

Lemma 5.6.14. *Let* $\varphi \in \operatorname{Hom}(\Gamma, G)$. *If the G-orbit of* φ *is open in* $\operatorname{Hom}(\Gamma, G)$, *then* $[\varphi] \in \operatorname{Fix}(\Gamma, G)$.

Proof. The action of $\mathbb{R}_+^*$ is continuous. If $[\varphi]$ is an open point in $\operatorname{Hom}(\Gamma, G)/G$, then the stabilizer of $[\varphi]$ is an open subgroup of $\mathbb{R}_+^*$. □

We consider the subset

$$F'(\Gamma, G) := \{(\varphi_1, \ldots, \varphi_k) \in \operatorname{Hom}(\Gamma, G),\ \xi_l(\varphi_i) = 0 \text{ for all } i \text{ and } l\}, \tag{5.33}$$

and $F''(\Gamma, G)$ its image by the quotient map: $\operatorname{Hom}(\Gamma, G) \to \operatorname{Hom}(\Gamma, G)/G$.

Lemma 5.6.15. $\operatorname{Fix}(\Gamma, G) \subset F''(\Gamma, G)$.

Proof. Let $[\varphi] \in \operatorname{Fix}(\Gamma, G)$, then for all $t \in \mathbb{R}_+^*$ there exists an element $g_t \in G$, such that $t \cdot \varphi = g_t \cdot \varphi$. Thus, for all $l = 1, \ldots, n$ and $i = 1, \ldots, k$, we have

$$t\xi_l(\varphi_i) = \xi_l(g_t \varphi_i g_t^{-1}),$$

which is equivalent by (5.30) to

$$t\xi_l(\varphi_i) = r_l(g_t)^{-1} r_l(\varphi_i)\xi_l(g_t) + r_l(g_t)^{-1}\xi_l(\varphi_i) - r_l(g_t)^{-1}\xi_l(g_t). \tag{5.34}$$

If $\tau_l(\varphi_i) \in 2\pi\mathbb{Z}$, then $t\xi_l(\varphi_i) = r_l(g_t)^{-1}\xi_l(\varphi_i)$ for all t, which implies that $\xi_l(\varphi_i) = 0$. For $\tau_l(\varphi_i) \notin 2\pi\mathbb{Z}$, let

$$x_{\varphi,l,i} := (I - r_l(\varphi_i))^{-1}\xi_l(\varphi_i). \tag{5.35}$$

Then from (5.34), we get

$$\xi_l(g_t) = (I - tr_l(g_t))x_{\varphi,l,i}.$$

In particular, for all i, j such that $\tau_l(\varphi_i), \tau_l(\varphi_i) \notin 2\pi\mathbb{Z}$, we have the relation $x_{\varphi,l,i} = x_{\varphi,l,j}$. Let $g \in G$ be an element of G satisfying

$$\xi_l(g) = \begin{cases} x_{\varphi,l,i} & \text{if there exist } i \text{ such that } \tau_l(\varphi_i) \notin 2\pi\mathbb{Z}, \\ 0 & \text{otherwise.} \end{cases}$$

By a direct calculation, using (5.30), we show that $\xi_l(g\varphi_i g^{-1}) = 0$ for all l and i. Therefore, $g \cdot \varphi \in F'(\Gamma, G)$ and $[g \cdot \varphi] = [\varphi]$. □

Lemma 5.6.16. *The restriction of the quotient map $\pi : \mathrm{Hom}(\Gamma, G) \to \mathrm{Hom}(\Gamma, G)/G$ to $F'(\Gamma, G)$ is a continuous bijection from $F'(\Gamma, G)$ to $F''(\Gamma, G)$.*

Proof. The restriction is continuous and by definition of $F''(\Gamma, G)$ is surjective. To see that the restriction is injective, suppose that there exists $\varphi \in F'(\Gamma, G)$ and $g \in G$ such that $g \cdot \varphi \in F'(\Gamma, G)$, we have to prove that $\varphi = g \cdot \varphi$. By definition of $F'(\Gamma, G)$,

$$\xi_l(\varphi_i) = \xi_l(g\varphi_i g^{-1}) = 0 \quad \text{for all } l \text{ and } i. \tag{5.36}$$

If there exists i such that $\tau_l(\varphi_i) \notin 2\pi\mathbb{Z}$, the last equality implies that $\xi_l(g) = 0$. Then from equality (5.31), we conclude that $\zeta(g\varphi_i g^{-1}) = \zeta(\varphi_i)$. Therefore, $\varphi = g \cdot \varphi$. □

Observe that for all $\varphi \in F'(\Gamma, G)$, the image $\varphi(\Gamma)$ is an Abelian subgroup of the Abelian subgroup

$$\mathbb{R}^n \times Z(H_{2n+1}) = \{g \in G,\ \xi_l(g) = 0,\ l = 1, \dots, n\} \cong \mathbb{R}^{n+1}.$$

Then φ factors through the quotient map π :, $\Gamma \to \Gamma' = \Gamma/[\Gamma, \Gamma]$, where $[\Gamma, \Gamma]$ is the derivative group. We consider the map

$$\begin{array}{rcl} \xi : F'(\Gamma, G) & \longrightarrow & \mathrm{Hom}(\Gamma', \mathbb{R}^{n+1}) \\ \varphi & \longmapsto & \tilde{\varphi}, \end{array}$$

where $\tilde{\varphi}(\gamma[\Gamma, \Gamma]) = \varphi(\gamma)$.

Lemma 5.6.17. *The map ξ is a homeomorphism.*

Proof. The inverse of ξ is the map $\tilde{\varphi} \mapsto \tilde{\varphi} \circ \pi$, and the bicontinuity is clear. □

The Abelian group Γ' is finitely generated. Let $T(\Gamma')$ be the subgroup of the torsion elements in Γ'. Then the quotient group $\Gamma'' = \Gamma'/T(\Gamma')$ is an Abelian torsion-free finitely generated group. Thus, Γ'' is a free Abelian group isomorphic to $\mathbb{Z}^l$, where $l = \mathrm{rk}(\Gamma'')$.

Lemma 5.6.18. *The topological spaces* $\mathrm{Hom}(\Gamma', \mathbb{R}^{n+1})$ *and* $\mathrm{Hom}(\Gamma'', \mathbb{R}^{n+1})$ *are homeomorphic.*

Proof. Note that $T(\Gamma') \subset \ker(\psi)$ for all $\psi \in \mathrm{Hom}(\Gamma', \mathbb{R}^{n+1})$. Then as before, the map

$$\begin{array}{rcl} \xi' : \mathrm{Hom}(\Gamma', \mathbb{R}^{n+1}) & \longrightarrow & \mathrm{Hom}(\Gamma'', \mathbb{R}^{n+1}) \\ \psi & \longmapsto & \tilde{\psi}, \end{array}$$

defined by $\tilde{\psi}(\gamma T(\Gamma')) = \psi(\gamma)$, is a continuous bijection, its inverse is $\tilde{\psi} \mapsto \tilde{\psi} \circ \pi'$, where π' is the quotient map $\Gamma' \to \Gamma''$. □

Proof of Theorem 5.6.10. Suppose that $[\varphi]$ is open in $\mathrm{Hom}(\Gamma, G)/G$, then by Lemmas 5.6.14 and 5.6.15, $[\varphi]$ is an open point in $F''(\Gamma, G)$. Using Lemma 5.6.16, we see that the inverse image of $[\varphi]$ in $F'(\Gamma, G)$ is an isolated point. As a consequence of Lemmas 5.6.17 and 5.6.18, the space $\mathrm{Hom}(\Gamma'', R^{n+1})$ contains an isolated point. If the rank l of Γ'' is nonzero, then $\mathrm{Hom}(\Gamma'', R^{n+1})$ is homeomorphic to $M_{n,l}(\mathbb{R})$. Thus, $l = 0$ and the set $\mathrm{Hom}(\Gamma'', R^{n+1})$ is reduced to the trivial homomorphism. Using again Lemmas 5.6.18 and 5.6.17, we see that $F'(\Gamma, G) = \{0\}$. In particular, φ is the trivial homomorphism. In this case, φ is a fixed point by the action of G on $\mathrm{Hom}(\Gamma, G)$, and thus φ is an isolated point in $\mathrm{Hom}(\Gamma, G)$. But $\mathrm{Hom}(\Gamma, G)$ is closed, then for all $\psi \in \mathrm{Hom}(\Gamma, G)$, $\lim_{t\to 0} t \cdot \psi \in F'(\Gamma, G) = \{\varphi\}$. It comes out that φ is an isolated point if only if $\mathrm{Hom}(\Gamma, G)$ is reduced to the trivial homomorphism, which is impossible because the natural injection of Γ in G is not trivial. □

5.6.3 Proofs of Theorem 5.6.11 and Corollary 5.6.12

Let us prove first the following results.

Lemma 5.6.19. *If H is a dilation-invariant subgroup, then $\mathscr{R}(\Gamma, G, H)$ is an $\mathbb{R}_+^*$-stable subspace of* $\mathrm{Hom}(\Gamma, G)$. *In particular, $\mathbb{R}_+^*$ acts continuously on $\mathscr{T}(\Gamma, G, H)$.*

Proof. As $\mathbb{R}_+^*$ acts by a continuous automorphism, φ is injective and $\varphi(\Gamma)$ is discrete if and only if $(t \cdot \varphi)$ is also injective and $(t \cdot \varphi)(\Gamma)$ is discrete. Recall that G is simply connected, then the subgroup Γ is torsion-free. Thus, the action of Γ on G/H is proper only if it is free. By definition of the parameter space (3.1), we have to prove that the action of Γ on G/H via φ is proper if and only if the action of Γ on G/H via $t \cdot \varphi$ is proper. Let S be a compact set in G. Then

$$\begin{aligned}
(t \cdot \varphi)(\Gamma)_S &= \{\gamma \in (t \cdot \varphi)(\Gamma),\ \gamma SH \cap SH \neq \emptyset\} \\
&= t\{\gamma \in \varphi(\Gamma),\ (t \cdot \gamma)SH \cap SH \neq \emptyset\} \\
&= t\{\gamma \in \varphi(\Gamma),\ t^{-1}((t \cdot \gamma)SH \cap SH) \neq \emptyset\} \\
&= t\{\gamma \in \varphi(\Gamma),\ \gamma(t \cdot S)H \cap (t \cdot S)H \neq \emptyset\} \\
&= t\varphi(\Gamma)_{t \cdot S}.
\end{aligned}$$

Therefore, the sets $(t \cdot \varphi)(\Gamma)_S$ and $\varphi(\Gamma)_{t\cdot S}$ are homeomorphic. □

Lemma 5.6.20. *Let $\varphi \in \mathscr{R}(\Gamma, G, H)$. If φ is a locally rigid, then $G \cdot \varphi \cap F'(\Gamma, G) \neq \emptyset$. In particular, Γ is an Abelian subgroup.*

Proof. Assume that φ is locally rigid, then $[\varphi]$ is an open point in $\mathscr{T}(\Gamma, G, H)$. By continuity of the action of $\mathbb{R}_+^*$, we deduce that $[\varphi] \in \mathrm{Fix}(\Gamma, G)$ and we conclude by Lemma 5.6.15 that $G \cdot \varphi \cap F'(\Gamma, G) \neq \emptyset$. Let φ' be an element of the intersection, then $\varphi'(\Gamma)$ is a subgroup of the Abelian subgroup $\mathbb{R}^n \times Z(H_{2n+1})$. As φ' is injective, Γ is Abelian. □

Lemma 5.6.21. *Let $\varphi \in F'(\Gamma, G) \cap \mathscr{R}(\Gamma, G, H)$. Then for any real number $t \neq 0$, the element $\varphi_t \in F'(\Gamma, G)$ defined by*

$$\zeta((\varphi_t)_i) = t\zeta(\varphi_i) \quad and \quad \tau_l((\varphi_t)_i) = t\tau_l(\varphi_i) \quad for\ all\ l = 1, \dots, n,$$

is also an element of $\mathscr{R}(\Gamma, G, H)$.

Proof. The subgroup $\varphi(\Gamma)$ is a discrete subgroup of the connected simply connected Lie subgroup $\mathbb{R}^n \times Z(H_{2n+1})$. Thus, $\varphi(\Gamma)$ has a syndetic hull L_φ, which is the linear span of the generators $\varphi_1, \dots, \varphi_k$. Similarly, if $t \neq 0$ the subgroup $\varphi_t(\Gamma)$ is a discrete subgroup of $\mathbb{R}^n \times Z(H_{2n+1})$ and its syndetic hull is equal to L_φ. By Fact 2.1.8, the action of $\varphi(\Gamma)$ is proper if and only if the action of $\varphi_t(\Gamma)$ is proper. □

Let $\mathbb{R}^* := \mathbb{R} \setminus \{0\}$. From Lemma 5.6.21, for $\varphi \in F'(\Gamma, G) \cap \mathscr{R}(\Gamma, G, H)$, the map

$$\begin{aligned} c_\varphi : \mathbb{R}^* &\longrightarrow \mathscr{R}(\Gamma, G, H) \\ t &\longmapsto \varphi_t \end{aligned}$$

is a well-defined and continuous injection. As a consequence, we can state the following.

Lemma 5.6.22. *The map $\overline{c_\varphi} : \mathbb{R}^* \to \mathscr{T}(\Gamma, G, H)$, $t \mapsto [\varphi_t]$ is a continuous injection.*

Proof. The map $\overline{c_\varphi}$ is the composition of two continuous injective maps, c_φ and the restriction of the quotient map to $F'(\Gamma, G) \cap \mathscr{R}(\Gamma, G, H)$. □

Proof of the Theorem 5.6.11. Suppose that φ is locally rigid. By Lemma 5.6.20, we can assume that $\varphi \in F'(\Gamma, G)$. Then the map c_φ is well-defined and using Lemma 5.6.22, we conclude that $\overline{c_\varphi}^{-1}([\varphi])$ is an open point in $\mathbb{R}^*$, which is impossible. □

Proof of Corollary 5.6.12. Assume that U is connected. By Proposition 5.6.5, the subgroup $K = U \ltimes L(H_0)$ is a syndetic hull of H, where $L(H_0)$ is the syndetic hull of H_0. As $\varphi(\Gamma)$ is a torsion-free, $\mathscr{R}(\Gamma, G, H) = \mathscr{R}(\Gamma, G, L(H))$ by Lemma 3.1.2. Proposition 5.6.2 induces the existence of $g \in G$, such that $\mathrm{Ad}_g(\mathfrak{k})$ is a subgraded subalgebra, where $\mathfrak{k}$ is the Lie subalgebra of K. Then gKg^{-1} is a dilation-invariant subgroup and by Lemma 3.1.1 we conclude that $\mathscr{R}(\Gamma, G, H) = \mathscr{R}(\Gamma, G, gKg^{-1})$.

Assume now $H \cap Z(H_{2n+1})$ is nontrivial. As before, we have $\mathscr{R}(\Gamma, G, H) = \mathscr{R}(\Gamma, G, K)$. By hypothesis, $L(H_0) \cap Z(H_{2n+1})$ is a nontrivial connected subgroup of $Z(H_{2n+1})$, which is one-dimensional, thus $Z(H_{2n+1}) \subset L(H_0)$. Then the Lie subalgebra of $L(H_0)$ is graded and from Lemma 5.6.9 we conclude that K is dilation-invariant. □

5.7 A local rigidity theorem for finite actions

Let G be a Lie group and Γ a finite group. We show here that the space $\mathrm{Hom}(\Gamma, G)/G$ is discrete and in addition finite if G has finitely many connected components. This means that in the case where Γ is a discontinuous group for the homogeneous space G/H, where H is a closed subgroup of G. All of the elements of the parameter space are locally rigid. Equivalently, any Clifford–Klein form of the finite fundamental group does not admit nontrivial continuous deformations. We have the following.

Theorem 5.7.1. *Let G be a Lie group and Γ a finite discontinuous group for a homogeneous space G/H, where H is a closed subgroup of G. Then any homomorphism in $\mathscr{R}(\Gamma, G, H)$ is strongly locally rigid.*

Theorem 5.7.1 is obtained as a direct consequence from a more general result (cf. Theorem 5.7.4 below). The first goal is to show that the quotient topology on the space $\mathrm{Hom}(\Gamma, G)/G$ turns out to be discrete for any Lie group G and for any finite group Γ. In this setting, we show that $\mathrm{Hom}(\Gamma, G)/G$ and G/G_0 share some common properties, where G_0 denotes the connected component of the identity. Namely, $\mathrm{Hom}(\Gamma, G)/G$ is discrete and countable as it is the case for G/G_0. Likewise $\mathrm{Hom}(\Gamma, G)/G$ is finite whenever G/G_0 is.

Compact subgroups of G are of paramount importance, and we first quote the following result.

Fact 5.7.2 ([82, Theorem 14.1.3], [83, Theorem 3.1] and [17]). *Any Lie group G with finitely many connected components contains a compact subgroup C with the property that for every other compact subgroup U of G, there exists $g \in G$ such that $gUg^{-1} \subset C$. Let G_0 designate as above the identity connected component of G. This subgroup has the following properties:*
- (i) *C is a maximal compact subgroup of G.*
- (ii) *$C \cap G_0$ is connected and C intersects each connected component of G.*
- (iii) *Any other maximal compact subgroup of G is conjugate to C by an element of G_0.*
- (iv) *$C_0 := C \cap G_0$ is a maximal compact subgroup of G_0.*
- (v) *If $U \subset G$ is a compact subgroup intersecting with each connected component of G and for which $U \cap G_0$ is a maximal compact subgroup of G_0, then U is a maximal compact subgroup of G.*

Our second objective is to generate a first-step solution to a conjecture of Montgomery: If C is a compact Lie group operating on a compact manifold M, then there exist at most a finite number of inequivalent orbits. For a finite group H, let $\mathrm{ord}(H)$ denote its cardinality. We record the following.

Fact 5.7.3 ([112, Lemma 1]). *Let G be a compact Lie group and B an integer. Then there are at most a finite number of mutually nonconjugate subgroups H with $\mathrm{ord}(H) \leqslant B$.*

Recall that the group G acts on $\mathrm{Hom}(\Gamma, G)$ through inner conjugation, i. e. for $\varphi \in \mathrm{Hom}(\Gamma, G)$ and $g \in G$, we consider the element φ^g of $\mathrm{Hom}(\Gamma, G)$ defined by $\varphi^g(\gamma) =$

$g\varphi(\gamma)g^{-1}$, $\gamma \in \Gamma$. In this respect, we define an equivalence relation on $\mathrm{Hom}(\Gamma, G)$ by

$$\varphi_1 \sim \varphi_2 \mod (G) \quad \text{if there exists } g \in G \text{ such that } \varphi_1^g = \varphi_2. \tag{5.37}$$

To summarize, we prove the following.

Theorem 5.7.4. *Let G be a Lie group and Γ a finite group. Then the space* $\mathrm{Hom}(\Gamma, G)/G$ *is discrete and at most countable. This space is finite if in addition G has finitely many connected components.*

Proof. Assume first that G has finitely many connected components. By Fact 5.7.2, G contains a maximal compact subgroup C. Put $\Gamma := \{\gamma_1, \dots, \gamma_p\}$. We denote by $\mathscr{F}_p(C)$ (resp., $\mathscr{F}_p(G)$) the set of subgroups of C (resp., of G) of orders less or equal to p. By Fact 5.7.2, the natural map $\mathscr{F}_p(C)/C \longrightarrow \mathscr{F}_p(G)/G$ is surjective and by Fact 5.7.3, $\mathscr{F}_p(C)/C$ is finite. Hence $\mathscr{F}_p(G)/G$ is a finite set, say, $\{[\Delta_1], \dots, [\Delta_M]\}$, where $\Delta_j \subset G$ with $\#\Delta_j \leqslant p$. Then $\#(\mathrm{Hom}(\Gamma, G)/G) \leqslant \sum_{j=1}^{M} \#(\mathrm{Hom}(\Gamma, \Delta_j))$. This shows that $\mathrm{Hom}(\Gamma, G)/G$ is finite.

We now move to the general case. Let $\varphi \in \mathrm{Hom}(\Gamma, G)$ and we define a subgroup $\widetilde{G}(\varphi) := \pi^{-1}(\pi(\varphi(\Gamma)))$ of G, where $\pi : G \longrightarrow G/G_0$ is the natural quotient map. As $\varphi(\Gamma) \subset \widetilde{G}(\varphi)$ and

$$\widetilde{G}(\varphi) = \bigcup_{\gamma \in \Gamma} \varphi(\gamma)\, G_0 = \bigcup_{1 \leqslant i \leqslant p} \varphi(\gamma_i)\, G_0,$$

then $\mathrm{Hom}(\Gamma, \widetilde{G}(\varphi))/\widetilde{G}(\varphi)$ is finite. Since G/G_0 is countable, there exist $I \subset \mathbb{N}$ and some class representatives $\{g_i\}_{i \in I}$ such that $G = \bigcup_{i \in I} g_i\, G_0$. Therefore,

$$\Lambda := \left\{ \bigcup_{i \in F} g_i\, G_0 \,\middle|\, F \subset I,\ F \text{ is finite} \right\}$$

is a countable set (as a partition of sets) and so is the set $\Sigma := \{\widetilde{G}(\varphi) \mid \varphi \in \mathrm{Hom}(\Gamma, G)\}$. There exists a sequence $\{\varphi_k\}_{k \in \mathbb{N}}$ of $\mathrm{Hom}(\Gamma, G)$ such that $\Sigma = \bigcup_{k \in \mathbb{N}} \widetilde{G}_k$, where $\widetilde{G}_k := \pi^{-1}(\pi(\varphi_k(\Gamma)))$. As the natural map,

$$\bigcup_{k \in \mathbb{N}} \mathrm{Hom}(\Gamma, \widetilde{G}_k)/\widetilde{G}_k \longrightarrow \mathrm{Hom}(\Gamma, G)/G,$$

is surjective, it follows that $\mathrm{Hom}(\Gamma, G)/G$ is countable.

Let $\{\varphi_k\}_{k \in \mathbb{N}}$ be a sequence of $\mathrm{Hom}(\Gamma, G)$ that converges to $\varphi \in \mathrm{Hom}(\Gamma, G)$. We prove that $\varphi_k \sim \varphi \bmod(\widetilde{G}(\varphi))$, for k large enough, where the relation $\sim$ is as defined above in (5.37). Note that there exists $N \in \mathbb{N}$ such that for any $k \geqslant N$ and $i \in \{1, \dots, p\}$, we have $\varphi_k(\gamma_i)G_0 = \varphi(\gamma_i)G_0$ and then $\widetilde{G}(\varphi_k) = \widetilde{G}(\varphi)$. Hence $\{\varphi_k\}_{k \in \mathbb{N}}$ converges to φ in $\mathrm{Hom}(\Gamma, \widetilde{G}(\varphi))$. As $\widetilde{G}(\varphi)$ has finitely many connected components, then $\varphi_k \sim \varphi \bmod(\widetilde{G}(\varphi))$, for k large enough. This shows that $\mathrm{Hom}(\Gamma, G)/G$ is discrete. □

5.7.1 (Strong) local rigidity for $K \ltimes \mathbb{R}^n$

Let as in Section 2.5, K be a compact subgroup of $GL_n(\mathbb{R})$ and $G := K \ltimes \mathbb{R}^n$ the semidirect product of K (with respect to the Euclidean product $\langle \cdot, \cdot \rangle$ on $\mathbb{R}^n$) and $\mathbb{R}^n$. The goal here is to study the rigidity proprieties of deformation parameters of the natural action of a discontinuous subgroup $\Gamma \subset G$, on a homogeneous space G/H, where H stands for a closed subgroup of G. That is, we prove the following local (and global) rigidity theorem: The parameter space admits a locally rigid (equivalently a strongly locally rigid) point if and only if Γ is finite. In this last situation, we also show that $\mathscr{R}(\Gamma, G, H)$ is a finite union of G-orbits for which the corresponding subgroups act fixed point freely on G/H and the deformation space $\mathscr{T}(\Gamma, G, H)$ is a finite set.

Theorem 5.7.5. *Let $G = K \ltimes \mathbb{R}^n$ be a compact extension of $\mathbb{R}^n$, H is a closed subgroup of G and Γ a discontinuous group for G/H. Then the following assertions are equivalent:*
(1) *$\mathscr{R}(\Gamma, G, H)$ contains a locally rigid element.*
(2) *Γ is finite.*
(3) *Any element of $\mathscr{R}(\Gamma, G, H)$ is a strongly, locally, rigid element.*
(4) *Any element of $\mathscr{R}(\Gamma, G, H)$ is locally rigid.*

Proof. Let Γ be a discrete subgroup of G and as above Γ_∞ a subgroup of Γ fulfilling Theorem 1.2.15. As Γ/Γ_∞ is finite, Γ_∞ is cocompact in Γ and, therefore, Γ acts properly on G/H if and only if Γ_∞ does as in Fact 2.1.8.

For $\mu \in \mathbb{R}^*$ and Γ a discrete subgroup of G, we set

$$\Gamma_\mu = \{(A, \mu x) \mid (A, x) \in \Gamma\}.$$

It is readily seen that Γ_μ is also a discrete subgroup of G. We first prove the following.

Lemma 5.7.6. *For any $\varphi \in \mathrm{Hom}(\Gamma, G)$, there exist $\varphi_1 \in \mathrm{Hom}(\Gamma, K)$ and a map $\varphi_2 : \Gamma \to \mathbb{R}^n$ such that $\varphi(\gamma) = (\varphi_1(\gamma), \varphi_2(\gamma))$ and $\varphi_2(\gamma\gamma') = \varphi_2(\gamma) + \varphi_1(\gamma)\varphi_2(\gamma')$, for any γ and $\gamma' \in \Gamma$.*

Proof. The homomorphism condition says that

$$(\varphi_1(\gamma\gamma'), \varphi_2(\gamma\gamma')) = (\varphi_1(\gamma), \varphi_2(\gamma))(\varphi_1(\gamma'), \varphi_2(\gamma')),$$

for γ and $\gamma' \in \Gamma$. This gives in turn that

$$\varphi_1 \in \mathrm{Hom}(\Gamma, K) \quad \text{and} \quad \varphi_2(\gamma\gamma') = \varphi_2(\gamma) + \varphi_1(\gamma)\varphi_2(\gamma'). \qquad \square$$

We next remark the following.

Lemma 5.7.7. *Let Γ be a discrete subgroup of G and H a closed one. Then Γ acts properly on G/H if and only if Γ_μ does, for any $\mu \in \mathbb{R}^*$.*

Proof. As K is conjugate to a subgroup of $O_n(\mathbb{R})$, we can and do assume that $K \subset O_n(\mathbb{R})$. The statement is obvious whenever Γ is finite. For this, we will consider the case where

Γ is infinite. Thanks to Lemma 1.2.5, Γ contains an element of infinite order. It is then enough to show the result when Γ is Abelian. Going through a conjugation, we can assume, as in Lemma 1.2.18, that any element (A, x) of Γ satisfies the condition

$$x = P_A(x). \tag{5.38}$$

Let $(x_1, \dots, x_p)$ be a family of maximal rank in $p_2(\Gamma)$, where $p_2 : G \longrightarrow \mathbb{R}^n$ stands for the second projection. Fix $r > 0$ and set $T_r := [C_r \cdot H \cdot C_r] \cap \Gamma_\mu$, where $C_r = K \times B(0, r)$. Clearly, T_r is a subset of $K \times p_2(T_r)$. Consider $(A, \mu x) \in T_r$. There exist (S, u), (S', u') elements of C_r and $(B, y) \in H$ such that

$$(A, \mu x) = (S, u)(B, y)(S', u').$$

Then $\mu x = Sy + v$ with $v := u + SBu' \in B(0, 2r)$. Moreover, there exist $\alpha_1, \dots, \alpha_p \in \mathbb{R}$ such that $x = \sum_{i=1}^p \alpha_i x_i$. Consider now $\overline{x} = \sum_{i=1}^p [\mu\alpha_i] x_i$, where $[a]$ designates the floor of a given real number a. It is easy to observe that $\overline{x} \in p_2(\Gamma)$. Indeed, for any $1 \leqslant i \leqslant p$, as $x_i \in p_2(\Gamma)$, there exists $A_i \in K$ such that $(A_i, x_i) \in \Gamma$. As Γ is Abelian satisfying (5.38), then thanks to Lemma 1.2.18,

$$\prod_{i=1}^p (A_i, x_i)^{[\mu\alpha_i]} = \left(\prod_{i=1}^p A_i^{[\mu\alpha_i]}, \sum_{i=1}^p [\mu\alpha_i] x_i \right).$$

Besides, we have

$$\|\mu x - \overline{x}\| \leqslant \sum_{i=1}^p (\mu\alpha_i - [\mu\alpha_i]) \cdot \|x_i\| \leqslant c := \sum_{i=1}^p \|x_i\|.$$

The latter inequality means that there exists $v' \in B(0, c)$ such that $\mu x = \overline{x} + v'$. This gives in turn that $\overline{x} = Sy + w$ where $w := v - v' \in B(0, 2r + c)$. Since $\overline{x} \in p_2(\Gamma)$, there exists $\Omega \in K$ such that $(\Omega, \overline{x}) \in \Gamma$. Hence we get

$$(\Omega, \overline{x}) = (S, w)(B, y)(B^{-1}S^{-1}\Omega, 0).$$

This in turn yields $(\Omega, \overline{x}) \in [C_{2r+c} \cdot H \cdot C_{2r+c}] \cap \Gamma$. As $[C_{2r+c} \cdot H \cdot C_{2r+c}] \cap \Gamma$ is finite, there exists c' such that $\|\overline{x}\| \leqslant c'$ and, therefore,

$$\|\mu x\| \leqslant \|\mu x - \overline{x}\| + \|\overline{x}\| \leq 2r + c + c'.$$

This proves that $p_2(T_r)$ is bounded, which implies that T_r is finite, and hence the action is proper as in Lemma 4.3.23. □

Back to the proof of Theorem 5.7.5. The assertions (3) $\Longrightarrow$ (4) and (4) $\Longrightarrow$ (1) are obvious. The assertion (2) $\Longrightarrow$ (3) comes from Theorem 5.7.1. We only prove (1) $\Longrightarrow$ (2). It suffices to show that local rigidity fails whenever Γ is infinite. Assume that Γ is infinite. Take $\varphi \in \mathscr{R}(\Gamma, G, H)$ and $\mu \in \mathbb{R}^*$. For any $\gamma \in \Gamma$, set $\varphi(\gamma) = (\varphi_1(\gamma), \varphi_2(\gamma))$ and $\varphi_\mu(\gamma) = (\varphi_1(\gamma), \mu\varphi_2(\gamma))$. Notice that $\varphi_\mu(\Gamma)$ is also discrete since φ_μ is a homomorphism and $\varphi_\mu(\Gamma) = (\varphi(\Gamma))_\mu$. As $\varphi(\Gamma)$ acts properly on G/H, we get as in Lemma 5.7.7, $\varphi_\mu(\Gamma)$ does for any $\mu \in \mathbb{R}^*$. We shall now prove that $\varphi_\mu(\Gamma)$ acts freely on G/H. Assume that for some $(S, t) \in G$ and $\mu \in \mathbb{R}^*$, we have $(S, t)(B, y)(S, t)^{-1} = (A, \mu x)$, where $(A, x) \in \varphi(\Gamma)$ and $(B, y) \in H$. This means that $(A, \mu x)$, (A, x) and (B, y) are of finite orders because $P_A(\mu x) = \mu P_A(x) = 0$, as in Lemma 1.2.5. Hence A and B are similar and thanks to Lemma 1.2.6, (A, x) and (B, y) are conjugate, then $\varphi(\Gamma)$ does not act freely on G/H, which is absurd. Therefore, $\varphi_\mu \in \mathscr{R}(\Gamma, G, H)$ for any $\mu \in \mathbb{R}^*$.

Assume now that the orbit of φ is open in $\mathscr{R}(\Gamma, G, H)$, and there exists $\varepsilon > 0$ such that for any $\mu \in]1-\varepsilon, 1+\varepsilon[$, there exists $(M_\mu, t_\mu) \in G$ such that for any $\gamma \in \Gamma$,

$$\varphi_\mu(\gamma) = (M_\mu, t_\mu)\varphi(\gamma)(M_\mu, t_\mu)^{-1}.$$

This means already that M_μ commutes with $\varphi_1(\gamma)$ and

$$M_\mu\varphi_2(\gamma) = (\varphi_1(\gamma) - I)t_\mu + \mu\varphi_2(\gamma). \tag{5.39}$$

As Γ is infinite, there exists an element $\gamma_0 \in \Gamma$ of infinite order thanks to Lemma 1.2.5. Let $(\varphi_1(\gamma_0), \varphi_2(\gamma_0)) = (A_0, x_0)$, then equation (5.39) gives

$$M_\mu x_0 = (A_0 - I_n)t_\mu + \mu x_0.$$

Since $(A_0 - I)t_\mu \in \{\ker A_0 - I\}^\perp$, then $P(A_0)M_\mu x_0 = \mu P_{A_0}(x_0)$. As P_{A_0} is a polynomial of A_0, it commutes with M_μ. At the end,

$$M_\mu P_{A_0}(x_0) = \mu P_{A_0}(x_0),$$

and $P_{A_0}(x_0)$ is not zero by Lemma 1.2.5. This is absurd as M_μ is orthogonal. □

5.7.2 (Strong) local rigidity for Heisenberg motion groups

This section aims to study the local rigidity proprieties of deformation parameters of the natural action of a discontinuous group $\Gamma \subset G = G_n$ acting on a homogeneous space G_n/H, where H stands for a closed subgroup of the Heisenberg motion group $G_n := \mathbb{U}_n \ltimes \mathbb{H}_n$. That is, the parameter space admits a locally rigid (equivalently a strongly locally rigid) point if and only if Γ is finite. The results are mainly obtained thanks to basic upshots concerning closed subgroups of the group in question as in Section 1.3. We keep all the notation and definitions there.

For any $\mu \in \mathbb{R}_+^*$, define the map $\Theta_\mu : G \to G, \quad (A, z, t) \mapsto (A, \mu z, \mu^2 t)$. Clearly, Θ_μ is an automorphism of G. For any $\varphi \in \operatorname{Hom}(\Gamma, G)$, set $\varphi_\mu := \Theta_\mu \circ \varphi$.

Proposition 5.7.8. *For any $\mu \in \mathbb{R}_+^*$ and any $\varphi \in \mathscr{R}(\Gamma, G, H)$, $\varphi_\mu \in \mathscr{R}(\Gamma, G, H)$.*

Proof. Remark first that for any $\mu \in \mathbb{R}_+^*$ and any $\varphi \in \mathscr{R}(\Gamma, G, H)$, $\varphi_\mu(\Gamma) = \Theta_\mu \circ \varphi(\Gamma)$. Then φ_μ is injective and $\varphi_\mu(\Gamma)$ is discrete. Now for $\gamma := (A_\gamma, z_\gamma, t_\gamma) \in \varphi(\Gamma)$ of infinite order, $\gamma_\mu = (A_\gamma, \mu z_\gamma, \mu^2 t_\gamma)$ is also of infinite order. By Lemma 1.3.1, there exists $u_\gamma \in \mathbb{C}^n$ such that $\tau_{u_\gamma}\gamma\tau_{u_\gamma}^{-1} = (A_\gamma, z'_\gamma, t'_\gamma)$ where $A_\gamma z'_\gamma = z'_\gamma$, $t'_\gamma z'_\gamma = 0$ and $\|z'_\gamma\| + |t'_\gamma| \neq 0$. A direct computation shows that

$$\tau_{\mu u_\gamma}\gamma_\mu\tau_{\mu u_\gamma}^{-1} = (A_\gamma, \mu z'_\gamma, \mu^2 t'_\gamma). \tag{5.40}$$

Equation (5.40) shows that for any $\mu \in \mathbb{R}_+^*$, $\varphi_\mu(\Gamma)$ and $\varphi(\Gamma)$ are of the same type. Moreover, $\operatorname{pr}(\varphi_\mu(\Gamma))$ is discrete if and only if $\operatorname{pr}(\varphi(\Gamma))$ is. Indeed, for any positive real ρ,

$$\begin{aligned}
\operatorname{Card}[\operatorname{pr}(\varphi_\mu(\Gamma)) \cap (\mathbb{U}_n \times B(0,\rho))] &= \operatorname{Card}\{(A, \mu z) \in \operatorname{pr}(\varphi_\mu(\Gamma)), \|\mu z\| \leq \rho\} \\
&= \operatorname{Card}\left\{(A, \mu z) \in \operatorname{pr}(\varphi_\mu(\Gamma)), \|z\| \leq \frac{\rho}{\mu}\right\} \\
&= \operatorname{Card}\left\{(A, z) \in \operatorname{pr}(\varphi(\Gamma)), \|z\| \leq \frac{\rho}{\mu}\right\},
\end{aligned}$$

where the symbol "Card" means the cardinality. Hence $\varphi_\mu(\Gamma)$ acts properly on G/H whenever $\varphi(\Gamma)$ does.

It remains to see that $\varphi_\mu(\Gamma)$ acts freely on G/H. Assume that one can find $\lambda_\mu \in g\varphi_\mu(\Gamma)g^{-1} \cap H$. Since the Γ-action on G/H is proper, λ_μ is of finite order and so is $\delta_\mu = g^{-1}\lambda_\mu g$. Note that $\delta_\mu = (B, \mu v, \mu^2 s)$ for some $\delta = (B, v, s) \in \varphi(\Gamma)$. By the arguments of the proofs of Lemma 1.3.1 and Proposition 1.3.3, this leads us to find $u_1, u_2 \in \mathbb{C}^n$ such that

$$\tau_{u_1}\delta\tau_{u_1}^{-1} = \tau_{u_2}\delta_\mu\tau_{u_2}^{-1} = (B, 0, 0).$$

In particular, δ and δ_μ are G-conjugate to each other. Hence δ_μ is in $g'\varphi(\Gamma)(g')^{-1} \cap H$ for some $g' \in G$. This implies that δ_μ is the identity element of G since $\varphi(\Gamma)$-action on G/H is free. Therefore, $g\varphi_\mu(\Gamma)g^{-1} \cap H = \{(I, 0, 0)\}$ for any $g \in G$, and thus the $\varphi_\mu(\Gamma)$-action on G/H is free. This completes the proof. □

The main result in this section is the following.

Theorem 5.7.9. *Let G be the Heisenberg motion group, H is a closed subgroup of G and Γ a discontinuous group for G/H. Then the statement of Theorem 5.7.5 holds.*

Proof. Thanks to Theorem 5.7.4, it suffices to show the assertion (3) $\Rightarrow$ (4), which means that the local rigidity fails whenever Γ is infinite. Take $\varphi \in \mathscr{R}(\Gamma, G, H)$ and $\mu \in \mathbb{R}_+^*$. Thanks to Proposition 5.7.8, $\varphi_\mu \in \mathscr{R}(\Gamma, G, H)$. Now assume that the G-orbit of φ is an open set in $\mathscr{R}(\Gamma, G, H)$. There exists $\varepsilon > 0$ such that for any $\mu \in]1-\varepsilon, 1+\varepsilon[$, there exists $(M_\mu, z_\mu, 0) \in G$ such that for any $\gamma \in \Gamma$,

$$\varphi_\mu(\gamma) = (M_\mu, z_\mu, 0)\varphi(\gamma)(M_\mu, z_\mu, 0)^{-1}. \tag{5.41}$$

For any $\gamma \in \Gamma$, write $\varphi(\gamma) = (A(\gamma), z(\gamma), t(\gamma))$ and then equation (5.41) reads

$$M_\mu \in \bigcap_{\gamma \in \Gamma} \mathscr{C}(A(\gamma)), \tag{5.42}$$

$$\mu z(\gamma) = M_\mu (I - A(\gamma)) M_\mu^{-1} z_\mu + M_\mu z(\gamma) \tag{5.43}$$

and

$$\mu^2 t(\gamma) = t(\gamma) - \frac{1}{2} \operatorname{Im}[\langle z_\mu, M_\mu z(\gamma)\rangle) - \langle z_\mu + M_\mu z(\gamma), M_\mu A(\gamma) M_\mu^{-1} z_\mu \rangle]. \tag{5.44}$$

Here, $\mathscr{C}(A(\gamma))$ denotes the set of commutators of $A(\gamma)$. By Proposition 1.3.8, as Γ is infinite, there exists $\gamma_0 \in \Gamma$ of infinite order. Let $\varphi(\gamma_0) = (A_0, z_0, t_0)$. One can assume thanks to Lemma 1.3.1 that $A_0 z_0 = z_0$, $t_0 z_0 = 0$ and $\|z_0\| + |t_0| \neq 0$. If $z_0 = 0$ and $t_0 \neq 0$, then (5.44) leads to $\mu^2 t_0 = t_0$ for any $\mu \in]1 - \varepsilon, 1 + \varepsilon[$, which is absurd.

If $z_0 \neq 0$ and $t_0 = 0$, then

$$\begin{aligned} \mu z_0 &= \mu P_{A_0}(z_0) \\ &= P_{A_0}(\mu z_0) \\ &\overset{\text{by (5.43)}}{=} P_{A_0} M_\mu (I - A_0) M_\mu^{-1} z_\mu + P_{A_0} M_\mu z_0 \\ &\overset{\text{by (5.42)}}{=} \underbrace{P_{A_0}(I - A_0)}_{=0} z_\mu + M_\mu P_{A_0}(z_0) \\ &= M_\mu z_0, \end{aligned}$$

which is also absurd since $M_\mu \in \mathbb{U}_n$ and $|\mu| \neq 1$. □

5.7.3 A variant of the local rigidity conjecture

As a direct consequence of Theorem 5.7.4, for any finite subgroup Γ of a connected nilpotent Lie group G, any $\varphi \in \mathscr{R}(\Gamma, G, H)$ is locally rigid and, therefore, the deformation space $\mathscr{T}(\Gamma, G, H)$ is discrete. This gives some evidence to the indirect statement of the following conjecture.

Conjecture 5.7.10. *Let G be a connected nilpotent Lie group, H a connected subgroup of G and Γ a nontrivial discontinuous group for G/H. Then the local rigidity holds if and only if Γ is a finite group.*

In this case, finite subgroups of G turn out to be central and, therefore, trivial when G is in addition simply connected. This finally shows that Conjectures 5.1.1 and 5.7.10 are equivalent in this setup.

On the other hand, it will be interesting to figure out whether Conjecture 5.1.1 may hold for compact extensions of nilpotent Lie groups as already shown in Theorems 5.7.5 and 5.7.9.

6 Stability concepts and Calabi–Markus phenomenon

Let G be a Lie group, H a closed subgroup of G and Γ a discrete subgroup of G. The concept of stability of an element $\varphi \in \operatorname{Hom}(\Gamma, G)$ measures in general the fact that in a neighborhood of φ, the properness property of the action on G/H is preserved. This concept may be one fundamental genesis to understand the local structure of the deformation space. Many open problems and related issues can be found in [32, 96, 99] and [100].

The stability fails for the $\mathbb{Z}$-action on $\mathbb{R}$ in $\operatorname{Aff}(\mathbb{R})$ [98, Example 1.1.1], whereas stability holds for the standard compact Lorentzian space form (cf. [95]). When the set $\mathscr{R}(\Gamma, G, H)$ is an open subset of $\operatorname{Hom}(\Gamma, G)$, then obviously each of its elements is stable, which is the case for any irreducible Riemannian symmetric space with the assumption that Γ is a torsion-free uniform lattice of G ([98] and [130]).

The determination of stable points is a very difficult problem in general, which is our subject of deal in this chapter. This reduces in fact to describe explicitly the interior of the subset of $\operatorname{Hom}(\Gamma, G)$ of injective homomorphisms with discrete image. This set fails to be open in $\operatorname{Hom}(\Gamma, G)$ but nevertheless the stability property may hold as reveals Example 6.4.19 below. This arouses our attention to consider some other variants of stability, such as geometric and near stability. These variants may help us in understanding the local geometric features of the deformations in question. We are then led to investigate several kinds of questions of geometric nature related to the structure of the deformation space and as a result, many stability theorems will be established in the nilpotent and exponential cases and also in the context of some compact extensions.

On the other hand, it may then happen that there does not exist an infinite discrete subgroup Γ of G, which acts properly discontinuously on G/H. This phenomenon was first discovered by E. Calabi and L. Markus [52] for $(G, H) = (SO(n, 1), SO(n-1, 1))$, and is called the Calabi–Markus phenomenon. For instance, T. Kobayashi proved in [94] that the Calabi–Markus phenomenon occurs if and only if $\operatorname{rank}_{\mathbb{R}} G = \operatorname{rank}_{\mathbb{R}} H$ when G is a reductive linear Lie group, and H a closed reductive subgroup of G. Besides, he showed in [97] that for a proper closed subgroup H of a solvable Lie group G, there exists a discontinuous group Γ for G/H such that the fundamental group $\pi_1(\Gamma\backslash G/H)$ is infinite, showing that the Calabi–Markus phenomenon does not occur in this context.

We here deal with Calabi–Markus's phenomenon and the question of existence of compact Clifford–Klein forms in the context of some compact extensions of nilpotent Lie groups. The obtained results are mainly based on several upshots proved in previous chapters.

https://doi.org/10.1515/9783110765304-006

6.1 Stability concepts

We keep all our notation and settings. For a discontinuous group Γ for a homogeneous space G/H, we pose

$$\mathrm{Hom}^0(\Gamma, G) = \{\varphi \in \mathrm{Hom}(\Gamma, G) : \varphi \text{ is injective}\}$$

and

$$\mathrm{Hom}^0_d(\Gamma, G) = \{\varphi \in \mathrm{Hom}^0(\Gamma, G) : \varphi(\Gamma) \text{ is discrete}\}.$$

Then clearly $\mathrm{Hom}^0(\Gamma, G)$ and $\mathrm{Hom}^0_d(\Gamma, G)$ coincide whenever Γ is finite.

Definition 6.1.1 (The concept of stability). Let us come back for a while to a general locally compact group G. Let Γ be a discrete subgroup of G. We focus attention here on the Γ-action on G/H when H is a noncompact closed subgroup of G. The homomorphism φ is said to be *topologically stable* or merely *stable* in the sense of Kobayashi–Nasrin [98], if there is an open set in $\mathrm{Hom}(\Gamma, G)$, which contains φ and is contained in $\mathscr{R}(\Gamma, G, H)$. When the set $\mathscr{R}(\Gamma, G, H)$ is an open subset of $\mathrm{Hom}(\Gamma, G)$, then obviously each of its elements is stable, which is the case for any irreducible Riemannian symmetric spaces with the assumption that Γ is a torsion-free uniform lattice of G (cf. [98] and [130]).

The following general fact allows to establish a relationship between local rigidity and stability properties.

Proposition 6.1.2. *A point in $\mathscr{R}(\Gamma, G, H)$ is rigid if and only if it is locally rigid and stable.*

Proof. Let U be a neighborhood in $\mathrm{Hom}(\Gamma, G)$ of a locally rigid point $u \in \mathscr{R}(\Gamma, G, H)$ and let it be contained in $\mathscr{R}(\Gamma, G, H)$. Then $G \cdot u$ is an open set of $\mathscr{R}(\Gamma, G, H)$ and is contained in $G \cdot U$, which is also open in $\mathrm{Hom}(\Gamma, G)$. Thus, $G \cdot u$ is an turn open in $\mathrm{Hom}(\Gamma, G)$; this conclusively shows that u is rigid. The converse is trivial. □

Definition 6.1.3 (Geometric stability, cf. [15]). A homomorphism $\varphi \in \mathscr{R}(\Gamma, G, H)$ is said *geometrically stable* if there exist a neighborhood $\mathscr{V}_\varphi \subset \mathrm{Hom}(\Gamma, G)$ such that for any $\psi \in \mathscr{V}_\varphi$, $\overline{\psi(\Gamma)}$, the closure of $\psi(\Gamma)$ in G acts properly on G/H.

The following remark is then immediate.

Remark 6.1.4. Any point in $\mathscr{R}(\Gamma, G, H)$ is geometrically stable if Γ is finite or H is compact.

Definition 6.1.5 (Near stability, cf. [15]). A homomorphism $\varphi \in \mathscr{R}(\Gamma, G, H)$ is said to be *nearly stable* if there is an open set in $\mathrm{Hom}^0_d(\Gamma, G)$, which contains φ and is contained in $\mathscr{R}(\Gamma, G, H)$.

The following is also immediate.

Fact 6.1.6. *Assume the* Γ *is torsion-free and* H *is compact. Then any homomorphism in* $\mathscr{R}(\Gamma, G, H)$ *is nearly stable.*

Recall now the set $\pitchfork(\Gamma : G)$ as in Definition 2.1.1, consisting of subsets H for which $SHS^{-1} \cap \Gamma$ is compact for any compact set S in G and $\pitchfork_{gp}(\Gamma : G)$ the set of all closed connected subgroups belonging to $\pitchfork(\Gamma : G)$.

Definition 6.1.7 (Stability of discrete subgroups).
(1) Let Γ be a discrete subgroup of G. We set $\mathrm{Stab}(\Gamma : G)$ the set of all $H \in \pitchfork_{gp}(\Gamma : G)$ for which the parameter space $\mathscr{R}(\Gamma, G, H)$ is open.
(2) A discrete subgroup Γ of G is said to be topologically stable (or merely stable), if $\mathrm{Stab}(\Gamma : G) = \pitchfork_{gp}(\Gamma : G)$.

Remark 6.1.8. The notion of stability is defined for discrete subgroups. For a discrete subgroup $\Gamma \subset G$, Γ becomes a discontinuous group for G/H for any $H \in \pitchfork_{gp}(\Gamma : G)$.

6.2 Stability of nilmanifold actions

6.2.1 Case of Heisenberg groups

We now come back to the Heisenberg case. We refer to Section 4.1 for notation and definitions. The aim is to provide a necessary and sufficient condition for the stability property. Such a result is fundamental when we treat the question whether the deformation space is (or is not) equipped with a smooth manifold structure. We first record the following direct consequence.

Proposition 6.2.1. *Retain the same assumptions as in Theorem 4.1.28 (or also in Theorem 4.1.33). Then the stability property holds and* G *acts on the parameter space* $\mathscr{R}(\mathfrak{l}, \mathfrak{g}, \mathfrak{h})$ *with constant dimension orbits.*

We now look at the case where Γ is not maximal in G according to Definition 4.1.38.

Proposition 6.2.2. *Let* H *be a connected subgroup of the Heisenberg group* G*, which does not meet the center of* G *and* Γ *a nonmaximal discontinuous group of* G *for the homogeneous space* G/H*. Then:*
(1) $\mathscr{T}_0(\mathfrak{l}, \mathfrak{g}, \mathfrak{h})$ *is an open dense smooth manifold of* $\mathscr{T}(\mathfrak{l}, \mathfrak{g}, \mathfrak{h})$ *and the set of stable points precisely coincides with* $R_0(\mathfrak{l}, \mathfrak{g}, \mathfrak{h})$*, which in turn consists of points whose orbits are of maximal dimension among orbits in* $R(\mathfrak{l}, \mathfrak{g}, \mathfrak{h})$.
(2) $\mathscr{T}(\mathfrak{l}, \mathfrak{g}, \mathfrak{h})$ *fails to be a Hausdorff space.*
(3) *The* G*-orbits of* $R(\mathfrak{l}, \mathfrak{g}, \mathfrak{h})$ *are not of constant dimensions.*

Proof. Let $i \in \{1, \dots, k\}$ and let $M = M(x, A, B) \in R^i(\mathfrak{l}, \mathfrak{g}, \mathfrak{h}) = \bigcup_{\theta \in I_{2n-s}^{k-1}} R^i_\theta(\mathfrak{l}, \mathfrak{g}, \mathfrak{h})$. Then

$$\operatorname{rk}\begin{pmatrix} A \\ B \end{pmatrix} = \operatorname{rk}\begin{pmatrix} A^1 \cdots A^{i-1} A^{i+1} \cdots A^k \\ B^1 \cdots B^{i-1} B^{i+1} \cdots B^k \end{pmatrix} = k - 1$$

and ${}^tB^pA^q - {}^tA^pB^q = 0$ for all $p, q = 1, \dots, k$. We stick to the notation $\mathfrak{a}$ and $\mathfrak{b}$ to designate the subspaces of $\mathbb{R}^{2n}$ spanned with the column vectors $\{M^j(0, A, B),\ j = 1, \dots, k\}$ and $\{M^j(0, -B, A),\ j = 1, \dots, k\}$, respectively. Then $\dim(\mathfrak{b}) = \dim(\mathfrak{a}) = k-1$ and as above, the homomorphism condition is being equivalent to $\mathfrak{b} \subset \mathfrak{a}^\perp$ but in turn $\mathfrak{b} \subsetneq \mathfrak{a}^\perp$ as Γ is not maximal. Fix $T = {}^t(-T^2, T^1) \in \mathfrak{a}^\perp \setminus \mathfrak{b}$ and let $M_n(x, A, B)$ be the matrix obtained when replacing in $M(x, A, B)$, the columns ${}^t(A^i, B^i)$ by

$${}^t\left(A^i + \frac{1}{n}T^1, B^i + \frac{1}{n}T^2\right).$$

Then clearly $M_n \in R_0(\mathfrak{l}, \mathfrak{g}, \mathfrak{h})$ and $(M_n)_n$ converges to M, and finally no point inside the layers $R^i_\theta(\mathfrak{l}, \mathfrak{g}, \mathfrak{h})$ ($\theta \in I_{2n-s}^{k-1}$ and $i \in \{1, \dots, k\}$) is stable. This also shows that $\mathscr{T}_0(\mathfrak{l}, \mathfrak{g}, \mathfrak{h})$ is an open, dense, smooth manifold of $\mathscr{T}(\mathfrak{l}, \mathfrak{g}, \mathfrak{h})$ as the canonical surjection is open and continuous. This completes the proof of the first point.

We now show that $\mathscr{T}(\Gamma, G, H)$ fails to be a Hausdorff space in this case. In fact, let $M_i = M(x_i, 0, C)$, $i = 1, 2$ where $x_i = (t_i, 0, \dots, 0)$ and $t_1 \neq t_2$, $A_s = 0$ and

$$C = \begin{pmatrix} 0 & 0 \\ 0 & I_{k-1} \end{pmatrix}.$$

So, obviously, $M_i \in R^1_\theta(\mathfrak{l}, \mathfrak{g}, \mathfrak{h})$ where $\theta = (2n - s - k + 2, \dots, 2n - s)$ and $[M_1] \neq [M_2]$ in $\mathscr{T}(\Gamma, G, H)$. Given any open neighborhoods V_i of M_i in $R(\mathfrak{l}, \mathfrak{g}, \mathfrak{h})$, for ε small enough, the matrix $M_i^\varepsilon = M(x_i, 0, C_\varepsilon)$ belongs to V_i for $i = 1, 2$ where

$$C_\varepsilon = \begin{pmatrix} 0 & 0 \\ \varepsilon & 0 \\ 0 & I_{k-1} \end{pmatrix}.$$

This leads to the existence of $g_\varepsilon \in G$ such that $\mathrm{Ad}_{g_\varepsilon} M_1^\varepsilon = M_2^\varepsilon$, which is enough to conclude. This proves the second point. The proof of point 3 is obvious. □

We now state our main result in this section.

Theorem 6.2.3. *Let $H = \exp(\mathfrak{h})$ be a connected subgroup of the Heisenberg group $G = \exp(\mathfrak{g})$ and Γ a discontinuous group of G for the homogeneous space G/H with a syndetic hull $L = \exp(\mathfrak{l})$. Then the following assertions are equivalent:*
1. *The space $\mathscr{T}(\mathfrak{l}, \mathfrak{g}, \mathfrak{h})$ is equipped with a smooth manifold structure.*
2. *The space $\mathscr{T}(\mathfrak{l}, \mathfrak{g}, \mathfrak{h})$ is a Hausdorff space.*
3. *$\dim G \cdot \psi$ is constant for any $\psi \in \mathscr{R}(\mathfrak{l}, \mathfrak{g}, \mathfrak{h})$.*

(4) *The stability holds.*
(5) $\mathfrak{z}(\mathfrak{g}) \subset \psi(\mathfrak{l}) + \mathfrak{h}$ *for any* $\psi \in \mathscr{R}(\mathfrak{l}, \mathfrak{g}, \mathfrak{h})$.

More generally, the space $\mathscr{T}(\mathfrak{l}, \mathfrak{g}, \mathfrak{h})$ *admits a smooth manifold as its dense open subset whose preimage consists of topologically stable and maximal-dimensional orbit points.*

Proof. The proof will be divided into several steps. Under the condition $\mathfrak{z}(\mathfrak{g}) \subset [\mathfrak{l}, \mathfrak{l}] + \mathfrak{h}$, which means that either $\mathfrak{l}$ is non-Abelian or $\mathfrak{h}$ meets the center $\mathfrak{z}(\mathfrak{g})$, any deformation space $\mathscr{T}(\mathfrak{l}, \mathfrak{g}, \mathfrak{h})$ is endowed with a smooth manifold structure as a direct consequence of Theorems 4.1.28 and 4.1.33. Beyond these cases, Γ turns out to be Abelian. With Proposition 6.2.1 in hand, we only have to prove that 5 implies 1. We first remark that Lemma 4.1.1 asserts that $\exp \mathfrak{l}$ acts properly on G/H if and only if $\mathfrak{z}(\mathfrak{g}) \not\subset \mathfrak{h}$, $\mathfrak{l} \cap \mathfrak{h} = \{0\}$ and $\mathfrak{z}(\mathfrak{g}) \cap (\mathfrak{h} \oplus \mathfrak{l}) = \mathfrak{l} \cap \mathfrak{z}(\mathfrak{g})$. This entails that this condition means that either $\mathfrak{z}(\mathfrak{g}) \subset \mathfrak{h}$ or $\mathfrak{z}(\mathfrak{g}) \subset \psi(\mathfrak{l})$ for any $\psi \in R(\mathfrak{l}, \mathfrak{g}, \mathfrak{h})$. This last fact means then that $\mathfrak{l}$ is maximal and, therefore, $R(\mathfrak{l}, \mathfrak{g}, \mathfrak{h})$ is open in $\mathrm{Hom}^{\circ}(\mathfrak{l}, \mathfrak{g})$, so that the stability property holds. Proposition 4.1.35 allows to conclude. □

Remark 6.2.4. We showed in other words that the statements of Theorem 6.2.3 only hold in the cases where Γ is not Abelian, Γ is Abelian and maximal (according to Definition 4.1.38) or Γ is Abelian and $\mathfrak{h}$ meets the center of $\mathfrak{g}$. This is also equivalent to the fact that $\mathfrak{z}(\mathfrak{g}) \subset [\mathfrak{l}, \mathfrak{l}] + \mathfrak{h}$ or Γ is Abelian and maximal, which in turn encompasses the setting of compact Clifford–Klein forms.

6.2.2 From H_{2n+1} to $H_{2n+1} \times H_{2n+1}/\Delta$

We study in this section the context of Subsection 4.1.9. Let Γ be a nontrivial discontinuous group for the homogenous space P/Δ, where $P = \exp(\mathfrak{p})$ and $\mathfrak{p} = \mathfrak{h}_{2n+1} \times \mathfrak{h}_{2n+1}$.

In this contest, as Γ is a discontinuous group for the homogeneous space P/Δ, $\mathfrak{l} \cap \mathfrak{D} = \{0\}$, then $\dim \mathfrak{l} \leq 2n + 1$. On the other hand, the vector space spanned by $\{(Z, 0), (X_i, 0), (0, X_i),\ i = 1, \dots, n\}$ is a $(2n + 1)$-dimensional Abelian subalgebra of $\mathfrak{p}$. That is, $\mathfrak{l}$ is maximal Abelian (as in Definition 4.1.37) if and only if $\dim \mathfrak{l} = 2n + 1$.

Theorem 6.2.5. *Let* Γ *be a discontinuous group for the homogeneous space* P/Δ *with the syndetic hull* $L = \exp(\mathfrak{l})$. *Then the following assertions are equivalent:*
(i) $\mathfrak{z}(\mathfrak{p}) \subset \mathfrak{D} \oplus \psi(\mathfrak{l})$ *for any* $\psi \in R(\mathfrak{l}, \mathfrak{p}, \mathfrak{D})$.
(ii) *The subalgebra* $\mathfrak{l}$ *is non-Abelian or maximal Abelian.*
(iii) *The stability holds. Namely,* $R(\Gamma, P; P/\Delta)$ *is open in* $\mathrm{Hom}(\Gamma, P)$.

Proof. Let ψ be an element of $R(\mathfrak{l}, \mathfrak{p}, \mathfrak{D})$ and $M = M(x, y, A, B)$ its matrix as above. Then $\mathfrak{z}(\mathfrak{p}) \subset \mathfrak{D} \oplus \psi(\mathfrak{l})$ if and only if there exist two vectors $S \in \mathfrak{l}$ and $T \in \mathfrak{D}$ satisfying

$$MS = (Z, 0) + T. \tag{6.1}$$

We prove first that (i) ⇔ (ii). If $\mathfrak{l}$ is non-Abelian, the matrix M equals to

$$M((x_0,\tilde{x}),(y_0,\tilde{y}),(0,A_0),(0,B_0)), \quad \text{where } x_0 \neq 0 \text{ and } \operatorname{rank}(B_0) = k-1.$$

Equality (6.1) holds by taking $S = \frac{1}{x_0}(Z,0)$ and $T = \frac{y_0}{x_0}(Z,Z)$. Suppose that $\mathfrak{l}$ is maximal Abelian. Then $\dim(\mathfrak{l}) = 2n+1$ and the matrix

$$\begin{pmatrix} x \\ B \end{pmatrix} \in \mathcal{M}_{2n+1}(\mathbb{R})$$

is invertible according to Proposition 4.1.43. Equality (6.1) holds for

$$S = \begin{pmatrix} x \\ B \end{pmatrix}^{-1} (Z,0) \quad \text{and} \quad T = \begin{pmatrix} y \\ A \end{pmatrix} S.$$

Conversely, if $\mathfrak{l}$ is Abelian and nonmaximal, let $\psi \in R(\mathfrak{l},\mathfrak{p},\mathfrak{D})$ such that its matrix $M = M(x,y,0,B)$ with

$$B = \begin{pmatrix} I_k \\ 0 \end{pmatrix}.$$

Equation (6.1) gives that $BS = 0$ and so $S = 0$, which is impossible as $(Z,0) \notin \mathfrak{D}$.

We prove now that (ii) ⇔ (iii). When $\mathfrak{l}$ is non-Abelian, Corollary 4.1.44 enables us to conclude that any element of $R(\mathfrak{l},\mathfrak{p},\mathfrak{D})$ is stable. In case $\mathfrak{l}$ is maximal Abelian, $\dim(\mathfrak{l}) = 2n+1$. Therefore, the Clifford–Klein form $\Gamma\backslash G/H$ is compact and the parameter space is open in $\operatorname{Hom}(\mathfrak{l},\mathfrak{p})$. Conversely, if $\mathfrak{l}$ is Abelian and nonmaximal to see that $R(\mathfrak{l},\mathfrak{p},\mathfrak{D})$ is not open in $\operatorname{Hom}(\mathfrak{l},\mathfrak{p})$, choose $M_0 = M(x_0,y_0,0,B_0) \in R(\mathfrak{l},\mathfrak{p},\mathfrak{D})$ such that $x_0 = (1,0,\dots,0)$ and

$$B_0 = \begin{pmatrix} I_{k-1} \\ 0 \end{pmatrix}.$$

Let $M_\varepsilon = M(x_0,y_0,A_\varepsilon,B_0)$, with

$$A_\varepsilon = \begin{pmatrix} A_{1,\varepsilon} \\ 0 \end{pmatrix}$$

such that $A_{1,\varepsilon} = (a_{i,j}) \in \mathcal{M}_{n,k}(\mathbb{R})$, $a_{1,1} = \varepsilon$ and $a_{i,j} = 0$ elsewhere. It is easy to check that the matrix M_ε satisfies the homomorphism conditions as here $K = 0$, and hence M_ε is in $\mathcal{E}$. For $w_\varepsilon = (\frac{1}{\varepsilon},0,\dots,0) \in \mathbb{R}^{2n}$, we have

$$\operatorname{rk}\begin{pmatrix} x_0 - w_\varepsilon A_\varepsilon \\ B \end{pmatrix} = k-1.$$

So, for all $\varepsilon > 0$, $M_\varepsilon \notin R(\mathfrak{l},\mathfrak{p},\mathfrak{D})$ according to Proposition 4.1.43. □

Hausdorffness of the deformation space

Theorem 6.2.6. *Let Γ be a discontinuous group for the homogeneous space P/Δ. Then the deformation space $\mathscr{T}(\Gamma, P; P/\Delta)$ is a Hausdorff space if and only if Γ is maximal Abelian.*

Proof. When the subgroup Γ is maximal Abelian, then by Proposition 6.2.8, the deformation space is a Hausdorff space. We first study the case where $\mathfrak{l}$ is Abelian and nonmaximal. In this case, $\dim(\mathfrak{l}) \leq 2n$. Let $M = M(x,y,A,B)$ and $M' = M(x,y',A,B)$ with $y = (y_1, 0, \dots, 0)$, $y' = (y_1', 0, \dots, 0)$, $y_1 \neq y_1'$,

$$A = \begin{pmatrix} A_1 \\ A_2 \end{pmatrix} \quad \text{and} \quad B = \begin{pmatrix} B_1 \\ B_2 \end{pmatrix}.$$

First, assume that $\mathrm{rk}(\Gamma) = k \leq n$ and take here $A = 0$ and

$$B = \begin{pmatrix} I_k \\ 0 \end{pmatrix}.$$

For $\varepsilon > 0$, let also $M_\varepsilon = M(x,y,A_\varepsilon,B)$ and $M'_\varepsilon = M(x,y',A_\varepsilon,B)$ be such that $A_\varepsilon = (a_{i,j})$ where $a_{2n,1} = \varepsilon$ and $a_{i,j} = 0$ otherwise. It is clear that M, M', M_ε and M'_ε are in $R(\mathfrak{l}, \mathfrak{p}, \mathfrak{D})$ and that $[M] \neq [M']$. For

$$u_\varepsilon = \left(\frac{y_1' - y_1}{\varepsilon}, 0, \dots, 0\right), \quad v_\varepsilon = \left(0, \dots, 0, \frac{y_1' - y_1}{\varepsilon}\right)$$

and $g_\varepsilon \in P$ parameterized by $(u_\varepsilon, v_\varepsilon)$, one has $g_\varepsilon \cdot M_\varepsilon = M'_\varepsilon$ and so $[M_\varepsilon] = [M'_\varepsilon]$. Let $V_{[M]}$ and $V_{[M']}$ be two open sets in $\mathscr{T}(\mathfrak{l}, \mathfrak{p}, \mathfrak{D})$ containing $[M]$ and $[M']$, respectively. If σ denotes the projection map from $R(\mathfrak{l}, \mathfrak{p}, \mathfrak{D})$ to $\mathscr{T}(\mathfrak{l}, \mathfrak{p}, \mathfrak{D})$, then $\sigma^{-1}(V_{[M]}) = V_M$ and $\sigma^{-1}(V_{[M']}) = V_{M'}$ are two open sets in $R(\mathfrak{l}, \mathfrak{p}, \mathfrak{D})$ containing M and M', respectively. There exists therefore $\varepsilon > 0$ such that $M_\varepsilon \in V_M$ and $M'_\varepsilon \in V_{M'}$, which shows that $V_{[M]} \cap V_{[M']} \neq \emptyset$. Thus, the space $\mathscr{T}(\mathfrak{l}, \mathfrak{p}, \mathfrak{D})$ fails to be a Hausdorff space.

Suppose now that $n < k < 2n$ and choose A and B in $\mathscr{M}_{2n,k}(\mathbb{R})$ such that

$$B_1 = \begin{pmatrix} I_n & 0 \end{pmatrix}, \quad B_2 = \begin{pmatrix} 0 & I_{k-n} \\ 0 & 0 \end{pmatrix}, \quad A_1 = 0, \quad A_2 = -B_2.$$

Let $M_\varepsilon = M(x,y,A_\varepsilon,B_\varepsilon)$ and $M'_\varepsilon = M(x,y',A_\varepsilon,B_\varepsilon)$ such that $A_{1\varepsilon} = 0$, $B_{1,\varepsilon} = B_1$,

$$B_{2,\varepsilon} = -A_{2\varepsilon} = \begin{pmatrix} 0 & I_{k-n} \\ \nabla_\varepsilon & 0 \end{pmatrix},$$

where $\nabla_\varepsilon = (\mu_{i,j}) \in \mathscr{M}_{2n-k,n}(\mathbb{R})$ with $\mu_{1,1} = \varepsilon$ and $\mu_{i,j} = 0$ otherwise. Then the same conclusion holds when we take $u_\varepsilon = (u_1, \dots, u_{2n})$ and $v_\varepsilon = (v_1, \dots, v_{2n})$ in $\mathbb{R}^{2n}$ such that $u_1 = v_{k+1} = \frac{y_1 - y_1'}{\varepsilon}$ and $u_i = v_j = 0$ otherwise.

Finally, if $k = 2n$, let $B_1 = (I_n \quad 0)$,

$$B_2 = \begin{pmatrix} 0_{\mathscr{M}_{n-1,1}(\mathbb{R})} & I_{n-1} & 0_{\mathscr{M}_{n-1,n}(\mathbb{R})} \\ 0 & 0_{\mathscr{M}_{1,n-1}(\mathbb{R})} & 0_{\mathscr{M}_{1,n}(\mathbb{R})} \end{pmatrix},$$

$A_1 = 0$ and $A_2 = -B_2$. Choose $M_\varepsilon = M(x, y, A_\varepsilon, B)$ and $M'_\varepsilon = M(x, y', A_\varepsilon, B)$ such that

$$B_{1,\varepsilon} = B_1, \quad B_{2,\varepsilon} = \begin{pmatrix} 0_{\mathscr{M}_{n-1,1}(\mathbb{R})} & I_{n-1} & 0_{\mathscr{M}_{n-1,n}(\mathbb{R})} \\ \varepsilon & 0_{\mathscr{M}_{1,n-1}(\mathbb{R})} & 0_{\mathscr{M}_{1,n}(\mathbb{R})} \end{pmatrix}, \quad A_{1\varepsilon} = 0 \quad \text{and} \quad A_{2\varepsilon} = -B_{2\varepsilon}.$$

Suppose here that $x = (0, \dots, 0, 1)$. This gives that $M, M', M_\varepsilon, M'_\varepsilon$ are all in $R(\mathfrak{l}, \mathfrak{p}, \mathfrak{D})$ and $[M] \neq [M']$ as $\operatorname{rank}(B) = k - 1$ in these circumstances. However, for $u_\varepsilon = (u_1, \dots, u_{2n})$ and $v_\varepsilon = (v_1, \dots, v_{2n})$ in $\mathbb{R}^{2n}$ such that $u_1 = v_{k+1} = \frac{y_1 - y'_1}{\varepsilon}$ and $u_i = v_j = 0$ otherwise. We get the result.

Let now $\mathfrak{l}$ be a non-Abelian subalgebra of $\mathfrak{p}$ of dimension k spanned by $\{e_1, \dots, e_k\}$ such that $\mathfrak{l} \cap \mathfrak{z} = \mathbb{R}e_1$. Suppose first that $k \leq n + 1$. For $i = 1, 2$, let

$$M_i = M((1, \tilde{x}), (0, \tilde{y}_i), (0, 0), (0, B))$$

such that $\tilde{x}$, $\tilde{y}_i \in \mathbb{R}^{k-1}$, $\tilde{y}_1 \neq \tilde{y}_2$ and

$$B = \begin{pmatrix} B_1 \\ B_2 \end{pmatrix}, \quad B_1 = \begin{pmatrix} I_{k-1} \\ 0 \end{pmatrix}, \quad B_2 = \begin{pmatrix} -\frac{1}{2}K_0 \\ 0 \end{pmatrix}.$$

So, M_1 and M_2 are in $R(\mathfrak{l}, \mathfrak{p}, \mathfrak{D})$ and $[M_1] \neq [M_2]$. Moreover, for $\varepsilon > 0$, let

$$M_{i\varepsilon} = M((1 + \varepsilon, \tilde{x}), (0, \tilde{y}_i), (0, A_\varepsilon), (0, B))$$

such that

$$A_\varepsilon = \begin{pmatrix} A_{1\varepsilon} \\ A_{2\varepsilon} \end{pmatrix} \quad \text{with } A_{1\varepsilon} = \begin{pmatrix} \varepsilon I_{k-1} \\ 0 \end{pmatrix} \text{ and } A_{2\varepsilon} = 0.$$

It is clear that $M_{i\varepsilon}$ are in $R(\mathfrak{l}, \mathfrak{p}, \mathfrak{D})$ and as $\operatorname{rk}(A_\varepsilon) = \operatorname{rk}(B) = \operatorname{rk}(A_\varepsilon + B) = k - 1$, $[M_{1\varepsilon}] = [M_{2\varepsilon}]$, which is enough to conclude that the deformation space fails to be a Hausdorff space.

Assume now that $\mathfrak{l}$ is non-Abelian and $n + 2 \leq \dim(\mathfrak{l}) \leq 2n + 1$ and take a basis $\{U, e_2, \dots, e_n, e_{n+1}, \dots, e_k\}$ of $\mathfrak{l}$. We can and do assume that $[e_i, e_j] = 0$ for all $i, j \geq n + 1$. In these circumstances, the matrix K_0 can be written in this basis as

$$K_0 = \begin{pmatrix} K_1 & K_2 \\ -{}^tK_2 & 0_{\mathscr{M}_{k-1-n}(\mathbb{R})} \end{pmatrix},$$

where $K_1 \in \mathscr{M}_n(\mathbb{R})$ is a skew symmetric matrix and K_2 is in $\mathscr{M}_{n,k-1-n}(\mathbb{R})$. Choose then two matrices

$$M_i = M((1,\tilde{x}),(0,\tilde{y}_i),(0,A),(0,B)), \quad i = 1,2,$$

such that $\tilde{x}$, $\tilde{y}_i \in \mathbb{R}^{k-1}$, $\tilde{y}_1 = (a,0,\ldots,0)$, $\tilde{y}_2 = (b,0,\ldots,0)$, $a \neq b$,

$$B = \begin{pmatrix} B_1 \\ B_2 \end{pmatrix} = \begin{pmatrix} I_{k-1} \\ 0 \end{pmatrix}$$

with

$$B_1 = \begin{pmatrix} I_n \\ 0 \end{pmatrix} \quad \text{and} \quad A = \begin{pmatrix} 0 \\ A_2 \end{pmatrix}, \quad A_2 \in \mathscr{M}_{n,k-1}(\mathbb{R}).$$

Let $E = K_0 - {}^tB_2B_1 + {}^tB_1B_2$. Then

$$E = \begin{pmatrix} E_1 & E_2 \\ -{}^tE_2 & 0_{\mathscr{M}_{n,k-1-n}(\mathbb{R})} \end{pmatrix},$$

where $E_1 \in \mathscr{M}_n(\mathbb{R})$ is a skew symmetric matrix and E_2 is in $\mathscr{M}_{n,k-1-n}(\mathbb{R})$. Then the homomorphism conditions (4.23) hold when

$${}^tA_2B_1 - {}^tB_1A_2 = E.$$

Putting $A_2 = (A' \quad A'')$ for $A' \in \mathscr{M}_n(\mathbb{R})$ and $A'' \in \mathscr{M}_{n,k-1-n}(\mathbb{R})$, the last equation holds when $A'' = E_2$ and A' is an upper triangular matrix such that ${}^tA' - A' = E_1$. So, M_1 and M_2 are in $R(\mathfrak{l},\mathfrak{p},\mathfrak{D})$ and as the first column of A vanishes, for all $v \in \mathbb{R}^{2n}$ such that $vA \neq \tilde{y}_1 - \tilde{y}_2$, that is, $[M_1] \neq [M_2]$. Moreover, for $\varepsilon > 0$, let

$$M_{i\varepsilon} = M((1,\tilde{x}),(0,\tilde{y}_i),(0,A_\varepsilon),(0,B)), \quad i = 1,2,$$

where

$$A_\varepsilon = \begin{pmatrix} A_{1\varepsilon} \\ A_{2\varepsilon} \end{pmatrix} \quad \text{and} \quad A_{1\varepsilon} = (a_{ij}) \in \mathscr{M}_{n,k-1}(\mathbb{R})$$

such that $a_{n1} = \varepsilon$ and $a_{ij} = 0$ elsewhere. We can easily check as earlier that $M_{i\varepsilon}$ are in $R(\mathfrak{l},\mathfrak{p},\mathfrak{D})$ and $[M_{1\varepsilon}] = [M_{2\varepsilon}]$, which is enough to conclude that the deformation space fails to be a Hausdorff space. □

Remark 6.2.7.

(1) In [39], S. Barmeier, studied the case where $n = 1$ and Γ is the discrete Heisenberg group where the parameter space is defined modulo the kernel $G \to \mathrm{Diff}(G/H)$. He proved that in these circumstances the deformation space is homeomorphic to $GL_2(\mathbb{R}) \times \mathbb{R}^\times \times \mathbb{R}^3$.

(2) In the case of the Heisenberg group H_{2n+1} and H and Γ are arbitrary, Theorem 6.2.3 says that there is equivalence between the fact that $\mathscr{T}(\Gamma, H_{2n+1}; H_{2n+1}/H)$ is a Hausdorff space and the fact that the stability holds. We remark that here it is no longer the case $P = H_{2n+1} \times H_{2n+1}$. The reader can consult the reference [60] for more details.

6.2.3 Case of 2-step nilpotent Lie groups

Coming back to our setting where G is nilpotent and 2-step, we refer to Section 4.2 for all of the materials. From the fact that the pair (G, H) have the Lipsman property (cf. Definition 2.1.10), if $\mathfrak{l}$ is an ideal of $\mathfrak{g}$, then $\exp(\mathfrak{l})$ acts properly on G/H if and only if $\mathfrak{l} \cap \mathfrak{h} = \{0\}$ (cf. Lemma 2.1.11). As a consequence of this observation, we get the following result concerning the case where $\mathfrak{l}$ is a maximal Abelian subalgebra (as in Definition 4.1.37) of $\mathfrak{g}$.

Proposition 6.2.8. *If $\mathfrak{l}$ is a maximal Abelian subalgebra of $\mathfrak{g}$, then the stability property holds.*

Proof. As $\mathfrak{l}$ is maximal and Abelian, then $\varphi(\mathfrak{l})$ is also maximal and Abelian for any $\varphi \in \mathrm{Hom}^0(\mathfrak{l}, \mathfrak{g})$. It follows that $\mathfrak{z} \subset \varphi(\mathfrak{l})$ for all $\varphi \in \mathrm{Hom}^0(\mathfrak{l}, \mathfrak{g})$ and, therefore,

$$\begin{aligned}\mathscr{R}(\mathfrak{l}, \mathfrak{g}, \mathfrak{h}) &= \{\varphi \in \mathrm{Hom}^0(\mathfrak{l}, \mathfrak{g}) : \ \mathrm{Ad}_g\, \varphi(\mathfrak{l}) \cap \mathfrak{h} = \{0\}\} \\ &= \{\varphi \in \mathrm{Hom}^0(\mathfrak{l}, \mathfrak{g}) : \ \varphi(\mathfrak{l}) \cap \mathfrak{h} = \{0\}\},\end{aligned}$$

which is an open set of $\mathrm{Hom}(\mathfrak{l}, \mathfrak{g})$. □

As a direct consequence of Theorem 4.2.3, we get the following.

Proposition 6.2.9. *If $\dim \mathfrak{z}' = \dim[\mathfrak{l}, \mathfrak{l}]$, then the stability property holds.*

Proof. Put $s = \dim([\mathfrak{l}, \mathfrak{l}])$, then we have $A_2 \in M_s(\mathbb{R})$. Thus, the inequation $\mathrm{rk}(B_4) < \dim \mathfrak{l}'$ implies

$$\mathrm{rk}\begin{pmatrix} A_2 & \star \\ 0 & B_4 \end{pmatrix} \leq s + \mathrm{rk}(B_4) < \dim \mathfrak{l}.$$

This means $\mathscr{R}_2$ is empty and $\mathscr{R}(\mathfrak{l}, \mathfrak{g}, \mathfrak{h}) = \mathscr{R}_1$, which is open in $\mathrm{Hom}(\mathfrak{l}, \mathfrak{g})$. □

Remark 6.2.10. Note that when $\mathfrak{l}$ is Abelian, the hypothesis of Proposition 6.2.9 holds if and only if $\mathfrak{z} \subset \mathfrak{h}$, which means in particular that $\mathfrak{h}$ is an ideal of $\mathfrak{g}$. So the proposition is a consequence from a more general result, easily derived from Lemma 2.1.11: If $\mathfrak{h}$ is an ideal of $\mathfrak{g}$, then the parameter space is open in $\mathrm{Hom}(\mathfrak{l}, \mathfrak{g})$.

A stability theorem

In the rest of this section, we give a new sufficient criterion of stability, which is useful when Γ is Abelian as we show in some examples.

Definition 6.2.11. Let $\mathfrak{l}$, $\mathfrak{g}$ and $\mathfrak{h}$ be as above, the subalgebra $\mathfrak{l}$ is said to satisfy $(\star)$ for $\mathfrak{g}/\mathfrak{h}$ if there is a decomposition of $\mathfrak{h} = (\mathfrak{z}\cap\mathfrak{h})\oplus\mathfrak{h}'\oplus\mathfrak{h}''$ and $\mathfrak{g} = (\mathfrak{z}\cap\mathfrak{h})\oplus\mathfrak{z}'\oplus\mathfrak{h}'\oplus\mathfrak{h}''\oplus V = \mathfrak{h}\oplus\mathfrak{z}'\oplus V$. Here, $\mathfrak{h}''$ and V are some subspaces of $\mathfrak{g}$ such that:

$(\star 1)$ $(\mathfrak{z}\cap\mathfrak{h})\oplus\mathfrak{h}'$ is an ideal of $\mathfrak{g}$.

$(\star 2)$ $\operatorname{rk}\begin{pmatrix} B_3'' \\ B_4 \end{pmatrix} = \operatorname{rk}(B_4)$ or $B_4 = 0$, where $B_3 = \begin{pmatrix} B_3' \\ B_3'' \end{pmatrix}$.

Example 6.2.12. Let $\mathfrak{g} = \mathbb{R}\text{-span}\{X, Y_1, \dots, Y_n, Z_1, \dots, Z_n\}$ such that $[X, Y_i] = Z_i$ for $1 \le i \le n$. For all $r > 0$ and $t \ge 0$, let $\mathfrak{h} = \mathfrak{h}_{r,t} = \mathbb{R}\text{-span}\{X, Y_1, \dots, Y_{r-t-1}, Z_1, \dots, Z_{r-1}\}$ and $\mathfrak{l} = \mathfrak{l}_{r,t,q} = \mathbb{R}\text{-span}\{Y_{r+t+q}, \dots, Y_n, Z_r, \dots, Z_n\}$ with $0 \le q \le n-r-t$. Then $\mathfrak{h}_{r,t}$ and $\mathfrak{l}_{r,t,q}$ are subalgebras of $\mathfrak{g}$ and the subalgebras

$$(\mathfrak{z}\cap\mathfrak{h})\oplus\mathfrak{h}' = \mathbb{R}\text{-span}\{Y_1, \dots, Y_{r+t-1}, Z_1, \dots, Z_{r-1}\}$$

of $\mathfrak{h}$ is an ideal of $\mathfrak{g}$ and $\mathfrak{h}'' = \mathbb{R}\text{-span}\{X\}$.

$$B_1 = (u_{i,j})_{\substack{1\le i\le r-1,\\ 1\le j\le k}}, \quad B_2 = (u_{i,j})_{\substack{r\le i\le n,\\ 1\le j\le k}}, \quad B_3' = (v_{i,j})_{\substack{1\le i\le r+t-1,\\ 1\le j\le k}}, \quad B_3'' = (x_1, \dots, x_k)$$

and $B_4 = (v_{i,j})_{\substack{r+t\le i\le n,\\ 1\le j\le k}}$. Then

$$\begin{aligned}\operatorname{Hom}(\mathfrak{l}, \mathfrak{g}) &= \left\{ \begin{pmatrix} (u_{i,j})_{1\le i\le n, 1\le j\le k} \\ (v_{i,j})_{1\le i\le r+t-1, 1\le j\le k} \\ (x_j)_{1\le j\le k} \\ (v_{i,j})_{r+t\le i\le n, 1\le j\le k} \end{pmatrix} \middle| \begin{array}{l} x_i v_{l,j} - x_j v_{l,i} = 0,\ 1 \le l, i \le n \\ \text{and } 1 \le j \le k \end{array} \right\} \\ &= \{\varphi(B) \in M_{2n+1,k}(\mathbb{R}) : B_3'' = 0\} \\ &\quad \cup \left\{\varphi(B) \in M_{2n+1,k}(\mathbb{R}) : \operatorname{rk}\begin{pmatrix} B_3'' \\ B_4 \end{pmatrix} \le 1\right\},\end{aligned}$$

and $\mathfrak{l}$ satisfies $(\star)$ for $\mathfrak{g}/\mathfrak{h}$.

Example 6.2.13. Let $\mathfrak{g} = \mathbb{R}\text{-span}\{X_1, \dots, X_n, (Z_{i,j})_{1\le i<j\le n}\}$ such that $[X_i, X_j] = Z_{i,j}$ for $i \ne j$. Let $\mathfrak{h} = \mathbb{R}\text{-span}\{X_1, \dots, X_q, (Z_{i,j})_{1\le i<j\le q}\}$, $\mathfrak{h}'' = \mathbb{R}\text{-span}\{X_1, \dots, X_q\}$, $\mathfrak{h}' = \{0\}$, $\mathfrak{z}\cap\mathfrak{h} = \mathbb{R}\text{-span}\{(Z_{i,j})_{1\le i<j\le q}\}$ of dimension r, $\mathfrak{l} = \mathfrak{z}' \oplus \mathbb{R}X_n$ an Abelian subalgebra of $\mathfrak{g}$ with $\mathfrak{z}' = \mathbb{R}\text{-span}\{(Z_{i,j})_{q\le i<j\le n}\}$ and $V = \mathbb{R}\text{-span}\{X_{q+1}, \dots, X_n\}$. Then we have

$$\operatorname{Hom}(\mathfrak{l}, \mathfrak{g}) = \left\{\varphi(B) \in M_{r+m'+n,k}(\mathbb{R}) : \operatorname{rk}\begin{pmatrix} B_3'' \\ B_4 \end{pmatrix} \le 1\right\},$$

and $\mathfrak{l}$ satisfies $(\star)$ for $\mathfrak{g}/\mathfrak{h}$.

Example 6.2.14. Let $\mathfrak{g}$ be a 2-step nilpotent Lie group, $\mathfrak{l}$ an subalgebra of $\mathfrak{g}$ and $\mathfrak{h}$ an ideal of $\mathfrak{g}$, then we can write $\mathfrak{h} = (\mathfrak{z} \cap \mathfrak{h}) \oplus \mathfrak{h}'$ and, therefore, $\mathfrak{h}'' = \{0\}$, so as in Definition 6.2.11, for all $\varphi(A, B) \in \mathrm{Hom}(\mathfrak{l}, \mathfrak{g})$ we have B_3'' is a trivial matrix. Then $\mathfrak{l}$ satisfies the property $(\star)$ for $\mathfrak{g}/\mathfrak{h}$.

Example 6.2.15. Let $\mathfrak{g} = \mathbb{R}\text{-span}\{X, T, Y_1, \ldots, Y_4, Z_1, \ldots, Z_4, U_0, U_3, U_4\}$ such that $[X, Y_i] = Z_i$, $i = 1, \ldots, 4$, $[X, T] = U_0$, $[T, Y_3] = U_3$ and $[T, Y_4] = U_4$. Take $\mathfrak{h} = \mathbb{R}\text{-span}\{X, Y_4, Z_4, U_4\}$ and $\mathfrak{l} = \mathbb{R}\text{-span}\{Y_1, Y_2, Y_3, Z_1, Z_2, Z_3\}$. Let

$$\mathscr{B} = \{Z_4, U_0, U_3, U_4, Z_1, Z_2, Z_3, Y_4, X, T, Y_1, Y_2, Y_3\}$$

be a basis of $\mathfrak{g}$ and

$$\varphi = \begin{pmatrix} B_1 \\ B_2 \\ B_3' \\ B_3'' \\ B_4 \end{pmatrix}$$

with

$$B_1 = \begin{pmatrix} z_{4,1} & \cdots & z_{4,6} \\ u_{0,1} & \cdots & u_{0,6} \\ u_{3,1} & \cdots & u_{3,6} \\ u_{4,1} & \cdots & u_{4,6} \end{pmatrix}, \quad B_2 = \begin{pmatrix} z_{1,1} & \cdots & z_{1,6} \\ z_{2,1} & \cdots & z_{2,6} \\ z_{3,1} & \cdots & z_{3,6} \end{pmatrix},$$

$B_3' = (y_{4,1}, \ldots, y_{4,6})$, $B_3'' = (x_1, \ldots, x_6)$ and

$$B_4 = \begin{pmatrix} t_1 & \cdots & t_6 \\ y_{1,1} & \cdots & y_{1,6} \\ y_{2,1} & \cdots & y_{2,6} \\ y_{3,1} & \cdots & y_{3,6} \end{pmatrix}.$$

Then

$$\begin{aligned} \mathrm{Hom}(\mathfrak{l}, \mathfrak{g}) &= \left\{ \begin{pmatrix} B_1 \\ B_2 \\ B_3' \\ B_3'' \\ B_4 \end{pmatrix} \in M_{13,6}(\mathbb{R}) \;\middle|\; \begin{array}{l} x_j t_k - x_k t_j = 0, \\ x_j y_{i,k} - x_k y_{i,j} = 0,\ i = 1, \ldots, 4, \\ t_j y_{i,k} - t_k y_{i,j} = 0,\ i = 3, 4, \\ 1 \leq j, k \leq 6 \end{array} \right\} \\ &= \left\{ \begin{pmatrix} B_1 \\ B_2 \\ B_3' \\ B_3'' \\ B_4 \end{pmatrix} \in M_{13,6}(\mathbb{R}) \;\middle|\; \mathrm{rk}\begin{pmatrix} B_3'' \\ B_4 \end{pmatrix} \leq 1 \right\} \end{aligned}$$

$$\cup \left\{ \begin{pmatrix} B_1 \\ B_2 \\ B_3' \\ 0 \\ B_4 \end{pmatrix} \in M_{13,6}(\mathbb{R}) \;\middle|\; \begin{matrix} t_j y_{i,k} - t_k y_{i,j} = 0, \ i = 3, 4, \\ 1 \le j, k \le 6 \end{matrix} \right\}.$$

Therefore, $\mathfrak{l}$ satisfies $(\star)$ for $\mathfrak{g}/\mathfrak{h}$.

Remark 6.2.16. As it is the case of Heisenberg groups, the Hausdorff property of the deformation space is equivalent to the fact that the stability property holds (cf. Theorem 6.2.3). The following example shows that such a phenomenon may fail for general 2-step nilpotent Lie groups.

Example 6.2.17. Let us resume Example 6.2.15 above. In this situation, we have $\mathfrak{z}' = \mathbb{R}\text{-span}\{Z_1, Z_2, Z_3, U_0, U_3\}$ and $\dim(\mathfrak{z}') < \dim(\mathfrak{l})$, so the hypothesis of Theorem 6.2.18 is satisfied. This shows that $\mathscr{R}(\mathfrak{l}, \mathfrak{g}, \mathfrak{h})$ is open in $\mathrm{Hom}(\mathfrak{l}, \mathfrak{g})$. We now write $\mathfrak{g} = \mathfrak{z}_1 \oplus \mathfrak{z}_2 \oplus \mathfrak{z}_3 \oplus \mathfrak{h}' \oplus \mathfrak{l}' \oplus V$, where $\mathfrak{z} \cap \mathfrak{l} = \mathbb{R}\text{-span}\{Z_1, Z_2, Z_3\} = \mathfrak{z}_1$, $\mathfrak{l}' = \mathbb{R}\text{-span}\{Y_1, Y_2, Y_3\}$, $\mathfrak{z} \cap \mathfrak{h} = \mathfrak{z}_2 = \mathbb{R}\text{-span}\{U_4, Z_4\}$, $\mathfrak{h}' = \mathbb{R}\text{-span}\{X, Y_4\}$, $\mathfrak{z}_3 = \mathbb{R}\text{-span}\{U_0, U_3\}$ and $V = \mathbb{R}\text{-span}\{T\}$. If we choose $\mathfrak{l}_1 = \mathbb{R}\text{-span}\{Y_1, Y_2, Z_1, Z_2, Z_3\}$ as a subalgebra of $\mathfrak{l}$ of codimension one, then we obtain $\mathfrak{l}_1^{\perp} = \mathbb{R}\text{-span}\{T, Y_1, Y_2, Y_3, Y_4, Z_1, Z_2, Z_3, Z_4, U_0, U_3, U_4\}$. Then $[\mathfrak{l}_1^{\perp}, \mathfrak{l}] \ni U_3 \notin \mathfrak{h} \oplus \mathfrak{l}$ and $\mathfrak{z}(\mathfrak{l}) = \mathfrak{l}$, which means that $\mathfrak{l}_1 + \mathfrak{z}(\mathfrak{l}) = \mathfrak{l}$. So by Proposition 4.2.10, $\mathscr{T}(\mathfrak{l}, \mathfrak{g}, \mathfrak{h})$ fails to be a Hausdorff space.

We now state the main result of this section.

Theorem 6.2.18. *Let $\mathfrak{g}$ be a 2-step nilpotent Lie algebra, $\mathfrak{h}$ and $\mathfrak{l}$ some subalgebras of $\mathfrak{g}$. Assume that $\mathfrak{l}$ satisfies the property $(\star)$ for $\mathfrak{g}/\mathfrak{h}$ and that $\dim(\mathfrak{z}') < \dim(\mathfrak{l})$. Then the stability property holds. More precisely,*

$$\mathscr{R}(\mathfrak{l}, \mathfrak{g}, \mathfrak{h}) = \left\{ \varphi(A, B) \in \mathrm{Hom}(\mathfrak{l}, \mathfrak{g}) : \mathrm{rk} \begin{pmatrix} A_2 & B_2 \\ 0 & B_4 \end{pmatrix} = \dim(\mathfrak{l}) \right\},$$

where $\varphi(A, B)$ is as in (4.32).

Proof. For a $n \times k$ matrix X, we have

$$\mathrm{rk}\, X = k \iff \ker X = \{0\}.$$

Thus, it is sufficient to show

$$\ker \begin{pmatrix} A_2 & B_2 \\ 0 & B_4 \end{pmatrix} = \{0\} \iff (\forall g \in G) \ker \begin{pmatrix} A_2 & B_2 + \sigma'(g) B' \\ 0 & B_4 \end{pmatrix} = \{0\}.$$

Thus, it is enough to show

$$\ker \begin{pmatrix} A_2 & B_2 + \sigma'(g) B' \\ 0 & B_4 \end{pmatrix} = \ker \begin{pmatrix} A_2 & B_2 \\ 0 & B_4 \end{pmatrix}$$

for any $g \in G$. Here, fix $g \in G$. Then this equation follows from

$$\ker B_4 \subset \ker \sigma'(g)B'. \tag{6.2}$$

Here, we decompose

$$\sigma'(g) = (\sigma_2'(g) \quad \sigma_2''(g) \quad \delta_2(g)), \quad B' = \begin{pmatrix} B_3' \\ B_3'' \\ B_4 \end{pmatrix}.$$

To show the inclusion (6.2), we use two claims.

Claim 1. $\sigma_2'(g) = 0$.

Proof of Claim 1. By the assumption that $I := (\mathfrak{z} \cap \mathfrak{h}) \oplus \mathfrak{h}'$ is an ideal of $\mathfrak{g}$, we obtain

$$[\mathfrak{g}, I] \subset \mathfrak{z} \cap \mathfrak{h} \tag{6.3}$$

because we have

$$[\mathfrak{g}, I] \subset [\mathfrak{g}, \mathfrak{g}] \subset \mathfrak{z} \quad \text{and} \quad [\mathfrak{g}, I] \subset I \subset \mathfrak{h}.$$

The inclusion (6.3) means that $\mathrm{ad}(X)Y \in \mathfrak{z} \cap \mathfrak{h}$ for any $X \in \mathfrak{g}$ and any $Y \in I$. Thus, $\sigma_2'(g) = 0$. □

Claim 2. $\ker B_4 \subset \ker B_3''$.

Proof of Claim 2. By the assumption $\dim(\mathfrak{z}') < \dim(\mathfrak{l})$, the condition

$$\mathrm{rk}\begin{pmatrix} A_2 & B_2 + \sigma(g)B' \\ 0 & B_4 \end{pmatrix} = k$$

implies $B_4 \neq 0$. Thus, as $\mathfrak{l}$ *satisfies the property* $(\star)$ for $\mathfrak{g}/\mathfrak{h}$, we obtain

$$\mathrm{rk}\begin{pmatrix} B_3'' \\ B_4 \end{pmatrix} = \mathrm{rk}(B_4).$$

This means

$$\ker B_4 \subset \ker B_3''.$$

Thus, Claim 1 and Claim 2 imply the inclusion (6.2). □

6.2.4 The threadlike case

Unlike the context of Heisenberg groups, the stability may fail to hold even generically on $\mathscr{R}(\Gamma, G, H)$. The setup of threadlike Lie groups appears as an instance to ascertain such a fact. In this case, the proper action of any connected Lie subgroup on a homogeneous space is equivalent to its free action (cf. Theorem 2.3.2). This fact greatly contributes to simplify the explicit determination of the parameter and the deformation spaces and, therefore, to seek the stability feature. The readers can consult the references [37] and [38] for more comprehensive details.

We resume all of the notation and definitions in Section 4.4: $G = \exp(\mathfrak{g})$ the thread-like nilpotent Lie group as defined in Subsection 1.1.3, with its one-codimensional Abelian ideal $\mathfrak{g}_0 = \mathbb{R}\text{-span}\{Y_1, \ldots, Y_n\}$. Let $G_0 = \exp(\mathfrak{g}_0)$ and let $\mathfrak{z}(\mathfrak{g})$ be the center of $\mathfrak{g}$.

We first have the following:

Proposition 6.2.19. *Assume that $H \not\subset G_0$ and $\Gamma \simeq \mathbb{Z}^k$ a discrete subgroup of G. Then the stability holds in the following cases:*
(i) $k > 2$.
(ii) $k = 2$ *and* $q < n$.
(iii) $k = 2$ *and* $q = n = 2$
(iv) $k = 1$ *and* $q = 1$.

Proof. As $\text{Hom}^0(\mathfrak{l}, \mathfrak{g})$ is an open set in $\text{Hom}(\mathfrak{l}, \mathfrak{g})$, $M \in \mathscr{R}(\Gamma, G, H)$ is stable if and only if there exists an open subset V of $\text{Hom}^0(\mathfrak{l}, \mathfrak{g})$, such that $M \in V \subset \mathscr{R}(\Gamma, G, H)$. In the case when $k > 2$, the result stems directly from equations (4.63) and (4.65). Assume now that $k = 2$ and $q < n$. It follows in view of Proposition 4.4.9 that $R(\mathfrak{l}, \mathfrak{g}, \mathfrak{h}) = R_{0,2}$. It suffices then to see that

$$\text{Hom}^0(\mathfrak{l}, \mathfrak{g}) \setminus R_{0,2} = \coprod_{j=0}^{3} (K_{j,2} \setminus R_{j,2}) = \left(\coprod_{j=1}^{3} K_{j,2} \right) \coprod (K_{0,2} \setminus R_{0,2})$$

is closed in $\text{Hom}^0(\mathfrak{l}, \mathfrak{g})$. Let then $M \in \overline{\text{Hom}^0(\mathfrak{l}, \mathfrak{g}) \setminus R_{0,2}}^{\text{Hom}^0(\mathfrak{l},\mathfrak{g})}$. There exists therefore a sequence $(M_i)_{i \in \mathbb{N}}$ assumed to belong to $\text{Hom}^0(\mathfrak{l}, \mathfrak{g}) \setminus R_{0,2}$, which converges to M. So, we can extract from $(M_i)_{i \in \mathbb{N}}$ a subsequence $(M_{i_s})_{s \in \mathbb{N}}$ of elements in $K_{j,2} \setminus R_{j,2}$ for some $j \in \{0, 1, 2, 3\}$. If

$$M_{i_s} = \begin{pmatrix} {}^t\vec{0} \\ N_1^s \\ N_2^s \end{pmatrix} \in K_{0,2} \setminus R_{0,2},$$

then obviously $\text{rank}\, N_1^s < 2$, which gives us

$$M = \begin{pmatrix} {}^t\vec{0} \\ N_1 \\ N_2 \end{pmatrix} \in K_{0,2} \setminus R_{0,2}.$$

Suppose now that

$$M_{i_s} = \begin{pmatrix} x_s & \lambda_s x_s \\ \vec{y_s} & \lambda_s \vec{y_s} \\ z_s & \lambda_s z_s + \beta_s \end{pmatrix} \in K_{1,2} \coprod K_{3,2}.$$

If $(\lambda_s)_s$ goes to infinity as s goes to $+\infty$, then we can easily check that $M \in K_{2,2} \coprod (K_{0,2} \setminus R_{0,2})$. Otherwise, $M \in K_{1,2} \coprod K_{3,2} \coprod (K_{0,2} \setminus R_{0,2})$. Finally, if

$$M_{i_s} = \begin{pmatrix} 0 & x_s \\ \vec{0} & \vec{y_s} \\ z_1^s & z_2^s \end{pmatrix} \in K_{2,2},$$

then its limit M belongs to $K_{2,2} \coprod (K_{0,2} \setminus R_{0,2})$. We get therefore that $M \in \operatorname{Hom}^0(\mathfrak{l}, \mathfrak{g}) \setminus R_{0,2}$, which completes the proof in this case. The case where $k = 2$ and $q = n = 2$ corresponds to the case where G is the three-dimensional Heisenberg group and the stability holds everywhere on the parameter space (cf. Theorem 6.2.3). Suppose finally that $k = 1$ and $q = 1$, then the result is immediate since we have $\operatorname{Hom}^0(\mathfrak{l}, \mathfrak{g}) = K_{0,1} \coprod K_{1,1} \simeq \mathbb{R}^{n+1} \setminus \{0\}$ and

$$\mathscr{R}(\Gamma, G, H) = R_{0,1} \coprod R_{1,1} = \left\{ \begin{pmatrix} x \\ y_1 \\ \vec{y} \end{pmatrix} \in M_{n+1,1}(\mathbb{R}) : y_1 \in \mathbb{R}^\times \right\} \simeq \mathbb{R} \times \mathbb{R}^\times \times \mathbb{R}^{n-1},$$

which is an open set in $(\mathbb{R}^{n+1})^\times$. □

Proposition 6.2.20. *We keep the hypotheses and notation as in Proposition* 6.2.19.
(i) *If $k = 2$ and $q = n$, then the set of nonstable points is the set:*

$$R_{0,2}^1 = \left\{ \begin{pmatrix} 0 & 0 \\ 0 & 0 \\ \vec{y} & \lambda\vec{y} \\ z_1 & z_2 \end{pmatrix} \in R_{0,2} : z_1, z_2, \lambda \in \mathbb{R}, {}^t\vec{y} \in \mathbb{R}^{n-2} \right\}$$
$$\cup \left\{ \begin{pmatrix} 0 & 0 \\ 0 & 0 \\ \vec{0} & \vec{y} \\ z_1 & z_2 \end{pmatrix} \in R_{0,2} : z_1, z_2 \in \mathbb{R}, {}^t\vec{y} \in \mathbb{R}^{n-2} \right\}.$$

(ii) *If $k = 1$ and $q \geq 2$, then the set of nonstable points is the following:*

$$R_{0,1}^1 = \left\{ \begin{pmatrix} 0 \\ 0 \\ \vec{y_1} \\ \vec{y_2} \end{pmatrix} \in M_{n+1,1}(\mathbb{R}) : {}^t\vec{y_1} \in \mathbb{R}^{q-1} \setminus \{0\}, {}^t\vec{y_2} \in \mathbb{R}^{n-q} \right\}.$$

Proof. First of all, we prove in the case when $k = 2$ and $q = n$ that the set $\coprod_{j=1}^{3} R_{j,2}$ is stable. Remark that $\coprod_{j=1}^{3} R_{j,2}$ is an open set in $\coprod_{j=1}^{3} K_{j,2}$. Hence we only have to show that $\coprod_{j=1}^{3} K_{j,2}$ is open in $\mathrm{Hom}^0(\mathfrak{l}, \mathfrak{g})$. We argue similarly as in Proposition 6.2.19 to prove that $K_{0,2}$ is closed in $\mathrm{Hom}^0(\mathfrak{l}, \mathfrak{g})$. Let now M be a nonstable element of $\mathscr{R}(\Gamma, G, H)$. Then $M \in R_{0,2}$ and any open set, which contains M meets $\mathrm{Hom}^0(\mathfrak{l}, \mathfrak{g}) \setminus \mathscr{R}(\Gamma, G, H)$. It follows that

$$M = \begin{pmatrix} {}^t\vec{0} \\ N \end{pmatrix} \in \overline{\mathrm{Hom}^0(\mathfrak{l}, \mathfrak{g}) \setminus \mathscr{R}(\Gamma, G, H)}^{\mathrm{Hom}^0(\mathfrak{l},\mathfrak{g})} \cap R_{0,2}.$$

So, we can find a sequence $(M_i)_{i\in\mathbb{N}}$ in $\mathrm{Hom}^0(\mathfrak{l}, \mathfrak{g}) \setminus \mathscr{R}(\Gamma, G, H)$, which converges to M. We therefore extract a subsequence of $(M_i)_{i\in\mathbb{N}}$ lying in $(K_{j,2} \setminus R_{j,2})$ for some $j \in \{1, 2, 3\}$, which leads to the fact that $M \in R_{0,2}^1$ as was to be shown. Conversely, let

$$M = \begin{pmatrix} 0 & 0 \\ 0 & 0 \\ \vec{y} & \lambda\vec{y} \\ z_1 & z_2 \end{pmatrix} \in R_{0,2}^1$$

$$\left(\text{resp., } M = \begin{pmatrix} 0 & 0 \\ 0 & 0 \\ \vec{0} & \vec{y} \\ z_1 & z_2 \end{pmatrix} \right).$$

We can therefore see that the sequence $(M_i)_{i\in\mathbb{N}^\times}$ of elements in $\mathrm{Hom}^0(\mathfrak{l}, \mathfrak{g}) \setminus \mathscr{R}(\Gamma, G, H)$ defined by

$$(M_i)_{i\in\mathbb{N}^\times} = \left(\begin{pmatrix} \frac{1}{i} & \frac{\lambda}{i} \\ 0 & 0 \\ \vec{y} & \lambda\vec{y} \\ z_1 & z_2 \end{pmatrix} \right)_{i\in\mathbb{N}^\times}$$

$$\left(\text{resp., } (M_i)_{i\in\mathbb{N}^\times} = \left(\begin{pmatrix} 0 & \frac{1}{i} \\ 0 & 0 \\ \vec{0} & \vec{y} \\ z_1 & z_2 \end{pmatrix} \right)_{i\in\mathbb{N}^\times} \right),$$

converges to M. This gives that M is not an inner point of $\mathscr{R}(\Gamma, G, H)$.

We now look at the situation where $k = 1$ and $q \geq 2$. We show that $R_{1,1}$ is stable. Remark that $R_{1,1}$ is open in $K_{1,1}$. Whence, we only have to prove that $K_{1,1}$ is open in $\mathrm{Hom}^0(\mathfrak{l}, \mathfrak{g})$. Let then $M \in \overline{K_{0,1}}^{\mathrm{Hom}^0(\mathfrak{l},\mathfrak{g})}$. There exists therefore a sequence $(M_i)_{i\in\mathbb{N}}$ of elements in $K_{0,1}$, which converges to M. Writing

$$M_i = \begin{pmatrix} 0 \\ \vec{y_i} \end{pmatrix},$$

we get that $M \in K_{0,1}$. Remark now that every

$$M = \begin{pmatrix} 0 \\ 0 \\ \vec{y_1} \\ \vec{y_2} \end{pmatrix} \in R^1_{0,1}$$

cannot be an inner point as being the limit of the sequence

$$(M_i)_{i \in \mathbb{N}^\times} = \left(\begin{pmatrix} \frac{1}{i} \\ 0 \\ \vec{y_1} \\ \vec{y_2} \end{pmatrix} \right)_{i \in \mathbb{N}^\times} \in \operatorname{Hom}^0(\mathfrak{l}, \mathfrak{g}) \setminus \mathscr{R}(\Gamma, G, H).$$

It follows therefore that any $M \in R^1_{0,1}$ is nonstable. Let

$$M = \begin{pmatrix} 0 \\ \vec{y_1} \\ \vec{y_2} \end{pmatrix}$$

be a nonstable point in $R_{0,1}$, then $M \in \overline{\operatorname{Hom}^0(\mathfrak{l}, \mathfrak{g}) \setminus \mathscr{R}(\Gamma, G, H)}^{\operatorname{Hom}^0(\mathfrak{l},\mathfrak{g})}$. Thus, there exists a sequence $(M_i)_{i \in \mathbb{N}}$ of elements in $\operatorname{Hom}^0(\mathfrak{l}, \mathfrak{g}) \setminus R(\Gamma, G, H) = (K_{0,1} \setminus R_{0,1}) \coprod (K_{1,1} \setminus R_{1,1})$, which converges to M and which belongs to $K_{j,1} \setminus R_{j,1}$ for some $j \in \{0, 1\}$. Suppose first that

$$M_i = \begin{pmatrix} 0 \\ \vec{0} \\ \vec{y_i} \end{pmatrix} \in K_{0,1} \setminus R_{0,1}$$

for all $i \in \mathbb{N}$, it comes out then that $M \in R^1_{0,1}$. Finally, if

$$M_i = \begin{pmatrix} x_i \\ 0 \\ \vec{y_i} \end{pmatrix} \in K_{1,1} \setminus R_{1,1}, \quad i \in \mathbb{N},$$

then necessarily $\lim_{i \to +\infty} x_i = 0$, which gives that $M \in R^1_{0,1}$. This completes the proof of Proposition 6.2.20. □

We now tackle at the case where $H \subset G_0$. We prove the following.

Proposition 6.2.21. *We keep the same hypotheses and notation. Assume that $H \subset G_0$, then the set $\mathscr{R}(\Gamma, G, H) \setminus R_{0,k}$ is stable for any $k \in \{1, \dots, n\}$.*

Proof. Recall that if $k = 2$ and $Y_n \in \mathfrak{h}$, then $\mathscr{R}(\Gamma, G, H) \setminus R_{0,2} = \emptyset$. Moreover, if $k = 1$ then $\mathscr{R}(\Gamma, G, H) \setminus R_{0,1} = K_{1,1}$, which is an open set of $\mathrm{Hom}^0(\mathfrak{l}, \mathfrak{g})$ as in the proof of Proposition 6.2.19. In addition, in the situation where $k = 2$ and $Y_n \notin \mathfrak{h}$, we have $\mathscr{R}(\Gamma, G, H) \setminus R_{0,k} = \coprod_{j=1}^{3} K_{j,2}$, which is open in $\mathrm{Hom}^0(\mathfrak{l}, \mathfrak{g})$ as provided in the proof of Proposition 6.2.20.

In the rest of the proof, we shall study the topological stability of elements of $R_{0,k}$ when $\mathfrak{h} \subset \mathfrak{g}_0$. Recall that for every $k \in \{1, \dots, n\}$, we have

$$R_{0,k} = \left\{ \begin{pmatrix} {}^t\vec{0} \\ N \end{pmatrix} \in H_{0,k} : \operatorname{rank}(A(t)N ⋒ M_{\mathfrak{h},\mathscr{B}_0}) = k + p \text{ for any } t \in \mathbb{R} \right\}.$$

Define for any $k \in \{1, \dots, n\}$, the set $A_k = \{(i_1, \dots, i_{p+k}) \in \mathbb{N}^{k+p} : 1 \le i_1 < \cdots < i_{p+k} \le n\}$, and we denote for any $N \in \mathrm{Hom}^0(\mathfrak{l}, \mathfrak{g})$, by $\Delta_{(i_1,\dots,i_{p+k})}(N, t)$ the relative minor of order $k + p$ obtained by considering the lines $i_1, \dots, i_{p+k}$ of the matrix $A(t)N ⋒ M_{\mathfrak{h},\mathscr{B}^0}$. It appears then clear that

$$\begin{pmatrix} {}^t\vec{0} \\ N \end{pmatrix} \in R_{0,k}$$

if and only if $\sum_{a \in A_k} \Delta_a^2(N, t) \neq 0$ for any $t \in \mathbb{R}$. There exists therefore a one variable polynomial function P_N, $N \in \mathrm{Hom}^0(\mathfrak{l}, \mathfrak{g})$ such that

$$R_{0,k} = \left\{ \begin{pmatrix} {}^t\vec{0} \\ N \end{pmatrix} \in H_{0,k} : P_N(t) \neq 0 \text{ for any } t \in \mathbb{R} \right\}.$$

Writing now $P_N(t) = b_d(N)t^d + \cdots + b_0(N)$ where $(b_j)_{0 \le j \le d}$ are polynomial functions on the coefficients of N and $b_d(N) \neq 0$. Let

$$m = \max\{d \in \mathbb{N} : b_d \text{ is not identically zero on } H_{0,k}\}.$$

Suppose first that m is odd, then

$$R_{0,k} = \left\{ \begin{pmatrix} {}^t\vec{0} \\ N \end{pmatrix} \in H_{0,k} : b_m(N) = 0,\ b_{m-1}(N)t^{m-1} + \cdots + b_0(N) \neq 0 \text{ for any } t \in \mathbb{R} \right\},$$

which is not an open set of $K_{0,k}$. The stability fails therefore at this stage. Suppose now that m is even, and we define subsets of $R_{0,k}$ by

$$R'_{0,k} = \left\{ \begin{pmatrix} {}^t\vec{0} \\ N \end{pmatrix} \in H_{0,k} : b_m(N) \neq 0, P_N(t) \neq 0 \text{ for any } t \in \mathbb{R} \right\},$$

and $R^1_{0,k} = R_{0,k} \setminus R'_{0,k}$.

We first show that $R'_{0,k}$ is stable. For $N \in \mathrm{Hom}^0(\mathfrak{l}, \mathfrak{g})$, such that $b_m(N) \neq 0$, we can decompose P_N as $P_N(t) = b_m(N) \prod_{i=1}^{\frac{m}{2}}(t^2 + \alpha_i(N)t + \beta_i(N))$ for some nontrivial functions α_i, β_i, which depend continuously upon the coefficients of N, when restricted if needed to a smaller set, still denoted by $R'_{0,k}$. We get then that

$$R'_{0,k} = \left\{ \begin{pmatrix} {}^t\vec{0} \\ N \end{pmatrix} \in H_{0,k} : b_m(N) \neq 0, \alpha_i^2(N) - 4\beta_i(N) < 0, 1 \leq i \leq \frac{m}{2} \right\},$$

which is an open set of $H_{0,k}$. This means therefore that the subset of nonstable elements of $R(\Gamma, G, H)$ is included in $R^1_{0,k}$, which completes the proof of the theorem. □

Case of compact Clifford–Klein forms

This section is devoted to the study of the deformation space when the Clifford–Klein form $\Gamma\backslash G/H$ is compact, which means that $\varphi(\Gamma)\backslash G/H$ is a compact manifold for any $\varphi \in \mathscr{R}(\Gamma, G, H)$. Unlike the case of Heisenberg groups, the deformation space fails in this case to be endowed with a manifold structure. We have the following.

Theorem 6.2.22. *Let $G = \exp \mathfrak{g}$ be a threadlike Lie group, H a closed connected subgroup of G, $\Gamma \simeq \mathbb{Z}^k$ a discrete subgroup of G such that the Clifford–Klein form $\Gamma\backslash G/H$ is compact. Let $L = \exp \mathfrak{l}$ be the syndetic hull of Γ. Then the parameter space is semialgebraic and open in* $\mathrm{Hom}(\mathfrak{l}, \mathfrak{g})$. *More precisely, the deformation space $\mathscr{T}(\Gamma, G, H)$ is described as follows:*
1. *If $k > 2$, then $\mathfrak{h} \not\subset \mathfrak{g}_0$ and $\mathscr{T}(\Gamma, G, H) \simeq \mathscr{T}_{0,k}$.*
2. *If $k = 2$ and $\mathfrak{h} \not\subset \mathfrak{g}_0$, then $\mathscr{T}(\Gamma, G, H) \simeq \mathscr{T}_{0,k}$.*
3. *If $k = 2$ and $\mathfrak{h} \subset \mathfrak{g}_0$, then $\mathscr{T}(\Gamma, G, H) \simeq \coprod_{i=1}^{3} \mathscr{T}_{i,2}$.*
4. *If $k = 1$ and $\mathfrak{h} \not\subset \mathfrak{g}_0$, then $\mathscr{T}(\Gamma, G, H) \simeq \mathscr{T}_{0,1} \coprod \mathscr{T}_{1,1}$.*
5. *If $k = 1$ and $\mathfrak{h} \subset \mathfrak{g}_0$, then $\mathscr{T}(\Gamma, G, H) = \mathscr{T}_{1,1}$.*

Furthermore, except the case where $k = 1$ and $\mathfrak{h} \not\subset \mathfrak{g}_0$, the deformation space is a Hausdorff space. In this situation, points in the deformation space, which cannot be separated belong to the same action of $\mathbb{Z}$.

Proof. Suppose first that $k > 2$, then for any $\varphi \in \mathscr{R}(\Gamma, G, H)$ we have that $\varphi(\Gamma) \subset G_0$. Moreover, since the Clifford–Klein form $\varphi(\Gamma)\backslash G/H$ is compact as mentioned above, we can deduce that $\mathfrak{h} \not\subset \mathfrak{g}_0$ for dimension reasons, and then according to Proposition 6.2.19, the parameter space is stable and has a structure of a semialgebraic set. Suppose now that $k = 2$ and $\mathfrak{h} \not\subset \mathfrak{g}_0$, then $q = 2 < n$. It follows, in view of Proposition 4.4.9 that $\mathscr{R}(\Gamma, G, H) = R_{0,2}$, which is stable according to Proposition 6.2.19 and semialgebraic. As for the case where $k = 2$ and $\mathfrak{h} \subset \mathfrak{g}_0$, we have $\varphi(\Gamma) \not\subset G_0$ for any $\varphi \in \mathscr{R}(\Gamma, G, H)$ and we get that $R_{0,2} = \emptyset$. One gets then $\mathscr{R}(\Gamma, G, H) = \coprod_{i=1}^{3} R_{i,2}$, which is stable according to Proposition 6.2.21, and on the other hand semialgebraic as being a finite union of semialgebraic sets.

We now study the situation where $k = 1$ and $\mathfrak{h} \not\subset \mathfrak{g}_0$. In this case, $q = 1$. Then by Proposition 6.2.19, the parameter space is stable and semialgebraic. Finally, if $\mathfrak{h} \subset \mathfrak{g}_0$, then $R_{0,1} = \emptyset$ and $\mathscr{R}(\Gamma, G, H) = R_{1,1}$, which is stable as shown in Proposition 6.2.21 and also semialgebraic.

From the characterization of the parameter space given in Propositions 4.4.8 and 4.4.9, it is easy to check that in the cases where $k \geq 2$ and $\mathfrak{h} \not\subset \mathfrak{g}_0$, the space $\mathscr{T}(\Gamma, G, H)$ is a Hausdorff space. The case 4 is also immediate. We now pay attention to the case when $k = 2$ and $\mathfrak{h} \subset \mathfrak{g}_0$. Let $[M_1], [M_2] \in \mathscr{T}(\Gamma, G, H)$ such that $[M_1] \neq [M_2]$. We designate by

$$M_j = \begin{pmatrix} x_j & x_j' \\ y_j & y_j' \\ \vec{0} & \vec{0} \\ \beta_j & \beta_j' \end{pmatrix} \quad (j = 1, 2).$$

Provided that

$$\begin{pmatrix} x_j & x_j' \\ y_j & y_j' \end{pmatrix} \neq \begin{pmatrix} x_j & x_j' \\ y_j & y_j' \end{pmatrix},$$

we can obviously separate $G \cdot M_1$ and $G \cdot M_2$ by open neighborhoods. In addition, elements in $R_{1,2}$ and $R_{2,2}$ are immediately separated by disjoint open sets. We only have to treat the case where

$$M_j = \begin{pmatrix} x_j & \lambda x_j \\ y_j & \lambda y_j \\ \vec{0} & \vec{0} \\ z_j & z_j' \end{pmatrix} \quad (j = 1, 2)$$

and

$$\begin{pmatrix} x_1 & \lambda x_1 \\ y_1 & \lambda y_1 \end{pmatrix} = \begin{pmatrix} x_2 & \lambda x_2 \\ y_2 & \lambda y_2 \end{pmatrix}$$

for $\lambda \neq 0$. Remark that $G \cdot M_1 \neq G \cdot M_2$ if and only if $z_1' - \lambda z_1 \neq z_2' - \lambda z_2$, and that

$$G \cdot M_i = \left\{ \begin{pmatrix} x_j & \lambda x_j \\ y_j & \lambda y_j \\ \vec{u_j} & \lambda \vec{u_j} \\ v & \lambda v + z_j' - \lambda z_j \end{pmatrix} : {}^t\vec{u_j} \in \mathbb{R}^{n-1}, v \in \mathbb{R} \right\}.$$

When we project the orbit through the partial coordinates system (Y_n, Y_n), we get the closed line $(v, \lambda v) + (0, z_1' - \lambda z_1) \neq (v, \lambda v) + (0, z_2' - \lambda z_2)$. So, one can obviously separate the orbits in question at this level. Finally, we show that the deformation space is not a Hausdorff space in case 4. Indeed, let $[M_1], [M_2] \in \mathscr{T}_{0,1}$ such that $[M_1] \neq [M_2]$. Noting

$$M_j = \begin{pmatrix} 0 \\ y_j \\ 0 \\ \vec{a_j} \end{pmatrix} \quad (j = 1, 2), \tag{6.4}$$

where ${}^t a_j = (a_{j,3}, \ldots, a_{j,n}) \in \mathbb{R}^{n-2}$ and we suppose that $y_1 = y_2$. Let now $\mathscr{V}_j$ be an open neighborhood of $[M_j]$ for $j = 1, 2$. We get that $\pi^{-1}(\mathscr{V}_j)$ is an open neighborhood of M_j in $\mathscr{R}(\Gamma, G, H)$. There exists then $\varepsilon > 0$ such that

$$M_j^{\varepsilon} = \begin{pmatrix} \varepsilon \\ y_j \\ \varepsilon \\ \vec{a_j} \end{pmatrix} \in \pi^{-1}(\mathscr{V}_j)$$

for $j = 1, 2$, and on the other hand $g_\varepsilon \in G$ such that $g_\varepsilon \cdot M_1^{\varepsilon} = M_2^{\varepsilon}$, which conclusively shows that $\mathscr{V}_1 \cap \mathscr{V}_2 \neq \emptyset$. This proves that the deformation space is not a Hausdorff space in this case. Finally, it is clear that two points $[M_j]$, $j = 1, 2$ in $\mathscr{T}(\Gamma, G, H)$, which are not separated belong to $\mathscr{T}_{0,1}$ and of the form (6.4) with $y_1 = y_2$. Let φ_j be the Lie algebras homomorphisms associated to M_j, $j = 1, 2$. Now any $g \in G$ is written as $g = \exp(xY_1) \cdot g'$ for some $x \in \mathbb{R}$ and $g' \in H$ as G is solvable and H is simply connected (cf. Theorem 1.1.6). Then clearly $\exp(\mathbb{Z}\varphi_1(T))gH = \exp(\mathbb{Z}\varphi_2(T))gH$, where $\exp(T)$ is any generator of Γ. This achieves the proof of the theorem. □

Remark 6.2.23. In case 4 of Theorem 6.2.22, the refined deformation space defined in Section 3.1.2 also fails to be a Hausdorff space. Indeed, thanks to Lemma 2.1.11 and the fact that H is normal, one easily sees that

$$\widehat{\mathscr{R}}(\Gamma, G, H) = \left\{ \begin{pmatrix} x \\ y_1 \\ \vec{0} \end{pmatrix} \in M_{n+1,1}(\mathbb{R}) : x \in \{0, 1\}, y_1 \in \mathbb{R}^{\times} \right\},$$

endowed with the quotient topology. The result follows as G acts on $\widehat{\mathscr{R}}(\Gamma, G, H)$ trivially.

Smoothness of the deformation space

It is an interesting question to know whether the deformation space can be endowed with a manifold structure in the case of compact Clifford–Klein forms provided that it is a Hausdorff space. In [98], it is shown that the deformation space contains a smooth manifold as its open dense subset. Using Theorem 6.2.22, we prove the following.

Theorem 6.2.24. *Let G be a threadlike Lie group, H a closed connected subgroup of G, Γ an Abelian discontinuous group for the homogeneous space G/H such that the Clifford–Klein form $\Gamma\backslash G/H$ is compact. Then the associated deformation space contains a smooth manifold as its open dense subset, which is contained in the set of maximal-dimensional orbits.*

Proof. We make use of Theorem 6.2.22. Denote as before $\pi : \mathscr{R}(\Gamma, G, H) \to \mathscr{T}(\Gamma, G, H)$ the canonical projection, then the image of any open dense subset of $\mathscr{R}(\Gamma, G, H)$ is open and dense in $\mathscr{T}(\Gamma, G, H)$. When $k > 2$ (or $k = 2$ and $H \not\subset G_0$), take

$$R' = \left\{ \begin{pmatrix} {}^t\vec{0} \\ N \end{pmatrix} \in R_{0,k}, N = (x_{i,j})_{1\leq i\leq n, 1\leq j\leq k};\ x_{1,1} \neq 0 \right\},$$

which is open and dense in $\mathscr{R}(\Gamma, G, H)$. In addition, no point in R' is fixed through the action of G. So all G-orbits of R' are of maximal dimension. We note $\mathscr{T}' = \pi(R')$ which, by Proposition 4.4.12, is homeomorphic to the set $\mathscr{T}_{0,k}(1, 1)$. We now look at the situation where $k = 2$ and $H \subset G_0$. We prove that $R' = R_{3,2}$ is an open and dense subset in $\mathscr{R}(\Gamma, G, H)$. Let $M \in \mathscr{R}(\Gamma, G, H) \setminus R'$. So, we can find a sequence $(M_i)_{i\in\mathbb{N}}$ in $\mathscr{R}(\Gamma, G, H) \setminus R' = R_{1,2} \cup R_{2,2}$, which converges to M. This leads immediately to the fact that $M \in R_{1,2} \cup R_{2,2}$ which shows that $\mathscr{R}(\Gamma, G, H) \setminus R'$ is closed in $\mathscr{R}(\Gamma, G, H)$. We now prove that R' is dense in $\mathscr{R}(\Gamma, G, H)$. Let

$$M = \begin{pmatrix} 0 & x \\ \vec{0} & \vec{y} \\ z_1 & z_2 \end{pmatrix} \in R_{2,2},$$

there exists then $n_0 \in \mathbb{N}$ such that the sequence

$$M_n = \begin{pmatrix} \frac{x}{n} & x \\ \frac{1}{n}\vec{y} & \vec{y} \\ z_1 & z_2 \end{pmatrix}, \quad n > n_0,$$

belongs to R' and converges to M. Suppose now that

$$M = \begin{pmatrix} x & 0 \\ \vec{y} & \vec{0} \\ z_1 & z_2 \end{pmatrix} \in R_{1,2},$$

then obviously the sequence

$$M_n = \begin{pmatrix} x & \frac{x}{n} \\ \vec{y} & \frac{1}{n}\vec{y} \\ z_1 & z_2 + \frac{z_1}{n} \end{pmatrix}, \quad n > 0$$

belongs to R' and converges to M. Finally, by Proposition 4.4.13 we have $\mathscr{T}' = \pi(R')$ is homeomorphic to $\mathscr{T}_{3,2}$ and G-orbits of elements of R' are of maximal dimension as reveals equation (4.70). This achieves the proof in this case. We now treat the case where $k = 1$. If $H \subset G_0$, then the result is immediate by Proposition 4.4.14. Otherwise, we note $R' = R_{1,1}$ and we argue similarly as previously to see that R' is open and dense in $\mathscr{R}(\Gamma, G, H)$. Clearly, all G-orbits inside R' are of maximal dimension. □

More significantly, the phenomenon of Hausdorffness of the deformation space is strongly linked to the feature of adjoint orbits of the basis group G on $\mathscr{R}(\Gamma, G, H)$, specifically to their dimensions. We have the following.

Theorem 6.2.25. *Let G be a threadlike Lie group, H a closed connected subgroup of G and Γ a discontinuous subgroup for G/H. If G acts on the parameter space $\mathscr{R}(\Gamma, G, H)$ with constant dimension orbits, then the deformation space $\mathscr{T}(\Gamma, G, H)$ is a Hausdorff space. When the Clifford–Klein form $\Gamma\backslash G/H$ is compact, this implication becomes an equivalence.*

Proof. When Γ is non-Abelian, we can immediately see that $\dim G \star M = n$ for any $M \in \mathscr{R}(\mathfrak{l}, \mathfrak{g}, \mathfrak{h})$ and Theorem 4.4.22 allows us to conclude. Suppose now that Γ is Abelian. Recall that in this situation we have an analogue decomposition of the parameter space as in the non-Abelian case, and accordingly the dimension of all orbits in the parameter space is constant if and only if $k > 2$. Proposition 4.4.12 is then sufficient to conclude.

Suppose now that the Clifford–Klein form $\Gamma\backslash G/H$ is compact. Recall from Theorem 6.2.22 that if $k = 1$ and $\mathfrak{h} \not\subset \mathfrak{g}_0$, the deformation space is not a Hausdorff space and the dimension of the G-orbits is not constant. Except for this previous case, the deformation space is a Hausdorff space and the dimension of all G-orbits is constant. This achieves the proof of the theorem. □

Remark 6.2.26.

(1) The deformation space is not connected, in general, in the nilpotent context. Though it is not clear that $\mathscr{R}(\Gamma, G, H)$ is a semialgebraic set in this case, it appears that it is so in most of the cases when restricted to the threadlike setting (or to the case of compact nilpotent Clifford–Klein forms). This means that the parameter space splits to a finite number of connected components with respect to the Euclidean topology, which gives in turn that the deformation space does, when endowed with the quotient topology. The number of the connected components is actually respected through the G-action. Looking at case 4 in Theorem 6.2.22 above for the simple setting when $k = 1$ and $\mathfrak{h} \subset \mathfrak{g}_0$, the deformation space $\mathscr{T}(\Gamma, G, H)$ turns out to be homeomorphic to $\mathscr{T}_{1,1}$, which admits two connected components. In [98], it is shown that the parameter space $\mathscr{R}(\Gamma, G, H)$ of the action of $\mathbb{Z}^k$ on $\mathbb{R}^{k+1}$ consists of two disconnected semialgebraic sets of different dimensions, but topologically connected.

(2) The points in the deformation space, which cannot be separated correspond to a faithful translation action on $G/H \simeq \mathbb{R}$ in the compact case. Beyond the compact case, this phenomenon fails to hold. Indeed, we can for instance consider the case where $\mathfrak{h} = \mathbb{R}\text{-span}\{X, Y_3, \dots, Y_n\}$ and $\Gamma = \exp(\mathbb{Z}Y_1)$.

6.2.5 Stability of discrete subgroups

Stable discrete subgroups as in Definition 6.1.7 have important impact on local geometric and differential structures of the related parameter and deformation spaces as reveal many studied cases (cf. Chapter 4). For more details, the readers can consult the reference [29]. Our next upshot about stable discrete subgroups of threadlike Lie groups is as follows.

Theorem 6.2.27. *Let G be a threadlike Lie group and Γ a non-Abelian discrete subgroup of G. Then Γ is stable. In this case, for any $H \in \pitchfork_{gp}(\Gamma : G)$, the deformation parameter space $\mathscr{R}(\Gamma, G, H)$ is a semialgebraic smooth manifold of dimension $n+k$ if $k > 3$ and $n+4$ otherwise.*

Proof. We take throughout the proof, a non-Abelian discrete subgroup Γ and let H be a fixed subgroup of G, which belongs to $\pitchfork_{gp}(\Gamma : G)$. We will show that the parameter space $\mathscr{R}(\Gamma, G, H)$ is open and semialgebraic. In light of Theorem 3.2.4, this space is homeomorphic to

$$\mathscr{R}(\mathfrak{l}, \mathfrak{g}, \mathfrak{h}) = \{M \in \mathscr{U} : \operatorname{rank}(g \star M ⋒ M_{\mathfrak{h},\mathscr{B}}) = k + \dim \mathfrak{h} \text{ for all } g \in G\},$$

where $\mathscr{B} = \{X, Y_1, \dots, Y_n\}$ and the symbol ⋒ merely means the concatenation of matrices written through $\mathscr{B}$. We also assume that $3 \leq k \leq n + 1 - \dim \mathfrak{h}$, otherwise $\mathscr{R}(\mathfrak{l}, \mathfrak{g}, \mathfrak{h})$ is empty. We note for all $j \in \{0, \dots, k\}$, $R_{j,k} = \mathscr{R}(\mathfrak{l}, \mathfrak{g}, \mathfrak{h}) \cap K_{j,k}$. Then according to Proposition 4.4.17, we have that $\mathscr{R}(\mathfrak{l}, \mathfrak{g}, \mathfrak{h}) = R_{0,k}$ whenever $k > 3$. Otherwise, $\mathscr{R}(\mathfrak{l}, \mathfrak{g}, \mathfrak{h}) = \coprod_{j=0}^{2} R_{j,3}$. We denote by

$$\mathscr{I}_{\mathscr{B}}^{\mathfrak{h}} = \{i_1 < \dots < i_q\} \quad (q = \dim \mathfrak{h})$$

the set of indices i $(1 \leq i \leq n+1)$ such that $\mathfrak{h} \cap \mathfrak{g}^i \neq \mathfrak{h} \cap \mathfrak{g}^{i-1}$, where $\mathfrak{g}^i = \{Y_n, \dots, Y_i\}$, $i = 1, \dots, n$, $\mathfrak{g}^0 = \mathfrak{g}$, and $\mathfrak{g}^{n+1} = \{0\}$. We first prove the following.

Proposition 6.2.28. *We keep the same hypotheses and notation as before. Then:*
(i) *If $\mathfrak{h} \not\subset \mathfrak{g}_0$, then $\mathfrak{h} = \mathbb{R}\text{-span}\{X + h_1 Y_1 + \dots + h_n Y_n\}$ for some $h_1, \dots, h_n \in \mathbb{R}$ and*

$$R_{0,k} = \{M_0(U, V) \in K_{0,k} : (u_1 - h_1 u) \in \mathbb{R}^{\times}\}. \tag{6.5}$$

(ii) *If $\mathfrak{h} \subset \mathfrak{g}_0$, then $\mathscr{I}_{\mathscr{B}}^{\mathfrak{h}} \subset \{1, \dots, p-1\}$ and $R_{0,k} = K_{0,k}$.*

Proof. Suppose first that $\mathfrak{h} \not\subset \mathfrak{g}_0$, then obviously $\mathfrak{h} = \mathbb{R}\text{-span}\{\widetilde{X}\}$ for some $\widetilde{X} = X + h_1 Y_1 + \cdots + h_n Y_n$ where $h_1, \dots, h_n \in \mathbb{R}$. Let $M = M_0(U, V) \in K_{0,k}$. Then clearly,

$$\begin{aligned} M \in \mathscr{R}(\mathfrak{l}, \mathfrak{g}, \mathfrak{h}) &\Leftrightarrow \operatorname{rank}(M ⋒ g \star M_{\mathfrak{h},\mathscr{B}}) = k+1 \quad \text{for all } g \in G \\ &\Leftrightarrow \operatorname{rank}\lfloor {}^t(u, u_1, \dots, u_n), {}^t(1, h_1, \alpha_2, \dots, \alpha_n)\rfloor = 2 \quad \text{for all } \alpha_2, \dots, \alpha_n \in \mathbb{R} \\ &\Leftrightarrow u_1 - u h_1 \in \mathbb{R}^{\times}. \end{aligned}$$

Suppose now that $\mathfrak{h} \subset \mathfrak{g}_0$. If $\mathscr{I}_{\mathscr{B}}^{\mathfrak{h}} \cap \{p, \dots, n\} \neq \emptyset$, then there exists $i_0 \in \{p, \dots, n\}$ and $\widetilde{Y} = Y_{i_0} + h_{i_0+1} Y_{i_0+1} + \cdots + h_n Y_n \in \mathfrak{h}$ for some $h_{i_0+1}, \dots, h_n \in \mathbb{R}$, which is impossible as L acts on G/H properly. So, $\mathscr{I}_{\mathscr{B}}^{\operatorname{Ad}_g(\mathfrak{h})} \cap \{p, \dots, n\} = \emptyset$ for all $g \in G$, which gives that $\operatorname{rank}(g \star M ⋒ M_{\mathfrak{h},\mathscr{B}}) = k + q$ for all $M \in K_{0,k}$. □

We argue similarly as in the previous proposition to treat the case where $k = 3$.

Proposition 6.2.29. *We keep the same hypotheses and notation as before. Then:*
(i) *If $\mathfrak{h} \not\subset \mathfrak{g}_0$, then $\mathfrak{h} = \mathbb{R}\text{-span}\{X + h_1 Y_1 + \cdots + h_n Y_n\}$, for some $h_1, \dots, h_n \in \mathbb{R}$, then*

$$R_{1,3} = \{M_1(U, V, \lambda) \in K_{1,3} : (u_1 - h_1 u) \in \mathbb{R}^{\times}\} \tag{6.6}$$

and

$$R_{2,3} = \{M_2(U, V) \in K_{2,3} : (v_1 - h_1 v) \in \mathbb{R}^{\times}\}. \tag{6.7}$$

(ii) *If $\mathfrak{h} \subset \mathfrak{g}_0$, then $\mathscr{I}_{\mathscr{B}}^{\mathfrak{h}} \subset \{1, \dots, n-2\}$ and $R_{j,3} = K_{j,3}, j = 0, 1, 2$.*

Now according to Propositions 6.2.28 and 6.2.29, the parameter space has a structure of a semialgebraic set. It remains therefore to show that it is open in $\operatorname{Hom}(\mathfrak{l}, \mathfrak{g})$. As $\operatorname{Hom}^0(\mathfrak{l}, \mathfrak{g})$ is an open set in $\operatorname{Hom}(\mathfrak{l}, \mathfrak{g})$, $M \in \mathscr{R}(\mathfrak{l}, \mathfrak{g}, \mathfrak{h})$ is stable if and only if there exists an open subset $\mathscr{V}$ of $\operatorname{Hom}^0(\mathfrak{l}, \mathfrak{g})$, such that $M \in \mathscr{V} \subset \mathscr{R}(\mathfrak{l}, \mathfrak{g}, \mathfrak{h})$. In the case when $k > 3$, the result stems directly from Proposition 6.2.28 and equation (6.5). Assume now that $k = 3$. The result is immediate when $\mathfrak{h} \subset \mathfrak{g}_0$. Suppose now that $\mathfrak{h} \not\subset \mathfrak{g}_0$. It suffices then to see that $\operatorname{Hom}^0(\mathfrak{l}, \mathfrak{g}) \setminus \mathscr{R}(\mathfrak{l}, \mathfrak{g}, \mathfrak{h}) = \coprod_{j=0}^{2} (K_{j,3} \setminus R_{j,3})$ is closed in $\operatorname{Hom}^0(\mathfrak{l}, \mathfrak{g})$. Let then

$$M \in \overline{\operatorname{Hom}^0(\mathfrak{l}, \mathfrak{g}) \setminus \mathscr{R}(\mathfrak{l}, \mathfrak{g}, \mathfrak{h})}^{\operatorname{Hom}^0(\mathfrak{l}, \mathfrak{g})},$$

there exists therefore a sequence $(M_i)_{i \in \mathbb{N}}$ assumed to belong to $\operatorname{Hom}^0(\mathfrak{l}, \mathfrak{g}) \setminus \mathscr{R}(\mathfrak{l}, \mathfrak{g}, \mathfrak{h})$, which converges to M. So we can extract from $(M_i)_{i \in \mathbb{N}}$ a subsequence $(M_{i_s})_{s \in \mathbb{N}}$ of elements in $K_{j,3} \setminus R_{j,3}$ for some $j \in \{0, 1, 2\}$. If $M_{i_s} \in K_{j,3} \setminus R_{j,3}$ for $j \in \{0, 2\}$, then obviously its limit M belongs to $K_{j,3} \setminus R_{j,3}$.

Suppose now that

$$M_{i_s} = \begin{pmatrix} u^s & \lambda_s u^s & 0 \\ u_1^s & \lambda_s u_1^s & 0 \\ \vdots & \vdots & \vdots \\ u_{n-2}^s & \lambda_s u_{n-2}^s & 0 \\ u_{n-1}^s & v_{n-1}^s & 0 \\ u_n^s & v_n^s & u^s(v_{n-1}^s - \lambda_s u_{n-1}^s) \end{pmatrix} \in K_{1,3} \setminus R_{1,3}$$

for some real sequence $(\lambda_s)_{s\in\mathbb{N}}$. If $(\lambda_s)_s$ goes to infinity as s goes to $+\infty$, then we can easily check that $M \in K_{2,3} \setminus R_{2,3}$. Otherwise, $M \in (K_{0,3} \setminus R_{0,3}) \coprod (K_{1,3} \setminus R_{1,3})$. We get therefore that $M \in \mathrm{Hom}^0(\mathfrak{l}, \mathfrak{g}) \setminus \mathscr{R}(\mathfrak{l}, \mathfrak{g}, \mathfrak{h})$. The rest of the proof follows from Theorem 4.4.16. □

Corollary 6.2.30. *Let Γ be a discrete subgroup and $H \in \mathfrak{M}_{gp}(\Gamma : G)$. The parameter space $\mathscr{R}(\Gamma, G, H)$ is endowed with a smooth manifold structure of dimension $n + k$ whenever $k > 3$. For $k = 3$, it is a disjoint union of an open, dense, smooth manifold of dimension $n + 4$ and a closed, smooth manifold of dimension $n + 3$.*

Proof. The result is immediate when $k > 3$. If $k = 3$, it is not hard to see that $\widetilde{R_{1,3}} = R_{0,3} \coprod R_{1,3}$ and $R_{2,3}$ are respectively open sets in $\widetilde{K_{1,3}}$ and $K_{2,3}$ and that $\widetilde{R_{1,3}}$ is an open dense set of $\mathscr{R}(\mathfrak{l}, \mathfrak{g}, \mathfrak{h})$. □

6.3 Stability in the exponential setting

We first look back at the case of connected, simply connected, nilpotent Lie groups. We prove the following result and provide in Remark 6.3.4 below a counterexample showing the failure of its validity in the general context of exponential Lie groups.

6.3.1 Case of general compact Clifford–Klein forms

Proposition 6.3.1. *Let $G = \exp \mathfrak{g}$ be a connected, simply connected, nilpotent Lie group, $H = \exp \mathfrak{h}$ a connected subgroup of G and Γ a discontinuous group for G/H of syndetic hull $L = \exp \mathfrak{l}$. If the Clifford–Klein form $\Gamma\backslash G/H$ is compact, then the stability holds everywhere. That is, the parameter space $\mathscr{R}(\mathfrak{l}, \mathfrak{g}, \mathfrak{h})$ is semialgebraic and open.*

Proof. The following lemma dealing with proper actions when the Clifford–Klein form is compact is proved in [135].

Lemma 6.3.2. *Let G be a connected, simply connected, nilpotent Lie group, L and H be its connected, closed subgroups. If $L\backslash G/H$ is compact, then the following three conditions are equivalent:*

(i) *The L-action on G/H is proper.*

(ii) *The L-action on G/H is free.*
(iii) $L \cap H = \{0\}$.

This result enables us to see that $\mathscr{R}(\mathfrak{l}, \mathfrak{g}, \mathfrak{h})$ is homeomorphic to the set

$$\left\{\psi \in \operatorname{Hom}(\mathfrak{l}, \mathfrak{g}) \,\middle|\, \begin{array}{l} \dim \psi(\mathfrak{l}) = \dim \mathfrak{l} \text{ and} \\ \psi(\mathfrak{l}) \cap \mathfrak{h} = \{0\} \end{array}\right\},$$

which is open in $\operatorname{Hom}(\mathfrak{l}, \mathfrak{g})$ and semialgebraic. □

We now find back our settings. G designates an exponential solvable Lie group, H a nonnormal connected maximal subgroup of G and Γ a discontinuous group for G/H. The most important problem in the study of the deformation space of discontinuous groups is the description of the parameter space $\mathscr{R}(\Gamma, G, H)$ as in equation (3.1), which is homeomorphic to $\mathscr{R}(\mathfrak{l}, \mathfrak{g}, \mathfrak{h})$ as in Theorem 3.2.4. Our main result in this section is the following.

Theorem 6.3.3. *Let G be an exponential Lie group, H a nonnormal, connected maximal subgroup of G and Γ a discontinuous group for G/H. Then the parameter space $\mathscr{R}(\Gamma, G, H)$ is semialgebraic. Furthermore, the stability property holds if and only if Γ is of rank two.*

Proof. We make use of Propositions 5.3.6 and 5.3.7 concerning the determination of the parameter space $\mathscr{R}(\mathfrak{l}, \mathfrak{g}, \mathfrak{h})$, which are the key points in the proof. Clearly, $\mathscr{R}(\mathfrak{l}, \mathfrak{g}, \mathfrak{h})$ is semialgebraic. For $k = 1$, the parameter space is a smooth manifold of dimension $n-1$ as reveals Proposition 5.3.6, so no stable points can be found. For $k = 2$, the parameter space is the intersection of $\operatorname{Hom}(\mathfrak{l}, \mathfrak{g})$ and the Zariski open set defined by

$$R = \left\{ \begin{pmatrix} a_1 & a_2 \\ x_1 & x_2 \\ y_1 & y_2 \\ \vec{z_1} & \vec{z_2} \end{pmatrix} \in \mathscr{M}_{n,2}(\mathbb{R}),\ x_1 y_2 - x_2 y_1 \neq 0 \right\},$$

which is enough to conclude. □

Remark 6.3.4. We now provide a counterexample to the result of Proposition 6.3.1 in the context of exponential Lie groups. Remark that the first example in Lemma 5.3.8 produces a parameter space $\mathscr{R}^2_{1,1}$ homeomorphic to $\mathbb{R}^\times$, which is not open in $\operatorname{Hom}(\mathfrak{l}, \mathfrak{g}) \simeq \mathbb{R}^2$. This case corresponds indeed to a situation where the Clifford–Klein form in question is compact.

6.4 Stability for Euclidean motion groups

We first prove some elementary results, which will be of use throughout the section. The following is a first observation in linear algebra; it states that there are at most finitely many $O_n(\mathbb{R})$-conjugacy classes $[X]$ for which $X^n = I$ for fixed $N \in \mathbb{N}^*$.

Fact 6.4.1. *Fix $N \in \mathbb{N}^*$. The set of solutions of the equation $(\mathscr{E}) : X^N = I$ in $O_n(\mathbb{R})$ coincides with the set*

$$\mathscr{S}_N = \bigsqcup_{1\le i\le M_N} \{S^{-1}A_i S \mid S \in O_n(\mathbb{R})\},$$

where $M_N \in \mathbb{N}^*$ and for $1 \le i \le M_N$, $A_i^N = I$ and A_i and A_j belong to different classes of conjugacy for $1 \le i \ne j \le M_N$.

Corollary 6.4.2. *Fix $N \in \mathbb{N}^*$ and let $\{A_p\}_{p\in\mathbb{N}}$ be a sequence of orthogonal matrices satisfying for any $p \in \mathbb{N}, A_p^N = I$. If $\{A_p\}_{p\in\mathbb{N}}$ converges to an orthogonal matrix A, say, there exist $p_0 \in \mathbb{N}$ such that for any $p \ge p_0$, $A_p = S_p^{-1} A S_p$ for some $S_p \in O_n(\mathbb{R})$.*

Proof. Thanks to Fact 6.4.1, $\{A_p\}_{p\in\mathbb{N}} \subset \mathscr{S}_N$. We remark that there exists $j \in \{1, \dots, M_N\}$ such that the orbit $\Theta_j = \{S^{-1}B_j S \mid S \in O_n(\mathbb{R})\}$, being a closed subset of $O_n(\mathbb{R})$, contains an infinite term of the sequence $\{A_p\}_{p\in\mathbb{N}}$. Then there exists $p_0 \in \mathbb{N}$ such that for any $p \ge p_0$, A_p and A are sitting inside Θ_j. This justifies the statement. □

We now recall a classical structural result for finitely generated Abelian groups.

Fact 6.4.3 ([2, Theorem 7.22]). *Any finitely generated Abelian group M is isomorphic to $\mathbb{Z}^r \oplus M_1$ where $|M_1| < \infty$. The integer r is an invariant of M. Any finite Abelian group is a direct sum of cyclic groups of prime power order and these prime power orders, counted with multiplicity, completely characterize the finite Abelian group up to isomorphism. Also, any finite Abelian group is uniquely isomorphic to a group $\mathbb{Z}/(s_1) \times \cdots \times \mathbb{Z}/(s_m)$ where s_i divides s_{i+1} for any $i \in \{1, \dots, m-1\}$.*

As a direct consequence, we get the following.

Proposition 6.4.4. *Let $\mathscr{L}$ be a discrete subgroup of $\mathbb{R}^n$. For any subgroup $\mathscr{L}'$ of $\mathscr{L}$, there exists a unique sequence of positive integers $s_1, \dots, s_m$ satisfying s_i divides s_{i+1} for any $1 \le i \le m-1$, and such that*

$$\mathscr{L}/\mathscr{L}' \simeq \mathbb{Z}^{k-k'} \times \mathbb{Z}/(s_1) \times \cdots \times \mathbb{Z}/(s_m),$$

where s_i divides s_{i+1} for any $i \in \{1, \dots, m-1\}$ and k, k' designate the rank of $\mathscr{L}$ and $\mathscr{L}'$, respectively.

Let us mention that for a given group K with an Abelian normal and finitely generated subgroup K_a of finite index in K, there exists by Fact 6.4.3 an integer k and a unique sequence of positive integers $s_1, \dots, s_m$ satisfying s_i divides s_{i+1} for any $1 \leq i \leq m-1$, and such that

$$K_a \simeq \mathbb{Z}^k \times \mathbb{Z}/(s_1) \times \cdots \times \mathbb{Z}/(s_m). \tag{6.8}$$

We easily check that the integer k is an invariant of K.

Definition 6.4.5. Let K be a group with an Abelian normal and finitely generated subgroup K_a of finite index in K. The integer k given in equation (6.8) is said to be the effective rank of K and denoted by $r_e(K)$.

We now come back to our group G. For a subgroup K of G, set

$$\mathscr{F}_K := \{P_A(x) : (A, x) \in K\} \subset \mathbb{R}^n. \tag{6.9}$$

We have the following.

Lemma 6.4.6. *For any discrete subgroup Γ of G, the rank of the family $\mathscr{F}_\Gamma$ coincides with the effective rank of Γ.*

Proof. Remark first that for any $g = (S_g, t_g) \in G$,

$$\Gamma^g = \{(S_g^{-1} A S_g, S_g^{-1}[x + (A - I)t_g]) : (A, x) \in \Gamma\}.$$

As for any $(A, x) \in \Gamma$,

$$\begin{aligned} P_{S_g^{-1}AS_g}(S_g^{-1}[x + (A - I)t_g]) &= S_g^{-1} P_A S_g (S_g^{-1}[x + (A - I)t_g]) \\ &= S_g^{-1} P_A(x) + S_g^{-1} \underbrace{P_A(A - I)(t_g)}_{0}, \end{aligned}$$

we get $\mathscr{F}_{\Gamma^g} = S_g^{-1} \mathscr{F}_\Gamma$. Take g as described in Theorem 1.2.24 and keep the same notation. Obviously, the rank of $\mathscr{F}_{\Gamma^g}$ is exactly the rank of the family $\{y_{k,i}\}_{1 \leq i \leq k}$. Since $\{y_i\}_{1 \leq i \leq k_0}$ generates an Abelian normal subgroup Γ_a^g of finite index in Γ^g, satisfying $\langle y_1, \dots, y_k \rangle \simeq \mathbb{Z}^k$ and $\langle y_{k+1}, \dots, y_{k_0} \rangle$ is finite, we get $r_e(\Gamma) = k$. □

The following upshot, will be of use later.

Proposition 6.4.7. *Let Γ be a discrete subgroup of G and $\mathscr{K}_\Gamma$ the set of elements of Γ of finite order. Then $\mathrm{pr}_1(\mathscr{K}_\Gamma)$ is finite.*

Proof. We keep the same notation as in Theorem 1.2.24. Let $\Gamma_0 = \{a_0, \dots, a_{N-1}\}$ as a finite set. Any $\gamma \in \Gamma$ is written as $\gamma = b_\gamma a_{i_\gamma} \delta_{j_\gamma}$ where $b_\gamma \in \Gamma_\infty$, $i_\gamma \in \{0, \dots, N-1\}$ and $j_\gamma \in \{0, \dots, q-1\}$ with $a_0 = \gamma_0 = e$. Let $\gamma \in \mathscr{K}_\Gamma$, then $\gamma^q \in \Gamma_a$ and, therefore, $\gamma^q \in \Gamma_0$, which entails that $\gamma^{qN} = e$. Assume that $\mathscr{K}_\Gamma$ is infinite. Then

$$\mathscr{K}_\Gamma = \bigcup_{\substack{0 \leq i \leq N-1 \\ 0 \leq j \leq q-1}} \mathscr{K}_{i,j},$$

where $\mathscr{K}_{i,j} = \{\gamma \in \mathscr{K}_\Gamma : \gamma = b_\gamma a_i \delta_j\}$. Then any $\gamma \in \mathscr{K}_{i,j}$ can be written as

$$\gamma = (A_\gamma, x_\gamma) = (B_\gamma, y_\gamma)(C_i, 0)(S_j, z_j) = (B_\gamma C_i S_j, y_\gamma + B_\gamma C_i z_j),$$

where $a_i = (C_i, 0)$, $b_\gamma = (B_\gamma, y_\gamma)$ and $\delta_j = (S_j, z_j)$; hence $C_i^N = I$ and $B_\gamma C_i z_j = z_j$. Assume that for a given $j \in \{0, \dots, q-1\}$ and $i \in \{0, \dots, N-1\}$, $\mathrm{pr}_1(\mathscr{K}_{i,j})$ is infinite. Then one can find an infinite sequence $\{A_p\}_{p\in\mathbb{N}}$ of $p_1(\mathscr{K}_{i,j})$ converging to some $A \in O_n(\mathbb{R})$. Then for $p \in \mathbb{N}$, we can write $A_p = B_p C_i S_j$. Besides, let $(\delta_{j_{(s,\gamma)}})_{s\in\mathbb{N}}$ be the sequence such that

$$b_\gamma a_i \delta_j = \delta_{j_{(1,\gamma)}} b_\gamma a_i \quad \text{and} \quad b_\gamma a_i \delta_{j_{(s,\gamma)}} = \delta_{j_{(s+1,\gamma)}} b_\gamma a_i.$$

This gives likewise a sequence $(S_{j_{(s,p)}})_{s\in\mathbb{N}}$ such that

$$B_p C_i S_j = S_{j_{(1,p)}} B_p C_i \quad \text{and} \quad B_p C_i S_{j_{(s,p)}} = S_{j_{(s+1,p)}} B_p C_i.$$

As the sequence $\{B_p\}_{p\in\mathbb{N}}$ converges, the sequences $\{S_{j_{(s,p)}}\}_{p\in\mathbb{N}}$ also do and as $S_{j_{(s,p)}}$ is of finite values, there exists $p'' \geq p'$ such that for any $p \geq p''$, $S_{j_{(s,p)}} = S_{j_{(s,p'')}}$. As $A_p^{qN} = I$, we deduce that $(B_p C_i S_j)^{qN} = I$ and then

$$(B_p)^{qN} \underbrace{(C_i)^{qN}}_{I} S_{j_{(qN-1,p'')}} \cdots S_{j_{(1,p'')}} S_j = I.$$

This gives that $(B_p)^{qN} = (S_{j_{(qN-1,p'')}} \cdots S_{j_{(1,p'')}} S_j)^{-1}$, which does not depend upon p. Furthermore, the matrix form of B_p reads

$$B_p = \begin{pmatrix} I_{m_+} & & & & & \\ & \varepsilon_p I_{m_-} & & & & \\ & & d_p(-1,1) & & & \\ & & & r(\theta_1(p)) & & \\ & & & & \ddots & \\ & & & & & r(\theta_l(p)) \end{pmatrix},$$

where $\varepsilon_p \in \{-1, 1\}$ and $d_p(-1,1)$ is a diagonal matrix of diagonal values in $\{-1, 1\}$. Since $(B_p)^{qN}$ does not depend upon p, the sequence $\{B_p\}_{p\in\mathbb{N}}$ turns out to be stationary and, therefore, $\{A_p\}_{p\in\mathbb{N}}$ must be finite, which is absurd and, therefore, $\mathrm{pr}_1(\mathscr{K}_{i,j})$ cannot be infinite. □

Thanks to Proposition 6.4.7, one can write

$$\mathscr{K}_\Gamma = \bigsqcup_{1\leq i\leq M_\Gamma} \mathscr{K}_i,$$

where $\mathscr{K}_i = \{(A_i, x) \in \mathscr{K}_\Gamma\}$ and M_Γ designates the cardinality of $\mathrm{pr}_1(\mathscr{K}_\Gamma)$. Fix $i \in \{1, \dots, M_\Gamma\}$ and $(A_i, x_0) \in \mathscr{K}_i$. Remark that $\mathscr{K}_i(A_i, x_0)^{-1} = \{(I, x - x_0), x \in \mathrm{pr}_2(\mathscr{K}_i)\} := \Lambda_i$

is a discrete subgroup because $x \in \ker(A_i - I)^{\perp}$ for any $(A_i, x) \in \mathscr{K}_i$. This entails that $\mathscr{F}_i := \mathrm{pr}_2(\Lambda_i)$ is a discrete subgroup of $\ker(A_i - I)^{\perp}$. Let Λ_i act by conjugation on the set $\mathscr{K}_i$. We finally prove the following.

Proposition 6.4.8. *$\mathscr{K}_i/\Lambda_i$ is finite.*

Proof. Since $A_i - I$ induces an isomorphism on $\ker(A_i - I)^{\perp}$, its restriction on $\mathscr{F}_i$ is injective. Furthermore,

$$(A_i, x)^2 (A_i, x_0)^{-1} = (A_i, x + A_i(x - x_0)) \in \mathscr{K}_i,$$

which entails that $A_i(x - x_0) \in \mathscr{F}_i$ and also $(A_i - I)(x - x_0) \in \mathscr{F}_i$. Then $(A_i - I)(\mathscr{F}_i) \subset \mathscr{F}_i$. Obviously, $(A_i - I)(\mathscr{F}_i)$ and $\mathscr{F}_i$ are of the same rank. Thanks to Proposition 6.4.4 there exist $s_1, \ldots, s_m \in \mathbb{Z}$ satisfying s_i divides s_{i+1} for any $1 \leq i \leq m-1$, and such that

$$\mathscr{F}_i/(A_i - I)(\mathscr{F}_i) \simeq \mathbb{Z}/(s_1) \times \cdots \times \mathbb{Z}/(s_m).$$

Hence $\mathscr{F}_i/(A_i - I)(\mathscr{F}_i)$ is a finite group of cardinality $s = \prod_{j=1}^{m} s_j$. Denote

$$\mathscr{F}_i/(A_i - I)(\mathscr{F}_i) = \{\overline{0}, \overline{x_1 - x_0}, \ldots, \overline{x_{s-1} - x_0}\}.$$

Fix $(A_i, x) \in \mathscr{K}_i$, then $x - x_0 \in \mathscr{F}_i$, and so there exists $y \in \mathscr{F}_i$, $j \in \{0, \ldots, s-1\}$ such that $x - x_0 = x_j - x_0 + (A_i - I)(y)$, which is in turn equivalent to the fact that $x = x_j + (A_i - I)(y)$, and finally

$$(A_i, x) = (I, -y)(A_i, x_j)(I, y).$$

At the end, $\mathscr{K}_i = \bigsqcup_{0 \leq j \leq d-1} \Theta_{i,j}$ where $\Theta_{i,j}$ is the orbit of (A_i, x_j) under the action of Λ_i by conjugation. □

6.4.1 Geometric stability

We now study the above variants of stability in the context of Euclidean motion groups. As an immediate and important consequence of Proposition 2.5.1, we get the following description of the parameter space of the action of any discontinuous group acting on a homogeneous space G/H, where G stands for the Euclidean motion group.

Corollary 6.4.9. *Let Γ be a discrete subgroup of the Euclidean motion group G acting properly on G/H, then*

$$\mathscr{R}(\Gamma, G, H) = \{\varphi \in \mathrm{Hom}_d^0(\Gamma, G) : \varphi(\Gamma) \text{ acts freely on } G/H\}.$$

One immediate aftermath concerning geometric properties of Clifford–Klein forms is when the fundamental group is torsion-free for which we have the following.

Proposition 6.4.10. *$\mathscr{R}(\Gamma, G, H)$ and $\mathrm{Hom}_d^0(\Gamma, G)$ coincide whenever Γ is torsion-free.*

Proof. It suffices to remark that thanks to Proposition 2.5.1, H is compact whenever Γ is infinite. As Γ is torsion-free, $\varphi(\Gamma) \cap gHg^{-1}$ is trivial for any $g \in G$ and any $\varphi \in \mathrm{Hom}_d^0(\Gamma, G)$, and then Corollary 6.4.9 completes the proof. □

The following is then an immediate consequence of Remarks 6.1.4 and Proposition 2.5.1.

Proposition 6.4.11. *Let G be the Euclidean motion group, H a closed subgroup of G and Γ a discontinuous group for G/H. Then any point in $\mathscr{R}(\Gamma, G, H)$ is geometrically stable.*

6.4.2 Near stability

Our main result in this section is to characterize the set of nearly stable points in the context of Euclidean motion groups. We will prove the following.

Theorem 6.4.12. *Let G be the Euclidean motion group, H a closed subgroup of G and Γ a discontinuous group for G/H. Then any point in $\mathscr{R}(\Gamma, G, H)$ is nearly stable. That is, $\mathscr{R}(\Gamma, G, H)$ is an open set of $\mathrm{Hom}_d^0(\Gamma, G)$.*

Proof. Thanks to Theorem 5.7.4, the parameter space $\mathscr{R}(\Gamma, G, H)$ is a finite union of open G-orbits whenever Γ is finite, which proves the statement in this case. Assume then that Γ is infinite, then H is compact by Proposition 2.5.1. Thanks to Fact 6.1.6, we assume right away that Γ is not torsion-free. Let $\{\varphi_p\}_{p\in\mathbb{N}}$ be a sequence of $\mathrm{Hom}_d^0(\Gamma, G)$ converging to $\varphi \in \mathrm{Hom}_d^0(\Gamma, G)$ for which $\varphi_p(\Gamma)$ does not act freely on G/H. There exists then $\gamma_p \in \Gamma \setminus \{e\}$ such that $\varphi_p(\gamma_p) \in g_p^{-1}Hg_p \setminus \{e\}$ for some $g_p \in G$. As φ_p is injective, $\varphi_p(\gamma_p)$ and γ_p have the same finite order. Then $\gamma_p \in \mathscr{K}_\Gamma$, and since

$$\mathscr{K}_\Gamma = \bigsqcup_{1\le i\le M_\Gamma} \mathscr{K}_i,$$

one can find $i \in \{1, \dots, M_\Gamma\}$ and a subsequence $\{\gamma_{\alpha(p)}\}_{p\in\mathbb{N}}$ in such a way that $\gamma_{\alpha(p)} \in \mathscr{K}_i$ for any $\alpha(p)$. Furthermore, thanks to Proposition 6.4.8,

$$\mathscr{K}_i = \bigsqcup_{0\le j\le d-1} \Theta_{i,j}$$

and we can therefore assume that there exists $j \in \{0, \dots, d-1\}$ such that $\gamma_{\alpha(p)} \in \Theta_{i,j}$. Hence

$$\gamma_{\alpha(p)} = (I, -y_{\alpha(p)})(A_i, x_j)(I, y_{\alpha(p)}).$$

Define $B_{\alpha(p)} \in O_n(\mathbb{R})$, $z_{\alpha(p)} \in \mathbb{R}^n$ for $p \in \mathbb{N}$ by the identity $(B_{\alpha(p)}, z_{\alpha(p)}) = \varphi_{\alpha(p)}(\gamma_{\alpha(p)})$. Thanks to Corollary 6.4.2, we can extract a convergent subsequence $\{B_{\beta(p)}\}_{p\in\mathbb{N}}$ sitting in the orbit of a matrix $B \neq I$, say, in such a way that $(B_{\beta(p)}, z_{\beta(p)}) \in g_{\beta(p)}^{-1} H g_{\beta(p)} \backslash \{e\}$ for some $y_{\beta(p)} \in G$. This gives in turn that

$$\varphi_{\beta(p)}((A_i, x_j)) = \varphi_{\beta(p)}((I, y_{\beta(p)}))(B_{\beta(p)}, z_{\beta(p)})\varphi_{\beta(p)}((I, y_{\beta(p)}))^{-1} =: (B_{\beta(p)}, z'_{\beta(p)}).$$

Finally, $\varphi_{\beta(p)}((A_i, x_j)) \in (g'_{\beta(p)})^{-1} H g'_{\beta(p)}$ for $g'_{\beta(p)} = g_{\beta(p)}\varphi_{\beta(p)}((I, y_{\beta(p)}))^{-1}$. Furthermore, the convergence of the sequence $(\varphi_{\beta(p)}((A_i, x_j)))_p$ to $\varphi((A_i, x_j))$ entails the convergence of $((B_{\beta(p)}, z'_{\beta(p)}))_p$ to some (B, z). Thanks to Corollary 6.4.2, $B_{\beta(p)} = S_p B S_p^{-1}$ for a given sequence $\{S_p\}_{p\in\mathbb{N}} \subset O_n(\mathbb{R})$. We can hence assume that $\{S_p\}_{p\in\mathbb{N}}$ converges to a matrix S and then $B \in S^{-1}\operatorname{pr}_1(H)S$. Thanks to Lemma 1.2.6, there exists $g \in G$ such that $(B, z) \in g^{-1}Hg$, which gives that $\varphi \notin \mathscr{R}(\Gamma, G, H)$ and the proof is complete. □

6.4.3 Case of crystallographic discontinuous groups

We study in this section the set of stable points. This simply reduces to describe the set of interior points of $\operatorname{Hom}_d^0(\Gamma, G)$ inside $\operatorname{Hom}(\Gamma, G)$. We first prove some preliminary useful results, which will be fundamental ingredients to reach our objectives. Let us start with the following.

Proposition 6.4.13. *Let Γ be a discrete subgroup of G and Γ_a a finitely generated Abelian normal subgroup of finite index in Γ. Take any $\varphi \in \operatorname{Hom}_d^0(\Gamma, G)$. If $\varphi_{|\Gamma_a}$ is an interior point of $\operatorname{Hom}_d^0(\Gamma_a, G)$ in $\operatorname{Hom}(\Gamma_a, G)$, then φ is an interior point of $\operatorname{Hom}_d^0(\Gamma, G)$ in $\operatorname{Hom}(\Gamma, G)$.*

Proof. We first prove the following.

Lemma 6.4.14. *Assume the same setting of Proposition 6.4.13. For $\varphi \in \operatorname{Hom}(\Gamma, G)$, the condition $\varphi_{|\Gamma_a} \in \operatorname{Hom}_d(\Gamma_a, G)$ implies $\varphi \in \operatorname{Hom}_d(\Gamma, G)$.*

Let $\varphi \in \operatorname{Hom}_d^0(\Gamma, G)$ be such that $\varphi_{|\Gamma_a}$ is an interior point of $\operatorname{Hom}_d^0(\Gamma_a, G)$ in $\operatorname{Hom}(\Gamma_a, G)$, and φ is not an interior point of $\operatorname{Hom}_d^0(\Gamma, G)$ in $\operatorname{Hom}(\Gamma, G)$. Take a sequence $\{\varphi_p\}_{p\in\mathbb{N}}$ of $\operatorname{Hom}(\Gamma, G)$ converging to φ such that $\varphi_p \notin \operatorname{Hom}_d^0(\Gamma, G)$. Since $\{\varphi_{p|\Gamma_a}\}_{p\in\mathbb{N}}$ converges to $\varphi_{|\Gamma_a}$ in $\operatorname{Hom}(\Gamma_a, G)$, there exists $p_0 \in \mathbb{N}$ such that for any $p \geq p_0$, $\varphi_{p|\Gamma_a} \in \operatorname{Hom}_d^0(\Gamma_a, G)$. By Lemma 6.4.14, for any $p \geq p_0$, $\varphi_p \in \operatorname{Hom}_d(\Gamma, G)$. This leads us to take a sequence $\{g_p\}_{p\geq p_0}$ in $\Gamma\backslash\{e\}$ such as $\varphi_p(g_p) = e$. Take $\overline{\delta}_0, \overline{\delta}_1, \dots, \overline{\delta}_{q-1}$ for the complete representative of the finite set Γ/Γ_a, where $q = \sharp\Gamma/\Gamma_a$. Then we can take $\alpha_p \in \Gamma_a$ such that

$$g_p = \alpha_p \delta_{j_p},$$

where $j_p \in \{0, 1, \dots, q-1\}$, $p \geq p_0$. As $\varphi_{p|\Gamma_a}$ is injective, then $\delta_{j_p} \notin \Gamma_a$. Hence $\varphi_p(\alpha_p) = \varphi_p(\delta_{j_p}^{-1})$, which entails that $\varphi_p(\alpha_p^q) = \varphi_p(\delta_{j_p}^{-q})$. Therefore,

$$\alpha_p^q = (\delta_{j_p}^{-q}).$$

This means that the set $\{\alpha_p^q\}_{p\geq p_0}$ is finite. Therefore, the set $\{\alpha_p\}_{p\geq p_0}$ is also finite because Γ_a is finitely generated Abelian group. Then the set $\{g_p\}_{p\geq p_0}$ is finite in Γ. This entails that there exists $\tilde{g} \in \{g_p\}_{p\geq p_0}$ such that for p large enough, $\varphi_p(\tilde{g}) = e$, and then $\varphi(\tilde{g}) = e$, which is impossible. This completes the proof. □

Lemma 6.4.15. *Let $\{\gamma_p\}_{p\in\mathbb{N}}$ and $\{\gamma'_p\}_{p\in\mathbb{N}}$ be two sequences of $SO_2(\mathbb{R}) \ltimes \mathbb{R}^2$ converging to (I_2, y) and (I_2, y'), respectively. Assume that both $\{\gamma_p\}_{p\in\mathbb{N}}$ and $\{\gamma'_p\}_{p\in\mathbb{N}}$ are not translations of $\mathbb{R}^2$ and commute. Then y and y' are linearly dependent.*

Proof. Consider the following equation:

$$\lim_{\theta\to 0}\frac{1}{\theta}(r(\theta) - I_2) = \frac{dr}{d\theta}(0) = \begin{pmatrix} 0 & -1 \\ 1 & 0 \end{pmatrix}. \tag{6.10}$$

Suppose that $\gamma_p = (r(\theta_p), y_p)$ and $\gamma'_p = (r(\theta'_p), y'_p)$. The commutativity condition implies

$$(r(\theta_p) - I_2)y'_p = (r(\theta'_p) - I_2)y_p. \tag{6.11}$$

Since γ_p and γ'_p are not translations, then for any $p \in \mathbb{N}$, $\theta_p, \theta'_p \neq 0$. Then we have

$$\frac{1}{\theta_p}(r(\theta_p) - I_2)y'_p = \rho_p\frac{1}{\theta'_p}(r(\theta'_p) - I_2)y_p,$$

where $\rho_p = \frac{\theta'_p}{\theta_p}$. By the convergence of $\{y'_p\}_{p\in\mathbb{N}}$ to y', $\{y_p\}_{p\in\mathbb{N}}$ to y and equation (6.10), $\{\rho_p\}_{p\in\mathbb{N}}$ converges to some real number ρ and we obtain $y' = \rho y$. □

Remark 6.4.16. The last lemma allows us to see that the converse of Proposition 6.4.13 is not true in general, as reveals the following example.

Example 6.4.17. Set $G = I(3)$ and Γ the discrete subgroup generated by

$$\gamma = \left(I_3, \begin{pmatrix} e_1 \\ 0 \end{pmatrix}\right)$$

and

$$\delta = \left(\begin{pmatrix} r(\frac{2\pi}{3}) & \\ & 1 \end{pmatrix}, \begin{pmatrix} 0_2 \\ 0 \end{pmatrix}\right),$$

where $e_1 = \binom{1}{0}$. The subgroup Γ_a generated by γ and

$$\gamma' = \left(I_3, \begin{pmatrix} r(\frac{2\pi}{3})e_1 \\ 0 \end{pmatrix}\right)$$

is an Abelian normal subgroup of index 3 in Γ and such that $\mathrm{Hom}_d^0(\Gamma_a, G)$ is not open in $\mathrm{Hom}(\Gamma_a, G)$, but $\mathrm{Hom}_d^0(\Gamma, G)$ is open in $\mathrm{Hom}(\Gamma, G)$. Indeed, for any positive integer p, we define $\varphi_p \in \mathrm{Hom}(\Gamma_a, G)$ as follows:

$$\varphi_p(\gamma) = \gamma \quad \text{and} \quad \varphi_p(\gamma') = \left(\begin{pmatrix} 1 & \\ & r(\frac{1}{p}) \end{pmatrix}, \begin{pmatrix} \frac{-1}{2} \\ \frac{\sqrt{3}}{2} e_1 \end{pmatrix} \right).$$

Then $\{\varphi_p\}$ converges to id_{Γ_a} and $\varphi_p \notin \mathrm{Hom}_d^0(\Gamma_a, G)$; otherwise, $r_e(\varphi_p(\Gamma_a)) = 1 \neq r_e(\Gamma) = 2$, which is absurd. Furthermore, let $\{\psi_p\}_{p\in\mathbb{N}}$ be a sequence of homomorphisms, which converges to some $\psi \in \mathrm{Hom}_d^0(\Gamma, G)$ and such that for any $p \in \mathbb{N}$, $\psi_p \notin \mathrm{Hom}_d^0(\Gamma, G)$. Due to Corollary 6.4.2, for p large enough, there exists $S_p \in O_3(\mathbb{R})$ such that

$$(S_p, 0_3)^{-1}\psi_p(\delta)(S_p, 0_3) = \left(\begin{pmatrix} r(\frac{m\pi}{3}) & \\ & 1 \end{pmatrix}, \begin{pmatrix} y_2(p) \\ 0 \end{pmatrix} \right),$$

where $m \in \{2, 4\}$ and thanks to Lemma 1.2.1,

$$(S_p, 0_3)^{-1}\psi_p(\gamma)(S_p, 0_3) = \left(\begin{pmatrix} r(\theta(p)) & \\ & \varepsilon_1 \end{pmatrix}, \begin{pmatrix} x_2(p) \\ \alpha_p \end{pmatrix} \right)$$

and

$$(S_p, 0_3)^{-1}\psi_p(\gamma')(S_p, 0_3) = \left(\begin{pmatrix} r(\theta'(p)) & \\ & \varepsilon_2 \end{pmatrix}, \begin{pmatrix} x_2'(p) \\ \beta_p \end{pmatrix} \right),$$

where $\varepsilon_i \in \{-1, 1\}$. We can assume that the sequence $\{(S_p, 0_3)\}_{p\in\mathbb{N}}$ converges to some $g \in G$. By Lemma 6.4.15,

$$g^{-1}\psi(\gamma)g = \left(\begin{pmatrix} I_2 & \\ & \varepsilon_1 \end{pmatrix}, \begin{pmatrix} x_2 \\ \alpha \end{pmatrix} \right)$$

and

$$g^{-1}\psi(\gamma')g = \left(\begin{pmatrix} I_2 & \\ & \varepsilon_2 \end{pmatrix}, \begin{pmatrix} \rho x_2 \\ \beta \end{pmatrix} \right).$$

It happens therefore that the subgroup of $\mathbb{R}^2$ generated by x_2 and ρx_2 is not stable by the subgroup generated by $r(\frac{m\pi}{3})$, which contradicts Theorem 1.2.24. Hence for p large enough, $\psi_p \in \mathrm{Hom}_d^0(\Gamma, G)$ and then $\mathrm{Hom}_d^0(\Gamma, G)$ is open in $\mathrm{Hom}(\Gamma, G)$.

We now prove the following the main result.

Theorem 6.4.18. *Let Γ be a discontinuous crystallographic group for a homogenous space G/H, then any $\varphi \in \mathscr{R}(\Gamma, G, H)$ is stable. That is, $\mathscr{R}(\Gamma, G, H) = \mathrm{Hom}_d^0(\Gamma, G)$ is open in* $\mathrm{Hom}(\Gamma, G)$.

Proof. Let Γ be a crystallographic subgroup of G acting discontinuously on G/H, then H is compact due to Proposition 2.5.1. Let $\psi \in \mathrm{Hom}_d^0(\Gamma, G)$ such that $\psi(\Gamma)$ does not act freely on G/H. There exists therefore $\gamma = (C_\gamma, x_\gamma) \in \Gamma$ of finite order such that $\psi(\gamma) \in sHs^{-1}$ for some $s = (S, t_s) \in G$. Since $\psi(\gamma) = g\gamma g^{-1}$ for some $g = (A, t) \in GL_n(\mathbb{R}) \ltimes \mathbb{R}^n$ by Lemma 1.2.20, we get $g\gamma g^{-1} = shs^{-1}$ for some $h = (C_h, t_h) \in H$. This implies that C_γ and C_h being of finite order, are conjugate in $GL_n(\mathbb{R})$ and then conjugate in $O_n(\mathbb{R})$. Thanks to Lemma 1.2.6, γ and h are conjugate in G, which is absurd. Therefore, $\mathscr{R}(\Gamma, G, H) = \operatorname{Hom}^0_d(\Gamma, G)$.

On the other hand, Γ contains an Abelian normal subgroup Γ_a, say, which is generated by n free translations and of finite index in Γ. Hence Γ_a admits generators $\{\gamma_1, \dots, \gamma_n\}$ such that for any $i \in \{1, \dots, n\}$, $\gamma_i = (I, x_i)$ where the family $\{x_1, \dots, x_n\}$ is of rank n. For any $\varphi \in \operatorname{Hom}^0_d(\Gamma, G)$, $\varphi_{|\Gamma_a}$ is an isomorphism from Γ_a into $\varphi(\Gamma_a)$, by Lemma 1.2.20, there exists $g_\varphi := (A_\varphi, t_\varphi) \in GL_n(\mathbb{R}) \ltimes \mathbb{R}^n$ such that $\varphi(\gamma_i) = (I, A_\varphi x_i)(1 \le i \le n)$. Let $\{\varphi_p\}_{p\in\mathbb{N}}$ be a sequence of homomorphisms, which converges to φ. For any $i \in \{1, \dots, n\}$ and any $p \in \mathbb{N}$, set $\varphi_p(\gamma_i) = (A_i(p), x_i(p)) \in G$. Making use of Fact 1.2.1, there exists $g_p := (S_p, 0) \in G$ for which $\phi_p := g_p^{-1}\varphi_p g_p$ is such that for any $i \in \{1, \dots, n\}$,

$$\phi_p(\gamma_i) = \left(\begin{pmatrix} D_i(p) & & & \\ & r(\theta_{1,i}(p)) & & \\ & & \ddots & \\ & & & r(\theta_{l(p),i}(p)) \end{pmatrix}, \begin{pmatrix} y'_i(p) \\ x_{1,i}(p) \\ \vdots \\ x_{l(p),i}(p) \end{pmatrix} \right), \tag{6.12}$$

where for any $i \in \{1, \dots, n\}$, $D_i(p)$ is a diagonal matrix of $O_{m(p)}(\mathbb{R})$ where $m(p) = n - 2l(p)$, and where

$$\begin{pmatrix} y'_i(p) \\ x_{1,i}(p) \\ \vdots \\ x_{l(p),i}(p) \end{pmatrix} = S_p^{-1} x_i(p)$$

for some $x_{s,i}(p) \in \mathbb{R}^2$ and $y'_i(p) \in \mathbb{R}^{m(p)}$. Since for any $i \in \{1, \dots, n\}$, the sequence $\{A_i(p)\}_{p\in\mathbb{N}}$ converges to I, then for p large enough $D_i(p) = I_{m(p)}$. On the other hand, we can find a subsequence $\{\phi_{\alpha(p)}\}_{p\in\mathbb{N}}$ such that $l(\alpha(p)) = l$. If for instance $l \ge 1$, then equation (6.12) reads

$$\phi_{\alpha(p)}(\gamma_i) = \left(\begin{pmatrix} I_m & & & \\ & r(\theta_{1,i}(\alpha(p))) & & \\ & & \ddots & \\ & & & r(\theta_{l,i}(\alpha(p))) \end{pmatrix}, \begin{pmatrix} y'_i(\alpha(p)) \\ x_{1,i}(\alpha(p)) \\ \vdots \\ x_{l,i}(\alpha(p)) \end{pmatrix} \right). \tag{6.13}$$

For any $i \in \{1, \dots, n\}$ and any $s \in \{1, \dots, l\}$, the sequence $\{r(\theta_{s,i}(\alpha(p)))\}_{p\in\mathbb{N}}$ tends to I_2. If there exists $i_0 \in \{1, \dots, n\}$ and $s_0 \in \{1, \dots, l\}$ such that the set $J_{i_0,s_0} := \{p \in \mathbb{N} \mid$

$r(\theta_{s_0,i_0}(\alpha(p))) \neq I_2\}$ is infinite, then by Lemma 6.4.15, the sequence $\{\phi_{\alpha(p)}\}_{p\in\mathbb{N}}$ converges to ϕ such that

$$\phi(\gamma_i) = \left(\begin{pmatrix} I_m & & & \\ & I_2 & & \\ & & \ddots & \\ & & & I_2 \end{pmatrix}, \begin{pmatrix} y_i' \\ x_{1,i} \\ \vdots \\ x_{l,i} \end{pmatrix}\right), \tag{6.14}$$

and for which the family $\{x_{s_0,i}\}_{1\leq i\leq n}$ is of rank 1. Therefore, the family

$$\left\{\begin{pmatrix} y_i' \\ x_{1,i} \\ \vdots \\ x_{l,i} \end{pmatrix}\right\}_{1\leq i\leq n}$$

is of rank less or than equal to $n-1$, which is absurd. This leads us to conclude that for sufficiently large $p, l = 0$. Equivalently, for any $i \in \{1,\ldots,n\}$ $A_i(p) = I$, and hence $\varphi_p(\gamma_i) = (I, x_i(p))$ for p large enough. Now the convergence of the sequence $\{\varphi_p\}_{p\in\mathbb{N}}$ to φ is equivalent to the convergence of the sequence of matrices M_p, say, of column $x_i(p)$ $(1 \leq i \leq n)$ to the invertible matrix $M(\varphi)$, say, of column $A_\varphi x_i$ $(1 \leq i \leq n)$. As $GL_n(\mathbb{R})$ is an open set of $M_n(\mathbb{R})$, there exists p_0 such that for $p \geq p_0$ M_p is invertible, then the family $\{x_i(p)\}_{1\leq i\leq n}$ is of rank n, and thus $\varphi_{p|\Gamma_a} \in \mathrm{Hom}_d^0(\Gamma_a, G)$. This justifies that $\varphi_{|\Gamma_a}$ is an interior point of $\mathrm{Hom}_d^0(\Gamma_a, G)$ in $\mathrm{Hom}(\Gamma_a, G)$. By Proposition 6.4.13, we get that $\mathrm{Hom}_d^0(\Gamma, G)$ is open in $\mathrm{Hom}(\Gamma, G)$ and this completes the proof of the theorem. □

6.4.4 Further remarks

I. In the setting where $G = O_n(\mathbb{R}) \ltimes \mathbb{R}^n$, the geometric stability is evidently verified. Indeed whenever Γ acts properly on G/H, the pair (Γ, H) is such that H is compact or Γ is finite, and hence any $\varphi \in \mathrm{Hom}(\Gamma, G)$ is such that $\overline{\varphi(\Gamma)}$ acts properly on G/H. When Γ is finite, the parameter space $\mathscr{R}(\Gamma, G, H)$ turns out to be a finite union of G-orbits for which the corresponding subgroups act fixed point freely on G/H as shown in [13]. Then $\mathscr{R}(\Gamma, G, H)$ is open in $\mathrm{Hom}(\Gamma, G)$ (thus any homomorphism is stable).

The following example shows that the parameter space $\mathscr{R}(\Gamma, G, H)$ may be open in $\mathrm{Hom}(\Gamma, G)$, but $\mathrm{Hom}_d^0(\Gamma, G)$ fails in general to be.

Example 6.4.19. Let Γ be the discrete subgroup of $I(5)$ generated by

$$\gamma_1 = \left(\begin{pmatrix} 1 & & \\ & I_2 & \\ & & I_2 \end{pmatrix}, \begin{pmatrix} 1 \\ 0_2 \\ 0_2 \end{pmatrix}\right) \quad \text{and} \quad \gamma_2 = \left(\begin{pmatrix} 1 & & \\ & r(\frac{2\pi}{3}) & \\ & & r(\frac{2\pi}{3}) \end{pmatrix}, \begin{pmatrix} 0 \\ 0_2 \\ 0_2 \end{pmatrix}\right)$$

and let

$$H = \left\{ \left(\begin{pmatrix} I_3 & \\ & A \end{pmatrix}, 0_5 \right) \middle| A \in SO_2(\mathbb{R}) \right\}.$$

Notice that $\mathrm{Hom}_d^0(\Gamma, I(5))$ is not open in $\mathrm{Hom}(\Gamma, I(5))$. Indeed, for any $p \in \mathbb{N}^*$ define the homomorphism φ_p by

$$\varphi_p(\gamma_1) = \left(\begin{pmatrix} r(\frac{1}{p}) & & \\ & 1 & \\ & & I_2 \end{pmatrix}, \begin{pmatrix} e_1 \\ 0 \\ 0_3 \end{pmatrix} \right)$$

and

$$\varphi_p(\gamma_2) = \left(\begin{pmatrix} 1 & & \\ & I_2 & \\ & & r(\frac{2\pi}{3}) \end{pmatrix}, \begin{pmatrix} 0 \\ 0_2 \\ 0_2 \end{pmatrix} \right),$$

where

$$e_1 = \begin{pmatrix} 1 \\ 0 \end{pmatrix}.$$

If $\varphi_p \in \mathrm{Hom}_d^0(\Gamma, I(5))$, then $r_e(\varphi(\Gamma)) = 0$, which is absurd. We easily check that the limit φ of the sequence $\{\varphi_p\}_{p\in\mathbb{N}}$ satisfies $\varphi \in \mathrm{Hom}_d^0(\Gamma, I(5))$. On the other hand, $\mathscr{R}(\Gamma, I(5), H)$ is open in $\mathrm{Hom}(\Gamma, I(5))$. In fact, any homomorphism $\psi \in \mathscr{R}(\Gamma, I(5), H)$ is such that $\psi(\gamma_2) \notin g^{-1}Hg$ for any $g \in G$, and thus there exists $g_\psi \in G$ such that

$$\widetilde{\psi}(\gamma_2) \in \left\{ \left(\begin{pmatrix} 1 & & \\ & r(\frac{2i\pi}{3}) & \\ & & r(\frac{2j\pi}{3}) \end{pmatrix}, \begin{pmatrix} 0 \\ 0_2 \\ 0_2 \end{pmatrix} \right) \middle| i, j \in \{1, 2\} \right\},$$

where $\widetilde{\psi} := g_\psi^{-1}\psi g_\psi$. Let $\{\psi_p\}_{p\in\mathbb{N}}$ be a sequence of homomorphism, which converges to some $\psi \in \mathscr{R}(\Gamma, I(5), H)$. For p large enough $\widetilde{\psi}_p(\gamma_2) = \widetilde{\psi}(\gamma_2)$, thanks to Corollary 6.4.2. Then

$$\widetilde{\psi}_p(\gamma_1) = \left(\begin{pmatrix} 1 & & \\ & r(\theta_p) & \\ & & r(\theta'_p) \end{pmatrix}, \begin{pmatrix} \alpha_p \\ 0_2 \\ 0_2 \end{pmatrix} \right).$$

Necessarily, $\{\alpha_p\}_{p\in\mathbb{N}}$ converges to some $\alpha \in \mathbb{R}^*$; otherwise $r_e(\psi(\Gamma)) = 0$, which is absurd. Therefore, for sufficiently large p, $\alpha_p \neq 0$ and then $\psi_p \in \mathscr{R}(\Gamma, I(5), H)$.

For an infinite discrete subgroup Γ of G, the following example shows that the parameter space may admits both stable and nonstable points.

Example 6.4.20. Let Γ be the discrete subgroup of $I(5)$ generated by

$$\gamma_1 = \left(\begin{pmatrix} 1 & & \\ & I_2 & \\ & & I_2 \end{pmatrix}, \begin{pmatrix} 1 \\ 0_2 \\ 0_2 \end{pmatrix}\right) \quad \text{and} \quad \gamma_2 = \left(\begin{pmatrix} 1 & & \\ & I_2 & \\ & & r(\frac{2\pi}{3}) \end{pmatrix}, \begin{pmatrix} 0 \\ 0_2 \\ 0_2 \end{pmatrix}\right)$$

and let

$$H = \left\{ \left(\begin{pmatrix} 1 & & \\ & A & \\ & & A \end{pmatrix}, 0_5\right) \middle| A \in SO_2(\mathbb{R}) \right\}.$$

Notice that id_Γ is not stable. Indeed, for any $p \in \mathbb{N}^*$ define the homomorphism φ_p by

$$\varphi_p(\gamma_1) = \left(\begin{pmatrix} r(\frac{1}{p}) & & \\ & 1 & \\ & & I_2 \end{pmatrix}, \begin{pmatrix} e_1 \\ 0 \\ 0_3 \end{pmatrix}\right)$$

and

$$\varphi_p(\gamma_2) = \left(\begin{pmatrix} 1 & & \\ & I_2 & \\ & & r(\frac{2\pi}{3}) \end{pmatrix}, \begin{pmatrix} 0 \\ 0_2 \\ 0_2 \end{pmatrix}\right),$$

where

$$e_1 = \begin{pmatrix} 1 \\ 0 \end{pmatrix}.$$

If $\varphi_p \in \mathrm{Hom}_d^0(\Gamma, I(5))$, then $r_e(\varphi(\Gamma)) = 0$, which is absurd. We easily check that the sequence $(\varphi_p)_{p\in\mathbb{N}}$ converges to id_Γ, and thus id_Γ is not stable point of $\mathscr{R}(\Gamma, I(5), H)$. On the other hand, let φ be the homomorphism such that

$$\varphi(\gamma_1) = \gamma_1 \quad \text{and} \quad \varphi(\gamma_2) = \left(\begin{pmatrix} 1 & & \\ & r(\frac{4\pi}{3}) & \\ & & r(\frac{2\pi}{3}) \end{pmatrix}, \begin{pmatrix} 0 \\ 0_2 \\ 0_2 \end{pmatrix}\right).$$

Then φ is a stable point. Indeed, as described in Example 6.4.19, If a sequence of homomorphisms $(\varphi_p)_{p\in\mathbb{N}}$ converges to φ, then for p large enough φ_p belongs to $\mathrm{Hom}_d^0(\Gamma, I(5))$ and by Theorem 6.4.12 we deduce that for sufficiently large p, $\varphi_p \in \mathscr{R}(\Gamma, I(5), H)$.

II. For general compact extensions of $\mathbb{R}^n$, the geometric and near stability do not hold as it is shown in Example 6.4.22 below. The tangential homogeneous spaces of reductive homogeneous spaces are typical examples. Remark first the following immediate fact.

Fact 6.4.21. *Suppose that $G := G_1 \times G_2$ is a Lie group, $\Gamma \subset G_1 \times \{e'\}$ is a discrete subgroup and $H \subset \{e\} \times G_2$ is a closed subgroup where e and e' designate the unity elements of G_1 and G_2, respectively. Then Γ acts on G/H properly discontinuously and freely.*

Example 6.4.22. Let $G = K \ltimes \mathbb{R}^4$, where $K = O_2(\mathbb{R}) \times O_2(\mathbb{R})$. Any $g \in G$ is written as

$$g = (A_g, x_g) = \left(\begin{pmatrix} A_{1,g} & \\ & A_{2,g} \end{pmatrix}, \begin{pmatrix} x_{1,g} \\ x_{2,g} \end{pmatrix} \right), \tag{6.15}$$

where $A_{1,g}, A_{2,g} \in O_2(\mathbb{R})$ and $x_{1,g}, x_{2,g} \in \mathbb{R}^2$. Let $(e_i)_{1\leq i\leq 4}$ be the standard basis of $\mathbb{R}^4$. Let Γ be generated by $\gamma_1 = (I_4, e_1)$ and

$$H = \left\{ \left(I_4, \begin{pmatrix} 0_2 \\ x \end{pmatrix} \right) \,\middle|\, x \in \mathbb{R}^2 \right\}.$$

Thanks to Fact 6.4.21, Γ acts properly and freely on G/H. We show next that id_Γ is neither geometrically nor nearly stable. Set $C_r := K \times B(0, \rho)$, where $B(0, \rho)$ is the closed ball of center 0 and radius ρ. For any $p \in \mathbb{N}^*$, the homomorphism φ_p defined as

$$\varphi_p(\gamma_1) = \left(\begin{pmatrix} r(\frac{1}{p}) & \\ & I_2 \end{pmatrix}, \begin{pmatrix} x_0 \\ \frac{1}{p} x_0 \end{pmatrix} \right) =: \mathfrak{s}_p,$$

where

$$x_0 = \begin{pmatrix} 1 \\ 0 \end{pmatrix}.$$

By Theorem 1.2.24, $\varphi_p(\Gamma)$ is discrete, and hence $\varphi_p \in \mathrm{Hom}_d^0(\Gamma, G)$. Obviously, $\{\varphi_p\}_{p\in\mathbb{N}^*}$ converges to the identity of Γ. On the other hand, any $\gamma \in \varphi_p(\Gamma)$ is such that $\gamma = \sigma_p^m$ for some $m \in \mathbb{Z}$, so it is of the matrix form

$$\gamma = \sigma_p^m = \left(\begin{pmatrix} r(\frac{m}{p}) & \\ & I_2 \end{pmatrix}, \begin{pmatrix} v_m \\ \frac{m}{p} x_0 \end{pmatrix} \right),$$

where $v_m = (I_2 - r(\frac{1}{p}))^{-1}(I_2 - r(\frac{m}{p}))x_0$ as in equation (6.11). Remark that

$$\|v_m\| \leq \frac{2}{|\cos(\frac{1}{2p})|} =: \rho_p$$

and $\lim_{m\to\infty} \|\frac{m}{p} x_0\| = +\infty$; hence $\overline{\varphi_p(\Gamma)}$ is not compact and we have

$$\sigma_p^m = \underbrace{\left(\begin{pmatrix} I_2 & \\ & I_2 \end{pmatrix}, \begin{pmatrix} v_m \\ 0_2 \end{pmatrix} \right)}_{\in C_{\rho_p}} \underbrace{\left(\begin{pmatrix} I_2 & \\ & I_2 \end{pmatrix}, \begin{pmatrix} 0_2 \\ \frac{m}{p} x_0 \end{pmatrix} \right)}_{\in H} \underbrace{\left(\begin{pmatrix} r(\frac{m}{p}) & \\ & I_2 \end{pmatrix}, \begin{pmatrix} 0_2 \\ 0_2 \end{pmatrix} \right)}_{\in C_{\rho_p}},$$

and $\varphi_p(\Gamma) \subset C_{\rho_p} H C_{\rho_p}$ so that $\varphi_p \notin \mathscr{R}(\Gamma, G, H)$. This allows us to conclude that for any neighborhood $\mathscr{V}$ of id_Γ, there exists $p \in \mathbb{N}$ such that $\varphi_p \in \mathscr{V} \cap \mathrm{Hom}_d^0(\Gamma, G)$ and, therefore, id_Γ is not geometrically stable and not nearly stable.

III. Table 6.1 summarizes the obtained results concerning the different stability variants in the setting of Euclidean motion groups.

Table 6.1: Summarizing various stability results.

	Near Stability	Geometric Stability	Stability
Γ finite	holds	holds	holds
Γ infinite	holds	holds	fails in general
Γ crystallographic	holds	holds	holds

6.5 The Calabi–Markus phenomenon

Let (G, H) be in full generality. The problem of finding discontinuous groups for G/H is trivial when H is compact. So, the interesting cases consist of constructing infinite discontinuous groups for G/H when H and G/H are noncompact and are the subject of our inquiry in the present subsection. When H is noncompact, it may happen that there does not exist an infinite discrete subgroup Γ of G, which acts properly discontinuously on G/H. This phenomenon is called the Calabi–Markus phenomenon. We here focus attention on this issue in the context of some compact extensions of nilpotent Lie groups.

6.5.1 Case of Euclidean motion groups

In the context of Euclidean motion groups, we proved that when Γ is infinite (this is for instance the case for crystallographic groups), the proper action obliges the subgroup H not supposed to be connected, to be compact (cf. Proposition 2.5.1). This phenomenon makes it easier the study of the parameter and the deformation spaces and their topological and geometrical features. As such, this important fact allows us to prove that the Calabi–Markus occurs in the setting of Euclidean motion groups. That is, if H is a closed noncompact subgroup of G, then G/H does not admit a compact Clifford–Klein form, unless G/H itself is compact. Let us record first the following questions posed by T. Kobayashi in [95] for general Lie groups.

Question 1. *Does G/H admit a Clifford–Klein form of infinite fundamental group?*

Question 2. *Does G/H admit a noncommutative-free group as a discontinuous group?*

We first give complete answers to these questions in the context of Euclidean motion based on prior results. More precisely, we prove the following.

Theorem 6.5.1. *Let G be the Euclidean motion group. Then:*
(1) *The Calabi–Markus phenomenon occurs for G. That is, if H is a closed, noncompact subgroup of G, then G/H does not admit a compact Clifford–Klein form, unless G/H itself is compact.*
(2) *If H is a closed subgroup of G, then G/H admits a Clifford–Klein form of infinite fundamental group if and only if H is compact.*
(3) *G/H admits no noncommutative-free group as a discontinuous group. More precisely, G itself admits no noncommutative-free discrete subgroups.*

Proof. The first point of the theorem immediately follows from Proposition 2.5.1. This proposition also shows that it is not possible to get a Clifford–Klein form $\Gamma\backslash G/H$ of infinite fundamental group unless H is compact, which proves the second point. The third point is an immediate consequence of Lemma 1.2.22. □

Remark 6.5.2.
(1) Given a closed subgroup H of G, one important question is to find a nontrivial discrete subgroup Γ of G in such a way that $\Gamma\backslash G/H$ is a Clifford–Klein form. We show hereafter that this is not true in general in the setting of Euclidean motion groups. When for instance G/H is compact, it fails in general to admit a compact Clifford–Klein form $\Gamma\backslash G/H$ with a nontrivial discontinuous group Γ for G/H. Indeed, take $n = 4$, $G = I(4)$ and

$$H = \left\{ \left(\begin{pmatrix} A & 0 \\ 0 & B \end{pmatrix}, \begin{pmatrix} x \\ x' \end{pmatrix} \right) \,\middle|\, A, B \in O_2(\mathbb{R}) \text{ and } x, x' \in \mathbb{R}^2 \right\}.$$

Then H is not compact and, therefore, the proper action holds if and only if Γ is finite as in Proposition 2.5.1. Let $g = (A, x)$ be an element of G of finite order. Then A itself is of finite order and according to equation (1.15), A is conjugate to an element of the form

$$\begin{pmatrix} A_1 & 0 \\ 0 & A_2 \end{pmatrix},$$

where A_i belongs to $O_2(\mathbb{R})$, $i = 1, 2$. Using Lemma 1.2.6, the element g is conjugate to an element of H. Hence any element of G of finite order is conjugate to an element of H. This entails that the unique Clifford–Klein form is G/H since the fixed-point free action holds uniquely when Γ is trivial.

(2) For the same context where $G = I(4)$, we show that one can find nontrivial discontinuous groups for a compact homogeneous space G/H. Let

$$H = \left\{ \left(\begin{pmatrix} I_2 & 0 \\ 0 & A \end{pmatrix}, \begin{pmatrix} x_1 \\ x_2 \end{pmatrix} \right) \middle| \ A \in O_2(\mathbb{R}), x_1, x_2 \in \mathbb{R}^2 \right\}.$$

Then G/H is compact and the proper action holds if and only if Γ is finite. Let

$$\gamma = \left(\begin{pmatrix} r(\frac{2\pi}{3}) & 0 \\ 0 & r(\frac{2\pi}{3}) \end{pmatrix}, 0 \right)$$

and $\Gamma = \langle \gamma \rangle$. Then clearly $\Gamma \backslash G/H$ is a compact Clifford–Klein form.

In several contexts (G is nilpotent connected and simply connected, G is exponential and Γ is Abelian, G is the reduced Heisenberg group, etc.), it is shown that whenever Γ is infinite, $\operatorname{Hom}_d^0(\Gamma, G)$ is an open set in $\operatorname{Hom}(\Gamma, G)$ (cf. [25, 31, 34]). In the context of Euclidean motion groups, we shall show in contrast that it is no longer the case. Let us start by studying an example, which is analogue to [98, Example 1.1.1] for the action of $\operatorname{Aff}(\mathbb{R})$ on $\mathbb{R}$.

Example 6.5.3. Let $G = I(2)$, $(e_i)_{1 \leq i \leq 2}$ the standard basis of $\mathbb{R}^2$ and $\Gamma = \langle \gamma \rangle$, where $\gamma = (I_2, e_1)$. Γ is a discrete subgroup of G. Consider the sequence of noninjective homomorphisms $\{\varphi_p\}_{p \in \mathbb{N}^*}$ such that $\varphi_p(\gamma) = (r(\frac{\pi}{p}), e_1)$. Clearly, $\{\varphi_p\}$ converges to id_Γ which is in $\operatorname{Hom}_d^0(\Gamma, G)$. This shows that this set is not open in $\operatorname{Hom}(\Gamma, G)$. Else, let ψ_p such that $\psi_p(\gamma) = (r(\frac{1}{p}), e_1)$, ψ_p is not of discrete image, and converges to id_Γ.

We now proceed to prove our first rigidity result in this context.

Proposition 6.5.4. *Let G be the Euclidean motion group and Γ a finite subgroup of G. Then $\operatorname{Hom}^0(\Gamma, G)/G$ is finite.*

Proof. Let $\{\varphi_s\}_{s \in \mathbb{N}}$ be a sequence in $\operatorname{Hom}^0(\Gamma, G)$ converging to φ. Denote $\varphi_s = (\varphi_{1,s}, \varphi_{2,s})$ and $\varphi = (\varphi_1, \varphi_2)$. Remark that by Lemma 5.7.6, the sequence $\{\varphi_{1,s}\}_{s \in \mathbb{N}}$ belongs to $\operatorname{Hom}(\Gamma, O_n(\mathbb{R}))$. Let $\gamma \in \Gamma$ such that $\varphi_{1,s}(\gamma) = I$, then $\varphi_s(\gamma)$ is an element of infinite order whenever $\varphi_{2,s}(\gamma) \neq 0$. This means conclusively that $\{\varphi_{1,s}\}_{s \in \mathbb{N}}$ is in $\operatorname{Hom}^0(\Gamma, O_n(\mathbb{R}))$ and converges to φ_1. Therefore, one of the orbits $[\psi_i]$'s contains an infinite number of elements of this sequence and, therefore, due to its convergence there exists $i_0 \in \{1, \ldots, m\}$, $s_0 \in \mathbb{N}$ such that for any $s \geq s_0$, $\varphi_{1,s} \in [\psi_{i_0}]$ and so $\varphi_1 \in [\psi_{i_0}]$. Equivalently, for $s \geq s_0$ there exists $A_s \in O_n(\mathbb{R})$ such that $\varphi_{1,s} = A_s \varphi_1 A_s^{-1}$. Hence $p_1(\varphi_s(\Gamma)) = A_s p_1(\varphi(\Gamma)) A_s^{-1}$. On the other hand, by Fact 1.2.10, there exist t_s and $t \in \mathbb{R}^n$ such that

$$(I, t)\varphi(\Gamma)(I, -t) = p_1(\varphi(\Gamma)) \times \{0\} \quad \text{and} \quad (I, t_s)\varphi_s(\Gamma)(I, -t_s) = p_1(\varphi_p(\Gamma)) \times \{0\},$$

and then one can easily check that $\varphi_s = (A_s, t - t_s)(\varphi_1, 0)(A_s, t - t_s)^{-1}$, which completes the proof. □

Remark 6.5.5. Corollary 6.5.4 shows immediately that for any finite subgroup Γ, the parameter space $\mathscr{R}(\Gamma, G, H)$ is a finite union of G-orbits for which the corresponding subgroups act fixed point freely on G/H. Then $\mathscr{R}(\Gamma, G, H)$ is open in $\mathrm{Hom}(\Gamma, G)$ (thus any homomorphism is stable) and the deformation space $\mathscr{T}(\Gamma, G, H)$ is a finite set.

6.5.2 The case of $SO_n(\mathbb{R}) \ltimes \mathbb{R}^n$

Let $M(n) := SO_n(\mathbb{R}) \ltimes \mathbb{R}^n$ be the semidirect product of the rotation group $SO_n(\mathbb{R})$ (with respect to the canonical Euclidean product on $\mathbb{R}^n$) and $\mathbb{R}^n$. Then, obviously, one can remark that the statement of Proposition 2.5.1 holds for $M(n)$ and so does that of Corollary 6.4.9. This allows us to affirm the following.

Theorem 6.5.6. *The conclusions of Theorems* 5.1.5 *and* 6.5.1 *hold for the group $M(n)$ for any $n \geq 2$.*

When more generally the compact component $SO_n(\mathbb{R})$ is replaced by a general compact subgroup of $\mathrm{Aut}(\mathbb{R}^n)$, the statement of Proposition 2.5.1 clearly fails to hold as many examples reveal. So, there is no hope that the statements of Theorem 6.5.1 hold for general compact extensions of $\mathbb{R}^n$ as in the coming subsection.

6.5.3 Case of the semidirect product $K \ltimes \mathbb{R}^n$

We now prove the following.

Theorem 6.5.7. *Let H be a closed, connected subgroup of $G := K \ltimes \mathbb{R}^n$. Then the Calabi–Markus phenomenon occurs if and only if, for any linear subspace $V \neq \{0\}$ of $\mathbb{R}^n$, there exists $k \in K$ such that $V \cap [k \cdot E_H] \neq \{0\}$.*

Proof. If there exists some nontrivial linear subspace V of $\mathbb{R}^n$ such that $K \cdot V \cap E_H \neq \{0\}$, then for a nonzero vector u of V the discrete subgroup $\Gamma = \{(I, pu), p \in \mathbb{Z}\}$ acts properly on $G/\mathscr{L}_H$ thanks to Theorem 2.5.7. Note here that Γ is torsion-free and then it acts freely on $G/\mathscr{L}_H$ whenever it acts properly on $G/\mathscr{L}_H$. Conversely, if there exists an infinite discrete subgroup Γ, which acts discontinuously on $G/\mathscr{L}_H$, then $\mathscr{L}_G$ acts properly $G/\mathscr{L}_H$, and hence $K \cdot E_\Gamma \cap E_H \neq \{0\}$, which completes the proof. □

6.5.4 Case of Heisenberg motion groups

We here refer back to Section 1.3 with all the notation and results, and prove the following.

Theorem 6.5.8. *Let G be a Heisenberg motion group and H a closed subgroup of G. Then G/H admits an infinite discontinuous group if and only if H is conjugate to a subgroup of G^1 or for any $r > 0$, $H \cap (\mathbb{U}_n \times B(0,r) \times \mathbb{R})$ is compact.*

Proof. As in Proposition 1.3.8, an infinite discrete subgroup contains an element of infinite order γ, say, and by Lemma 1.3.1, there exists $g \in G$ such that $g\gamma g^{-1} = (A, z, t)$ where $Az = z$, $tz = 0$ and $\|z\| + |t| \neq 0$. If $t = 0$, then by Proposition 2.6.3 H is conjugate to a subgroup of G^1. Otherwise, $H \cap (\mathbb{U}_n \times B(0,r) \times \mathbb{R})$ is compact for any $r > 0$ thanks to Proposition 2.6.2.

Conversely, if H is conjugate to a subgroup of G^1, take Γ the subgroup of type (B) generated by $(I, z, 0)$ for some $z \in \mathbb{C}^n \setminus \{0\}$. As $\mathrm{pr}(\Gamma)$ is discrete, Proposition 2.6.3 asserts that Γ acts properly on G/H and, therefore, discontinuously on G/H. Finally, if $H \cap (\mathbb{U}_n \times B(0,r) \times \mathbb{R})$ is compact for any $r > 0$ one can consider Γ to be the subgroup of type (A) generated by $(I, 0, 1)$ and the result follows from Proposition 2.6.2. □

Remark 6.5.9. When H is conjugate to a subgroup of G^1 and satisfies $H \cap (\mathbb{U}_n \times B(0,r) \times \mathbb{R})$ is compact for any $r > 0$, then obviously H is compact and, therefore, any infinite torsion- free discrete subgroup of G acts discontinuously on G/H. The readers can consult the reference [16] for more comprehensive details.

The following is a direct consequence of Theorem 6.5.8.

Corollary 6.5.10. *The Calabi–Markus phenomenon occurs for a homogenous space G/H if and only if H contains an infinite discrete subgroup conjugate to a subgroup of $\mathbb{U}_n \times \mathbb{C}^n \times \{0\}$ and $H \cap (\mathbb{U}_n \times B(0,r) \times \mathbb{R})$ is not compact for some $r > 0$.*

Indeed, if H contains an infinite discrete subgroup conjugate to a subgroup of $\mathbb{U}_n \times \mathbb{C}^n \times \{0\}$, it then contains an element (B, z, t) for which $P_B(z) = z \neq 0$, and Proposition 1.3.14 allows to conclude.

6.5.5 Existence of compact Clifford–Klein forms

This section aims to study the existence of a compact Clifford–Klein form $\Gamma \backslash G/H$, where H stands for a closed, connected subgroup of a Lie group G and Γ is an infinite discontinuous group for G/H. Our first result is the following.

Proposition 6.5.11. *Let H be a closed connected subgroup of G and R its solvable radical. The following are equivalent:*
(i) *G/H admits a compact Clifford–Klein form.*
(ii) *G/R admits a compact Clifford–Klein form.*

Proof. Suppose that G/H admits a compact Clifford–Klein form $\Gamma \backslash G/H$, where Γ designates a discontinuous group for G/H. Equivalently, there exists a compact set C of G

such as $G = \Gamma CH$. Clearly, Γ acts discontinuously on G/R. $\Gamma CH = \Gamma(CS)R$, which entails that $\Gamma\backslash G/R$ is compact.

Conversely, assume that there exists a discontinuous group Γ for G/R such that $\Gamma\backslash G/R$ is compact. Through Theorem 1.2.24, we easily get that Γ contains an Abelian torsion-free group Γ_a of finite index in Γ, which entails that $\Gamma_a\backslash G/R$ is compact. Write $H = SR$ for some Levi complement S of R. For any compact set K of G, $\Gamma_a \cap KHK^{-1} = \Gamma_a \cap (KS)RK^{-1}$ is compact by means of Lemma 2.5.5. Therefore, Γ_a acts properly on G/H and gives in turn that for any $g \in G$, $\Gamma_a \cap gHg^{-1}$ is a finite group and then trivial. Hence Γ_a acts freely on G/H. It suffices now to consider a compact set C such that $G = \Gamma_a CR$. Then $G = \Gamma_a CSR$, as $C \subset CS$, and hence $\Gamma_a\backslash G/H$ is compact. □

We next prove the following.

Theorem 6.5.12. *Let $G := K \ltimes \mathbb{R}^n$ and H a closed, connected subgroup of G. Then G/H admits a compact Clifford–Klein form, if and only if, there exists a linear subspace V of $\mathbb{R}^n$ such that for all $k \in K$, $\mathbb{R}^n = V \oplus [k \cdot E_H]$.*

Proof. By Proposition 6.5.11 and Proposition 1.2.38, we can assume that H is a solvable subgroup of G. Assume first that for any linear subspace V of $\mathbb{R}^n$, there exists $k \in K$ such that $V \cap [k \cdot E_H] \neq \{0\}$ or $V + [k \cdot E_H] \neq \mathbb{R}^n$.

Let now Γ be a discontinuous group for G/H and take $V = E_\Gamma$. By Theorem 2.5.7, there exists $k \in K$ such that $E_\Gamma \oplus [k \cdot E_H] \neq \mathbb{R}^n$, which is in turn equivalent to the fact that $E_\Gamma \oplus E_H \neq \mathbb{R}^n$, as $\dim(E_H) = \dim(k \cdot E_H)$. Let S be the orthogonal complement of $E_\Gamma \oplus E_H$ in $\mathbb{R}^n$. From Corollary 1.2.41, there exist $\tau := (I, t) \in G$ and $\tau' := (I, t') \in G$ such that $E_H = \mathrm{pr}_2(H^\tau)$ and $\mathrm{pr}_2(\Gamma^{\tau'}) \subset E_\Gamma$. Let (I, s) and (I, s') be two elements of S such that $\Gamma^{\tau'}(I, s)H^\tau = \Gamma^{\tau'}(I, s')H^\tau$. There exist $(A, v) \in \Gamma^{\tau'}$ and $(B, y) \in H^\tau$ such that $(I, s') = (A, v)(I, s)(B, y)$. It follows that $B = A^{-1}$ and $s' = v + As + Ay$. From Lemma 1.2.28, E_Γ and E_H are stable by A and so is S. Hence $As \in S$, $v + Ay = 0$ and $s' = As$. This entails that the set $\{\Gamma^{\tau'}(I, \alpha s)H^\tau \mid \alpha \in \mathbb{R}\}$ is not compact and lies in $\Gamma^{\tau'}\backslash G/H^\tau$.

Conversely, let $g = (A, x) \in G$. As $V \oplus [A \cdot E_H] = \mathbb{R}^n$, there exist $x' \in V$ and $x'' \in E_H$ such that $x = x' + Ax''$. Let now $(v_1, \dots, v_p)$ be a basis of V and Γ the discrete subgroup of G spanned by the set $(v_1, \dots, v_p)$. Then Γ acts properly on G/H, as $E_\Gamma = V$ by Theorem 2.5.7. In addition, it is clear that Γ acts freely on G/H. Further by Corollary 1.2.37, there exists $\tau := (I, t) \in G$ such that $E_H = \mathrm{pr}_2(H^\tau)$. Hence there exists $A'' \in K$ such that $(A'', x'') \in H^\tau$. From Proposition 1.2.34, we get $x'' \in \ker(A'' - I)$. Indeed, it is enough to observe that $P_{A''}(x'') = x''$, as $x'' \in E_H$. Therefore,

$$\begin{aligned}
\Gamma g H^\tau &= \Gamma(A, x' + x'')H^\tau = \Gamma(A, x')(I, x'')H^\tau \\
&= \Gamma(A, x')(A''^{-1}, 0)(A'', x'')H^\tau \\
&= \Gamma(AA''^{-1}, x')H^\tau.
\end{aligned}$$

As V/Γ is compact, there exists a compact subset C of $\mathbb{R}^n$ such that $V = \Gamma \cdot C$. Write $x' = x'_\Gamma + x'_C$, where $x'_\Gamma \in \Gamma$ and $x'_C \in C$. As a consequence, we get

$$\Gamma g H^\tau = \Gamma(AA''^{-1}, x'_\Gamma + x'_C)H^\tau = \Gamma(I, x'_\Gamma)(AA''^{-1}, x'_C)H^\tau = \Gamma(AA''^{-1}, x'_C)H^\tau. \tag{6.16}$$

Consider now the natural continuous mapping $\pi : G \longrightarrow \Gamma\backslash G/H^\tau$. It follows from equation (6.16) that $\Gamma\backslash G/H^\tau = \pi(G) \subset \pi(K \times C)$. This shows that $\Gamma\backslash G/H$ is compact. □

Remark 6.5.13. The relation between the existence of compact Clifford–Klein forms and the Calabi–Markus phenomenon depends upon the choice of the closed subgroup H. Note that in general, the Calabi–Markus phenomenon (i. e., the nonexistence of infinite discontinuous group for the homogenous space G/H) entails the nonexistence of compact Clifford–Klein forms unless the homogenous space G/H itself is compact. As proved in [13], both properties are equivalent in the setting of Euclidean motion groups. However, this is no longer the case as revealed in the coming examples.

Example 6.5.14. Set

$$K = \left\{ \begin{pmatrix} A & \\ & B \end{pmatrix} \middle| \; A, B \in SO_2 \right\}.$$

For any vector $x = \binom{x_1}{x_2} \in \mathbb{R}^4$, where $x_1, x_2 \in \mathbb{R}^2$ and any $k = \begin{pmatrix} A & \\ & B \end{pmatrix} \in K$, one has

$$k \cdot x = \begin{pmatrix} Ax_1 \\ Bx_2 \end{pmatrix}.$$

Put $G = K \ltimes \mathbb{R}^4$ and

$$H_1 = \left\{ \begin{pmatrix} 0 \\ t \\ 0 \\ t' \end{pmatrix} \middle| \; t, t' \in \mathbb{R} \right\}.$$

Then G/H_1 is not compact and there exist no infinite discrete discontinuous group for G/H. Indeed, for any nonzero vector $u = \binom{u_1}{u_2}$, $u_1, u_2 \in \mathbb{R}^2$ there exists $A_1, B_1 \in SO_2$ such that

$$A_1 u_1 = \begin{pmatrix} 0 \\ \|u_1\| \end{pmatrix} \quad \text{and} \quad B_1 u_2 = \begin{pmatrix} 0 \\ \|u_2\| \end{pmatrix}.$$

Hence $\begin{pmatrix} A_1 & \\ & B_1 \end{pmatrix} \cdot u \in H_1$, which is enough to conclude thanks to Theorem 6.5.7.

Take now

$$H_2 = \left\{ \begin{pmatrix} 0 \\ t \\ 0 \\ t \end{pmatrix} \middle| t \in \mathbb{R} \right\}.$$

There exists an infinite discontinuous group Γ for the homogenous space G/H_2. It suffices to consider the discrete subgroup

$$\Gamma = \left\{ \left(I_4, \begin{pmatrix} p \\ 0 \\ 0 \\ 0 \end{pmatrix} \right) \middle| p \in \mathbb{Z} \right\}.$$

Indeed, for $k \in K$ and $u \in E_H$ such that $ku \in E_\Gamma$, we easily get that $u = 0$, which entails that Γ acts properly on G/H_2 thanks to Theorem 2.5.7. As Γ is torsion-free, then evidently Γ is discontinuous for G/H_2. Moreover, for a given linear subspace V such that for any $k \in K$, $V \cap k \cdot E_{H_2} = \{0\}$, V does not contain vectors of the form $\binom{a}{b}$ for $a, b \in \mathbb{R}^2$ such that $\|a\| = \|b\|$. Hence

$$V \cap \mathbb{R}\text{-span} \left\{ \begin{pmatrix} 1 \\ 0 \\ 1 \\ 0 \end{pmatrix}, \begin{pmatrix} 0 \\ 1 \\ 0 \\ 1 \end{pmatrix} \right\} = \{0\},$$

and then the dimension of V is less than 2, which gives in turn that for any $k \in K$, $V \oplus k \cdot E_{H_2} \neq \mathbb{R}^4$. Thanks to Theorem 6.5.12, G/H_2 does not admit a compact Clifford–Klein form.

Case of Heisenberg motion groups

The following theorem provides a criterion for the existence of compact Clifford–Klein forms as an important consequence from Theorem 6.5.8 in the case of Heisenberg motion groups.

Theorem 6.5.15. *Let H be a closed subgroup of G. The homogenous space G/H admits a compact Clifford–Klein form if and only if either H or G/H is compact.*

Proof. There is nothing to do in the setting where G/H is compact. Assume now that H is compact. Then any torsion-free group is a discontinuous group for G/H. Consider $\{e_s\}_{1 \leq s \leq n}$ the standard basis of $\mathbb{R}^n$ and define for any $s \in \{1, \ldots, n\}$ $u_s = e_s$ and $u_{n+s} = ie_s$. Consider the subgroup L of $\mathbb{C}^n$ of generator $\{u_s\}_{1 \leq s \leq 2n}$. Take Γ to be the subgroup of G

generated by $\{\gamma_s\}_{1\le s\le 2n}$ where, for any $s \in \{1,\dots,2n\}$,

$$\gamma_s = (I, u_s, 0).$$

A direct computation shows that

$$[\gamma_{s+n}, \gamma_s] = (I, 0, 1).$$

Then Γ is not Abelian and for any $v \in L$ and any $p \in \mathbb{Z}$, $(I, v, \frac{p}{2}) \in \Gamma$. Hence Γ coincides with the set $\{I\} \times L \times \frac{1}{2}\mathbb{Z}$ and, therefore, Γ is a discrete subgroup of G. On the other hand, we easily check that $G = \Gamma \cdot C_1$, for some compact set C_1 of G. Thus, $G = \Gamma \cdot C_1 \cdot H$, which gives in turn that $\Gamma\backslash G/H$ is compact.

Conversely, assume that both H and G/H are not compact. Since $\Gamma\backslash G/H$ is compact but not for G/H, the discrete group Γ should be infinite. Let Γ be an infinite discontinuous group for G/H such that $\Gamma\backslash G/H$ is compact. Therefore, Γ is not of the type (C), by Proposition 2.6.4. Assume first that Γ is of the type (A), then Γ is conjugate to a subgroup of G^1. By Fact 1.3.6, Γ contains some torsion-free subgroup Γ^* of Γ of finite index in Γ and isomorphic to a central subgroup of $\mathbb{H}_n$. Then Γ^* is of rank 1 and generated by some $(A_0, 0, t_0)$. Write $\Gamma = \Gamma^* \cdot \Lambda$, for some finite set Λ, say, and remark that the compactness property of $\Gamma\backslash G/H$ is equivalent to that of $\Gamma^*\backslash G/H$. On the other hand, $\Gamma^* \subset \{(I, 0, pt_0) : p \in \mathbb{Z}\}\cdot\mathscr{C}$ where $\mathscr{C}$ designates the closure of the set $\{(A_0^p, 0, 0) : p \in \mathbb{Z}\}$. If for some compact set C of G, $G = \Gamma \cdot C \cdot H$ then $G = \{(I, 0, pt_0) : p \in \mathbb{Z}\} \cdot \mathscr{C} \cdot C' \cdot H$, for some other compact set C' of G. Therefore, $\{(I, 0, pt_0) : p \in \mathbb{Z}\}\backslash G/H$ is compact. Hence we can take $(A_0, 0, t_0) = (I, 0, 1)$.

By Proposition 2.6.2, for any $r > 0$, $H \cap (\mathbb{U}_n \times B(0, r) \times \mathbb{R})$ is compact, which says that $\mathrm{pr}(H)$ is a closed subgroup of $\mathbb{U}_n \ltimes \mathbb{C}^n$ thanks to Lemma 1.3.15. Assume for a while that $\mathbb{U}_n \ltimes \mathbb{C}^n/\mathrm{pr}(H)$ is not compact, then there exists some noncompact cross-section $\mathscr{X}$ for the $\mathrm{pr}(H)$-cosets of $\mathbb{U}_n \ltimes \mathbb{C}^n$ for which $\mathbb{U}_n \ltimes \mathbb{C}^n/\mathrm{pr}(H)$ is homeomorphic to $\mathscr{X}$. Write $\mathbb{U}_n \ltimes \mathbb{C}^n = \mathscr{X} \cdot \mathrm{pr}(H)$, then

$$G = (\mathscr{X} \cdot \mathrm{pr}(H)) \times \mathbb{R}. \tag{6.17}$$

Remark further that if there exist $(B, z, t), (B, z, t') \in H$, then necessarily $t = t'$. Otherwise, $(I, 0, t - t') \in H$, which contradicts the hypothesis. Hence one can consider the mapping $\mathfrak{s}_1 : \mathrm{pr}(H) \to \mathbb{R}, (B, u) \mapsto s$, for which (B, u, s) is the unique element of H such that $(B, u) \in \mathrm{pr}(H)$.

With respect to decomposition (6.17), any $g \in G$ can be written as

$$g = (A, z, t) = (\underbrace{(A_{\mathscr{X}}, z_{\mathscr{X}}) \cdot (A_H, z_H)}_{\in \mathbb{U}_n \ltimes \mathbb{C}^n}, t)$$

and then

$$g = \Big(\underbrace{A_{\mathscr{X}}, z_{\mathscr{X}}}_{\in \mathscr{X}}, t - \mathfrak{s}_1((A_H, z_H)) + \frac{1}{2}\operatorname{Im}\langle z_{\mathscr{X}}, A_{\mathscr{X}} z_H\rangle\Big) \cdot (\underbrace{A_H, z_H}_{\in \mathrm{pr}(H)}, \mathfrak{s}_1((A_H, z_H))). \quad (6.18)$$

Besides, for any $m := (A_{\mathscr{X}}, z_{\mathscr{X}}) \in \mathscr{X}$ and for any $p \in \mathbb{Z}$, there exists some $t_p \in \mathbb{R}$ such that

$$t_p - \mathfrak{s}_1((A_H, z_H)) + \frac{1}{2}\operatorname{Im}\langle z_H, A_H z_{\mathscr{X}}\rangle = p.$$

Thus, by equation (6.18), G/H contains a subset $\Sigma = \{\overline{(m,p)} : m \in \mathscr{X}, p \in \mathbb{Z}\}$, where $\overline{(m,p)}$ designates the class of (m,p) for the H-coset in G. The action of Γ on G/H induces an action on Σ given as

$$(I, 0, q) \cdot \overline{(m,p)} = \overline{(m, p+q)} \quad \text{for each } q \in \mathbb{Z}.$$

Recall the covering map $\pi : G/H \to \Gamma\backslash G/H$, then

$$\widetilde{\Sigma} := \pi(\Sigma) = \pi(\Sigma'),$$

where $\Sigma' = \{\overline{(m,0)} : m \in \mathscr{X}\}$ is the noncompact set of G/H. Now $\pi_{|\Sigma'}$ is injective and $\pi(\Sigma')$ is a noncompact set lying in $\Gamma\backslash G/H$, and hence a contradiction.

Suppose now that $\mathbb{U}_n \ltimes \mathbb{C}^n/\mathrm{pr}(H)$ is compact. Necessarily, there exists $(A, z) \in \mathrm{pr}(H)$ such that $P_A(z) \neq 0$, we can assume that $Az = z \neq 0$. If for some $\alpha \in \mathbb{R}$, $(A, \alpha i z) \in \mathrm{pr}(H)$ then for $t, t' \in \mathbb{R}$ such that $h = (A, z, t) \in H$ and $h' = (A, \alpha i z, t') \in H$, a direct computation shows that $[h, h'] = (I, 0, -\alpha) \in H$, which is absurd. This already gives that for any $\alpha, \alpha' \in \mathbb{R}$, the fact that $\alpha \neq \alpha'$ is equivalent to the fact $(A, \alpha' i z) \notin (A, \alpha i z)\,\mathrm{pr}(H)$. Then $\{\overline{(A, \alpha i z)} : \alpha \in \mathbb{R}\}$ is a noncompact, closed subset sitting inside $\mathbb{U}_n \ltimes \mathbb{C}^n/\mathrm{pr}(H)$, which is also absurd. Therefore, Γ is necessarily of type (B) and $\mathrm{pr}(\Gamma)$ is discrete thanks to Proposition 2.6.3. Moreover, for any $(B, u, s) \in H$, $P_B(u) = 0$. Hence H is conjugate to a subgroup of G^1. As we may assume that $H \subset G^1$, and clearly H contains some $(A, 0, t)$ with $t \neq 0$. Therefore, the subgroup Γ_H generated by $(A, 0, t)$ is a discrete cocompact subgroup of H. Remark that the compactness of $\Gamma\backslash G/H$ is equivalent to that of $\Gamma_H\backslash G/\Gamma$. Since Γ_H is of type (A), then it suffices to substitute Γ_H for Γ and Γ for H (as neither Γ nor G/Γ is compact) and use the same arguments as above to meet a contradiction. □

6.5.6 Concluding remarks

As in Theorem 6.5.1, there is a relationship between the existence of a compact Clifford–Klein forms and the Calabi–Markus phenomenon in the context of Euclidean motion groups. That is, there is no infinite discontinuous groups for a noncompact homogenous space $I(n)/H$ whenever H is a closed subgroup, which is not compact.

However, this is not longer the case for the Heisenberg motion groups. We present the following examples to illustrate this issue.

Example 6.5.16. Let H be the subgroup of G generated by $h_1 = (I, z, 0)$ and $h_2 = (I, \sqrt{2}z, 1)$, for some $z \in \mathbb{C}^n$ with $\|z\| = 1$. Remark that for any $h \in H$ there exist $p, q \in \mathbb{Z}$ such that $h = (I, (p + q\sqrt{2})z, q)$, and subsequently H is discrete. For any $(p, q) \in \mathbb{Z}^2 \backslash \{(0, 0)\}$, we have

$$\tau_{u_{p,q}} h \tau_{u_{p,q}}^{-1} = (I, (p + q\sqrt{2})z, 0)$$

for $u_{p,q} = \frac{-iz}{|p+q\sqrt{2}|^2}$. On the other hand, there exist some infinite sequences $\{\alpha(p)\}_{p\in\mathbb{N}} \subset \mathbb{Z}$ and $\{\beta(p)\}_{p\in\mathbb{N}} \subset \mathbb{Z}$ such that $\{\alpha(p) + \sqrt{2}\beta(p)\}_{p\in\mathbb{N}}$ tends to 0. For p large enough, $|\alpha(p) + \sqrt{2}\beta(p)| < 1$ and then $(I, (\alpha(p) + \beta(p)\sqrt{2})z, \beta(p)) \in \mathbb{U}_n \times B(0, 1) \times \mathbb{R}$. This allows us to conclude that there is no infinite discontinuous groups for G/H. Subsequently, there is no compact Clifford–Klein forms for G/H.

For a noncompact homogenous space G/H, it is clear that in order to obtain a compact Clifford–Klein form, we necessarily need to find an infinite discontinuous group Γ for G/H, which is not the case as in Theorem 6.5.10. On the other hand, one can find a homogenous space admitting infinite discontinuous groups but without compact Clifford–Klein forms as shown in the following example.

Example 6.5.17. Set $n = 2$ and fix a vector $z \in \mathbb{C}^2 \backslash \{0\}$, $H = \{(I_2, tz, 0) : t \in \mathbb{R}\}$. The discrete subgroup Γ of generator $(I_2, 0, 1)$ is discontinuous for the homogenous space G/H and then the Calabi–Markus's phenomenon does not hold. On the other hand, G/H does not admit a compact Clifford–Klein forms by Theorem 6.5.15.

7 Discontinuous actions on reduced nilmanifolds

In the present chapter, the point is to remove the assumption on the groups in question to be simply connected. We refer to the background from Chapter 1 for the general theory. That is, G denotes an exponential connected Lie group, $\widetilde{G}$ the universal covering of G and $\mathfrak{g}$ the real exponential solvable Lie algebra of both G and $\widetilde{G}$. This means that $\mathfrak{g}$ is solvable, the exponential mapping of G is surjective and the associated exponential map $\exp_{\widetilde{G}} : \mathfrak{g} \to \widetilde{G}$ is a C^∞-diffeomorphism.

The attention is first focused on the reduced Heisenberg group H^r_{2n+1} for which the universal covering is H_{2n+1}, defined in Subsection 1.1.2. Unlike the context of H_{2n+1} (cf. Theorem 6.2.3), we show here that the deformation space $\mathscr{T}(\Gamma, G, H)$ is a Hausdorff space and even endowed with a smooth manifold structure for any arbitrary connected subgroup H of G and any arbitrary discontinuous group Γ for G/H. Indeed, we will provide a disjoint decomposition of $\mathscr{T}(\Gamma, G, H)$ into open smooth manifolds of a common dimension. On the other hand, we show that the stability property holds for any deformation parameter, giving evidence that in some small neighborhood V_φ of any element φ of the parameter space. The proper action of the discrete subgroup $\psi(\Gamma)$, $\psi \in V_\varphi$ on G/H is preserved.

Moving to the product Lie group $G = H^r_{2n+1} \times H^r_{2n+1}$ and $\Delta_G = \{(x, x) \in G : x \in H^r_{2n+1}\}$ the diagonal subgroup of G, we provide given any discontinuous group $\Gamma \subset G$ for G/Δ_G, a layering of the parameter space $\mathscr{R}(\Gamma, G, \Delta_G)$, which is shown to be endowed with a smooth manifold structure. We also show that the stability property holds. On the other hand, a (strong) local rigidity theorem is obtained for both H^r_{2n+1} and $H^r_{2n+1} \times H^r_{2n+1}$. That is, the parameter space admits a locally rigid point if and only if Γ is finite (giving an affirmative answer to Conjecture 5.7.10) and this is also equivalent to the fact that the deformation space is a Hausdorff space.

We also consider the setting of reduced threadlike groups and tackle similar questions. We show that Conjecture 5.7.10 holds for Abelian discontinuous groups and that non-Abelian discontinuous groups are stable. We also single out the notion of stability on layers and show that any Abelian discontinuous group is stable on layers. More detailed results on this subject could be found in [27] and [28].

7.1 Reduced Heisenberg groups

7.1.1 Backgrounds

Let $\mathfrak{g} := \mathfrak{h}_{2n+1}$ designate the Heisenberg Lie algebra of dimension $2n + 1$ as defined in Section 1.1.2 and H_{2n+1} the corresponding connected and simply connected Lie group. Let $G := H^r_{2n+1}$ be the reduced Heisenberg Lie group, which is defined as the quotient of H_{2n+1} by the central discrete subgroup $\exp_{\widetilde{G}}(\mathbb{Z}Z)$, where Z designates a nonzero generator of the center of $\mathfrak{g}$ as in equation (1.6). The center $Z(G)$ of G is compact and is

https://doi.org/10.1515/9783110765304-007

identified to the torus $\mathbb{T}$, the group of complex numbers of modulus 1. The group G is therefore identified to $\mathbb{R}^{2n} \ltimes \mathbb{T}$. As the exponential mapping $\exp := \exp_G$ is given by

$$\exp(U + \lambda Z) = (U, e^{2i\pi\lambda}), \quad U \in \mathbb{R}^{2n} \text{ and } \lambda \in \mathbb{R},$$

G can be equipped with the following law:

$$(X, Y, e^{2i\pi t}) * (X', Y', e^{2i\pi s}) = (X + X', Y + Y', e^{2i\pi(t+s+\frac{1}{2}(<X',Y>-<X,Y'>))}),$$

where $X, Y, X', Y' \in \mathbb{R}^n$, $t, s \in \mathbb{R}$ and $<, >$ denotes the usual Euclidian scalar product. According to Proposition 1.4.11, the exponential map $\exp : \mathfrak{g} \to G$ is surjective. Recall that the Lie algebra $\mathfrak{g}$ acts on itself by the adjoint representation ad, that is,

$$\mathrm{ad}_T(Y) = [T, Y], \quad T, Y \in \mathfrak{g}.$$

The group G also acts on $\mathfrak{g}$ by the adjoint representation Ad, defined by

$$\mathrm{Ad}_g = \exp \circ \mathrm{ad}_T, \quad g = \exp T \in G.$$

For $\overrightarrow{w} \in \mathbb{R}^{2n}$ and $c \in \mathbb{R}$, we adopt the notation

$$\exp(\overrightarrow{w} + cZ) = \begin{pmatrix} {}^t\overrightarrow{w} \\ e^{2i\pi c} \end{pmatrix}.$$

7.1.2 Discrete subgroups of H_{2n+1}^r

Let Γ be a discrete subgroup of G. We first pose the following.

Definition 7.1.1. Let ε_Γ be the integer given by $\varepsilon_\Gamma = 0$ if Γ is torsion-free and $\varepsilon_\Gamma = 1$ otherwise. Let also r_Γ be the rank of Γ, which is the cardinality of a minimal generating set. The nonnegative integer $l_\Gamma := r_\Gamma - \varepsilon_\Gamma$ is called the length of the subgroup Γ.

We next prove the following structure result.

Proposition 7.1.2. *For a discrete subgroup Γ of G, there exist a unique nonnegative integer l_Γ and a linearly independent family of vectors $\{\overrightarrow{w}_1, \ldots, \overrightarrow{w}_{l_\Gamma}\}$ of $\mathbb{R}^{2n}$ such that*

(1) *If Γ is torsion-free, then*

$$\Gamma = \left\{ \begin{pmatrix} {}^t\overrightarrow{w}_1 \\ e^{2i\pi c_1} \end{pmatrix}^{n_1} \cdots \begin{pmatrix} {}^t\overrightarrow{w}_{l_\Gamma} \\ e^{2i\pi c_{l_\Gamma}} \end{pmatrix}^{n_{l_\Gamma}} ; n_1, \ldots, n_{l_\Gamma} \in \mathbb{Z} \right\},$$

for some $c_1, \ldots, c_{l_\Gamma} \in \mathbb{R}$.

(2) *Otherwise, let* $q \in \mathbb{N}^*$ *be the order of* $\Gamma \cap Z(G)$. *Then*

$$\Gamma = \left\{ \begin{pmatrix} {}^t\vec{w}_1 \\ e^{2i\pi c_1} \end{pmatrix}^{n_1} \cdots \begin{pmatrix} {}^t\vec{w}_{l_\Gamma} \\ e^{2i\pi c_{l_\Gamma}} \end{pmatrix}^{n_{l_\Gamma}} \begin{pmatrix} {}^t\vec{0} \\ e^{2i\pi \frac{1}{q}} \end{pmatrix}^{s}; \ n_1, \dots, n_{l_\Gamma}, s \in \mathbb{Z} \right\},$$

for some $c_1, \dots, c_{l_\Gamma} \in \mathbb{R}$.

Proof. We first consider the surjective projection

$$\begin{aligned} \pi_1 : G &\longrightarrow G/Z(G) \\ \begin{pmatrix} {}^t\vec{w} \\ e^{2i\pi c} \end{pmatrix} &\longmapsto {}^t\vec{w}. \end{aligned}$$

Then $\pi_1(\Gamma)$ is a discrete subgroup of $\mathbb{R}^{2n}$. This gives that there exist a nonnegative integer l_Γ and a family $\{\vec{w}_1, \dots, \vec{w}_{l_\Gamma}\}$ of linearly independent vectors of $\mathbb{R}^{2n}$ such that

$$\pi_1(\Gamma) = \{n_1 {}^t\vec{w}_1 + \cdots + n_{l_\Gamma} {}^t\vec{w}_{l_\Gamma}; \ n_1, \dots, n_{l_\Gamma} \in \mathbb{Z}\}$$

is a discrete subgroup of $\mathbb{R}^{2n}$. As for all $j \in \{1, \dots, l_\Gamma\}$, $\vec{w}_j \in \pi_1(\Gamma)$, there exists $c_j \in \mathbb{R}$ such that $\gamma_j = \begin{pmatrix} {}^t\vec{w}_j \\ e^{2i\pi c_j} \end{pmatrix} \in \Gamma$ and $\pi_1(\gamma_j) = {}^t\vec{w}_j$. Then

$$\begin{aligned} \pi_1(\Gamma) &= \{n_1 \pi_{1|_\Gamma}(\gamma_1) + \cdots + n_{l_\Gamma} \pi_{1|_\Gamma}(\gamma_{l_\Gamma}); \ n_1, \dots, n_{l_\Gamma} \in \mathbb{Z}\} \\ &= \{\pi_{1|_\Gamma}(\gamma_1^{n_1}) + \cdots + \pi_{1|_\Gamma}(\gamma_{l_\Gamma}^{n_{l_\Gamma}}); \ n_1, \dots, n_{l_\Gamma} \in \mathbb{Z}\} \\ &= \{\pi_{1|_\Gamma}(\gamma_1^{n_1} \cdots \gamma_{l_\Gamma}^{n_{l_\Gamma}}); \ n_1, \dots, n_{l_\Gamma} \in \mathbb{Z}\} \\ &= \pi_{1|_\Gamma}(\{\gamma_1^{n_1} \cdots \gamma_{l_\Gamma}^{n_{l_\Gamma}}; \ n_1, \dots, n_{l_\Gamma} \in \mathbb{Z}\}). \end{aligned}$$

Hence

$$\Gamma = \{\gamma_1^{n_1} \cdots \gamma_{l_\Gamma}^{n_{l_\Gamma}}; \ n_1, \dots, n_{l_\Gamma} \in \mathbb{Z}\} \cdot (\Gamma \cap Z(G)).$$

We now show that l_Γ is unique. Indeed, if l_Γ and l'_Γ are two distinct such integers with $l_\Gamma < l'_\Gamma$, say, there exist two linearly independent families of vectors $\{\vec{w}_1, \dots, \vec{w}_{l_\Gamma}\}$ and $\{\vec{w}'_1, \dots, \vec{w}'_{l'_\Gamma}\}$ of $\mathbb{R}^{2n}$ such that

$$\begin{aligned} \Gamma &= \left\{ \begin{pmatrix} {}^t\vec{w}_1 \\ e^{2i\pi c_1} \end{pmatrix}^{n_1} \cdots \begin{pmatrix} {}^t\vec{w}_{l_\Gamma} \\ e^{2i\pi c_{l_\Gamma}} \end{pmatrix}^{n_{l_\Gamma}}; \ n_1, \dots, n_{l_\Gamma} \in \mathbb{Z} \right\} \cdot (\Gamma \cap Z(G)) \\ &= \left\{ \begin{pmatrix} {}^t\vec{w}'_1 \\ e^{2i\pi c'_1} \end{pmatrix}^{m_1} \cdots \begin{pmatrix} {}^t\vec{w}'_{l'_\Gamma} \\ e^{2i\pi c'_{l'_\Gamma}} \end{pmatrix}^{m_{l'_\Gamma}}; \ m_1, \dots, m_{l'_\Gamma} \in \mathbb{Z} \right\} \cdot (\Gamma \cap Z(G)) \end{aligned}$$

for some $c_1, \dots, c_{l_\Gamma}, c'_1, \dots, c'_{l'_\Gamma} \in \mathbb{R}$. There exist then for all $j \in \{1, \dots, l'_\Gamma\}$ some integers $(n_i^j)_{\substack{1\leqslant i\leqslant l_\Gamma \\ 1\leqslant j\leqslant l'_\Gamma}} \in \mathbb{Z}$ such that $\overrightarrow{w}'_j = \sum_{i=1}^{l_\Gamma} n_i^j \overrightarrow{w}_i$. This is impossible given $l_\Gamma < l'_\Gamma$. □

The following is an immediate consequence of the last proposition.

Corollary 7.1.3. *Any discrete subgroup of G is finitely generated.*

Remark 7.1.4. The integer l_Γ is indeed the length of Γ.

7.1.3 A matrix-like writing of elements of Hom(Γ, G)

This subsection aims to describe the set $\mathrm{Hom}(\Gamma, G)$ of homomorphisms from Γ to G. Let $\mathcal{M}_{r,s}(\mathbb{C})$ be the vector space of matrices of r rows and s columns. When $r = s$, we adopt the notation $\mathcal{M}_r(\mathbb{C})$ instead of $\mathcal{M}_{r,r}(\mathbb{C})$. Let now $\{\gamma_1, \dots, \gamma_k\}$ be a set of generators of Γ. Thanks to the injective map

$$\mathrm{Hom}(\Gamma, G) \to G \times \cdots \times G, \quad \varphi \mapsto (\varphi(\gamma_1), \dots, \varphi(\gamma_k))$$

to equip $\mathrm{Hom}(\Gamma, G)$ with the relative topology induced from the direct product $G \times \cdots \times G$ and the identification of $G \times \cdots \times G$ to the space $\mathcal{M}_{2n+1,k}(\mathbb{C})$, it appears clear that the map

$$\Psi : \mathrm{Hom}(\Gamma, G) \longrightarrow \mathcal{M}_{2n+1,k}(\mathbb{C}), \tag{7.1}$$

which associates to any element $\varphi \in \mathrm{Hom}(\Gamma, G)$, its matrix

$$M_\varphi(A, B, z) = \begin{pmatrix} A \\ B \\ e^{2i\pi z} \end{pmatrix} = \begin{pmatrix} C \\ e^{2i\pi z} \end{pmatrix} \in \mathcal{M}_{2n+1,k}(\mathbb{C}), \quad C = \begin{pmatrix} A \\ B \end{pmatrix}, \tag{7.2}$$

where A and $B \in \mathcal{M}_{n,k}(\mathbb{R})$ and $z := (z_1, \dots, z_k) \in \mathbb{R}^k$, with

$$e^{2i\pi z} := (e^{2i\pi z_1} \quad \cdots \quad e^{2i\pi z_k}) \in \mathcal{M}_{1,k}(\mathbb{C})$$

is a homeomorphism on its range. Let us write $C = \lfloor C^1, \dots, C^k \rfloor$, where this symbol merely designs the matrix constituted of the columns $C^1, \dots, C^k$. This means indeed that

$$\varphi(\gamma_j) := \exp(C^j + z_j Z)$$

for any $1 \leq j \leq k$. Let $\mathscr{E}$ denote the subset of $\mathscr{M}_{2n+1,k}(\mathbb{C})$ consisting of the totality of matrices as in (7.2), which is homeomorphic to the set $\mathscr{M}_{2n,k}(\mathbb{R}) \times \mathbb{T}^k$. Through the next coming sections, Γ will serve as a discontinuous group for a homogeneous space G/H. Recall the definitions:

$$\operatorname{Hom}^0(\Gamma, G) = \{\varphi \in \operatorname{Hom}(\Gamma, G) : \varphi \text{ is injective}\}$$

and

$$\operatorname{Hom}^0_d(\Gamma, G) = \{\varphi \in \operatorname{Hom}^0(\Gamma, G) : \varphi(\Gamma) \text{ is discrete}\}.$$

The set $\operatorname{Hom}(\Gamma, G)$ is homeomorphically identified to a subset $\mathscr{U}$ of $\mathscr{E}$ and $\operatorname{Hom}^0_d(\Gamma, G)$ to a subset $\mathscr{U}^0_d$ of $\mathscr{U}$. The group G acts on $\mathscr{E}$ through the law: For $g = \exp X$, with $X \in \mathfrak{g}$ with coordinates ${}^t(\alpha, \beta, \gamma)$, $\alpha, \beta \in \mathscr{M}_{1,n}(\mathbb{R}), \gamma \in \mathbb{R}$,

$$g \star \begin{pmatrix} C = \lfloor C^1, \dots, C^k \rfloor \\ e^{2i\pi z} \end{pmatrix} = \begin{pmatrix} \lfloor g \cdot C^1 \cdot g^{-1}, \dots, g \cdot C^k \cdot g^{-1} \rfloor \\ e^{2i\pi(z_1 + \alpha C^1_1 - \beta C^1_2)} \dots e^{2i\pi(z_k + \alpha C^k_1 - \beta C^k_2)} \end{pmatrix}, \quad C^i = \begin{pmatrix} C^i_1 \\ C^i_2 \end{pmatrix},$$

where $C^i_1, C^i_2 \in \mathscr{M}_{n,1}(\mathbb{R})$, $i \in \{1, \dots, k\}$. The map $\Psi : \operatorname{Hom}(\Gamma, G) \longrightarrow \mathscr{E}$ given in equation (7.1) turns out to be G-equivariant.

For

$$M = M(A, B, z) = \begin{pmatrix} A \\ B \\ e^{2i\pi z} \end{pmatrix} \in \mathscr{M}_{2n+1,k}(\mathbb{C}),$$

$$g \star M = \operatorname{Ad}_{\exp X} \cdot M = \begin{pmatrix} A \\ B \\ e^{2i\pi(z - \beta A + \alpha B)} \end{pmatrix}. \tag{7.3}$$

For all $j \in \{1, \dots, l = l_\Gamma\}$, we consider the notation

$${}^t\overrightarrow{w}_j = \begin{pmatrix} {}^t\overrightarrow{w}^1_j \\ {}^t\overrightarrow{w}^2_j \end{pmatrix} \in \mathbb{R}^{2n},$$

where $\overrightarrow{w}^1_j, \overrightarrow{w}^2_j \in \mathbb{R}^n$. Let $p_{ij} = 0$ if Γ is torsion-free and

$$p_{ij} = q(\prec \overrightarrow{w}^1_j, \overrightarrow{w}^2_i \succ - \prec \overrightarrow{w}^1_i, \overrightarrow{w}^2_j \succ)$$

otherwise. Let also

$$P(\Gamma) = \begin{pmatrix} 0 & p_{12} & \cdots & \cdots & p_{1l} \\ -p_{12} & \ddots & \ddots & & \vdots \\ \vdots & \ddots & \ddots & \ddots & \vdots \\ \vdots & & \ddots & \ddots & p_{l-1\,l} \\ -p_{1l} & \cdots & \cdots & -p_{l-1\,l} & 0 \end{pmatrix} \in \mathscr{M}_l(\mathbb{R}),$$

$\mathscr{A}(l,\mathbb{R})$ the subspace of $\mathscr{M}_l(\mathbb{R})$ of skew-symmetric matrices and $\mathscr{A}(l,\mathbb{Z})$ the subset of $\mathscr{A}(l,\mathbb{R})$ with entries in $\mathbb{Z}$. We now prove the following.

Proposition 7.1.5. *We keep the same notation and hypotheses. Let $M(A,B,z)$ be as in (7.2), where $A,B \in \mathscr{M}_{n,k}(\mathbb{R})$. We have:*

(1) *If Γ is torsion-free, then $k = l$ and*

$$\mathscr{U} = \{M(A,B,z) \in \mathscr{E} : {}^tAB - {}^tBA \in \mathscr{A}(l,\mathbb{Z})\}.$$

(2) *Otherwise, $k = l+1$ and*

$$\mathscr{U} = \left\{ M(A,B,z) \in \mathscr{E} \;\middle|\; \begin{array}{l} A = \begin{pmatrix} A' & {}^t\vec{0} \end{pmatrix}, B = \begin{pmatrix} B' & {}^t\vec{0} \end{pmatrix},\ A',B' \in \mathscr{M}_{n,l}(\mathbb{R}), \\ z = \left(z_1,\dots,z_l,\frac{p}{q}\right),\ p \in \{0,\dots,q-1\}, z_1,\dots,z_l \in \mathbb{R} \text{ and} \\ {}^tA'B' - {}^tB'A' \in \frac{p}{q}P(\Gamma) + \mathscr{A}(l,\mathbb{Z}) \end{array} \right\}.$$

Proof. It is sufficient to prove the proposition when Γ is not torsion-free. Indeed otherwise, $P(\Gamma) = 0$ and the same arguments work. For $\varphi \in \mathrm{Hom}(\Gamma,G)$, $M_\varphi(A,B,z) \in \mathscr{U}$ and

$$\gamma_{l+1} = \begin{pmatrix} {}^t\vec{0} \\ e^{2i\pi\frac{1}{q}} \end{pmatrix},$$

we have $\varphi(\gamma_{l+1}) = \gamma_{l+1}^p$ for some $p \in \{0,\dots,q-1\}$. Now, let $r,j \in \{1,\dots,l\}$. Then

$$\varphi(\gamma_r\gamma_j\gamma_r^{-1}\gamma_j^{-1}) = \varphi(\gamma_r)\varphi(\gamma_j)\varphi(\gamma_r)^{-1}\varphi(\gamma_j)^{-1} = \begin{pmatrix} {}^t\vec{0} \\ e^{2i\pi({}^tA^jB^r - {}^tB^jA^r)} \end{pmatrix}.$$

On the other hand, we have

$$\gamma_r\gamma_j\gamma_r^{-1}\gamma_j^{-1} = \begin{pmatrix} {}^t\vec{0} \\ e^{2i\pi(<\vec{W}_j^1,\vec{W}_r^2> - <\vec{W}_r^1,\vec{W}_j^2>)} \end{pmatrix} = \begin{pmatrix} {}^t\vec{0} \\ e^{2i\pi\frac{p_{rj}}{q}} \end{pmatrix} = \gamma_{l+1}^{p_{rj}}.$$

As $\varphi \in \mathrm{Hom}(\Gamma, G)$, then

$$\varphi(\gamma_r\gamma_j\gamma_r^{-1}\gamma_j^{-1}) = (\varphi(\gamma_{l+1}))^{p_{rj}} = \gamma_{l+1}^{pp_{rj}} = \begin{pmatrix} {}^t\vec{0} \\ e^{2i\pi\frac{pp_{rj}}{q}} \end{pmatrix}.$$

This gives ${}^tA^jB^r - {}^tB^jA^r \in \frac{p}{q}p_{rj} + \mathbb{Z}$ for some $p \in \{0,\ldots,q-1\}$. Let now $M(A,B,z) = \lfloor g_1,\ldots,g_{l+1}\rfloor \in \mathscr{E}$ such that $g_j = {}^t(C^j, e^{2i\pi z_j})$ for $j \in \{1,\ldots,l\}$ and $g_{l+1} = {}^t(0, e^{2i\pi\frac{p}{q}})$ for some $p \in \{0,\ldots,q-1\}$ with the convention that $g_{l+1} = e$ if Γ is torsion-free, which satisfies the required conditions. Let φ be the map defined by

$$\begin{aligned} \varphi : \Gamma \quad & \to \quad G \\ \gamma_1^{n_1}\cdots\gamma_l^{n_l}\gamma_{l+1}^{n_{l+1}} \quad & \mapsto \quad g_1^{n_1}\cdots g_l^{n_l}g_{l+1}^{n_{l+1}}. \end{aligned}$$

We need to show that $\varphi \in \mathrm{Hom}(\Gamma, G)$. Let $\gamma = \gamma_1^{n_1}\cdots\gamma_l^{n_l}\gamma_{l+1}^{n_{l+1}}$ and $\gamma' = \gamma_1^{m_1}\cdots\gamma_l^{m_l}\gamma_{l+1}^{m_{l+1}}$ in Γ. Therefore,

$$\begin{aligned} \varphi(\gamma\gamma') &= \varphi(\gamma_1^{n_1}\cdots\gamma_l^{n_l}\gamma_{l+1}^{n_{l+1}}\gamma_1^{m_1}\cdots\gamma_l^{m_l}\gamma_{l+1}^{m_{l+1}}) \\ &= \varphi(\gamma_1^{n_1+m_1}\cdots\gamma_l^{n_l+m_l}\gamma_{l+1}^{m}), \end{aligned}$$

where

$$m = n_{l+1} + m_{l+1} - \sum_{1\leqslant j<i\leqslant l} n_i m_j p_{ij}.$$

Then

$$\begin{aligned} \varphi(\gamma\gamma') &= g_1^{n_1+m_1}\cdots g_l^{n_l+m_l}g_{l+1}^{m} \\ &= g_1^{n_1}\cdots g_l^{n_l}g_{l+1}^{n_{l+1}}g_1^{m_1}\cdots g_l^{m_l}g_{l+1}^{m_{l+1}} \\ &= \varphi(\gamma_1^{n_1}\cdots\gamma_l^{n_l}\gamma_{l+1}^{n_{l+1}})\varphi(\gamma_1^{m_1}\cdots\gamma_l^{m_l}\gamma_{l+1}^{m_{l+1}}) \\ &= \varphi(\gamma)\varphi(\gamma'). \end{aligned}$$

This shows that $\lfloor g_1,\ldots,g_{l+1}\rfloor \in \mathscr{U}$, which is enough to conclude. □

Any information concerning the structures of the spaces $\mathrm{Hom}(\Gamma, G)$ and $\mathscr{R}(\Gamma, G, H)$ may help to understand the properties and the structure of the deformation space $\mathscr{T}(\Gamma, G, H)$. The sets $\mathrm{Hom}(\Gamma, G)$ and $\mathscr{R}(\Gamma, G, H)$ may have some singularities and there is no clear reason to say that the parameter space $\mathscr{R}(\Gamma, G, H)$ is an analytic or algebraic or smooth manifold. For instance, when the parameter space is a semialgebraic set, it has certainly a finite number of connected components, which means in turn that the deformation space itself enjoys this feature. Corollary 7.1.15 below will be set toward such a purpose. Up to this step, let $\mathscr{U}$ and $\mathscr{E}$ be as in Subsection 7.1.3. We have the following.

Corollary 7.1.6. *For a discrete subgroup Γ of G, the set* $\mathrm{Hom}(\Gamma, G)$ *is homeomorphic to a disjoint union of open (and hence closed) algebraic sets in* $\mathscr{U}$. *(Disjoint means here with empty pairwise intersection.)*

Proof. Recall that $\mathrm{Hom}(\Gamma, G)$ is homeomorphically identified to a subset $\mathscr{U}$ of $\mathscr{E}$. It suffices then to show that $\mathscr{U}$ splits to a disjoint union of open algebraic sets in $\mathscr{U}$. We only treat the case where Γ is torsion-free, the other case is handled similarly. For $D \in \mathscr{A}(l, \mathbb{Z})$, let

$$\mathscr{U}_D = \{M(A,B,z) \in \mathscr{E} : {}^tAB - {}^tBA = D\}.$$

We have

$$\mathscr{U} = \coprod_{D \in \mathscr{A}(l,\mathbb{Z})} \mathscr{U}_D.$$

Clearly, the sets $\mathscr{U}_D$ are algebraic in $\mathscr{U}$ and $\mathscr{U}_D \cap \mathscr{U}_{D'} \neq \emptyset$ for $D \neq D' \in \mathscr{A}(l, \mathbb{Z})$. We only need to show that $\mathscr{U}_D$ is open in $\mathscr{U}$ for all $D \in \mathscr{A}(l, \mathbb{Z})$. Let $(A_j)_{j \in \mathbb{N}}, (B_j)_{j \in \mathbb{N}}$ be some sequences of $\mathscr{M}_{n,k}(\mathbb{R})$ and $(z_j)_{j \in \mathbb{N}}$ a sequence in $\mathbb{R}^k$ such that $(M(A_j, B_j, z_j))_{j \in \mathbb{N}}$ is a sequence in ${}^c\mathscr{U}_D$, which converges to $M(A,B,z)$ in $\mathscr{E}$. This means that there exists a sequence $(D_j)_{j \in \mathbb{N}} \subset \mathscr{A}(l, \mathbb{Z}) \backslash \{D\}$ such that ${}^tA_jB_j - {}^tB_jA_j = D_j$ for all $j \in \mathbb{N}$ and $({}^tA_jB_j - {}^tB_jA_j)_{j \in \mathbb{N}}$ converges to ${}^tAB - {}^tBA$. Then $(D_j)_{j \in \mathbb{N}}$ is stationary and $M(A,B,z) \in {}^c\mathscr{U}_D$. □

We next show the following.

Proposition 7.1.7. *Let G be the reduced Heisenberg Lie group and Γ a discrete subgroup of G. Then:*

(1) *If Γ is torsion-free, then*

$$\mathscr{U}_d^0 = \{M(A,B,z) \in \mathscr{U} : \mathrm{rk}(C) = l\}.$$

(2) *Otherwise, if the symbol $\wedge$ means the greatest common divisor, the set $\mathscr{U}_d^0$ reads*

$$\left\{ M(A,B,z) \in \mathscr{U} \;\middle|\; \begin{array}{l} A = \begin{pmatrix} A' & {}^t\vec{0} \end{pmatrix}, B = \begin{pmatrix} B' & {}^t\vec{0} \end{pmatrix},\ A', B' \in \mathscr{M}_{n,l}(\mathbb{R}), \\ z = \left(z_1, \dots, z_l, \dfrac{p}{q}\right),\ p \in \{1, \dots, q-1\},\ p \wedge q = 1, z_1, \dots, z_l \in \mathbb{R}, \\ \mathrm{rk}\begin{pmatrix} A' \\ B' \end{pmatrix} = l \end{array} \right\}.$$

Proof. As in Proposition 7.1.5, it is sufficient to consider the case where Γ is not torsion-free. Let us first recall the following well-known result.

Lemma 7.1.8 ([48], Corollary TG VII.3). *Let $(\vec{a}_i)_{1 \leqslant i \leqslant p}$ be a linearly independent family of p vectors of $\mathbb{R}^n$ and $\vec{b} = \sum_{i=1}^p t_i \vec{a}_i$ a linear combination of real coefficients t_i. Then the subgroup of $\mathbb{R}^n$ generated by $\{\vec{a}_1, \dots, \vec{a}_p, \vec{b}\}$ is discrete if and only if t_i are rational.*

Let $\varphi \in \mathrm{Hom}_d^0(\Gamma, G)$ and $M_\varphi(A,B,z) \in \mathscr{U}_d^0$, then $\varphi(\Gamma \cap Z(G)) = \Gamma \cap Z(G)$. Therefore,

$$A = \begin{pmatrix} A' & {}^t\vec{0} \end{pmatrix}, \quad B = \begin{pmatrix} B' & {}^t\vec{0} \end{pmatrix}, \quad A', B' \in \mathscr{M}_{n,l}(\mathbb{R}),$$

$z = (z_1, \ldots, z_l, \frac{p}{q})$, $z_1, \ldots, z_l \in \mathbb{R}$, $p \in \{1, \ldots, q-1\}$ and $p \wedge q = 1$. We now show that

$$\mathrm{rk}\left(C' = \begin{pmatrix} A' \\ B' \end{pmatrix}\right) = l.$$

As Γ is not torsion-free, then according to Proposition 7.1.2,

$$\Gamma = \left\{ \begin{pmatrix} {}^t\vec{w}_1 \\ e^{2i\pi c_1} \end{pmatrix}^{n_1} \cdots \begin{pmatrix} {}^t\vec{w}_l \\ e^{2i\pi c_l} \end{pmatrix}^{n_l} \begin{pmatrix} {}^t\vec{0} \\ e^{2i\pi \frac{1}{q}} \end{pmatrix}^{n} ; n_1, \ldots, n_l, n \in \mathbb{Z} \right\},$$

where $\{\vec{w}_1, \ldots, \vec{w}_l\}$ is a linearly independent family of $\mathbb{R}^{2n}$ and $c_1, \ldots, c_l \in \mathbb{R}$. The columns of the matrix C' generate a discrete subgroup of $\mathbb{R}^{2n}$. According to Lemma 7.1.8, if $\mathrm{rk}(C') = l' < l$, then the columns of C' are $\mathbb{Q}$-linearly dependent. We can and do assume that $\mathrm{rk}\lfloor C'^1, \ldots, C'^{l'} \rfloor = l'$. We denote by $I = \{1, \ldots, l'\}$. Let $j_0 \in \{1, \ldots, l\} \setminus I$ such that $C'^{j_0} = \sum_{j \in I} \lambda_j C'^j$, where $\lambda_j \in \mathbb{Q}$ for $j \in I$. We denote by

$$\lambda_j = \frac{p_j}{q_j}, \quad Q = \prod_{j \in I} q_j, \quad Q_j = \frac{Q}{q_j}$$

and $\gamma = \gamma_1^{Q_1 p_1} \cdots \gamma_{l'}^{Q_{l'} p_{l'}} \gamma_{j_0}^{-Q}$. As $\mathrm{rk}\lfloor {}^t\vec{w}_1, \ldots, {}^t\vec{w}_{l'}, {}^t\vec{w}_{j_0} \rfloor = l' + 1$, we have $\gamma \neq e$. Moreover, it is not hard to see that

$$\exp\left(Q\left(-\sum_{j \in I} \lambda_j z_j + z_{j_0} \right) Z \right) \in \varphi(\Gamma) \cap Z(G).$$

This gives that $-\sum_{j \in I} \lambda_j z_j + z_{j_0} \in \mathbb{Q}$, which contradicts the fact that φ is injective. Conversely, let $\varphi \in \mathrm{Hom}(\Gamma, G)$ and $M_\varphi(A,B,z) \in \mathscr{U}$ such that

$$A = \begin{pmatrix} A' & {}^t\vec{0} \end{pmatrix}, \quad B = \begin{pmatrix} B' & {}^t\vec{0} \end{pmatrix}, \quad A', B' \in \mathscr{M}_{n,l}(\mathbb{R}), \quad \mathrm{rk}(C') = l$$

$z = (z_1, \ldots, z_l, \frac{p}{q})$, $p \in \{1, \ldots, q-1\}$, $p \wedge q = 1$ and $z_1, \ldots, z_l \in \mathbb{R}$. Let us show that φ is injective. Let $\gamma \in \ker\varphi$, then $\gamma \in \Gamma \cap Z(G)$. Therefore, $\gamma = \exp(\frac{p'}{q} Z)$ for some $p' \in \mathbb{Z}$. Hence

$$\ker\varphi = \left\{ \exp\left(\frac{p'}{q} Z\right) \in \Gamma : \exp\left(\frac{pp'}{q} Z\right) = e \right\} = \{e\},$$

which entails that φ is injective. We now show that $\varphi(\Gamma)$ is discrete. For $A' = (a_{ij})_{\substack{1 \leqslant i \leqslant n \\ 1 \leqslant j \leqslant l}}$ and $B' = (b_{ij})_{\substack{1 \leqslant i \leqslant n \\ 1 \leqslant j \leqslant l}}$, let $(m_j^1)_{j \in \mathbb{N}}, \ldots, (m_j^l)_{j \in \mathbb{N}}$ and $(m_j)_{j \in \mathbb{N}}$ be some integers sequences such that the sequence $(u_j)_{j\in\mathbb{N}}$ of $\varphi(\Gamma)$ defined by

$$u_j = \exp\left(m_j^1\left(\sum_{i=1}^{n}(a_{i1}X_i+b_{i1}Y_i)+z_1Z\right)\right)\cdots\exp\left(m_j^l\left(\sum_{i=1}^{n}(a_{il}X_i+b_{il}Y_i)+z_lZ\right)\right)\exp\left(m_j\frac{p}{q}Z\right)$$

converges. Hence, the sequence

$$\exp\left(\sum_{i=1}^{n}(m_j^1a_{i1}+\cdots+m_j^la_{il})X_i+(m_j^1b_{i1}+\cdots+m_j^lb_{il})Y_i\right)$$

converges, which implies that $m_j^1a_{i1}+\cdots+m_j^la_{il}$ and $m_j^1b_{i1}+\cdots+m_j^lb_{il}$ converge for all $i\in\{1,\dots,n\}$. As $\operatorname{rk}(C')=l$, these sequences converge. Therefore, $(u_j)_{j\in\mathbb{N}}$ is stationary. □

The last proposition shows that $\mathscr{U}_d^0$ is open in $\mathscr{U}$. So the following becomes clear.

Corollary 7.1.9. *Let G be the reduced Heisenberg Lie group and Γ a discrete subgroup of G. Then the set $\operatorname{Hom}_d^0(\Gamma,G)$ is open in $\operatorname{Hom}(\Gamma,G)$.*

We now show the following result.

Theorem 7.1.10. *Let G be the reduced Heisenberg Lie group and Γ a discrete subgroup of G of length l. Then:*
1. *$\operatorname{Hom}_d^0(\Gamma,G)$ is homeomorphic to a disjoint union of semialgebraic and open smooth manifolds in $\mathscr{U}_d^0$ of a common dimension equals to $(2n+1)l-\frac{1}{2}l(l-1)$.*
2. *$\operatorname{Hom}_d^0(\Gamma,G)$ and $\operatorname{Hom}_d^0(\Gamma,G)/G$ are endowed with smooth manifold structures of dimensions $(2n+1)l-\frac{1}{2}l(l-1)$ and $2nl-\frac{1}{2}l(l-1)$, respectively.*

Proof. We only treat the case where Γ is torsion-free, the other case is handled similarly. For $D\in\mathscr{A}(l,\mathbb{Z})$, let

$$\mathscr{U}_{d,D}^0=\mathscr{U}_d^0\cap\{M(A,B,z)\in\mathscr{E}:\ {}^tAB-{}^tBA=D\}.$$

We have

$$\mathscr{U}_d^0=\coprod_{D\in\mathscr{A}(l,\mathbb{Z})}\mathscr{U}_{d,D}^0. \tag{7.4}$$

Clearly, the sets $\mathscr{U}_{d,D}^0$ are semialgebraic and open in $\mathscr{U}_d^0$, and we only need to show that $\mathscr{U}_{d,D}^0$ is endowed with a smooth manifold structure for all $D\in\mathscr{A}(l,\mathbb{Z})$. Let

$$\mathcal{V}=\left\{M(A,B,z)\in\mathscr{E}:\ \operatorname{rk}\begin{pmatrix}A\\B\end{pmatrix}=l\right\}$$

and ψ_L the smooth map

$$\begin{array}{rcl} \psi_D : \mathcal{V} & \to & \mathscr{A}(l, \mathbb{R}) \\ M & \mapsto & {}^tAB - {}^tBA - D. \end{array}$$

Clearly, $\mathscr{U}^0_{d,D} = \psi_D^{-1}(\{0\})$. The goal now is to show that zero is a regular value of the map ψ_D. The derivative of ψ_D at a point $M = M(A, B, z) \in \mathscr{U}^0_{d,D}$ is given by

$$\begin{array}{rcl} d(\psi_D)_M : \mathscr{E} & \to & \mathscr{A}(l, \mathbb{R}) \\ X = M(H, K, h) & \mapsto & {}^tHB - {}^tBH + {}^tAK - {}^tKA. \end{array}$$

So, clearly we have

$$d(\psi_D)_M(X) = {}^t\begin{pmatrix} H \\ -K \end{pmatrix}\begin{pmatrix} B \\ A \end{pmatrix} - {}^t\begin{pmatrix} B \\ A \end{pmatrix}\begin{pmatrix} H \\ -K \end{pmatrix},$$

which is enough to conclude thanks to Lemma 4.1.29. This shows the first point.

For the second point, the set $\mathscr{U}^0_d$ splits to a disjoint union of open smooth manifolds of dimension $(2n+1)l - \frac{1}{2}l(l-1)$ and, therefore, endowed with a smooth manifold structure. Lemma 4.1.30 allows conclude that $\mathrm{Hom}^0_d(\Gamma, G)$ is endowed with a smooth manifold structure.

Now, we focus attention to the space $\mathrm{Hom}^0_d(\Gamma, G)/G$. For any $X = {}^t(\alpha, \beta, \gamma) \in \mathfrak{g}$ and $M(A, B, z) \in \mathscr{U}^0_d$, we have as in equation (7.3),

$$\mathrm{Ad}_{\exp X} \cdot M(A, B, z) = M(A, B, z - \beta A + \alpha B).$$

Here, $\gamma \in \mathbb{R}$, α and β are in $\mathbb{R}^n$. For $M(A, B, z) \in \mathscr{U}^d_0$, we can easily see that the matrix through the canonical basis of $\mathbb{R}^{2n}$ and $\mathbb{R}^l$ of the map $\Phi_{A,B} : \mathbb{R}^n \times \mathbb{R}^n \to \mathbb{R}^l, (\alpha, \beta) \mapsto \beta A - \alpha B$ is $M(\Phi_{A,B}) = (-{}^tB \quad {}^tA)$, which means that $\mathrm{rk}(M(\Phi_{A,B})) = l$ and that $\Phi_{A,B}$ is surjective. Let

$$\widetilde{\mathscr{U}^0_d} = \{M(A, B, 0) \in \mathscr{U} : \mathrm{rk}(C) = l\},$$

which as above, is endowed with a smooth manifold structure. Then the mapping

$$\tilde{\pi} : \mathscr{U}^0_d/G \to \widetilde{\mathscr{U}^0_d}; [M(A, B, z)] \mapsto M(A, B, 0)$$

is a continuous bijection. In addition, its inverse coincides with the restriction of the canonical quotient surjection to $\widetilde{\mathscr{U}^0_d}$ regarded as a subset of $\mathscr{U}^0_d$. This shows that $\mathrm{Hom}^0_d(\Gamma, G)/G$ is endowed with a smooth manifold structure. □

7.1.4 Proper action on reduced homogeneous spaces

Let $H = \exp \mathfrak{h}$ be a closed, connected subgroup of the reduced Heisenberg group G, Γ a discrete subgroup and L the syndetic hull of Γ (as in Theorem 1.4.12). We need to characterize the proper action of the closed, connected subgroup L on the homogeneous space G/H. As L contains the center of G, our goal is to prove the following.

Proposition 7.1.11. *Let $H = \exp \mathfrak{h}$ and $L = \exp \mathfrak{l}$ be closed, connected subgroups of G such that L contains the center of G. We have the following:*
(1) *The action of L on G/H is free if and only if $\mathfrak{l} \cap \mathfrak{h} = \{0\}$.*
(2) *If $\mathfrak{z}(\mathfrak{g}) \subsetneq \mathfrak{h}$, then L acts properly on G/H if and only if $\mathfrak{l} \cap \mathfrak{h} = \mathfrak{z}(\mathfrak{g})$.*
(3) *If $\mathfrak{z}(\mathfrak{g}) \not\subset \mathfrak{h}$, then the action of L on G/H is proper if and only if the action of L on G/H is free.*

Proof. (1) We first prove that $\exp(\mathfrak{l} \cap \mathfrak{h}) = L \cap H$. Let $t \in L \cap H$. Then they exist $T_1 \in \mathfrak{l}$ and $T_2 \in \mathfrak{h}$ such that $t = \exp T_1 = \exp T_2$. This implies that $T_1 = T_2 \operatorname{mod}(\mathfrak{z}(\mathfrak{g}))$. As $\mathfrak{z}(\mathfrak{g}) \subset \mathfrak{l}$, then $T_2 \in \mathfrak{h} \cap \mathfrak{l}$, which implies that $t \in \exp(\mathfrak{l} \cap \mathfrak{h})$. Now suppose that L acts on G/H freely. As $\mathfrak{z}(\mathfrak{g}) \subset \mathfrak{l}$, then $\mathfrak{z}(\mathfrak{g}) \not\subset \mathfrak{h}$ and, therefore, H is nilpotent and simply connected. As $\exp(\mathfrak{l} \cap \mathfrak{h}) = L \cap H = \{e\}$, we get $\mathfrak{l} \cap \mathfrak{h} = \{0\}$. The converse implication is trivial as $\exp(\mathfrak{h} \cap \mathfrak{l}) = H \cap L$ and $Z(G) \subset L$.

(2) Suppose that L acts properly on G/H, which implies that the triplet (L, G, H) is (CI). Then $\mathfrak{h} \cap \mathfrak{l} \subsetneq \mathfrak{z}(\mathfrak{g})$ and, therefore, $\mathfrak{z}(\mathfrak{g}) = \mathfrak{h} \cap \mathfrak{l}$. Conversely, let us assume that L and H are not compact; otherwise, our assertion is clear. We consider the norm $\|g\| = \inf\{\|X\|, \exp X = g\}$, for $g \in G$. Suppose that the action of L on G/H is not proper, then there exists a compact set $S \subset G$ such that $SHS^{-1} \cap L$ is not relatively compact. Hence one can find sequences $V_j \in \mathfrak{h}$, $W_j \in \mathfrak{l}$, A_j and $B_j \in \mathfrak{g}$ such that:
(a) $\exp A_j \in S$ and $\exp B_j \in S$,
(b) $\lim_{j\to+\infty} \|\exp V_j\| = \lim_{j\to+\infty} \|\exp W_j\| = +\infty$,
(c) $\exp W_j = \exp A_j \exp V_j \exp(-B_j)$.

Moreover, G is a 2-step nilpotent Lie group. Then the last equation gives

$$W_j = V_j + (A_j - B_j) \operatorname{mod}(\mathfrak{z}(\mathfrak{g})). \tag{7.5}$$

Let $W_j = W_j' \operatorname{mod}(\mathfrak{z}(\mathfrak{g}))$, $V_j = V_j' \operatorname{mod}(\mathfrak{z}(\mathfrak{g}))$, $A_j = A_j' \operatorname{mod}(\mathfrak{z}(\mathfrak{g}))$ and $B_j = B_j' \operatorname{mod}(\mathfrak{z}(\mathfrak{g}))$. Obviously assertion (b) gives

$$\lim_{j\to+\infty} \|V_j'\| = \lim_{j\to+\infty} \|W_j'\| = +\infty.$$

Then we can assume that

$$\lim_{j\to+\infty} \frac{V_j'}{\|V_j'\|} = V', \quad \lim_{j\to+\infty} \frac{W_j'}{\|W_j'\|} = W', \quad \text{where } V' \in \mathfrak{h},\ W' \in \mathfrak{l},\ \|V'\| = \|W'\| = 1.$$

Let $\alpha_j' = \frac{\|V_j'\|}{\|W_j'\|}$. Then equation (7.5) gives

$$\frac{W_j'}{\|W_j'\|} = \alpha_j \frac{V_j'}{\|V_j'\|} + \frac{A_j' - B_j'}{\|W_j'\|}.$$

Thus, $(\alpha_j')_j$ converges to $\alpha' \in \mathbb{R}^*$. Then $W' \in \mathfrak{h} \cap \mathfrak{l} = \mathfrak{z}(\mathfrak{g})$, which is impossible as $W' \notin \mathfrak{z}(\mathfrak{g})$.

(3) As $\mathfrak{h}$ does not contain the center of $\mathfrak{g}$, then H is simply connected. If the action of L on G/H is proper, we have for all $g \in G$, $gLg^{-1} \cap H$ is a compact subgroup of H and, therefore, $gLg^{-1} \cap H = \{e\}$. Then the action of L on G/H is free. Conversely, we can and do assume that L and H are not compact; otherwise, our assertion is clear. Suppose that the action of L on G/H is not proper, then there exists a compact $S \subset G$ such that $SHS^{-1} \cap L$ is not relatively compact. Hence one can find sequences $V_j \in \mathfrak{h}$, $W_j \in \mathfrak{l}$, A_j and $B_j \in \mathfrak{g}$ meeting conditions (a)–(c) of assertion (2). Moreover, G is 2-step nilpotent Lie group then the equation (c) gives the equation (7.5). We have $\mathfrak{z}(\mathfrak{g}) \not\subset \mathfrak{h}$. Then according to Proposition 1.1.12, there exists a basis $\{X_1, \dots, X_n, Y_1, \dots, Y_n, Z\}$ of $\mathfrak{g}$ satisfying $[X_i, Y_j] = \delta_{ij} Z$ and $\mathfrak{h} = \mathbb{R}\text{-span}(X_1, \dots, X_s)$, where $s = \dim \mathfrak{h}$. Let $W_j = W_j' \bmod(\mathfrak{z}(\mathfrak{g}))$ where $W_j' \in \mathfrak{l}$, $A_j = A_j' \bmod(\mathfrak{z}(\mathfrak{g}))$ and $B_j = B_j' \bmod(\mathfrak{z}(\mathfrak{g}))$ where $A_j', B_j' \in \mathfrak{g}$. Hence the same procedure as in the proof of assertion (2) gives a contradiction. □

7.1.5 The parameter space

Let G be the reduced Heisenberg Lie group, $H = \exp \mathfrak{h}$ be a closed, connected subgroup of G and Γ be a discontinuous group for the homogeneous space G/H. This section aims to study the parameter space (3.1) and to study the stability property. For $\varphi \in \mathrm{Hom}_d^0(\Gamma, G)$, let $L_\varphi = \exp_G \mathfrak{l}_\varphi$ be the syndetic hull of $\varphi(\Gamma)$. We first prove the following.

Lemma 7.1.12. *We keep the notation of Section 7.1.2. For any $M_\varphi(A, B, z) \in \mathscr{U}_d^0$, we have $\mathfrak{l}_\varphi = \mathbb{R}\text{-span}(Z, C^1, \dots, C^l)$.*

Proof. We still adopt the notation of the proof of Theorem 1.4.12. The closed subgroup $\widetilde{\varphi(\Gamma)}$ coincides with the closed subgroup $\exp_{\widetilde{G}}(\mathbb{Z}(C^1 + z_1 Z)) \cdots \exp_{\widetilde{G}}(\mathbb{Z}(C^l + z_l Z)) \exp_{\widetilde{G}}(\mathbb{Z}Z)$. It is then clear that the Lie algebra of the syndetic hull $\widetilde{L_\varphi}$ of $\widetilde{\varphi(\Gamma)}$ is the Lie subalgebra $\mathfrak{l}_\varphi = \mathbb{R}\text{-span}\{C^1, \dots, C^l, Z\}$. As $L_\varphi = \exp_G \mathfrak{l}_\varphi$, we are done. □

As a direct consequence of Proposition 7.1.11, we get the following description of the parameter space $\mathscr{R}(\Gamma, G, H)$.

Proposition 7.1.13. *Let G be the reduced Heisenberg Lie group, $H = \exp \mathfrak{h}$ be a closed, connected subgroup of G and Γ be a discontinuous group for the homogeneous space G/H. Then*

$$\mathscr{R}(\Gamma, G, H) = \{\varphi \in \mathrm{Hom}_d^0(\Gamma, G) : \mathfrak{h} \cap \mathfrak{l}_\varphi \subseteq \mathfrak{z}(\mathfrak{g})\}.$$

More precisely:
(1) *If* $\mathfrak{z}(\mathfrak{g}) \not\subset \mathfrak{h}$, *then* $\mathscr{R}(\Gamma, G, H) = \{\varphi \in \mathrm{Hom}_d^0(\Gamma, G) : \mathfrak{h} \cap \mathfrak{l}_\varphi = \{0\}\}$.
(2) *Otherwise,* $\mathscr{R}(\Gamma, G, H) = \{\varphi \in \mathrm{Hom}_d^0(\Gamma, G) : \mathfrak{h} \cap \mathfrak{l}_\varphi = \mathfrak{z}(\mathfrak{g})\}$.

Proof. Let $\varphi \in \mathscr{R}(\Gamma, G, H)$. We first show that the proper action of $\varphi(\Gamma)$ on G/H implies its free action. It is clear that the proper action implies that the triplet $(G, H, \varphi(\Gamma))$ is (CI), which gives that for all $g \in G$, the subgroup $K := \varphi(\Gamma) \cap gHg^{-1}$ is central and then finite as $\varphi(\Gamma)$ is discrete. As the map $\varphi : \Gamma \to \varphi(\Gamma)$ is a group isomorphism and K is finite and cyclic, we get that $\varphi^{-1}(K) = K$. Therefore, $K \subset \Gamma \cap H = \{e\}$. Thus, the action of $\varphi(\Gamma)$ on G/H is free. As L_φ contains $\varphi(\Gamma)$ cocompactly,

$$\mathscr{R}(\Gamma, G, H) = \{\varphi \in \mathrm{Hom}_d^0(\Gamma, G) : L_\varphi \text{ acts properly on } G/H\}.$$

Now, Proposition 7.1.11 allows us to conclude. □

7.1.6 Stability of discrete subgroups

The question whether it is possible to characterize all stable discrete subgroups of connected nilpotent Lie groups is a difficult question (posed in [12]). We now provide an answer in our context, and also give an affirmative answer to Conjecture 5.1.1.

Theorem 7.1.14. *Let* $G := H_{2n+1}^r$ *be the reduced Heisenberg Lie group and* Γ *a discontinuous group of length* l_Γ *for the homogeneous space* G/H *where* H *is a connected, closed subgroup of* G. *Then we have:*
(1) *The stability property holds. That is, any discrete subgroup of G is stable (in the sense of Definition 2.1.1).*
(2) *The parameter space* $\mathscr{R}(\Gamma, G, H)$ *and the deformation space* $\mathscr{T}(\Gamma, G, H)$ *are endowed with smooth manifold structures of dimensions* $(2n+1)l_\Gamma - \frac{1}{2}l_\Gamma(l_\Gamma - 1)$ *and* $2nl_\Gamma - \frac{1}{2}l_\Gamma(l_\Gamma - 1)$, *respectively.*
(3) *The G-orbits of* $\mathscr{R}(\Gamma, G, H)$ *have a common dimension equals to* l_Γ.
(4) *The parameter space* $\mathscr{R}(\Gamma, G, H)$ *admits a locally rigid point if and only if* Γ *is a finite group.*

Proof. (1) Let $\{T_1, \dots, T_r\}$ be a basis of $\mathfrak{h}$ and $\mathfrak{l}_\varphi = \mathbb{R}\text{-span}\{C^1, \dots, C^l, Z\}$ as in Lemma 7.1.12. Thanks to Proposition 7.1.13, it is clear that $\mathscr{R}(\Gamma, G, H)$ is homeomorphic to the set

$$\{M_\varphi(A, B, z) \in \mathscr{U}_d^0 : \mathrm{rk}\lfloor T_1, \dots, T_r, C^1, \dots, C^l, Z \rfloor = r + l + 1 - \dim(\mathfrak{h} \cap \mathfrak{l}_\varphi)\},$$

which is a Zariski open set in $\mathscr{U}_d^0$. This completes the proof as $\mathscr{U}_d^0$ is open in $\mathscr{U}$ as in Corollary 7.1.9. This also entails that the parameter space is open in $\mathrm{Hom}(\Gamma, G)$ and that any discrete subgroup of G is stable.

(2) The parameter and the deformation spaces are open in $\mathrm{Hom}(\Gamma, G)$ and $\mathrm{Hom}^0_d(\Gamma, G)/G$, respectively. They are therefore endowed with a smooth manifold structure with the mentioned dimensions thanks to Theorem 7.1.10.

(3) We immediately see that $\dim G \star M(A, B, z) = l$ for any $M(A, B, z) \in \mathscr{R}(\Gamma, G, H)$.

(4) Assume first that Γ is a finite group, then it is a central and cyclic group. As such, we have the following:

$$\mathscr{T}(\Gamma, G, H) = \mathscr{R}(\Gamma, G, H) = \mathrm{Hom}^0(\Gamma, G) = \mathrm{Aut}(\Gamma),$$

where the last means the automorphism group of Γ, which is a finite group. So, the strong local rigidity property holds.

Let now Γ be infinite. As $\mathscr{R}(\Gamma, G, H)$ is endowed with a smooth manifold structure and $\dim G \star M(A, B, z) \lneq \dim \mathscr{R}(\Gamma, G, H)$ for any $M(A, B, z) \in \mathscr{R}(\Gamma, G, H)$, then the local rigidity fails to hold. □

As a direct consequence of the decomposition (7.4) and the proof of Theorem 7.1.14, we set the following.

Corollary 7.1.15. *The parameter and the deformation spaces split into semialgebraic smooth manifolds.*

Corollary 7.1.16. *Let G be the reduced Heisenberg Lie group, $H = \exp \mathfrak{h}$ be a closed, connected subgroup of G and Γ be a discontinuous group for the homogeneous space G/H of length l_Γ. Then the deformation space $\mathscr{T}(\Gamma, G, H)$ splits into open smooth manifolds of common dimension equals to $2nl_\Gamma - \frac{1}{2}l_\Gamma(l_\Gamma - 1)$.*

7.2 From H^r_{2n+1} to $(H^r_{2n+1} \times H^r_{2n+1})/\Delta$

Let $\mathfrak{g} := \mathfrak{h}_{2n+1} \times \mathfrak{h}_{2n+1}$ be the Lie algebra of dimension $4n + 2$ and $\widetilde{G}$ the corresponding connected and simply connected Lie group. Let

$$\exp_{\widetilde{G}} : \mathfrak{g} \to \widetilde{G}$$

be the associated exponential map, which is a global C^∞-diffeomorphism from $\mathfrak{g}$ onto $\widetilde{G}$. We assume henceforth that $G := H^r_{2n+1} \times H^r_{2n+1}$, which we can identify to $\mathbb{R}^{2n} \ltimes \mathbb{T}^2$. Indeed, G is the quotient of $\widetilde{G}$ by the central discrete subgroup $\Lambda = \exp_{\widetilde{G}}(\mathbb{Z}(Z, 0)) \exp_{\widetilde{G}}(\mathbb{Z}(0, Z))$. Let

$$\exp := \exp_G : \mathfrak{g} \to G$$

be the associated exponential map, which is surjective.

For $(X, Y), (X', Y') \in \mathfrak{g}$, the Lie bracket is given by

$$[(X, Y), (X', Y')] = ([X, X'], [Y, Y']).$$

It is then obvious that the group G is a 2-step nilpotent Lie group. The group G acts on $\mathfrak{g}$ by the adjoint action Ad_G such that for all $(X,X'),(Y,Y') \in \mathfrak{g}$,

$$\begin{aligned}\mathrm{Ad}_{\exp(X,X')}(Y,Y') &= e^{\mathrm{ad}_{(X,X')}}(Y,Y') \\ &= (Y,Y') + \mathrm{ad}_{(X,X')}(Y,Y') \\ &= (Y,Y') + ([X,Y],[X',Y']).\end{aligned}$$

From now on, we fix

$$\mathscr{B} = \{(X_i,X_i),(Y_i,Y_i),(X_i,0),(Y_i,0),(Z,Z),(Z,0),\ 1 \leqslant i \leqslant n\}$$

a basis of $\mathfrak{g}$ where $\{X_i,Y_i,Z,\ 1 \leqslant i \leqslant n\}$ is a basis of $\mathfrak{h}_{2n+1}$ for which $[X_i,Y_j] = \delta_{ij}Z$. The corresponding exponential mapping exp of G is given by

$$\begin{aligned}&\exp\left(\sum_{j=1}^{n}[r_{j1}(X_j,X_j) + r_{j2}(Y_j,Y_j) + s_{j1}(X_j,0) + s_{j2}(Y_j,0)] + c_1(Z,Z) + c_2(Z,0)\right) \\ &:= (\mathbf{r}_1,\mathbf{r}_2,\mathbf{s}_1,\mathbf{s}_2,e^{2i\pi c_1},e^{2i\pi c_2}),\end{aligned}$$

where $\mathbf{r}_j = (r_{1j},\dots,r_{nj})$ and $\mathbf{s}_j = (s_{1j},\dots,s_{nj})$, $j \in \{1,2\}$.
G is equipped with the following law:

$$\begin{aligned}&(\mathbf{r}_1,\mathbf{r}_2,\mathbf{s}_1,\mathbf{s}_2,e^{2i\pi c_1},e^{2i\pi c_2}) * (\mathbf{r}_1',\mathbf{r}_2',\mathbf{s}_1',\mathbf{s}_2',e^{2i\pi c_1'},e^{2i\pi c_2'}) \\ &= (\mathbf{r}_1+\mathbf{r}_1',\mathbf{r}_2+\mathbf{r}_2',\mathbf{s}_1+\mathbf{s}_1',\mathbf{s}_2+\mathbf{s}_2',e^{2i\pi(c_1+c_1'+\frac{1}{2}R)},e^{2i\pi(c_2+c_2'+\frac{1}{2}S)})\end{aligned}$$

with

$$\begin{aligned}R &= \prec \mathbf{r}_1,\mathbf{r}_2' \succ - \prec \mathbf{r}_1',\mathbf{r}_2 \succ, \\ S &= \prec \mathbf{r}_1,\mathbf{s}_2' \succ - \prec \mathbf{s}_2,\mathbf{r}_1' \succ + \prec \mathbf{s}_1,\mathbf{r}_2' \succ - \prec \mathbf{r}_2,\mathbf{s}_1' \succ + \prec \mathbf{s}_1,\mathbf{s}_2' \succ - \prec \mathbf{s}_2,\mathbf{s}_1' \succ,\end{aligned}$$

$\mathbf{r}_j,\mathbf{s}_j,\mathbf{r}_j',\mathbf{s}_j' \in \mathbb{R}^n$, $c_j,c_j' \in \mathbb{R}$, $j \in \{1,2\}$ and $\prec,\succ$ denotes the usual Euclidean scalar product.

7.2.1 Posed problems and main results

Let $G = H_{2n+1}^r \times H_{2n+1}^r$ be the $(4n+2)$-dimensional Lie group and $\Delta_G = \{(x,x) \in G : x \in H_{2n+1}^r\}$ the diagonal subgroup of G. Given any discontinuous group $\Gamma \subset G$ for G/Δ_G, we now provide a layering of the parameter space $\mathscr{R}(\Gamma,G,\Delta_G)$, which is shown to be endowed with a smooth manifold structure. we also show in this section that the stability property holds. On the other hand, a (strong) local rigidity theorem is obtained. That is, the parameter space $\mathscr{R}(\Gamma,G,\Delta_G)$ admits a rigid point if and only if Γ is finite (giving an affirmative answer to Conjecture 5.7.10) and this is also equivalent to the fact that the deformation space is a Hausdorff space.

The point consists in finding a unique integer l_Γ, said to be *the effective rank of* Γ that depends only upon Γ and a family of infinite order generators of Γ of cardinality l_Γ, which generates the group $\Gamma/T(\Gamma)$, where $T(\Gamma)$ designates the normal torsion subgroup of Γ. In this context, the following main result will be proved.

Theorem 7.2.1. *Let $G := H^r_{2n+1} \times H^r_{2n+1}$ and Γ a discontinuous group for the homogeneous space G/Δ_G of effective rank l_Γ. Then we have:*
(1) *The stability property holds.*
(2) *The parameter space $\mathscr{R}(\Gamma, G, \Delta_G)$ is endowed with smooth manifold structure of dimension $(4n+2)l_\Gamma - \frac{1}{2}l_\Gamma(l_\Gamma - 1)$.*
(3) *When $n \geqslant 2l_\Gamma$, the space $\mathscr{T}(\Gamma, G, \Delta_G)$ admits an open smooth manifold as its dense subset whose preimage by the canonical surjection*

$$\pi : \mathscr{R}(\Gamma, G, \Delta_G) \to \mathscr{T}(\Gamma, G, \Delta_G)$$

consists of maximal dimensional orbit points.

A second objective is to prove the following main result.

Theorem 7.2.2. *Let G and Γ be as in Theorem 7.2.1. Then the following assertions are equivalent:*
(1) *The deformation space $\mathscr{T}(\Gamma, G, \Delta_G)$ is a Hausdorff space.*
(2) *Γ is finite.*
(3) *$\mathscr{R}(\Gamma, G, \Delta_G)$ admits a locally rigid homomorphism.*
(4) *$\mathscr{R}(\Gamma, G, \Delta_G)$ admits a rigid homomorphism.*
(5) *The G-orbits of $\mathscr{R}(\Gamma, G, \Delta_G)$ have a common dimension.*

7.2.2 Discontinuous groups for $(H^r_{2n+1} \times H^r_{2n+1})/\Delta$

This subsection aims to describe the discrete subgroups of G. Let then Γ be a discrete infinite subgroup of G. We will construct a family of infinite order generators of Γ denoted by $\{\gamma_1, \dots, \gamma_{l_\Gamma}\}$. More precisely,

$$\Gamma = \{\gamma_1^{m_1} \cdots \gamma_{l_\Gamma}^{m_{l_\Gamma}};\ m_1, \dots, m_{l_\Gamma} \in \mathbb{Z}\} \cdot \Gamma \cap Z(G).$$

Let us first give in the following lemma the explicit expression of the torsion-free part of Γ. The same arguments as in Proposition 7.1.2 give us following result.

Lemma 7.2.3. *For a discrete infinite subgroup Γ of G, there exist a unique nonnegative integer l_Γ, a linearly independent family of vectors $\{\mathbf{w}_1, \dots, \mathbf{w}_{l_\Gamma}\}$ of $\mathbb{R}^{4n}$ and some nonnegative integers $c_1, \dots, c_{l_\Gamma}, c'_1, \dots, c'_{l_\Gamma} \in \mathbb{R}$ such that for all $j \in \{1, \dots, l_\Gamma\}$,*

$$\gamma_j = (\mathbf{w}_j, e^{2i\pi c_j}, e^{2i\pi c'_j}).$$

Definition 7.2.4. The integer l_Γ is called the *effective rank of* Γ.

Now using the fact that $\Gamma \cap Z(G)$ is finite central subgroup of G, we get immediately that $\Gamma \cap Z(G) = \{\gamma_{l_\Gamma+1}^m : m \in \mathbb{Z}\}$, where

$$\gamma_{l_\Gamma+1} \in \{(\mathbf{0}_{4n}, 1, e^{2i\pi\frac{1}{q}}), (\mathbf{0}_{4n}, e^{2i\pi\frac{1}{q}}, 1)\}, \quad q \in \mathbb{N}^*$$

or $\Gamma \cap Z(G) = \{\gamma_{l_\Gamma+1}^{m_1}\gamma_{l_\Gamma+2}^{m_2} : m_1, m_2 \in \mathbb{Z}\}$ where

$$\gamma_{l_\Gamma+1} = (\mathbf{0}_{4n}, 1, e^{2i\pi\frac{1}{q}}) \quad \text{and} \quad \gamma_{l_\Gamma+2} = (\mathbf{0}_{4n}, e^{2i\pi\frac{1}{q'}}, 1), \quad q, q' \in \mathbb{N}^*.$$

From now on, we denote by $l := l_\Gamma$ and

$$k = \begin{cases} l & \text{if } \Gamma \text{ is torsion-free,} \\ l+1 & \text{if } \Gamma \cap Z(G) \text{ is a cyclic subgroup,} \\ l+2 & \text{otherwise.} \end{cases}$$

We now prove the following.

Proposition 7.2.5. *Let* $G = H_{2n+1}^r \times H_{2n+1}^r$, $\Delta = \exp \mathfrak{d}$ *the diagonal subgroup of* G *and* $L = \exp \mathfrak{l}$ *a closed, connected subgroup of* G *such that* L *contains the center of* G. *Then* L *acts properly on* G/Δ *if and only if* $\mathfrak{l} \cap \mathfrak{d} = \mathbb{R}(Z, Z)$.

Proof. Suppose that L acts properly on G/Δ, which implies that the triplet (G, Δ, L) is (CI). Then $\mathfrak{d} \cap \mathfrak{l} \subseteq \mathfrak{z}(\mathfrak{g})$ and as $\mathfrak{d} \cap \mathfrak{z}(\mathfrak{g}) = \mathbb{R}(Z, Z)$ we get $\mathfrak{l} \cap \mathfrak{d} = \mathbb{R}(Z, Z)$. Conversely, let us assume that L is not compact; otherwise, our assertion is clear. We consider the norm $\|(g, g')\| = \inf\{\|(X, X')\|, \exp(X, X') = (g, g')\}$ for $(g, g') \in G$. Suppose that the action of L on G/Δ is not proper. There exists then a compact set $S \subset G$ such that $S\Delta S^{-1} \cap L$ is not compact. Hence one can find sequences $(V_j, V_j)_{j\in\mathbb{N}}, (W_j^1, W_j^2)_{j\in\mathbb{N}}$ and $((A_j^1, A_j^2)_{j\in\mathbb{N}}, (B_j^1, B_j^2)_{j\in\mathbb{N}})$ in $\mathfrak{d}$, $\mathfrak{l}$ and $\mathfrak{g}^2$, respectively, such that:

(a) $\exp(A_j^1, A_j^2) \in S$ and $\exp(B_j^1, B_j^2) \in S$,

(b) $\lim_{j\to+\infty} \|\exp(V_j, V_j)\| = \lim_{j\to+\infty} \|\exp(W_j^1, W_j^2)\| = +\infty$,

(c) $\exp(W_j^1, W_j^2) = \exp(A_j^1, A_j^2)\exp(V_j, V_j)\exp(-B_j^1, -B_j^2)$.

Moreover, as G is a 2-step nilpotent, the last equation gives

$$(W_j^1, W_j^2) = (V_j, V_j) + ((A_j^1, A_j^2) - (B_j^1, B_j^2)) \operatorname{mod}(\mathfrak{z}(\mathfrak{g})). \tag{7.6}$$

Let

$$(W_j^1, W_j^2) = (W_j'^1, W_j'^2) \operatorname{mod}(\mathfrak{z}(\mathfrak{g})), \quad (V_j, V_j) = (V_j', V_j') \operatorname{mod}(\mathfrak{z}(\mathfrak{g})),$$
$$(A_j^1, A_j^2) = (A_j'^1, A_j'^2) \operatorname{mod}(\mathfrak{z}(\mathfrak{g})), \quad (B_j^1, B_j^2) = (B_j'^1, B_j'^2) \operatorname{mod}(\mathfrak{z}(\mathfrak{g})).$$

Obviously assertion (b) gives

$$\lim_{j\to+\infty} \|(V'_j, V'_j)\| = \lim_{j\to+\infty} \|(W'^1_j, W'^2_j)\| = +\infty.$$

Then we can assume that

$$\lim_{j\to+\infty} \frac{(V'_j, V'_j)}{\|(V'_j, V'_j)\|} = (V', V'), \quad \lim_{j\to+\infty} \frac{(W'^1_j, W'^2_j)}{\|(W'^1_j, W'^2_j)\|} = (W'^1, W'^2),$$

where

$$(V', V') \in \mathfrak{d}, \quad (W'^1, W'^2) \in \mathfrak{l}, \quad \|(V', V')\| = \|(W'^1, W'^2)\| = 1.$$

Let $\alpha_j = \frac{\|(V'_j, V'_j)\|}{\|(W'^1_j, W'^2_j)\|}$, then equation (7.6) gives

$$\frac{(W'^1_j, W'^2_j)}{\|(W'^1_j, W'^2_j)\|} = \alpha_j \frac{(V'_j, V'_j)}{\|(V'_j, V'_j)\|} + \frac{(A'^1_j, A'^2_j) - (B'^1_j, B'^2_j)}{\|(W'^1_j, W'^2_j)\|}.$$

Thus, $(\alpha_j)_j$ converges to $\alpha \in \mathbb{R}^*$. Then $(W'^1, W'^2) \in \mathfrak{d} \cap \mathfrak{l} = \mathbb{R}(Z, Z)$, which is impossible as $(W'^1, W'^2) \notin \mathbb{R}(Z, Z)$. □

7.2.3 The deformation parameters set Hom(Γ, G)

Let Γ be a discrete subgroup of G and Hom(Γ, G) the set of homomorphisms from Γ to G. Let $\mathscr{M}_{r,s}(\mathbb{C})$ be the vector space of matrices of r rows and s columns. When $r = s$, we adopt the notation $\mathscr{M}_r(\mathbb{C})$ instead of $\mathscr{M}_{r,r}(\mathbb{C})$. Let now $\{\gamma_1, \ldots, \gamma_k\}$ be a set of generators of Γ given as in previous section. Recall that Hom(Γ, G) is endowed with the pointwise convergence topology. The same topology is obtained by using the injective map

$$\mathrm{Hom}(\Gamma, G) \to G \times \cdots \times G, \quad \varphi \mapsto (\varphi(\gamma_1), \ldots, \varphi(\gamma_k))$$

to equip Hom(Γ, G) with the relative topology induced from the direct product $G^k = G \times \cdots \times G$. Now, we identify any element $(\mathbf{r}, e^{2i\pi z}, e^{2i\pi z'})$, $\mathbf{r} \in \mathbb{R}^{4n}$, $z, z' \in \mathbb{R}$, of G by the associated column vector and the product set G^k by the set of matrices in $\mathscr{M}_{4n+2,k}(\mathbb{C})$ defined by

$$\mathscr{E} = \left\{ \begin{pmatrix} C \\ e^{2i\pi \mathbf{z}} \\ e^{2i\pi \mathbf{z}'} \end{pmatrix} \in \mathscr{M}_{4n+2,k}(\mathbb{C}) : C \in \mathscr{M}_{4n,k}(\mathbb{R}), \mathbf{z}, \mathbf{z}' \in \mathbb{R}^k \right\},$$

where, for $\mathbf{z} = (z_1, \ldots, z_k) \in \mathbb{R}^k$,

$$e^{2i\pi \mathbf{z}} := \begin{pmatrix} e^{2i\pi z_1} & \cdots & e^{2i\pi z_k} \end{pmatrix} \in \mathbb{T}^k.$$

Then it appears clear that the map

$$\Psi : \operatorname{Hom}(\Gamma, G) \longrightarrow \mathscr{M}_{4n+2,k}(\mathbb{C}), \tag{7.7}$$

which associates to any element $\varphi \in \operatorname{Hom}(\Gamma, G)$ its matrix $M_\varphi := \lfloor \varphi(\gamma_1), \ldots, \varphi(\gamma_k) \rfloor$, where this symbol merely designates the matrix constituted of the columns $\varphi(\gamma_1)$, $\ldots, \varphi(\gamma_k)$, is a homeomorphism on its range. In order to give a description of $\Psi(\operatorname{Hom}(\Gamma, G))$, we consider the following notation. For all $u \in \{1, \ldots, l\}$, we have $\gamma_u = {}^t(\mathbf{w}_u, e^{2i\pi c_u}, e^{2i\pi c'_u})$. Then let

$$\mathbf{w}_u = (\mathbf{w}_u^1, \mathbf{w}_u^2, \mathbf{w}_u^3, \mathbf{w}_u^4) \in \mathbb{R}^{4n},$$

where $\mathbf{w}_u^s \in \mathbb{R}^n, s \in \{1, \ldots, 4\}$ for all $v \in \{1, \ldots, l\}$,

$$\begin{aligned} p_{uv} = q(&\prec \mathbf{w}_u^1, \mathbf{w}_v^4 \succ - \prec \mathbf{w}_u^4, \mathbf{w}_v^1 \succ + \prec \mathbf{w}_u^3, \mathbf{w}_v^2 \succ - \prec \mathbf{w}_u^2, \mathbf{w}_v^3 \succ \\ &+ \prec \mathbf{w}_u^3, \mathbf{w}_v^4 \succ - \prec \mathbf{w}_u^4, \mathbf{w}_v^3 \succ) \end{aligned}$$

when Γ is not torsion-free, and 0 otherwise.

Let also

$$P(\Gamma) = \begin{pmatrix} 0 & p_{12} & \cdots & \cdots & p_{1l} \\ -p_{12} & \ddots & \ddots & & \vdots \\ \vdots & \ddots & \ddots & \ddots & \vdots \\ \vdots & & \ddots & \ddots & p_{l-1\,l} \\ -p_{1l} & \cdots & \cdots & -p_{l-1\,l} & 0 \end{pmatrix} \in \mathscr{M}_l(\mathbb{R})$$

and $\mathscr{A}_l(\mathbb{Z})$ the subspace of skew-symmetric matrices of $\mathscr{M}_l(\mathbb{Z})$.

Moreover, for

$$C = \begin{pmatrix} C_1 \\ \vdots \\ C_4 \end{pmatrix} \in \mathscr{M}_{4n,l}(\mathbb{R}),$$

we have the following.

Proposition 7.2.6. *We keep the same notation and hypotheses. Let Γ be a discontinuous group for G/Δ. We have:*

(1) *If* Γ *is torsion-free, then* $\mathrm{Hom}(\Gamma, G)$ *is homeomorphic to*

$$\mathcal{U} = \left\{ \begin{pmatrix} C \\ e^{2i\pi z} \\ e^{2i\pi z'} \end{pmatrix} \in \mathcal{E} \;\middle|\; \begin{matrix} {}^tC_1C_4 - {}^tC_4C_1 + {}^tC_3C_2 - {}^tC_2C_3 + {}^tC_3C_4 - {}^tC_4C_3 \in \mathscr{A}_l(\mathbb{Z}), \\ {}^tC_1C_2 - {}^tC_2C_1 \in \mathscr{A}_l(\mathbb{Z}) \end{matrix} \right\}.$$

(2) *Otherwise,* $\Gamma \cap Z(G) = \langle \gamma_{l+1} \rangle$ *is a cyclic subgroup and* $\mathrm{Hom}(\Gamma, G)$ *is homeomorphic to*

$$\mathcal{U} = \left\{ \begin{pmatrix} C & {}^t\mathbf{0}_{4n} \\ e^{2i\pi z} & e^{2i\pi \frac{a}{b}} \\ e^{2i\pi z'} & e^{2i\pi \frac{a'}{b'}} \end{pmatrix} \in \mathcal{E} \;\middle|\; \begin{matrix} {}^tC_1C_4 - {}^tC_4C_1 + {}^tC_3C_2 - {}^tC_2C_3 \\ + {}^tC_3C_4 - {}^tC_4C_3 \in \frac{a'}{b'}\mathrm{P}(\Gamma) + \mathscr{A}_l(\mathbb{Z}), \\ {}^tC_1C_2 - {}^tC_2C_1 \in \frac{a}{b}\mathrm{P}(\Gamma) + \mathscr{A}_l(\mathbb{Z}), \\ a, b, a', b' \in \mathbb{Z}, a \wedge b = a' \wedge b' = 1, \\ b \vee b' \text{ divides } q \end{matrix} \right\},$$

where the symbols $\wedge$ *and* $\vee$ *mean the greatest common divisor and the least common multiple, respectively.*

Proof. It is sufficient to prove the proposition when Γ is not torsion-free. Indeed otherwise, $P(\Gamma) = 0$ and the same arguments work. We prove that $\Psi(\mathrm{Hom}(\Gamma, G)) = \mathcal{U}$. For $\varphi \in \mathrm{Hom}(\Gamma, G)$, we have $\varphi(\gamma_{l+1}) = {}^t(\mathbf{0}_{4n}, e^{2i\pi \frac{a}{b}}, e^{2i\pi \frac{a'}{b'}})$ where $a, b, a', b' \in \mathbb{Z}$ such that $a \wedge b = a' \wedge b' = 1$, $b \vee b'$ divides q. Now, let $u, v \in \{1, \dots, l\}$. Then it is not hard to see that

$$\gamma_u\gamma_v\gamma_u^{-1}\gamma_v^{-1} = {}^t(\mathbf{0}_{4n}, 1, e^{2i\pi \frac{p_{uv}}{q}}) = \gamma_{l+1}^{p_{uv}}.$$

As $\varphi \in \mathrm{Hom}(\Gamma, G)$, then

$$\varphi(\gamma_u\gamma_v\gamma_u^{-1}\gamma_v^{-1}) = (\varphi(\gamma_{l+1}))^{p_{uv}} = {}^t(\mathbf{0}_{4n}, e^{2i\pi p_{uv}\frac{a}{b}}, e^{2i\pi p_{uv}\frac{a'}{b'}}).$$

On the other hand, for the same reason, if we denote by $C_j = \lfloor C_j^1, \dots, C_j^l \rfloor$, we have

$$\begin{aligned} \varphi(\gamma_u\gamma_v\gamma_u^{-1}\gamma_v^{-1}) &= \varphi(\gamma_u)\varphi(\gamma_v)\varphi(\gamma_u)^{-1}\varphi(\gamma_v)^{-1} \\ &= {}^t(\mathbf{0}_{4n}, e^{2i\pi({}^tC_1^uC_2^v - {}^tC_2^uC_1^v)}, e^{2i\pi({}^tC_1^uC_4^v - {}^tC_4^uC_1^v + {}^tC_3^uC_2^v - {}^tC_2^uC_3^v + {}^tC_3^uC_4^v - {}^tC_4^uC_3^v)}). \end{aligned}$$

This gives that

$$\begin{cases} {}^tC_1^uC_2^v - {}^tC_2^uC_1^v \in p_{uv}\frac{a}{b} + \mathbb{Z}, \\ {}^tC_1^uC_4^v - {}^tC_4^uC_1^v + {}^tC_3^uC_2^v - {}^tC_2^uC_3^v + {}^tC_3^uC_4^v - {}^tC_4^uC_3^v \in p_{uv}\frac{a'}{b'} + \mathbb{Z}. \end{cases}$$

Conversely, let

$$M = \begin{pmatrix} C & {}^t\mathbf{0}_{4n} \\ e^{2i\pi \mathbf{z}} & e^{2i\pi \frac{a}{b}} \\ e^{2i\pi \mathbf{z}'} & e^{2i\pi \frac{a'}{b'}} \end{pmatrix} \in \mathscr{U}.$$

Noting $C = \lfloor C^1, \ldots, C^l \rfloor$, $g_j = {}^t({}^tC^j, e^{2i\pi z_j}, e^{2i\pi z_j'})$ and $g_{l+1} = {}^t(\mathbf{0}_{4n}, e^{2i\pi \frac{a}{b}}, e^{2i\pi \frac{a'}{b'}})$, for $j \in \{1, \ldots, l\}$, we construct the map φ defined by

$$\begin{aligned} \varphi : \Gamma \quad & \rightarrow \quad G \\ \gamma_1^{n_1} \cdots \gamma_{l+1}^{n_{l+1}} & \mapsto \quad g_1^{n_1} \cdots g_{l+1}^{n_{l+1}}. \end{aligned}$$

We show that $\varphi \in \operatorname{Hom}(\Gamma, G)$. Let $\gamma = \gamma_1^{n_1} \cdots \gamma_{l+1}^{n_{l+1}}$ and $\gamma' = \gamma_1^{m_1} \cdots \gamma_{l+1}^{m_{l+1}}$ in Γ. It is not hard to see that

$$\gamma\gamma' = \gamma_1^{n_1+m_1} \cdots \gamma_l^{n_l+m_l} \gamma_{l+1}^{m},$$

where

$$m = n_{l+1} + m_{l+1} - \sum_{1 \leqslant v < u \leqslant l} n_u m_v p_{vu}.$$

Then

$$\begin{aligned} \varphi(\gamma\gamma') &= g_1^{n_1+m_1} \cdots g_l^{n_l+m_l} g_{l+1}^{m} \\ &= g_1^{n_1} \cdots g_l^{n_l} g_{l+1}^{n_{l+1}} g_1^{m_1} \cdots g_l^{m_l} g_{l+1}^{m_{l+1}} \\ &= \varphi(\gamma_1^{n_1} \cdots \gamma_l^{n_l} \gamma_{l+1}^{n_{l+1}}) \varphi(\gamma_1^{m_1} \cdots \gamma_l^{m_l} \gamma_{l+1}^{m_{l+1}}) \\ &= \varphi(\gamma)\varphi(\gamma'). \end{aligned}$$

This shows that $M = \Psi(\varphi)$, which is enough to conclude. □

Now, we have the following.

Corollary 7.2.7. *We keep the same notation and hypotheses. The set* $\operatorname{Hom}(\Gamma, G)$ *is homeomorphic to a disjoint union of open algebraic sets in* $\mathscr{U}$.

Proof. Recall that $\operatorname{Hom}(\Gamma, G)$ is homeomorphically identified to a subset $\mathscr{U}$ of $\mathscr{E}$. It suffices then to show that $\mathscr{U}$ splits to a disjoint union of open algebraic sets in $\mathscr{U}$. We only treat the case where Γ is torsion-free as the other case is handled similarly. For $D = (D_1, D_2) \in \mathscr{A}_l(\mathbb{Z})^2$, let

$$\mathscr{U}_D = \left\{ \begin{pmatrix} C \\ e^{2i\pi \mathbf{z}} \\ e^{2i\pi \mathbf{z}'} \end{pmatrix} \in \mathscr{E} \;\middle|\; \begin{aligned} & {}^tC_1C_4 - {}^tC_4C_1 + {}^tC_3C_2 - {}^tC_2C_3 + {}^tC_3C_4 - {}^tC_4C_3 = D_1, \\ & {}^tC_1C_2 - {}^tC_2C_1 = D_2 \end{aligned} \right\}.$$

We have

$$\mathscr{U} = \coprod_{D \in \mathscr{A}_l(\mathbb{Z})^2} \mathscr{U}_D.$$

Clearly, the set $\mathscr{U}_D$ are algebraic in $\mathscr{U}$ and we show that $\mathscr{U}_D$ is open in $\mathscr{U}$. Let $M \in \mathscr{U}$ and $(M_j)_{j\in\mathbb{N}}$ a sequence of ${}^c\mathscr{U}_D$, which converges to M. Let

$$M = \begin{pmatrix} C \\ e^{2i\pi z} \\ e^{2i\pi z'} \end{pmatrix}, \quad M_j = \begin{pmatrix} C^j \\ e^{2i\pi z_j} \\ e^{2i\pi z'_j} \end{pmatrix} \quad \text{and} \quad C^j = \begin{pmatrix} C_{1j} \\ \vdots \\ C_{4j} \end{pmatrix}.$$

The sequences:

$${}^tC_{1j}C_{4j} - {}^tC_{4j}C_{1j} + {}^tC_{3j}C_{2j} - {}^tC_{2j}C_{3j} + {}^tC_{3j}C_{4j} - {}^tC_{4j}C_{3j} \quad \text{and} \quad {}^tC_{1j}C_{2j} - {}^tC_{2j}C_{1j}$$

of $\mathscr{A}_l(\mathbb{Z})$ converge to

$$D'_1 := {}^tC_1C_4 - {}^tC_4C_1 + {}^tC_3C_2 - {}^tC_2C_3 + {}^tC_3C_4 - {}^tC_4C_3 \quad \text{and} \quad D'_2 := {}^tC_1C_2 - {}^tC_2C_1,$$

respectively, such that $(D'_1, D'_2) \neq (D_1, D_2)$. Then they are stationary, which gives that $M \in {}^c\mathscr{U}_D$. □

Now, our goal is the determination of the set of all injective group homomorphisms of discrete images denoted by $\mathrm{Hom}^0_d(\Gamma, G)$. The same arguments as in Proposition 7.1.7 give us the following result.

Proposition 7.2.8. *We keep the same notation and hypotheses. Then:*
(1) *If Γ is torsion-free, then $\mathrm{Hom}^0_d(\Gamma, G)$ is homeomorphic to*

$$\mathscr{U}^0_d = \left\{ \begin{pmatrix} C \\ e^{2i\pi z} \\ e^{2i\pi z'} \end{pmatrix} \in \mathscr{U} : \mathrm{rk}(C) = l \right\}.$$

(2) *Otherwise, $\mathrm{Hom}^0_d(\Gamma, G)$ is homeomorphic to*

$$\mathscr{U}^0_d = \left\{ \begin{pmatrix} C & {}^t\mathbf{0}_{4n} \\ e^{2i\pi z} & e^{2i\pi \frac{a}{b}} \\ e^{2i\pi z'} & e^{2i\pi \frac{a'}{b'}} \end{pmatrix} \in \mathscr{U} \;\middle|\; \begin{array}{l} \mathrm{rk}(C) = l, \\ a^2 + a'^2 \neq 0 \text{ and } b \vee b' = q \end{array} \right\}.$$

The last proposition shows that $\mathscr{U}^0_d$ is open in $\mathscr{U}$. So the following becomes clear.

Corollary 7.2.9. *The set $\mathrm{Hom}^0_d(\Gamma, G)$ is open in $\mathrm{Hom}(\Gamma, G)$.*

We now show the following result.

Corollary 7.2.10. *We keep the same notation and hypotheses. Then* $\mathrm{Hom}_d^0(\Gamma, G)$ *is homeomorphic to a disjoint union of open semialgebraic sets in* $\mathscr{U}_d^0$.

Proof. When Γ is torsion-free, it suffices to see that

$$\mathscr{U}_d^0 = \coprod_{D \in \mathscr{A}_l(\mathbb{Z})^2} (\mathscr{U}_d^0 \cap \mathscr{U}_D) \tag{7.8}$$

and that the sets $\mathscr{U}_d^0 \cap \mathscr{U}_D$, $D \in \mathscr{A}_l(\mathbb{Z})^2$ are semialgebraic and open in $\mathscr{U}_d^0$. □

7.2.4 The parameter space

Let Γ be a discontinuous group for the homogeneous space G/Δ. This subsection aims to give the explicit description of the parameter space. For $\varphi \in \mathrm{Hom}_d^0(\Gamma, G)$, let $L_\varphi = \exp \mathfrak{l}_\varphi$ be the syndetic hull of $\varphi(\Gamma)$.

Proposition 7.2.11. *Let* Γ *be a discontinuous group for the homogeneous space* G/Δ. *Then:*

(1) *If* Γ *is torsion-free, then* $\mathscr{R}(\Gamma, G, \Delta)$ *is homeomorphic to*

$$\mathscr{R} = \left\{ \begin{pmatrix} C \\ e^{2i\pi \mathbf{z}} \\ e^{2i\pi \mathbf{z}'} \end{pmatrix} \in \mathscr{U} : \operatorname{rk}\begin{pmatrix} C_3 \\ C_4 \end{pmatrix} = l \right\}.$$

(2) *Otherwise,* $\mathscr{R}(\Gamma, G, \Delta)$ *is homeomorphic to*

$$\mathscr{R} = \left\{ \begin{pmatrix} C & {}^t\mathbf{0}_{4n} \\ e^{2i\pi \mathbf{z}} & e^{2i\pi \frac{a}{b}} \\ e^{2i\pi \mathbf{z}'} & e^{2i\pi \frac{a'}{b'}} \end{pmatrix} \in \mathscr{U}_d^0 \;\middle|\; \begin{matrix} \operatorname{rk}\begin{pmatrix} C_3 \\ C_4 \end{pmatrix} = l \\ \text{and } b \text{ divides } b' \end{matrix} \right\}.$$

Proof. It is sufficient to prove the proposition when Γ is not torsion-free. Indeed otherwise, the same arguments work. Let

$$M_\varphi = \begin{pmatrix} C & {}^t\mathbf{0}_{4n} \\ e^{2i\pi \mathbf{z}} & e^{2i\pi \frac{a}{b}} \\ e^{2i\pi \mathbf{z}'} & e^{2i\pi \frac{a'}{b'}} \end{pmatrix} = \Psi(\varphi) \in \mathscr{U}_d^0$$

such that $C = \lfloor C^1, \dots, C^l \rfloor$ and $L^j \in \mathfrak{g}$ such that ${}^t({}^tC^j, 1, 1) = \exp L^j$, $j \in \{1, \dots, l\}$. It is not hard to see that $\mathfrak{l}_\varphi = \mathbb{R}\text{-span}\{L^1, \dots, L^l, (Z, Z), (Z, 0)\}$. Thanks to Proposition 7.2.5, it is clear that $\mathscr{R}(\Gamma, G, \Delta)$ is homeomorphic to the set

$$\left\{ \begin{pmatrix} C & {}^t\mathbf{0}_{4n} \\ e^{2i\pi\mathbf{z}} & e^{2i\pi\frac{a}{b}} \\ e^{2i\pi\mathbf{z}'} & e^{2i\pi\frac{a'}{b'}} \end{pmatrix} \in \mathscr{U}_d^0 \;\middle|\; \begin{array}{l} \operatorname{rk}\left(\begin{pmatrix} I_{2n} \\ 0 \end{pmatrix} ⋒ C \right) = 2n + l \\ \text{and } \varphi(\Gamma) \text{ acts freely on } G/\Delta \end{array} \right\},$$

where the symbol ⋒ merely means the concatenation. Hence

$$\operatorname{rk} \begin{pmatrix} C_3 \\ C_4 \end{pmatrix} = l.$$

Now, show that the free action of $\varphi(\Gamma)$ on G/Δ is equivalent to the fact that b divides b'. Let $g \in G$ and

$$u = \begin{pmatrix} C^1 \\ e^{2i\pi z_1} \\ e^{2i\pi z_1'} \end{pmatrix}^{m_1} \cdots \begin{pmatrix} C^l \\ e^{2i\pi z_l} \\ e^{2i\pi z_l'} \end{pmatrix}^{m_l} \begin{pmatrix} {}^t\mathbf{0}_{4n} \\ e^{2i\pi\frac{a}{b}} \\ e^{2i\pi\frac{a'}{b'}} \end{pmatrix}^{m} \in \varphi(\Gamma) \cap g\Delta g^{-1}$$

for some integers $m_1, \ldots, m_l$ and m. Write $C = (c_{uv})_{\substack{1 \leq u \leq 4n \\ 1 \leq v \leq l}}$. Then

$$\sum_{v=1}^{l} m_v c_{uv} = 0 \quad \text{for all } u \in \{2n + 1, \ldots, 4n\}.$$

Moreover, as $\operatorname{rk} \binom{C_3}{C_4} = l$ then $m_v = 0$ for all $v \in \{1, \ldots, l\}$ and $m\frac{a'}{b'} \in \mathbb{Z}$. Hence

$$\begin{aligned} \varphi(\Gamma) \cap g\Delta g^{-1} &= \left\{ \begin{pmatrix} {}^t\mathbf{0}_{4n} \\ e^{2i\pi m\frac{a}{b}} \\ 1 \end{pmatrix} : m = kb', \ k \in \mathbb{Z} \right\} \\ &= \left\{ \begin{pmatrix} {}^t\mathbf{0}_{4n} \\ e^{2i\pi kb'\frac{a}{b}} \\ 1 \end{pmatrix} : k \in \mathbb{Z} \right\}. \end{aligned}$$

It is clear then that if b does not divide b', then $e \neq \exp(\frac{b'a}{b}(Z, Z)) \in \varphi(\Gamma) \cap g\Delta g^{-1}$. Therefore, the action of $\varphi(\Gamma)$ on G/Δ is not free. □

7.2.5 Proof of Theorem 7.2.1

(1) It is clear that the parameter space is open in $\operatorname{Hom}(\Gamma, G)$.

(2) We only treat the case where Γ is torsion-free, the other case is handled similarly. For $D \in \mathscr{A}_l(\mathbb{Z})^2$, let

$$\mathscr{R}_D = \mathscr{U}_D \cap \mathscr{R}.$$

Therefore,

$$\mathscr{R} = \coprod_{D \in \mathscr{A}_l(\mathbb{Z})^2} \mathscr{R}_D.$$

Clearly, the sets $\mathscr{R}_D, D \in \mathscr{A}_l(\mathbb{Z})^2$ are semialgebraic and open in $\mathscr{R}$, and we only need to show that $\mathscr{R}_D$ is endowed with a smooth manifold structure for all $D \in \mathscr{A}_l(\mathbb{Z})^2$. Let $D = (D_1, D_2) \in \mathscr{A}_l(\mathbb{Z})^2$. It is clear that the set

$$\mathscr{F}_{D_2} = \left\{ M = \begin{pmatrix} C \\ e^{2i\pi \mathbf{z}} \\ e^{2i\pi \mathbf{z}'} \end{pmatrix} \in \mathscr{E} \;\middle|\; \begin{array}{l} {}^tC_1C_2 - {}^tC_2C_1 = D_2, \\ \operatorname{rk}\begin{pmatrix} C_3 \\ C_4 \end{pmatrix} = l \end{array} \right\}$$

is open in $\mathscr{E}$. Now let ψ_{D_1} the smooth map

$$\begin{array}{llll} \psi_{D_1} : \mathscr{F}_{D_2} & \to & \mathscr{A}_l(\mathbb{R}) \\ M & \mapsto & {}^tC_1C_4 - {}^tC_4C_1 + {}^tC_3C_2 - {}^tC_2C_3 + {}^tC_3C_4 - {}^tC_4C_3 - D_1. \end{array}$$

Clearly, $\mathscr{R}_D = \psi_{D_1}^{-1}(\{0\})$. The goal now is to show that zero is a regular value of the map ψ_{D_1}. The derivative of ψ_{D_1} at a point

$$M = \begin{pmatrix} C \\ e^{2i\pi \mathbf{z}} \\ e^{2i\pi \mathbf{z}'} \end{pmatrix} \in \mathscr{R}_D,$$

is given by

$$\begin{array}{lll} d(\psi_{D_1})_M : \mathscr{E} & \to & \mathscr{A}_l(\mathbb{R}) \\ X = \begin{pmatrix} H \\ e^{2i\pi \mathbf{h}} \\ e^{2i\pi \mathbf{h}'} \end{pmatrix} & \mapsto & {}^t\begin{pmatrix} H_1 \\ H_3 \\ H_4 \\ -H_4 \\ -H_2 \\ H_3 \end{pmatrix} \begin{pmatrix} C_4 \\ C_2 \\ -C_3 \\ C_1 \\ C_3 \\ C_4 \end{pmatrix} - {}^t\begin{pmatrix} C_4 \\ C_2 \\ -C_3 \\ C_1 \\ C_3 \\ C_4 \end{pmatrix} \begin{pmatrix} H_1 \\ H_3 \\ H_4 \\ -H_4 \\ -H_2 \\ H_3 \end{pmatrix} \end{array}$$

for

$$H = \begin{pmatrix} H_1 \\ \vdots \\ H_4 \end{pmatrix}.$$

This is enough to conclude thanks to Lemma 4.1.29 that $\mathscr{R}_D$ is endowed with a smooth manifold structure of dimension $(4n + 2)l - \frac{1}{2}l(l - 1)$. Then the set $\mathscr{R}$ splits to a disjoint union of open smooth manifolds of dimension $(4n+2)l - \frac{1}{2}l(l-1)$ and, therefore, endowed with a smooth manifold structure.

The group G acts on $\mathscr{E}$ through the law:

$$g \star \begin{pmatrix} C \\ e^{2i\pi\mathbf{z}} \\ e^{2i\pi\mathbf{z}'} \end{pmatrix} = \mathrm{Ad}_g \cdot \begin{pmatrix} C \\ e^{2i\pi\mathbf{z}} \\ e^{2i\pi\mathbf{z}'} \end{pmatrix} = \begin{pmatrix} C \\ e^{2i\pi(\mathbf{z}-\boldsymbol{\beta}C_1+\boldsymbol{\alpha}C_2)} \\ e^{2i\pi(\mathbf{z}'-\boldsymbol{\delta}C_1+\boldsymbol{\gamma}C_2-(\boldsymbol{\beta}+\boldsymbol{\delta})C_3+(\boldsymbol{\alpha}+\boldsymbol{\gamma})C_4)} \end{pmatrix} \tag{7.9}$$

for $g = {}^t(\boldsymbol{\alpha}, \boldsymbol{\beta}, \boldsymbol{\gamma}, \boldsymbol{\delta}, e^{2i\pi d}, e^{2i\pi d'}) \in G$ with $\boldsymbol{\alpha}, \boldsymbol{\beta}, \boldsymbol{\gamma}, \boldsymbol{\delta} \in \mathbb{R}^n, d, d' \in \mathbb{R}$.

The map $\Psi : \mathrm{Hom}(\Gamma, G) \longrightarrow \mathscr{E}$ given in equation (7.7) turns out to be G-equivariant. So the deformation space is homeomorphic to $\mathscr{T} = \mathscr{R}/G$.

(3) We now show that in the case where $n \geqslant 2l$, the space $\mathscr{T}$ admits an open smooth manifold as its dense subset. Let then

$$\mathscr{R}_0 = \left\{ \begin{pmatrix} C \\ e^{2i\pi\mathbf{z}} \\ e^{2i\pi\mathbf{z}'} \end{pmatrix} \in \mathscr{R} : \mathrm{rk}\begin{pmatrix} C_1 \\ C_2 \end{pmatrix} = \mathrm{rk}\begin{pmatrix} C_1 + C_3 \\ C_2 + C_4 \end{pmatrix} = l \right\}$$

and

$$\mathscr{T}_0 = \left\{ \begin{pmatrix} C \\ \mathbf{1}_l \\ \mathbf{1}_l \end{pmatrix} \in \mathscr{R}_0 \right\}.$$

We show that $\mathscr{T}_0$ is a cross-section of all orbits of $\mathscr{R}_0$. Let

$$M = \begin{pmatrix} C \\ e^{2i\pi\mathbf{z}} \\ e^{2i\pi\mathbf{z}'} \end{pmatrix} \in \mathscr{R}_0.$$

It is not hard to see that

$$G \star M = \left\{ \begin{pmatrix} C \\ e^{2i\pi\left(\mathbf{z}+\mathbf{v}t\left(\begin{smallmatrix} C_1 \\ C_2 \end{smallmatrix}\right)\right)} \\ e^{2i\pi\left(\mathbf{z}'+\left(\mathbf{u}\left(\begin{smallmatrix} C_1+C_3 \\ C_2+C_4 \end{smallmatrix}\right)-\mathbf{v}\left(\begin{smallmatrix} C_1 \\ C_2 \end{smallmatrix}\right)\right)\right)} \end{pmatrix} : \mathbf{u}, \mathbf{v} \in \mathbb{R}^{2n} \right\}.$$

As $M \in \mathscr{R}_0$, then

$$\begin{pmatrix} C \\ \mathbf{1}_l \\ \mathbf{1}_l \end{pmatrix} \in G \star M \cap \mathscr{T}_0.$$

The other inclusion is clear. The next step consists in showing that the map

$$\begin{aligned}\Phi : \mathscr{R}_0/G &\to \mathscr{T}_0\\ [M] &\mapsto \begin{pmatrix} C \\ \mathbf{1}_l \\ \mathbf{1}_l \end{pmatrix}\end{aligned}$$

is a homeomorphism. First of all, it is clear that Φ is well-defined. In fact, let $M_1, M_2 \in \mathscr{R}_0$ such that $[M_1] = [M_2]$. Then $\Phi([M_1]) = G \star M_1 \cap \mathscr{T}_0 = G \star M_2 \cap \mathscr{T}_0 = \Phi([M_2])$. For

$$M_1 = \begin{pmatrix} C \\ e^{2i\pi\mathbf{z}} \\ e^{2i\pi\mathbf{z}'} \end{pmatrix}, \quad M_2 = \begin{pmatrix} D \\ e^{2i\pi\mathbf{t}} \\ e^{2i\pi\mathbf{t}'} \end{pmatrix} \in \mathscr{R}_0$$

such that $\Phi([M_1]) = \Phi([M_2])$, we have

$$\begin{pmatrix} C \\ \mathbf{1}_l \\ \mathbf{1}_l \end{pmatrix} = \begin{pmatrix} D \\ \mathbf{1}_l \\ \mathbf{1}_l \end{pmatrix} \in G \star M_1 \cap G \star M_2.$$

It follows that $[M_1] = [M_2]$, which leads to the injectivity of Φ. Now, to see that Φ is surjective, it is sufficient to verify that for all $M \in \mathscr{T}_0$, we have $\Phi([M]) = M$ as $G \star M \cap \mathscr{T}_0 = \{M\}$. Let π the canonical surjection $\pi : \mathscr{R}_0 \to \mathscr{R}_0/G$. Thus, we can easily see the continuity of $\widetilde{\Phi} = \Phi \circ \pi$, which is equivalent to the continuity of Φ. Also, it is clear that $\Phi^{-1} = \pi_{|\mathscr{T}_0}$, so the bicontinuity follows. It is not hard to see that $\mathscr{T}_0$ is an open smooth manifold. Now, we show that the space $\mathscr{T}_0$ is a dense subset of $\mathscr{T}$. Let

$$M = \begin{pmatrix} C \\ e^{2i\pi\mathbf{z}} \\ e^{2i\pi\mathbf{z}'} \end{pmatrix} \in \mathscr{R} \backslash \mathscr{R}_0.$$

Suppose first that

$$\operatorname{rk}\begin{pmatrix} C_1 \\ C_2 \end{pmatrix} < l.$$

Then we can assume that $\operatorname{rk}(C_1) = r < l$. We denote by $C_i = \lfloor C_i^1, \dots, C_i^l \rfloor$ for $i = 1, \dots, 4$. Let $\mathfrak{a}$ designate the subspace of $\mathbb{R}^n$ spanned by the column vectors $\{C_1^1, \dots, C_1^l, C_3^1, \dots, C_3^l\}$. It is not hard to see that $\dim \mathfrak{a}^\perp = n - \dim \mathfrak{a} \geqslant l - r$, as $n \geqslant 2l$. Fix $\{T_1, \dots, T_{l-r}\}$ a linearly independent family of vectors of $\mathfrak{a}^\perp$. Without loss of generality, we can suppose that

$$\operatorname{rk}(C_1) = \operatorname{rk}\lfloor C_1^1 \cdots C_1^r \rfloor.$$

Let then for $j \in \mathbb{N}^*$, $C_{2j} = \lfloor C^1_{2j}, \dots, C^l_{2j} \rfloor$ such that

$$C^i_{2j} = \begin{cases} C^i_2 & \text{if } i = 1, \dots, r, \\ C^i_2 + \frac{1}{j} T_{i-r} & \text{if } i = r+1, \dots, l. \end{cases}$$

The result is then immediate if we can extract a subsequence $M_{\varphi(j)}$ of

$$M_j = \begin{pmatrix} C_1 \\ C_{2j} \\ C_3 \\ C_4 \\ e^{2i\pi z} \\ e^{2i\pi z'} \end{pmatrix}$$

such that

$$\operatorname{rk}\begin{pmatrix} C_1 + C_3 \\ C_{2\varphi(j)} + C_4 \end{pmatrix} = l.$$

Otherwise, we can assume that $\operatorname{rk}(C_1 + C_3) < \operatorname{rk}(C_1)$. Let $\{S_1, \dots, S_p\}$ span an orthogonal vector subspace, supplement to $\mathbb{R}\text{-span}\{C^1_1 + C^1_3, \dots, C^l_1 + C^l_3\}$ inside the space $\mathbb{R}\text{-span}\{C^1_1, \dots, C^l_1, C^1_3, \dots, C^l_3\}$. Then $p = \operatorname{rk}(C_1, C_3) - \operatorname{rk}(C_1 + C_3) \geq \operatorname{rk}(C_1) - \operatorname{rk}(C_1 + C_3)$. Now, we are in position to form a new sequence $(C_{4m})_{m \in \mathbb{N}^*}$ as above, from the matrix C_4 using the vectors $S_1, \dots, S_p$. Let $s = \operatorname{rk}(C_1 + C_3)$ and replace ith column of C_4 by $C^i_4 + \frac{1}{m} S_{i-s}$ for $s + 1 \leqslant i \leqslant r$. Let $M_{j,m}$ be the matrix obtained when replacing in M the matrix C_2 and C_4 by C_{2j} and C_{4m}, respectively. Now if

$$\operatorname{rk}\begin{pmatrix} C_1 \\ C_2 \end{pmatrix} = l,$$

it suffices to take $C_{2j} = C_2$ and to construct C_{4m} as above. Then clearly $M_{j,m} \in \mathscr{R}_0$ and $(M_{j,m})_{j,m \in \mathbb{N}^*}$ converge to M. This shows that $\mathscr{T}_0 = \Phi(\mathscr{R}_0)$ is a dense set of $\mathscr{T}$ as Φ is a homeomorphism. □

For the convenience of readers, we give the following examples to illustrate the last point of Theorem 7.2.1.

Example 7.2.12. Let $n = 4$ and $l = 2$. We take

$$C_1 = \begin{pmatrix} 1 & 0 \\ 0 & 0 \\ 0 & 0 \\ 0 & 0 \end{pmatrix}, \quad C_2 = \begin{pmatrix} 0 & 0 \\ 0 & 0 \\ 0 & 0 \\ 0 & 0 \end{pmatrix}, \quad C_3 = \begin{pmatrix} -1 & 0 \\ 0 & 0 \\ 0 & 0 \\ 0 & 0 \end{pmatrix} \quad \text{and} \quad C_4 = \begin{pmatrix} 0 & 0 \\ 0 & 0 \\ 0 & 0 \\ 0 & 1 \end{pmatrix}.$$

Then $r = 1$ and $l - r = 1$. Since

$$\mathfrak{a}^{\perp} = \mathbb{R}\text{-span}\left\{\begin{pmatrix}0\\1\\0\\0\end{pmatrix}, \begin{pmatrix}0\\0\\1\\0\end{pmatrix}, \begin{pmatrix}0\\0\\0\\1\end{pmatrix}\right\},$$

we have

$$\begin{pmatrix}C_1\\C_{2j}\end{pmatrix} = \begin{pmatrix}1 & 0\\0 & 0\\0 & 0\\0 & 0\\0 & 0\\0 & \frac{1}{j}\\0 & 0\\0 & 0\end{pmatrix} \quad \text{and} \quad \begin{pmatrix}C_1 + C_3\\C_{2j} + C_{4m}\end{pmatrix} = \begin{pmatrix}0 & 0\\0 & 0\\0 & 0\\0 & 0\\\frac{1}{m} & 0\\0 & \frac{1}{j}\\0 & 0\\0 & 1\end{pmatrix},$$

for m and $j \in \mathbb{N}^*$.

Example 7.2.13. Let $n = 4$ and $l = 2$. We take

$$C_1 = \begin{pmatrix}1 & 0\\0 & 0\\0 & 0\\0 & 0\end{pmatrix}, \quad C_2 = \begin{pmatrix}0 & 0\\0 & 0\\0 & 0\\0 & 1\end{pmatrix}, \quad C_3 = \begin{pmatrix}-1 & 0\\0 & 0\\0 & 0\\0 & 0\end{pmatrix} \quad \text{and} \quad C_4 = \begin{pmatrix}0 & 0\\0 & 0\\0 & 0\\0 & 1\end{pmatrix}.$$

Then $r = 1$ and $l - r = 1$. Since

$$\mathfrak{a}^{\perp} = \mathbb{R}\text{-span}\left\{\begin{pmatrix}0\\1\\0\\0\end{pmatrix}, \begin{pmatrix}0\\0\\1\\0\end{pmatrix}, \begin{pmatrix}0\\0\\0\\1\end{pmatrix}\right\},$$

we have

$$\begin{pmatrix}C_1\\C_2\end{pmatrix} = \begin{pmatrix}1 & 0\\0 & 0\\0 & 0\\0 & 0\\0 & 0\\0 & 0\\0 & 0\\0 & 1\end{pmatrix} \quad \text{and} \quad \begin{pmatrix}C_1 + C_3\\C_2 + C_{4m}\end{pmatrix} = \begin{pmatrix}0 & 0\\0 & 0\\0 & 0\\0 & 0\\\frac{1}{m} & 0\\0 & 0\\0 & 0\\0 & 2\end{pmatrix},$$

for $m \in \mathbb{N}^*$.

7.2.6 Proof of Theorem 7.2.2

Assume first that Γ is a finite group. Then

$$\mathscr{T} = \left\{ \begin{pmatrix} {}^t\mathbf{0}_{4n} \\ e^{2i\pi \frac{a}{b}} \\ e^{2i\pi \frac{a'}{b'}} \end{pmatrix} \;\middle|\; \begin{array}{l} a, b, a', b' \in \mathbb{Z}, a^2 + a'^2 \neq 0, \\ a \wedge b = a' \wedge b' = 1, b \vee b' = q, b \text{ divides } b' \end{array} \right\},$$

which is a finite group. So the strong local rigidity property holds. Suppose now that Γ is infinite. As $\mathscr{R}$ is endowed with a smooth manifold structure and $\dim G \star M \lneq \dim \mathscr{R}$ for any $M \in \mathscr{R}$ then the local rigidity fails to hold. We now study the Hausdorffness of the deformation space. In the case where Γ is finite, it is not hard to see that the deformation space is Hausdorff. Assume that Γ is infinite. It is sufficient to look at the case where Γ is torsion-free. Let $[M_1], [M_2] \in \mathscr{T}$ such that $[M_1] \neq [M_2]$. We designate by

$$M_j = \begin{pmatrix} C \\ e^{2i\pi \mathbf{z}_j} \\ e^{2i\pi \mathbf{z}'} \end{pmatrix}, \quad j \in \{1, 2\}$$

with

$$\begin{pmatrix} C_1 \\ C_2 \end{pmatrix} = 0, \quad \begin{pmatrix} C_3 \\ C_4 \end{pmatrix} = \begin{pmatrix} I_l \\ 0 \end{pmatrix}$$

and $\mathbf{z}_j = (z_j, 0, \ldots, 0)$ such that $z_1 - z_2 \notin \mathbb{Z}$. Let now $\mathscr{V}_j$ be an open neighborhood of $[M_j]$ for $j = 1, 2$. If π denotes the projection map from $\mathscr{R}$ to $\mathscr{T}$, we get that $\pi^{-1}(\mathscr{V}_j)$ is an open neighborhood of M_j in $\mathscr{R}$. Suppose first that $l \leqslant n$. There exists then $\varepsilon > 0$ such that

$$M_j^\varepsilon = \begin{pmatrix} C^\varepsilon \\ e^{2i\pi \mathbf{z}_j} \\ e^{2i\pi \mathbf{z}'} \end{pmatrix} \in \pi^{-1}(\mathscr{V}_j), \quad j \in \{1, 2\},$$

where

$$C^\varepsilon = \begin{pmatrix} 0 \\ C_2^\varepsilon \\ I_l \\ 0 \end{pmatrix} \quad \text{and} \quad C_2^\varepsilon = \begin{pmatrix} 0 & 0 \\ \varepsilon & 0 \end{pmatrix}.$$

On the other hand, for

$$g_\varepsilon = {}^t\left(\mathbf{0}_{n-1}, \frac{z_2 - z_1}{\varepsilon}, \mathbf{0}_{2n-1}, \frac{z_1 - z_2}{\varepsilon}, z_1 - z_2, \mathbf{0}_{n-1}, e^{2i\pi d}, e^{2i\pi d'}\right) \in G,$$

one has $g_\varepsilon \star M_1^\varepsilon = M_2^\varepsilon$, which conclusively shows that $\mathscr{V}_1 \cap \mathscr{V}_2 \neq \emptyset$. This proves that the deformation space is not a Hausdorff space in this case. Suppose now that $n < l < 2n$. There exists then $\varepsilon > 0$ such that

$$M_j^\varepsilon = \begin{pmatrix} C^\varepsilon \\ e^{2i\pi \mathbf{z}_j} \\ e^{2i\pi \mathbf{z}'} \end{pmatrix} \in \pi^{-1}(\mathscr{V}_j), \quad j \in \{1,2\},$$

where

$$C^\varepsilon = \begin{pmatrix} 0 \\ C_2^\varepsilon \\ I_l \\ 0 \end{pmatrix} \quad \text{and} \quad C_2^\varepsilon = \begin{pmatrix} 0 & 0 \\ \nabla_\varepsilon & 0 \end{pmatrix},$$

and $\nabla_\varepsilon = (\mu_{ij})_{i,j} \in \mathscr{M}_{2n-l,n}(\mathbb{R})$ with $\mu_{11} = \varepsilon$ and $\mu_{ij} = 0$ otherwise. Then the same conclusion holds when we take

$$g_\varepsilon = {}^t\left(\mathbf{0}_{l-n}, \frac{z_2 - z_1}{\varepsilon}, \mathbf{0}_{2n-1}, \frac{z_1 - z_2}{\varepsilon}, \mathbf{0}_{2n-l-1}, z_1 - z_2, \mathbf{0}_{n-1}, e^{2i\pi d}, e^{2i\pi d'}\right) \in G.$$

Finally, suppose that $l = 2n$. There exists then $\varepsilon > 0$ such that

$$M_j^\varepsilon = \begin{pmatrix} C^\varepsilon \\ e^{2i\pi \mathbf{z}_j} \\ e^{2i\pi \mathbf{z}'} \end{pmatrix} \in \pi^{-1}(\mathscr{V}_j), \quad j \in \{1,2\},$$

where

$$C^\varepsilon = \begin{pmatrix} 0 \\ C_2^\varepsilon \\ I_l \end{pmatrix} \quad \text{and} \quad C_2^\varepsilon = \begin{pmatrix} 0 & 0 \\ \varepsilon & 0 \end{pmatrix}.$$

Then the same conclusion holds when we take

$$g_\varepsilon = {}^t\left(\mathbf{0}_{n-1}, \frac{z_2 - z_1}{\varepsilon}, \mathbf{0}_{2n-1}, \frac{z_1 - z_2}{\varepsilon}, z_1 - z_2, \mathbf{0}_{n-1}, e^{2i\pi d}, e^{2i\pi d'}\right) \in G.$$

7.3 Reduced threadlike groups

From now on and unless a specific mention $\mathfrak{g}$ designates the threadlike Lie algebra of dimension $n+1, n \geq 2$ as defined in Subsection 1.1.3. Recall the subspace $\mathfrak{g}_0 = \mathbb{R}\text{-span}\{Y_1, \dots, Y_n\}$, which is a one-codimensional Abelian ideal of $\mathfrak{g}$. The center $\mathfrak{z}(\mathfrak{g})$ of $\mathfrak{g}$ is however one-dimensional and it is the space $\mathbb{R}\text{-span}\{Y_n\}$. The reduced threadlike Lie group G is the quotient of $\widetilde{G}$ by the central discrete subgroup $\exp_{\widetilde{G}}(\mathbb{Z}Y_n)$. The corresponding exponential map $\exp := \exp_G$ is given by

$$\exp(xX + y_1Y_1 + \cdots + y_nY_n) = (x, y_1, \ldots, y_{n-1}, e^{2i\pi y_n}), \quad x, y_j \in \mathbb{R}, 1 \leq j \leq n.$$

For the sake of simplicity, we identify from now on this element by the associated column vector.

7.3.1 Proper action of closed subgroups of G

We look first at the case of closed connected subgroup of G. Thus, we only have to study the structure of the associated Lie subalgebra. Let then $\mathfrak{h}$ be a p-dimensional subalgebra of $\mathfrak{g}$. We are going to construct a strong Malcev basis $\mathscr{B}_{\mathfrak{h}}$ of $\mathfrak{h}$ extracted from $\mathscr{B}$. Recall that a family of vectors $\{Z_1, \ldots, Z_m\}$ is said to be a strong Malcev basis of a Lie algebra $\mathfrak{l}$ ($m = \dim \mathfrak{l}$) if $\mathfrak{l}_s = \mathbb{R}\text{-span}\{Z_1, \ldots, Z_s\}$ is an ideal of $\mathfrak{l}$ for all $s \in \{1, \ldots, m\}$.

We denote by $\mathfrak{h}_0 = \mathfrak{h} \cap \mathfrak{g}_0$ and $\mathscr{I}_{\mathfrak{g}_0}^{\mathfrak{h}_0} = \{i_1 < \cdots < i_{p_0}\}$ the set of indices $i \in \{1, \ldots, n\}$ such that $\mathfrak{h}_0 + \mathfrak{g}^i = \mathfrak{h}_0 + \mathfrak{g}^{i+1}$, where $(\mathfrak{g}^i)_{1 \leq i \leq n+1}$ is the decreasing sequence of ideals of $\mathfrak{g}$ given by

$$\mathfrak{g}^i = \mathbb{R}\text{-span}\{Y_i, \ldots, Y_n\}, \quad i = 1, \ldots, n \quad \text{and} \quad \mathfrak{g}^{n+1} = \{0\}. \tag{7.10}$$

We note for all $i_s \in \mathscr{I}_{\mathfrak{g}_0}^{\mathfrak{h}_0}$, $\tilde{Y}_s = Y_{i_s} + \sum_{r=i_s+1}^{n} \alpha_{r,s} Y_r \in \mathfrak{h}$. We get therefore that, if $\mathfrak{h} \subset \mathfrak{g}_0$, then $\mathscr{B}_{\mathfrak{h}} = \{\tilde{Y}_1, \ldots, \tilde{Y}_{p_0}\}$. Otherwise, there exists $\tilde{X} = X + \sum_{r=1}^{n} x_r Y_r \in \mathfrak{h}$. Hence $\mathscr{B}_{\mathfrak{h}} = \{\tilde{X}, \tilde{Y}_1, \ldots, \tilde{Y}_{p_0}\}$. We denote by $M_{\mathfrak{h},\mathscr{B}} \in M_{n+1,p}(\mathbb{R})$ the matrix of $\mathscr{B}_{\mathfrak{h}}$ written in the basis $\mathscr{B}$. Then Lemma 1.1.13 asserts that there exists a basis $\mathscr{B} = \{X, X_1, \ldots, X_n\}$ of $\mathfrak{g}$ such that

$$[X, X_i] = X_{i+1}, \quad i = 1, \ldots, n-1,$$
$$[X_i, X_j] = 0, \quad i, j = 1, \ldots, n$$

and $\mathfrak{h} = \mathbb{R}\text{-span}\{X, X_{n-p+2}, \ldots, X_n\}$. This will be of use in the sequel and permits to have a particular form of the matrix $M_{\mathfrak{h},\mathscr{B}}$.

We also opt for the notation $G_0 = \exp(\mathfrak{g}_0)$ and $Z(G) = \exp(\mathfrak{z}(\mathfrak{g}))$. The following immediate result will be of several uses.

Lemma 7.3.1. *Let Γ be a closed Abelian subgroup of G. Then for any* $\exp T_1, \exp T_2 \in \Gamma$, *we have*

$$[T_1, T_2] \in \mathbb{Z}Y_n.$$

Proof. Using the universal covering $\widetilde{G}$ of G, we can easily check that

$$\exp_{\widetilde{G}} T_1 \exp_{\widetilde{G}} T_2 = \exp_{\widetilde{G}} T_2 \exp_{\widetilde{G}} T_1 \exp_{\widetilde{G}}(sY_n)$$

for some $s \in \mathbb{Z}$. We get therefore that

$$T_1 - e^{\mathrm{ad}_{T_2}(T_1)} = -\left([T_2, T_1] + \cdots + \frac{1}{(n-1)!}[T_2, [\ldots, [T_2, T_1], \ldots]]\right) = sY_n.$$

Fixing a sequence of ideals as in equation (7.10) and suppose that $[T_2, T_1] \in \mathfrak{g}^i \setminus \mathfrak{g}^{i+1}$, we get that $T_1 - e^{\mathrm{ad}_{T_2}(T_1)} \in \mathfrak{g}^i \setminus \mathfrak{g}^{i+1}$. This is enough to conclude. □

We now look at the point to establish a criterion on the triple (G, H, L) such that the action of L on G/H is proper or free. Here, H and L designate closed connected subgroups of G. The main result of this subsection is the following.

Theorem 7.3.2. *Let $G = \exp \mathfrak{g}$ be a reduced threadlike Lie group, $H = \exp \mathfrak{h}$ and $L = \exp \mathfrak{l}$ some closed, connected subgroups of G. We have the following:*
1. *If the action of L on G/H is free, then $\mathrm{Ad}_g\, \mathfrak{h} \cap \mathfrak{l} = \{0\}$ for any $g \in G$.*
2. *Assume that one of the subgroups H or L contains the center of G. Then:*
   (i) *The action of L on G/H is free if and only if $\mathrm{Ad}_g\, \mathfrak{h} \cap \mathfrak{l} = \{0\}$ for any $g \in G$.*
   (ii) *The action of L on G/H is proper if and only if $\mathfrak{l} \cap \mathrm{Ad}_g\, \mathfrak{h} \subseteq \mathfrak{z}(\mathfrak{g})$ for any $g \in G$.*

Proof. (1) Assume that the action of L on G/H is free and suppose that there exist $V \in \mathfrak{l} \setminus \{0\}$, $U \in \mathfrak{h}$ and $g \in G$ such that $V = \mathrm{Ad}_g\, U$. Then

$$\exp V = g(\exp U)g^{-1} = e.$$

This leads to the fact that $U = V = pY_n$ for some $p \in \mathbb{Z}^*$. Then $Z(G) \subset H \cap L$, which contradicts the free action of L on G/H.

(2) Suppose now that $Z(G) \subset L$. To prove assertion (i), it suffices to show that for all $g \in G$, we have $\exp(\mathfrak{l} \cap \mathrm{Ad}_g\, \mathfrak{h}) = L \cap gHg^{-1}$. Let $g \in G$. It is clear that $\exp(\mathfrak{l} \cap \mathrm{Ad}_g\, \mathfrak{h}) \subset L \cap gHg^{-1}$. Let now $a \in L \cap gHg^{-1}$. Then there exist $T \in \mathfrak{g}$, $V \in \mathfrak{l}$ and $U \in \mathfrak{h}$ such that $a = \exp V = \exp T \exp U \exp(-T)$. This implies that $V = \mathrm{Ad}_{\exp T}\, U \bmod(\mathfrak{z}(\mathfrak{g}))$. As $\mathfrak{z}(\mathfrak{g}) \subset \mathfrak{l}$, then $\mathrm{Ad}_{\exp T}\, U \in \mathrm{Ad}_{\exp T}\, \mathfrak{h} \cap \mathfrak{l}$, which implies that $a \in \exp(\mathfrak{l} \cap \mathrm{Ad}_{\exp T}\, \mathfrak{h})$.

The proof of assertion (ii) will be divided through the following lemmas.

Lemma 7.3.3. *Let G be a reduced threadlike Lie group. The surjective projection defined by*

$$\begin{array}{rcl} \pi_G : G & \longrightarrow & G/Z(G) = \overline{G} \\ {}^t(\mathbf{w}, e^{2i\pi c}) & \longmapsto & {}^t\mathbf{w} \end{array} \tag{7.11}$$

is closed and proper.

Proof. Let F be a closed subset of G, then

$$\pi_G^{-1}(\pi_G(F)) = FZ(G).$$

This gives that $\pi_G(F)$ is closed in $\overline{G}$. Let now $\overline{S}$ be a compact set of $\overline{G}$ and ${}^t(\mathbf{u}_j, e^{2i\pi z_j})_{j\in\mathbb{N}}$ a sequence of $\pi_G^{-1}(\overline{S})$. Then $(\mathbf{u}_j)_{j\in\mathbb{N}}$ is a sequence of $\overline{S}$. Therefore, we can extract a subsequence $(\mathbf{u}_{j_s})_{s\in\mathbb{N}}$, which converges to $\mathbf{u} \in \overline{S}$. Moreover, we can extract from ${}^t(\mathbf{0}_n, e^{2i\pi z_{j_s}})_{s\in\mathbb{N}}$ a sequence, which converges to ${}^t(\mathbf{0}_n, e^{2i\pi z})$ and which we denote ${}^t(\mathbf{0}_n, e^{2i\pi z_{j_s}})_{s\in\mathbb{N}}$. The resulting sequence ${}^t(\mathbf{u}_{j_s}, e^{2i\pi z_{j_s}})_{s\in\mathbb{N}}$ converges to ${}^t(\mathbf{u}, e^{2i\pi z}) \in \pi_G^{-1}(\overline{S})$. □

Lemma 7.3.4. *Let G be a reduced threadlike Lie group, H and L some closed, connected subgroups of G. Assume that one of the subgroups H or L contains the center of G, then L acts properly on G/H if and only if $\overline{L}$ acts properly on $\overline{G}/\overline{H}$, where $\overline{L} = \pi_G(L)$ and $\overline{H} = \pi_G(H)$.*

Proof. Assume, for example, that $Z(G) \subset L$. Suppose first that L acts properly on G/H and let $\overline{S}$ be a compact set of $\overline{G}$. Then $\pi_G^{-1}(\overline{S})$ is a compact set of G and then $\pi_G^{-1}(\overline{S})H(\pi_G^{-1}(\overline{S}))^{-1} \cap L$ is compact in G. As $Z(G) \subset L$, we can easily see that $\overline{SHS}^{-1} \cap \overline{L}$ is compact in $\overline{G}$. Conversely, let S be a compact set of G. It is not hard to see that

$$\begin{aligned} SHS^{-1} \cap L \subset SHS^{-1}Z(G) \cap L &= \pi_G^{-1}\big(\pi_G(S)\overline{H}(\pi_G(S))^{-1}\big) \cap \pi_G^{-1}(\overline{L}) \\ &= \pi_G^{-1}\big(\pi_G(S)\overline{H}(\pi_G(S))^{-1} \cap \overline{L}\big), \end{aligned}$$

which is compact in G. Hence $SHS^{-1} \cap L$ is compact. □

To conclude, it suffices to use Proposition 2.3.1. □

7.3.2 Hom(Γ, *G*) for an Abelian discrete subgroup Γ

Let G be the reduced threadlike Lie group and Γ be a discrete Abelian nontrivial subgroup of rank $k \in \mathbb{N}$ of G. Recall the set of all injective group homomorphisms $\operatorname{Hom}_d^0(\Gamma, G)$ from Γ to G, which will be of interest in the next section, merely because it is involved in deformations and it is viewed as a starting means to study the parameter and the deformation spaces.

We will next construct a family of infinite order generators of Γ, which we denote by $\{\gamma_1, \dots, \gamma_k\}$ whenever $k \geq 1$, and γ_{k+1} a finite order (central) element of Γ if Γ is not torsion-free. More precisely, if Γ is torsion-free, then

$$\Gamma = \langle \gamma_1, \dots, \gamma_k \rangle.$$

This notation means that $\gamma \in \Gamma$ if and only if $\gamma = \gamma_1^{n_1} \cdots \gamma_k^{n_k}$ for some $n_1, \dots, n_k \in \mathbb{Z}$. In the case where Γ is not torsion-free, we get

$$\Gamma = \langle \gamma_1, \dots, \gamma_k \rangle \oplus \langle \gamma_{k+1} \rangle.$$

Let us first give in the following two lemmas the explicit the expression of the torsion-free part of Γ.

Lemma 7.3.5. *If $\Gamma \subset G_0$, then there exist a linearly independent family $\{\mathbf{w}_1, \dots, \mathbf{w}_k\}$ of $\mathbb{R}^{n-1}$ and $v_1, \dots, v_k \in \mathbb{R}$ such that for any $j \in \{1, \dots, k\}$,*

$$\gamma_j = {}^t(0, \mathbf{w}_j, e^{2i\pi v_j}).$$

Proof. We consider the surjective projection

$$\begin{aligned} \pi_{G_0} : G_0 \qquad & \longrightarrow \quad G_0/Z(G) \\ {}^t(0, \mathbf{w}, e^{2i\pi v}) & \longmapsto \quad {}^t(0, \mathbf{w}). \end{aligned}$$

Then $\pi_{G_0}(\Gamma)$ is a discrete subgroup of $G_0/Z(G) \simeq \mathbb{R}^{n-1}$. This gives that there exist a unique nonnegative integer k_Γ and a family $\{\mathbf{w}_1, \dots, \mathbf{w}_{k_\Gamma}\}$ of linearly independent vectors of $\mathbb{R}^{n-1}$ such that

$$\pi_{G_0}(\Gamma) = \left\{ \begin{pmatrix} 0 \\ {}^t\mathbf{w}_1 \end{pmatrix}^{n_1} \cdots \begin{pmatrix} 0 \\ {}^t\mathbf{w}_{k_\Gamma} \end{pmatrix}^{n_{k_\Gamma}} : n_1, \dots, n_{k_\Gamma} \in \mathbb{Z} \right\}.$$

The integer k_Γ is nothing but the rank k of Γ. Then, for all $j \in \{1, \dots, k\}$, there exists $v_j \in \mathbb{R}$ such that $\gamma_j = {}^t(0, \mathbf{w}_j, e^{2i\pi v_j}) \in \Gamma$. Hence we have that

$$\Gamma = \{\gamma_1^{n_1} \cdots \gamma_k^{n_k} : n_1, \dots, n_k \in \mathbb{Z}\} \cdot (\Gamma \cap Z(G)).$$

□

Lemma 7.3.6. *If $\Gamma \not\subset G_0$, then $k \leq 2$ and we can take γ_1 as*

$$\gamma_1 = {}^t(a, \mathbf{w}, e^{2i\pi v}),$$

where $\mathbf{w} \in \mathbb{R}^{n-1}$, $a \in \mathbb{R}^*$ and $v \in \mathbb{R}$. Moreover, if $k = 2$, we can take γ_2 as

$$\gamma_2 = {}^t(\mathbf{0}_{n-1}, \alpha, e^{2i\pi v'}),$$

with $\mathbf{0}_{n-1} = (0, \dots, 0) \in \mathbb{R}^{n-1}$, $\alpha \in \mathbb{R}^*$ and $v' \in \mathbb{R}$.

Proof. We consider the surjective projection defined in equation (7.11). First of all, it is clear that $\overline{G}$ is a connected and simply connected threadlike Lie group and its associated Lie algebra is given by $\overline{\mathfrak{g}} := \mathfrak{g}/\mathfrak{z}(\mathfrak{g})$. We denote by $\rho_{\mathfrak{g}} : \mathfrak{g} \to \mathfrak{g}/\mathfrak{z}(\mathfrak{g})$ the canonical surjection, then clearly $\exp_{\overline{G}} \circ \rho_{\mathfrak{g}} = \pi_G \circ \exp_G$. On the other hand, we denote by $\overline{\Gamma} = \pi_G(\Gamma)$ which is a discrete Abelian subgroup of $\overline{G}$ and $\overline{L} = \exp_{\overline{G}} \overline{\mathfrak{l}}$ its corresponding syndetic hull. Since $\Gamma \not\subset G_0$, there exists $T \in \mathfrak{g} \setminus \mathfrak{g}_0$ such that $\exp T \in \Gamma$. We get that $\overline{T} := \rho_{\mathfrak{g}}(T) \in \overline{\mathfrak{l}}$. As $\mathfrak{l}$ is Abelian, there exists $\alpha \in \mathbb{R}$ such that

$$\overline{\mathfrak{l}} = \mathbb{R}\overline{T} \oplus \mathbb{R}(\alpha \overline{Y}_{n-1}).$$

We get therefore that

$$\overline{\Gamma} = \exp_{\overline{G}}(\mathbb{Z}\overline{T})\exp_{\overline{G}}(\mathbb{Z}(\alpha\overline{Y}_{n-1}))$$

and then that

$$\Gamma = \exp(\mathbb{Z}T)\exp(\mathbb{Z}(\alpha Y_{n-1} + \nu Y_n)) \cdot (\Gamma \cap Z(G))$$

for some $\nu \in \mathbb{R}$. We obtain $k_\Gamma = \dim \overline{\Gamma}$, which is unique according to Theorem 1.4.12. □

Now using the fact that $\Gamma \cap Z(G)$ is a finite central subgroup of G, we get immediately the following.

Lemma 7.3.7. *If Γ is not torsion-free, then*

$$\gamma_{k+1} = {}^t(\mathbf{0}_n, e^{\frac{2i\pi}{q}}),$$

where $q \in \mathbb{N}^*$ is the cardinality of $\Gamma \cap Z(G)$.

Suppose from now on that Γ is an Abelian discrete subgroup of rank k and having a family of generators $\gamma_1, \dots, \gamma_l$ defined as in Lemmas 7.3.5, 7.3.6 and 7.3.7. Here,

$$l = \begin{cases} k & \text{if } \Gamma \text{ is torsion-free,} \\ k+1 & \text{otherwise.} \end{cases}$$

Now, we are ready to give an explicit description of $\mathrm{Hom}(\Gamma, G)$ and $\mathrm{Hom}_d^0(\Gamma, G)$. Recall first that $\mathrm{Hom}(\Gamma, G)$ is endowed with the pointwise convergence topology. The same topology is obtained by using the injective map

$$\mathrm{Hom}(\Gamma, G) \to G \times \cdots \times G, \quad \varphi \mapsto (\varphi(\gamma_1), \dots, \varphi(\gamma_l))$$

to equip $\mathrm{Hom}(\Gamma, G)$ with the relative topology induced from the direct product $G^l = G \times \cdots \times G$. Having in mind the identification of any element $\exp(xX + \sum_{i=1}^n y_i Y_i) \in G$ by the column vector ${}^t(x, y_1, \dots, y_{n-1}, e^{2i\pi y_n})$, we identify the product set G^l by the set of matrices in $M_{n+1,l}(\mathbb{C})$ defined by

$$\mathscr{M}_{n+1,l} = \left\{ \begin{pmatrix} C \\ e^{2i\pi \mathbf{z}} \end{pmatrix} \in M_{n+1,l}(\mathbb{C}) : C \in M_{n,l}(\mathbb{R}), \mathbf{z} \in \mathbb{R}^l \right\}$$

for $\mathbf{z} = (z_1, \dots, z_l)$, $e^{2i\pi \mathbf{z}} := (e^{2i\pi z_1}, \dots, e^{2i\pi z_l}) \in \mathbb{T}^l$. Then it appears clear that the following map:

$$\begin{aligned} \Psi : \mathrm{Hom}(\Gamma, G) &\to \mathscr{M}_{n+1,l} \\ \varphi &\mapsto (\varphi(\gamma_1) \quad \cdots \quad \varphi(\gamma_l)) \end{aligned} \tag{7.12}$$

is a homeomorphism on its image. Our task is reduced to give the description of $\Psi(\mathrm{Hom}(\Gamma, G))$ and $\Psi(\mathrm{Hom}_0^d(\Gamma, G))$, which will be denoted respectively by $\mathscr{H}$ and $\mathscr{K}$. Toward such a purpose, we consider the following discussion.

The torsion-free case

Throughout the present subsection, Γ denotes a torsion-free Abelian discrete subgroup of G. A preliminary algebraic interpretation of $\mathscr{H}$ is given by the following:

$$\mathscr{H} = \left\{ \begin{pmatrix} C \\ e^{2i\pi \mathbf{z}} \end{pmatrix} \in \mathscr{M}_{n+1,k} \;\middle|\; \begin{array}{l} C = (C_1 \quad \cdots \quad C_k), \\ \begin{pmatrix} C_u \\ e^{2i\pi z_u} \end{pmatrix}\begin{pmatrix} C_v \\ e^{2i\pi z_v} \end{pmatrix} = \begin{pmatrix} C_v \\ e^{2i\pi z_v} \end{pmatrix}\begin{pmatrix} C_u \\ e^{2i\pi z_u} \end{pmatrix}, 1 \le u, v \le k \end{array} \right\}.$$

Indeed, for $\varphi \in \mathrm{Hom}(\Gamma, G)$ we have

$$\varphi(\gamma_u)\varphi(\gamma_v) = \varphi(\gamma_v)\varphi(\gamma_u), \quad 1 \le u, v \le k.$$

Thus, we get $\Psi(\mathrm{Hom}(\Gamma, G)) \subset \mathscr{H}$. Let conversely

$$M = \begin{pmatrix} C_1 & \cdots & C_k \\ e^{2i\pi z_1} & \cdots & e^{2i\pi z_k} \end{pmatrix} \in \mathscr{H},$$

we can define a group homomorphism: $\varphi : \Gamma \to G$ satisfying

$$\varphi(\gamma_j) = \begin{pmatrix} C_j \\ e^{2i\pi z_j} \end{pmatrix}.$$

The next step consists in giving an explicit description of $\mathscr{H}$. Toward such a purpose, we define the sets $H_{j,k}, j \in \{0, \dots, k\}$ as follows:

$$H_{0,k} = \left\{ \begin{pmatrix} \mathbf{0}_k \\ N \\ e^{2i\pi \mathbf{z}} \end{pmatrix} \in \mathscr{M}_{n+1,k} : N \in M_{n-1,k}(\mathbb{R}), \mathbf{z} \in \mathbb{R}^k \right\}$$

and for any $j \in \{1, \dots, k\}$,

$$H_{j,k} = \left\{ \begin{pmatrix} \lambda_1{}^t\mathbf{c} & \cdots & \lambda_k{}^t\mathbf{c} \\ y_1 & \cdots & y_k \\ e^{2i\pi z_1} & \cdots & e^{2i\pi z_k} \end{pmatrix} \in \mathscr{M}_{n+1,k} \;\middle|\; \begin{array}{l} \lambda_j = 1, \mathbf{c} \in \mathbb{R}^* \times \mathbb{R}^{n-2}, \\ \lambda_s, y_s, z_s \in \mathbb{R}, s \in \{1, \dots, k\}, \\ (\lambda_u y_v - \lambda_v y_u)\rho_1(\mathbf{c}) \in \mathbb{Z}, u, v \in \{1, \dots, k\} \end{array} \right\}.$$

Here, ρ_j designates the projection map from $\mathbb{R}^{n-1}$ on $\mathbb{R}$ given by

$$\rho_j(c_1, \dots, c_{n-1}) = c_j, \quad 1 \le j \le n-1.$$

Proposition 7.3.8. *With the same notation and hypotheses, we have*

$$\mathscr{H} = \bigcup_{j=0}^{k} H_{j,k}.$$

Proof. First, it is clear that $\bigcup_{j=0}^{k} H_{j,k} \subset \mathscr{H}$. Conversely, let

$$M = \begin{pmatrix} C_1 \cdots C_k \\ e^{2i\pi \mathbb{Z}} \end{pmatrix} \in \mathscr{H}.$$

Let for all $1 \leq j \leq k$, ${}^tC_j = (c_j, c_{1,j}, \ldots, c_{n-1,j})$. Then for any $u, v \in \{1, \ldots, k\}$, we get by Lemma 7.3.1,

$$\left[c_u X + \sum_{j=1}^{n-1} c_{j,u} Y_j, c_v X + \sum_{j=1}^{n-1} c_{j,v} Y_j\right] \in \mathbb{Z} Y_n,$$

which gives rise to the following equations:

$$\begin{cases} c_u c_{j,v} - c_v c_{j,u} = 0, & u, v \in \{1, \ldots, k\},\ j \in \{1, \ldots, n-2\}, \\ c_u c_{n-1,v} - c_v c_{n-1,u} \in \mathbb{Z}, & u, v \in \{1, \ldots, k\}. \end{cases} \tag{7.13}$$

If $(c_1, \ldots, c_k) = \mathbf{0}_k$, then $M \in H_{0,k}$. Otherwise, there exists $j_0 \in \{1, \ldots, k\}$ such that $c_{j_0} \neq 0$. Then by equations (7.13), we have that $M \in H_{j_0,k}$. □

Next, we give the description of $\mathscr{K}$ according to the rank of Γ. We denote for all $j \in \{0, \ldots, k\}$ by $K_{j,k} := \Psi(\mathrm{Hom}_d^0(\Gamma, G)) \cap H_{j,k}$. Then

$$\mathscr{K} = \bigcup_{j=0}^{k} K_{j,k}.$$

Our task is then reduced to the determination of those layers according to the value of k.

Case 1: $k \geq 3$

Proposition 7.3.9. *Assume that Γ is a torsion-free Abelian discrete subgroup of G of rank $k \geq 3$. Under the above notation, we have*

$$\mathscr{K} = K_{0,k} = \left\{ \begin{pmatrix} \mathbf{0}_k \\ N \\ e^{2i\pi \mathbb{Z}} \end{pmatrix} \in H_{0,k} : N \in M_{n-1,k}^0(\mathbb{R}) \right\},$$

with $M_{n,m}^0(\mathbb{R})$ denotes the set of all (n, m) matrices of maximal rank.

Proof. Let $\{\mathbf{u}_1, \dots, \mathbf{u}_k\}$ be a linearly independent family of $\mathbb{R}^n$ and $v_1, \dots, v_k \in \mathbb{R}$ such that $\gamma_j = {}^t(\mathbf{u}_j, e^{2i\pi v_j})$. Let

$$\begin{pmatrix} \mathbf{0}_k \\ N \\ e^{2i\pi \mathbf{z}} \end{pmatrix} = \Psi(\varphi) \in K_{0,k}$$

such that $N = (N_1 \quad \cdots \quad N_k)$. Suppose that $\operatorname{rk} N < k$, as $\varphi(\Gamma)$ is discrete, there exist a set of distinct integers $J = \{j_0, j_1, \dots, j_{k'}\} \subsetneq \{1, \dots, k\}$ and $\eta_1, \dots, \eta_{k'} \in \mathbb{Q}^*$ such that $N_{j_0} = \sum_{s=1}^{k'} \eta_s N_{j_s}$. We denote for $s \in \{1, \dots, k'\}$,

$$\eta_s = \frac{\alpha_s}{\beta_s}, \quad b = \prod_{s=1}^{k'} \beta_s, \quad b_s = \frac{b}{\beta_s},$$

and

$$\gamma = \gamma_{j_0}^{-b} \gamma_{j_1}^{b_1 \alpha_1} \cdots \gamma_{j_{k'}}^{b_{k'} \alpha_{k'}}.$$

As $\operatorname{rk}(\mathbf{u}_{j_0} \quad \cdots \quad \mathbf{u}_{j_{k'}}) = k' + 1$, we have $\gamma \neq e$. Moreover, it is not hard to see that

$$\varphi(\gamma) = {}^t\left(\mathbf{0}_n, e^{2i\pi b(\eta_1 z_{j_1} + \dots + \eta_s z_{j_s} - z_{j_0})}\right) \in \varphi(\Gamma) \cap Z(G).$$

This gives that

$$\eta_1 z_{j_1} + \dots + \eta_s z_{j_s} - z_{j_0} \in \mathbb{Q}.$$

Let $(u, v) \in \mathbb{Z} \times \mathbb{Z}^*$ such that

$$\eta_1 z_{j_1} + \dots + \eta_s z_{j_s} - z_{j_0} = \frac{u}{v},$$

we have $\varphi(\gamma^v) = e$, which contradicts the fact that φ is injective. Suppose now that there exists $j \in \{1, \dots, k\}$ such that $K_{j,k} \neq \emptyset$ and let

$$\begin{pmatrix} \lambda_1 {}^t\mathbf{c} & \cdots & \lambda_k {}^t\mathbf{c} \\ y_1 & \cdots & y_k \\ e^{2i\pi z_1} & \cdots & e^{2i\pi z_k} \end{pmatrix} = \Psi(\varphi) \in K_{j,k}.$$

Suppose that there exist s such that $y_s - \lambda_s y_j = 0$. If $\lambda_s \notin \mathbb{Q}$, then there exists $a \in \mathbb{R} \setminus (\mathbb{Z} + \lambda_s \mathbb{Z})$ and a sequence $(m_u + \lambda_s m'_u)_{u \in \mathbb{N}}$ in $\mathbb{Z} + \lambda_s \mathbb{Z}$, which converges to a. Thus, we can extract from $(\varphi(\gamma_j^{m_u} \gamma_s^{m'_u}))_{u \in \mathbb{N}}$ a convergent (hence a stationary as $\varphi(\Gamma)$ is discrete) subsequence. This shows that $a = m_{u_0} + \lambda_s m'_{u_0}$, for some $u_0 \in \mathbb{N}$, which contradicts the fact that $a \notin \mathbb{Z} + \lambda_s \mathbb{Z}$. It follows therefore that $\lambda_s \in \mathbb{Q}$. Let $(\alpha, \beta) \in \mathbb{Z} \times \mathbb{Z}^*$ be such that $\lambda_s = \frac{\alpha}{\beta}$. By an easy computation, we get

$$\varphi(y_j^{\alpha} y_s^{\beta}) = {}^t(\mathbf{0}_n, e^{2i\pi\beta(\lambda z_j + z_s)}) \in \varphi(\Gamma) \cap Z(G)$$

and conclusively $\lambda_s z_j - z_s \in \mathbb{Q}$. Let $(\alpha', \beta') \in \mathbb{Z} \times \mathbb{Z}^*$ such that $\lambda_s z_j - z_s = \frac{\alpha'}{\beta'}$, then for $m_1 = -\alpha\beta'$ and $m_2 = \beta\beta' \neq 0$, we have

$$e \neq y_j^{m_1} y_s^{m_2} \in \operatorname{Ker} \varphi,$$

which contradicts the injectivity of φ. This gives that $\lambda_s y_j - y_s \neq 0, s \in \{1, \dots, k\} \setminus \{j\}$. We denote by $p_s = (y_s - \lambda_s y_j)\rho_1(\mathbf{c}) \in \mathbb{Z}^*$, then we get

$$(\lambda_s y_{s'} - \lambda_{s'} y_s)\rho_1(\mathbf{c}) = \lambda_s p_{s'} - \lambda_{s'} p_s \in \mathbb{Z}^*.$$

A routine computation shows that for s, s′ $\in \{1, \dots, k\} \setminus \{j\}$ such that $s \neq s'$, we have

$$\varphi(y_{s'}^{p_s} y_s^{-p_{s'}} y_j^{(\lambda_s p_{s'} - \lambda_{s'} p_s)}) \in \varphi(\Gamma) \cap Z(G)$$

this gives rise to

$$p_s z_{s'} - p_{s'} z_s + (\lambda_s p_{s'} - \lambda_{s'} p_s) z_j = \frac{\alpha}{\beta}$$

for some $(\alpha, \beta) \in \mathbb{Z} \times \mathbb{Z}^*$. Hence

$$(y_{s'}^{p_s} y_s^{-p_{s'}} y_j^{(\lambda_s p_{s'} - \lambda_{s'} p_s)})^{\beta} \in \operatorname{Ker} \varphi.$$

Thus, we have $K_{j,k} = \emptyset$ for all $j \in \{1, \dots, k\}$. □

Case 2: $k = 2$
We argue similarly as in the previous proposition to prove the following result.

Proposition 7.3.10. *Let G be a reduced threadlike Lie group and Γ a torsion-free Abelian discrete subgroup of G of rank 2. The sets $K_{j,2}, j = 0, 1, 2$, are given by*

$$K_{0,2} = \left\{ \begin{pmatrix} \mathbf{0}_2 \\ N \\ e^{2i\pi \mathbf{z}} \end{pmatrix} \in H_{0,2} : N \in M_{n-1,2}^0(\mathbb{R}) \right\}$$

and for all $j \in \{1, 2\}$,

$$K_{j,2} = \left\{ \begin{pmatrix} \lambda_1 {}^t\mathbf{c} & \lambda_2 {}^t\mathbf{c} \\ y_1 & y_2 \\ e^{2i\pi z_1} & e^{2i\pi z_2} \end{pmatrix} \in H_{j,2} : (\lambda_1 y_2 - \lambda_2 y_1) \in \mathbb{R}^* \right\}.$$

Case 3: $k = 1$

The description of $K_{j,1}$ is similar to the previous case. Using the same analysis as in Proposition 7.3.9, we obtain the following result.

Proposition 7.3.11. *We keep the same notation. The sets $K_{j,1}, j = 0, 1$, are given by*

$$K_{0,1} = \left\{ \begin{pmatrix} 0 \\ {}^t\mathbf{n} \\ e^{2i\pi z} \end{pmatrix} \in H_{0,1} : \mathbf{n} \in (\mathbb{R}^{n-1})^* \right\}$$

and

$$K_{1,1} = H_{1,1}.$$

The torsion case

Suppose that Γ is an Abelian discrete subgroup of G, which admits a finite order element. Using the injective map Ψ defined in equation (7.12), the representative matrix of any element of $\mathrm{Hom}(\Gamma, G)$ admits an additional column, which represents the finite order element. It appears immediate that when restricted to this case, the same analysis as in Subsection 7.3.2 takes place to describe the sets $\mathrm{Hom}(\Gamma, G)$ just by adding the aforementioned column. In order to avoid the redundance, let us define the sets $\widetilde{H}_{j,k}, j = 0, \dots, k$, as follows:

$$\widetilde{H}_{0,k} = \left\{ \begin{pmatrix} \mathbf{0}_k & 0 \\ N & {}^t\mathbf{0}_{n-1} \\ e^{2i\pi \mathbf{z}} & e^{\frac{2i\pi r}{q}} \end{pmatrix} \in \mathcal{M}_{n+1,k+1} : N \in M_{n-1,k}(\mathbb{R}), \mathbf{z} \in \mathbb{R}^k, r \in \mathbb{Z} \right\}$$

and for any $j \in \{1, \dots, k\}$,

$$\widetilde{H}_{j,k} = \left\{ \begin{pmatrix} \lambda_1 {}^t\mathbf{c} & \cdots & \lambda_k {}^t\mathbf{c} & {}^t\mathbf{0}_{n-1} \\ y_1 & \cdots & y_k & 0 \\ e^{2i\pi z_1} & \cdots & e^{2i\pi z_k} & e^{\frac{2i\pi r}{q}} \end{pmatrix} \in \mathcal{M}_{n+1,k+1} \;\middle|\; \begin{array}{l} \lambda_j = 1, \mathbf{c} \in \mathbb{R}^* \times \mathbb{R}^{n-2}, \\ \lambda_s, y_s, z_s \in \mathbb{R}, r \in \mathbb{Z}, \\ (\lambda_u y_v - \lambda_v y_u)\rho_1(\mathbf{c}) \in \mathbb{Z}, \\ s, u, v \in \{1, \dots, k\} \end{array} \right\}.$$

The following proposition stems immediately from Proposition 7.3.8, which describes the structure of $\mathrm{Hom}(\Gamma, G)$. Indeed, the last added vector of $\Psi(\varphi)$, $\varphi \in \mathrm{Hom}(\Gamma, G)$, takes the following form:

$$\varphi(\gamma_{k+1}) = \begin{pmatrix} {}^t\mathbf{0}_n \\ e^{\frac{2i\pi r}{q}} \end{pmatrix} = \gamma_{k+1}^r \in Z(G),$$

for some $r \in \mathbb{Z}$, being the image by a group homomorphism of the finite order element γ_{k+1} of $\Gamma \cap Z(G)$.

Proposition 7.3.12. *Let G be a reduced threadlike Lie group and Γ an Abelian discrete subgroup of G of rank k with a finite order element. The set* $\mathrm{Hom}(\Gamma, G)$ *is homeomorphic to* $\bigcup_{j=0}^{k} \widetilde{H}_{j,k}$.

Take now $\varphi \in \mathrm{Hom}_d^0(\Gamma, G)$, we get that $q \wedge r = 1$ where the symbol $\wedge$ means the greatest common divisor. It appears clear that when $k = 0$, Γ is a central and cyclic subgroup and then $\mathrm{Hom}(\Gamma, G)$ is homeomorphic to Γ. Furthermore,

$$\mathrm{Hom}_d^0(\Gamma, G) = \mathrm{Aut}(\Gamma), \tag{7.14}$$

where the last means the automorphism group of Γ, which is a finite group. Suppose now that $k > 0$. First, it is easy to see that the description of $\mathscr{K}$ is similar to the torsion-free case, using the same analysis as in previous subsection. The analogues of the objects $K_{j,k}$, $j \in \{0, \ldots, k\}$, given by $\widetilde{K}_{j,k} := \Psi(\mathrm{Hom}_d^0(\Gamma, G)) \cap \widetilde{H}_{j,k}$ are as follows. For $j = 0$, we have

$$\widetilde{K}_{0,k} = \left\{ \begin{pmatrix} \mathbf{0}_k & 0 \\ N & {}^t\mathbf{0}_{n-1} \\ e^{2i\pi \mathbf{z}} & e^{\frac{2i\pi r}{q}} \end{pmatrix} \in \widetilde{H}_{0,k} : N \in M_{n-1,k}^0(\mathbb{R}), 0 < r < q,\ r \wedge q = 1 \right\}.$$

Let now $j \in \{1, \ldots, k\}$. If $k \geq 3$, then

$$\widetilde{K}_{j,k} = \emptyset.$$

Otherwise, we get the following discussion. When $k = 2$, we have for $j \in \{1, 2\}$:

$$\widetilde{K}_{j,2} = \left\{ \begin{pmatrix} \lambda_1 {}^t\mathbf{c} & \lambda_2 {}^t\mathbf{c} & {}^t\mathbf{0}_{n-1} \\ y_1 & y_2 & 0 \\ e^{2i\pi z_1} & e^{2i\pi z_2} & e^{\frac{2i\pi r}{q}} \end{pmatrix} \in \widetilde{H}_{j,2} : (\lambda_1 y_2 - \lambda_2 y_1) \in \mathbb{R}^*,\ 0 < r < q,\ r \wedge q = 1 \right\}.$$

Finally, when $k = 1$, we have

$$\widetilde{K}_{1,1} = \left\{ \begin{pmatrix} {}^t\mathbf{c} & {}^t\mathbf{0}_n \\ e^{2i\pi z} & e^{\frac{2i\pi r}{q}} \end{pmatrix} \in \widetilde{H}_{1,1} : 0 < r < q,\ r \wedge q = 1 \right\}.$$

Now we can state the following proposition, which will be used later.

Proposition 7.3.13. *Let G be a reduced threadlike Lie group and Γ a discrete Abelian subgroup of G. The set* $\mathrm{Hom}_d^0(\Gamma, G)$ *is open in* $\mathrm{Hom}(\Gamma, G)$.

Proof. We only treat the case where Γ is torsion-free, the other case is handled similarly. Whenever $k \leq 2$, the result is immediate. In fact, it suffices to see that the set

$$\mathscr{M}^0 = \left\{ \begin{pmatrix} C \\ e^{2i\pi \mathbf{z}} \end{pmatrix} \in \mathscr{M}_{n+1,k} : C \in M_{n,k}^0(\mathbb{R}) \right\}$$

is open in

$$\mathscr{M} = \left\{ \begin{pmatrix} C \\ e^{2i\pi\mathbf{z}} \end{pmatrix} \in \mathscr{M}_{n+1,k} : C \in M_{n,k}(\mathbb{R}) \right\}$$

and, therefore, $\mathscr{K} = \mathscr{H} \cap \mathscr{M}^0$ is open in $\mathscr{H} \cap \mathscr{M} = \mathscr{H}$. Thus, we only have to treat the case when $k > 2$. We prove that $\mathscr{H} \setminus \mathscr{K}$ is closed in $\mathscr{H}$. Let $M \in \overline{\mathscr{H} \setminus \mathscr{K}}^{\mathscr{H}}$, there exists therefore a sequence $(M_s)_{s\in\mathbb{N}}$ belongs to $\mathscr{H} \setminus \mathscr{K}$ and converges to M. Then we can extract from $(M_s)_{s\in\mathbb{N}}$ a subsequence $(M_{\sigma(s)})_{s\in\mathbb{N}}$ of elements in $H_{j,k} \setminus K_{j,k}$ for some $j \in \{0, \dots, k\}$. Suppose first that $j = 0$, writing

$$M_{\sigma(s)} = \begin{pmatrix} \mathbf{0}_k \\ N_s \\ e^{2i\pi\mathbf{z}_s} \end{pmatrix},$$

we get that

$$M = \begin{pmatrix} \mathbf{0}_k \\ N \\ e^{2i\pi\mathbf{z}} \end{pmatrix} \in H_{0,k} \setminus K_{0,k}$$

as $\operatorname{rk}(N) \leq \operatorname{rk} N_s < k$. Suppose now that $j \in \{1, \dots, k\}$. We get therefore

$$M_{\sigma(s)} = \begin{pmatrix} \lambda_{1,s}{}^t\mathbf{c}_s & \cdots & \lambda_{k,s}{}^t\mathbf{c}_s \\ y_{1,s} & \cdots & y_{k,s} \\ e^{2i\pi z_{1,s}} & \cdots & e^{2i\pi z_{k,s}} \end{pmatrix}.$$

If there exists $j_0 \in \{1, \dots, k\}$ such that $\lim_{s\to+\infty} \rho_1(\lambda_{j_0,s}\mathbf{c}_s)$ is not zero, then $M \in H_{j_0,k}$. Otherwise, $M \in H_{0,k}$. Let then $N = (N_1 \quad \cdots \quad N_k) \in M_{n-1,k}(\mathbb{R})$ and $\mathbf{z} \in \mathbb{R}^k$ such that

$$M = \begin{pmatrix} \mathbf{0}_k \\ N \\ e^{2i\pi\mathbf{z}} \end{pmatrix}.$$

If there exists $j_0 \in \{1, \dots, k\}$ such that $\lambda_{j_0,s}$ goes to infinity as s goes to $+\infty$, then we can easily check that $\lim_{s\to+\infty} \mathbf{c}_s = \mathbf{0}_{n-1}$. Moreover, we can easily check that the sequence $\rho_1(\mathbf{c}_s)(\lambda_{j_0,s}y_{j,s} - y_{j_0,s})$ of $\mathbb{Z}$ is stationary and equals to zero. We get therefore that $\lim_{s\to+\infty} y_{j,s} = 0$ and then $N_j = {}^t\mathbf{0}_{n-1}$, which is enough to conclude. Suppose finally that for all $j' \in \{1, \dots, k\}$, $\lambda_{j',s}$ goes to $\lambda_{j'} \in \mathbb{R}$ as s goes to $+\infty$. Thus, $\lambda_{j',s}y_{j,s} = y_{j',s}$. It follows therefore that $\operatorname{rk}(N) \leq 1$. □

7.3.3 Parameter and deformation spaces

Let $H = \exp\mathfrak{h}$ be a closed connected subgroup of G and Γ an Abelian discontinuous subgroup for G/H. As shows the title, the purpose of this section is to give an explicit description of the parameter and the deformation spaces of the action of Γ on G/H. We start by the following result.

Proposition 7.3.14. *Let G be a reduced threadlike Lie group, $H = \exp\mathfrak{h}$ a closed, connected group of G and Γ an Abelian discontinuous subgroup for G/H. We have*

$$\mathscr{R}(\Gamma, G, H) = \{\varphi \in \operatorname{Hom}_d^0(\Gamma, G) : \mathfrak{l}_\varphi \cap \operatorname{Ad}_g \mathfrak{h} \subseteq \mathfrak{z}(\mathfrak{g}) \textit{ for any } g \in G\},$$

where $\mathfrak{l}_\varphi$ is the Lie subalgebra associated to the synthetic hull of $\varphi(\Gamma)$.

Proof. We first show that the proper action of $\varphi(\Gamma)$ on G/H implies its free action. It is clear that the proper action implies that the triplet $(G, H, \varphi(\Gamma))$ is (CI), which gives that for all $g \in G$, the subgroup $K := \varphi(\Gamma) \cap gHg^{-1}$ is central and then finite as $\varphi(\Gamma)$ is discrete. As the map $\varphi : \Gamma \to \varphi(\Gamma)$ is a group isomorphism and K is finite and cyclic, we get that $\varphi^{-1}(K) = K$. Therefore, $K \subset \Gamma \cap gHg^{-1} = \{e\}$, for all $g \in G$. Thus, the action of $\varphi(\Gamma)$ on G/H is free. As $L_\varphi = \exp\mathfrak{l}_\varphi$ contains $\varphi(\Gamma)$ cocompactly, we get that

$$\mathscr{R}(\Gamma, G, H) = \{\varphi \in \operatorname{Hom}_d^0(\Gamma, G) : L_\varphi \text{ acts properly on } G/H\}.$$

Now, Theorem 7.3.2 allows us to conclude. □

Thanks to Theorem 7.3.2, one sees that if Γ acts properly on G/H then $p \leq n - k + \dim(\mathfrak{h} \cap \mathfrak{z}(\mathfrak{g}))$. We assume therefore that this previous condition is satisfied. Moreover, we fix from now on a basis $\mathscr{B} = \{X, Y_1, \dots, Y_n\}$ of $\mathfrak{g}$ with nontrivial Lie brackets defined in (1.7) and supposed to be adapted to $\mathfrak{h}$ when $\mathfrak{h} \not\subset \mathfrak{g}_0$ as in Lemma 1.1.13. Consider the following notation:

$$M_{\mathfrak{h},\mathscr{B}} = \begin{pmatrix} M^1_{\mathfrak{h},\mathscr{B}} \\ \mathbf{z}_\mathfrak{h} \end{pmatrix}, \quad \mathbf{z}_\mathfrak{h} \in \mathbb{R}^p$$

and

$$M^1_{\mathfrak{h},\mathscr{B}} = \begin{pmatrix} \mathbf{x}_\mathfrak{h} \\ M^0_{\mathfrak{h},\mathscr{B}} \end{pmatrix}, \quad \mathbf{x}_\mathfrak{h} \in \mathbb{R}^p.$$

Now, we consider the action of G on $\mathscr{M}_{n+1,l}$ defined as follows. For $g \in G, {}^tC_j \in \mathbb{R}^n, z_j \in \mathbb{R}, j \in \{1, \dots, l\}$:

$$g \star \begin{pmatrix} C_1 & \cdots & C_l \\ e^{2i\pi z_1} & \cdots & e^{2i\pi z_l} \end{pmatrix} = \left(g \begin{pmatrix} C_1 \\ e^{2i\pi z_1} \end{pmatrix} g^{-1} \quad \cdots \quad g \begin{pmatrix} C_l \\ e^{2i\pi z_l} \end{pmatrix} g^{-1} \right). \tag{7.15}$$

Lemma 7.3.15. *The map Ψ defined in (7.12) is G-equivariant. That is, for any $\varphi \in \operatorname{Hom}(\Gamma, G)$ and $g \in G$, we have $\Psi(g \cdot \varphi) = g \star \Psi(\varphi)$.*

Proof. It is sufficient to see that for any $\varphi \in \operatorname{Hom}(\Gamma, G)$ and $g \in G$:

$$\begin{aligned} g \star \Psi(\varphi) &= (g\varphi(\gamma_1)g^{-1}, \ldots, g\varphi(\gamma_l)g^{-1}) \\ &= (g \cdot \varphi(\gamma_1), \ldots, g \cdot \varphi(\gamma_l)) \\ &= \Psi(g \cdot \varphi). \end{aligned}$$

□

We consider then the orbit space

$$\mathscr{T} = \mathscr{R}/G.$$

In light of Proposition 7.3.14 and Lemma 7.3.15, the following result is immediate.

Proposition 7.3.16. *Let G be a reduced threadlike Lie group, $H = \exp \mathfrak{h}$ a closed, connected group of G and Γ an Abelian discontinuous subgroup for G/H. The parameter space is homeomorphic to*

$$\mathscr{R} = \left\{ \begin{pmatrix} C \\ e^{2i\pi \mathbf{z}} \end{pmatrix} \in \mathscr{K} : \operatorname{rk}(C ⋒ M^1_{\mathrm{Ad}_g \mathfrak{h}, \mathscr{B}}) = k + p - \dim(\mathfrak{h} \cap \mathfrak{z}(\mathfrak{g})) \text{ for any } g \in G \right\},$$

where the symbol ⋒ merely means the concatenation of matrices. Moreover, the deformation space $\mathscr{T}(\Gamma, G, H)$ is homeomorphic to $\mathscr{T}$.

In the sequel, we fix some notation and we define the necessary ingredients, which will be used in the description of $\mathscr{T}$. First, we give the explicit expression of the action of G on $\mathscr{M}_{n+1,l}$. Let then $g = {}^t(x, x_1, \ldots, x_{n-1}, e^{2i\pi x_n}) \in G$, $C \in M_{n,l}(\mathbb{R})$ and $\mathbf{z} \in \mathbb{R}^l$, we have

$$g \star \begin{pmatrix} C \\ e^{2i\pi \mathbf{z}} \end{pmatrix} = \begin{pmatrix} B_n(g)C \\ e^{2i\pi(\mathbf{z} + P_n(g)C)} \end{pmatrix}, \tag{7.16}$$

where $P_n(g) = (P_{n,1}(g) \cdots P_{n,n}(g))$ satisfies

$$\begin{cases} P_{n,1}(g) = -\sum_{s=1}^{n-1} \frac{x^{s-1}}{s!} x_{n-s}, \\ P_{n,j}(g) = \frac{x^{n-j+1}}{(n-j+1)!}, \quad j = 2, \ldots, n \end{cases} \tag{7.17}$$

and

$$B_n(g) = \begin{pmatrix} 1 & 0 \\ Q_n(g) & A_{n-1}(x) \end{pmatrix} \in M_{n,n}(\mathbb{R}). \tag{7.18}$$

Here, $Q_n(g) = {}^t(Q_{1,n}(g) \cdots Q_{n-1,n}(g)) \in M_{n-1,1}(\mathbb{R})$ is such that

$$\begin{cases} Q_{1,n}(g) = 0, \\ Q_{j,n}(g) = -\sum_{s=1}^{j-1} \frac{x^{s-1}}{s!} x_{j-s}, \quad j = 2, \ldots, n-1 \end{cases} \tag{7.19}$$

and

$$A_{n-1}(x) = \begin{pmatrix} 1 & 0 & \cdots & \cdots & 0 \\ x & 1 & \ddots & & \vdots \\ \frac{x^2}{2} & x & 1 & \ddots & \vdots \\ \vdots & \ddots & \ddots & \ddots & 0 \\ \frac{x^{n-2}}{(n-2)!} & \cdots & \frac{x^2}{2} & x & 1 \end{pmatrix} \in M_{n-1,n-1}(\mathbb{R}). \tag{7.20}$$

Now, we define the following sets. For $u, v \in \mathbb{N}^*$ satisfying $1 \leq u \leq n$ and $1 \leq v \leq l$,

$$\mathscr{M}_{n+1,l}(u,v) = \left\{ \begin{pmatrix} C \\ e^{2i\pi \mathbf{z}} \end{pmatrix} \in \mathscr{M}_{n+1,l}(\mathbb{C}) \;\middle|\; \begin{array}{l} C = (c_{r,s}) \in M_{n,l}(\mathbb{R}), \mathbf{z} \in \mathbb{R}^l, \\ c_{r,s} = 0, 1 \leq r \leq u-1, 1 \leq s \leq l, \\ c_{u,s} = 0, 1 \leq s \leq v-1, \\ c_{u,v} \neq 0 \end{array} \right\}.$$

If $u \leq n-1$, we put

$$\mathscr{M}'_{n+1,l}(u,v) = \left\{ \begin{pmatrix} C \\ e^{2i\pi \mathbf{z}} \end{pmatrix} \in \mathscr{M}_{n+1,l}(u,v) : c_{(u+1),v} = 0 \right\}.$$

Otherwise,

$$\mathscr{M}'_{n+1,l}(n,v) = \left\{ \begin{pmatrix} C \\ e^{2i\pi \mathbf{z}} \end{pmatrix} \in \mathscr{M}_{n+1,l}(n,v) : \rho_v(\mathbf{z}) = 0 \right\}.$$

The next step consists in giving an explicit description of $\mathscr{R}$ and $\mathscr{T}$ in order to have comprehensive details about the parameter and the deformation spaces. For such a purpose, we can divide the task into two parts as in the previous section.

The torsion-free case

We consider the following notation. For all $k \in \{1, \ldots, n\}$ and $j \in \{0, \ldots, k\}$,

$$R_{j,k} := \Psi(\mathscr{R}(\Gamma, G, H)) \cap K_{j,k},$$
$$R_{0,k}(u,v) := R_{0,k} \cap \mathscr{M}_{n+1,k}(u,v)$$

and

$$J_k = \{(u, v) \in \{1, \dots, n\} \times \{1, \dots, k\} : R_{0,k}(u, v) \neq \emptyset\}.$$

Case 1: $k \geq 3$

Proposition 7.3.17. *Assume that Γ is a torsion-free Abelian discrete subgroup of G of rank $k \geq 3$. Under the above notation, we have:*

(1) *The parameter space is homeomorphic to $\mathscr{R} = R_{0,k}$ where:*
 (i) *If $H \not\subset G_0$, then $R_{0,k} = K_{0,k}$ if $p = 1$. Otherwise,*

$$R_{0,k} = \left\{ \begin{pmatrix} \mathbf{0}_k \\ N \\ e^{2i\pi \mathbf{z}} \end{pmatrix} \in K_{0,k} : N = \begin{pmatrix} N_1 \\ N_2 \end{pmatrix}, N_1 \in M^0_{n-p+1,k}(\mathbb{R}) \right\}.$$

 (ii) *If $H \subset G_0$, then*

$$R_{0,k} = \left\{ \begin{pmatrix} \mathbf{0}_k \\ N \\ e^{2i\pi \mathbf{z}} \end{pmatrix} \in K_{0,k} \;\middle|\; \begin{array}{l} \mathrm{rk}(A_{n-1}(x) N \Cap M^0_{\mathfrak{h},\mathscr{B}}) = k + p - \dim(\mathfrak{h} \cap \mathfrak{z}(\mathfrak{g})) \\ \text{for any } x \in \mathbb{R} \end{array} \right\}.$$

(2) *For any $(u, v) \in J_k$, the set $R_{0,k}(u, v)$ is G-invariant and $R_{0,k}(u, v)/G$ is homeomorphic to*

$$\mathscr{T}_{0,k}(u, v) := R_{0,k}(u, v) \cap \mathscr{M}'_{n+1,k}(u, v). \tag{7.21}$$

Proof. (1) As a direct consequence from Proposition 7.3.9, we have $R_{j,k} = \emptyset$ for all $j \in \{1, \dots, k\}$. Now, in order to find the description of $R_{0,k}$, we shall discuss the two following cases. Assume first that $H \not\subset G_0$. If $p = 1$, then $H = \exp(\mathbb{R}X)$ and $\mathrm{Ad}_G\, \mathfrak{h} = \mathbb{R}(X + \mathfrak{g}^2)$, where $(\mathfrak{g}^j)_{1 \leq j \leq n+1}$ is the sequence defined as in equation (7.10). Hence, for any

$$\begin{pmatrix} \mathbf{0}_k \\ N \\ e^{2i\pi \mathbf{z}} \end{pmatrix} \in K_{0,k},$$

we have

$$\mathrm{rk} \begin{pmatrix} \mathbf{0}_k & 1 \\ N & {}^t\mathbf{a} \end{pmatrix} = k + 1$$

for all $\mathbf{a} \in \{0\} \times \mathbb{R}^{n-2}$. Then $R_{0,k} = K_{0,k}$. Suppose now that $p \geq 2$, then for all $g \in G$, the matrix $M^1_{\mathrm{Ad}_g\, \mathfrak{h},\mathscr{B}}$ is of the form

$$\begin{pmatrix} 1 & 0 & 0 & 0 & \cdots & 0 \\ 0 & 0 & 0 & 0 & \cdots & 0 \\ * & {}^t\mathbf{0}_{n-p} & \vdots & \vdots & & \\ * & 1 & 0 & \vdots & & \vdots \\ \vdots & * & 1 & 0 & & \\ \vdots & \vdots & \ddots & \ddots & \ddots & \vdots \\ * & * & \cdots & * & 1 & 0 \end{pmatrix} \in M_{n,p}(\mathbb{R}). \tag{7.22}$$

Now, let

$$M = \begin{pmatrix} \mathbf{0}_k \\ N \\ e^{2i\pi \mathbf{z}} \end{pmatrix} \in K_{0,k}$$

and write

$$N = \begin{pmatrix} N_1 \\ N_2 \end{pmatrix}, \tag{7.23}$$

where $N_1 \in M_{n-p+1,k}(\mathbb{R})$ and $N_2 \in M_{p-2,k}(\mathbb{R})$. We get then

$$M \in \mathscr{R} \Leftrightarrow N_1 \in M^0_{n-p+1,k}(\mathbb{R}).$$

Finally, when $H \subset G_0$, the result follows immediately from Proposition 7.3.16.

(2) It suffices to prove that $R_{0,k}$ is G-stable. Thus, the G-stability of $R_{0,k}(u,v)$ follows. First, it is not hard to see that the action of G on $R_{0,k}$ is reduced to the action of $\exp(\mathbb{R}X)$. Indeed,

$$G \star \begin{pmatrix} \mathbf{0}_k \\ N \\ e^{2i\pi \mathbf{z}} \end{pmatrix} = \left\{ \begin{pmatrix} \mathbf{0}_k \\ A_{n-1}(x)N \\ e^{2i\pi(\mathbf{z}+T_n(x)N)} \end{pmatrix} : x \in \mathbb{R} \right\}.$$

Here, $A_{n-1}(x) \in M_{n-1,n-1}(\mathbb{R})$ is given by equation (7.20) and

$$T_n(x) = \begin{pmatrix} \frac{1}{(n-1)!}x^{n-1} & \cdots & x \end{pmatrix} \in M_{1,n-1}(\mathbb{R}).$$

The fact that N is of maximal rank is equivalent to $A_{n-1}(x)N$ is also of maximal rank for any $x \in \mathbb{R}$. If furthermore we write N as in (7.23), where $N_1 \in M_{n-p+1,k}(\mathbb{R})$ whenever

$p > 1$, we get

$$A_{n-1}(x)\begin{pmatrix} N_1 \\ N_2 \end{pmatrix} = \begin{pmatrix} A_{n-p+1}(x)N_1 \\ N_2(x) \end{pmatrix},$$

for some $N_2(x) \in M_{p-2,k}(\mathbb{R})$. Therefore,

$$\operatorname{rk}(A_{n-p+1}(x)N_1) = \operatorname{rk} N_1, \quad x \in \mathbb{R}.$$

To conclude, it suffices to see that

$$A_{n-1}(x)A_{n-1}(x') = A_{n-1}(x + x'), \quad x, x' \in \mathbb{R}.$$

Now, let $(u, v) \in J_k$, we show that $\mathscr{T}_{0,k}(u, v)$ is a cross-section of all adjoint orbits of $R_{0,k}(u, v)$. Let

$$M = \begin{pmatrix} \mathbf{0}_k \\ N \\ e^{2i\pi \mathbf{z}} \end{pmatrix} \in R_{0,k}(u, v).$$

Noting

$$N = \{(a_{r,s}), 1 \le r \le n-1,\ 1 \le s \le k\},$$

we get $a_{u-1,v} \neq 0$. Let

$$t_{N,\mathbf{z}} = \begin{cases} -\frac{a_{u,v}}{a_{u-1,v}} & \text{if } u < n, \\ -\frac{\rho_v(\mathbf{z})}{a_{n-1,v}} & \text{if } u = n. \end{cases}$$

By an easy computation, we show that

$$G \star M \cap \mathscr{T}_{0,k}(u, v) = \exp(t_{N,\mathbf{z}}X) \star M.$$

The next step consists in proving that the following map:

$$\begin{array}{rccl} (\Phi_{0,k})_{(u,v)} : R_{0,k}(u, v)/G & \to & \mathscr{T}_{0,k}(u, v) \\ {[M]} & \mapsto & \exp(t_{N,\mathbf{z}}X) \star M \end{array}$$

is an homeomorphism. First, it is not hard to see that $(\Phi_{0,k})_{(u,v)}$ is well-defined. In fact, let $M_j \in R_{0,k}(u, v), j = 1, 2$ such that $[M_1] = [M_2]$. Then

$$(\Phi_{0,k})_{(u,v)}([M_1]) = G \star M_1 \cap \mathscr{T}_{0,k}(u, v) = G \star M_2 \cap \mathscr{T}_{0,k}(u, v) = (\Phi_{0,k})_{(u,v)}([M_2]).$$

The injectivity of $(\Phi_{0,k})_{(u,v)}$ is immediate. Indeed, for $M_j \in R_{0,k}(u,v), j = 1,2$ such that

$$(\Phi_{0,k})_{(u,v)}([M_1]) = (\Phi_{0,k})_{(u,v)}([M_2]),$$

we have

$$G \star M_1 \cap \mathscr{T}_{0,k}(u,v) = G \star M_2 \cap \mathscr{T}_{0,k}(u,v) \in G \star M_1 \cap G \star M_2,$$

which gives that $[M_1] = [M_2]$. Now, to see that $(\Phi_{0,k})_{(u,v)}$ is surjective, it suffices to verify that for all $M \in \mathscr{T}_{0,k}(u,v)$, we have

$$(\Phi_{0,k})_{(u,v)}([M]) = M.$$

To achieve the proof, we prove that $(\Phi_{0,k})_{(u,v)}$ is bicontinuous. Let $(\pi_{0,k})_{(r,s)}$ be the canonical surjection

$$(\pi_{0,k})_{(r,s)} : R_{0,k}(r,s) \to R_{0,k}(r,s)/G.$$

Thus, we can easily see the continuity of $(\widetilde{\Phi}_{0,k})_{(u,v)} = (\Phi_{0,k})_{(u,v)} \circ (\pi_{0,k})_{(u,v)}$, which is equivalent to the continuity of $(\Phi_{0,k})_{(u,v)}$. Finally, it is clear that $((\Phi_{0,k})_{(u,v)})^{-1} = ((\pi_{0,k})_{(u,v)})_{|\mathscr{T}_{0,k}(u,v)}$, then the bicontinuity follows. □

Case 2: $k = 2$

Proposition 7.3.18. *Let G be a reduced threadlike Lie group and Γ a torsion-free Abelian discrete subgroup of G.*

(1) *The parameter space is homeomorphic to $\mathscr{R} = \bigcup_{j=0}^{2} R_{j,2}$ where:*
   (i) *If $H \not\subset G_0$, then $R_{0,2} = K_{0,2}$ if $p = 1$. Otherwise,*

$$R_{0,2} = \left\{ \begin{pmatrix} \mathbf{0}_2 \\ N \\ e^{2i\pi z} \end{pmatrix} \in K_{0,2} : N = \begin{pmatrix} N_1 \\ N_2 \end{pmatrix}, N_1 \in M^0_{n-p+1,2}(\mathbb{R}) \right\}$$

   and for $j \in \{1,2\}$:

$$R_{j,2} = \left\{ \begin{pmatrix} \lambda_1{}^t\mathbf{c} & \lambda_2{}^t\mathbf{c} \\ y_1 & y_2 \\ e^{2i\pi z_1} & e^{2i\pi z_2} \end{pmatrix} \in K_{j,2} : \rho_2(\mathbf{c}) \neq 0 \right\}.$$

   (ii) *If $H \subset G_0$, then*

$$R_{0,2} = \left\{ \begin{pmatrix} \mathbf{0}_2 \\ N \\ e^{2i\pi z} \end{pmatrix} \in K_{0,2} \;\middle|\; \begin{array}{l} \mathrm{rk}(A_{n-1}(x)N \Cap M^0_{\mathfrak{h},\mathscr{B}}) = p + 2 - \dim(\mathfrak{h} \cap \mathfrak{z}(\mathfrak{g})) \\ \text{for any } x \in \mathbb{R} \end{array} \right\}.$$

Moreover, for $j \in \{1,2\}$*:*

$$R_{j,2} = \begin{cases} \emptyset & \text{if } n-1 \in \mathscr{I}_{\mathfrak{g}_0}^{\mathfrak{h}_0}, \\ K_{j,2} & \text{otherwise.} \end{cases}$$

(2) *For any* $(u,v) \in J_2$*, the set* $R_{0,2}(u,v)$ *is* G*-invariant and* $R_{0,2}(u,v)/G$ *is homeomorphic to* $\mathscr{T}_{0,2}(u,v)$ *given as in equation (7.21) and the components* $R_{j,2}$*,* $j = 1,2$ *are* G*-invariant and homeomorphic to*

$$\mathscr{T}_{j,2} = \left\{ \begin{pmatrix} \lambda_1{}^t\mathbf{c} & \lambda_2{}^t\mathbf{c} \\ 0 & 0 \\ 1 & 1 \end{pmatrix} \in R_{j,2} : \rho_s(\mathbf{c}) = 0, s = 3,\dots,n-1 \right\}.$$

Proof. We opt for the same arguments as in Proposition 7.3.17 to have the description of $R_{0,k}$. Now, let us give the description of $R_{j,2}$, $j = 1,2$, according to the position of H inside G. Let

$$M = \begin{pmatrix} \lambda_1{}^t\mathbf{c} & \lambda_2{}^t\mathbf{c} \\ y_1 & y_2 \\ e^{2i\pi z_1} & e^{2i\pi z_2} \end{pmatrix} \in K_{j,2}.$$

In the case where $H \not\subset G_0$, we have $\mathbb{R}(X + \mathfrak{g}^2) \subset \mathrm{Ad}_G\,\mathfrak{h}$ and then for all $g \in G$, the matrix $M^1_{\mathrm{Ad}_g\,\mathfrak{h},\mathscr{B}}$ is of the form ${}^t(1,0,*,\dots,*)$ if $p = 1$. Otherwise, such a matrix is given as equation (7.22). Hence the assertion

$$\mathrm{rk}\left(\begin{pmatrix} \lambda_1{}^t\mathbf{c} & \lambda_2{}^t\mathbf{c} \\ y_1 & y_2 \end{pmatrix} ⋒ M^1_{\mathfrak{h},\mathscr{B}}\right) = 2 + p - \dim(\mathfrak{h} \cap \mathfrak{z}(\mathfrak{g}))$$

for any $g \in G$, is equivalent to

$$\mathrm{rk}\left(\begin{pmatrix} \lambda_1{}^t\mathbf{c} & \lambda_2{}^t\mathbf{c} \\ y_1 & y_2 \end{pmatrix} ⋒ \begin{pmatrix} 1 \\ 0 \\ {}^t\mathbf{a} \end{pmatrix}\right) = 3,$$

for $a \in \mathbb{R}^{n-2}$, which is in turn equivalent to $\rho_2(\mathbf{c}) \in \mathbb{R}^*$.

We now treat the case where $H \subset G_0$. Suppose first that $n-1 \in \mathscr{I}_{\mathfrak{g}_0}^{\mathfrak{h}_0}$, then there exists $\alpha_n \in \mathbb{R}$ such that $Y_{n-1} + \alpha_n Y_n \in \mathfrak{h}$. Then $M_{\mathfrak{h},\mathscr{B}}$ is of the form

$$\begin{pmatrix} A & {}^t\mathbf{0}_{n-1} \\ a & 1 \\ b & \alpha_n \end{pmatrix}.$$

As

$$\operatorname{rk}\begin{pmatrix} \lambda_1{}^t\mathbf{c} & \lambda_2{}^t\mathbf{c} & {}^t\mathbf{0}_{n-1} \\ y_1 & y_2 & 1 \end{pmatrix} = 2,$$

we get

$$\operatorname{rk}\left(\begin{pmatrix} \lambda_1{}^t\mathbf{c} & \lambda_2{}^t\mathbf{c} \\ y_1 & y_2 \end{pmatrix} ⋒ M^1_{\mathfrak{h},\mathscr{B}}\right) < p + 2 - \dim(\mathfrak{h} \cap \mathfrak{z}(\mathfrak{g})).$$

Suppose finally that $n - 1 \notin \mathscr{I}^{\mathfrak{h}_0}_{\mathfrak{g}_0}$. Then it is not hard to see that

$$\operatorname{rk}\left(\begin{pmatrix} \lambda_1{}^t\mathbf{c} & \lambda_2{}^t\mathbf{c} \\ y_1 & y_2 \end{pmatrix} ⋒ M^1_{\mathfrak{h},\mathscr{B}}\right) = p + 2 - \dim(\mathfrak{h} \cap \mathfrak{z}(\mathfrak{g})).$$

This gives that $R_{j,2} = K_{j,2}, j = 1, 2$.

(2) We only have to show the G-invariance of $R_{j,k}, j = 1, 2$, as for the cross-sections, we argue as in previous proposition. Using equation (7.16), we have

$$G \star \begin{pmatrix} \lambda_1{}^t\mathbf{c} & \lambda_2{}^t\mathbf{c} \\ y_1 & y_2 \\ e^{2i\pi z_1} & e^{2i\pi z_2} \end{pmatrix} = \left\{ \begin{pmatrix} \lambda_1 B_{n-1}(g)^t\mathbf{c} & \lambda_2 B_{n-1}(g)^t\mathbf{c} \\ y_1 + \lambda_1 P_{n-1}(g)^t\mathbf{c} & y_2 + \lambda_2 P_{n-1}(g)^t\mathbf{c} \\ e^{2i\pi(z_1 + P_n(g)^t(\lambda_1\mathbf{c}, y_1))} & e^{2i\pi(z_2 + P_n(g)^t(\lambda_2\mathbf{c}, y_2))} \end{pmatrix} : g \in G \right\},$$

where $B_j(g)$ and $P_j(g), j \in \{n-1, n\}$ are defined in equations (7.17) and (7.18), respectively. Thus, we can easily see that for $j = 1, 2$,

$$\rho_j(\mathbf{c}) = \rho_j({}^t(B_{n-1}(g)^t\mathbf{c})). \qquad \square$$

Case 3: $k = 1$

The description of the parameter and the deformation spaces is similar to the previous case. Using the same analysis as in Propositions 7.3.17 and 7.3.18, we obtain the following result.

Proposition 7.3.19. *We keep the same notation and hypothesis and suppose that $k = 1$.*

(1) *The parameter space is homeomorphic to $\mathscr{R} = R_{0,1} \cup R_{1,1}$ where:*
   (i) *If $H \not\subset G_0$, then $R_{0,1} = K_{0,1}$ if $p = 1$. Otherwise,*

$$R_{0,1} = \left\{ \begin{pmatrix} 0 \\ {}^t\mathbf{n}_1 \\ {}^t\mathbf{n}_2 \\ e^{2i\pi z} \end{pmatrix} \in K_{0,1} : {}^t\mathbf{n}_1 \in (\mathbb{R}^{n-p+1})^* \right\}.$$

Moreover,

$$R_{1,1} = \left\{ \begin{pmatrix} {}^t\mathbf{c} \\ e^{2i\pi z} \end{pmatrix} \in K_{1,1} : \rho_2(\mathbf{c}) \neq 0 \right\}.$$

(ii) *If $H \subset G_0$, then*

$$R_{0,1} = \left\{ \begin{pmatrix} 0 \\ {}^t\mathbf{n} \\ e^{2i\pi z} \end{pmatrix} \in K_{0,1} \;\middle|\; \begin{array}{l} \mathrm{rk}(A_{n-1}(x){}^t\mathbf{n} \circledcirc M^0_{\mathfrak{h},\mathscr{B}}) = p + 1 - \dim(\mathfrak{h} \cap \mathfrak{z}(\mathfrak{g})) \\ \textit{for any } x \in \mathbb{R} \end{array} \right\}$$

and

$$R_{1,1} = K_{1,1}.$$

(2) *The components $R_{0,1}(u,v)$ and $R_{1,1}$ are G-invariant. Their corresponding G-orbit sets are homeomorphic to $\mathscr{T}_{0,1}(u,v)$, given as in equation (7.21) and*

$$\mathscr{T}_{1,1} = \left\{ \begin{pmatrix} {}^t\mathbf{c} \\ 1 \end{pmatrix} \in R_{1,1} : \rho_s(\mathbf{c}) = 0, s = 3, \ldots, n \right\},$$

respectively.

The torsion case

Suppose first that $k = 0$, then we have

$$\mathscr{R}(\Gamma, G, H) = \mathrm{Hom}^0_d(\Gamma, G). \tag{7.24}$$

Moreover, as $\Gamma = \langle \gamma_1 \rangle \subset Z(G)$, we get for all $g \in G$ and $\varphi \in \mathscr{R}(\Gamma, G, H)$, $g \cdot \varphi(\gamma_1) = \varphi(\gamma_1)$ and then

$$\mathscr{T}(\Gamma, G, H) = \mathscr{R}(\Gamma, G, H). \tag{7.25}$$

We now treat the case where $k > 0$. First, using the same analysis as in Propositions 7.3.9, 7.3.10 and 7.3.11, it is easy to see that the description of $\mathscr{R}$ and $\mathscr{T}$ is similar to the torsion-free case. The analogues of the objects $R_{j,k}$, given by $\widetilde{R}_{j,k} := \Psi(\mathscr{R}(\Gamma, G, H)) \cap \widetilde{K}_{j,k}$, $j \in \{0, \ldots, k\}$, are as follows. For $j = 0$, if $H \not\subset G_0$, then

$$\widetilde{R}_{0,k} = \widetilde{K}_{0,k}.$$

Otherwise,

$$\widetilde{R}_{0,k} = \left\{ \begin{pmatrix} \mathbf{0}_k & 0 \\ N & {}^t\mathbf{0}_{n-1} \\ e^{2i\pi \mathbf{z}} & e^{\frac{2i\pi}{q}} \end{pmatrix} \in \widetilde{K}_{0,k} \;\middle|\; \begin{array}{l} \mathrm{rk}(A_{n-1}(x)N \circledcirc M^0_{\mathfrak{h},\mathscr{B}}) = k + p - \dim(\mathfrak{h} \cap \mathfrak{z}(\mathfrak{g})) \\ \text{for any } x \in \mathbb{R} \end{array} \right\}.$$

Let now $j \in \{1,\dots,k\}$. If $k \geq 3$, then

$$\widetilde{R}_{j,k} = \emptyset.$$

Otherwise, we get the following discussion. When $k = 2$, we have for $j \in \{1,2\}$,

$$\widetilde{R}_{j,2} = \left\{ \begin{pmatrix} \lambda_1{}^t\mathbf{c} & \lambda_2{}^t\mathbf{c} & {}^t\mathbf{0}_{n-1} \\ y_1 & y_2 & 0 \\ e^{2i\pi z_1} & e^{2i\pi z_2} & e^{\frac{2i\pi r}{q}} \end{pmatrix} \in \widetilde{K}_{j,2} : \rho_2(\mathbf{c}) \neq 0 \right\}$$

if $H \not\subset G_0$. Otherwise,

$$\widetilde{R}_{j,2} = \begin{cases} \emptyset & \text{if } n-1 \in \mathscr{I}_{\mathfrak{g}_0}^{\mathfrak{h}_0}, \\ \widetilde{K}_{j,2} & \text{otherwise.} \end{cases}$$

Finally, for $k = 1$, if $H \not\subset G_0$ then

$$\widetilde{R}_{1,1} = \left\{ \begin{pmatrix} {}^t\mathbf{c} & {}^t\mathbf{0}_{n-1} \\ y & 0 \\ e^{2i\pi z} & e^{\frac{2i\pi r}{q}} \end{pmatrix} \in \widetilde{K}_{1,1} : \rho_2(\mathbf{c}) \neq 0 \right\}.$$

Otherwise,

$$\widetilde{R}_{1,1} = \widetilde{K}_{1,1}.$$

We remark that we have the same description of the sets $(\widetilde{R}_{j,k})_{0 \leq j \leq k}$ as in torsion-free subcases except for the case where $H \not\subset G_0$ for which $\widetilde{R}_{0,k}$ admits a slightly different expression. In fact, using the free action of Γ on G/H, we get that $H = \exp(\mathbb{R}X)$. That is $p = 1$. We get then $\widetilde{R}_{0,k} = \widetilde{K}_{0,k}$. Last but not least, using the same analysis of the previous subsection, if we denote by J_{k+1} be the set of all $(u,v) \in \{1,\dots,n\} \times \{1,\dots,k+1\}$ for which $\widetilde{R}_{0,k}(u,v) := \widetilde{R}_{0,k} \cap \mathscr{M}_{n+1,k+1}(u,v) \neq \emptyset$, then the G-invariance of the components $\widetilde{R}_{0,k}(u,v)$ and $(\widetilde{R}_{j,k}), j = 1,\dots,k$ is also required. We get then that $\widetilde{R}_{0,k}(u,v)/G$ is homeomorphic to

$$\mathscr{T}_{0,k}(u,v) := \widetilde{R}_{0,k}(u,v) \cap \mathscr{M}'_{n+1,k+1}(u,v). \tag{7.26}$$

Moreover, for $k = 2$, the set $\widetilde{R}_{j,2}/G$ is homeomorphic to

$$\mathscr{T}_{j,2} = \left\{ \begin{pmatrix} \lambda_1{}^t\mathbf{c} & \lambda_2{}^t\mathbf{c} & {}^t\mathbf{0} \\ 0 & 0 & 0 \\ 1 & 1 & e^{\frac{2i\pi r}{q}} \end{pmatrix} \in \widetilde{R}_{j,2} : \rho_s(\mathbf{c}) = 0, s = 3,\dots,n-1 \right\}.$$

Finally, when $k = 1$, the set $\widetilde{R}_{1,1}/G$ is homeomorphic to

$$\mathscr{T}_{1,1} = \left\{ \begin{pmatrix} {}^t\mathbf{c} & {}^t\mathbf{0}_n \\ 1 & e^{\frac{2i\pi r}{q}} \end{pmatrix} \in \widetilde{R}_{1,1} : \rho_s(\mathbf{c}) = 0, s = 3, \ldots, n \right\}.$$

7.3.4 The local rigidity problem

Theorem 7.3.20. *Let G be a threadlike Lie group, H a connected Lie subgroup of G and Γ an Abelian discontinuous group for G/H. Then Conjecture 5.7.10 holds.*

Proof. When G is simply connected, a positive answer to Conjecture 5.1.1 is the subject of Theorem 5.1.5. Indeed, G admits no nontrivial finite discrete subgroups. Suppose now that G is a reduced threadlike Lie group. The result is immediate when Γ is finite by equations (7.14), (7.24) and (7.25). Suppose now that Γ is infinite, that is, $k \geq 1$. We have then to show that the G-orbit of any $M \in \mathscr{R}$ is not open in the parameter space. Assume first that $k \geq 3$ and let

$$M = \begin{pmatrix} \mathbf{0}_k \\ N \\ e^{2i\pi\mathbf{z}} \end{pmatrix} \in R_{0,k}.$$

The sequence $(M_s)_{s\in\mathbb{N}^*}$ given by

$$M_s = \begin{pmatrix} \mathbf{0}_k \\ A_{n-1}(\frac{1}{s})N \\ e^{2i\pi\mathbf{z}} \end{pmatrix}$$

belongs to $\mathscr{R} \setminus G \star M$ and converges to M. Now for $k = 2$, thus either we are in the context where M belongs to $R_{0,2}$ and then the last arguments apply or

$$M = \begin{pmatrix} \lambda_1{}^t\mathbf{c} & \lambda_2{}^t\mathbf{c} \\ y_1 & y_2 \\ e^{2i\pi z_1} & e^{2i\pi z_2} \end{pmatrix} \in R_{j,2}.$$

Then it suffices to consider the sequence $(M_s)_{s\in\mathbb{N}^*}$ given by

$$M_s = \begin{pmatrix} \lambda_1{}^t\mathbf{c}_s & \lambda_2{}^t\mathbf{c} \\ y_1 & y_2 \\ e^{2i\pi z_1} & e^{2i\pi z_2} \end{pmatrix},$$

where $\mathbf{c}_s = (c_{1,s}, \ldots, c_{n-1,s})$ satisfies $\rho_j(\mathbf{c}_s) = \rho_j(\mathbf{c})$ for all $j \in \{1, \ldots, n-1\} \setminus \{2\}$ and

$$\rho_2(\mathbf{c}_s) = \begin{cases} \rho_2(\mathbf{c}) + \frac{1}{s} & \text{if } \rho_2(\mathbf{c}) > 0, \\ \rho_2(\mathbf{c}) - \frac{1}{s} & \text{otherwise.} \end{cases}$$

Finally, for $k = 1$, the same argument as in case $k = 2$ applies. □

7.3.5 The stability problem

As shown in the previous section, the description of the parameter space in the case when $H \subset G_0$ is not so explicit. Then we are led to treat separately the cases $H \not\subset G_0$ and $H \subset G_0$.

Assume first that $H \not\subset G_0$ and let us move on to the following discussion.

Case 1: $k \geq 2$

Proposition 7.3.21. *Let G be a reduced threadlike Lie group, H a closed connected subgroup of G and Γ an Abelian discontinuous group for the homogeneous space G/H of rank $k \geq 2$. Assume that $H \not\subset G_0$, then the stability holds.*

Proof. According to Proposition 7.3.13, it suffices to show that $\mathscr{R}(\Gamma, G, H)$ is an open set in $\mathrm{Hom}_d^0(\Gamma, G)$, which is equivalent to prove that $\mathscr{R}$ is open in $\mathscr{K}$. By Propositions 7.3.9 and 7.3.17, this result is immediate when $k > 2$. We tackle now the case where $k = 2$. Let $M \in \overline{\mathscr{K} \setminus \mathscr{R}}^{\mathscr{K}}$. Assume first that $M \in K_{0,2}$. Then we can find a sequence $(M_s)_{s\in\mathbb{N}}$ in $\mathscr{K} \setminus \mathscr{R}$, which converges to M. Suppose that we can extract a subsequence $(M_{\sigma(s)})_{s\in\mathbb{N}}$ lying in $(K_{j,2} \setminus R_{j,2})$ for some $j \in \{1, 2\}$. Without loss of generality, we can suppose that $j = 1$. Note that

$$M_{\sigma(s)} = \begin{pmatrix} {}^t\mathbf{c}_s & \lambda_s {}^t\mathbf{c}_s \\ y_{1,s} & y_{2,s} \\ e^{2i\pi z_{1,s}} & e^{2i\pi z_{2,s}} \end{pmatrix}.$$

We get then that the sequence $\rho_1(\mathbf{c}_s)(y_{2,s} - \lambda_s y_{1,s})$ of $\mathbb{Z}$ converges to 0 and is stationary, which contradicts the fact that $M_s \in \mathscr{K}$ for all $s \in \mathbb{N}$. We get therefore that for all subsequence $(M_{\sigma(s)})_{s\in\mathbb{N}}$ of $(M_s)_{s\in\mathbb{N}}$ there exists $s_0 \in \mathbb{N}$ such that for all $s \geq s_0$ we have $M_{\sigma(s)} \in K_{0,2} \setminus R_{0,2}$, which gives that $M \in K_{0,2} \setminus R_{0,2}$. Assume now that there exists $j \in \{1, 2\}$ such that $M \in K_{j,2}$. Then we can find a sequence in $\mathscr{K} \setminus \mathscr{R}$, which converges to M and for which we can extract a subsequence lying in $(K_{j',2} \setminus R_{j',2})$ for some $j' \in \{1, 2\}$, which leads to the fact that $M \in (K_{1,2} \cup K_{2,2}) \setminus R_{j,2}$. □

Case 2: $k = 1$

Proposition 7.3.22. *Let G be a reduced threadlike Lie group, H a closed, connected subgroup of G such that $H \not\subset G_0$ and Γ an Abelian discontinuous group for the homogeneous space G/H of rank 1. If $n \in \{2, p\}$, then the stability holds in the parameter space. Otherwise, the set of nonstable parameters is given by*

$$\mathscr{R}' = \left\{ \begin{pmatrix} \mathbf{0}_l \\ N \\ e^{2i\pi z} \end{pmatrix} \in \mathscr{R} : N = \begin{pmatrix} \mathbf{0}_l \\ N' \end{pmatrix} \right\}.$$

Proof. First of all, it is not hard to see that when $n = p$ or $n = 2$, the result is immediate as

$$\mathscr{K} = \left\{ \begin{pmatrix} {}^t\mathbf{a} \\ e^{2i\pi z} \end{pmatrix} \in \mathscr{H} : \mathbf{a} \in (\mathbb{R}^n)^* \right\}$$

and

$$\mathscr{R} = \left\{ \begin{pmatrix} {}^t\mathbf{a} \\ e^{2i\pi z} \end{pmatrix} \in \mathscr{K} : \mathbf{a} = (a_1, \dots, a_n) \in \mathbb{R}^n, a_2 \in \mathbb{R}^* \right\}.$$

Suppose finally that $n > 2$ and $p < n$. First of all, it is not hard to see that $R_{1,1}$ is open in $\mathscr{R}$. Then let

$$M = \begin{pmatrix} 0 \\ {}^t\mathbf{n} \\ e^{2i\pi z} \end{pmatrix} \in \mathscr{R}$$

be a nonstable point. This is equivalent to $M \in \overline{\mathscr{K} \setminus \mathscr{R}}^{\mathscr{K}}$. There exists therefore a sequence $(M_s)_{s \in \mathbb{N}} \subset \mathscr{K} \setminus \mathscr{R}$, which converges to M. Then we can extract a subsequence lying in $(K_{1,1} \setminus R_{1,1})$, which leads to the fact that $M \in \mathscr{R}'$. Conversely, it is clear that any element

$$M = \begin{pmatrix} 0 \\ {}^t\mathbf{n} \\ e^{2i\pi z} \end{pmatrix} \in \mathscr{R}'$$

is a limit of a sequence

$$(M_s)_{s \in \mathbb{N}^*}$$

given by

$$M_s = \begin{pmatrix} \frac{1}{s} \\ {}^t\mathbf{n} \\ e^{2i\pi z} \end{pmatrix},$$

which gives that $M \in \overline{\mathscr{K} \setminus \mathscr{R}}^{\mathscr{K}}$. □

We tackle now the case where $H \subset G_0$ and $k \in \{1, 2\}$.

Proposition 7.3.23. *Let G be a reduced threadlike Lie group, H a closed, connected subgroup of G such that $H \subset G_0$ and Γ an Abelian discontinuous group for the homogeneous space G/H. Then the set $\mathscr{R} \setminus R_{0,k}$ is stable for any $k \in \{1, 2\}$.*

Proof. Suppose first that $k = 2$. If $n - 1 \in \mathscr{I}_{\mathfrak{g}_0}^{\mathfrak{h}_0}$, then $\mathscr{R} \setminus R_{0,2} = \emptyset$. Otherwise, it is not hard to see that $R_{1,2} \cup R_{2,2} = K_{1,2} \cup K_{2,2}$ is open in $\mathscr{K}$. Now, if $k = 1$, then $K_{0,1}$ is closed in $\mathscr{K}$, which gives that $K_{1,1} = R_{1,1}$ is open in $\mathscr{K}$. □

We finally study the stability of elements of the parameter space when $H \subset G_0$ and $k \geq 3$. For such a purpose, we consider the following notation. Let $p_1 := p - \dim(\mathfrak{h} \cap \mathfrak{z}(\mathfrak{g}))$ and for any $k \in \{1, \dots, n - p_1 - 1\}$, we set

$$S_k = \{(i_1, \dots, i_{p_1+k}) \in \mathbb{N}^{p_1+k} : 1 \leq i_1 < \dots < i_{p_1+k} \leq n - 1\}.$$

We denote for any $N \in M_{n-1,k}(\mathbb{R}), x \in \mathbb{R}$, by $\Delta_{(i_1,\dots,i_{p_1+k})}(N, x)$ the relative minor of order $k + p_1$ obtained by considering the rows $i_1, \dots, i_{p_1+k}$ of the matrix $A_{n-1}(x)N \Cap M^0_{\mathfrak{h},\mathscr{B}}$ and

$$P_N(x) = \sum_{\mathbf{a} \in S_k} \Delta^2_{\mathbf{a}}(N, x),$$

we get that

$$R_{0,k} = \left\{ \begin{pmatrix} \mathbf{0}_k \\ N \\ e^{2i\pi \mathbf{z}} \end{pmatrix} \in K_{0,k} : P_N(x) \neq 0 \text{ for any } x \in \mathbb{R} \right\}.$$

We denote by $d(N)$ the degree of P_N and

$$d = \max\{d(N) \in \mathbb{N} : N \in M^0_{n-1,k}(\mathbb{R})\}.$$

Let then

$$P_N(x) = a_{d(N)}(N)x^{d(N)} + \dots + a_0(N),$$

where $(a_j)_{0\leq j\leq d(N)}$ are polynomial functions on the coefficients of N and $a_{d(N)}(N) \neq 0$. We define the following subset of $R_{0,k}$ by

$$(R_{0,k})_d = \left\{ \begin{pmatrix} \mathbf{0}_k \\ N \\ e^{2i\pi \mathbf{z}} \end{pmatrix} \in K_{0,k} : a_d(N) \neq 0, P_N(x) \neq 0 \text{ for any } x \in \mathbb{R} \right\}.$$

We prove that $(R_{0,k})_d$ is open in $\mathscr{K}$. Let

$$\begin{pmatrix} \mathbf{0}_k \\ N \\ e^{2i\pi \mathbf{z}} \end{pmatrix} \in K_{0,k}$$

such that $a_d(N) \neq 0$. We decompose P_N as

$$P_N(x) = a_d(N) \prod_{j=1}^{\frac{d}{2}} (x^2 + \alpha_j(N)x + \beta_j(N))$$

for some nontrivial functions α_j, β_j, which depend continuously upon the coefficients of N, when restricted if needed to a smaller set, still denoted by $(R_{0,k})_d$. We get then that

$$(R_{0,k})_d = \left\{ \begin{pmatrix} \mathbf{0}_k \\ N \\ e^{2i\pi \mathbf{z}} \end{pmatrix} \in K_{0,k} : a_d(N) \neq 0, \alpha_j^2(N) - 4\beta_j(N) < 0, 1 \leq j \leq \frac{d}{2} \right\},$$

is an open set of $K_{0,k}$. In the case where $k \geq 3$, $K_{0,k} = \mathscr{K}$. This means therefore that the subset of nonstable elements of $\mathscr{R} = R_{0,k}$ is included in $R_{0,k} \setminus (R_{0,k})_d$.

As we saw, it is in general difficult to characterize the set of nonstable parameters. Hence it would be very interesting to study the local stability relatively to the different layers.

Definition 7.3.24. Let $\mathscr{L} = (H_i)_{i\in I}$ be a finite covering of sets of $\mathrm{Hom}(\Gamma, G)$. A homomorphism $\varphi \in R_i := H_i \cap \mathscr{R}(\Gamma, G, H)$ is said to be *stable on layers with respect to the layering* $\mathscr{L}$, if there is an open set in H_i (which is not necessarily open in $\mathrm{Hom}(\Gamma, G)$), which contains φ and is contained in $\mathscr{R}(\Gamma, G, H)$.

The parameter space $\mathscr{R}(\Gamma, G, H)$ is said to be stable on layers with respect to the layering $\mathscr{L}$ if each of its elements is stable on layers with respect to $\mathscr{L}$ (cf. [90] for more details). Remark that the stability of the parameter space implies the stability on layers, but the converse fails to hold in general. Note also that the stability on layers holds in the case of Heisenberg groups (cf. [20]). We here provide an explicit decomposition of $\mathrm{Hom}(\Gamma, G)$ and we prove that the parameter space is stable on layers with respect to such decomposition. We now prove the following main result.

Theorem 7.3.25. *Let G be a threadlike Lie group, H a closed, connected subgroup of G and Γ an Abelian discontinuous subgroup for G/H. There exists a covering of sets of* $\mathrm{Hom}(\Gamma, G)$, *which permits the parameter space to be stable on layers.*

Proof. Recall first that when G is simply connected, we have an analogue decomposition $\mathrm{Hom}(\Gamma, G)$ and $\mathscr{R}(\Gamma, G, H)$ (cf. Theorem 4.4.1 and Proposition 4.4.7). We can argue exactly with the same way as in reduced situation to prove that the stability on layers holds on the parameter space with respect to a adequate layering. Suppose then that G is a reduced threadlike Lie group and Γ a torsion-free Abelian discontinuous subgroup of G for G/H. From Propositions 7.3.17, 7.3.18 and 7.3.19, it is easy to check that in the cases where $H \not\subset G_0$ the parameter space is stable on layers with respect to the layering $(H_{j,k})_{0\le j\le k}$ given in Proposition 7.3.8. We now pay attention to the case when $H \subset G_0$. We are going to construct a refined covering of $\mathrm{Hom}(\Gamma, G)$ obtained from the decomposition given in Proposition 7.3.8. For such purpose, we keep all our notation as above and we put

$$H_j = \left\{ \begin{pmatrix} \mathbf{0}_k \\ N \\ e^{2i\pi\mathbf{z}} \end{pmatrix} \in H_{0,k} : d(N) = j \right\}.$$

We prove then that the parameter space is stable on layers with respect to the layering given by

$$\mathrm{Hom}(\Gamma, G) = \left(\bigcup_{j=0}^{d} H_j \right) \cup \left(\bigcup_{j=1}^{k} H_{j,k} \right). \tag{7.27}$$

Remark first that H_j is G-invariant. Indeed, it suffices to see that $P_{A(x')N}(x) = P_N(x + x')$ for all $N \in M_{n-1,k}(\mathbb{R})$ and $x, x' \in \mathbb{R}$ and then $d(A(x')N) = d(N)$. Let now

$$P_N(x) = a_{d(N)}(N) \prod_{s=0}^{\frac{d(N)}{2}} (x^2 + \alpha_{d(N),s}(N)x + \beta_{d(N),s}(N))$$

be the decomposition of P_N where $a_{d(N)}, \alpha_{d(N),s}$ and $\beta_{d(N),s}$ are some polynomial functions depending upon the coefficients of N. We get therefore that for any $j \in \{1, \dots, m\}$,

$$R_j = \mathscr{R} \cap H_j = \left\{ \begin{pmatrix} \mathbf{0}_k \\ N \\ e^{2i\pi\mathbf{z}} \end{pmatrix} \in H_j : (\alpha_{d(N),s})^2 - 4\beta_{d(N),s} < 0, s = 1, \dots, \frac{j}{2} \right\}.$$

It follows therefore that R_j is open in H_j. Finally, it is not hard to see that $R_{j,k}$ is open in $K_{j,k}$ for any $j \in \{1, \dots, k\}$. □

7.4 A stability theorem for non-Abelian actions

We still consider in this section the setting of a reduced threadlike Lie group G. Let H be an arbitrary closed subgroup of G and $\Gamma \subset G$ a non-Abelian discontinuous group for G/H. Unlike the setting where Γ is Abelian, we show that the stability property holds.

Our main target now is the following.

Theorem 7.4.1. *Let G be a threadlike group, then any non-Abelian discrete subgroup of G is stable.*

In the case where G is simply connected, the proof of Theorem 7.4.1 is subject of Theorem 6.2.27. We thus only treat the nonsimply connected case. In the case where Γ is Abelian, the property of stability fails to hold in general and depends upon the structure and the position of H and Γ inside G (cf. Subsection 7.3.5).

The proof of Theorem 7.4.1 will be provided later after introducing several other results. The first is the following.

Proposition 7.4.2. *Let G be a reduced threadlike Lie group and Γ a discrete non-Abelian subgroup of G such that* $\operatorname{rank}(\Gamma) = k = p + \varepsilon$ *with $\varepsilon = 0$ if Γ is torsion-free and $\varepsilon = 1$ otherwise. Then $p \geqslant 3$ and there exist $\gamma, \gamma_q, \dots, \gamma_{n-1} \in G$ such that*

$$\Gamma = \{\gamma^m \gamma_q^{m_q} \cdots \gamma_{n-1}^{m_{n-1}},\ m, m_j \in \mathbb{Z}\}(\Gamma \cap Z(G))$$

with $q = n - p + 1$, $\gamma\gamma_j\gamma^{-1}\gamma_j^{-1} = \gamma_{j+1}$; $q \leqslant j \leqslant n-2$, $\gamma\gamma_{n-1}\gamma^{-1}\gamma_{n-1}^{-1} = e$ and $\gamma_i\gamma_j\gamma_i^{-1}\gamma_j^{-1} = e$; $q \leqslant i, j \leqslant n-1$.

Proof. We consider the surjective projection

$$\begin{aligned} \pi : G &\longrightarrow G/Z(G) = G' \\ {}^t(\omega, e^{2i\pi v}) &\longmapsto {}^t\omega. \end{aligned}$$

First of all, it is clear that G' is a connected and simply connected threadlike Lie group and its associated Lie algebra is given by $\mathfrak{g}' := \mathfrak{g}/\mathfrak{z}(\mathfrak{g})$. We denote by $\exp_{G'} : \mathfrak{g}' \to G'$ the associated exponential map and $\rho : \mathfrak{g} \to \mathfrak{g}/\mathfrak{z}(\mathfrak{g})$ the canonical surjection, then clearly $\exp_{G'} \circ \rho = \pi \circ \exp$. On the other hand, we denote by $\Gamma' = \pi(\Gamma)$, which is a discrete subgroup of G' and $L' = \exp_{G'} \mathfrak{l}'$ its corresponding syndetic hull. Since $\Gamma \not\subset G_0$, there exists $T \in \mathfrak{g} \backslash \mathfrak{g}_0$ such that $\exp T \in \Gamma$. This gives that $\rho(T) \in \mathfrak{l}' \backslash \rho(\mathfrak{g}_0)$. Therefore, $\mathfrak{l}' \not\subset \rho(\mathfrak{g}_0)$. Let $\overline{W} = \rho(W)$ for $W \in \mathfrak{g}$. According to Lemma 1.1.13, there exists a strong Malcev basis $\mathscr{B}' = \{\overline{X}, \overline{Y}_1, \dots, \overline{Y}_{n-1}\}$ of $\mathfrak{g}'$ such that

$$\begin{aligned} [\overline{X}, \overline{Y}_i] &= \overline{Y}_{i+1}, \quad i = 1, \dots, n-2, \\ [\overline{Y}_i, \overline{Y}_j] &= \overline{0}, \quad i, j = 1, \dots, n-1 \end{aligned}$$

and $\mathfrak{l}' = \mathbb{R}\text{-span}\{\overline{X}, \overline{Y}_{n-p+1}, \dots, \overline{Y}_{n-1}\}$ with $p = \dim \mathfrak{l}'$. We now need the following lemma.

Lemma 7.4.3. *Γ is non-Abelian if and only if Γ' is.*

Proof. Assume that Γ' is Abelian, then so is $\mathfrak{l}'$. Therefore, $\mathfrak{l}' = \mathbb{R}\text{-span}\,\{\overline{X}\}$ or $\mathfrak{l}' = \mathbb{R}\text{-span}\,\{\overline{X}, \overline{Y}_{n-1}\}$. In the case where $\mathfrak{l}' = \mathbb{R}\text{-span}\,\{\overline{X}\}$, it is clear that Γ is Abelian. Otherwise, $\Gamma' = \exp_{G'}(\mathbb{Z}\overline{X})\exp_{G'}(\mathbb{Z}\overline{Y}_{n-1})$, which implies that $\Gamma = \exp(\mathbb{Z}(X + \alpha Y_n))\exp(\mathbb{Z}(Y_{n-1} + \alpha_{n-1}Y_n))(\Gamma \cap Z(G))$ for some $\alpha, \alpha_{n-1} \in \mathbb{R}$, and finally Γ is Abelian. The inverse implication is trivial. □

According to Lemma 7.4.3, we have that $p \geqslant 3$. We can assume that

$$\mathfrak{l}' = \mathbb{R}\text{-span}\{\overline{X}, \overline{Z}_{n-p+1}, \overline{Z}_{n-p+2}, \dots, \overline{Z}_{n-2}, \overline{Z}_{n-1}\},$$

where
(i) $Z_{n-p+1} = Y_{n-p+1}$,
(ii) $Z_j = Y_j + \sum_{i=j+1}^{n-1} u_{i,j} Y_i$, $j = n-p+2, \dots, n-2$,
(iii) $Z_{n-1} = Y_{n-1}$

such that $u_{i,j} \in \mathbb{R}$ and satisfying

$$\exp_{G'}(\overline{X})\exp_{G'}(\overline{Z}_j)\exp_{G'}(-\overline{X})\exp_{G'}(-\overline{Z}_j) = \exp_{G'}(\overline{Z}_{j+1}), \quad j = n-p+1, \dots, n-2.$$

This gives that

$$L' = \exp_{G'}(\mathbb{R}\overline{X})\exp_{G'}(\mathbb{R}\overline{Z}_{n-p+1})\cdots\exp_{G'}(\mathbb{R}\overline{Z}_{n-1}).$$

Then

$$\Gamma' = \exp_{G'}(\mathbb{Z}\overline{X})\exp_{G'}(\mathbb{Z}\overline{Z}_{n-p+1})\cdots\exp_{G'}(\mathbb{Z}\overline{Z}_{n-1}),$$

and there exist $\alpha, \alpha_j \in \mathbb{R}$ for $n-p+1 \leqslant j \leqslant n-1$ such that

$$\Gamma = \exp(\mathbb{Z}(X + \alpha Y_n))\exp(\mathbb{Z}(Z_{n-p+1} + \alpha_{n-p+1}Y_n))\cdots\exp(\mathbb{Z}(Z_{n-1} + \alpha_{n-1}Y_n))(\Gamma \cap Z(G)).$$

□

7.4.1 Description of Hom(Γ, G)

Our main result in this subsection consists in giving an explicit description of $\mathrm{Hom}(\Gamma, G)$. We regard the product set $G^k = G \times \cdots \times G$ as a set of matrices in $\mathscr{M}_{n+1,k}(\mathbb{C})$ defined by

$$\mathscr{E}_k = \left\{ \begin{pmatrix} C \\ e^{2i\pi z} \end{pmatrix} \in \mathscr{M}_{n+1,k}(\mathbb{C}) \;\middle|\; \begin{array}{l} C \in \mathscr{M}_{n,k}(\mathbb{R}), \\ z := (z_1, \dots, z_k) \in \mathbb{R}^k, \\ e^{2i\pi z} := (e^{2i\pi z_1} \quad \dots \quad e^{2i\pi z_k}) \end{array} \right\}.$$

Recall that

$$\varepsilon = \begin{cases} 0 & \text{if } \Gamma \text{ is torsion-free,} \\ 1 & \text{otherwise.} \end{cases}$$

In the case where Γ is not torsion-free, we assume that $s \in \mathbb{N}^*$ is the cardinality of $\Gamma \cap Z(G)$ and

$$\gamma_n = \begin{pmatrix} 0 \\ e^{2i\pi\frac{1}{s}} \end{pmatrix}$$

one generator. Let $\{\gamma, \gamma_q, \dots, \gamma_{n-1+\varepsilon}\}$ be a family of generators of Γ, we consider the injective map

$$\begin{aligned} \Psi : \operatorname{Hom}(\Gamma, G) &\longrightarrow \mathscr{M}_{n+1,k}(\mathbb{C}) \\ \varphi &\longmapsto \lfloor \varphi(\gamma), \varphi(\gamma_q), \dots, \varphi(\gamma_{n-1+\varepsilon}) \rfloor. \end{aligned} \tag{7.28}$$

It is not hard to check that Ψ is a homeomorphism on its range. So, our task is reduced to give an explicit description of $\Psi(\operatorname{Hom}(\Gamma, G))$. Toward such a purpose, we define some matrices in $\mathscr{E}_p$:

$$M_u = \begin{pmatrix} u & 0 & 0 & \cdots & 0 \\ u_1 & 0 & 0 & \cdots & 0 \\ \vdots & \vdots & \vdots & \vdots & \vdots \\ u_{q-1} & 0 & 0 & \cdots & 0 \\ u_q & v_{q,q} & 0 & \cdots & 0 \\ u_{q+1} & v_{q+1,q} & u v_{q,q} & \ddots & \vdots \\ \vdots & \vdots & v_{q+2,q+1} & \ddots & \vdots \\ \vdots & \vdots & \vdots & & 0 \\ u_{n-1} & v_{n-1,q} & v_{n-1,q+1} & \cdots & u^{n-q-1} v_{q,q} \\ e^{2i\pi u_n} & e^{2i\pi v_{n,q}} & e^{2i\pi v_{n,q+1}} & \cdots & e^{2i\pi v_{n,n-1}} \end{pmatrix},$$

$$M_{u,\lambda} = \begin{pmatrix} u & \lambda u & 0 & 0 & 0 & \cdots & 0 \\ u_1 & \lambda u_1 & 0 & 0 & 0 & \cdots & 0 \\ \vdots & \vdots & \vdots & \vdots & \vdots & & \vdots \\ u_{n-3} & \lambda u_{n-3} & 0 & 0 & 0 & \cdots & 0 \\ u_{n-2} & v_{n-2,q} & 0 & 0 & 0 & \cdots & 0 \\ u_{n-1} & v_{n-1,q} & u(v_{n-2,q} - \lambda u_{n-2}) & 0 & 0 & \cdots & 0 \\ e^{2i\pi u_n} & e^{2i\pi v_{n,q}} & e^{2i\pi v_{n,q+1}} & e^{2i\pi v_{n,q+2}} & 1 & \cdots & 1 \end{pmatrix},$$

$$M_{v_q} = \begin{pmatrix} 0 & v_q & 0 & 0 & \cdots & 0 \\ 0 & v_{1,q} & 0 & 0 & \cdots & 0 \\ \vdots & \vdots & \vdots & \vdots & & \vdots \\ 0 & v_{n-3,q} & 0 & 0 & \cdots & 0 \\ u_{n-2} & v_{n-2,q} & 0 & 0 & \cdots & 0 \\ u_{n-1} & v_{n-1,q} & -u_{n-2}v_q & 0 & \cdots & 0 \\ e^{2i\pi u_n} & e^{2i\pi v_{n,q}} & e^{2i\pi v_{n,q+1}} & 1 & \cdots & 1 \end{pmatrix}$$

and

$$M_0 = \begin{pmatrix} 0 & 0 & 0 & \cdots & 0 \\ u_1 & v_{1,q} & 0 & \cdots & 0 \\ \vdots & \vdots & \vdots & & \vdots \\ u_{n-1} & v_{n-1,q} & 0 & \cdots & 0 \\ e^{2i\pi u_n} & e^{2i\pi v_{n,q}} & 1 & \cdots & 1 \end{pmatrix}.$$

Finally, we define the sets:

$$\begin{aligned} \mathscr{E}^0_{0,p} &= \{M_u \in \mathscr{E}_p : u \in \mathbb{R}^*\}, \\ \mathscr{E}^0_{1,p} &= \{M_{u,\lambda} \in \mathscr{E}_p : \lambda u \in \mathbb{R}^*\}, \\ \mathscr{E}^0_{2,p} &= \{M_{v_q} \in \mathscr{E}_p : v_q \in \mathbb{R}^*\}, \\ \mathscr{E}^0_{3,p} &= \{M_0 \in \mathscr{E}_p\}, \\ \mathscr{E}^1_{0,p} &= \{M_u \Cap {}^t(0_n, e^{2i\pi \frac{r}{s}}) \in \mathscr{E}_{p+1} : M_u \in \mathscr{E}^0_{0,p},\ r \in \{0,\dots,s-1\}\}, \\ \mathscr{E}^1_{1,p} &= \{M_{u,\lambda} \Cap {}^t(0_n, e^{2i\pi \frac{r}{s}}) \in \mathscr{E}_{p+1} : M_{u,\lambda} \in \mathscr{E}^0_{1,p},\ r \in \{0,\dots,s-1\}\}, \\ \mathscr{E}^1_{2,p} &= \{M_{v_q} \Cap {}^t(0_n, e^{2i\pi \frac{r}{s}}) \in \mathscr{E}_{p+1} : M_{v_q} \in \mathscr{E}^0_{2,p},\ r \in \{0,\dots,s-1\}\}, \\ \mathscr{E}^1_{3,p} &= \{M_0 \Cap {}^t(0_n, e^{2i\pi \frac{r}{s}}) \in \mathscr{E}_{p+1} : r \in \{0,\dots,s-1\}\} \end{aligned}$$

and

$$\mathscr{H}^\varepsilon_{j,p} = \mathscr{E}^\varepsilon_{j,p} \cap \Psi(\mathrm{Hom}(\Gamma, G)), \quad j \in \{0,1,2,3\},$$

where $\Cap$ merely means the concatenation of matrices.

The following proposition describes the structure of the set $\mathrm{Hom}(\Gamma, G)$.

Proposition 7.4.4. *Let G be a reduced threadlike Lie group and Γ a non-Abelian discrete subgroup of G. The set $\mathrm{Hom}(\Gamma, G)$ is homeomorphic to $\coprod_{j=0}^3 \mathscr{H}^0_{j,p}$ if Γ is torsion-free and to $\coprod_{j=0}^3 \mathscr{H}^1_{j,p}$ otherwise.*

Proof. According to Proposition 7.4.2, when Γ is torsion-free we have

$$\Gamma = \{\gamma^m \gamma_q^{m_q} \cdots \gamma_{n-1}^{m_{n-1}},\ m, m_j \in \mathbb{Z}\}$$

such that $q = n - p + 1$, $\gamma\gamma_{n-1}\gamma^{-1}\gamma_{n-1}^{-1} = e$, $\gamma\gamma_j\gamma^{-1}\gamma_j^{-1} = \gamma_{j+1}$; $q \leqslant j \leqslant n-2$ and $\gamma_i\gamma_j\gamma_i^{-1}\gamma_j^{-1} = e$; $q \leqslant i, j \leqslant n-1$. Let $\varphi \in \operatorname{Hom}(\Gamma, G)$ and

$$\Psi(\varphi) := M_\varphi = \begin{pmatrix} u & v_q & \cdots & v_{n-1} \\ u_1 & v_{1,q} & \cdots & v_{1,n-1} \\ \vdots & \vdots & & \vdots \\ u_{n-1} & v_{n-1,q} & \cdots & v_{n-1,n-1} \\ e^{2i\pi u_n} & e^{2i\pi v_{n,q}} & \cdots & e^{2i\pi v_{n,n-1}} \end{pmatrix}.$$

As $\gamma\gamma_{n-1}\gamma^{-1}\gamma_{n-1}^{-1} = e$, we get $\varphi(\gamma)\varphi(\gamma_{n-1})\varphi(\gamma)^{-1}\varphi(\gamma_{n-1})^{-1} = e$ and, therefore,

$$uv_{1,n-1} - u_1 v_{n-1} = 0.$$

Let $j \in \{q, \ldots, n-2\}$. We have $\gamma\gamma_j\gamma^{-1}\gamma_j^{-1} = \gamma_{j+1}$ and then $\varphi(\gamma)\varphi(\gamma_j)\varphi(\gamma)^{-1}\varphi(\gamma_j)^{-1} = \varphi(\gamma_{j+1})$. This gives that

$$\begin{cases} v_{j+1} = v_{1,j+1} = 0, \\ v_{2,j+1} = uv_{1,j} - u_1 v_j. \end{cases}$$

Likewise, for $i, j \in \{q, \ldots, n-1\}$, $\gamma_i\gamma_j\gamma_i^{-1}\gamma_j^{-1} = e$ and then $\varphi(\gamma_i)\varphi(\gamma_j)\varphi(\gamma_i)^{-1}\varphi(\gamma_j)^{-1} = e$. This also gives

$$v_i v_{1,j} - v_{1,i} v_j = 0.$$

Finally, we obtain the following:

$$\begin{cases} v_{j+1} = v_{1,j+1} = 0, & q \leqslant j \leqslant n-2, \quad (1) \\ v_{2,j+1} = uv_{1,j} - u_1 v_j, & q \leqslant j \leqslant n-2. \quad (2) \end{cases} \tag{7.29}$$

Equations (1) and (2) in (7.29) give $v_{2,j} = 0$ for all $j \in \{q+2, \ldots, n-1\}$. Therefore,

$$M_\varphi = \begin{pmatrix} u & v_q & 0 & 0 & \cdots & 0 \\ u_1 & v_{1,q} & 0 & 0 & \cdots & 0 \\ u_2 & v_{2,q} & uv_{1,q} - u_1 v_q & 0 & \cdots & 0 \\ u_3 & v_{3,q} & v_{3,q+1} & v_{3,q+2} & \cdots & v_{3,n-1} \\ \vdots & \vdots & \vdots & \vdots & & \vdots \\ u_{n-1} & v_{n-1,q} & v_{n-1,q+1} & v_{n-1,q+2} & \cdots & v_{n-1,n-1} \\ e^{2i\pi u_n} & e^{2i\pi v_{n,q}} & e^{2i\pi v_{n,q+1}} & e^{2i\pi v_{n,q+2}} & \cdots & e^{2i\pi v_{n,n-1}} \end{pmatrix}.$$

We are conclusively led to the following discussions:

Case 1: If $u \neq 0$, assume for a while that $v_q = 0$. For all $j \in \{q, \dots, n-2\}$, we have $\gamma\gamma_j\gamma^{-1}\gamma_j^{-1} = \gamma_{j+1}$ and then $\varphi(\gamma)\varphi(\gamma_j)\varphi(\gamma)^{-1}\varphi(\gamma_j)^{-1} = \varphi(\gamma_{j+1})$. Hence we get that

$$M_\varphi = \begin{pmatrix} u & 0 & 0 & \cdots & \cdots & 0 \\ u_1 & v_{1,q} & 0 & \cdots & \cdots & 0 \\ u_2 & v_{2,q} & uv_{1,q} & \ddots & & 0 \\ u_3 & v_{3,q} & v_{3,q+1} & \ddots & \ddots & \vdots \\ \vdots & \vdots & \vdots & & \ddots & 0 \\ u_{n-q} & v_{n-q,q} & v_{n-q,q+1} & \cdots & \cdots & u^{n-q-1}v_{1,q} \\ u_{n-q+1} & v_{n-q+1,q} & v_{n-q+1,q+1} & \cdots & \cdots & v_{n-q+1,n-1} \\ \vdots & \vdots & \vdots & & & \vdots \\ u_{n-1} & v_{n-1,q} & v_{n-1,q+1} & \cdots & \cdots & v_{n-1,n-1} \\ e^{2i\pi u_n} & e^{2i\pi v_{n,q}} & e^{2i\pi v_{n,q+1}} & \cdots & \cdots & e^{2i\pi v_{n,n-1}} \end{pmatrix}$$

with

$$\begin{cases} v_{i,j} = \sum_{k=1}^{i-1} \frac{u^k}{k!} v_{i-k,j-1}, & q+1 \leqslant j \leqslant n-1,\ 2 \leqslant i \leqslant n-1, \\ v_{nj} - \sum_{k=1}^{n-1} \frac{u^k}{k!} v_{n-k,j-1} \in \mathbb{Z}, & q+1 \leqslant j \leqslant n-1. \end{cases}$$

As $\gamma\gamma_{n-1}\gamma^{-1}\gamma_{n-1}^{-1} = e$, we get $\varphi(\gamma)\varphi(\gamma_{n-1})\varphi(\gamma)^{-1}\varphi(\gamma_{n-1})^{-1} = e$, which gives in turn that

$$\begin{cases} v_{i,q} = 0, & 1 \leqslant i \leqslant q-1, \\ v_{i,n-1} = 0, & 1 \leqslant i \leqslant n-2, \\ u^{n-q} v_{q,q} \in \mathbb{Z}. \end{cases}$$

Therefore,

$$M_\varphi = \begin{pmatrix} u & 0 & 0 & \cdots & 0 \\ u_1 & 0 & 0 & \cdots & 0 \\ \vdots & \vdots & \vdots & & \vdots \\ u_{q-1} & 0 & 0 & \cdots & 0 \\ u_q & v_{q,q} & 0 & \cdots & 0 \\ u_{q+1} & v_{q+1,q} & uv_{q,q} & \ddots & \vdots \\ \vdots & \vdots & v_{q+2,q+1} & \ddots & \vdots \\ \vdots & \vdots & \vdots & & 0 \\ u_{n-1} & v_{n-1,q} & v_{n-1,q+1} & \cdots & u^{n-q-1}v_{q,q} \\ e^{2i\pi u_n} & e^{2i\pi v_{n,q}} & e^{2i\pi v_{n,q+1}} & \cdots & e^{2i\pi v_{n,n-1}} \end{pmatrix} \in \mathscr{E}^0_{0,p}.$$

Let now $v_q \neq 0$. There exists $\lambda \in \mathbb{R}^*$ such that $v_q = \lambda u$. For all $j \in \{q, \dots, n-2\}$, we have $\gamma\gamma_j\gamma^{-1}\gamma_j^{-1} = \gamma_{j+1}$ and then $\varphi(\gamma)\varphi(\gamma_j)\varphi(\gamma)^{-1}\varphi(\gamma_j)^{-1} = \varphi(\gamma_{j+1})$. This entails in turn that

$$M_\varphi = \begin{pmatrix} u & \lambda u & 0 & \cdots & \cdots & 0 \\ u_1 & v_{1,q} & 0 & \cdots & \cdots & 0 \\ u_2 & v_{2,q} & u(v_{1,q} - \lambda u_1) & \ddots & & 0 \\ u_3 & v_{3,q} & v_{3,q+1} & \ddots & \ddots & \vdots \\ \vdots & \vdots & \vdots & & \ddots & 0 \\ u_{n-q} & v_{n-q,q} & v_{n-q,q+1} & \cdots & \cdots & u^{n-q-1}(v_{1,q} - \lambda u_1) \\ u_{n-q+1} & v_{n-q+1,q} & v_{n-q+1,q+1} & \cdots & \cdots & v_{n-q+1,n-1} \\ \vdots & \vdots & \vdots & & & \vdots \\ u_{n-1} & v_{n-1,q} & v_{n-1,q+1} & \cdots & \cdots & v_{n-1,n-1} \\ e^{2i\pi u_n} & e^{2i\pi v_{n,q}} & e^{2i\pi v_{n,q+1}} & \cdots & \cdots & e^{2i\pi v_{n,n-1}} \end{pmatrix}$$

with

$$\begin{cases} v_{i,j} = \sum_{k=1}^{i-1} \frac{u^k}{k!} v_{i-k,j-1}, & q+2 \leqslant j \leqslant n-1,\ 3 \leqslant i \leqslant n-1, \\ v_{nj} - \sum_{k=1}^{n-1} \frac{u^k}{k!} v_{n-k,j-1} \in \mathbb{Z}, & q+2 \leqslant j \leqslant n-1. \end{cases}$$

Besides, $\varphi(\gamma_q)\varphi(\gamma_{q+1})\varphi(\gamma_q)^{-1}\varphi(\gamma_{q+1})^{-1} = e$ and $\varphi(\gamma)\varphi(\gamma_q)\varphi(\gamma)^{-1}\varphi(\gamma_q)^{-1} = \varphi(\gamma_{q+1})$ and this allows us to write

$$\begin{cases} v_{j,q} = \lambda u_j, & 1 \leqslant j \leqslant n-3, \\ v_{i,q+1} = 0, & 1 \leqslant i \leqslant n-2, \\ v_{n-1,q+1} = u(v_{n-2,q} - \lambda u_{n-2}), \\ \lambda u^2 (v_{n-2,q} - \lambda u_{n-2}) \in \mathbb{Z}. \end{cases}$$

We get finally that $M_\varphi \in \mathscr{E}_{1,p}^0$.

Case 2: If $u = 0$ and $v_q \neq 0$, we have for all $j \in \{q, \dots, n-2\}$, $\gamma\gamma_j\gamma^{-1}\gamma_j^{-1} = \gamma_{j+1}$ and then $\varphi(\gamma)\varphi(\gamma_j)\varphi(\gamma)^{-1}\varphi(\gamma_j)^{-1} = \varphi(\gamma_{j+1})$. Hence

$$M_\varphi = \begin{pmatrix} 0 & v_q & 0 & 0 & \cdots & 0 \\ u_1 & v_{1,q} & 0 & 0 & \cdots & 0 \\ u_2 & v_{2,q} & v_{2,q+1} & 0 & \cdots & 0 \\ \vdots & \vdots & \vdots & \vdots & & \vdots \\ u_{n-3} & v_{n-3,q} & v_{n-3,q+1} & 0 & \cdots & 0 \\ u_{n-2} & v_{n-2,q} & v_{n-2,q+1} & 0 & \cdots & 0 \\ u_{n-1} & v_{n-1,q} & v_{n-1,q+1} & 0 & \cdots & 0 \\ e^{2i\pi u_n} & e^{2i\pi v_{n,q}} & e^{2i\pi v_{n,q+1}} & 1 & \cdots & 1 \end{pmatrix}$$

with

$$\begin{cases} v_{i,q+1} = -\sum_{k=1}^{i-1} \frac{v_q{}^k}{k!} u_{i-k}, & 2 \leqslant i \leqslant n-1, \\ v_{n,q+1} + \sum_{k=1}^{n-1} \frac{v_q{}^k}{k!} u_{n-k} \in \mathbb{Z}. \end{cases}$$

We have $\varphi(\gamma_q)\varphi(\gamma_{q+1})\varphi(\gamma_q)^{-1}\varphi(\gamma_{q+1})^{-1} = e$ and then $u_i = 0$ for all $i \in \{1, \ldots, n-3\}$, $u_{n-2}v_q^2 \in \mathbb{Z}$ and $M_\varphi \in \mathscr{E}_{2,p}^0$; otherwise, $M_\varphi \in \mathscr{E}_{3,p}^0$. Now, when Γ is not torsion-free, we get an additional column at the representative matrices of homomorphisms, which reads as $\varphi(\gamma_n) = \gamma_n^r$ such that $r \in \{0, \ldots, s-1\}$. □

From now on, we identify any homomorphism $\varphi \in \operatorname{Hom}(\Gamma, G)$ with its corresponding matrix $\Psi(\varphi) \in \coprod_{j=0}^{3} \mathscr{H}_{j,p}^{\varepsilon}$. Let

$$\mathscr{K}_{j,p}^{\varepsilon} = \Psi(\operatorname{Hom}_{\mathrm{d}}^0(\Gamma, G)) \cap \mathscr{H}_{j,p}^{\varepsilon}, \quad j \in \{0, 1, 2, 3\}.$$

The following two propositions accurately determine the structure of $\operatorname{Hom}_{\mathrm{d}}^0(\Gamma, G)$.

Proposition 7.4.5. *Keep the same notation, and assume that Γ is torsion-free. Then $\operatorname{Hom}_d^0(\Gamma, G)$ is homeomorphic to $\mathscr{K}$ such that:*
(1) *If $p \geqslant 4$, then $\mathscr{K} = \mathscr{K}_{0,p}^0 = \{M \in \mathscr{H}_{0,p}^0 : v_{q,q} \neq 0\}$.*
(2) *If $p = 3$, then $\mathscr{K} = \coprod_{j=0}^{2} \mathscr{K}_{j,3}^0$, where*

$$\mathscr{K}_{0,3}^0 = \{M \in \mathscr{H}_{0,3}^0 : v_{n-2,n-2} \neq 0\},$$
$$\mathscr{K}_{1,3}^0 = \{M \in \mathscr{H}_{1,3}^0 : v_{n-2,n-2} - \lambda u_{n-2} \neq 0\}$$

and

$$\mathscr{K}_{2,3}^0 = \{M \in \mathscr{H}_{2,3}^0 : u_{n-2} \neq 0\}.$$

Proof. We have

$$\Psi(\operatorname{Hom}_{\mathrm{d}}^0(\Gamma, G)) \cap \mathscr{H}_{3,p}^0 = \emptyset.$$

In fact, let $\varphi \in \mathscr{H}_{3,p}^0$, then we have $\varphi(\gamma)\varphi(\gamma_q)(\varphi(\gamma))^{-1}(\varphi(\gamma_q))^{-1} = e$. This implies that $\varphi(\gamma\gamma_q\gamma^{-1}\gamma_q^{-1}) = e$. If φ is injective, then $\gamma\gamma_q\gamma^{-1}\gamma_q^{-1} = e$ and, therefore, Γ is Abelian, which is impossible.

We begin by the case where $p \geqslant 4$, we have

$$\Psi(\operatorname{Hom}_{\mathrm{d}}^0(\Gamma, G)) \cap \left(\mathscr{H}_{1,p}^0 \coprod \mathscr{H}_{2,p}^0 \coprod \mathscr{H}_{3,p}^0\right) = \emptyset.$$

In fact, let $\varphi \in \mathscr{H}_{j,p}^0$, $j = 1$. Then $\varphi(\gamma_{n-p+3}) = \exp(v_{n,n-p+3}Y_n) \in \varphi(\Gamma) \cap Z(G)$. Suppose that $\varphi(\Gamma)$ is discrete, then $v_{n,n-p+3} = \frac{a}{b} \in \mathbb{Q}$ for some $b \neq 0$. Therefore, $\varphi(\gamma_{n-p+3}^b) = e$ and φ is not injective. Let now $\varphi \in \operatorname{Hom}(\Gamma, G)$ and $M_\varphi \in \mathscr{K}_{0,p}^0$. It is not hard to show that φ is injective and $\varphi(\Gamma)$ is discrete. Assume that $M_\varphi \in \mathscr{H}^0_{0,p}$ with $v_{q,q} = 0$, then $\varphi(y_{n-1}) = \exp(v_{n,n-1}Y_n) \in \varphi(\Gamma) \cap Z(G)$, which is discrete if $\varphi(\Gamma)$ is discrete and then $v_{n,n-1} = \frac{a}{b} \in \mathbb{Q}$, $b \neq 0$. Therefore, $\varphi(y^b_{n-1}) = e$ and φ is not injective. We treat similarly the case where $p = 3$. □

The case where Γ is not torsion-free is straightforward.

Proposition 7.4.6. *With the same notation, when Γ is not torsion-free, $\mathrm{Hom}^0_{\mathrm{d}}(\Gamma, G)$ is homeomorphic to $\mathscr{K}$ such that:*

(1) *If $p \geqslant 4$, then*

$$\mathscr{K} = \mathscr{K}^1_{0,p} = \{M \in \mathscr{H}^1_{0,p} : v_{q,q} \neq 0,\ r \in \{1,\dots,s-1\} \text{ and } r \wedge s = 1\}.$$

(2) *If $p = 3$, then $\mathscr{K} = \coprod_{j=0}^{2} \mathscr{K}^1_{j,3}$, where*

$$\mathscr{K}^1_{0,3} = \{M \in \mathscr{H}^1_{0,3} : v_{n-2,n-2} \neq 0,\ r \in \{1,\dots,s-1\} \text{ and } r \wedge s = 1\},$$
$$\mathscr{K}^1_{1,3} = \{M \in \mathscr{H}^1_{1,3} : v_{n-2,n-2} - \lambda u_{n-2} \neq 0,\ r \in \{1,\dots,s-1\} \text{ and } r \wedge s = 1\}$$

and

$$\mathscr{K}^1_{2,3} = \{M \in \mathscr{H}^1_{2,3} : u_{n-2} \neq 0,\ r \in \{1,\dots,s-1\} \text{ and } r \wedge s = 1\}.$$

7.4.2 Explicit determination of the parameter space

The following result follows from the Theorem 7.3.2.

Proposition 7.4.7. *Let G be a reduced threadlike Lie group, $H = \exp \mathfrak{h}$ a closed, connected subgroup of G and Γ a non-Abelian discontinuous group for G/H. We have*

$$\mathscr{R}(\Gamma, G, H) = \{\varphi \in \mathrm{Hom}^0_{\mathrm{d}}(\Gamma, G) : \mathrm{Ad}_g\, \mathfrak{h} \cap \mathfrak{l}_\varphi \subseteq \mathfrak{z}(\mathfrak{g}) \text{ for any } g \in G\},$$

where $\mathfrak{l}_\varphi$ is the Lie subalgebra associated to the syndetic hull of $\varphi(\Gamma)$ (as in Theorem 1.4.12).

Proof. Let $\varphi \in \mathrm{Hom}^0_{\mathrm{d}}(\Gamma, G)$. We first show that the proper action of $\varphi(\Gamma)$ on G/H implies its free action. It is clear that the proper action implies that the triplet $(G, H, \varphi(\Gamma))$ is (CI), which gives that for all $g \in G$, the subgroup $K := \varphi(\Gamma) \cap gHg^{-1}$ is central and then finite as $\varphi(\Gamma)$ is discrete. As the map $\varphi : \Gamma \to \varphi(\Gamma)$ is a group isomorphism and K is finite and cyclic, we get that $\varphi^{-1}(K) = K$. Therefore, $K \subset \Gamma \cap H = \{e\}$. Thus, the action of $\varphi(\Gamma)$ on G/H is free. As L_φ contains $\varphi(\Gamma)$ cocompactly,

$$\mathscr{R}(\Gamma, G, H) = \{\varphi \in \mathrm{Hom}^0_{\mathrm{d}}(\Gamma, G) : L_\varphi \text{ acts properly on } G/H\}.$$

Now, Theorem 7.9 allows us to conclude. □

Using Theorem 7.9, the proper action of Γ on G/H implies that $r \leqslant n-p+\dim(\mathfrak{h}\cap\mathfrak{z}(\mathfrak{g}))$ where $r = \dim \mathfrak{h}$. We assume from now on that this previous condition is satisfied. Let $g \in G$, $\mathscr{B}_{\mathrm{Ad}_g \mathfrak{h}}$ a strong Malcev basis of $\mathrm{Ad}_g \mathfrak{h}$ extracted from $\mathscr{B}$ and $M_{\mathrm{Ad}_g \mathfrak{h},\mathscr{B}}$ the matrix of $\mathrm{Ad}_g \mathfrak{h}$ written in the basis $\mathscr{B} = \{X, Y_1, \dots, Y_n\}$. We put

$$M_{\mathrm{Ad}_g \mathfrak{h},\mathscr{B}} = \begin{pmatrix} M^1_{\mathrm{Ad}_g \mathfrak{h},\mathscr{B}} \\ z_{\mathrm{Ad}_g \mathfrak{h}} \end{pmatrix}, \quad z_{\mathrm{Ad}_g \mathfrak{h}} \in \mathbb{R}^r.$$

In light of Proposition 7.4.7, the following result is immediate.

Lemma 7.4.8. *Let G be a reduced threadlike Lie group, $H = \exp \mathfrak{h}$ a nontrivial, closed, connected subgroup of G and Γ a non-Abelian discontinuous group for G/H. The parameter space is homeomorphic to*

$$\mathscr{R} = \left\{ \begin{pmatrix} C \\ e^{2i\pi z} \end{pmatrix} \in \mathscr{K} : \mathrm{rk}(C \Cap M^1_{\mathrm{Ad}_g \mathfrak{h},\mathscr{B}}) = p + r - \dim(\mathfrak{h} \cap \mathfrak{z}(\mathfrak{g}))\ \textit{for any}\ g \in G \right\}.$$

We denote by $\mathscr{R}^{\varepsilon}_{j,p} = \Psi(\mathscr{R}(\Gamma, G, H)) \cap \mathscr{K}^{\varepsilon}_{j,p}$. Then according to Propositions 7.4.5 and 7.4.6, we have that $\mathscr{R} = \mathscr{R}^{\varepsilon}_{0,p}$ whenever $p \geqslant 4$. Otherwise, $\mathscr{R} = \coprod_{j=0}^{2} \mathscr{R}^{\varepsilon}_{j,3}$. Let first $p \geqslant 4$. We denote by

$$\mathscr{I}^{\mathfrak{h}}_{\mathscr{B}} = \{i_1 < \cdots < i_r\} \quad (r = \dim \mathfrak{h})$$

the set of indices $i \in \{1, \dots, n\}$ such that $\mathfrak{h}\cap\mathfrak{g}^i \neq \mathfrak{h}\cap\mathfrak{g}^{i+1}$, where $\mathfrak{g}^i = \mathbb{R}\text{-span}\{Y_i, \dots, Y_n\}$, $i = 1, \dots, n$, $\mathfrak{g}^0 = \mathfrak{g}$, and $\mathfrak{g}^{n+1} = \{0\}$.

Proposition 7.4.9. *We keep the same hypotheses and notation as before. Suppose that $p \geqslant 4$ and $\mathfrak{h} \not\subset \mathfrak{g}_0$. Then there exist $h_1, \dots, h_n \in \mathbb{R}$ such that*

$$\mathfrak{h} = \mathbb{R}\text{-span}\{X + h_1 Y_1 + \cdots + h_n Y_n\} \oplus \mathfrak{h} \cap \mathfrak{z}(\mathfrak{g})$$

and

$$\mathscr{R}^{\varepsilon}_{0,p} = \{M \in \mathscr{K}^{\varepsilon}_{0,p} : (u_1 - h_1 u) \in \mathbb{R}^*\}.$$

Proof. If $\mathfrak{z}(\mathfrak{g}) \not\subset \mathfrak{h}$, then it is clear that $\mathfrak{h} = \mathbb{R}\text{-span}\{\widetilde{X}\}$ for some $\widetilde{X} = X + h_1 Y_1 + \cdots + h_n Y_n$ where $h_1, \dots, h_n \in \mathbb{R}$. Otherwise, as Γ acts freely on G/H, we have $\mathfrak{h} = \mathbb{R}\text{-span}\{X + h_1 Y_1 + \cdots + h_n Y_n, Y_n\}$ for some $h_1, \dots, h_n \in \mathbb{R}$.

Let $M \in \mathscr{K}^{\varepsilon}_{0,p}$. Clearly, we have

$$\begin{aligned} M \in \mathscr{R} &\Leftrightarrow \mathrm{rk}(C \Cap M^1_{\mathrm{Ad}_g \mathfrak{h},\mathscr{B}}) = p + r - \dim(\mathfrak{h} \cap \mathfrak{z}(\mathfrak{g})) \quad \text{for all } g \in G \\ &\Leftrightarrow \mathrm{rk}\lfloor {}^t(u, u_1, \dots, u_{n-1}), {}^t(1, h_1, \alpha_2, \dots, \alpha_{n-1}) \rfloor = 2 \quad \text{for all } \alpha_2, \dots, \alpha_{n-1} \in \mathbb{R} \\ &\Leftrightarrow u_1 - uh_1 \in \mathbb{R}^*. \end{aligned}$$

□

Proposition 7.4.10. *We keep the same hypotheses and notation as before. If $p \geqslant 4$ and $\mathfrak{h} \subset \mathfrak{g}_0$, then $\mathscr{I}_{\mathscr{B}}^{\mathfrak{h}} \subset \{1, \dots, q-1, n\}$ and $\mathscr{R}_{0,p}^{\varepsilon} = \mathscr{K}_{0,p}^{\varepsilon}$.*

Proof. Suppose that $\mathscr{I}_{\mathscr{B}}^{\mathfrak{h}} \cap \{q, \dots, n-1\} \neq \emptyset$, then there exists $i_0 \in \{q, \dots, n-1\}$ and $\widetilde{Y} = Y_{i_0} + h_{i_0+1}Y_{i_0+1} + \cdots + h_n Y_n \in \mathfrak{h}$ for some $h_{i_0+1}, \dots, h_n \in \mathbb{R}$, which is impossible as L acts on G/H properly. So, $\mathscr{I}_{\mathscr{B}}^{\mathrm{Ad}_g \mathfrak{h}} \cap \{q, \dots, n-1\} = \emptyset$ for all $g \in G$. This gives that $\mathrm{rk}(C ⋒ M^1_{\mathrm{Ad}_g \mathfrak{h}, \mathscr{B}}) = r + p - \dim(\mathfrak{h} \cap \mathfrak{z}(\mathfrak{g}))$ for all

$$M = \begin{pmatrix} C \\ e^{2i\pi z} \end{pmatrix} \in \mathscr{K}_{0,p}^{\varepsilon}, \quad g \in G.$$ □

We argue similarly as in the previous propositions to treat the case where $p = 3$.

Proposition 7.4.11. *We keep the same hypotheses and notation as before. Suppose that $p = 3$ and $\mathfrak{h} \not\subset \mathfrak{g}_0$. Then there exist $h_1, \dots, h_n \in \mathbb{R}$ such that*

$$\mathfrak{h} = \mathbb{R}\text{-span}\{X + h_1 Y_1 + \cdots + h_n Y_n\} \oplus \mathfrak{h} \cap \mathfrak{z}(\mathfrak{g}).$$

Moreover, we have

$$\mathscr{R}_{j,3}^{\varepsilon} = \{M \in \mathscr{K}_{j,3}^{\varepsilon} : (u_1 - h_1 u) \in \mathbb{R}^*\}, \quad j \in \{0, 1\}$$

and

$$\mathscr{R}_{2,3}^{\varepsilon} = \{M \in \mathscr{K}_{2,3}^{\varepsilon} : (v_{1,q} - h_1 v_q) \in \mathbb{R}^*\}.$$

Proposition 7.4.12. *We keep the same hypotheses and notation as before. If $p = 3$ and $\mathfrak{h} \subset \mathfrak{g}_0$, then $\mathscr{I}_{\mathscr{B}}^{\mathfrak{h}} \subset \{1, \dots, q-1, n\}$ and $\mathscr{R}_{j,3}^{\varepsilon} = \mathscr{K}_{j,3}^{\varepsilon}$, $j \in \{0, 1, 2\}$.*

Proposition 7.4.13. *Let $G = \mathbb{G}_3^r$ be the 4-dimensional reduced threadlike Lie group, $H = \exp \mathfrak{h}$ a nontrivial closed connected Lie subgroup of G and Γ a non-Abelian discontinuous group for G/H. We have $p = 3$, $\mathfrak{h} = \mathfrak{z}(\mathfrak{g})$ and $\mathscr{R}_{j,3}^{0} = \mathscr{K}_{j,3}^{0}$, $j \in \{0, 1, 2\}$.*

Proof. As $r \leqslant 3 - p + \dim(\mathfrak{h} \cap \mathfrak{z}(\mathfrak{g}))$, we have that $r \leqslant \dim(\mathfrak{h} \cap \mathfrak{z}(\mathfrak{g}))$. If $\mathfrak{z}(\mathfrak{g}) \not\subset \mathfrak{h}$, then H is trivial, which is impossible. Now, when $\mathfrak{z}(\mathfrak{g}) \subset \mathfrak{h}$, we get $r = 1$ and $\mathfrak{h} = \mathfrak{z}(\mathfrak{g})$. Hence, $\mathscr{R}_{j,3}^{0} = \mathscr{K}_{j,3}^{0}$, $j \in \{0, 1, 2\}$. □

7.4.3 Proof of Theorem 7.4.1

We start by proving this lemma that will be used later.

Lemma 7.4.14. *The disjoint components $\mathscr{R}_{j,p}^{\varepsilon}$, $j \in \{0, 1, 2\}$ are G-invariant.*

Proof. The group G acts on $\mathrm{Hom}(\mathfrak{l}, \mathfrak{g})$ through the law $g \star \varphi = \mathrm{Ad}_g \circ \varphi$. Let

$$M = \begin{pmatrix} u & & \vec{0} & \\ u_1 & & & \\ U_2 & & N & \\ e^{2i\pi u_n} & e^{2i\pi v_{n,q}} & \dots & e^{2i\pi v_{n,n-1}} \end{pmatrix} \in \mathscr{R}^0_{0,p},$$

where $U_2 = {}^t(u_2, \dots, u_{n-1}) \in \mathbb{R}^{n-2}$ and $N = \lfloor W_q, \dots, W_{n-1} \rfloor \in \mathscr{M}_{n-2,n-q}(\mathbb{R})$ with

$$\begin{cases} W_q = {}^t(v_{q,q}, \dots, v_{n-1,q}) \in \mathbb{R}^{n-2}, \\ W_{q+1} = {}^t(0, uv_{q,q}, v_{q+2,q+1}, \dots, v_{n-1,q+1}) \in \mathbb{R}^{n-2}, \\ W_{q+2} = {}^t(0, 0, u^2 v_{q,q}, v_{q+3,q+2}, \dots, v_{n-1,q+2}) \in \mathbb{R}^{n-2}, \\ \vdots \\ W_{n-2} = {}^t(0, \dots, 0, u^{n-q-2} v_{q,q}, v_{n-1,n-2}) \in \mathbb{R}^{n-2}, \\ W_{n-1} = {}^t(0, \dots, 0, u^{n-q-1} v_{q,q}) \in \mathbb{R}^{n-2}. \end{cases}$$

Let $g = \exp(xX + y_1 Y_1 + \cdots + y_n Y_n) \in G$ with $x, y_1, \dots, y_n \in \mathbb{R}$. We have

$$g \star M = \begin{pmatrix} u & & \vec{0} & \\ u_1 & & & \\ U_2' & & A_{n-3}(x)N & \\ e^{2i\pi u_n'} & e^{2i\pi v_{n,q}'} & \dots & e^{2i\pi v_{n,n-1}'} \end{pmatrix},$$

where

$$A_n(x) = \begin{pmatrix} 1 & 0 & \cdots & \cdots & 0 \\ x & 1 & \ddots & & \vdots \\ \frac{x^2}{2} & x & 1 & \ddots & \vdots \\ \vdots & \ddots & \ddots & \ddots & 0 \\ \frac{x^n}{(n)!} & \cdots & \frac{x^2}{2} & x & 1 \end{pmatrix} \in \mathscr{M}_{n+1}(\mathbb{R}),$$

$${}^t(U_2', e^{2i\pi u_n'}) = g \star U_1 - u_1 Y_1, \quad \text{with } U_1 = {}^t(u_1, U_2, e^{2i\pi u_n})$$

and

$$\begin{cases} v_{n,q}' = v_{n,q} + \sum_{j=1}^{n-q} \frac{x^j}{j!} v_{n-j,q}, \\ v_{n,q+1}' = v_{n,q+1} + \sum_{j=1}^{n-q-2} \frac{x^j}{j!} v_{n-j,q+1} + \frac{x^{n-q-1}}{(n-q-1)!} uv_{q,q}, \\ \vdots \\ v_{n,n-2}' = v_{n,n-2} + xv_{n-1,n-2} + \frac{x^2}{2} u^{n-q-2} v_{q,q}, \\ v_{n,n-1}' = v_{n,n-1} + xu^{n-q-1} v_{q,q}. \end{cases}$$

It is then clear that $g * M \in \mathscr{R}^0_{0,p}$. Likewise, we obtain that $\mathscr{R}^1_{0,p}$ is G-invariant. Suppose now that $p = 3$. Let

$$M = \begin{pmatrix} u & \lambda u & 0 \\ u_1 & \lambda u_1 & 0 \\ \vdots & \vdots & \vdots \\ u_{n-3} & \lambda u_{n-3} & 0 \\ u_{n-2} & v_{n-2,q} & 0 \\ u_{n-1} & v_{n-1,q} & u(v_{n-2,q} - \lambda u_{n-2}) \\ e^{2i\pi u_n} & e^{2i\pi v_{n,q}} & e^{2i\pi v_{n,q+1}} \end{pmatrix} \in \mathscr{R}^0_{1,3}.$$

We have

$$g \star M = \begin{pmatrix} u & \lambda u & 0 \\ u_1 & \lambda u_1 & 0 \\ u'_2 & \lambda u'_2 & 0 \\ \vdots & \vdots & \vdots \\ u'_{n-3} & \lambda u'_{n-3} & 0 \\ u'_{n-2} & v'_{n-2,q} & 0 \\ u'_{n-1} & v'_{n-1,q} & u(v_{n-2,q} - \lambda u_{n-2}) \\ e^{2i\pi u'_n} & e^{2i\pi v'_{n,q}} & e^{2i\pi v'_{n,q+1}} \end{pmatrix},$$

where

$$\begin{cases} u'_i = u_i + \sum_{j=1}^{i-1} \frac{x^{j-1}}{j!}(xu_{i-j} - uy_{i-j}), \quad i \in \{2, \dots, n\}, \\ v'_{n-2,q} = v_{n-2,q} + \lambda \sum_{j=1}^{n-3} \frac{x^{j-1}}{j!}(xu_{n-j-2} - uy_{n-j-2}), \\ v'_{n-1,q} = v_{n-1,q} + (xv_{n-2,q} - \lambda uy_{n-2}) + \lambda \sum_{j=2}^{n-2} \frac{x^{j-1}}{j!}(xu_{n-j-1} - uy_{n-j-1}), \\ v'_{n,q} = v_{n,q} + (xv_{n-1,q} - \lambda uy_{n-1}) + \frac{x}{2}(xv_{n-2,q} - \lambda uy_{n-2}) + \lambda \sum_{j=3}^{n-1} \frac{x^{j-1}}{j!}(xu_{n-j-1} - uy_{n-j-1}), \\ v'_{n,q+1} = v_{n,q+1} + xu(v_{n-2,q} - \lambda u_{n-2}). \end{cases}$$

The result follows in this case. Similarly, we obtain that $\mathscr{R}^1_{1,3}$ is G-invariant.

Let

$$M = \begin{pmatrix} 0 & v_q & 0 \\ 0 & v_{1,q} & 0 \\ \vdots & \vdots & \vdots \\ 0 & v_{n-3,q} & 0 \\ u_{n-2} & v_{n-2,q} & 0 \\ u_{n-1} & v_{n-1,q} & -u_{n-2}v_q \\ e^{2i\pi u_n} & e^{2i\pi v_{n,q}} & e^{2i\pi v_{n,q+1}} \end{pmatrix} \in \mathscr{R}^0_{2,3}.$$

We have

$$g * M = \begin{pmatrix} 0 & v_q & 0 \\ 0 & v_{1,q} & 0 \\ 0 & v'_{2,q} & 0 \\ \vdots & \vdots & \vdots \\ 0 & v'_{n-3,q} & 0 \\ u_{n-2} & v'_{n-2,q} & 0 \\ u'_{n-1} & v'_{n-1,q} & -u_{n-2}v_q \\ e^{2i\pi u'_n} & e^{2i\pi v'_{n,q}} & e^{2i\pi v'_{n,q+1}} \end{pmatrix},$$

where

$$\begin{cases} u'_{n-1} = u_{n-1} + xu_{n-2}, \\ u'_n = u_n + xu_{n-1} + \frac{x^2}{2}u_{n-2}, \\ v'_{i,q} = v_{i,q} + \sum_{j=1}^{i-1} \frac{x^{j-1}}{j!}(xv_{i-j,q} - v_q y_{i-j}), \quad i \in \{2, \dots, n\}, \\ v'_{n,q+1} = v_{n,q+1} - xu_{n-2}v_q. \end{cases}$$

The result also follows in this case. □

It is sufficient to prove Theorem 7.4.1 when Γ is torsion-free. We first prove that $\mathscr{R}$ is open in $\mathscr{K}$. In the case when $p \geqslant 4$, the result stems directly from Propositions 7.4.9 and 7.4.10. Suppose now that $p = 3$. It suffices then to see that

$$\mathscr{K} \setminus \mathscr{R} = \coprod_{j=0}^{2} (\mathscr{K}^0_{j,3} \setminus \mathscr{R}^0_{j,3})$$

is closed in $\mathscr{K}$. Let then $M \in \overline{\mathscr{K} \setminus \mathscr{R}}^{\mathscr{K}}$. There exists therefore a sequence $(M_m)_{m\in\mathbb{N}}$ assumed to belong to $\mathscr{K} \setminus \mathscr{R}$, which converges to M. So we can extract from $(M_m)_{m\in\mathbb{N}}$ a subsequence $(M_{m_s})_{s\in\mathbb{N}}$ of elements in $\mathscr{K}^0_{j,3} \setminus \mathscr{R}^0_{j,3}$ for some $j \in \{0, 1, 2\}$. If $M_{m_s} \in \mathscr{K}^0_{j,3} \setminus \mathscr{R}^0_{j,3}$ for $j \in \{0, 2\}$, then obviously its limit M belongs to $\mathscr{K}^0_{j,3} \setminus \mathscr{R}^0_{j,3}$.

Suppose now that

$$M_{m_s} = \begin{pmatrix} u^s & \lambda_s u^s & 0 \\ u^s_1 & \lambda_s u^s_1 & 0 \\ \vdots & \vdots & \vdots \\ u^s_{n-3} & \lambda_s u^s_{n-3} & 0 \\ u^s_{n-2} & v^s_{n-2,n-2} & 0 \\ u^s_{n-1} & v^s_{n-1,n-2} & u^s(v^s_{n-2,n-2} - \lambda_s u^s_{n-2}) \\ e^{2i\pi u^s_n} & e^{2i\pi v^s_{n,n-2}} & e^{2i\pi v^s_{n,n-1}} \end{pmatrix} \in \mathscr{K}^0_{1,3} \setminus \mathscr{R}^0_{1,3}$$

for some real sequence $(\lambda_s)_{s\in\mathbb{N}}$. If $(\lambda_s)_s$ goes to infinity as s goes to $+\infty$, then we can easily check that $M \in \mathscr{K}^0_{2,3} \setminus \mathscr{R}^0_{2,3}$. Otherwise, $M \in (\mathscr{K}^0_{0,3} \setminus \mathscr{R}^0_{0,3}) \coprod (\mathscr{K}^0_{1,3} \setminus \mathscr{R}^0_{1,3})$. We get therefore that $M \in \mathscr{K} \setminus \mathscr{R}$. We argue similarly to obtain that $\mathscr{K}$ is open in $\coprod_{j=0}^{3} \mathscr{H}_{j,p}^{0}$. This completes the proof of our theorem. □

7.4.4 A concluding remark

Table 7.1 summarizes the results concerning local rigidity, stability and Hausdorfness in the setting of (double) Heisenberg and threadlike Lie groups.

Table 7.1: Summarizing the rigidity and stability results.

	Rigidity	**Stability**	**Hausdorffness**
$G = H_{2n+1}$, H and Γ are arbitrary	Local rigidity fails to hold	Stability holds $\Longleftrightarrow$ G-orbits have a common dimension	The deformation space is equipped with a smooth manifold structure $\Longleftrightarrow$ The deformation space is a Hausdorff space $\Longleftrightarrow$ G-orbits have a constant dimension
$G = H_{2n+1}^{r}$, H and Γ are arbitrary	Local rigidity holds $\Longleftrightarrow$ Strong local rigidity holds $\Longleftrightarrow$ Γ is finite	Stability holds	The parameter and the deformation spaces are endowed with smooth manifold structures
$G = H_{2n+1} \times H_{2n+1}$, $H = \Delta_G$ and Γ are arbitrary	Local rigidity fails to hold	Stability holds $\Longleftrightarrow$ Γ is non-Abelian or maximal Abelian	The deformation space is a Hausdorff space $\Longleftrightarrow$ Γ is maximal Abelian
$G = H_{2n+1}^{r} \times H_{2n+1}^{r}$, $H = \Delta_G$ and Γ are arbitrary	Local rigidity holds $\Longleftrightarrow$ Strong local rigidity holds $\Longleftrightarrow$ Γ is finite	Stability holds	The deformation space is a Hausdorff space $\Longleftrightarrow$ Γ is finite $\Longleftrightarrow$ The parameter space admits a rigid homomorphism $\Longleftrightarrow$ The G-orbits have a common dimension
$G = \mathbb{G}_n^{r}$, H and Γ are arbitrary	Local rigidity holds $\Longleftrightarrow$ Γ is finite (Γ is Abelian)	Stability holds for Abelian Γ	The deformation space is a Hausdorff space for non-Abelian Γ

8 Deformation of topological modules

The purpose here is to describe a dequantization procedure for topological modules over a deformed algebra. We first define the characteristic variety of a topological module as the common zeroes of the annihilator of the representation obtained by setting the deformation parameter to zero. On the other hand, the Poisson characteristic variety is defined as the common zeroes of the ideal obtained by considering the annihilator of the deformed representation, and then setting the deformation parameter to zero.

An involutive (or coisotropic) submanifold of a Poisson manifold V is a submanifold W of V such that the ideal of functions that vanish out on W is a Poisson subalgebra of $C^\infty(V)$. In the context of the quantization by deformation, several authors [46, 54] have proposed some methods to associate to an involutive submanifold, a left ideal of the deformed algebra $(C^\infty(V)[\![\nu]\!], *)$, where the star-product $*$ on V comes from the constructions of M. Kontsevich [102] or of D. Tamarkin [126].

The aim here is to propose a reverse step. It consists in describing a method for dequantizing some modules on a deformed algebra, which means to associate to such module, an involutive submanifold and a Poisson submanifold of the underlying Poisson manifold. The C^∞ case is poorly adapted, but the analytic or algebraic cases turn to be appropriate. We first consider the complex field and then propose a natural involution on the deformed algebra where the objects thus define hold on the real field.

These are (strongly pseudo-unitary) modules, which means modules endowed with bilinear Hermitian nondegenerate forms compatible with the involution, taking values in $\mathbb{C}[\![\nu]\!]$ (where ν is the parameter of deformation), and such that the quotient bilinear form obtained where $\nu = 0$ is still nondegenerate.

We here exclusively limit to the algebraic framework. We introduce the notion of divisible ideals of a deformed algebra $\mathscr{A}$. Inspired by [86], we shall also define the notion of the characteristic and Poisson characteristic manifolds for a topologically free $\mathscr{A}$-module. These manifolds are not necessarily conic (unlike what happens in [86] or [79]). Thanks to the introduction of the deformation parameter, the notion of strongly pseudo-unitarity allows to define these objects on the real field.

Using Gabber's theorem, we show the involutivity of the characteristic variety. The Poisson characteristic variety is indeed a Poisson subvariety of the underlying Poisson manifold. We compute explicitly the characteristic variety in several examples in the Poisson linear case, including the dual of any exponential solvable Lie algebra. In the nilpotent case, we show that any coadjoint orbit appears as the Poisson characteristic variety of a well-chosen topological module.

In the case of an exponential solvable Lie group $G = \exp \mathfrak{g}$, we substantiate the Zariski closure conjecture claiming that for an irreducible unitary representation π of G, associated to a coadjoint orbit Ω via the Kirillov orbit method, the Poisson characteristic variety associated to a topological module with an adequate way coincides with the Zariski closure in $\mathfrak{g}^*$ of the orbit Ω. We compute the characteristic varieties in

https://doi.org/10.1515/9783110765304-008

some fundamental examples where we show that the conjecture holds. We also prove the conjecture in some restrictive cases.

8.1 Geometric objects associated with deformed algebras

Let $(V, \{,\})$ be a real analytic (resp., algebraic) Poisson manifold, which means a real analytic (resp., algebraic) manifold endowed with a 2-tensor P with analytic (resp., regular) coefficients such that the Schouten bracket $[P, P]$ vanishes. The Poisson bracket endows the structural sheaf $\mathcal{O}$ of analytic (resp., regular) functions germs with a Poisson structure. We will limit to the flat case $V = \mathbb{R}^d$. So by complexification, one obtains a complex analytic (resp., algebraic) Poisson manifold structure on $V^{\mathbb{C}} = \mathbb{C}^d$.

Let A be the Poisson algebra $\mathcal{O}_V$ of analytic (resp., polynomials) functions on $\mathbb{C}^d$. M. Kontsevich (cf. [102]) built a star-product # on A:

$$f \# g = \sum_{k \geq 0} \nu^k C_k(f, g), \tag{8.1}$$

where the coefficients C_k are the bidifferential operators described using completely explicit formulas only involving partial differentials of constants of the Poisson 2-tensor structure (see again [8] and [110]). In particular, if the Poisson tensor is of analytic coefficients (resp., polynomials), then so is the star-product. In other words, the star-product (8.1) endows

$$\mathscr{A} = A[\![\nu]\!] \tag{8.2}$$

with an associative topologically free algebra structure on $\mathbb{C}[\![\nu]\!]$. This algebra is naturally filtered by $\mathscr{A}_n = \nu^n \mathscr{A}$, and its associated graduation is naturally isomorphic to the algebra of polynomials $A[\nu]$, endowed with the commutative product of A extended through $\mathbb{C}[\nu]$-linearity. Note that the star-product (8.1) is of real coefficients: if f and g are of real values on V, then so is $f \# g$.

The star-product (8.1) is equivalent to another star-product $*$ (called of Duflo–Kontsevich [6]) (cf. [55, 87] and [110]) having the same properties, and satisfying furthermore the following property: for all f, g in the center of $(\mathscr{A}, *)$, we have

$$f * g = fg.$$

In addition, the two algebras $(\mathscr{A}, \#)$ and $(\mathscr{A}, *)$ are isomorphic.

Let us restrict to the algebraic frame. We shall define the characteristic manifold $V(\mathscr{M}) \subset V^{\mathbb{C}}$ of a topologically free $\mathscr{A}$-module $\mathscr{M}$ as well as its Poisson characteristic manifold $VA(\mathscr{M})$ by adapting the definitions of [86] as follows: Consider the annihilator $\operatorname{Ann} \mathscr{M}$ of the $\mathscr{A}$-module $\mathscr{M}$ and define $V(\mathscr{M})$ as the common zeroes set of the annihilator of the A-module $M = \mathscr{M}/\nu\mathscr{M}$, and $VA(\mathscr{M})$ as the common zeroes set of $\operatorname{Ann}\mathscr{M}/(\operatorname{Ann}\mathscr{M} \cap \nu\mathscr{A})$. These two objects are affine submanifolds of $\mathbb{C}^n$ (i. e., defined by the annihilator of finite number of polynomials). We show (using [72]) that $V(\mathscr{M})$ is an involutive submanifold of $V^{\mathbb{C}}$, and $VA(\mathscr{M})$ is a Poisson submanifold of $V^{\mathbb{C}}$ (what justifies the name), and that we always have the inclusion

$$V(\mathscr{M}) \subset VA(\mathscr{M}).$$

After showing (using [53]) that the involution $f \longmapsto f^*$ defined by

$$f^*(\xi) = \overline{f(\overline{\xi})}$$

is an antiautomorphism of the algebra $(\mathscr{A}, *)$, we introduce the notion of a strongly pseudo-unitary $\mathscr{A}$-module, and we show that the associated characteristic manifold is defined on the real field, as well as its Poisson characteristic manifold. More precisely, we first extend the complex conjugation to an automorphism of $\mathbb{C}[\![\nu]\!]$ when fixing $\nu = i\hbar$ is purely imaginary, which means $\overline{\nu} = -\nu$. The module $\mathscr{M}$ is said to be pseudo-unitary (or equivalently that the associate representation is $*$-representation [50]), if there exists a sesquilinear nondegenerated form $\langle -, - \rangle_\nu$ on $\mathscr{M}$, taking value in $\mathbb{C}[\![\nu]\!]$, which is Hermitian, that is,

$$\langle m, n \rangle_\nu = \overline{\langle n, m \rangle_\nu}, \quad m, n \in \mathscr{M}$$

and compatible with the involution, that is, satisfying for all $a \in \mathscr{A}$:

$$\langle am, n \rangle_\nu = \langle m, a^* n \rangle_\nu.$$

The form $\langle -, - \rangle_\nu$ induced by passage to the quotient reveals an Hermitian form $\langle -, - \rangle_0$ on M. If this form is nondegenerate, we say that the module $\mathscr{M}$ is strongly pseudo-unitary (or equivalently, the associated representation is a strongly nondegenerate $*$-representation). Unlike [50], we do not necessarily assume positivity on the sesquilinear form. A strongly pseudo-unitary module endowed with a well-defined positive form will said to be (strongly unitary).

8.1.1 Topological modules on the ring of formal series

We refer in this subsection to [53, 87] and [69]. Let $k[\![\nu]\!]$ be the ring of formal series on any field k endowed with the ν-adic topology, defined by the ultrametric distance:

$$d(a, b) = 2^{-\operatorname{val}(a-b)},$$

with $\operatorname{val} a = \sup\{j, a \in \nu^j k[\![\nu]\!]\}$. This distance makes of $k[\![\nu]\!]$ a complete topological ring. Further, on a $k[\![\nu]\!]$-module $\mathscr{M}$, we put an invariant topology by translation by deciding that the family $v^j\mathcal{M}, j \in \mathbb{N}$ forms a neighborhood basis of zero. This topology is separated if and only if the intersection of all $v^j\mathcal{M}$ is reduced to $\{0\}$. In this case, we can define the valuation:

$$\operatorname{val} m = \sup\{j, m \in v^j\mathcal{M}\}$$

and also the topology defined by the ultrametric distance

$$d(m, m') = 2^{-\operatorname{val}(m-m')}.$$

One says that a $k[\![v]\!]$-module $\mathcal{M}$ is torsion-free, if the action of v is an injection of $\mathcal{M}$. A topologically free $k[\![v]\!]$-module $\mathcal{M}$ is isomorphic to $M[\![v]\!]$ for some vector space M.

Proposition 8.1.1 ([87, Proposition XVI.2.4], [47] and [50, Lemma A1]). *A $k[\![v]\!]$-module $\mathcal{M}$ is topologically free if and only if it is separated, complete and torsion-free.*

Definition 8.1.2. Let $\mathcal{M}$ be a $k[\![v]\!]$-module, and let $\mathcal{N}$ be a sub-$k[\![v]\!]$-module of $\mathcal{M}$. We consider the k-vector spaces $M = \mathcal{M}/v\mathcal{M}$ and $N = \mathcal{N}/v\mathcal{N}$. The inclusion $i : \mathcal{N} \hookrightarrow \mathcal{M}$ induces a k-linear map:

$$i_0 : N \longrightarrow M.$$

We say (cf. [69]) that the sub-$k[\![v]\!]$-module $\mathcal{N}$ is divisible if the map i_0 is injective.

Remark 8.1.3.
(1) For a topologically free module $\mathcal{M} = M[\![v]\!]$, the vector space M can also be seen as the quotient $\mathcal{M}/v\mathcal{M}$.
(2) The sub-$k[\![v]\!]$-module $\mathcal{N}$ is divisible if and only if $v\mathcal{N} = \mathcal{N} \cap v\mathcal{M}$. A simple example of nondivisible sub-$k[\![v]\!]$-module is $vM[\![v]\!] \subset M[\![v]\!]$.
(3) For any $k[\![v]\!]$-module $\mathcal{M}$, we write again "$m = O(v^j)$ in $\mathcal{M}$" for $m \in v^j\mathcal{M}$. A sub-$k[\![v]\!]$-module $\mathcal{N}$ of $\mathcal{M}$ is then divisible if and only if, the fact that for all $m \in \mathcal{N}$, $m = O(v)$ in $\mathcal{M}$ implies that $m = O(v)$ in $\mathcal{N}$. The associative algebra $\mathcal{A}$ defined in (8.2) is by construction a $\mathbb{C}[\![v]\!]$-module topologically free.

Definition 8.1.4.
(1) A topological $\mathcal{A}$-module is a $\mathbb{C}[\![v]\!]$-module $\mathcal{M}$ endowed with a $\mathbb{C}[\![v]\!]$-bilinear map:

$$\begin{aligned} \Phi : \mathcal{A} \times \mathcal{M} &\longrightarrow \mathcal{M} \\ (\varphi, m) &\longmapsto \pi_v(\varphi)m, \end{aligned} \tag{8.3}$$

making of $\mathcal{M}$ an $\mathcal{A}$-module.
(2) Let $\mathcal{M}_1$ and $\mathcal{M}_2$ be two topological $\mathcal{A}$-modules. A morphism of topological $\mathcal{A}$-modules (or an intertwining operator) of $\mathcal{M}_1$ on $\mathcal{M}_2$ is a $\mathbb{C}[\![v]\!]$-linear continuous map commuting with the actions of elements of $\mathcal{A}$. We say that $\mathcal{M}_1$ and $\mathcal{M}_2$

are equivalent, if there exists an intertwining bijective bicontinuous operator of $\mathcal{M}_1$ on $\mathcal{M}_2$.

Proposition 8.1.5. *Let $\mathcal{M}$ be a topological $\mathcal{A}$-module. Then the $\mathbb{C}[\![v]\!]$-bilinear map Φ defined by equation (8.3) is continuous for the v-adic topologies of $\mathcal{A}$ and of $\mathcal{M}$.*

Proof. Let $m \in \mathcal{M}$ and $a \in A$. The family $\{\mathcal{W}_j = \Phi(a,m) + v^j\mathcal{M}\}_j$ forms a neighborhood basis of $\Phi(a,m) = \pi_v(a)m$. Consider the neighborhood $\mathcal{U}_j = a + v^j\mathcal{A}$ and $\mathcal{V}_j = m + v^j\mathcal{M}$ of a and of m, respectively. It is then clear that the image by Φ of the product $\mathcal{U}_i \times \mathcal{V}_j$ is included in $\mathcal{W}_j$, which shows the continuity. In particular, the multiplication of $\mathcal{A} \times \mathcal{A}$ in $\mathcal{A}$ is $\mathbb{C}[\![v]\!]$-bilinear and continuous for the v-adic topology, which makes of $\mathcal{A}$ a topological algebra. □

8.1.2 Divisible ideals

Let $\mathcal{A} = A[\![v]\!]$ be a free topologically module, where we identify A with $\mathcal{A}/v\mathcal{A}$. Let $\mathcal{J}$ be a left ideal of $\mathcal{A}$. It is immediate to see that $J = \mathcal{J}/(\mathcal{J} \cap v\mathcal{A})$ is an ideal of the commutative algebra A. The same happens in the context of right ideals.

Proposition 8.1.6. *Let $\mathcal{J}$ be a divisible bilateral ideal of $\mathcal{A}$. Then $J = \mathcal{J}/(\mathcal{J} \cap v\mathcal{A})$ is a Poisson ideal of A.*

Proof. For $a_0 \in J$ and $b_0 \in A$, let $a = a_0 + va_1 + \cdots \in \mathcal{J}$ and $b = b_0 + vb_1 + \cdots \in \mathcal{A}$. By divisibility of $\mathcal{J}$, we have

$$\frac{1}{v}(a * b - b * a) \in \mathcal{J},$$

which means by considering the constant term that $\{a_0, b_0\}$ belongs to J. □

We also remark that as $\mathcal{J}$ is divisible, we can also identify J to $\mathcal{J}/v\mathcal{J}$.

8.1.3 Cancelations

Let $\mathcal{M}$ be a topological $\mathcal{A}$-module. We define the annihilator $\operatorname{Ann}\mathcal{M}$ of the module $\mathcal{M}$ as the set of $\varphi \in \mathcal{A}$ such that $\pi_v(\varphi)m = 0$ for all $m \in \mathcal{M}$. We see immediately that $\operatorname{Ann}\mathcal{M}$ is a bilateral ideal of $\mathcal{A}$, and that $\operatorname{Ann}\mathcal{M}$ is divisible if $\mathcal{M}$ is torsion-free. The following is then immediate.

Proposition 8.1.7. *Let $\mathcal{M}$ be an $\mathcal{A}$-module. Then $M = \mathcal{M}/v\mathcal{M}$ is a module on the commutative algebra $A = \mathcal{A}/v\mathcal{A}$.*

We note by π_0 the representation of A on the associated module M.

Proposition 8.1.8. *Let $\mathscr{M}$ be a topological $\mathscr{A}$-module. Then the annihilator of the A-module $M = \mathscr{M}/\nu\mathscr{M}$ is stable by the Poisson bracket of A.*

Proof. We note by π_ν the representation of $\mathscr{A}$ in the module $\mathscr{M}$. The annihilator of M can be seen as the set of $f \in A$ such that $\pi_\nu(f)u = O(\nu)$ for all $u \in \mathscr{M}$. For all $f, g \in \operatorname{Ann} M$, we have

$$\pi_\nu(f * g)u = \pi_\nu(f)\pi_\nu(g)u = O(\nu^2),$$

which gives

$$\begin{aligned}\pi_\nu(\{f, g\})u &= \frac{1}{\nu}\pi_\nu(f * g - g * f).u + O(\nu)\\ &= O(\nu),\end{aligned}$$

and hence the result. □

Remark 8.1.9. The ring A being Noetherian (cf. [47]), the annihilator $\operatorname{Ann} M$ is finitely generated.

8.1.4 Involutivity

Definition 8.1.10. Let $V = \mathbb{R}^d$ be a real algebraic flat Poisson variety, and let $W \subset V^{\mathbb{C}}$ be an affine subvariety. Let $I(W)$ be the ideal of A consisting of functions that vanish on W. We say that W is involutive or coisotropic if the ideal $I(W)$ is stable by the Poisson bracket. This is equivalent to the cancellation of the Poisson 2-tensor outside the conormal fiber of W (cf. [46, 54]).

Proposition 8.1.11. *Let W be an affine submanifold of $V^{\mathbb{C}}$. Let W^r be the nonsingular part of W. If W is involutive, then for all symplectic leaf S of $V^{\mathbb{C}}$ meeting W^r, the intersection $W^r \cap S$ is a coisotropic subvariety of S, which means that for all $x \in W^r \cap S$ we have*

$$(T_x(W^r \cap S))^\omega \subset T_x(W^r \cap S), \tag{8.4}$$

where the exponent ω designates the orthogonal in T_xS with respect to the symplectic form.

Proof. Let $x \in W^r \cap S$ and let P_x be the Poisson 2-tensor in x. Let also $\tilde{P}_x : T_x^* V^{\mathbb{C}} \to T_x V^{\mathbb{C}}$ be the skew-symmetric associated linear map, defined by

$$\langle \eta, \tilde{P}_x(\xi) \rangle = P_x(\xi, \eta). \tag{8.5}$$

Then the image of $\tilde{P}_x$ is precisely T_xS. □

Lemma 8.1.12. *We have the equality,*

$$T_x(W^r \cap S)^\omega = \tilde{P}_x(T_x(W^r \cap S)^\perp), \tag{8.6}$$

where the symbol $\perp$ designates the orthogonal of a subspace in the dual.

Proof. Let $\xi \in T_x(W^r \cap S)^\perp$. Then for all $Y \in T_x(W^r \cap S)$, we have

$$\omega(\tilde{P}_x(\xi), Y) = \langle \xi, Y \rangle = 0, \tag{8.7}$$

which shows the inclusion

$$\tilde{P}_x(T_x(W^r \cap S)^\perp) \subset T_x(W^r \cap S)^\omega.$$

To show the opposite inclusion, we consider X in $T_x(W^r \cap S)^\omega$, image by $\tilde{P}_x$ of an element ξ of $T_x^* V$, and it is immediate from (8.7) that ξ belongs to $T_x(W^r \cap S)^\perp$. □

Proof. Let us go back to the proof of Proposition 8.1.11. Let $X \in T_x S$, by definition of a symplectic leaf, there exists $\varphi \in A$ such that X coincides with the Hamiltonian field $H_\varphi(x)$. Then X belongs to $(T_x(W^r \cap S))^\omega$ if and only if for all $Y \in T_x(W^r \cap S)$ we have

$$\omega(X, Y) = Y \cdot \varphi(x) = d\varphi(x)(Y) = 0.$$

Now, the space B_x of $d\varphi(x)$ where $\varphi \in I(W)$ is the orthogonal of $T_x W^r$. Thanks to the transversality condition, we can therefore write

$$T_x(W^r \cap S)^\perp = (T_x W^r \cap T_x S)^\perp = B_x + (T_x S)^\perp. \tag{8.8}$$

Let then $X, Y \in T_x(W^r \cap S)^\omega$. From (8.8) and Lemma 8.1.12, there exists $\varphi, \psi \in I(W)$ such that $X = \tilde{P}_x(d\varphi(x)) = H_\varphi(x)$ and $Y = H_\psi(x)$. We have then thanks to the involutivity,

$$\omega(X, Y) = \{\varphi, \psi\}(x) = 0,$$

which ends the demonstration. □

Remark 8.1.13. We have the equality

$$T_x(W^r \cap S)^\omega = \tilde{P}_x((T_x W^r)^\perp). \tag{8.9}$$

The following complements Proposition 8.1.11.

Proposition 8.1.14. *Let W be an affine submanifold of $V^{\mathbb{C}}$. Let W^r be the nonsingular part of W. Suppose that there exists a Zariski dense set $\mathcal{U}$ of W^r such that:*

(1) *For all* $x \in \mathscr{U}$, *the intersection of* W^r *with the symplectic leaf* S_x *going through* x *is transverse.*
(2) $W^r \cap S_x$ *is coisotropic in* S_x.

Then W *is involutive.*

Proof. It is easy to see that under the hypotheses of the proposition, if f and g belong to $I(W)$, then $\{f,g\}$ vanishes on the dense Zariski open set $\mathscr{U}$ of W^r, and so on W in whole. Recall [127] that even in the algebraic case, the symplectic leaves are not generally algebraic subvarieties of $V^{\mathbb{C}}$. An example of this situation is given later in Subsection 8.3.4. □

8.1.5 Characteristic manifolds

Let $\mathscr{M}$ be a topological $\mathscr{A}$-module. The characteristic variety $V(\mathscr{M})$ of $\mathscr{M}$ is defined as the set of common zeroes of the annihilator of the A-module $M = \mathscr{M}/\nu\mathscr{M}$. If $\mathscr{M}$ is torsion-free, it is called the Poisson characteristic variety of $\mathscr{M}$, and as in [86] will be noted $VA(\mathscr{M})$, being the set of common zeroes of the ideal $\operatorname{Ann}\mathscr{M}/(\operatorname{Ann}\mathscr{M} \cap \nu\mathscr{A})$ of A. This terminology is justified by Proposition 8.1.8 below.

As A is a commutative Noetherian ring, the associated graduation $A[\nu]$ of $\mathscr{A}$ is also Noetherian (cf. [47]). We record the following theorem (cf. [72, Theorem I]).

Theorem 8.1.15 (Integrability of characteristic varieties: O. Gabber). *Suppose that* $\mathscr{M}$ *is a finitely generated* $\mathscr{A}$*-module and let* $M = \mathscr{M}/\nu\mathscr{M}$. *Then the radical* $J(\mathscr{M})$ *of* $\operatorname{Ann} M$ *is stable by the Poisson bracket.*

Remark 8.1.16. $J(\mathscr{M})$ is the ideal of elements of A that vanishes on the characteristic variety $V(\mathscr{M})$. Theorem 8.1.15 therefore says that $V(\mathscr{M})$ is involutive.

Theorem 8.1.17. *The set of common zeroes of a Poisson ideal is a Poisson submanifold of* $V^{\mathbb{C}}$.

Proof. Let J be a Poisson ideal of A, and let $f \in J$. Let x be canceling all elements of J and let $y = \phi_t(x)$, where $(\phi_t)_{|t|<\varepsilon}$ is the flow of the Hamiltonian vector field $H_g, g \in A$. By the analyticity of the flow, we have for $|t|$ small enough:

$$(f \circ \phi_t)(x) = \sum_{k=0}^{\infty} \frac{1}{k!} (\operatorname{ad}^k g.f)(x) t^k, \tag{8.10}$$

where the right member is convergent. But the term

$$(\operatorname{ad}^k g.f)(x) = \{g, \{g, \dots, \{g, f\}, \dots\}\}(x)$$

vanishes, as J is a Poisson ideal. Then $(f \circ \phi_t)(x)$ vanishes for t enough small. It follows that the symplectic leaf passing through x is entirely contained in the set of common zeros of J. □

Corollary 8.1.18. *The Poisson characteristic variety $VA(\mathscr{M})$ of a free torsion topological $\mathscr{A}$-module $\mathscr{M}$ is a Poisson submanifold of $V^{\mathbb{C}}$.*

Proof. As $\mathscr{M}$ is torsion-free, the ideal $\operatorname{Ann}\mathscr{M}$ of $\mathscr{A}$ is divisible, and so from Proposition 8.1.6, $\operatorname{Ann}\mathscr{M}/(\operatorname{Ann}\mathscr{M} \cap \nu\mathscr{A})$ is a Poisson ideal of A. □

Proposition 8.1.19. *For any torsion-free topological $\mathscr{A}$-module $\mathscr{M}$, we have the inclusion:*

$$V(\mathscr{M}) \subset VA(\mathscr{M}). \tag{8.11}$$

Proof. This result naturally comes from the following characterizations:

$$\begin{aligned} \operatorname{Ann} M &= \operatorname{Ann}(\mathscr{M}/\nu\mathscr{M}) = \{\varphi_0 \in A / \pi_\nu(\varphi_0) = O(\nu)\}, \\ \operatorname{Ann}\mathscr{M}/(\operatorname{Ann}\mathscr{M} \cap \nu\mathscr{A}) &= \{\varphi_0 \in A / \exists \varphi = \varphi_0 + \nu\varphi_1 + \cdots \in \mathscr{A} / \pi_\nu(\varphi) = 0\}. \end{aligned} \tag{8.12}$$

Clearly, the second ideal is contained in the first and then the converse inclusion happens for characteristic varieties. □

Remark 8.1.20. The annihilators of two topologically free equivalent modules coincide. Therefore, the characteristic Poisson manifold $VA(\mathscr{M})$ only depends upon the equivalence class of $\mathscr{M}$, according to Definition 8.1.4. The same does not hold for the characteristic variety $V(\mathscr{M})$.

8.1.6 Involution and $*$-representations

A. Cattaneo and G. Felder remarked in [53] that the algebra $(\mathscr{A}, \#)$ defined in (8.1) is naturally equipped with an involution. We adapt their arguments here, and we show that this involution is also an involution for the Duflo–Kontsevich star-product $*$. We extend the complex conjugation to an involutive automorphism of $\mathbb{C}[\![\nu]\!]$ by simply taking $\overline{\nu} = -\nu$. We consider then ν as a purely imaginary complex number and we pose again $\nu = i\hbar$, where $\hbar = \overline{\hbar}$ can be considered as a real.

Proposition 8.1.21. *The semilinear involution $f \mapsto f^*$ of $\mathscr{A}$ defined by*

$$f^*(\xi) = \overline{f(\overline{\xi})} \tag{8.13}$$

is an antiautomorphism of the algebra $(\mathscr{A}, \#)$.

Proof. In the expression of the star-product, we can replace the parameter ν by any formal series without constant term. In particular, we can replace ν by $\hbar = -i\nu$ (which then verifies $\overline{\hbar} = \hbar$). The star-product (8.1) being defined by bidifferential operators with real coefficients, we can easily check that we have

$$(f \#_\hbar g)^* = f^* \#_\hbar g^*, \tag{8.14}$$

from where we deduce

$$(f \#_\nu g)^* = f^* \#_{-\nu} g^*. \tag{8.15}$$

Finally, we highlight (see also the remark in the end of Section 2 of [53]) that the Kontsevich star-product satisfies the parity property of its coefficients, which means

$$f \#_{-\nu} g = g \#_\nu f. \tag{8.16}$$

From (8.15), we have then

$$(f \#_\nu g)^* = g^* \#_\nu f^*, \tag{8.17}$$

and hence the result. □

Proposition 8.1.22. *The involution $f \mapsto f^*$ is an antiautomorphism of the algebra $(\mathscr{A}, *)$ endowed with the star-product of Duflo–Kontsevich. Its restriction to A is an (anti)automorphism of the commutative algebra A.*

Proof. Let $D = I + \nu D_1 + \cdots$ be the formal differential operator that realizes the equivalence between the two star-products:

$$f * g = D(D^{-1} f \# D^{-1} g). \tag{8.18}$$

This equivalence operator D has real coefficients, and thus commutes with the involution $f \mapsto f^*$. The semilinear involution $f \mapsto f^0$ defined by

$$f^0 = D[(D^{-1} f)^*] \tag{8.19}$$

coincides then with $f \mapsto f^*$. In the case of linear Poisson varieties, the same equality between the two star-products holds. The involution is obviously restricted to A. Moreover, the involution respects $\nu\mathscr{A}$, and the involution thus defined on $\mathscr{A}/\nu\mathscr{A}$ corresponds to this restriction via the canonical isomorphism of A on $\mathscr{A}/\nu\mathscr{A}$. □

Let π_ν be a representation of $(\mathscr{A}, *)$ in a topological module $\mathscr{M}$. As a result of [50], we say that π_ν is a $*$-representation of $\mathscr{A}$ in $\mathscr{M}$ if we have

$$\langle \pi_\nu(f) u, v \rangle_\nu = \langle u, \pi_\nu(f)^* v \rangle_\nu, \quad u, v \in \mathscr{M}, \quad f \in \mathscr{A} \tag{8.20}$$

for some Hermitian nondegenerate sesquilinear form $\langle -,-\rangle_\nu$ on $\mathscr{M}$ taking values in $\mathbb{C}[\![\nu]\!]$ (without hypothesis of positivity). We say that the $*$-representation is unitary if further this sesquilinear form is positively defined, which means that $\langle x,x\rangle_\nu > 0$ for all nonzero $x \in \mathscr{M}$, the order on $\mathbb{R}[\![\nu]\!]$ being the lexicography order.

This sesquilinear form defined by passing to the quotient is a complex sesquilinear form on $M = \mathscr{M}/\nu\mathscr{M}$. We will assume that the form $\langle -,-\rangle_\nu$ is strongly nondegenerate, that is to also assume that the quotient form is nondegenerate on M. In this case, we say that the $*$-representation is strongly nondegenerate. A unitary strongly nondegenerate representation is said to be strongly unitary.

A unitary $\mathscr{A}$-module (resp., pseudo-unitary) is by definition a topological module endowed with a unitary representation (resp., a $*$-representation) of $\mathscr{A}$. A strongly unitary $\mathscr{A}$-module (resp., strongly pseudo-unitary) is defined to be a topological module equipped with a strongly unitary representation (resp., with a strongly nondegenerate $*$-representation) of $\mathscr{A}$.

Proposition 8.1.23. *Let π_ν be a strongly nondegenerate $*$-representation of $(\mathscr{A}, *)$ in a topological module $\mathscr{M}$. Then the annihilator of $M = \mathscr{M}/\nu\mathscr{M}$ is generated by a finite number of self-adjoint elements of A. The same happens for* $\operatorname{Ann}\mathscr{M}/(\operatorname{Ann}\mathscr{M} \cap \nu\mathscr{A})$.

Proof. Any element f of A is written in a unique way:

$$f = f^+ + if^-, \tag{8.21}$$

where f^+ and f^- are self-adjoint. We have

$$f^+ = \frac{1}{2}(f + f^*), \quad f^- = \frac{1}{2i}(f - f^*). \tag{8.22}$$

Due to the fact that the scalar product on M is nondegenerate, we see that if f belongs to $\operatorname{Ann} M$, then so is f^*. Let $\{f_1,\dots,f_k\}$ be a system of generators of $\operatorname{Ann} M$. It is then clear that $\operatorname{Ann} M$ is generated by $\{f_1^+,\dots,f_k^+,f_1^-,\dots,f_k^-\}$. The same applies to $\operatorname{Ann}\mathscr{M}/(\operatorname{Ann}\mathscr{M} \cap \nu\mathscr{A})$, as we can see using the second equality (8.12) and the strong pseudo-unitarity of $\mathscr{M}$. □

Corollary 8.1.24. *Under the hypotheses below, the characteristic varieties $V(\mathscr{M})$ and $VA(\mathscr{M})$ of a strongly pseudo-unitary module are real, when $\mathscr{M}$ is torsion-free.*

Proof. The characteristic variety $V(\mathscr{M})$ is defined by the $2k$ equations:

$$f_j^+(\xi) = f_j^-(\xi) = 0, \quad j = 1,\dots,k. \tag{8.23}$$

Now the polynomials $f_j^\pm$ are well of real coefficients by definition of the involution. We have the same justification for $VA(\mathscr{M})$. □

Remark 8.1.25. Proposition 8.1.23 and Corollary 8.1.24 do not use the hypothesis of positivity on the scalar product. It is necessary to note that the notions of nondegeneracy introduced here are different from those introduced in [50], in which the unitary case is considered.

For the sake of brevity, note respectively $V(\mathcal{M}) = V(\pi_\nu)$ and $VA(\mathcal{M}) = VA(\pi_\nu)$ for the characteristic and the Poisson characteristic varieties.

8.1.7 Topologically convergent modules

We need later to specialize the parameter of deformation ν to be a nonzero complex number for the sake of convergence of the formal series. Let $\mathcal{M} = M[\![\nu]\!]$ be a topological free module on the deformed algebra $\mathcal{A} = A[\![\nu]\!]$, and let π_ν be the associated representation. We further assume that M is a locally convex topological space. Let $\mathcal{A}_0$ be the sub-$\mathbb{C}[\nu]$-algebra of $\mathcal{A}$ generated by A. This is the set of sums:

$$\sum_{j=0}^{N} \nu^j \alpha_j,$$

where $N \in \mathbb{N}$ and each α_j is a sum of terms of type $a_1 * \cdots * a_r$, with $a_1, \ldots, a_r \in A$. We say that $\mathcal{M} = M[\![\nu]\!]$ is weakly convergent if there is $R > 0$ such that for all $a \in \mathcal{A}_0$ and all $m \in M$, the entire series $\pi_\nu(a)m$ weakly converges for $\nu = \nu_0$ in the disk of radius R to a vector of M, which will be noted $\tilde{\pi_{\nu_0}}(a)m$. The uniqueness of this weak limit is ensured by the Hahn–Banach theorem. The radius of convergence $R_{\mathcal{M}}$ of the module is then defined as the upper limit of the radius R above.

Proposition 8.1.26. *A topological free weekly convergent module $\mathcal{M} = M[\![\nu]\!]$ of radius $R_{\mathcal{M}}$ induces a family of representations $(\tilde{\pi_{\nu_0}})_{\nu_0 \in D(0,R_{\mathcal{M}})}$ of $\mathcal{A}_0$ in M.*

Proof. For all $a, b \in A$, $m \in M$ and for all m' in the topological dual M', we have equality between the formal series:

$$\langle m', \pi_\nu(a * b)m \rangle = \langle m', \pi_\nu(a)\pi_\nu(b)m \rangle. \tag{8.24}$$

The equality still holds when a and b are in $\mathcal{A}_0$. We have convergence of these two integer series at $\nu = \nu_0 \in D(0, R_{\mathcal{M}})$, and the right member is also the limit of the entire series $\langle m', \pi_\nu(a)\tilde{\pi_{\nu_0}}(b)m \rangle$ at $\nu = \nu_0$. We have then for all $\nu_0 \in D(0, R_{\mathcal{M}})$ the equality:

$$\tilde{\pi_{\nu_0}}(a * b)m = \tilde{\pi_{\nu_0}}(a)\tilde{\pi_{\nu_0}}(b)m. \tag{8.25}$$

□

8.2 Case of linear Poisson manifolds

We now apply the previous results to the case of representations of Lie algebra. Let $V = \mathbb{R}^d$ be a linear Poisson manifold. Then the dual V^* of linear forms on V form a Lie subalgebra $\mathfrak{g}$ of the algebra $A = S(\mathfrak{g}^{\mathbb{C}})$ of polynomials on V endowed with the Poisson bracket. Let $\mathscr{A} = A[\![v]\!]$ be the associated deformed algebra. We then regard the Poisson manifold V as the dual $\mathfrak{g}^*$ of the Lie algebra $\mathfrak{g}$. The Poisson bracket of Kirillov–Kostant–Souriau is given for $\varphi, \psi \in A$ and $\ell \in \mathfrak{g}$ by the formula (cf. [61]):

$$\{\varphi, \psi\}(\ell) := \langle \xi, [d\varphi(\ell), d\psi(\ell)] \rangle. \tag{8.26}$$

Note that is possible to specialize the indeterminate v to a nonnull value in the expression of the star-product (cf. [68]): we introduce the family of complex parameter of Lie algebras $\mathfrak{g}^{\mathbb{C}}_{v_0}$, of the underlying same space $\mathfrak{g}^{\mathbb{C}}$ with the Lie bracket:

$$[X, Y]_{v_0} = v_0[X, Y].$$

The evaluation at $v = v_0$ provides a noncommutative law $*_{v_0}$ on $S(\mathfrak{g}^{\mathbb{C}})$, which is the multiplication of the enveloping algebra of $\mathfrak{g}^{\mathbb{C}}_{v_0}$ transported by Duflo isomorphism. We shall also introduce the notion of weakly convergent topologically free module, which allows us to specialize the indeterminate v to a nonzero value for representations. In a precise way, a weakly convergent topologically free module on $\mathscr{A}$ is a topologically free $\mathscr{A}$-module on $\mathscr{M} = M[\![v]\!]$, where M is a locally convex separated topological space, such that there exists $R > 0$ meeting the property that for all $a \in \mathscr{A}_0$ and all $m \in M$, the whole series $\pi_v(a)m$ is weakly convergent of radius R. Here, $\mathscr{A}_0$ designates the sub-$\mathbb{C}[v]$-algebra of $\mathscr{A}$ generated by A, and π_v the associate representation. The essential ingredient here is the specialization of the deformation parameter v, being any purely imaginary, which allows (cf. Proposition 8.1.26) to associate to a weakly convergent strongly unitary topologically free module on the deformed algebra, a family of unitary representations $(\rho_\hbar)$ of Lie algebras labeled by a real parameter $\hbar$.

Let then $(\mathfrak{g}_\hbar)_{\hbar\in\mathbb{R}}$ be the family of a real parameter of real Lie algebras of same underlying space $\mathfrak{g}$ with the bracket:

$$[X, Y]_\hbar = \hbar[X, Y].$$

We prove the equivalence between the notion of strongly unitary weakly convergent $\mathscr{A}$-module of radius R and the notion of one parameter family $(\rho_\hbar)_{\hbar\in]-R,R[}$ of unitary representations of $\mathfrak{g}_\hbar$ such that the associated family of representations of $\mathscr{U}(\mathfrak{g}_\hbar) \simeq (S(\mathfrak{g}), *_\hbar)$ depends weakly analytically of the parameter $\hbar$. The transition form is given by the formula:

$$\rho_\hbar(X) = -i\pi_v(X)|_{v=i\hbar},$$

for all $X \in \mathfrak{g}$. The factor i is explained by the fact that $\hbar$ is real while the indeterminate v is formally pure imaginary.

We then develop some examples in the linear Poisson setting, more precisely in the nilpotent and solvable cases, placed away from Verma modules treated in conclusion. In the last part, using the method of Kirillov orbits [91] and results of N. V. Pedersen [118, 119], we determine, in the case of exponential solvable Lie algebras the characteristic manifold of a strongly unitary module obtained by unitary induction of any real polarization. We also show that when V is the dual of a nilpotent Lie algebra, any symplectic leaf (i. e., any coadjoint orbit) can be seen as the characteristic Poisson manifold of a well-chosen strongly unitary $\mathscr{A}$-module $\mathscr{M}$.

8.2.1 The algebra $(\mathscr{A}, *)$

It results from the works of B. Shoikhet (cf. [70] and [72]), related to the annihilator of the associated weights on the "wheels" that the two star-products # and $*$ coincide. The algebra $(\mathscr{A}, *)$ is isomorphic to the enveloping formal complexified algebra:

$$\mathscr{U}_\nu(\mathfrak{g}^{\mathbb{C}}) = T(\mathfrak{g}^{\mathbb{C}})[\![\nu]\!]/\langle x \otimes y - y \otimes x - \nu[x,y]\rangle, \tag{8.27}$$

and we have precisely

$$f * g = \tau^{-1}(\tau f.\tau g). \tag{8.28}$$

Here, $\tau : \mathscr{A} \to U_\nu(\mathfrak{g}^{\mathbb{C}})$ is the Duflo isomorphism (cf. [64]):

$$\tau = \sigma \circ J(D)^{1/2},$$

where σ denotes the symmetrization map and $J(D)^{1/2}$ is the differential operator of infinite order with constant coefficients corresponding to the formal series

$$J(x)^{1/2} = \left(\det \frac{sh\,\mathrm{ad}\,\frac{\nu}{2}x}{\mathrm{ad}\,\frac{\nu}{2}x}\right)^{1/2}. \tag{8.29}$$

Note that $(S^n(\mathfrak{g}^{\mathbb{C}}))_{n\geq 0}$ is the increasing usual filtration of the symmetric algebra. We can then specialize the value of the deformation parameter ν. Indeed, for all f, g in A, the series at ν defining $f * g$ is polynomial in ν, and thus can be evaluated in any complex number ν. The star-product thus generates a family of noncommutative associative laws $(*_\nu)$ on A, the parameter ν varying through the field of complex numbers. Each of these algebras is identified via the Duflo isomorphism τ_ν to the enveloping Lie algebra $\mathfrak{g}_\nu^{\mathbb{C}}$ of the underlying vector space $\mathfrak{g}^{\mathbb{C}}$ but with the bracket defined by

$$[x,y]_\nu = \nu[x,y].$$

For a real parameter $\hbar$, it is worth to notice that the real Lie algebra $\mathfrak{g}_\hbar$ is nothing but the underlying vector space $\mathfrak{g}$ endowed with the bracket defined by $[x,y]_\hbar = \hbar[x,y]$.

8.2.2 Converging topological free modules on $(\mathscr{A}, *)$

Let as above $\mathscr{A}_0$ designate the $\mathbb{C}[v]$-subalgebra of $\mathscr{A}$ generated by A. As the entire series $a * b$ is a polynomial at v for all $a, b \in A$, we have

$$\mathscr{A}_0 = A[v].$$

For all $v_0 \in \mathbb{C}$, the evaluation in v_0,

$$\begin{array}{rcl} \mathrm{ev}_{v_0} : \mathscr{A}_0 & \longrightarrow & (A, *_{v_0}) \\ \sum_{k=0}^{n} v^k a_k & \longmapsto & \sum_{k=0}^{n} v_0^k a_k \end{array}$$

is a morphism of $\mathbb{C}$-algebras. We have the following.

Proposition 8.2.1. *Let $R > 0$, and let for all $v_0 \in D(0, R)$, π_{v_0} be a representation of the algebra $(A, *_{v_0})$ on a separated locally convex topological vector space M. We assume that for all $a \in A$, $m \in M$ and $v_0 \in D(0, R)$, the vector $\pi_{v_0}(a)m$ is given by the evaluation at v_0 of a weakly convergent entire series of radius $\geq R$. Then for v_0 in the disk of radius R is induced a representation $\tilde{\pi}_{v_0} = \pi_{v_0} \circ \mathrm{ev}_{v_0}$ of $\mathscr{A}_0$ in M, and $\mathscr{M} = M[\![v]\!]$ is then a topological weakly convergent-free module of radius $\geq R$. Conversely, a weakly convergent topological-free module of radius $\geq R$ induces for all $v_0 \in D(0, R)$ a representation π_{v_0} of $(A, *_{v_0})$ in M.*

Proof. Let $a, b \in A$ and $m \in M$. The equality,

$$\pi_{v_0}(a *_{v_0} b)m = \pi_{v_0}(a)\pi_{v_0}(b)m,$$

for all $v_0 \in D(0, R)$ implies the equality between the formal series:

$$\pi_v(a * b)m = \pi_v(a)\pi_v(b)m,$$

which makes of $\mathscr{M} = M[\![v]\!]$ a topological free module. It is by construction weakly convergent of radius $\geq R$.

Conversely, if M is a separated locally convex topological vector space and if $\mathscr{M} = M[\![v]\!]$ is a weakly convergent topological free $\mathscr{A}$-module of radius $\geq R$, consider for all $v_0 \in D(0, R)$ the representation $\tilde{\pi}_{v_0}$ of $\mathscr{A}_0$ in M as given in Proposition 8.1.26. If we note π_{v_0} the restriction of $\tilde{\pi}_{v_0}$ to $A \subset \mathscr{A}_0$, we have immediately for all $m \in M$:

$$\begin{aligned} \pi_{v_0}(a)\pi_{v_0}(b)m &= \tilde{\pi}_{v_0}(a)\tilde{\pi}_{v_0}(b)m \\ &= \tilde{\pi}_{v_0}(a * b)m \\ &= \pi_{v_0}(a *_{v_0} b)m. \end{aligned}$$

□

8.2.3 Unitarity

A unitary representation of a Lie algebra $\mathfrak{g}$ is a representation ρ realized on a pre-Hilbertian space such that the operators $\rho(X)$ are anti-Hermitian for all $X \in \mathfrak{g}$. The following theorem shows how to connect a strongly unitary topological free module on $\mathscr{A}$ to a family of unitary representations $\rho_\hbar$ of the Lie algebra $\mathfrak{g}_\hbar$, when $\hbar$ takes real values.

Theorem 8.2.2. *Let $R > 0$, and let M be a separated pre-Hilbertian locally convex topological vector space M. Assume that $\mathscr{M} = M[[\nu]]$ is a weakly convergent topological free module of radius $\geq R$. For all $\nu_0 \in D(0, R)$, let π_{ν_0} be the representation of the algebra $(A, *_{\nu_0})$ in M associated to $\mathscr{M}$ by Proposition 8.1.26. Then:*
(1) *The identity*

$$\rho_\hbar(X) = -i\pi_{i\hbar}(X)$$

for $X \in \mathfrak{g}_\hbar$, defines a representation $\rho_\hbar$ of the Lie algebra $\mathfrak{g}_\hbar$ in M for all $\hbar \in D(0, R)$.
(2) *The following two assertions are equivalent:*
  (a) *For all $\hbar \in]-R, R[$, the representation $\rho_\hbar$ is unitary.*
  (b) *The topological free module $\mathscr{M} = M[[\nu]]$, endowed with the scalar product taking values in $\mathbb{C}[[\nu]]$ naturally extending that of M, is strongly unitary.*

Proof. We have for $X, Y \in \mathfrak{g}$:

$$\begin{aligned}
[\rho_\hbar(X), \rho_\hbar(Y)] &= -[\pi_{i\hbar}(X), \pi_{i\hbar}(Y)] \\
&= -i\hbar\pi_{i\hbar}([X, Y]) \\
&= \hbar\rho_\hbar([X, Y]) \\
&= \rho_\hbar([X, Y]_\hbar),
\end{aligned}$$

hence the first part of the theorem.

The unitarity of the topological module $\mathscr{M} = M[[\nu]]$ allows to claim that for all $a \in \mathscr{A}$ and $u, v \in \mathscr{M}$, the equality between the formal series holds:

$$\langle \pi_\nu(a)u, v \rangle = \langle u, \pi_\nu(a^*)v \rangle.$$

As we have assumed that the indeterminate ν is purely imaginary, this equality is specialized (in view of Proposition 8.1.26) in all parameters $\nu_0 \in i]-R, R[$: for all $a \in A$ and $u, v \in M$, we have

$$\langle \pi_{\nu_0}(a)u, v \rangle = \langle u, \pi_{\nu_0}(a^*)v \rangle.$$

As $X^* = X$ for all $X \in \mathfrak{g}$, we see immediately that the operators $\pi_{i\hbar}(X)$ are Hermitian for all $\hbar \in]-R, R[$. The operator $\rho_\hbar(X)$ is thus anti-Hermitian, hence the implication (b) $\Rightarrow$ (a). The converse is immediate. □

8.3 The Zariski closure conjecture

We focus in this section on the setting of exponential solvable Lie groups. We first determine the characteristic variety of a representation π_ν associated in a natural way to an arbitrary unitary monomial representation induced by a real polarization (using [119]). and then substantiate the Zariski closure conjecture.

8.3.1 Kirillov–Bernat theory for exponential groups

Let G be an exponential solvable real Lie group G with Lie algebra $\mathfrak{g}$. Any $\xi \in \mathfrak{g}^*$ admits a *Pukánszky polarization*, that is, a maximal vector subspace $\mathfrak{h}$ of $\mathfrak{g}$ such that $\langle \xi, [\mathfrak{h}, \mathfrak{h}] \rangle = \{0\}$, which is moreover a Lie subalgebra of $\mathfrak{g}$ such that Pukánszky's condition:

$$\mathrm{Ad}^* H.\xi = \xi + \mathfrak{h}^\perp \tag{8.30}$$

holds, where $H = \exp \mathfrak{h}$ is the analytic subgroup of G with Lie algebra $\mathfrak{h}$. Consider the unitary induced representation:

$$\pi^{\xi,\mathfrak{h}} = \widetilde{\mathrm{Ind}}_H^G \chi_\xi, \tag{8.31}$$

where χ_ξ denotes the unitary character of H defined by

$$\chi_\xi(\exp X) = e^{-i\langle \xi, X \rangle}. \tag{8.32}$$

The representation $\pi^{\xi,\mathfrak{h}}$ is unitary and irreducible; its class does not depend on the choice of the Pukánszky polarization $\mathfrak{h}$, and two such representations $\pi^{\xi,\mathfrak{h}}$ and $\pi^{\eta,\mathfrak{h}'}$ are equivalent if and only if ξ and η are in the same coadjoint orbit. Moreover, any unitary irreducible representation of G arises this way. The *Kirillov map*,

$$\begin{aligned} \kappa : \mathfrak{g}^*/G \quad &\longrightarrow \quad \widehat{G} \\ \mathrm{Ad}^* G.\xi \quad &\longmapsto \quad [\pi^{\xi,\mathfrak{h}}], \end{aligned}$$

thus obtained is then a bijection. There are natural topologies on both sides of the Kirillov map: the *quotient topology* on the space $\mathfrak{g}^*/G$ of coadjoint orbits, and the *Fell topology* on the unitary dual $\widehat{G}$.

Theorem 8.3.1. *The Kirillov map κ is a homeomorphism.*

This theorem has been proved by I. D. Brown in 1973 (cf. [49]) for nilpotent, simply connected groups, and by H. Leptin and J. Ludwig in 1994 (cf. [106]) for general exponential groups (see also [26] for several details). Let O_f be the coadjoint orbit of

$f \in \mathfrak{g}^*$ and $d = \dim O_f$. Let $\mathrm{pr} : O_f \to G/H$ be the surjection defined by:

$$\mathrm{pr}(g.f) = gH. \tag{8.33}$$

This is well-defined because H contains the stabilizer G_f of f. We consider the restriction of the Poisson bracket from $\mathfrak{g}^*$ to the symplectic leaf O_f. According to [119], we designate by $\mathcal{E}^0(O_f)$ the space of functions $\varphi = \psi \circ \mathrm{pr}$, where $\psi \in C^\infty(G/H)$, and we designate by $\mathcal{E}^1(O_f)$ the normalizer de $\mathcal{E}^0(O_f)$ in $C^\infty(O_f)$ with respect to the Poisson bracket. Any $X \in \mathfrak{g}$ can be considered as a linear function on $\mathfrak{g}^*$. It induces by restriction a C^∞ function on O_f that we denote always X.

Lemma 8.3.2. *For all $X \in \mathfrak{g}$, the function X belongs to $\mathcal{E}^1(O_f)$.*

Proof. The Hamiltonian field H_X associated to the function X is equal to the fundamental field given by the coadjoint action:

$$H_X\varphi(\eta) = \frac{d}{dt}\Big|_{t=0} \varphi(\exp(-tX).\eta), \quad \eta \in O_f. \tag{8.34}$$

Note that $\varphi \in \mathcal{E}^0(O_f)$ if and only if $\varphi(gh.f) = \varphi(g.f)$ for all $g \in G$ and $h \in H$, or if $H_X\varphi = 0$ for all $X \in \mathfrak{h}$. Hence it is clear that for all $\varphi \in \mathcal{E}^0(O_f)$ and all $t \in \mathbb{R}$, $\eta \mapsto \varphi(\exp -tX.\eta)$ belongs also to $\mathcal{E}^0(O_f)$. We deduce that $H_X.\varphi = \{X, \varphi\}$ belongs also to $\mathcal{E}^0(O_f)$. □

Let $\tau = (\tau_1, \ldots, \tau_{d/2}) : G/H \xrightarrow{\sim} \Omega \subset \mathbb{R}^{d/2}$ a global chart. We will suppose here that the open set Ω is equal to the whole $\mathbb{R}^{d/2}$. It is always possible in the case of a solvable exponential group choosing a coexponential basis $(X_1, \ldots, X_{d/2})$ to $\mathfrak{h}$ in $\mathfrak{g}$ as in Theorem 1.1.5, and considering

$$\tau^{-1}(x_1, \ldots, x_{d/2}) = \exp x_1X_1 \cdots \exp x_{d/2}X_{d/2}H. \tag{8.35}$$

Assume that the global chart τ is defined as above. Let $(q_1, \ldots, q_{d/2})$ be the functions in $\mathcal{E}^0(O_f)$ defined by

$$q_j = \tau_j \circ \mathrm{pr}. \tag{8.36}$$

We first record the following two results [119, Theorems 2.2.2 and 3.2.3].

Theorem 8.3.3. *There exists a family $(p_1, \ldots, p_{d/2})$ in $\mathcal{E}^1(O_f)$ such that*

$$(p_1, \ldots, p_{d/2}, q_1, \ldots, q_{d/2})$$

forms a Darboux global chart,

$$\Phi : O_f \xrightarrow{\sim} W_f \times \mathbb{R}^{d/2},$$

where W_f is an open set in $\mathbb{R}^{d/2}$, which means that

$$\{p_i, p_j\} = \{q_i, q_j\} = 0, \quad \{p_i, q_j\} = \delta_i^j. \tag{8.37}$$

This open set is the whole $\mathbb{R}^{d/2}$ if and only if the polarization $\mathfrak{h}$ satisfies the Pukanszky condition (8.30).

Theorem 8.3.4. *Let (p, q) be a global Darboux coordinate system: $O_f \xrightarrow{\sim} W_f \times \mathbb{R}^{d/2} \subset \mathbb{R}^d$ as in Theorem* 8.3.3. *Then:*

(1) *With respect to these coordinates, the space $\mathcal{E}^0(O_f)$ is identified to the space of functions depending only on q, and the space $\mathcal{E}^1(O_f)$ is identified to the space of functions:*

$$\varphi(p, q) = \sum_{u=1}^{d/2} a_u(q)p_u + a_0(q), \tag{8.38}$$

where $a_0, a_1, \ldots, a_{d/2} \in C^\infty(\mathbb{R}^{d/2})$. In particular, for $X \in \mathfrak{g}$ one has

$$X(p, q) = \sum_{u=1}^{d/2} a_{X,u}(q)p_u + a_{X,0}(q), \tag{8.39}$$

where the $a_{X,u}$, $u = 0, \ldots, d/2$ are functions in $C^\infty(\mathbb{R}^{d/2})$.

(2) *There exists a unique unitary strongly continuous representation ρ of G in $L^2(\mathbb{R}^{d/2})$ such that the space $\mathcal{H}_\rho^\infty$ of its C^∞-vectors contains $C_c^\infty(\mathbb{R}^{d/2})$, and such that for all $\xi \in C^\infty(\mathbb{R}^{d/2})$ we have*

$$\rho(X)\xi(t) = \sum_{u=1}^{d/2} a_{X,u}(t)\frac{\partial \xi(t)}{\partial t_u} - ia_{X,0}(t)\xi(t) + \frac{1}{2}\left(\sum_{u=1}^{d/2} \frac{\partial a_{X,u}}{\partial t_u}(t)\right)\xi(t). \tag{8.40}$$

Remark 8.3.5. This representation is equivalent to the induced representation $\mathrm{Ind}_H^G \chi_f$. The expression of the functions $X(p, q)$ follows immediately from Lemma 8.3.2. In addition, the functions $a_{X,u}$ are entire with at most an exponential growth (cf. [7, Theorem 1.6]).

We can also deduce an information about the value at $q = 0$ of the functions $a_{X,u}$.

Lemma 8.3.6.

(1) *For any $X \in \mathfrak{g}$, we have $a_{X,0}(0) = \langle f, X\rangle$.*
(2) *For any $X \in \mathfrak{h}$ and any $u = 1, \ldots, d/2$, we have $a_{X,u}(0) = 0$.*
(3) *$a_{X_j,u}(0) = -\delta_j^u$, for any $j = 1, \ldots, d/2$ and any $u = 1, \ldots, d/2$.*

Proof. The first statement follows immediately from the fact that the point of O_f of coordinates $(0, 0)$ is f. The second follows from the fact that the set of points of coordinates (p,q) with $q = 0$ and $H.f$, is also included in $f + \mathfrak{h}^{\perp}$. As to the third assertion, it comes from a direct computation:

$$\begin{aligned}
a_{X_j,u}(0) &= \frac{\partial}{\partial p_u} X_j(p,0) \\
&= -\{q_u, X_j\}(p,0) \\
&= \{X_j, q_u\}(p,0) \\
&= \frac{d}{dt}\bigg|_{t=0} q_u(\exp -tX_j.(p,0)) \\
&= \frac{d}{dt}\bigg|_{t=0} q_u(p,(0,\dots,-t,\dots,0)) \quad (-t \text{ in the } j\text{th position}) \\
&= -\delta_j^u.
\end{aligned}$$

□

8.3.2 Construction of a representation π_ν of $\mathcal{A}$

We apply the previous construction to a family $(G_\hbar)_{\hbar \in \mathbb{R}-\{0\}}$ of exponential groups. More precisely, let $(\mathfrak{g}_\hbar)_{\hbar\in\mathbb{R}}$ be the family of solvable Lie algebras defined by the same underlying vector-space $\mathfrak{g}$, with the bracket

$$[X,Y]_\hbar = \hbar[X,Y].$$

We will denote by $\exp_\hbar$ the exponential of $\mathfrak{g}_\hbar$ in $G_\hbar$. Let $f \in \mathfrak{g}^*$, the coadjoint orbit $O_{f,\hbar} = G_\hbar.f \subset \mathfrak{g}^*$ is the same for all $\hbar \neq 0$, but the Poisson structure depends on $\hbar$: for all $\eta \in \mathfrak{g}$ and $\varphi,\psi \in C^\infty(\mathfrak{g}^*)$, we have

$$\{\varphi,\psi\}_\hbar(\eta) = \hbar\{\varphi,\psi\}_1(\eta) = \hbar\langle\eta, [d\varphi(\eta), d\psi(\eta)]\rangle. \tag{8.41}$$

We denote always O_f this common orbit, when it is not necessary to precise the Poisson structure. However, the orbit $O_{f,0}$ degenerates and is reduced to the point f, since the group G_0 is Abelian. There exists a subspace $\mathfrak{h}$ of $\mathfrak{g}$, which is a real polarization of f in $\mathfrak{g}_\hbar$ for all $\hbar \neq 0$. Let $H_\hbar = \exp_\hbar \mathfrak{h} \subset G_\hbar$. Let $\chi_{f,\hbar}$ be the character of $H_\hbar$ defined by

$$\chi_{f,\hbar}(\exp_\hbar X) = e^{-i\langle f,X\rangle}. \tag{8.42}$$

We use the results recorded in Subsection 8.3.1 to construct a simultaneous realization of all the induced $\operatorname{Ind}_{H_\hbar}^{G_\hbar}\chi_{f,\hbar}$ in the same $L^2(\mathbb{R}^{d/2})$. In fact, if (p,q) is the global Darboux coordinate system of Theorem 8.3.4 for $O_{f,1}$ (corresponding to $\hbar = 1$), then for all $\hbar \neq 0$, a global Darboux coordinate system for $O_{f,\hbar}$ is given by $(p,q') = (p,\hbar^{-1}q) = (p_1,\dots,p_{d/2},\hbar^{-1}q_1,\dots,\hbar^{-1}q_{d/2})$. For all $X \in \mathfrak{g}$, the corresponding function writes

$$\begin{aligned} X(p,q') &= \sum_{u=1}^{d/2} a_{X,u}(q)p_u + a_{X,0}(q) \\ &= \sum_{u=1}^{d/2} a_{X,u}(\hbar q')p_u + a_{X,0}(\hbar q'). \end{aligned}$$

Thus, by Theorem 8.3.4, there exists for all $\hbar \neq 0$ a unique unitary strongly continuous representation $\rho_\hbar$ from $G_\hbar$ to $L^2(\mathbb{R}^{d/2})$ such that the space $\mathcal{H}^\infty_{\rho_\hbar}$ of its C^∞ vectors contains $C^\infty_c(\mathbb{R}^{d/2})$, and verifying

$$\rho_\hbar(X)\xi(t) = \sum_{u=1}^{d/2} a_{X,u}(\hbar t)\frac{\partial \xi(t)}{\partial t_u} - ia_{X,0}(\hbar t)\xi(t) + \frac{\hbar}{2}\left(\sum_{u=1}^{d/2} \frac{\partial a_{X,u}}{\partial t_u}(\hbar t)\right)\xi(t). \tag{8.43}$$

This representation is equivalent to the induced representation $\operatorname{Ind}_{H_\hbar}^{G_\hbar} \chi_{f,\hbar}$. This representation is irreducible if and only if the polarization $\mathfrak{h}$ verifies the Pukanszky condition $H_\hbar.f = f + \mathfrak{h}^\perp$ for an arbitrary $\hbar \neq 0$. Otherwise, one can easily realize the induced representation $\operatorname{Ind}_{H_0}^{G_0} \chi_{f,0}$ in $L^2(\mathbb{R}^{d/2})$, and we obtain that

$$\rho_0(\exp tX_j).\varphi(t_1,\dots,t_{d/2}) = \varphi(t_1,\dots,t_{j-1},t_j - t, t_{j+1},\dots,t_{d/2}),$$

and for any $X \in \mathfrak{h}h$,

$$\rho_0(\exp tX).\varphi(t_1,\dots,t_{d/2}) = e^{-it\langle f,X\rangle}\varphi(t_1,\dots,t_{d/2}).$$

By differentiating in $t = 0$, we obtain

$$\rho_0(X_j).\varphi(t_1,\dots,t_{d/2}) = -\frac{\partial}{\partial t_j}\varphi(t_1,\dots,t_{d/2})$$

and for all $X \in \mathfrak{h}$,

$$\rho_0(X).\varphi(t_1,\dots,t_{d/2}) = -i\langle f,X\rangle.\varphi(t_1,\dots,t_{d/2}). \tag{8.44}$$

Thanks to Lemma 8.3.6, the representation ρ_0 of G_0 in $C^\infty_c(\mathbb{R}^{d/2})$ is obtained by extending to $\hbar = 0$ the formula (8.43). Let now K be a fixed compact in $\mathbb{R}^{d/2}$ with a nonempty interior, and let $M = C^\infty_K(\mathbb{R}^{d/2})$ be the space of C^∞-functions supported in K. This space is for all real numbers $\hbar$, a submodule of the C^∞-vectors module of $\rho_\hbar$. M is endowed with the Fréchet topology defined by the seminorms:

$$N_k(\varphi) = \sup_{|\alpha|\leq k} \sup_{t\in K} |D^\alpha \varphi(t)|.$$

The topological dual of M is the space of distributions supported in K. Let $\varphi \in M$ and $T \in M'$. Since the functions $a_{X,u}$ are entire functions, applying Lemma 8.3.6, we see that the expression $\langle T, \rho_\hbar(w)\varphi\rangle$ is entire as a function of the variable $\hbar$ for all $w \in S(\mathfrak{g})$, by identifying $S(\mathfrak{g})$ and $\mathcal{U}(\mathfrak{g}_\hbar)$ via the Duflo isomorphism. We consider

$$\pi_{v_0}(X) = i\rho_{-iv_0}(X), \tag{8.45}$$

and we obtain a family (π_{v_0}) of representations $(A, *_{v_0})$ in M satisfying the conditions of Proposition 8.2.1 (with $R = +\infty$). As above, one obtains a structure of weakly convergent and strongly unitary topologically free module $\mathcal{M} = M[\![v]\!]$. The expression of the representation π_v of the deformed algebra $\mathcal{A}$ in $\mathcal{M}$ is obtained as follows: for all $X \in \mathfrak{g}$, one has

$$\pi_v(X)\xi(t) = \sum_{u=1}^{\frac{d}{2}} ia_{X,u}(-ivt)\frac{\partial \xi}{\partial t_u}(t) + a_{X,0}(-ivt)\xi(t) + \frac{v}{2}\sum_{u=1}^{\frac{d}{2}} \frac{\partial}{\partial t_u}a_{X,u}(-ivt)\xi(t), \tag{8.46}$$

and the action of an element of $\mathcal{A}$ is obtained via the identification $\mathcal{A} = \mathcal{U}_v(\mathfrak{g}^{\mathbb{C}})$ given by the Duflo isomorphism.

8.3.3 The characteristic variety

Theorem 8.3.7. *Let $\mathfrak{g}$ be a exponential solvable Lie algebra, f an element of $\mathfrak{g}^*$ and $\mathfrak{h}$ a real polarization at f. Let d be the dimension of the coadjoint orbit of f, and let K be a compact in $\mathbb{R}^{d/2}$ with a nonempty interior. Let π_v be the representation of the deformed algebra $\mathcal{A}$ on the convergent topologically free module $M = C_K^\infty(\mathbb{R}^{d/2})[\![v]\!]$ constructed as in Subsection* 8.3.2. *Then*

$$V(\pi_v) = f + \mathfrak{h}^\perp. \tag{8.47}$$

Particularly, if $\mathfrak{h}$ satisfies the Pukanszky condition (8.30), *then we get that $V(\pi_v) = H.f$.*

Proof. Taking $v = 0$ in (8.46), we obtain

$$\pi_0(X) = \sum_{u=1}^{\frac{d}{2}} ia_{X,u}(0)\frac{\partial}{\partial t_u} + a_{X,0}(0).$$

By Lemma 8.3.6, we have

$$\begin{pmatrix} \pi_0(X_j) = -i\frac{\partial}{\partial t_j}, & j = 1, \dots, d/2, \\ \pi_0(X) = \langle f, X\rangle & \text{for any } X \in \mathfrak{h}. \end{pmatrix}$$

The annihilator of π_0 is the ideal of $S(\mathfrak{g})$ generated by $X - \langle f, X, X \in \mathfrak{h}$. As a consequence, we obtain the equality

$$V(\pi_\nu) = f + \mathfrak{h}^\perp. \qquad \square$$

8.3.4 Fundamental examples

We compute in this section the characteristic and the Poisson characteristic varieties for some examples of Poisson linear manifolds. The following lemma will be of use along the section.

Lemma 8.3.8. *Let K be a compact of $\mathbb{R}^n$ of nonempty interior, m a nonzero integer, $R > 0$, and let $(P_\nu)_{\nu \in D(0,R)}$ be a family of differential operators of order $\leq m$ whose coefficients restricted to K analytically depends upon ν. Then for all $\varphi \in C_c^\infty(\mathbb{R}^n)$ and all distribution T with supports included in K, the function $\nu \mapsto \langle T, P_\nu(\varphi)\rangle$ is analytic on $D(0,R)$.*

Proof. Let $C_K^\infty(\mathbb{R}^n)$ be the space of smooth functions on $\mathbb{R}^n$ of support included in K. We endow this space with the Frechet topology defined by seminorms:

$$N_k(\varphi) = \sup_{|\alpha| \leq k} \sup_{x \in K} |D^\alpha \varphi(x)|.$$

We consider the Laplacian Δ on $\mathbb{R}^n$. There exists an integer L such that the distribution $(1-\Delta)^{-L}T$ is a continuous function on K, since T is of finite order. We have then by integration by parts:

$$\langle T, P_\nu(\varphi)\rangle = \int_K (1-\Delta)^{-L} T(x)(1-\Delta)^L P_\nu(\varphi)(x) dx.$$

On the other hand,

$$(1-\Delta)^L P_\nu(\varphi)(x) = P'_\nu(\varphi)(x) = \sum_{k \geq 0} \nu^k Q_k(\varphi)(x),$$

where Q_k are differential operators of order $\leq m + 2L$. The coefficients of P'_ν being given on K by convergent entire series on $D(0,R)$, there exists for all $r < R$ a constant C such that the absolute values of the coefficients of Q_k are all dominated on K by Cr^{-k}. We have then

$$\sup_{x \in K} |Q_k(\varphi)(x)| \leq C' r^{-k} N_{m+2L}(\varphi),$$

from where the domination

$$\left| \int_K (1-\Delta)^{-L} T(x) Q_k \varphi(x)\, dx \right| \leq C'(\mathrm{Vol}\, K) N_{m+2L}(\varphi) \sup_{x \in K} |(1-\Delta)^{-L} T(x)| r^{-k}$$

holds on the disk of radius r. The expression $\langle T, P_\nu(\varphi)\rangle$ is thus given by the convergent entire series

$$\langle T, P_\nu(\varphi)\rangle = \sum_{k\geq 0} \nu^k \int_K (1-\Delta)^{-L} T(x) Q_k \varphi(x)\, dx.$$

This being true for all $r < R$, we have convergence of this entire series on $D(0,R)$. □

8.3.5 The Heisenberg groups

We here refer to Subsection 1.1.2 and consider the Heisenberg group $H_{2n+1} = \mathbb{R}^n \times \mathbb{R}^n \times \mathbb{R}$ with the product

$$(x,y,z)\cdot(x',y',z') = \left(x+x', y+y', z+z' + \frac{1}{2}(xy'-x'y)\right),$$

its Lie algebra $\mathfrak{h}_{2n+1}$ is generated by the family of vector fields $\{X_1,\dots,X_n, Y_1,\dots,Y_n, Z\}$ whose Lie brackets are given by

$$[X_i, X_j] = \delta_{ij} Z, \quad i,j = 1,\dots,n,$$

δ_{ij} being the Kronecker symbol. For

$$f_\lambda = \lambda Z^* + \sum_{i=1}^n a_i X_i^* + \sum_{i=1}^n b_i Y_i^* \in \mathfrak{h}_{2n+1}^*,$$

consider the representation ρ^λ associated to f_λ by the orbit method. We now discuss the following two cases.

Case 1: If $\lambda \neq 0$, we can take $f_\lambda = \lambda Z^*$. Indeed, the dimension of ρ^λ is infinite and the associated orbit of f_λ under the coadjoint action is $\Omega_{f_\lambda} = \{(\lambda, u, v),\ u, v \in \mathbb{R}^n\}$. If one realises ρ^λ using the polarization $\mathfrak{b} = \mathbb{R}\text{-span}\{Y_1, Y_2, \dots, Y_n, Z\}$, then ρ^λ acts on the space $L^2(\mathbb{R}^n)$. Also note ρ^λ its differential, which is realized on the Schwartz space $\mathscr{S}(\mathbb{R}^n)$, the Frechet space of the C^∞-vectors of ρ^λ (cf. [91]). It is given by

$$\begin{cases} \rho^\lambda(Z) = -i\lambda, \\ \rho^\lambda(X_i) = -\frac{\partial}{\partial t_i}, \quad i = 1,\dots,n, \\ \rho^\lambda(Y_j) = it_j, \quad j = 1,\dots,n. \end{cases}$$

Thus, the representation $\rho_\hbar^\lambda$ is defined by the relations:

$$\begin{cases} \rho_\hbar^\lambda(Z) = -i\lambda, \\ \rho_\hbar^\lambda(X_i) = -\frac{\partial}{\partial t_i}, \quad i = 1, \dots, n, \\ \rho_\hbar^\lambda(Y_j) = i\hbar t_j, \quad j = 1, \dots, n. \end{cases}$$

These expressions depend polynomially upon $\hbar$; this allows to use Proposition 8.1.26 and Theorem 8.2.2, which makes of $\mathscr{M} = \mathscr{S}(\mathbb{R}^n)[\![\nu]\!]$ a weakly convergent strongly unitary topological free module (of infinite radius). The associated representation of $\mathscr{A}$ is given by

$$\begin{cases} \pi_\nu^\lambda(Z) = \lambda, \\ \pi_\nu^\lambda(X_i) = -i\frac{\partial}{\partial t_i}, \quad i = 1, \dots, n, \\ \pi_\nu^\lambda(Y_j) = -i\nu t_j, \quad j = 1, \dots, n. \end{cases}$$

The annihilator of π_ν^λ is therefore generated by $Z - \lambda$, and thus

$$VA(\pi_\nu^\lambda) = \Omega_{f_\lambda}.$$

It immediately results that

$$\begin{cases} \pi_0^\lambda(Z) = \lambda, \\ \pi_0^\lambda(X_i) = -i\frac{\partial}{\partial t_i}, \quad i = 1, \dots, n, \\ \pi_0^\lambda(Y_j) = 0, \quad j = 1, \dots, n. \end{cases}$$

We then deduce that $\mathrm{Ann}(\pi_0^\lambda)$ is generated by the family $\{Z - \lambda,\ Y_j, j = 1, \dots, n\}$. It comes then

$$\begin{aligned} V(\pi_\nu^\lambda) &= \{l \in \mathfrak{h}_{2n+1}^* : l(Z - \lambda) = 0 \text{ and } l(Y_j) = 0,\ j = 1, \dots, n\} \\ &= \lambda Z^* \bigoplus_{i=1}^{n} \mathbb{R}X_i^* = f_\lambda + \mathfrak{b}^\perp. \end{aligned}$$

A similar calculation shows that $V(\pi_\nu^\lambda) = f_\lambda + \mathfrak{b}'^\perp$ if we realize ρ^λ using the polarization $\mathfrak{b}' = \mathbb{R}\text{-span}\{X_1, \dots, X_n, Z\}$. Therefore, we can clearly observe that $V(\pi_\nu^\lambda)$ depends upon the realization of π^λ.

Case 2: If $\lambda = 0$, then the orbit Ω_{f_0} reduces to $\{f_0\}$ and the associated polarization to f_0 is the Lie algebra $\mathfrak{h}_{2n+1}$. Hence,

$$\rho^0(X) = -if_0(X) \quad \text{for all } X \in \mathfrak{h}_{2n+1}.$$

In this case, we have $\pi_v^0(X) = \pi_0^0(X)$, for all $X \in \mathfrak{h}_{2n+1}$. We then deduce that the annihilator of π_v^0 is the ideal generated by the family $\{X - f_0(X),\ X \in \mathfrak{h}_{2n+1}\}$. It follows that

$$\begin{aligned} V(\pi_v^0) &= \{l \in \mathfrak{h}_{2n+1}^* / l(X - f_0(X)) = 0,\ X \in \mathfrak{h}_{2n+1}\} \\ &= \{f_0\}, \end{aligned}$$

and the same, $VA(\pi_v^0) = \{f_0\}$.

8.3.6 The *n*-step threadlike Lie algebras

We now refer to Subsection 1.1.3. Consider the n-step threadlike Lie algebras, G_n of Lie algebra $\mathfrak{g}_n$ of dimension $n + 1$ and endowed with a basis $\{X_1, \dots, X_{n+1}\}$ with the nonvanishing brackets:

$$[X_{n+1}, X_j] = X_{j-1}, \quad j = 2, \dots, n.$$

Let $\{X_1^*, \dots, X_{n+1}^*\}$ be the dual basis of $\mathfrak{g}^*$. The center of $\mathfrak{g}$ is generated by the vector X_1. Let $l = l_1 X_1^* + \cdots + l_{n+1} X_{n+1}^* \in \mathfrak{g}^*$ with $l_1 \neq 0$. Then $\mathfrak{b}(l) = \mathbb{R}\text{-span}\{X_1, \dots, X_n\}$ is an Abelian ideal of $\mathfrak{g}_n$, which polarizes l.

As the Pukanszky indices set is $\{2, n + 1\}$, it can be assumed without loss of generalities, that $l_2 = l_{n+1} = 0$ (cf. [18]). The irreducible unitary representation $\rho = \rho_l$ associated to l is then realized on $L^2(\mathbb{R})$. Its differential is given by

$$\begin{cases} \rho(X_{n+1}) = -\frac{\partial}{\partial t}, \\ \rho(X_1) = -i l_1, \\ \rho(X_2) = i t l_1, \\ \rho(X_3) = -i(l_3 + \frac{1}{2} t^2 l_1), \\ \vdots \\ \rho(X_n) = -i(l_n - l_{n-1} t + \frac{1}{2} l_{n-2} t^2 + \cdots + (-1)^{n-3} \frac{l_3}{(n-3)!} t^{n-3} + (-1)^{n-1} \frac{l_1}{(n-1)!} t^{n-1}). \end{cases}$$

It follows therefore that the representation $\rho_\hbar$ is determined by

$$\begin{cases} \rho_\hbar(X_{n+1}) = -\frac{\partial}{\partial t}, \\ \rho_\hbar(X_1) = -i l_1, \\ \rho_\hbar(X_2) = +i\hbar t l_1, \\ \rho_\hbar(X_3) = -i(l_3 + \frac{1}{2} \hbar^2 t^2 l_1), \\ \vdots \\ \rho_\hbar(X_n) = -i(l_n - l_{n-1} \hbar t + \frac{1}{2} l_{n-2} \hbar^2 t^2 + \cdots + \frac{l_3}{(n-3)!} (-\hbar t)^{n-3} + \frac{l_1}{(n-1)!} (-\hbar t)^{n-1}). \end{cases}$$

These expressions being polynomials upon $\hbar$, we can also apply Proposition 8.1.26 and Theorem 8.2.2, which makes of $\mathcal{M} = \mathcal{S}(\mathbb{R})[\![v]\!]$ a strongly unitary weakly convergent module. We get

$$\begin{cases} \pi_v(X_{n+1}) = -i\frac{\partial}{\partial t}, \\ \pi_v(X_1) = l_1, \\ \pi_v(X_2) = +ivtl_1, \\ \pi_v(X_3) = l_3 - \frac{1}{2}v^2t^2l_1, \\ \vdots \\ \pi_v(X_n) = l_n + ivl_{n-1}t - \frac{1}{2}v^2l_{n-2}t^2 + \cdots + (iv)^{n-3}\frac{l_3}{(n-3)!}t^{n-3} + (iv)^{n-1}\frac{l_1}{(n-1)!}t^{n-1}. \end{cases}$$

Making $v = 0$, we get

$$\begin{cases} \pi_0(X_{n+1}) = -i\frac{\partial}{\partial t}, \\ \pi_0(X_1) = l_1, \\ \pi_0(X_2) = 0, \\ \pi_0(X_3) = l_3, \\ \vdots \\ \pi_0(X_n) = l_n. \end{cases}$$

Noticing that $\mathrm{Ann}(\pi_0)$ is generated by the family $\{X_1 - l_1, X_2, X_3 - l_3, \ldots, X_n - l_n\}$, it comes that for $f = (f_1, \ldots, f_{n+1}) \in \mathfrak{g}^*, f \in V(\pi_v)$ if and only if $f \in l + \mathfrak{b}(l)^{\perp}$. On the other hand, we see that the annihilator of π_v is generated by the family $v_k, k = 1, \ldots, n$, with

$$v_k = X_k - l_k - \frac{l_{k-1}}{l_1}X_2 - \frac{l_{k-2}}{2l_1^2}X_2^2 - \cdots - \frac{l_3}{(k-3)!l_1^{k-3}}X_2^{k-3} - \frac{1}{(k-1)!l_1^{k-2}}X_2^{k-1}.$$

The Poisson characteristic variety $VA(\pi_v)$ is therefore equal to the coadjoint orbit Ω_l.

8.3.7 The affine group of the real line

Let $G = \mathrm{Aff}(\mathbb{R})$, its Lie algebra admits two generators X and Y such that $[X, Y] = Y$. This group admits two unitary irreducible representations ρ_+ and ρ_- associated respectively to the linear functionals Y^* and $-Y^*$. The differentials of these representations are defined by

$$\begin{cases} \rho_+(X) = -\frac{d}{dx}, \\ \rho_+(Y) = -ie^{-x} \end{cases} \quad \text{and} \quad \begin{cases} \rho_-(X) = -\frac{d}{dx}, \\ \rho_-(Y) = -ie^{-x}. \end{cases}$$

Hence, the expressions of $\rho_{+,\hbar}$ and $\rho_{-,\hbar}$ are respectively given by

$$\begin{cases} \rho_{+}, \hbar(X) = -\frac{d}{dx}, \\ \rho_{+}, \hbar(Y) = -ie^{-\hbar x} \end{cases} \quad \text{and} \quad \begin{cases} \rho_{-}, \hbar(X) = -\frac{d}{dx}, \\ \rho_{-}, \hbar(Y) = -ie^{-\hbar x}. \end{cases}$$

We restrict these differentials to $M = C_K^{\infty}(\mathbb{R})$ where K is a compact set of nonempty interior. Applying first Lemma 8.3.8 then Proposition 8.2.1 and Theorem 8.2.2, we make of $\mathcal{M} = M[\![v]\!]$ a weakly convergent unitary strongly module (of infinite radius). The annihilators of the representations π_{+v} and π_{-v} are trivial, and consequently, the sets $VA(\pi_{+,v})$ and $VA(\pi_{-,v})$ are equal to $\mathfrak{g}^*$. Taking $v = 0$, it follows that the annihilators of $\pi_{+,0}$ and $\pi_{-,0}$ are respectively generated by $\langle Y - 1\rangle$ and $\langle Y + 1\rangle$. We then obtain that

$$\begin{aligned} V(\pi_{+,v}) &= \{l = xX^* + yY^* \in \mathfrak{g}^* / \forall \varphi \in \operatorname{Ann} \pi_{+,0}, \varphi(l) = 0\} \\ &= \{l = xX^* + yY^* \in \mathfrak{g}^* / y - 1 = 0\} = Y^* + \mathfrak{b}^{\perp}, \end{aligned}$$

where $\mathfrak{b} = \mathbb{R}Y$ is the polarization associated to the functionals Y^* and $-Y^*$. Similarly,

$$V(\pi_{-,v}) = \{l = xX^* + yY^* \in \mathfrak{g}^* / y + 1 = 0\} = -Y^* + \mathfrak{b}^{\perp}.$$

Remark 8.3.9. For these two representations, the Poisson characteristic manifold coincides with the Zariski closure of the associated coadjoint orbit.

8.3.8 A 3-dimensional exponential solvable Lie group

Let $\mathfrak{g}$ be the Lie algebra generated by the three vectors $\{A, X, Y\}$ whose Lie brackets are given by $[A, X] = X - Y$, $[A, Y] = X + Y$, and let $G = \exp \mathfrak{g}$. Hence, G is an exponential noncompletely solvable Lie group.

Let $f = xX^* + yY^* + aA^* \in \mathfrak{g}^*$. If $x^2 + y^2 = 0$, then the orbit of f is reduced to the unit set $\{f\}$. Thereby, the computation done in Example 8.3.7 shows that in this case, $V(\pi_{f,v}) = \{f\}$. In the case where $x^2 + y^2 \neq 0$, the subalgebra $\mathfrak{b} = \mathbb{R}\text{-span}\{X, Y\}$ is a polarization of f satisfying the Pukanszky condition. Then let χ_f be the character defined on $B = \exp \mathfrak{b}$ by $\chi_f(\exp U) = e^{-if(U)}$ and $\rho_f = \operatorname{Ind}_B^G \chi_f$. We know (cf. [5]) that there exists a unique $\theta \in [0, 2\pi[$ such that $\rho = \rho_\theta = \rho_{f_\theta}$ où $f_\theta = \cos\theta X^* + \sin\theta Y^*$. The orbit Ω associated to ρ is parameterized by

$$\Omega = \{sA^* + e^{-t}\cos(t + \theta)X^* + e^{-t}\sin(t + \theta)Y^*,\ s, t \in \mathbb{R}\}.$$

On the other hand, we have that

$$\begin{cases} \rho(A) = -\frac{d}{dt}, \\ \rho(X) = -ie^{t}\cos(\theta + t), \\ \rho(Y) = -ie^{-t}\sin(\theta + t) \end{cases} \quad \text{and} \quad \begin{cases} \rho_\hbar(A) = -\frac{d}{dt}, \\ \rho_\hbar(X) = -ie^{-\hbar t}\cos(\theta + \hbar t), \\ \rho_\hbar(Y) = -ie^{-\hbar t}\sin(\theta + \hbar t). \end{cases}$$

We restrict our study to $M = C_K^\infty(\mathbb{R})$, where K is a compact set with a nonempty interior. With the same arguments as above, $\mathcal{M} = M[\![\nu]\!]$ stands for a unitary weakly convergent module (of infinite radius). The representation π_ν of $\mathcal{A}$ writes

$$\begin{cases} \pi_\nu(A) = -i\frac{d}{dt}, \\ \pi_\nu(X) = e^{i\nu t}\cos(\theta - i\nu t), \\ \pi_\nu(Y) = e^{i\nu t}\sin(\theta - i\nu t) \end{cases} \quad \text{and} \quad \begin{cases} \pi_0(A) = -i\frac{d}{dt}, \\ \pi_0(X) = \cos\theta, \\ \pi_0(Y) = \sin\theta. \end{cases}$$

It follows that the annihilator of π_ν is reduced to $\{0\}$, therefore,

$$VA(\pi_\nu) = \mathfrak{g}^*.$$

The characteristic Poisson manifold coincides here also with the Zariski closure of the associated coadjoint orbit. To see this, it suffices to be convinced that the logarithmic spiral is Zariski dense in the 2-dimensional plan, remarking that every line passing through the origin intersects this spiral infinitely many times. Otherwise, the annihilator of π_0 is the ideal generated by the two generators $\{X - \cos\theta, Y - \sin\theta\}$. The representation ρ acts on the space $L^2(G/B)$, which is isomorphic to $L^2(\exp\mathbb{R}A)$. Then we have

$$\begin{aligned} V(\pi_\nu) &= \{l \in \mathfrak{g}^*/l(X - \cos\theta) = 0 \text{ and } l(Y - \sin\theta) = 0\} \\ &= f_\theta + \mathbb{R}A^* = f_\theta + \mathfrak{b}^\perp. \end{aligned}$$

8.3.9 The Zariski closure conjecture

For any exponential solvable Lie algebra $\mathfrak{g}$ and any coadjoint orbit $\Omega \subset \mathfrak{g}^*$, we constructed above a topological module $(\pi_\nu, \mathcal{M})$ over the formal enveloping algebra:

$$\widehat{\mathcal{U}_\nu}(\mathfrak{g}_\mathbb{C}) = T(\mathfrak{g})[\![\nu]\!]/\langle x \otimes y - y \otimes x - \nu[x, y]\rangle,$$

identified to a deformed algebra $\mathcal{A} = (S(\mathfrak{g}_\mathbb{C})[\![\nu]\!], *)$ via a $\mathbb{C}[\![\nu]\!]$-module isomorphism (e. g., the symmetrization map or the Duflo isomorphism). This module is obtained by considering the differential of the associated irreducible unitary representation ρ of the group $G = \exp\mathfrak{g}$. This representation is constructed by inducing a unitary character of a polarization $\mathfrak{h}$ of $\xi \in \Omega$, that is, a maximal isotropic Lie subalgebra of $\mathfrak{g}$ with respect to the bilinear form $B_\ell = \langle \ell, [-, -]\rangle$. Any polarization obviously contains the radical $\mathfrak{g}(\ell)$ of the bilinear form B_ℓ. The polarization $\mathfrak{h}$ satisfies Pukanszky's condition (8.30). Namely, this construction gives rise for any real number $\hbar$ to a unitary irreducible representation $\rho_\hbar$ of the group $G_\hbar = \exp\mathfrak{g}_\hbar$, which is irreducible for $\hbar \neq 0$. Here, $\mathfrak{g}_\hbar$ is the Lie algebra obtained from $\mathfrak{g}$ by multiplying the Lie bracket by $\hbar$. Differentiating each of these representations and setting $\nu = i\hbar$, we get the requested topological module by considering ν as an indeterminate $\pi_\nu = i\rho_{-i\nu}$ (cf. [21]).

Examples 8.3.7 and 8.3.8 show that Theorem 8.3.11 does not hold in the case of a nonnilpotent exponential solvable group. In these two examples, the characteristic Poisson manifold is equal to the Zariski closure of the orbit. More precisely, we state the following conjecture.

Conjecture 8.3.10 (cf. [23]). *Let G be an exponential solvable Lie group, with Lie algebra $\mathfrak{g}$, and let π be an irreducible unitary representation of G, associated to a coadjoint orbit $(\mathrm{Ad}^*\ G)\ell = \Omega$ via the Kirillov orbit method. Then the Poisson characteristic variety $VA(\pi_\nu)$ coincides with the Zariski closure in $\mathfrak{g}^*$ of the orbit Ω.*

8.3.10 The nilpotent case

In Examples 8.3.5 and 8.3.6, the characteristic Poisson variety coincides with the coadjoint orbit. This property can be generalized for all unitary irreducible representation of a nilpotent group. We have the following.

Theorem 8.3.11. *Under the same hypotheses, Conjecture* 8.3.10 *holds for nilpotent Lie groups.*

Proof. Let $\mathfrak{g}$ be an n-dimensional nilpotent real Lie algebra. Let $f \in \mathfrak{g}^*$ and $\mathfrak{h}$ a real polarization at f and let $\rho_\hbar$ be the unitary representation of the simply connected group $G_\hbar = \exp \mathfrak{g}_\hbar$ given by the construction of Subsection 8.3.2. This representation is irreducible since $\mathfrak{h}$ satisfies the Pukanszky condition. The coadjoint orbit $O_f \subset \mathfrak{g}^*$ is an algebraic submanifold, given by the annulation of real valued polynomials $(Q_j)_{j=1,\ldots,n-d}$ where d designates the dimension of the orbit. Let τ be the Duflo isomorphism, which is here reduced to the symmetrization. Based on Theorem 2.3.2 of [118] (adapted here to our sign conventions in the definition of the character χ_f), the annihilator of $\rho_\hbar$ in $\mathcal{U}(\mathfrak{g}_\hbar)$ is the ideal generated by the u_j's with $\tau^{-1}(u_j)(\eta) = Q_j(i\eta)$. Hence, we see that the annihilator of the representation π_ν of the deformed algebra $\mathcal{A}$ is generated by the Q_j's. Thus, the ideal $\mathrm{Ann}\,\pi_\Omega^\nu/(\mathrm{Ann}\,\pi_\Omega^\nu \cap \nu\mathcal{A})$ of $S(\mathfrak{g})$ is also generated by the Q_j's, and consequently the orbit Ω and the characteristic Poisson variety coincide. □

We prove Conjecture 8.3.10 in several other situations: in the case $\mathfrak{g} = [\mathfrak{g}, \mathfrak{g}] + \mathfrak{g}(\ell)$, the proof is similar to the one in the nilpotent case. The case when the chosen Pukanszky polarization is normal is also possible to handle. We also treat several interesting low-dimensional examples. We have not yet been able to prove the conjecture in the general case.

8.3.11 First approach to Conjecture 8.3.10

We keep the notation of the introduction. Let d be the dimension of the coadjoint orbit $(\mathrm{Ad}^*\ G)\ell$. We choose a coexponential basis $(X_1, \ldots, X_{d/2})$ of the polarization $\mathfrak{h}$ in $\mathfrak{g}$. The

Hilbert space of the representation $\rho_\hbar$ is thus identified to $L^2(\mathbb{R}^{d/2})$ via the diffeomorphism

$$\begin{array}{rcl}\Phi : \mathbb{R}^{d/2} & \longrightarrow & G_\hbar/H_\hbar \\ \hbar t = (\hbar t_1, \dots, \hbar t_{d/2}) & \longmapsto & \exp t_1 X_1 \cdots \exp t_{d/2} X_{d/2} H.\end{array}$$

Recall from [21] that I is the ideal of those polynomials Q_0 such that $\pi_\nu(Q_0) = O(\nu)$, whereas J is the ideal of the polynomials Q_0 such that there exists a sequence $(Q_j)_{j\geq 1}$ of polynomials such that

$$Q = Q_0 + \nu Q_1 + \nu^2 Q_2 + \cdots + \nu^k Q_k + \cdots \in \operatorname{Ker} \pi_\nu.$$

In other words, I is the set of polynomials Q_0 such that $\pi_\nu(Q_0)$ is "small" (i. e., vanishes for $\nu = 0$), and J is the set of polynomials Q_0, which can be deformed into an element of the annihilator. One clearly has the inclusion $I \subset J$. Conjecture 8.3.10 can be reformulated as follows.

Conjecture 8.3.12. *Any polynomial vanishing on the orbit can be deformed into an element of the annihilator of the associated topological module* π_ν.

Now let $Q_0 \in S(\mathfrak{g})$ such that Q_0 vanishes on the orbit $\Omega = (\mathrm{Ad}^* G)\ell$. Recall that we look for $Q_1, Q_2, \dots \in S(\mathfrak{g})$ such that $\pi_\nu(Q_0 + \nu Q_1 + \nu^2 Q_2 + \cdots) = 0$. We have $Q_0 \in \operatorname{Ker} \pi_0$, hence $\pi_\nu(Q_0) = O(\nu)$. Moreover, $\operatorname{ad} X \cdot Q_0$ vanishes on Ω for any $X \in \mathfrak{g}$, hence $\pi_\nu(\operatorname{ad} X \cdot Q_0) = O(\nu)$, that is,

$$[\pi_\nu(X), \pi_\nu(Q_0)] = \nu \pi_\nu([X, Q_0]) = O(\nu^2).$$

Applying this to $X = X_1, \dots, X_{d/2}$ and using $\pi_\nu(X_j) = -\partial_j + O(\nu)$ [21, Lemme 5.1.4], we end up with the fact that

$$\dot{\pi}_0(Q_0) = \left.\frac{d}{d\nu}\right|_{\nu=0} \pi_\nu(Q_0)$$

is a partial differential operator with constant coefficients. Using the same lemma from [21] again, there exists an element $Q_1 \in S(\mathfrak{g})$ such that $\dot{\pi}_0(Q_0) = -\pi_0(Q_1)$. Hence we get

$$\pi_\nu(Q_0 + \nu Q_1) = O(\nu^2), \tag{8.48}$$

thus making a first step toward deforming Q_0 as we would like. Unfortunately, the next steps are much more harder, and we have been able to carry out the process only in the nilpotent case.

Lemma 8.3.13. *Let G be an exponential solvable Lie group with lie algebra $\mathfrak{g}$. Let $Q_0 \in J_\Omega$. Then, the rth derivative $\pi_0^{(r)}(Q_0)$ is a partial differential operator with polynomial coefficients of degree at most $r - 1$.*

Proof. The result has been just proved for $r = 1$. We now detail the case $r = 2$. Let $j, k \in \{1, \ldots, d/2\}$. Starting from the fact that $\operatorname{ad} X_j \operatorname{ad} X_k Q_0$ vanishes on Ω, we get

$$[\pi_\nu(X_j), [\pi_\nu(X_k), \pi_\nu(Q_0)]] = O(\nu^3). \tag{8.49}$$

Vanishing of the coefficient of ν^2 in (8.49) yields

$$[\pi_0(X_j), [\pi_0(X_k), \ddot{\pi}_0(Q_0)]] + [\dot{\pi}_0(X_j), [\pi_0(X_k), \dot{\pi}_0(Q_0)]] + [\pi_0(X_j), [\dot{\pi}_0(X_k), \dot{\pi}_0(Q_0)]] = 0,$$

hence

$$[-\partial_j, [-\partial_k, \ddot{\pi}_0(Q_0)]] + [\dot{\pi}_0(X_j), [-\partial_k, \dot{\pi}_0(Q_0)]] + [-\partial_j, [\dot{\pi}_0(X_k), \dot{\pi}_0(Q_0)]] = 0. \tag{8.50}$$

The second term of the right-hand side vanishes because $\dot{\pi}_0(Q_0)$ is a constant coefficient partial differential operator. The last term also vanishes because of the following computation:

$$[-\partial_j, [\dot{\pi}_0(X_k), \dot{\pi}_0(Q_0)]] = [[-\partial_j, \dot{\pi}_0(X_k)], \dot{\pi}_0(Q_0)] + [\dot{\pi}_0(X_k), [-\partial_j, \dot{\pi}_0(Q_0)]].$$

The second term of this sum vanishes as $\dot{\pi}_0(Q_0)$ is a constant coefficient partial differential operator. The first term also vanishes because $\dot{\pi}_0(X_k)$ is a partial differential operator with coefficients of degree at most one: this comes from the formula

$$\pi_\nu(X_k) = \sum_{u=1}^{\frac{d}{2}} a_{X,u}(\nu t)\frac{\partial}{\partial t_u} + a_{X,0}(\nu t) + \frac{\nu}{2}\sum_{u=1}^{\frac{d}{2}} \frac{\partial a_{X,u}(\nu t)}{\partial t_u}, \tag{8.51}$$

where $a_{X,0}, \ldots, a_{X,d/2}$ are analytic functions on $\mathbb{R}^{d/2}$ (cf. [119]). Thus the first term of the sum (8.50) vanishes, which is equivalent to the fact that $\ddot{\pi}_0(Q_0)$ is a differential operator with affine coefficients.

The case $r \geq 3$ is treated similarly, by induction on r. From

$$[\pi_\nu(X_{j_1}), [\cdots[\pi_\nu(X_{j_r}), \pi_\nu(Q_0)]\cdots]], \tag{8.52}$$

the vanishing of the coefficient of ν^r in (8.52) implies that

$$[\partial_{j_1}, [\cdots[\partial_{j_r}, \pi_0^{(r)}(Q_0)]\cdots]] = O(\nu^{r+1})$$

is a finite sum of Lie brackets involving only the operators $\pi_0^{(u)}(Q_0)$, $u = 1, \ldots, r-1$ and $\pi_0^{(s)}(X_j)$, $s = 1, \ldots, r-1$, $j = 1, \ldots, d/2$. The operators of the first family are partial differential operators with coefficients of degree at most $u-1$ by the induction hypothesis; those of the second family are partial differential operators with coefficients of degree at most s by (8.51). More precisely, we have

$$
\begin{aligned}
&(-1)^r[\partial_{j_1},[\cdots[\partial_{j_r},\pi_0^{(r)}(Q_0)]\cdots]] \\
&= -\sum_{\substack{u+s_1+\cdots+s_r=r,\\ s_j\geq 0,\ 0<u\leq r-1}} [\pi_0^{(s_1)}(X_1),[\cdots[\pi_0^{(s_r)}(X_r),\pi_0^{(u)}(Q_0)]\cdots]].
\end{aligned}
\tag{8.53}
$$

All terms of the right-hand side are differential operators with polynomial coefficients. Computing the degree of such a coefficient, we find

$$(u-1)+(s_r-1)+\cdots+(s_1-1)=-1, \tag{8.54}$$

which means that the corresponding term vanishes. Hence both sides of (8.53) vanish, which proves the claim. □

8.3.12 Partial solution of Conjecture 8.3.10

The "nonsaturated" case

The following theorem provides a proof of Conjecture 8.3.10 for a class of representations, which contains all unitary irreducible representations, which are nonsaturated with respect to an ideal of codimension one in $\mathfrak{g}$. The nilpotent case can always be reduced to tha case. This approach uses, as in [75, 118], a particular set of generators of the kernel of the representation.

Theorem 8.3.14. *Let G be an exponential solvable Lie group with Lie algebra $\mathfrak{g}$. Let $\ell \in \mathfrak{g}^*$ be such that $\mathfrak{g}(\ell)+\mathfrak{n}$ for some nilpotent ideal $\mathfrak{n}$ containing $[\mathfrak{g},\mathfrak{g}]$. Then Conjecture 8.3.10 holds.*

Proof. Let $p:\mathfrak{g}^*\twoheadrightarrow\mathfrak{n}^*$ be the projection defined by restriction. Let N be the analytic subgroup of G with Lie algebra $\mathfrak{n}$, and let π' be the unitary irreducible representation of N associated to $\ell'=p(\ell)$. We have

$$\Omega=(\mathrm{Ad}^*\,G)\ell=(\mathrm{Ad}^*\,N)\ell. \tag{8.55}$$

Then $p(\Omega)=(\mathrm{Ad}^*\,N)\ell'$. Let $(X_1,\dots,X_n)$ be a Jordan–Hölder basis of $\mathfrak{g}$ such that $(X_1,\dots,X_k)$ is a basis of N, and $(X_{k+1},\dots,X_n)\in\mathfrak{g}(\ell)$. Then for any $k\leq r\leq n$, the subspace $\mathfrak{g}_r$ spanned by $(X_1,\dots,X_r)$ is an ideal of $\mathfrak{g}$. This leads to the fact that the restriction of π to N is irreducible and equivalent to π'. Let $S=\{j_1,\dots,j_d\}\subset\{1,\dots,n\}$ be the set of jump indices, defined by

$$i\in S\Leftrightarrow X_i\notin\mathfrak{g}(\ell)+\mathfrak{g}_{i-1}. \tag{8.56}$$

Then obviously S is included in $\{1,\dots,k\}$. Let $(X_1^*,\dots X_n^*)$ be the basis of $\mathfrak{g}^*$ dual to $(X_1,\dots,X_n)$. Then, according to [120], there exist polynomial functions q_j on $\mathbb{R}^d$ such that

$$\Omega = (\mathrm{Ad}^* N)\ell = \left\{ \sum_{j=1}^{n} q_j(y) X_j^*,\ y \in \mathbb{R}^d \right\}. \tag{8.57}$$

Let $Q_j \in S(\mathfrak{g})$ defined by $Q_j = q_j(X_{j_1}, \ldots, X_{j_d})$. For any $i \in \{1, \ldots d\}$, we have $Q_{j_i} = X_{j_i}$, and the elements $X_j - \sigma(Q_j)$ in $\mathcal{U}(\mathfrak{g})$ generate $\operatorname{Ker} \pi$, where $\sigma : S(\mathfrak{g}) \to \mathcal{U}(\mathfrak{g})$ is the symmetrization map [120, Remark 3.1.6]. The proof goes then exactly as in the nilpotent case [21, Théorème 5.4.1]. □

The "saturated" case unsolved

In order to prove Conjecture 8.3.10 by induction on the dimension of the Lie algebra, one also should tackle the case when there exists a one-codimensional ideal $\mathfrak{g}'$ of $\mathfrak{g}$ containing $\mathfrak{n} + \mathfrak{g}(\ell)$ (the "saturated" case). Let us try a proof, the failure of which will show the main difficulties arising here.

Let $p : \mathfrak{g}^* \twoheadrightarrow \mathfrak{g}'^*$ be the projection map defined by restriction. Let X be some nonzero vector in $\mathfrak{g} \setminus \mathfrak{g}'$, and let $\ell' = p(\ell)$, for some $\ell \in \Omega$. The orbit $\Omega = (\mathrm{Ad}^* G)\ell$ is saturated with respect to $\mathfrak{g}'$, that is, $\mathfrak{g}(\ell) \subset \mathfrak{g}'$. Let G' be the analytic (normal) subgroup of G with Lie algebra $\mathfrak{g}'$. Then we have

$$p(\Omega) = \coprod_{s \in \mathbb{R}} \Omega'_s, \tag{8.58}$$

where Ω'_s is the coadjoint orbit of $p(\mathrm{Ad}^* \exp sX.\Omega'_0)$. We can slightly abusively write

$$\Omega'_s = \mathrm{Ad}^* \exp sX.\Omega'_0. \tag{8.59}$$

Let π'_s be the unitary irreducible representation associated with the coadjoint orbit Ω'_s for $s \in \mathbb{R}$. We have immediately for any $g_0 \in G'$:

$$\pi'_s(g_0) = \pi'_0(\exp sX.g_0.\exp -sX), \tag{8.60}$$

and the restriction of π desintegrates as

$$\pi|_{G'} = \int_{\mathbb{R}}^{\oplus} \pi'_s\, ds. \tag{8.61}$$

Let us remark that if $Q_0 \in J_{\Omega'_0}$, then $Q_s := \mathrm{Ad} \exp -sX.Q_0$ belongs to $J_{\Omega'_s}$ for any $s \in \mathbb{R}$. Now any polynomial $Q \in S(\mathfrak{g})$ can be written as

$$Q = \sum_{\alpha=0}^{m} Q_\alpha X^\alpha \tag{8.62}$$

with $Q_\alpha \in \mathcal{S}(\mathfrak{g}')$. Suppose $Q \in J_\Omega$. Then by the hypothesis of saturation, Q_α belongs to $J_{\Omega'_s}$ for any $\alpha \in \{0, \dots, m\}$ and for any $s \in \mathbb{R}$. Now using the induction hypothesis for $\mathfrak{g}'$, the polynomial Q_α can be deformed to an element $\widetilde{Q}_{\alpha,0} \in \operatorname{Ker} \pi'_{\nu,0}$. The element

$$\begin{aligned}\widetilde{Q}_{\alpha,s} &:= \operatorname{Ad}_\nu \exp_\nu(-sX).\widetilde{Q}_{\alpha,0}\\ &= \sum_{k\geq 0} \frac{(-1)^k}{k!} s^k \operatorname{ad}_\nu^k X.\widetilde{Q}_{\alpha,0}\\ &= \sum_{k\geq 0} \frac{(-1)^k}{k!} \nu^k s^k \operatorname{ad}^k X.\widetilde{Q}_{\alpha,0}\end{aligned}$$

belongs to $\operatorname{Ker} \pi'_{\nu,s}$ for any $s \in \mathbb{R}$.

Under the assumption that there exists $T \in \mathcal{U}(\mathfrak{g}')$ such that

$$\pi'_{s,\nu}(T) = -is.\operatorname{Id} \tag{8.63}$$

for any $s \in \mathbb{R}$, the element

$$\widetilde{Q}_\alpha := \sum_k \frac{(-i)^k}{k!} \nu^k T^k \operatorname{ad}^k X.\widetilde{Q}_{\alpha,0} \tag{8.64}$$

belongs to $\operatorname{Ker} \pi'_{\nu,s}$ for any $s \in \mathbb{R}$. The element $\widetilde{Q} := \sum_{\alpha=0}^m \widetilde{Q}_\alpha X^\alpha$ is the needed deformation of Q. The problem here is that (8.63) is strongly linked to Kirillov's surjectivity theorem, which does not hold in general beyond the nilpotent case.

8.3.13 The normal polarization case

Theorem 8.3.15. *Let G be an exponential solvable Lie group with Lie algebra $\mathfrak{g}$. Let $\ell \in \mathfrak{g}^*$ be such that the representation $\pi_{\ell,\nu}$ associated to ℓ is induced from a normal polarizing subgroup. Then Conjecture* 8.3.10 *holds.*

Proof. We fix once for all a Jordan–Hölder basis $\mathcal{B} = (Y_1, \dots, Y_n)$ of $\mathfrak{g}$ passing through $\mathfrak{b} = \mathbb{R}\text{-span}(Y_1, \dots, Y_p)$, the Lie algebra of the normal polarizing subgroup B. We first show that for any $Q \in S(\mathfrak{b})$, $\pi_{\ell,\nu}(Q)$ is a multiplication operator by means of an analytic function. Indeed, it is sufficient to prove that for any $X \in \mathfrak{b}$. Let $X_1 = Y_{p+1}, \dots, X_{\frac{d}{2}} = Y_n$. We have for $t = \exp(t_1X_1)\cdots\exp(t_{\frac{d}{2}}X_{\frac{d}{2}})$ and ξ a C^∞ vector of π_ℓ:

$$\begin{aligned}\pi_\ell(X)\xi(t) &= \frac{d}{ds}\bigg|_{s=0} \xi(\exp(-sX)t) = \frac{d}{ds}\bigg|_{s=0} \xi(t \cdot t^{-1}\exp(-sX)t)\\ &= \frac{d}{ds}\bigg|_{s=0} \xi(t)\chi_\ell(t^{-1}\exp(-sX)t),\end{aligned}$$

as the modular functions Δ_B and Δ_G coincide. This gives already the result.

Let $Q_0 \in J_\Omega$. As B polarizes any element of the form $g \cdot \ell := \mathrm{Ad}^*(g)\ell$ ($g \in G$), we get

$$0 = Q_0(B \cdot g \cdot \ell) = Q_0(g \cdot \ell + \mathfrak{b}^\perp), \tag{8.65}$$

for any $g \in G$ making use of the Pukanszky condition of $\mathfrak{b}$. If $\mathfrak{b} = \mathfrak{g}(\ell)$, then $\mathfrak{g} = \mathfrak{g}(\ell)$ and there is nothing to argue. Otherwise, the set $\{(g \cdot \ell)|_{\mathfrak{b}}, g \in G\}$ turns out to be nontrivial. Equation 8.65 shows finally that the polynomial Q_0 belongs to $S(\mathfrak{b})$, with respect to the choice of the Jordan–Hölder basis $\mathscr{B}$. We now consider the diffeomorphism

$$\Theta : \mathbb{R}^d \longrightarrow \Omega, \quad (p,q) = (p_1, \ldots, p_{\frac{d}{2}}, q_1, \ldots, q_{\frac{d}{2}}) \longmapsto \Theta(p,q)$$

with

$$\langle \Theta(p,q), X \rangle = X(p,q) = \sum_{u=1}^{\frac{d}{2}} a_{X,u}(p) q_u + a_{X,0}(p)$$

for any $X \in \mathfrak{g}$, where $a_{X,u}$ are the analytic functions defined for a while before. Besides, for $X \in \mathfrak{b}$ we get $X(p,q) = a_{X,0}(p)$. Hence

$$0 = Q_0(X(p,q)) = Q_0(X(p,0)) = Q_0(a_{X,0}(p)),$$

and finally

$$0 = Q_0(a_{X,0}(-ivp)) = Q_0(\pi_v(X)) = \pi_v(Q_0).$$

This achieves the proof of the theorem. □

8.3.14 The nilpotent case revisited

A proof of Conjecture 8.3.10 in the nilpotent case (cf. Theorem 8.3.11) is provided using the explicit description of the annihilator of the representation ρ due to C. Godfrey (cf. [75]) and N. V. Pedersen (cf. [118]). In this case, the orbit is Zariski closed, and then $VA(\pi_v) = \Omega$. The method we used there relies on specific generators of the ideal J, which are at the same time via symmetrization, generators of the annihilator of ρ. We propose here a direct method, which does not call for specific generators. We develop here a generator-free approach, based on the following variant of the Kirillov surjectivity theorem (see [58]).

Proposition 8.3.16. *Let G be a connected, simply connected nilpotent Lie group with Lie algebra $\mathfrak{g}$. Let $\ell \in \mathfrak{g}^*$, and let $\pi_{\ell,v}$ be the associated deformed $*$-representation of $(S(\mathfrak{g})[\![v]\!], *)$. Then $\pi_{\ell,v}$ is a surjective map from $(S(\mathfrak{g})[\![v]\!], *)$ onto the algebra $\mathrm{DOP}(vx, \partial)[\![v]\!]$ of formal series whose coefficients are partial differential operators with polynomial coefficients with respect to the variables $vx_1, \ldots, vx_{d/2}$.*

Proof. The operators $\pi_{\ell,\nu}(Q)$ for $Q \in S(\mathfrak{g})$ are indeed elements of $\mathrm{DOP}(\nu x,\partial)[\nu]$. The Kirillov surjectivity theorem applied to $\pi_{\ell,\nu}$ for any fixed ν shows that for any $D \in \mathrm{DOP}(\nu x,\partial)$ there exists $Q \in S(\mathfrak{g})[\nu]$ such that $\pi_{\ell,\nu}(Q) = D$. The passage to a formal series is straightforward. □

Alternative proof of Conjecture 8.3.10 *in the nilpotent case.* According to Lemma 8.3.13, the term $\nu^{r-1}\pi_0^{(r)}(Q_0)$ belongs to $\mathrm{DOP}(\nu x,\partial)[\nu]$. Applying Proposition 8.3.16, there exists $Q_r \in S(\mathfrak{g})[\nu]$ such that

$$\nu^{r-1}\pi_0^{(r)}(Q_0) = \pi_\nu(Q_r). \tag{8.66}$$

Hence

$$\pi_\nu(Q_0) = \nu\pi_\nu\left(\sum_{r\geq 1}\frac{1}{r!}Q_r\right), \tag{8.67}$$

as was to be shown. The sum in the right-hand side is finite because $\pi_0^{(r)}(Q_0)$ vanishes for sufficiently large r. □

Example 8.3.17. Besides the nilpotent case, some low-dimensional solvable examples have been treated in [21]. We develop here another example.

Let $(P_j)_{j=1,\ldots,8}$ be the differential operators on $\mathbb{R}^2$ defined by

$$P_1 = \frac{\partial}{\partial t_1}, \quad P_2 = (e^{t_1}\cos t_1)\frac{\partial}{\partial t_2}, \quad P_3 = (e^{t_1}\sin t_1)\frac{\partial}{\partial t_2}, \quad P_4 = i(e^{t_1}\cos t_1)t_2,$$
$$P_5 = i(e^{t_1}\sin t_1)t_2, \quad P_6 = i\frac{1}{2}e^{2t_1}\cos 2t_1, \quad P_7 = i\frac{1}{2}e^{2t_1}\sin 2t_1, \quad P_8 = ie^{2t_1}.$$

They generate an eight-dimensional exponential solvable Lie algebra $\mathfrak{g}$ with basis $(X_1,\ldots,X_8)$ and brackets:

$$\begin{aligned}
&[X_1,X_2] = X_2 - X_3, \quad [X_1,X_4] = X_4 - X_5, \quad [X_1,X_6] = X_6 - X_7,\\
&[X_1,X_3] = X_2 + X_3, \quad [X_1,X_5] = X_4 + X_5, \quad [X_1,X_7] = X_6 + X_7,\\
&[X_2,X_4] = X_6 + \frac{1}{2}X_8, \quad [X_3,X_5] = -X_6 + \frac{1}{2}X_8, \quad [X_2,X_5] = [X_3,X_4] = X_7,
\end{aligned}$$

the other brackets being zero. There is a unitary irreducible representation π of the group $G = \exp\mathfrak{g}$ such that $d\pi(X_j) = P_j$. According to [119], the Kirillov coadjoint orbit associated to π is given by the parametric representation $(p_1,p_2,q_1,q_2) \mapsto \xi(p_1,p_2,q_1,q_2) \in \mathfrak{g}^*$, with the following coordinates in the dual basis $(X_1^*,\ldots,X_8^*)$:

$$\xi_1 = p_1, \quad \xi_2 = (e^{q_1}\cos q_1)p_2, \quad \xi_3 = (e^{q_1}\sin q_1)p_2, \quad \xi_4(e^{q_1}\cos q_1)q_2,$$
$$\xi_5 = (e^{q_1}\sin q_1)q_2, \quad \xi_6 = \frac{1}{2}(e^{2q_1}\cos 2q_1), \quad \xi_7 = \frac{1}{2}(e^{2q_1}\sin 2q_1), \quad \xi_8 = e^{2q_1}.$$

The ideal of polynomials vanishing on the orbit is generated by $Q_{1,0}$, $Q_{2,0}$ and $Q_{3,0}$ with

$$Q_{1,0} = \xi_2\xi_5 - \xi_3\xi_4, \quad Q_{2,0} = 4(\xi_6^2 + \xi_7^2) - \xi_8^2, \quad Q_{3,0} = \xi_6(\xi_3^2 - \xi_2^2) - 2\xi_2\xi_3\xi_7. \tag{8.68}$$

The representation π_ν writes on the generators:

$$\begin{aligned}
&\pi_\nu(X_1) = i\frac{\partial}{\partial t_1}, \quad \pi_\nu(X_2) = i(e^{-i\nu t_1}\cos -i\nu t_1)\frac{\partial}{\partial t_2},\\
&\pi_\nu(X_3) = i(e^{i\nu t_1}\sin -i\nu t_1)\frac{\partial}{\partial t_2}, \quad \pi_\nu(X_4) = -(e^{-i\nu t_1}\cos -i\nu t_1)t_2,\\
&\pi_\nu(X_5) = -(e^{-i\nu t_1}\sin -i\nu t_1)t_2, \quad \pi_\nu(X_6) = -\frac{1}{2}e^{-2i\nu t_1}\cos -i\nu 2t_1,\\
&\pi_\nu(X_7) = -\frac{1}{2}e^{-2i\nu t_1}\sin -2i\nu t_1, \quad \pi_\nu(X_8) = -e^{-2i\nu t_1}.
\end{aligned}$$

The annihilator of π_ν is generated by the three elements $Q_1, Q_2, Q_3 \in \mathcal{U}_\nu(\mathfrak{g})$ defined by

$$Q_1 = X_2X_5 - X_3X_4, \quad Q_2 = 4(X_6^2 + X_7^2) - X_8^2, \quad Q_3 = X_6(X_3^2 - X_2^2) - 2X_2X_3X_7. \tag{8.69}$$

Our conjecture holds for this example, as we can see by comparing (8.68) and (8.69): indeed Q_i is a deformation of $Q_{i,0}$ for any $i = 1, 2, 3$.

8.3.15 Dixmier and Kirillov maps

Let $\mathfrak{g}$ be a solvable Lie algebra over an algebraically closed field. Recall that the Dixmier map [63] is a bijection from $\mathfrak{g}^*/G_{\mathrm{alg}}$ onto Prim $\mathcal{U}(\mathfrak{g})$, where G_{alg} is the algebraic adjoint group of $\mathfrak{g}$. This bijection has been proved to be bicontinuous by Olivier Mathieu [111]. If Conjecture 8.3.10 holds, it would give a concrete description of the Dixmier map (more precisely of its inverse) for any complexified exponential solvable Lie algebra $\mathfrak{g}$, by choosing a unitary irreducible representation ρ, differentiating the associated family $\rho_\hbar$ and considering the corresponding Poisson characteristic variety. It would be interesting to investigate the bicontinuity of this map in the light of this dequantization process.

As in Subsection 8.3.1, the Kirillov–Bernat map for an exponential solvable Lie group is a bijection κ from $\mathfrak{g}^*/G$ onto $\widehat{G}$, which can be identified with Prim $C^*(G)$, as exponential Lie groups are of type I (see, e. g., [22]). This bijection has been proved to be bicontinuous by H. Leptin and J. Ludwig [106]. An analogue of the Poisson characteristic variety is available by considering the C^*-algebra of the tangent groupoid $\mathcal{G} = \coprod_{\hbar\in[0,1]} G_\hbar$ of the Lie group G: we can consider the evaluation morphism:

$$\mathrm{ev}_\hbar : C^*(\mathcal{G}) \longrightarrow C^*(G_\hbar)$$

for any $\hbar \in [0,1]$. Any irreducible unitary representation ρ generates a primitive ideal $J_{\rho,\hbar} = \operatorname{Ann}\rho_\hbar$ of $C^*(G_\hbar)$ for any $\hbar \in]0,1]$. The representation ρ_0 of the Abelian group $G_0 = \mathfrak{g}$ provides the ideal $J_{\rho,0} = \operatorname{Ann}\rho_0$ of $C^*(G_0) \sim C_0(\mathfrak{g}^*)$. The family $(\rho_\hbar)_{\hbar\in[0,1]}$ gives rise to a unitary representation $\widetilde{\rho}$ of the groupoid $\mathcal{G}$. Let $\widetilde{J}$ be the annihilator of $\widetilde{\rho}$ in $C^*(\mathcal{G})$. The characteristic variety $V(\widetilde{\rho})$ is the set of common zeroes of $J_{\rho,0}$, whereas the Poisson characteristic variety $VA(\widetilde{\rho})$ is the set of common zeroes of the evaluation $\mathrm{ev}_0(\widetilde{J})$ of $\widetilde{J}$ at $\hbar = 0$. The analogue of our Conjecture 8.3.10 can be stated as follows.

Conjecture 8.3.18. *For any exponential solvable group G and for any $\rho \in \widehat{G}$,*

$$VA(\widetilde{\rho}) = \overline{\kappa^{-1}(\rho)},$$

where $\overline{\kappa^{-1}(\rho)}$ stands for the closure of the coadjoint orbit $\kappa^{-1}(\rho)$ with respect to the ordinary topology of yg*.

It would be worth investigating the bicontinuity of this map in the light of this C^*-algebraic dequantization process.

8.4 Some nonexponential restrictive cases

8.4.1 The two-dimensional motion group

It is the 3-dimensional real Lie algebra $\mathfrak{g}$ of basis (H,P,Q) with the brackets

$$[H,P] = -Q, \quad [H,Q] = P;$$

the other brackets are null. We refer again to M. Vergne [42, Chapter VIII 1.3] and [3]. If $X = aH + bP + cQ$, we compute easily the matrix $\operatorname{ad} X$ in this basis

$$\operatorname{ad} X = \begin{pmatrix} 0 & 0 & 0 \\ -c & 0 & a \\ b & -a & 0 \end{pmatrix}.$$

We have also

$$(\operatorname{ad} X)^2 = \begin{pmatrix} 0 & 0 & 0 \\ ab & -a^2 & 0 \\ ac & 0 & -a^2 \end{pmatrix},$$

as well as the equality

$$(\operatorname{ad} X)^3 = -a^2(\operatorname{ad} X).$$

We deduce the same explicit expression as in Subsection 8.4.2,

$$\exp(\operatorname{ad} X) = I + \frac{\sin a}{a}\operatorname{ad} X + \frac{1-\cos a}{a^2}(\operatorname{ad} X)^2.$$

The matrix of $\exp(\operatorname{ad}^* X)$ in the dual basis,

$$\exp(\operatorname{ad}^* X) = \begin{pmatrix} 1 & c\frac{\sin a}{a} + b\frac{1-\cos a}{a} & -b\frac{\sin a}{a} + c\frac{1-\cos a}{a} \\ 0 & \cos a & \sin a \\ 0 & -\sin a & \cos a \end{pmatrix}.$$

The coadjoint action $\exp X$ on an element $\xi = \mu H^* + \beta P^* + \gamma Q^*$ writes

$$\begin{aligned} \operatorname{Ad}^*(\exp X).\xi = {} & \left(\mu + \left(c\frac{\sin a}{a} + b\frac{1-\cos a}{a}\right)\beta + \left(-b\frac{\sin a}{a} + c\frac{1-\cos a}{a}\right)\gamma\right)H^* \\ & + ((\cos a)\beta + (\sin a)\gamma)P^* \\ & + ((-\sin a)\beta + (\cos a)\gamma)Q^*. \end{aligned}$$

The coadjoint orbits are all the points of μH^* and the cylinders $\Omega_r, r > 0$ of axe H^* and radius r, defined by the equation $\xi_2^2 + \xi_3^2 = r^2$. The Duflo isomorphism is once again given by

$$\tau = \frac{\sin(\nu\partial_1/2)}{\nu\partial_1/2} \circ \sigma,$$

where ∂_1 designates the partial derivative operator $\frac{\partial}{\partial\xi_1}$ in $\mathfrak{g}^*$, and σ designates the symmetrization. The algebra of invariants has a unique generator $C\ :\ \xi \mapsto \xi_2^2 + \xi_3^2$, and the Duflo isomorphism still consists in applying the identity on this generator. The Casimir $\tau(C)$, still denoted by C, writes

$$C = P^2 + Q^2.$$

We consider for $r > 0$, $\hbar \neq 0$ and $0 \leq \lambda < 1$, the family $\rho_{r,\lambda;\hbar}$ of the following unitary irreducible representations of $G_\hbar = \exp \mathfrak{g}_\hbar$: The representation $\rho_{r,\lambda;\hbar}$ acts on the space $\mathcal{H}_\lambda$ of the 2π-pseudo-periodic functions classes φ, satisfying

$$\varphi(t + 2\pi) = e^{2i\pi\lambda}\varphi(t), \qquad \int_0^{2\pi} |\varphi(t)|^2\, dt < +\infty.$$

Its differential is defined by

$$\rho_{r,\lambda;\hbar}(H) = -\hbar\frac{d}{d\theta},$$

$$\rho_{r,\lambda;\hbar}(P) = ir\sin(\theta),$$
$$\rho_{r,\lambda;\hbar}(Q) = ir\cos(\theta).$$

On the space $\mathcal{H}_\lambda^\infty$ of C^∞-vectors, which is here the space of C^∞- functions of $\mathcal{H}_\lambda$. These expressions are polynomials of variable $\hbar$. Considering $\nu = i\hbar$ and applying the techniques above, one gets therefore a family of unitary representations of $\mathcal{A}$ in the topologically free weakly convergent and strongly unitary module $\mathcal{H}_\lambda^\infty[\![\nu]\!]$:

$$\pi_{r,\lambda;\nu}(H) = -\nu\frac{d}{d\theta},$$
$$\pi_{r,\lambda;\nu}(P) = -r\sin(\theta),$$
$$\pi_{r,\lambda;\nu}(Q) = -r\cos(\theta).$$

Assuming that ν is null, we see that the annihilator $\operatorname{Ann}\pi_{r,\lambda;0}$ is generated by H and by $P^2 + Q^2 - r^2$. The characteristic variety $V(\pi_{r,\lambda;\nu})$ turns out to be the circle in Ω_r defined by the equations $\xi_2^2 + \xi_3^2 = r^2$ and $\xi_1 = 0$. Then it is independent of the parameter λ. The annihilator $\operatorname{Ann}\pi_{r,\lambda;\nu}$ is generated in $\mathcal{A}$ by $C - r^2$. The ideal $\operatorname{Ann}\pi_{r,\lambda;\nu}/\nu\operatorname{Ann}\pi_{r,\lambda;\nu}$ is also generated by $C - r^2$ in $S(\mathfrak{g})$. The characteristic Poisson variety $VA(\pi_{r,\lambda;\nu})$ is therefore equal to the coadjoint orbit Ω_r.

We consider now the space $\mathcal{H}_\lambda^\omega$ of analytic vectors. These are the entire functions φ on $\mathbb{R}$ such that $\varphi(t + 2\pi) = e^{2i\pi\lambda}\varphi(t)$. Let R be the opertaor on $C^\omega(\mathbb{R})[\![\nu]\!]$ defined by

$$R\varphi(\theta) = \nu\varphi(\nu\theta).$$

This operator is injective. Let $\widetilde{C}_\lambda = R(\mathcal{H}_\lambda^\omega[\![\nu]\!])$, and let $\widetilde{\pi}_{r,\lambda;\nu}$ be the representation of $\mathcal{A}$ on $\widetilde{C}_\lambda$ defined by

$$\widetilde{\pi}_{r,\lambda;\nu}(X) = R\circ\pi_{r,\lambda;\nu}\circ R^{-1}.$$

The two representations are obviously equivalent, and we have

$$\widetilde{\pi}_{r,\lambda;\nu}(H) = -\frac{d}{d\theta},$$
$$\widetilde{\pi}_{r,\lambda;\nu}(P) = -r\sin(\nu\theta),$$
$$\widetilde{\pi}_{r,\lambda;\nu}(Q) = -r\cos(\nu\theta).$$

The annulator of $\widetilde{\pi}_{r,\lambda;0}$ is generated by P and $Q + r$. This time the characteristic variety $V(\widetilde{\pi}_{r,\lambda;\nu})$ is therefore the generator of the cylinder Ω_r defined by the equations $\xi_3 = -r$ et $\xi_2 = 0$.

8.4.2 The diamond groups

It is the 4-dimensional real Lie algebra $\mathfrak{g}$ with the basis (H, P, Q, E) and the brackets

$$[H, P] = -Q, \quad [H, Q] = P, \quad [P, Q] = E,$$

the other brackets are null. W refer to M. Vergne in [42]. If $X = aH + bP + cQ + dE$, we easily compute the matrix of $\operatorname{ad} X$ in this basis

$$\operatorname{ad} X = \begin{pmatrix} 0 & 0 & 0 & 0 \\ -c & 0 & a & 0 \\ b & -a & 0 & 0 \\ 0 & -c & b & 0 \end{pmatrix}.$$

We also have

$$(\operatorname{ad} X)^2 = \begin{pmatrix} 0 & 0 & 0 & 0 \\ ab & -a^2 & 0 & 0 \\ ac & 0 & -a^2 & 0 \\ b^2 + c^2 & -ab & -ac & 0 \end{pmatrix},$$

as well as the equality

$$(\operatorname{ad} X)^3 = -a^2(\operatorname{ad} X).$$

We deduce the explicit expression

$$\exp(\operatorname{ad} X) = I + \frac{\sin a}{a} \operatorname{ad} X + \frac{1 - \cos a}{a^2}(\operatorname{ad} X)^2.$$

After transposition and inversion, the matrix of $\exp(\operatorname{ad}^* X)$ in the dual basis reads

$$\exp(\operatorname{ad}^* X) = \begin{pmatrix} 1 & c\frac{\sin a}{a} + b\frac{1-\cos a}{a} & -b\frac{\sin a}{a} + c\frac{1-\cos a}{a} & (b^2 + c^2)\frac{1-\cos a}{a^2} \\ 0 & \cos a & \sin a & c\frac{\sin a}{a} - b\frac{1-\cos a}{a} \\ 0 & -\sin a & \cos a & -b\frac{\sin a}{a} - c\frac{1-\cos a}{a} \\ 0 & 0 & 0 & 1 \end{pmatrix}.$$

The coadjoint action of $\exp X$ on an element $\xi = \mu H^* + \beta P^* + \gamma Q^* + \lambda E^*$ writes

$$\begin{aligned} \operatorname{Ad}^*(\exp X).\xi = \Big(\mu &+ \Big(c\frac{\sin a}{a} + b\frac{1 - \cos a}{a}\Big)\beta + \Big(-b\frac{\sin a}{a} + c\frac{1 - \cos a}{a}\Big)\gamma \\ &+ (b^2 + c^2)\frac{1 - \cos a}{a^2}\lambda\Big)H^* \\ &+ \Big((\cos a)\beta + (\sin a)\gamma + \Big(c\frac{\sin a}{a} - b\frac{1 - \cos a}{a}\Big)\lambda\Big)P^* \end{aligned}$$

$$+\left((-\sin a)\beta + (\cos a)\gamma + \left(-b\frac{\sin a}{a} - c\frac{1-\cos a}{a}\right)\lambda\right)Q^*$$
$$+\lambda E^*.$$

Let ξ_i denote the ith component of $\mathrm{Ad}^*(\exp X).\xi$. The last component ξ_4 is invariant under the coadjoint action. Two cases are to be considered here:

- First case: $\xi_4 = \lambda = 0$. The expression $\xi_2^2 + \xi_3^2 = \beta^2 + \gamma^2$ is invariant on the subspace defined by $\xi_4 = 0$. Then we see that the corresponding coadjoint orbits are the cylinders of axe H^* and every point of this axe, depending on whether $\xi_2^2 + \xi_3^2$ is positive or null. They are exactly the coadjoint orbits of the quotient of γ by its center (the Lie algebra of the 2-dimensional motion group): we treat this example later in Subsection 8.4.1.
- Second case: $\xi_4 = \lambda \neq 0$. Making $\exp Y$ act on ξ with $Y = bP + cQ$, we obtain

$$\begin{aligned}\mathrm{Ad}^*(\exp Y).\xi = (\mu + c\beta + b\gamma + (b^2+c^2)\lambda)H^* \\ + (\beta + c\lambda)P^* \\ + (\gamma - b\lambda)Q^* \\ + \lambda E^*.\end{aligned}$$

By an adequate choice of b and c, it brings us back to the case where $\beta = \gamma = 0$, which we will assume. The explicit formula giving $\mathrm{Ad}^*(\exp X).\xi$ can be simplified:

$$\begin{aligned}\mathrm{Ad}^*(\exp X).\xi = \left(\mu + (b^2+c^2)\frac{1-\cos a}{a^2}\lambda\right)H^* \\ + \left(c\frac{\sin a}{a} - b\frac{1-\cos a}{a}\right)\lambda P^* \\ + \left(-b\frac{\sin a}{a} - c\frac{1-\cos a}{a}\right)\lambda Q^* \\ + \lambda E^*.\end{aligned}$$

We notice that we have

$$\xi_1 - \frac{\xi_2{}^2 + \xi_3{}^2}{2\xi_4} = \mu \quad \text{and} \quad \xi_4 = \lambda. \tag{8.70}$$

The coadjoint orbits in this case are the paraboloids of revolution $\Omega_{\lambda,\mu}$ and axe H^* given by the equations (8.70).

Let φ be the entire function defined by

$$\varphi(z) = \frac{\operatorname{sh} z/2}{z/2}.$$

It is clear that the matrix of $\varphi^{1/2}(\operatorname{ad} X)$ is of the form:

$$\begin{pmatrix} 1 & 0 & 0 & 0 \\ * & \varphi^{1/2}(ia) & 0 & 0 \\ * & * & \varphi^{1/2}(ia) & 0 \\ * & * & * & 1 \end{pmatrix},$$

which gives, with the notation of Subsection 8.2.1 that

$$J(X)^{1/2} = \frac{\sin(va/2)}{va/2}.$$

Hence, the Duflo isomorphism is given by

$$\tau = \frac{\sin(v\partial_1/2)}{v\partial_1/2} \circ \sigma,$$

where ∂_1 designates the partial derivative operator $\frac{\partial}{\partial \xi_1}$ in $\mathfrak{g}^*$, and σ designates the symmetrization. Taking into account the equations (8.70), we easily see that the algebra $S(\mathfrak{g})^{\mathfrak{g}}$ of invariant polynomials on $\mathfrak{g}^*$ is generated by C_1 and C_2, where $C_1(\xi) = \xi_4$ and $C_2(\xi) = \xi_2^2 + \xi_3^2 - 2\xi_1\xi_4$. The coadjoint orbit $\Omega_{\lambda,\mu}$ defined by (8.70) is given similarly by the equations:

$$C_1(\xi) = \lambda, \quad C_2(\xi) = -2\lambda\mu. \tag{8.71}$$

The operator $J(D)^{1/2}$ acts on these two generators by the identity, thus the Duflo isomorphism τ goes back to apply the symmetrization on these generators. We abusively denote C_1 and C_2, the two Casimirs $\tau(C_1)$ and $\tau(C_2)$. We have explicitly

$$C_1 = E, \quad C_2 = P^2 + Q^2 - 2EH.$$

We use now the notation of Section 8.2. For all real number $\hbar$, there exists a family of unitary irreducible representations $\rho_{\lambda,\mu;\hbar}$ de $G_\hbar = \exp \mathfrak{g}_\hbar$, whose differentials are defined on the Schwartz space $\mathcal{S}(\mathbb{R})$ (provided with a scalar product of $L^2(\mathbb{R})$) by

$$\begin{aligned} \rho_{\lambda,\mu;\hbar}(H) &= -i\left(-\frac{1}{2\lambda}\frac{d^2}{dx^2} + \frac{1}{2}\lambda\hbar^2 x^2 + \mu\right), \\ \rho_{\lambda,\mu;\hbar}(P) &= -\frac{d}{dx}, \\ \rho_{\lambda,\mu;\hbar}(Q) &= i\lambda\hbar x, \\ \rho_{\lambda,\mu;\hbar}(E) &= -i\lambda. \end{aligned}$$

These representations are unitarily equivalent to those obtained by holomorphic induction from the point $\mu H^* + \lambda E^*$ and the complex polarization $\mathfrak{h}$ generated by

$H, E, P+iQ$ [42, Sections V.4 and VIII.1.4.4]. The expressions here are polynomials of the variable $\hbar$. Considering $\nu = i\hbar$ and applying the techniques above, we get a unitary representation $\pi_{\lambda,\mu;\nu}$ of the deformed algebra $\mathcal{A}$ in the topologically free, weakly convergent and strongly unitary module $\mathcal{S}(\mathbb{R})[\![\nu]\!]$:

$$\begin{aligned}
\pi_{\lambda,\mu;\nu}(H) &= -\frac{1}{2\lambda}\frac{d^2}{dx^2} - \frac{1}{2}\lambda\nu^2x^2 + \mu,\\
\pi_{\lambda,\mu;\nu}(P) &= -i\frac{d}{dx},\\
\pi_{\lambda,\mu;\nu}(Q) &= i\lambda\nu x,\\
\pi_{\lambda,\mu;\nu}(E) &= \lambda.
\end{aligned}$$

Admitting that the parameter ν is null, we see that the annihilators $\operatorname{Ann}\pi_{\lambda,\mu;0}$ are the ideal of $S(\mathfrak{g})$ generated by par $E-\lambda$, Q and $C_2 + 2\lambda\mu$. We deduce that the characteristic variety $V(\pi_{\lambda,\mu;\nu})$ is defined by the equations:

$$\xi_2^2 + \xi_3^2 - 2\xi_1\xi_4 = -2\lambda\mu, \quad \xi_4 = \lambda, \quad \xi_3 = 0.$$

The characteristic variety $V(\pi_{\lambda,\mu;\nu})$ generates therefore the paraboloid of revolution $\Omega_{\lambda,\mu}$. The nonexistence of real polarizations can be seen by the fact that the characteristic variety is not an affine subspace (cf. Subsection 8.3.2). The annihilator $\operatorname{Ann}\pi_{\lambda,\mu;\nu}$ is generated in $\mathcal{A}$ by $C_1-\lambda$ and $C_2+2\lambda\mu$. The ideal $\operatorname{Ann}\pi_{\lambda,\mu;\nu}/\nu\operatorname{Ann}\pi_{\lambda,\mu;\nu}$ is therefore generated in $S(\mathfrak{g})$ by $C_1-\lambda$ and $C_2+2\lambda\mu$. We deduce that the characteristic Poisson manifold $VA(\pi_{\lambda,\mu;\nu})$ coincides with the coadjoint orbit $\Omega_{\lambda,\mu}$.

8.4.3 The semisimple case: Verma modules

Let $\mathfrak{g}_1$ be a complex semisimple Lie algebra. The Killing form is defined by

$$(X,Y) = \operatorname{Tr}(\operatorname{ad}X \circ \operatorname{ad}Y)$$

is nondegenerate and invariant under the adjoint action. Thus, it induces a linear isomorphism κ from $\mathfrak{g}_1$ to its dual $\mathfrak{g}_1^*$, defined by

$$\kappa(X) = (X,-),$$

which intertwines the adjoint and the coadjoint representations. Let H be a semisimple element of $\mathfrak{g}_1$, $\mathfrak{h}_1$ be a Cartan subalgebra containing H, Δ the associated root system and let Δ_+ designate the set of positive roots coming from the choice of an order $\mathfrak{h}_1^*$. Let also W denote the Weyl group associated in this context. The Killing form restricted to $\mathfrak{h}_1$ is nondegenerate. Let $\lambda = \kappa^{-1}(H) \in \mathfrak{g}_1^*$. The radical decomposition of $\mathfrak{g}_1$ writes

$$\mathfrak{g}_1 = \mathfrak{n}_{1-} \oplus \mathfrak{h}_1 \oplus \mathfrak{n}_{1+},$$

where $\mathfrak{n}_{1\pm} = \bigoplus_{\alpha \in \pm\Delta_+} \mathfrak{g}_1^\alpha$. Let $\delta \in \mathfrak{h}_1^*$ be the half-sum of positive roots. The Cartan subalgebra $\mathfrak{h}_1$ is orthogonal to $\mathfrak{n}_{1+}$ and $\mathfrak{n}_{1-}$ with respect to the Killing form, which allows us to prove that $\kappa^{-1}(H) \in \mathfrak{g}_1^*$ vanishes on $\mathfrak{n}_{1\pm}$. Thus, we will identify $\mathfrak{h}_1^*$ to the orthogonal of $\mathfrak{n}_{1-} \oplus \mathfrak{n}_{1+}$ in $\mathfrak{g}_1^*$ by taking the trivial prolongation to $\mathfrak{n}_{1-} \oplus \mathfrak{n}_{1+}$ of the linear forms on $\mathfrak{h}_1$. If H is regular, then the stabilizer of λ under the coadjoint action equals to $\mathfrak{h}_1$, and the solvable subalgebra $\mathfrak{b}_1 = \mathfrak{h}_1 \oplus \mathfrak{n}_{1+}$ is a solvable polarization on λ.

The context above applies to the Lie algebras $(\mathfrak{g}_\nu)_{\nu \in \mathbb{C}-\{0\}}$ (with the notation of Section 8.2), that have all the same underlying vector space, which will be denoted by $\mathfrak{g}$. We will always identify $\mathfrak{g}$ and its dual thanks to the Killing form of $\mathfrak{g} = \mathfrak{g}_1$ (independently of ν), not with the Killing form of $\mathfrak{g}_\nu$. The radical decomposition

$$\mathfrak{g}_\nu = \mathfrak{n}_{\nu-} \oplus \mathfrak{h}_\nu \oplus \mathfrak{n}_{\nu+}$$

is independent from ν in what concerns the underlying vector spaces, which will be denoted by $\mathfrak{n}_-$, $\mathfrak{h}$ and $\mathfrak{n}_+$. Similarly, we denote $\mathfrak{b}_+ = \mathfrak{h} \oplus \mathfrak{n}_+$ and

$$\mathfrak{g}^\alpha = \{X \in \mathfrak{g},\ \forall A \in \mathfrak{h}, [A, X]_\nu = \nu\alpha(A)X\}.$$

The root system associated $\mathfrak{g}_\nu$ is $\nu\Delta$, and $\mathfrak{g}^\alpha$ coincides for all ν with the radical subspace of $\mathfrak{g}_\nu$ corresponding to $\nu\alpha$. Let M_λ^ν be the Verma module associated to λ for $\mathfrak{g}_\nu$. We have

$$M_\lambda^\nu = \mathcal{U}(\mathfrak{g}_\nu) \bigotimes_{\mathcal{U}(\mathfrak{b}_{\nu+})} \mathbb{C},$$

where an element $A + Y$ of $\mathfrak{b}_{\nu+} = \mathfrak{h}_\nu \oplus \mathfrak{n}_{\nu+}$ acts on $\mathbb{C}$ via multiplication by $(\lambda - \nu\delta)(A)$, and where the action of $\mathcal{U}(\mathfrak{g}_\nu)$ is given by left multiplication. When we see ν as an indeterminate, this module is topologically free on $\mathcal{A} = \mathcal{U}_\nu(\mathfrak{g})$. In fact, we can identify it to $\mathcal{U}_\nu(\mathfrak{n}_-)$, and to $S(\mathfrak{n}_-)[\![\nu]\!]$ via symmetrization. We have the decomposition on weight subspaces under the action of $\mathfrak{h}_\nu$:

$$M_\lambda^\nu = \bigoplus_{\beta \in Q_+} (M_\lambda^\nu)_{\lambda - \nu\delta - \nu\beta},$$

where Q_+ designates the set of linear combinations with positive integer coefficients of elements of Δ_+. Supposing $\nu = 0$, we see that all $A \in \mathfrak{h}_0$ acts on M_λ^0 by multiplication by $\lambda(A)$. We see also that action de $\mathfrak{n}_{0+}$ is trivial on M_λ^0, and the action de $\mathfrak{n}_{0-}$ is faithful. The annulator of M_λ^0 in $S(\mathfrak{g})$ is therefore the ideal generated by $\mathfrak{n}_+$ and by $A - \lambda(A)$, $A \in \mathfrak{h}$. We deduce

$$V(M_\lambda^\nu) = \lambda + \mathfrak{b}_+^\perp.$$

Let $\chi_\lambda^\nu : Z(\mathcal{U}(\mathfrak{g}_\nu)) \to \mathbb{C}$ be the central character of M_λ^ν. Let also $\gamma_\nu : Z(\mathcal{U}(\mathfrak{g}_\nu)) \xrightarrow{\sim} S(\mathfrak{h})^W$ be the Harish–Chandra isomorphism of $\mathfrak{g}_\nu$ (cf. [63]). Considering $S(\mathfrak{h})^W$ as the set of W-invariant polynomials on $\mathfrak{h}^*$, we get (cf. [63]):

$$\chi_\lambda^\nu(u) = \gamma_\nu(u)(\lambda).$$

The Duflo isomorphism $\tau_\nu : S(\mathfrak{g})^{\mathfrak{g}_\nu} \to Z(\mathcal{U}(\mathfrak{g}_\nu))$ is obtained by left composition of γ_ν^{-1} by the restriction on $\mathfrak{h}^*$ (Chevalley isomorphism). For all $v \in S(\mathfrak{g})^{\mathfrak{g}_\nu}$, we get finally

$$\chi_\lambda^\nu(\tau_\nu(v)) = v(\lambda).$$

The annihilator of the Verma module M_λ^ν is the bilateral ideal of $\mathcal{U}(\mathfrak{g}_\nu)$ generated by $\operatorname{Ker}\chi_\lambda^\nu$ [63, Theorem 8.4.3]. Then it is the bilateral ideal of $(S(\mathfrak{g}), *_\nu)$ generated by $\{v - v(\lambda),\ v \in S(\mathfrak{g})^{\mathfrak{g}_\nu}\}$ via the Duflo isomorphism.

Consider now ν as an indeterminate and M_λ^ν as topologically free on $\mathcal{A}$. The ideal $\operatorname{Ann} M_\lambda^\nu / (\operatorname{Ann} M_\lambda^\nu \cap \nu\mathcal{A})$ of $S(\mathfrak{g})$ is therefore the ideal generated by $\{v - v(\lambda),\ v \in S(\mathfrak{g})^{\mathfrak{g}_\nu}\}$. We deduce the characteristic Poisson variety:

$$VA(M_\lambda^\nu) = \{\xi \in \mathfrak{g}^*,\ v(\xi) = v(\lambda) \text{ for any } v \in S(\mathfrak{g})^{\mathfrak{g}_\nu}\}.$$

When λ (which means H) is regular, this turns out to be the coadjoint orbit of λ. When $\lambda = 0$, this is precisely the nilpotent cone.

Verma modules and real forms

The Verma module M_λ^ν has a natural symmetric bilinear form with good covariance properties, the Shapovalov form (cf. [65, 124]). To transform it in an Hermitian sesquilinear form, we can apply the results of Section 8.1.6. We keep the notation of Subsection 8.4.3. Let $(X_{-\alpha}, H_\alpha, X_\alpha)_{\alpha\in\Delta_+}$ be a Chevalley basis of $\mathfrak{g}_1$. Then

$$\{-iX_{-\alpha}, -iH_\alpha, -iX_\alpha\}_{\alpha\in\Delta_+}$$

is a Chevalley basis of $\mathfrak{g}_i$ (where $i = \sqrt{-1}$). Let $\mathfrak{g}_{i,\mathbb{R}}$ be the associated deployed real form of $\mathfrak{g}_i$. It is defined as the real vector space generated by $-iX_{-\alpha}$, $-iH_\alpha$, $-iX_\alpha$. This defines a real form $\mathfrak{g}_\mathbb{R}$ of the underlying vector space $\mathfrak{g}$, which is automatically a deployed real form $\mathfrak{g}_{\nu,\mathbb{R}}$ of the Lie algebra $\mathfrak{g}_\nu$ for all imaginary number ν. The conjugation respects the radical subspaces, then necessarily the three decomposition components

$$\mathfrak{g}_\nu = \mathfrak{n}_{\nu-} \oplus \mathfrak{h}_\nu \oplus \mathfrak{n}_{\nu+}.$$

The conjugation on $\mathfrak{g}$ (resp., on $\mathfrak{h}$) with respect to this real form induces a conjugation on the dual $\mathfrak{g}^*$ (resp., $\mathfrak{h}^*$) thanks to the formula

$$\overline{\xi}(X) := \overline{\xi(\overline{X})}.$$

We extend the conjugation by multiplicativity to the enveloping algebra $\mathcal{U}(\mathfrak{g}_\nu)$. This enveloping algebra has (thanks to the Poincaré–Birkhof-f-Witt theorem) the decomposition

$$\mathcal{U}(\mathfrak{g}_\nu) = \mathcal{U}(\mathfrak{h}_\nu) \oplus (\mathfrak{n}_{\nu-}\mathcal{U}(\mathfrak{g}_\nu) + \mathcal{U}(\mathfrak{g}_\nu)\mathfrak{n}_{\nu+}).$$

Let $P_\nu : \mathcal{U}(\mathfrak{g}_\nu) \to \mathcal{U}(\mathfrak{h}_\nu)$ be the projection corresponding to this decomposition. From the above, this projection is compatible with the conjugation, which means that

$$P(\overline{u}) = \overline{P(u)} \quad \text{for any } u \in \mathcal{U}(\mathfrak{g}_\nu).$$

We define the *transposition* on $\mathcal{U}(\mathfrak{g}_\nu)$ as follows:

$${}^tX_\alpha = X_{-\alpha}, \quad {}^tX_{-\alpha} = X_\alpha, \quad {}^tH_\alpha = H_\alpha,$$

and we extend this transposition to an antiautomorphism of $\mathcal{U}(\mathfrak{g}_\nu)$. We see immediately that the transposition commutes with the conjugation. We consider on $\mathcal{U}(\mathfrak{g}_\nu)$ the involution (semilinear) defined by

$$a^* = {}^t\overline{a}.$$

This involution (restricted to $\mathfrak{g}$) is a new conjugation on $\mathfrak{g}$ and, therefore, a new real form $\mathfrak{g}^{\mathbb{R}}$, which is a *compact* real form of the Lie algebra $\mathfrak{g}_\hbar$ where $\hbar = -i\nu$ is a nonzero real number (cf. [81, Section III.6]). We notice that

$$\mathfrak{h} \cap \mathfrak{g}^{\mathbb{R}} = i\mathfrak{h} \cap \mathfrak{g}_{\mathbb{R}}.$$

Denote by $\mathfrak{h}^{\mathbb{R}}$ this intersection and denote by τ the conjugation (in $\mathfrak{g}$ or $\mathfrak{h}$ or in their duals, resp.) with respect to this new real form. Suppose then that $\lambda \in \mathfrak{h}^*$ is real, which means that $\tau(\lambda) = \lambda$. The involution $*$ on $\mathcal{U}(\mathfrak{g}_\nu)$ coincides, via the Duflo isomorphism, with the involution given in Subsection 8.1.6, the conjugation being τ.

Consider the Verma module $M_\nu^{\lambda+\nu\delta}$ (instead of considering M_ν^λ as in Subsection 8.4.3). Consider also ν as an indeterminate. As modules on the deformed algebra $\mathcal{A}$, they coincide up to $O(\nu)$ and, therefore, their characteristic and Poisson characteristic varieties coincide. Let $e_{\lambda+\nu\delta}^\nu$ be the highest weight vector of $M_{\lambda+\nu\delta}^\nu$.

Suppose now that λ is a real number. Define then a sesquilinear form on the Verma module $M_{\lambda+\nu\delta}^\nu$ by the formula

$$\langle m, n\rangle_\nu = \langle a.e_{\lambda+\nu\delta}^\nu, b.e_{\lambda+\nu\delta}^\nu\rangle_\nu = P_\nu(a^*b)(\lambda).$$

This is an Hermitian version of the Shapovalov form of $M_{\lambda+\nu\delta}^\nu$.

Proposition 8.4.1. *The representation π_ν of $\mathcal{A} = \mathcal{U}(\mathfrak{g}_\nu)$ in $M^\nu_{\lambda+\nu\delta}$ satisfies for all $m, n \in M^\nu_\lambda$ and all $u \in \mathcal{U}(\mathfrak{g}_\nu)$. We have the following:*

$$\langle um, n\rangle_\nu = \langle m, u^* n\rangle_\nu.$$

Proof. Let $a, b \in \mathcal{U}(\mathfrak{g}_\nu)$ verify $m = a.e^\nu_\lambda$ and $n = b.e^\nu_\lambda$. The proposition is an immediate consequence of the equality

$$\langle ua.e^\nu_{\lambda+\nu\delta}, b.e^\nu_{\lambda+\nu\delta}\rangle = \langle a.e^\nu_{\lambda+\nu\delta}, u^* b.e^\nu_{\lambda+\nu\delta}\rangle.$$ □

The associated representation is therefore a $*$-representation. Moreover, the quotient form defined par $\langle -, -\rangle_\nu$ in $\nu = 0$ is Hermitian nondegenerate. In other words, the Verma module $M^\nu_{\lambda+\nu\delta}$ is strongly pseudo-unitary. Hence the characteristic varieties $V(\pi_\nu)$ and $VA(\pi_\nu)$ are defined on the real field, and they are given by the same explicit formulas as in Subsection 8.4.3.

Remark 8.4.2. Let $\lambda \in \mathfrak{h}^*$. Using Theorem 7.6.24 of [65], it is easy to see that the Verma module $M^\nu_{\lambda+\nu\delta}$ is simple (consequently, the Shapovalov form is nondegenerate) except the possibly where the set of ν's is countable discrete. Otherwise, on the question of unitarizability of some modules of highest weight, see [68] and [85].

8.5 A deformation approach of the Kirillov map

We now outline an approach of the Kirillov map for exponential groups based on deformations, and suggest a possible way toward an alternative proof of the Leptin–Ludwig bicontinuity theorem [106] along these lines.

8.5.1 Poisson ideals

General setting

Let A be a commutative C^*-algebra, which identifies itself via the Gel'fand transform with the algebra $C_0(\widehat{A})$ of continuous functions on the spectrum of A, which vanish at infinity. The closed ideals of A are in bijection with the closed subsets of $\widehat{A}$ via $F \mapsto I_F = \{a \in A, a|_F = 0\}$.

Suppose now that the spectrum $X = \widehat{A}$ of A is endowed with a Poisson manifold structure, that is, a C^∞ manifold structure together with a Lie bracket $\{-,-\}$ on $C^\infty(X)$ (the Poisson bracket), which is a derivation with respect to both arguments (Leibniz rule). A *Poisson ideal* of A is a closed ideal $J = I_F$ such that the set F of its common zeroes is invariant under the flows of Hamiltonian vector fields. For any $f \in J \cap C^\infty(X)$ and any $g \in C^\infty(X)_c$, $\{f, g\} \in J$, but this property does not characterize Poisson ideals. A Poisson ideal $J = I_F$ is *primitive* if F is the closure of a symplectic leaf in X. Let us denote by $\mathrm{Prim}_{\mathcal{P}} A$ the set of primitive Poisson ideals of A.

Proposition 8.5.1. *Any maximal ideal J of A contains a unique Poisson ideal $p(J)$, which is maximal for the inclusion in J, and it is primitive in the above sense. The map $p : X = \hat{A} \to \mathrm{Prim}_{\mathcal{P}} A$ thus defined is surjective.*

Proof. Let J be a maximal ideal of A, which corresponds to a point $\xi \in X$. Let S_ξ be the symplectic leaf through ξ, and let $p(J)$ be the ideal of elements of A, which vanish on S_ξ. The primitive Poisson ideal thus obtained is clearly maximal among the Poisson ideals contained in J. Uniqueness is clear, as the sum of two Poisson ideals is a Poisson ideal. Surjectivity of p follows from the fact that any point of X is contained in a symplectic leaf. □

Set now a "Jacobson-like" topology on $\mathrm{Prim}_{\mathcal{P}} A$: for a subset T of $\mathrm{Prim}_{\mathcal{P}} A$ one defines $\overline{T}$ by

$$\overline{T} = \{J \in \mathrm{Prim}_{\mathcal{P}} A,\ I(T) \subset J\},$$

where the Poisson ideal $I(T)$ ie defined by

$$I(\emptyset) = A \quad \text{and} \quad I(T) = \bigcap_{I \in T} I \quad \text{for } T \neq \emptyset.$$

Proposition 8.5.2. *There exists a unique topology on $\mathrm{Prim}_{\mathcal{P}} A$ such that $T \mapsto \overline{T}$ is the closure map for this topology, which makes $\mathrm{Prim}_{\mathcal{P}} A$ a T_0-space.*

Proof. We clearly have $\overline{\emptyset} = \emptyset$, $T \subset \overline{T}$ and $\overline{\overline{T}} = \overline{T}$. We define the closed subsets as those writing $\overline{T}$, $T \subset \mathrm{Prim}_{\mathcal{P}} A$. Let $(T_j)_{j\in\Lambda}$ be a collection of closed subsets. We have then

$$\begin{aligned}\bigcap_{j\in\Lambda} T_j &= \{I \in \mathrm{Prim}_{\mathcal{P}} A,\ I(T_j) \subset I \text{ for any } j \in \Lambda\}\\ &= \Big\{I \in \mathrm{Prim}_{\mathcal{P}} A,\ \sum_{j\in\Lambda} I(T_j) \subset I\Big\}\end{aligned}$$

whereas

$$\overline{\bigcap_{j\in\Lambda} T_j} = \Big\{I \in \mathrm{Prim}_{\mathcal{P}} A,\ \bigcap_{K \in \bigcap_{j\in\Lambda} T_j} K \subset I\Big\}.$$

But the inclusion $I(T_j) = \bigcap_{K\in T_j} K \subset \bigcap_{K \in \bigcap_{r\in\Lambda} T_r} K$, verified for any $j \in \Lambda$, yields the inclusion $\sum_{j\in\Lambda} I(T_j) \subset \bigcap_{K\in\bigcap_{j\in\Lambda} T_j} K$. Hence $\overline{\bigcap_{j\in\Lambda} T_j} \subset \bigcap_{j\in\Lambda} T_j$, which shows that the intersection $\bigcap_{j\in\Lambda} T_j$ is closed. In order to show that the union of two closed subsets is closed, we need the following lemma.

Lemma 8.5.3. *Let I_1 and I_2 be two Poisson ideals. Then a primitive Poisson ideal, which contains $I_1 \cap I_2$ contains I_1 or I_2.*

Proof. Let V_{I_1} (resp., V_{I_2}) be the set of common zeroes of the elements of I_1 (resp., I_2). These are two closed sets of X, which are invariant under Hamiltonian flows. We have $I_1I_2 \subset I_1 \cap I_2$, hence $V_{I_1 \cap I_2} \subset V_{I_1I_2}$, and $V_{I_1} \cup V_{I_2} \subset V_{I_1 \cap I_2}$. On the other hand,

$$\begin{aligned} V_{I_1I_2} &= \{\xi \in X,\ \forall (a,b) \in I_1 \times I_2,\ a(\xi)b(\xi) = 0\} \\ &= V_{I_1} \cup V_{I_2}. \end{aligned}$$

We have then $V_{I_1 \cap I_2} \subset V_{I_1I_2} = V_{I_1} \cup V_{I_2} \subset V_{I_1 \cap I_2}$, which finally yields

$$V_{I_1 \cap I_2} = V_{I_1I_2} = V_{I_1} \cup V_{I_2}. \tag{8.72}$$

Let I be a primitive Poisson ideal containing $I_1 \cap I_2$. Its set of common zeroes V_I is the closure of a symplectic leaf contained in $V_{I_1} \cup V_{I_2}$, hence contained in V_{I_1} or V_{I_2}, whence the result. □

End of proof of the proposition. Let T_1 and T_2 be two closed subsets of $\operatorname{Prim}_{\mathcal{P}} A$, let $I_1 = I(T_1)$ and $I_2 = I(T_2)$. Then

$$\begin{aligned} \overline{T_1 \cup T_2} &= \{J \in \operatorname{Prim}_{\mathcal{P}} A,\ I(T_1 \cup T_2) \subset J\} \\ &= \{J \in \operatorname{Prim}_{\mathcal{P}} A,\ I_1 \cap I_2 \subset J\} \\ &= \{J \in \operatorname{Prim}_{\mathcal{P}} A,\ I_1 \subset J \text{ or } I_2 \subset J\} \quad \text{(according to Lemma 8.5.3)} \\ &= T_1 \cup T_2. \end{aligned}$$

Finally, $\operatorname{Prim}_{\mathcal{P}} A$ is a T_0-space: in fact (see [62, Paragraph 3.1.3] for a similar argument), if I_1 and I_2 are two distinct elements of $\operatorname{Prim}_{\mathcal{P}} A$, we have for instance $I_1 \not\subset I_2$. Let then F be the closed subset of the elements $I \in \operatorname{Prim}_{\mathcal{P}} A$, which contain I_1. We have $I_1 \in F$ and $I_2 \notin F$, hence the complement of F inside $\operatorname{Prim}_{\mathcal{P}} A$ is a neighborhood of I_2, which does not contain I_1. □

Consider the equivalence relation on $X = \widehat{A}$ defined by

$$I \sim J \Leftrightarrow p(I) = p(J).$$

Two points of X are equivalent if and only if the closures of their symplectic leaves coincide. We endow $\widehat{A}/\sim$ with the quotient topology.

Proposition 8.5.4. *The map $\widetilde{p} : \widehat{A}/\sim \longrightarrow \operatorname{Prim}_{\mathcal{P}} A$ derived from p is an homeomorphism.*

Proof. The map $\widetilde{p}$ is clearly bijective. Let us prove continuity first: Let T be a closed subset of $\operatorname{Prim}_{\mathcal{P}} A$, let $I(T) = \bigcap_{J \in T} J$ be the associated Poisson ideal. We have then

$$\begin{aligned} p^{-1}(T) &= \{K \in \widehat{A},\ I(T) \subset p(K)\} \\ &= \{K \in \widehat{A},\ I(T) \subset K\}. \end{aligned}$$

Hence $p^{-1}(T)$ is the set of common zeroes of $I(T)$, which is closed in $\widehat{A}$. Hence the map p is continuous, and $\widetilde{p}$ as well by definition of the quotient topology. Let us now prove that $\widetilde{p}$ is closed: Let $\widetilde{U}$ be a closed subset of $\widehat{A}/\sim$, and let $U \subset \widehat{A}$ be the inverse image of $\widetilde{U}$ by the canonical projection. Let $J(U) = \bigcap_{K \in U} K$. Then $U = \{I \in \widehat{A},\ J(U) \subset I\}$, and

$$\widetilde{p}(\widetilde{U}) = p(U) = \{p(I), I \in \widehat{A},\ J(U) \subset I\}.$$

But we have

$$\begin{aligned} J(U) &= \bigcap_{K \in U} K \\ &= \bigcap_{\widetilde{K} \in \widetilde{p}(\widetilde{U})} \Big(\bigcap_{p(K) = \widetilde{K}} K \Big) \\ &= \bigcap_{\widetilde{K} \in \widetilde{p}(\widetilde{U})} \widetilde{K} \\ &= I(\widetilde{p}(\widetilde{U})). \end{aligned}$$

We have then

$$\begin{aligned} \widetilde{p}(\widetilde{U}) &= \{p(I), I \in \widehat{A},\ J(U) \subset I\} \\ &= \{p(I), I \in \widehat{A},\ I(\widetilde{p}(\widetilde{U})) \subset I\} \\ &= \{p(I), I \in \widehat{A},\ I(\widetilde{p}(\widetilde{U})) \subset p(I)\} \\ &= \{J \in \operatorname{Prim}_{\mathcal{P}} A,\ I(\widetilde{p}(\widetilde{U})) \subset J\} \\ &= \overline{\widetilde{p}(\widetilde{U})}, \end{aligned}$$

which proves that $\widetilde{p}(\widetilde{U})$ is closed. Hence the map $\widetilde{p}$ is bicontinuous. □

8.6 Type-I-ness and consequences

O. Takenouchi proved in 1957 that exponential groups are *type I groups* [125]. It means that any unitary representation π of G is quasi-equivalent to a representation ρ, which is multiplicity-free, that is, such that the commutant $\rho(G)'$ is commutative. This has nice consequences on both domain and range of the Kirillov map.

8.6.1 Linear Poisson manifolds and primitive Poisson ideals

Recall that a linear Poisson manifold is nothing but the dual $\mathfrak{g}^*$ of a Lie algebra $\mathfrak{g}$ with the Kirillov–Kostant–Souriau bracket (8.26) The symplectic leaves coincide with the coadjoint orbits under the action of a connected Lie group G with Lie algebra $\mathfrak{g}$ [131].

Theorem 8.6.1. *Let $X = \mathfrak{g}^*$ be a linear Poisson manifold, and let $A = C_0(X)$. Suppose that $\mathfrak{g}$ is solvable and that the associated connected simply connected Lie group G is of type I. Then $\mathrm{Prim}_{\mathcal{P}} A$ is homeomorphic to the space of coadjoint orbits $\mathfrak{g}^*/G$ endowed with the quotient topology.*

Proof. It is enough to show, according to Proposition 8.5.4, that the equivalence classes for $\sim$ are the coadjoint orbits in $\widehat{A} = \mathfrak{g}^*$. Two points ξ and η of $\mathfrak{g}^*$ verify $\xi \sim \eta$ if and only if $\overline{\mathrm{Ad}^*\, G.\xi} = \overline{\mathrm{Ad}^*\, G.\eta}$. But G is supposed to be solvable and of type one; hence each coadjoint orbit is locally closed, that is, open in its closure [10]. Hence $\xi \sim \eta$ if and only if $\mathrm{Ad}^*\, G.\xi = \mathrm{Ad}^*\, G.\eta$. □

8.6.2 The unitary dual and the orbit space

(1) Any strongly continuous unitary representation of a type I group G is uniquely determined, up to isomorphism, by its kernel: the unitary dual $\widehat{G}$ is then homeomorphic to the primitive spectrum $\mathrm{Prim}\, C^*(G)$ endowed with the *Jacobson topology* defined as follows: the closure of any subset T is given by

$$\overline{T} := \Big\{ J \in \mathrm{Prim}\, C^*(G),\ \bigcap_{K \in T} K \subset J \Big\}. \tag{8.73}$$

This topology is T_0: for any $J, K \in \mathrm{Prim}\, C^*(G)$ there is a neighborhood of J or K, which does not contain the other element (see [62]).

(2) According to Theorem 8.6.1, the orbit space $\mathfrak{g}^*/G$ of an exponential group is homeomorphic to the primitive Poisson ideal space $\mathrm{Prim}_{\mathcal{P}}\, C^*(\mathfrak{g})$ endowed with the Jacobson-like topology defined in paragraph 8.5.1.

As a consequence, $\mathfrak{g}^*/G$ and $\widehat{G}$ have very similar structure when G is exponential.

8.6.3 The Tangent groupoid and Rieffel's strict quantization

Let G be an exponential Lie group. The idea consists in "interpolating" between $\mathfrak{g}^*/G$ and $\widehat{G}$, more precisely between $\mathrm{Prim}\, C^*(G)$ and $\mathrm{Prim}_{\mathcal{P}}\, C^*(\mathfrak{g})$. In order to do this, we interpolate between G and its Lie algebra $\mathfrak{g}$ by means of the continuous family $(G_t)_{t\in[0,1]}$, where G_t has Lie algebra $\mathfrak{g}_t = (\mathfrak{g}, t[-,-])$. The *tangent groupoid* of G is the union of these groups. More precisely, it is given by

$$\mathcal{G} = \mathfrak{g} \times [0, 1] \tag{8.74}$$

with the law

$$(X,t)(Y,t) = (X \underset{t}{.} Y, t), \tag{8.75}$$

where $X \underset{t}{.} Y = \frac{1}{t}\log(\exp tX.\exp tY)$ is given near $(0,0)$ by the Baker–Campbell–Hausdorff series:

$$X \underset{t}{.} Y = X + Y + \frac{t}{2}[X,Y] + \frac{t^2}{12}([X,[X,Y]] + [Y,[Y,X]]) + \cdots. \tag{8.76}$$

The groupoid C^*-algebra $\mathcal{A} := C^*(\mathcal{G})$ is the C^*-algebra associated with the continuous field of C^*-algebras $\mathcal{A}_t := C^*(G_t)$. In particular, we have a surjective evaluation morphism

$$\begin{aligned} E_t : \mathcal{A} &\longrightarrow \mathcal{A}_t \\ a &\longmapsto a_t \end{aligned}$$

for any $t \in [0,1]$. The $\mathcal{A}_t$'s are all isomorphic except for $t = 0$: $\mathcal{A}_0$ is commutative and isomorphic to $C_0(\mathfrak{g}^*)$ via the Fourier–Gel'fand transformation. Denote by $\underset{t}{*}$ the convolution on $C_c^\infty(G_t) \sim C_c^\infty(\mathfrak{g})$, and by $\underset{t}{\#}$ the corresponding Fourier transform convolution on $\mathcal{AS}(\mathfrak{g}^*) = \mathcal{F}(C_c^\infty(\mathfrak{g}))$. The algebra $\widetilde{\mathcal{A}} := \mathcal{C}^1([0,1], C_c^\infty(\mathfrak{g}))$ is a dense subalgebra of $\mathcal{A}$, isomorphic via Fourier transform to $\widetilde{\mathcal{B}} := \mathcal{C}^1([0,1], \mathcal{AS}(\mathfrak{g}^*))$.

Theorem 8.6.2 (M. A. Rieffel, N. P. Landsman, B. Ramazan). *For any $f, g \in \widetilde{\mathcal{B}}$, the following holds in $C_0(\mathfrak{g}^*)$:*

$$\lim_{t\to 0} \frac{i}{t}(f(t) \underset{t}{\#} g(t) - g(t) \underset{t}{\#} f(t)) = \{f(0), g(0)\}. \tag{8.77}$$

Here, $\{-,-\}$ is the usual Kirillov–Kostant–Poisson bracket on $\mathfrak{g}^*$.

8.6.4 One-parameter families of representations

Let $\xi \in \mathfrak{g}^*$ and let $\mathfrak{h}$ be a Pukánszky polarization at ξ. For any $t \in [0,1]$, consider as above the unitary induced representation:

$$\pi_t^{\xi,\mathfrak{h}} = \widetilde{\operatorname{Ind}}_{H_t}^{G_t} \chi_\xi. \tag{8.78}$$

The coadjoint orbits of ξ under the action of G_t are all the same $\mathcal{O}$ except for $t = 0$. The representation $\pi_t^{\xi,\mathfrak{h}}$ is a model for the image of $\mathcal{O}$ by the Kirillov map of G_t when $t \neq 0$, but is not irreducible in general for $t = 0$.

Theorem 8.6.3. *The family $(\pi_t^{\xi,\mathfrak{h}})_{t\in]0,1]}$ is a continuous field of representations, that is, the map $t \mapsto |||\pi_t^{\xi,\mathfrak{h}}(a_t)|||$ is continuous on $]0,1]$ for any $a \in \mathcal{A}$, and extends by continuity at $t = 0$ to a limit $L^{\mathcal{O}}(a)$.*

Proof. The representations $\pi_t^{\xi,\mathfrak{h}}$ admit a common realization in $L^2(\mathbb{R}^{d/2})$. The derivatives are given for any $X \in \mathfrak{g}$ by

$$d\pi_t^{\xi,\mathfrak{h}}(X)f(y) = \sum_{u=1}^{d/2} a_{X,u}(ty)\frac{\partial f}{\partial y_u}(y)$$

$$- ia_{X,0}(ty)f(y) + \frac{t}{2}\left(\sum_{u=1}^{d/2} \frac{\partial a_{X,u}}{\partial y_u}(ty)\right)f(y). \tag{8.79}$$

Here, the $a_{X,u}$'s are analytic [7, 21, 119]. The expression above is hence analytic w. r. t. the parameter t, and the functions $t \mapsto d\pi_t^{\xi,\mathfrak{h}}(X)$ are *strongly continuous in the generalized sense* [88]. Exponentiating we get that for any $X \in \mathfrak{g}$ the operator-valued function $t \mapsto \pi_t^{\xi,\mathfrak{h}}(\exp X)$ is strongly continuous. Strong continuity of $t \mapsto \pi_t^{\xi,\mathfrak{h}}(a_t)$ follows from integration and Lebesgue's dominated convergence theorem.

Now consider for $t \in]0,1]$ the ideal $I_t^{\mathcal{O}} := \operatorname{Ker}\pi_t^{\xi,\mathfrak{h}} \in \operatorname{Prim}\mathcal{A}_t$, with $\operatorname{Ad}^* G_t.\xi = \mathcal{O}$, and the following closed ideal in the C^*-algebra $\mathcal{A}$:

$$J^{\mathcal{O}} := \bigcap_{t\in]0,1]} J_t^{\mathcal{O}}, \tag{8.80}$$

where $J_t^{\mathcal{O}} = E_t^{-1}(I_t^{\mathcal{O}}) \in \operatorname{Prim}\mathcal{A}$. The quotient $\mathcal{A}/J^{\mathcal{O}}$ is then a continuous field of C^*-algebras (with $(\mathcal{A}/J^{\mathcal{O}})_t = \mathcal{A}_t/I_t^{\mathcal{O}} \sim \pi_t(\mathcal{A}_t)$) (cf. [45]), whence the result. □

The value at $t = 0$ is the norm of the class of a_0 in the quotient $\mathcal{A}_0/E_0(J^{\mathcal{O}})$. We try to compute it by means of the *semicharacter formula* [66, 117], which generalizes the Kirillov character formula for simply connected nilpotent Lie groups: let us formulate it directly in our "variable" context. For any $t \in]0,1]$, there are a finite number of $\operatorname{Ad}^* G_t$-invariant subsets $\Omega_j \subset \mathfrak{g}^*, j = 1,\dots,k$ (independent of t), a family $(\chi_j^t)_{j=1,\dots,k}$ of characters of G_t with positive real values, a family $(Q_j)_{j=1,\dots,k}$ of polynomials on $\mathfrak{g}^*$ (independent of t) and a family $(\alpha_j^t)_{j=1,\dots,k}$ of positive $\operatorname{Ad}^* G_t$-invariant analytic functions on $\mathfrak{g}$ such that

$$\Omega_j = \{\xi \in \mathfrak{g}^*,\ Q_j(\xi) \neq 0 \text{ and } Q_r(\xi) = 0 \text{ for } r < j\}, \tag{8.81}$$

the dual $\mathfrak{g}^*$ is the disjoint union of the Ω_j's, and for any $\varphi \in C_c^\infty(G_t)$ and $\xi \in \Omega_j$ we have (with $\mathcal{O} = \operatorname{Ad}^* G_1.\xi$ and $\dim\mathcal{O} = d$):

$$\operatorname{Tr}\pi_t^{\xi,\mathfrak{h}}(u_j^t \underset{t}{*} \varphi) = t^{-d/2}\int_{\mathcal{O}} \widehat{\alpha_j^t\varphi}(\eta)\, Q_j(\eta)d\beta_{\mathcal{O}}(\eta). \tag{8.82}$$

Here, u_j^t is the symmetrization of Q_j in $\mathcal{U}(\mathfrak{g}_t)$ and $d\pi_t^{\xi,\mathfrak{h}}(u_j)$ is positive self-adjoint χ_j^t-semiinvariant. The measure $Q_j d\beta_{\mathcal{O}}$ is tempered.[1] In view of the semicharacter formula, we introduce "twisted Schatten seminorms" on $\widetilde{\mathcal{A}}$:

$$S_{2p}^{\mathcal{O},t}(a) := \left(t^{\frac{d}{2}} \operatorname{Tr} d\pi_t^{\xi,\mathfrak{h}}(u_j^t)(\pi_t^{\xi,\mathfrak{h}}(a_t)^* \pi_t^{\xi,\mathfrak{h}}(a_t))^p\right)^{\frac{1}{2p}}. \tag{8.83}$$

Theorem 8.6.4. *We have the following diagram, which suggests a good candidate for the limit:*

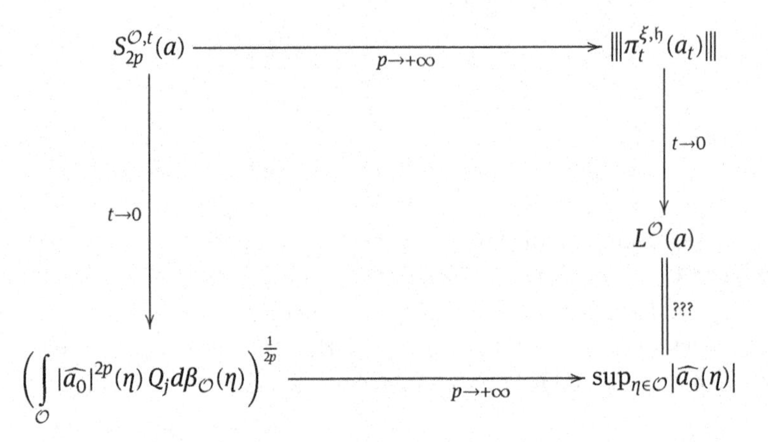

Proof. The right arrow is a direct consequence of Theorem 8.6.3, the left arrow comes from the semicharacter formula and the properties of the # product and the bottom arrow is a standard property of L^p norms. The upper arrow will show up as a consequence of the following lemma.

Lemma 8.6.5. *Let T be a bounded operator on a Hilbert space $\mathcal{H}$ and $\mathcal{D}$ a dense subspace of $\mathcal{H}$ containing the image of $\mathcal{H}$ by T and by its adjoint T^*. Let A be a nonnegative self-adjoint operator, which can be unbounded, defined on a domain containing $\mathcal{D}$, and such that $A|_{\mathcal{D}}$ is an isomorphism from $\mathcal{D}$ onto $\mathcal{D}$. Suppose moreover that $A(T^*T)^pA$ and $A^2(T^*T)^p$ are trace-class operators for any integer $p \geq 1$, and that the image of $A(T^*T)^pA$ is contained in $\mathcal{D}$ for any $p \geq 1$. Then:*

(1) *The following equality holds:*

$$\operatorname{Tr} A^2(T^*T)^p = \operatorname{Tr} A(T^*T)^p A = \operatorname{Tr}\left|A^2(T^*T)^p\right|, \tag{8.84}$$

where the notation $|U|$ stands for the operator $(U^*U)^{1/2}$.

(2) *We have*

$$|||T||| = \lim_{p\to\infty} (\operatorname{Tr} A(T^*T)^p A)^{\frac{1}{2p}}. \tag{8.85}$$

1 Notice the $t^{-d/2}$ factor, due to the normalization of the Liouville measure.

Proof. Let us denote by V the nonnegative self-adjoint trace-class operator $A(T^*T)^pA$ (with a fixed integer $p \geq 1$). Let (λ_i) be the sequence of positive eigenvalues of V (without multiplicities) decreasingly ordered. Consider the decomposition:

$$(\operatorname{Ker} V)^\perp = \bigoplus_i E_{\lambda_i}, \tag{8.86}$$

where E_{λ_i} is the eigenspace of V of eigenvalue λ_i (the direct sum is orthogonal). Every E_{λ_i} is finite-dimensional, V being a compact operator. The inclusion $E_{\lambda_i} \subset \mathcal{D}$ holds thanks to the hypotheses of the lemma. The subspace $F_{\lambda_i} := AE_{\lambda_i}$ then contains only eigenvectors of $U := A^2(T^*T)^p = AVA^{-1}$ for the eigenvalue λ_i. Reciprocally, if μ is a nonzero eigenvalue of U, the corresponding eigenspace F'_μ is in $\mathcal{D}$ (thanks again to the hypotheses of the lemma), and $E'_\mu := A^{-1}F'_\mu$ consists of eigenvectors of V for the eigenvalue μ. In particular, μ is one of the eigenvalues λ_i and we have $E'_\mu = E_{\lambda_i}, F'_\mu = F_{\lambda_i}$. The nonzero eigenvalues of U and of V are then positive, equal and occur with the same multiplicity, which proves the first assertion of the lemma.

The second assertion will be proved with the help of the first one: on the one hand, we have

$$\begin{aligned}(\operatorname{Tr} A(T^*T)^pA)^{\frac{1}{2p}} &= (\operatorname{Tr} A^2(T^*T)^p)^{\frac{1}{2p}} \\ &\leq (\operatorname{Tr}(A^2T^*T)|||(T^*T)^{p-1}|||)^{\frac{1}{2p}} \\ &\leq (\operatorname{Tr}(A^2T^*T))^{\frac{1}{2p}} |||T|||^{\frac{p-1}{p}},\end{aligned} \tag{8.87}$$

and using Assertion 1, we have on the other hand,

$$\operatorname{Tr} V = \operatorname{Tr} U = \operatorname{Tr} |U| \geq |||U|||. \tag{8.88}$$

We have then for any $\varepsilon > 0$,

$$\begin{aligned}(\operatorname{Tr} A(T^*T)^pA)^{\frac{1}{2p}} &\geq \sup_{v\in\mathcal{D}, \|v\|=1} \langle A(T^*T)^pA(v), v\rangle^{\frac{1}{2p}} \\ &\geq \sup_{v\in\mathcal{D}, \|v\|=1} \langle (T^*T)^pA(v), A(v)\rangle^{\frac{1}{2p}} \\ &\geq \sup_{v\in\mathcal{D}, \|v\|=1, \|Av\|\geq\varepsilon} \langle (T^*T)^pA(v), A(v)\rangle^{\frac{1}{2p}} \\ &\geq \sup_{w\in\mathcal{D}, \|A^{-1}w\|=1, \|w\|\geq\varepsilon} \langle (T^*T)^pw, w\rangle^{\frac{1}{2p}} \\ &\geq \sup_{w\in\mathcal{D}, \|A^{-1}w\|=1, \|w\|\geq\varepsilon} \langle (T^*T)^pw, w\rangle^{\frac{1}{2p}} \left(\frac{\varepsilon}{\|w\|}\right)^{\frac{1}{p}}\end{aligned}$$

hence finally,

$$(\mathrm{Tr}\, A(T^*T)^p A)^{\frac{1}{2p}} \geq \sup_{w \in \mathcal{D}, w \neq 0, \frac{\|w\|}{\|A^{-1}w\|} \geq \varepsilon} \left(\frac{\langle (T^*T)^p(w), w \rangle}{\|w\|^2} \right)^{\frac{1}{2p}} \varepsilon^{\frac{1}{p}}. \tag{8.89}$$

Let us now apply the spectral theorem for the nonnegative bounded self-adjoint operator T^*T:

$$(T^*T)^p = \int_0^{\|\|T\|\|^2} x^p \, dE(x) \tag{8.90}$$

for any positive integer p. According to the spectral theorem, one can write for any $w \in \mathcal{H} - \{0\}$ and for any integer $p \geq 1$:

$$\langle (T^*T)^p w, w \rangle = \|w\|^2 \|x\|_p^p,$$

where x stands for the identity map $x \mapsto x$ on $\mathbb{R}$, and where $\|-\|_p$ stands for the L^p norm with respect to the probability measure $\lambda_w = \|w\|^{-2} d\langle E_x w, w \rangle$. Inequality $\|x\|_p \geq \|x\|_1$ writes

$$\frac{\langle (T^*T)^p w, w \rangle}{\|w\|^2} \geq \left(\frac{\langle T^*Tw, w \rangle}{\|w\|^2} \right)^p. \tag{8.91}$$

For any $\alpha \in]0, 1[$, let us denote by C_α the (open nonempty) cone consisting of vectors $w \in \mathcal{H} - \{0\}$ such that

$$\frac{\langle T^*Tw, w \rangle}{\|w\|^2} > (1 - \alpha)^2 \|\|T\|\|^2.$$

According to (8.91), we have for any $w \in C_\alpha$,

$$\frac{\langle (T^*T)^p w, w \rangle}{\|w\|^2} > ((1 - \alpha)\|\|T\|\|)^{2p}. \tag{8.92}$$

Let us choose some $\varepsilon(\alpha) > 0$ such that

$$\varepsilon(\alpha) < \sup_{w \in C_\alpha \cap \mathcal{D}} \frac{\|w\|}{\|A^{-1}w\|}. \tag{8.93}$$

Starting from inequality (8.89) with $\varepsilon = \varepsilon(\alpha)$, one can then write using (8.93):

$$(\mathrm{Tr}\, A(T^*T)^p A)^{\frac{1}{2p}} \geq \sup_{w \in C_\alpha \cap \mathcal{D}, \frac{\|w\|}{\|A^{-1}w\|} \geq \varepsilon(\alpha)} \left(\frac{\langle (T^*T)^p w, w \rangle}{\|w\|^2} \right)^{\frac{1}{2p}} \varepsilon(\alpha)^{\frac{1}{p}}, \tag{8.94}$$

which yields, using (8.87) and (8.92),

$$(1-\alpha)|||T|||\,\varepsilon(\alpha)^{\frac{1}{p}} \le (\operatorname{Tr} A(T^*T)^p A)^{\frac{1}{2p}} \le (\operatorname{Tr}(A^2T^*T))^{\frac{1}{2p}}\,|||T|||^{\frac{p-1}{p}}. \tag{8.95}$$

The second assertion of the lemma follows by taking some arbitrarily small $\alpha > 0$ and by considering both sides of inequality (8.95) for $p \to +\infty$. □

Proof of Theorem 8.6.4 (*continued*). Fix any $t \in \,]0,1]$, and apply Lemma 8.6.5 to the operators $\widetilde{A} := (t^{\frac{d}{2}} d\pi_t^{\xi,\mathfrak{h}}(u_j^t))^{1/2}$ and $\widetilde{T} := \pi_t^{\xi,\mathfrak{h}}(a_t)$, which verify the hypotheses of Lemma 8.6.5 once we have identified all the representation spaces with $\mathcal{H} = L^2(\mathbb{R}^{d/2})$ as indicated above. The dense domain $\mathcal{D}$ can be taken as the intersection of the smooth vector spaces $\mathcal{H}^{\infty}_{\pi_t^{\xi,\mathfrak{h}}}$ for $t \in \,]0,1]$. □

The limit $L^{\mathcal{O}}(a)$ is indeed what we would expect.

Theorem 8.6.6. *The following equality holds:*

$$\lim_{t\to 0} |||\pi_t^{\xi,\mathfrak{h}}(a_t)||| = \sup_{\eta\in\mathcal{O}} |\widehat{a_0}(\eta)|. \tag{8.96}$$

Proof. This is actually a direct consequence of the general Leptin–Ludwig theorem for variable exponential groups. Indeed, if we see the groupoid $\mathcal{G}$ as a variable group with parameter space $[0,1]$, the orbit space Ω is $\mathfrak{g}^* \coprod (]0,1] \times \mathfrak{g}^*/G)$ endowed with the following topology: a point $(t,\lambda) \in \Omega$ is limit of the sequence (t_n,λ_n) if and only if $t_n \to t \in [0,1]$ and:

- either $t \neq 0$ and λ is limit of λ_n in $\mathfrak{g}^*/G$,
- or $t = 0$ and for any n there is $\xi_n \in \lambda_n$ such that $\lambda \in \mathfrak{g}^*$ is the limit of ξ_n.

In particular, the closure of $]0,1] \times \{\mathcal{O}\}$ in Ω is given by

$$\overline{]0,1] \times \{\mathcal{O}\}} = \mathcal{O} \coprod (]0,1] \times \{\mathcal{O}\}), \tag{8.97}$$

expressing the fact that a coadjoint orbit splits in the collection of all of its points when t reaches 0. The Leptin–Ludwig theorem then says that the closure of the set

$$E_{\mathcal{O}} := \{\pi_t^{\xi,\mathfrak{h}} \circ E_t,\ t \in \,]0,1]\} \subset \widehat{\mathcal{A}} \tag{8.98}$$

is $\{\chi_\xi \circ E_t,\ \xi \in \mathcal{O}\} \coprod E_{\mathcal{O}}$. The conclusion follows then from [71, Lemma I.9]. □

It would be of course very desirable to have an independent proof of this result: This could be reached by showing that the convergence indicated by the upper arrow of the diagram is uniform with respect to t (or, alternatively, by showing that the convergence indicated by left arrow of the diagram is uniform with respect to p), what we have been unable to prove. This in turn could then be a first step in an alternative proof of the Leptin–Ludwig bicontinuity theorem, the following conjecture being the second main ingredient.

Conjecture 8.6.7. *For any compact subset K of $\mathfrak{g}^*$ and for any $a \in \mathcal{A}$, the maps $t \mapsto |||\pi_t^{\xi,\mathfrak{h}}(a_t)|||$ are equicontinuous at $t = 0$ with respect to $\xi \in K$. Equivalently, for any quasi-compact subset T of $\mathfrak{g}^*/G$ and for any $a \in \mathcal{A}$, the maps $t \mapsto |||[\pi_t]^{\mathcal{O}}(a_t)|||$ are equicontinuous at $t = 0$ with respect to $\mathcal{O} \in T$.*

Remark that Conjecture 8.6.7 cannot be proved by simply looking at formulae (8.79): This comes from the fact that the analytic functions $a_{X,u}$, which depend on the initial point $\xi \in \mathfrak{g}^*$, do not always behave smoothly and can explode when ξ moves inside a stratum Ω_j and reaches its boudary. This phenomenon already occurs on the nilpotent Lie algebra $\mathfrak{g}_{5,4}$, defined by the basis $(X_1, X_2, X_3, X_4, X_5)$ and nonvanishing brackets $[X_5, X_3] = X_1$, $[X_4, X_3] = X_2$ and $[X_5, X_4] = X_3$.

8.6.5 Kirillov map revisited

We keep the previous notation: For $t \in]0,1]$ recall $I_t^{\mathcal{O}} := \operatorname{Ker} \pi_t^{\xi,\mathfrak{h}} \in \operatorname{Prim} \mathcal{A}_t$, with $\operatorname{Ad}^* G_t.\xi = \mathcal{O}$, and

$$J^{\mathcal{O}} := \bigcap_{t \in]0,1]} J_t^{\mathcal{O}}, \tag{8.99}$$

where $J_t^{\mathcal{O}} = E_t^{-1}(I_t^{\mathcal{O}}) \in \operatorname{Prim} \mathcal{A}$. For any closed ideal $I \subset \mathcal{A}$ and for any $t \in [0,1]$, we denote by $\operatorname{ev}_t(I)$ the closed ideal $E_t(I)$ of $\mathcal{A}_t$. Consider the set of closed ideals:

$$\mathcal{Q} := \{J^{\mathcal{O}}, \mathcal{O} \in \mathfrak{g}^*/G\}. \tag{8.100}$$

Theorem 8.6.8.
(1) *For any $t \in]0,1]$ the evaluation ev_t is a bijection from $\mathcal{Q}$ onto $\operatorname{Prim} \mathcal{A}_t$.*
(2) *There is a unique topology on $\mathcal{Q}$ such that ev_t is a homeomorphism for any $t \in]0,1]$.*
(3) *The evaluation ev_0 is a bijection from $\mathcal{Q}$ onto $\operatorname{Prim}_{\mathcal{P}} \mathcal{A}_0$, and the Kirillov map writes*

$$\kappa = \operatorname{ev}_1 \circ \operatorname{ev}_0^{-1}. \tag{8.101}$$

Proof. The first assertion comes from the fact that $E_t : J^{\mathcal{O}} \to I_t^{\mathcal{O}}$ is surjective and (2) is rather easy. It is enough to prove $\kappa(\overline{T}) = \overline{\kappa(T)}$ for any quasi-compact T of $\mathfrak{g}^*/G$ (by local quasi-compactness of both spaces involved). Let $\mathcal{T} \subset \mathfrak{g}^*$ be the union of the orbits in T. Set

$$J^{\mathcal{T}} := \bigcap_{\mathcal{O} \in \mathcal{T}} J^{\mathcal{O}}. \tag{8.102}$$

If $a \in J^{\mathcal{T}}$, then $\widehat{a_0}|_{\mathcal{T}} = 0$ as a consequence of Theorem 8.3.1. Hence we have

$$\mathcal{F}(\mathrm{ev}_0(J^{\mathcal{T}})) \subset \mathcal{I}_{\overline{\mathcal{T}}}. \tag{8.103}$$

We want the *reverse inclusion*. Suppose that Conjecture 8.6.7 is true. Then for any $a \in \mathcal{A}$ the map

$$t \mapsto \sup_{\mathcal{O}\in T} \| [\pi_t]^{\mathcal{O}}(a_t) \|$$

extends by continuity at $t = 0$ by

$$\sup_{\eta\in\mathcal{T}} |\widehat{a_0}(\eta)|$$

(using Theorems 8.6.3 and 8.6.6). Hence $\mathcal{A}/J^{\mathcal{T}}$ is a field of C^*-algebras, which is continuous at $t = 0$. Now let $a \in \mathcal{A}$ such that $\widehat{a_0}|_{\mathcal{T}} = 0$. It defines $\tilde{a} \in \mathcal{A}/J^{\mathcal{T}}$ such that $\|\tilde{a}_t\| \to 0$ when $t \to 0$. Hence we have, using Kasparov's notation,[2]

$$\tilde{a} \in (\mathcal{A}/J^{\mathcal{T}})|_{]0,1]} = \mathcal{A}|_{]0,1]}/(J^{\mathcal{T}} \cap \mathcal{A}|_{]0,1]}). \tag{8.104}$$

Now choose some element $a' \in \mathcal{A}|_{]0,1]}$ with image $\tilde{a}$ in the quotient. We obviously have $a' - a \in J^{\mathcal{T}}$ and

$$(a' - a)_0 = a_0, \tag{8.105}$$

which immediately yields the reverse inclusion $\mathcal{I}_{\overline{\mathcal{T}}} \subset \mathcal{F}(\mathrm{ev}_0(J^{\mathcal{T}}))$. We have then the equality

$$\mathcal{I}_{\overline{\mathcal{T}}} = \mathcal{F}(\mathrm{ev}_0(J^{\mathcal{T}})) \tag{8.106}$$

for any quasi-compact $T \subset \mathfrak{g}^*/G$, which means that $\mathrm{ev}_0 : \mathcal{Q} \to \mathrm{Prim}_{\mathcal{P}}(\mathcal{A}_0)$ is a homeomorphism. This proves Theorem 8.6.8. With Theorem 8.6.6 only, we can only repeat the proof for T containing a single orbit, which proves that ev_0 is a bijection. The first inclusion for quasi-compact T only shows that this bijection is open. □

Remark 8.6.9. Theorem 8.6.6 and Conjecture 8.6.7 together yield a proof of the bicontinuity of the Kirillov map. The pertinence of Theorem 8.6.8 is subordinated to finding a proof of Theorem 8.6.6 independent of the Leptin–Ludwig bicontinuity theorem, and a proof of Conjecture 8.6.7 would moreover yield an alternative proof of the latter.

---

2 According to which, for any open subset $U \subset [0,1]$, $\mathcal{A}|_U$ stands for the set of $a \in \mathcal{A}$ such that $E_t(a) = 0$ for any $t \notin U$.

Bibliography

[1] Abdelmoula. L, Baklouti. A, and Kédim. I, The Selberg–Weil–Kobayashi local rigidity theorem for exponential Lie groups, Int. Math. Res. Not. **2012**, No. 17, 4062–4084 (2012).

[2] Adkins. W. A and Weintraub. S. H, Algebra: An Approach via Module Theory, Graduate Texts in Mathematics, Springer-Verlag, New York, Berlin, Heidelberg (1992). ISBN 0-387-97839-9.

[3] Andler. M and Manchon. D, Opérateurs aux différences finies, calcul pseudo-différentiel et représentations des groupes de Lie, J. Geom. Phys. **27**, 1–29 (1998).

[4] Apanasov. B and Xie. X, Discrete actions on nilpotent Lie groups and negatively curved spaces, Differ. Geom. Appl. **20**, 11–29 (2004).

[5] Arnal. D, Baklouti. A, Ludwig. J, and Selmi. M, Separation of unitary representations of exponential Lie groups, J. Lie Theory **10**, 399–410 (2000).

[6] Arnal. D, Ben Amar. N, and Masmoudi. M, Cohomology of good graphs and Kontsevich linear star products, Lett. Math. Phys. **48**, 291–306 (1999).

[7] Arnal. D and Cortet. J-C, Représentations-$*$ des groupes exponentiels, J. Funct. Anal. **92**, No. 1, 103–135 (1990).

[8] Arnal. D, Manchon. D, and Masmoudi. M, Choix des signes pour la formalité de M. Kontsevich, Pac. J. Math. **203**, 23–66 (2002). arXiv:math.QA/0003003.

[9] Auslander. L, Bieberbach's theorem on space groups and discrete uniform subgroups of Lie groups II, Amer. J. Math. **83**, No. 2, 276–280 (1961).

[10] Auslander. L and Kostant. B, Polarization and unitary representations of solvable Lie groups, Invent. Math. **14**, 255–354 (1971).

[11] Baklouti. A, Deformation of discontinuous groups acting on some nilpotent homogeneous spaces, Proc. Jpn. Acad., Ser. A, Math. Sci. **85**, No. 4, 41–45 (2009).

[12] Baklouti. A, On discontinuous subgroups acting on solvable homogeneous spaces, Proc. Jpn. Acad., Ser. A, Math. Sci. **87**, 173–177 (2011).

[13] Baklouti. A and Bejar. S, On the Calabi–Markus phenomenon and a rigidity theorem for Euclidean motion groups, Kyoto J. Math. **56**, No. 2, 325–346 (2016).

[14] Baklouti. A and Bejar. S, On the Calabi–Markus phenomenon and a rigidity theorem for Euclidean motion groups. Kyoto. J. Math. **56**, No. 2, 325–346 (2016).

[15] Baklouti. A and Bejar. S, Variants of stability of discontinuous groups for Euclidean motion groups, Int. J. Math. **28**, No. 6, 26p (2017).

[16] Baklouti. A, Bejar. S and Dhahri. K, Deforming discontinuous groups for Heisenberg motion groups, Int. J. Math. **30**, No. 9, (2019).

[17] Baklouti. A, Bejar. S, and Fendri. R, A local rigidity theorem for finite actions on Lie groups and application to compact extensions of $\mathbb{R}^n$, Kyoto J. Math. **59**, No. 3, 607–618 (2019).

[18] Baklouti. A, Benson. C, and Ratcliff. G, Moment sets and the unitary dual of a nilpotent Lie group, J. Lie Theory **11**, 135–154 (2001).

[19] Baklouti. A, Bossofora. M, and Kedim. I, Deformation problems on three-step nilpotent Lie groups. Hiroshima Math. J. **49**, No. 2, 195–233 (2019).

[20] Baklouti. A, Boussoffara. M, and Kedim. I, Stability of Discontinuous Groups Acting on Homogeneous Spaces, Mathematical Notes. **103**, No. 4, 9–22 (2018).

[21] Baklouti. A, Dhieb. S, and Manchon. D, Orbites coadjointes et variétés caractéristiques, J. Geom. Phys. **54**, 1–41 (2005).

[22] Baklouti. A, Dhieb. S, and Manchon. D, A deformation approach of the Kirillov map for exponential groups, Adv. Pure Appl. Math. **2**, 421–436 (2011).

[23] Baklouti. A, Dhieb. S, and Manchon. D, The Poisson characteristic variety of unitary irreducible representations of exponential Lie groups, Springer Proceedings in Mathematics and Statistics. **290**, 207–217 (2019).

https://doi.org/10.1515/9783110765304-009

[24] Baklouti. A, Dhieb. S, and Tounsi. K, When is the deformation space $\mathscr{T}(\Gamma, H_{2n+1}, H)$ a smooth manifold?, Int. J. Math. **22**, No. 11, 1661–1681 (2011).
[25] Baklouti. A, ElAloui. N, and Kédim. I, A rigidity theorem and a stability theorem for two-step nilpotent Lie groups, J. Math. Sci. Univ. Tokyo **19**, 281–307 (2012).
[26] Baklouti. A, Fujiwara. H, and Ludwig. J, Representation Theory of Solvable Lie Groups and Related Topics, Springer Monographs in Mathematics (2021). eBook ISBN 978-3-030-82044-2.
[27] Baklouti. A, Ghaour. S, and Khlif. F, Deforming discontinuous subgroups of reduced Heisenberg groups, Kyoto J. Math. **55**, No. 1, 219–242 (2015).
[28] Baklouti. A, Ghaour. S, and Khlif. F, On discontinuous groups acting on $(\mathbb{H}^r_{2n+1} \times \mathbb{H}^r_{2n+1})/\Delta$, Adv. Pure Appl. Math. **6**, No. 2, 63–79 (2015).
[29] Baklouti. A, Ghaour. S, and Khlif. F, A stability theorem for non-abelian actions on threadlike homogeneous spaces, Springer Proceedings in Mathematics and Statistics. **207**, 117–135 (2017).
[30] Baklouti. A and Kédim. I, On the deformation space of Clifford–Klein forms of some exponential homogeneous spaces, Int. J. Math. **20**, No. 7, 817–839 (2009).
[31] Baklouti. A and Kédim. I, On non-Abelian discontinuous subgroups acting on exponential solvable homogeneous spaces, Int. Math. Res. Not. **2010**, No. 7, 1315–1345 (2010).
[32] Baklouti. A and Kédim. I, Open problems in deformation theory of discontinuous groups acting on homogeneous spaces, Int. J. Open Problems Comput. Math. **6**, No. 1, 115–131 (2013).
[33] Baklouti. A and Kédim. I, On the local rigidity of discontinuous groups for exponential solvable Lie groups, Adv. Pure. Appl. Maths. **4**, No. 1, 3–20 (2013).
[34] Baklouti. A, Kédim. I, and Yoshino. T, On the deformation space of Clifford–Klein forms of Heisenberg groups, Int. Math. Res. Not. **2008**, No. 16, Art. ID rnn066, 35p (2008). doi:10.1093/imrn/rnn066.
[35] Baklouti. A and Khlif. F, Proper actions on some exponential solvable homogeneous spaces, Int. J. Math. **16**, No. 9, 941–955 (2005).
[36] Baklouti. A and Khlif. F, Criterion of weak proper actions on solvable homogeneous spaces, Int. J. Math. **18** No. 8, 903–918 (2007).
[37] Baklouti. A, and Khlif. F, Deforming discontinuous subgroups for threadlike homogeneous spaces, Mathematical Notes. **146**, 117–140 (2010).
[38] Baklouti. A, Khlif. F, and Kooba. H, On the geometry of stable discontinuous subgroups acting on threadlike homogeneous spaces, Mathematical Notes. **89**, No. 5-6, 761–776 (2011).
[39] Barmeier. S, Deformations of the discrete Heisenberg group, Proc. Jpn. Acad., Ser. A, Math. Sci. **89**, No. 4, 55–59 (2013).
[40] Baues. O and Goldman. W, Is the deformation space of complete affine structures on the 2-torus smooth?, Geom. Dyn., Contemp. Math. **389**, 69–89 (2005).
[41] Benedetti. R and Risler. J. J, Real Algebraic and Semi-Algebraic Sets, Herman, 340p (1990).
[42] Bernat. P, Conze. N, Duflo. M, et al., Représentations des Groupes de Lie Résolubles, Monographies de la Société Mathématique de France, Vol. **4**, Dunod, Paris (1972).
[43] Bieberbach. L, Über die Bewegungsgruppen der Euklidischen Rame I, Math. Ann. **70**, 297–336 (1911).
[44] Bieberbach. L, Über die Bewegungsgruppen der Euklidischen Rame II, Math. Ann. **72**, 400–412 (1912).
[45] Blanchard. E, Déformations de C^*-algèbres de Hopf, Bull. Soc. Math. Fr. **124**, No. 1, 141–215 (1996).
[46] Bordemann. M, Ginot. G, Halbout. G, Herbig. H-C, and Waldmann. S, Star-représentations sur des sous-variétés co-isotropes, arXiv:math.QA/0309321 (2003).
[47] Bourbaki. N, Algèbre Commutative, Hermann, Paris (1961).

[48] Bourbaki. N, Éléments de Mathématiques Topologie Générale Chapitre 5 á 10, Springer (2007).
[49] Brown. I. D, Dual topology of a nilpotent Lie group, Ann. Sci. Éc. Norm. Supér. **6**, No. 3, 407–411 (1973).
[50] Bursztyn. H and Waldmann. S, $*$-ideals and formal Morita equivalence of $*$-algebras, Int. J. Math. **12** No 5, 555–577 (2001).
[51] Calabi. E, On compact, Riemannian manifolds with constant curvature, I, in Proceedings of Symposia in Pure Mathematics, Vol. **III**, Amer. Math. Soc., Providence, RI, 155–180 (1961).
[52] Calabi. E and Markus. L, Relativistic space forms, Ann. Math. **75**, 63–76 (1962).
[53] Cattaneo. A and Felder. G, A path integral approach to the Kontsevic quantization formula, Commun. Math. Phys. **212**, No. 3, 591–611 (2000).
[54] Cattaneo. A and Felder. G, Coisotropic submanifolds in Poisson geometry and branes in the Poisson sigma model, arXiv:math.QA/0309180 (2003).
[55] Cattaneo. A, Felder. G and Tomassini. L, From local to global deformation quantization of Poisson manifolds, arXiv:math.QA/0012228 (2000).
[56] Chevallier. D, Introduction à la Théorie des Groupes de Lie Réels, Ellipses (2006).
[57] Choi. S and Goldman. W, The classification of real projective structures on compact surfaces, Bull. Amer. Math. Soc. (N.S.) **34**, No. 2, 161–171 (1997).
[58] Corwin. L and Greenleaf. F, Representations of Nilpotent Lie Groups and Their Applications. Part 1, Cambridge Studies in Advanced Mathematics, Vol. **18**, Cambridge University Press (2004).
[59] Corwin. L, Greenleaf. F. P, and Grelaud. G, Direct integral decompositions and multiplicities for induced representations of nilpotent Lie groups, Trans. Amer. Math. Soc. **304**, 549–583 (1987).
[60] Dhieb. S, Deformation of discontinuous groups acting on $(H_{2n+1} \times H_{2n+1})/\Delta$, J. Lie Theory **26**, No. 2, 371–382 (2016).
[61] Dito. G, Kontsevich star product on the dual of a Lie algebra, Lett. Math. Phys. **48**, 307–322 (1999).
[62] Dixmier. J, Les C^*-Algèbres et Leurs Représentations, Gauthier-Villars, Paris (1964).
[63] Dixmier. J, Algèbres Enveloppantes, Gauthier-Villars, Paris (1974).
[64] Duflo. M, Opérateurs différentiels bi-invariants sur un groupe de Lie, Ann. Sci. Éc. Norm. Supér. (4) **10**, 265–288 (1977).
[65] Duflo. M, Sur la classification des idéaux primitifs dans l'algèbre enveloppante d'une algèbre de Lie semi-simple, Ann. Math. (2) **105**, No. 1, 107–120 (1977).
[66] Duflo. M and Raïs. M, Sur l'analyse harmonique sur les groupes de Lie résolubles, Ann. Sci. Éc. Norm. Supér. **9**, 107–144 (1976).
[67] Dunfield. N. M and Thurston. W. P, The virtual Haken conjecture: experiments and examples, Geom. Topol. **7**, 399–441 (2003).
[68] Enright. T, Howe. R, and Wallach. N, A classification of unitary highest weight modules, Prog. Math. **40**, 97–143 (1983).
[69] Etingof. P and Kazhdan. D, Quantization of Lie bialgebras, I, Sel. Math. **2**, 1–41 (1996).
[70] Felder. G and Shoikhet. B, Deformation quantization with traces, arXiv:math.QA/0002057 (2000).
[71] Fell. J. M. G, The structure of algebras of operator fields, Acta Math. **106**, 233–280 (1961).
[72] Gabber. O, The integrability of the characteristic variety, Am. J. Math. **103**, No. 3, 445–468 (1981).
[73] Ghys. É, Déformations des structures complexes sur les espaces homogènes de $SL(2, \mathbb{C})$, J. Reine Angew. Math. **468**, 113–138 (1995).
[74] Godement. R, Introduction à la Théorie des Groupes de Lie, Springer-Verlag, Berlin, Heidelberg (2004).

[75] Godfrey. C, Ideals of coadjoint orbits of nilpotent Lie algebras, Trans. Amer. Math. Soc. **233**, 295–307 (1977).

[76] Goldman. W, Non-standard Lorentz space forms, J. Differ. Geom. **21**, 301–308 (1985).

[77] Goldman. W and Millson. J. J, Local rigidity of discrete groups acting on complex hyperbolic space, Invent. Math. **88**, 495–520 (1987).

[78] Goldman. W. M, Locally homogeneous geometric manifolds, in Proceedings of the International Congress of Mathematicians, Hyderabad, India (2010).

[79] Granger. M and Maisonobe. Ph, A basic course on differential modules, in $\mathscr{D}$-Modules Cohérents et Holonomes, Les Cours du CIMPA, Hermann, Paris (1993).

[80] Hall. B. C, Lie Groups, Lie Algebras, and Representations, Springer (2004).

[81] Helgason. S, Differential Geometry and Symmetric Spaces, Academic Press (1962).

[82] Hilgert. J and Neeb. K-H, Structure and Geometry of Lie Groups, Springer Monographs in Mathematics, Springer Science+Business Media, LLC (2012). doi:10.1007/978-0-387-84794-8.

[83] Hochschild. G. P, The Structure of Lie Groups, Holden-Day Series in Mathematics, Holden-Day, Inc., San Francisco (1965).

[84] Hofmann. K. H and Neeb. K-H, The compact generation of closed subgroups of locally compact groups, J. Group Theory **12**, No. 4, 555–559 (2009).

[85] Jakobsen. H. P, Hermitian symmetric spaces and their unitary highest weight modules, J. Funct. Anal. **52** No. 3, 385–412 (1983).

[86] Joseph. A, On the classification of primitive ideals in the enveloping algebra of a semi-simple Lie algebra, in Lie Group Representations I, Lecture Notes in Mathematics, Vol. **1024**, Springer, 30–78 (1983).

[87] Kassel. Ch, Quantum Groups, Springer (1995).

[88] Kato. T, Perturbation Theory for Linear Operators, Springer (1966).

[89] Kédim. I, Rigidity of discontinuous actions on diamond homogeneous spaces. J. Math. Sci. Univ. Tokyo **23**, No. 2, 381–403 (2016).

[90] Khlif. F, Stability of discontinuous groups for reduced threadlike Lie groups, Int. J. Math. **26**, No. 08 (2015).

[91] Kirillov. A. A, Elements of the Theory of Representations, Springer (1976).

[92] Kobayashi. T, Proper action on homogeneous space of reductive type, Math. Ann. **285**, 249–263 (1989).

[93] Kobayashi. T, Discontinuous groups acting on homogeneous spaces of reductive type, in Proceeding of the Conference on Representation Theorie of Lie Groups and Lie Algebras held in 1990 August–September at Fuji-Kawaguchiko (ICM-90 Satellite Conference), Word Scientific, 59–75 (1992).

[94] Kobayashi. T, On discontinuous groups on homogeneous space with noncompact isotropy subgroups, J. Geom. Phys. **12**, 133–144 (1993).

[95] Kobayashi. T, Criterion of proper action on homogeneous space of reductive type, J. Lie Theory **6**, 147–163 (1996).

[96] Kobayashi. T, Discontinuous groups and Clifford–Klein forms of pseudo-Riemannian homogeneous manifolds, in Algebraic and Analytic Methods in Representation Theory, Perspectives in Mathematics, Vol. **17**, Academic Press, 99–165 (1996).

[97] Kobayashi. T, Deformation of compact Clifford–Klein forms of indefinite Riemannian homogeneous manifolds, Math. Ann. **310**, 394–408 (1998).

[98] Kobayashi. T, Discontinuous groups for non Riemannian homogeneous spaces, in Mathematics Unlimited—2001 and Beyond (B. Engquist and W. Schmid, eds), Springer-Verlag, 723–747 (2001).

[99] Kobayashi. T, On discontinuous group actions on non Riemannian homogeneous spaces, Sūgaku Expo. **22**, 1–19 (2009). arXiv:math.DG/0603319.

[100] Kobayashi. T and Nasrin. S, Deformation of properly discontinuous action of $\mathbb{Z}^k$ on $\mathbb{R}^{k+1}$, Int. J. Math. **17**, 1175–1193 (2006).
[101] Kobayashi. T and Yoshino. T, Compact Clifford–Klein forms of symmetric spaces-revisited, Pure Appl. Math. Q. **1**, No. 3, 591–663 (2005), special issue: in memory of Armand Borel, part 2 of 3.
[102] Kontsevich. M, Deformation quantization of Poisson manifolds I, arXiv:math.QA/9709040 (1997).
[103] Kulkarni. R. S, Proper actions and pseudo-Riemannian space forms, Adv. Math. **40**, No. 1, 10–51 (1981).
[104] Kulkarni. R. S, Lee. K. B, and Raymond. F, Deformation spaces for Seifert manifolds, in Geometry and Topology, Lecture Notes in Mathematics, Vol. **1167**, Springer, Berlin, 180–216 (1985).
[105] Labourie. F, Mozes. S, and Zimmer. R. J, On manifolds locally modelled on non-Riemannian homogeneous spaces, Geom. Funct. Anal. **5**, No. 6, 955–965 (1995).
[106] Leptin. H and Ludwig. J, Unitary Representation Theory of Exponential Lie Groups, De Gruyter Expositions in Mathematics, Vol. **18** (1994).
[107] Lipsman. R, Representations of exponential solvable Lie groups induced from maximal subgroups, Mich. Math. J. **40**, 299–320 (1993).
[108] Lipsman. R, Proper action and a compactness condition, J. Lie Theory **5**, 25–39 (1995).
[109] Magnin. L, Determination of 7-dimensional indecomposable nilpotent complex Lie algebras by adjoining a derivation to 6-dimensional Lie algebras, Algebr. Represent. Theory **13**, No. 6, 723–753 (2010).
[110] Manchon. D and Torossian. Ch, Cohomologie tangente et cup-produit pour la quantification de Kontsevich, arXiv:math.QA/0106205 (2001).
[111] Mathieu. O, Bicontinuity of the Dixmier map, J. Am. Math. Soc. **4**, No. 4, 837–863 (1991).
[112] Mostow. G. D, On a conjecture of Montgomery, Ann. Math. (2) **65**, No. 3, 513–516 (1957).
[113] Motzkin. T. S and Taussky. O, Pairs of matrices with property L, Trans. Amer. Math. Soc. **73**, 108–114 (1952).
[114] Nasrin. S, Criterion of proper actions for 2-step nilpotent Lie groups, Tokyo J. Math. **24**, No. 2, 535–543 (2001).
[115] Oliver. R. K, On Bieberbach's analysis of discrete Euclidean groups, Proc. Amer. Math. Soc. **80**, 15–21 (1980).
[116] Palais. R. S, On the existence of slices for actions of non-compact Lie groups, Ann. Math. **73**, 295–323 (1961).
[117] Pedersen. N. V, On the characters of exponential solvable Lie groups, Ann. Sci. Éc. Norm. Supér. **17**, No. 1, 1–29 (1984).
[118] Pedersen. N. V, On the infinitesimal kernel of irreducible representations of nilpotent Lie groups, Bull. Soc. Math. Fr. **112**, No. 42, 3–467 (1984).
[119] Pedersen. N. V, On the symplectic structure of coadjoint orbits of (solvable) Lie groups and applications. I, Math. Ann. **281**, 633–669 (1988).
[120] Pedersen. N. V, Orbits and primitive ideals of solvable Lie algebras, Math. Ann. **298**, No. 2, 275–326 (1994).
[121] Robinson. D. J, A Course in the Theory of Groups, 2nd ed., Springer-Verlag (1996).
[122] Saitô. M, Sur certain groupes de Lie résolubles I; II, Sci. Pap. Coll. Gen. Educ. Univ. Tokyo **7**, 1–11 (1957).
[123] Selberg. A, On discontinuous groups in higher-dimension symmetric spaces, in Contributions to Functional Theory, Tata Institute, Bombay, 147–164 (1960).
[124] Shapovalov. N. N, A certain bilinear form on the universal enveloping algebra of a complex semi-simple Lie algebra, Funct. Anal. Appl. **6**, No. 4, 307–312 (1972).

[125] Takenouchi. O, Sur la facteur-représentation d'un groupe de Lie résoluble de type (E), Math. J. Okayama Univ. **7**, 151–161 (1957).

[126] Tamarkin. D, Another proof of M. Kontsevich formality theorem, arXiv:math.QA/9803025 (1998).

[127] Vanhaecke. P, Integrable Systems in the Realm of Algebraic Geometry, Lecture Notes in Mathematics, Vol. **1638**, Springer (1996).

[128] Varadarajan. V. S, Lie Groups, Lie Algebras and Their Representations, Graduate Texts in Mathematics, Vol. **102**, Springer, New York, NY (1984).

[129] Weil. A, On discrete subgroups of Lie groups II, Ann. Math. **75**, 578–602 (1962).

[130] Weil. A, Remarks on the cohomology of groups, Ann. Math. **80**, 149–157 (1964).

[131] Weinstein. A, The local structure of Poisson manifolds, J. Differ. Geom. **18**, 523–557 (1983).

[132] Wüstner. M, A connected Lie group equals the square of the exponential image, J. Lie Theory **13**, 307–309 (2003).

[133] Yoshino. T, A counterexample to Lipsman's conjecture, Int. J. Math. **16**, 561–566 (2005).

[134] Yoshino. T, Criterion of proper actions for 3-step nilpotent Lie groups, Int. J. Math. **85**, 887–893 (2007).

[135] Yoshino. T, Deformation spaces of compact Clifford–Klein forms of homogeneous spaces of Heisenberg groups, in Representation Theory and Analysis on Homogeneous Spaces, RIMS Kôkyûroku Bessatsu B7, Res. Inst. Math. Sci. (RIMS), Kyoto, 45–55 (2008).

[136] Yoshino. T, A solution to Lipsman's conjecture for $\mathbb{R}^4$, Preprint.

Index

https://doi.org/10.1515/9783110765304-010

De Gruyter Expositions in Mathematics

Volume 71
Victor P. Maslov, Yurievich Oleg Shvedov†
The Canonical Operator in Multiparticle Problems and Quantum Field Theory, 2022
ISBN 978-3-11-076238-9, e-ISBN 978-3-11-076270-9

Volume 70
Aron Simis, Zaqueu Ramos
Graded Algebras in Algebraic Geometry, 2022
ISBN 978-3-11-063754-0, e-ISBN 978-3-11-064069-4

Volume 69/2
Mariusz Urbański, Mario Roy, Sara Munday
Non-Invertible Dynamical Systems. Volume 2: Finer Thermodynamic Formalism – Distance Expanding Maps and Countable State Subshifts of Finite Type, Conformal GDMSs, Lasota-Yorke Maps and Fractal Geometry, 2022
ISBN 978-3-11-070061-9, e-ISBN 978-3-11-070269-9

Volume 69/1
Mariusz Urbański, Mario Roy, Sara Munday
Non-Invertible Dynamical Systems. Volume 1: Ergodic Theory – Finite and Infinite, Thermodynamic Formalism, Symbolic Dynamics and Distance Expanding Maps, 2021
ISBN 978-3-11-070264-4, e-ISBN 978-3-11-070268-2

Volume 68
Yuri A. Bahturin
Identical Relations in Lie Algebras, 2021
ISBN 978-3-11-056557-7, e-ISBN 978-3-11-056665-9

Volume 34
Marek Jarnicki, Peter Pflug
Extension of Holomorphic Functions, 2nd Edition, 2020
ISBN 978-3-11-062766-4, e-ISBN 978-3-11-063027-5

Volume 67
Alcides Lins Neto, Bruno Scárdua
Complex Algebraic Foliations, 2020
ISBN 978-3-11-060107-7, e-ISBN 978-3-11-060205-0